Magnetic Resonance

Introduction, Advanced Topics and Applications to Fossil Energy

NATO ASI Series

Advanced Science Institutes Series

A series presenting the results of activities sponsored by the NATO Science Committee, which aims at the dissemination of advanced scientific and technological knowledge, with a view to strengthening links between scientific communities.

The series is published by an international board of publishers in conjunction with the NATO Scientific Affairs Division

A	Life Sciences	Plenum Publishing Corporation
B	Physics	London and New York
C	Mathematical and Physical Sciences	D. Reidel Publishing Company Dordrecht, Boston and Lancaster
D	Behavioural and Social Sciences	Martinus Nijhoff Publishers
E	Engineering and Materials Sciences	The Hague, Boston and Lancaster
F	Computer and Systems Sciences	Springer-Verlag
G	Ecological Sciences	Berlin, Heidelberg, New York and Tokyo

Series C: Mathematical and Physical Sciences Vol. 124

Magnetic Resonance
Introduction, Advanced Topics and Applications to Fossil Energy

edited by

Leonidas Petrakis

Gulf Research and Development Company and
University of Pittsburgh, Pittsburgh, Pennsylvania, U.S.A.

and

Jacques P. Fraissard

Université Pierre et Marie Curie
Laboratoire de Chimie des Surfaces, Paris, France

D. Reidel Publishing Company

Dordrecht / Boston / Lancaster

Published in cooperation with NATO Scientific Affairs Division

Proceedings of the NATO Advanced Study Institute on
Magnetic Resonance - Introduction, Advanced Topics and Applications to Fossil Energy
Maleme, Crete, Greece
July 3-15, 1983

Library of Congress Cataloging in Publication Data

NATO Advanced Study Institute on Magnetic Resonance (1983: Maleme, Crete)
Magnetic resonance.

(NATO ASI series. Series C, Mathematical and physical sciences; v. 124)
"Published in cooperation with NATO Scientific Affairs Division."
"Proceedings of the NATO Advanced Study Institute on Magnetic Resonance, Maleme, Crete, July 3–15, 1983"–T.p. verso.
Includes index.
1. Fossil fuels–Analysis–Congresses. 2. Magnetic resonance–Congresses. I. Petrakis, Leonidas, 1935– . II. Fraissard, Jacques P., 1934– . III. North Atlantic Treaty Organization. Scientific Affairs Division. IV. Title. V. Series.
TP321.N38 1984 662'.6 84-3393
ISBN-13: 978-94-009-6380-1 e-ISBN-13: 978-94-009-6378-8
DOI: 10.1007/978-94-009-6378-8

Published by D. Reidel Publishing Company
P.O. Box 17, 3300 AA Dordrecht, Holland

Sold and distributed in the U.S.A. and Canada
by Kluwer Academic Publishers,
190 Old Derby Street, Hingham, MA 02043, U.S.A.

In all other countries, sold and distributed
by Kluwer Academic Publishers Group,
P.O. Box 322, 3300 AH Dordrecht, Holland

D. Reidel Publishing Company is a member of the Kluwer Academic Publishers Group

---

TABLE OF CONTENTS

III - PARADIGMS OF MAGNETIC RESONANCE IN FOSSIL ENERGY STUDIES

Lectures

Papers

IV - RESEARCH PROSPECTS

Lectures

PREFACE

This volume contains the lectures presented at an Advanced Study Institute on "Magnetic Resonance Techniques in Fossil Energy Problems," which was held at the village of Maleme, Crete, in July of 1983. As of this writing, a different popular attitude prevails from that when the ASI was proposed as far as how critical the world energy picture is. In the popular press, a panglossian attitude (the "petroleum glut" of the 80's) has replaced the jeremiads of the 70's (a catastrophic "energy crisis"). Yet, there are certain important constants: (a) for the foreseeable future, fossil energy sources (petroleum, coal, oil shale, etc.) will continue to be of paramount importance; and (b) science and technology of the highest order are needed to extend the fossil energy resource base and to utilize it in a cost-effective manner that is also environmentally acceptable. It is precisely this second item that this volume addresses.

The volume introduces the phenomenology of magnetic resonance in a unified and detailed manner, starting from the behavior of spins in a magnetic field and proceeding to a density matrix consideration of the phenomena involved. The various important parameters of NMR and EPR are discussed in detail, as well as their significance and the information that may be obtained from their proper consideration. Advanced topics are then considered, including line broadening interactions, coherent averaging and multiple quantum methods in NMR, ENDOR, electron spin echoes, dynamic nuclear polarization, problems of quantification, 2-Dimensional NMR, Mossbauer, and instrumentation. This component of the volume can serve as an introduction to magnetic resonance and discussion of advanced topics, regardless of applications.

This volume also introduces the general fields of fossil energy, including resources, synthetic fuels processes, and structural aspects of fossil energy materials. Through the presentation of many specific recent results, the tremendous potentional of magnetic resonance for impacting fossil energy research and development is documented carefully. Finally, the issue of research problems and opportunities is discussed and specific recommendations are made. The volume is addressed both to research workers in the area of magnetic resonance as well as to those who are utilizing, or would like to utilize, the techniques for solving problems in fossil energy. In addition, it could serve as a textbook for a graduate course, and it should be of use to those who manage and/or fund research.

Many people contributed very generously the success of the Institute on which this volume is based. The members of the Organizing Committee are acknowledged on a separate page in this volume, and we take this opportunity to thank them again. We also acknowledge the special interest for this ASI of the following people: Dr. G. Caryotakis, Dr. S. W. Chun, Dr. M. DiLullo, Dr. C. Anderson Evans, Dr. M. Frodyma, Prof. G. J. Karabatsos, Mr. B. Knuttel, Dr. U. Merten, Dr. R. E. Roberts, Dr. H. Rosenwasser, Dr. R. E. Schwemm, Dr. J. Sharkey, Dr. C. Sinclair, Mr. P. V. Smith, Dr. R. Wainerdi, and Dr. T. Wirthlin.

In addition to the Scientific Affairs Division of the North Atlantic Treaty Organization, the Institute was sponsored by a number of other organizations: American Chemical Society, Fuel Chemistry Division; American Chemical Society, Petroleum Division; Bruker Analytische Messtechnik; Exxon Research and Engineering Company; Gulf Research & Development Company; IBM Instruments; JEOL; National Science Foundation; U.S. Department of Energy; U.S. Naval Air Systems Command; and Varian AG. Their support and sponsorship we acknowledge with thanks.

Finally, we acknowledge the efforts of all the people who contributed to organizational aspects of the Institute, administrative matters, and secretarial assistance and travel and local arrangements. We include in this list Lina Petrakis, Jean DeCroo, Pauline Devita, April Stevens, Ann Langerholc, Helen Antonopoulos, and Tom Balos.

Maroussi
July, 1983

LEONIDAS PETRAKIS
JACQUES FRAISSARD

ORGANIZING COMMITTEE

L. PETRAKIS, Director, Pittsburgh
J. FRAISSARD, Co-Director, Paris
T. AXENROD, New York
R. de BEER, Delft
G. BOTSARIS, Medford
W. BREY, Gainesville
P. CHICHE, Varneuil-en-Halotte
S. W. CHUN, Pittsburgh
J. J. DELPUECH, Nancy
J. J. DEREPPE, Louvain
G. GAVALAS, Pasadena
R. GERHARDS, Essen
B. GERSTEIN, Ames
M. GORBATY, Linden
G. J. KARABATSOS, East Lansing
W. R. LADNER, Cheltenham
G. MACIEL, Fort Collins
U. MERTEN, Pittsburgh
P. MONTANO, Morgantown
R. PELET, Rueil-Talmaison
A. PINES, Berkeley
D. PRATT, Pittsburgh
R. E. ROBERTS, Washington, D.C.
J. SMIDT, Delft
J. UEBERSFELD, Paris
J. WAUGH, Cambridge (Massachusetts)
I. WENDER, Pittsburgh
R. WIND, Delft
W. WISER, Salt Lake City

ΜΑΓΝΗΤΙΚΟΣ ΣΥΝΤΟΝΙΣΜΟΣ. ΕΙΣΑΓΩΓΗ.
(MAGNETIC RESONANCE. AN INTRODUCTION.)

David W. Pratt

Department of Chemistry, University of Pittsburgh
Pittsburgh, Pennsylvania 15260 USA

The basic principles common to all magnetic resonance experiments are reviewed, with special emphasis on the interpretation of NMR and EPR spectra in the solid state. The connection between experiments in the frequency and time domains is established, and a brief review of spin dynamics is given.

1. BASIC CONCEPTS.

1.1. Introduction.

The magnetic resonance spectrum of a nucleus, atom or molecule is just a plot of the probability that the system will absorb (or emit) a microwave or radiofrequency photon as a function of some external parameter. The external parameter in the nuclear magnetic resonance (NMR) experiment is usually frequency, whereas in the electron paramagnetic resonance (EPR) experiment, it is usually the magnetic field. Thus, the first task of the magnetic resonance spectroscopist is to measure the frequency of the absorbed or emitted radiation, the field, and the intensity of the transition as accurately as possible. The progressive development of experimental methods for making such measurements, in both the frequency and time domains, has placed this objective well within the reach of most practicing research scientists, even those who are not expert in the subject. However, once the data have been collected, one is faced with a second, more challenging task; to analyze the spectrum and interpret it correctly. This is a formidable objective because the energy levels giving rise to the spectrum, and their widths, represent the net effect of a large number of interactions which might be experienced by the nucleus, atom, or molecule under investigation. Yet it is also a rewarding

L. Petrakis and J. P. Fraissard (eds.), Magnetic Resonance. Introduction, Advanced Topics and Applications to Fossil Energy, 1–69.

task because the proper interpretation of an NMR or EPR spectrum leads to a greater understanding of the static and dynamic properties of that system, on a microscopic level. The purpose of this article is to provide the background necessary for making such an interpretation.

The different forces and interactions which manifest themselves in a spectroscopic experiment vary enormously according to the frequency range being studied. In this course we shall be concerned with nuclei, electrons, and their interactions with external fields; hence, the transitions we shall be interested in occur in the frequency range 10^6–10^{10} Hz, corresponding to wavelengths in the range 3×10^4–3 cm and energies in the range 3×10^{-5}–0.3 cm^{-1}. For comparison, we recall that the value of kT at room temperature is about 200 cm^{-1}. Thus, the energy level separations characteristic of the magnetic resonance experiment are very much less than thermal energies except at the very lowest temperatures. This has two important consequences. First, it is clear that the species under study is not strongly perturbed by the absorption of 1 cm^{-1} or less of energy; this is an important advantage of the technique when compared with other spectroscopic methods. But, at the same time, it is clear that the Boltzmann factors for the absorption process are very unfavorable; thus, magnetic resonance suffers in comparison with the sensitivity of higher frequency spectroscopies.

1.2. Quantum Mechanical Description of the Resonance Experiment.

For most chemical purposes, it is sufficient to regard the nucleus as having only mass (M), charge (+Ze), and position ($\underline{R}$), and similarly for the electron (m, −e, and $\underline{r}$). However, in order to interpret the low energy transitions in NMR and EPR, it is necessary to take into account three more properties of each "particle". These are, for the nucleus, the spin angular momentum $\underline{I}$, the magnetic moment $\underline{\mu}_n$ and the electric quadrupole moment eQ; and, for the electron, the spin angular momentum $\underline{S}$, the orbital angular momentum $\underline{L}$, and the magnetic moment $\underline{\mu}_e$. Consider now the general case of <u>either</u> a nucleus <u>or</u> an electron with angular momentum $\underline{F}$ and magnetic moment $\underline{\mu}_f$. $\underline{F}$ is a vector quantity, having both direction and magnitude. Quantum mechanics tell us that the magnitude of $\underline{F}$, $|\underline{F}|$, is restricted to the values $[F(F+1)\hbar^2]^{1/2}$, where F is the spin quantum number, having either integral or half-integral values. If $|\underline{F}|$ is different from zero, the vector $\underline{F}$ can take up any one of 2F+1 orientations with respect to a space-fixed axis, usually denoted by z. These orientations may be specified by the quantum number m_F, which represents the projection of $\underline{F}$ on the z axis (<u>i.e.</u>, $F_z = m_F\hbar$). For example, if F = 1, then $|\underline{F}| = \sqrt{2}\,\hbar$, $m_F = 0, \pm 1\,\hbar$, and there are three possible orientations of $\underline{F}$ with respect to z.

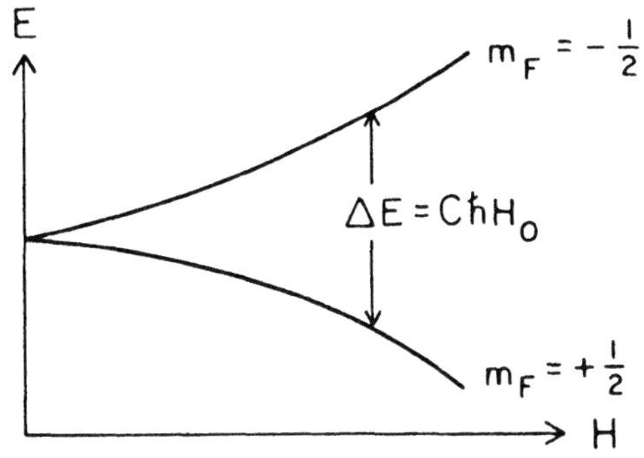

Fig. 1. Zeeman energy level diagram for an F=1/2 system.

In field-free space, each of the possible orientations of $\underline{F}$ is degenerate. (You may recall that a similar situation exists for the three 2p orbitals of the hydrogen atom.) But in the presence of a field, either electric or magnetic, this degeneracy is at least partially lifted. The field may be either internal or external in nature. Suppose we consider the case of an F = 1/2 particle in the presence of an external magnetic field, $\underline{H}_0$. The energy of interaction may be written as

$$\hat{H} = -\underline{\mu}_f \cdot \underline{H}_0 \tag{1}$$

If the field is directed along the z axis

$$\hat{H} = -\mu_f^z H_0 \tag{2}$$

Since $\underline{\mu}_f$ and $\underline{F}$ are co-linear, we may write

$$\underline{\mu}_f = C\underline{F}\hbar \tag{3}$$

$$\mu_f^z = CF_z\hbar = Cm_F\hbar \tag{4}$$

and

$$\hat{H} = -C\hbar m_F H_0 = E \tag{5}$$

where C is a proportionality constant. Since, for F = 1/2, $m_F = \pm 1/2$; there are two states which differ in energy by

$$\Delta E = E(m_F=-1/2) - E(m_F=1/2) = C\hbar H_0 \tag{6}$$

Since the energy difference is proportional to the field, an energy level diagram similar to that shown in Fig. 1 results.

In order for transitions between the two states in Fig. 1 to be observed, the condition

$$\Delta E = E_2 - E_1 = C\hbar H_0 = h\nu_0 \tag{7}$$

must be satisfied. Combining these relations, we have

$$C = \frac{\Delta E}{\hbar H_0} = \frac{2\pi\nu_0}{H_0} \quad \text{radians/sec Gauss} \tag{8}$$

For nuclei,

$$C\hbar = \gamma_n\hbar = g_n\beta_n \tag{9}$$

where γ_n is the nuclear magnetogyric ratio, g_n is the nuclear g factor, and β_n is the nuclear magneton ($\beta_n = e\hbar/2Mc = 5.050 \times 10^{-24}$ erg/G). In the case of protons, $g_n = 5.585$ and $\gamma_n = 2.675 \times 10^4$. Thus, the frequency required for a proton NMR experiment at 14 kG is

$$\nu_n = \frac{\gamma_n H_0}{2\pi} = \frac{g_n\beta_n H_0}{h} = 60 \times 10^6 \text{ Hz} \tag{10}$$

For electrons,

$$C\hbar = \gamma_e\hbar = g_e\beta_e \tag{11}$$

where γ_e is the electron magnetogyric ratio, g_e is the electronic g factor, and β_e is the Bohr magneton ($\beta_e = e\hbar/2mc = 9.274 \times 10^{-21}$ erg/G). In the case of electrons possessing only spin angular momentum, $g_e = 2.00232$ and $\gamma_e = 1.760 \times 10^7$. Thus, the frequency required for an EPR experiment at 14 kG is

$$\nu_e = \frac{\gamma_e H_0}{2\pi} = \frac{g_e\beta_e H_0}{h} = 39 \times 10^9 \text{ Hz} \tag{12}$$

Note that the principal source of difference in the values of ν_n and ν_e is the difference in the masses of the two "particles". Otherwise, the two experiments are identical in all respects.

1.3. Classical Description of the Resonance Experiment.

According to classical mechanics, the equation of motion of the angular momentum vector $\underline{F}$ in the presence of a field is

$$\frac{d\underline{F}}{dt} = \gamma_f \underline{F} \times \underline{H}_0 \tag{13}$$

Since $\underline{\mu}_f = \gamma_f \underline{F}$, we may eliminate $\underline{F}$, getting

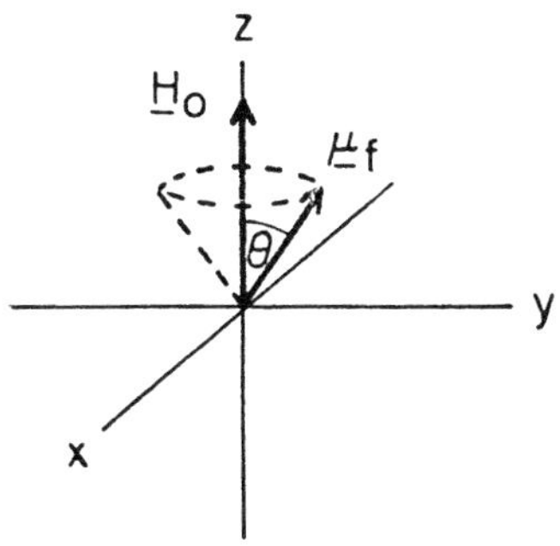

Fig. 2. Larmor precession in a fixed coordinate system.

$$\frac{d\underline{\mu}_f}{dt} = \underline{\mu}_f \times (\gamma_f \underline{H}_0) \tag{14}$$

This equation tells us that if $\underline{\mu}_f$ and $\underline{H}_0$ are initially pointing in different directions, changes in the orientation of $\underline{\mu}_f$ with time are perpendicular to both $\underline{\mu}_f$ and $\underline{H}_0$. In other words, $\underline{\mu}_f$ moves under the influence of the field, but in a manner which keeps the angle between the two vectors constant. If $\underline{H}_0$ is independent of time, $\underline{\mu}_f$ precesses about $\underline{H}_0$ with some angular frequency $\underline{\omega}_0$, like a gyroscope. This is called Larmor precession (cf., Fig. 2). Note that the precessional frequency, $\omega_0 = \gamma_f H_0$, is identical in magnitude with the angular frequency needed for magnetic resonance absorption, as found by elementary quantum theory.

Since Θ is fixed, Eq. (1) shows that the energy is constant in time. To change this energy, one must apply a perturbation which exerts a torque on $\underline{\mu}_f$. This is most conveniently viewed from the perspective of a coordinate system (x'y'z') rotating with the angular velocity $\underline{\omega}_0$ about the z'(=z) axis. The equation of motion then becomes

$$\frac{d\underline{\mu}_f}{dt} = \frac{\delta\underline{\mu}_f}{\delta t} + (\underline{\omega}_0 \times \underline{\mu}_f) = \underline{\mu}_f \times \gamma_f \underline{H}_0 \tag{15}$$

where $\delta\underline{\mu}_f/\delta t$ is the time derivative in the rotating frame. Rearranging Eq. (15), we have

$$\frac{\delta\underline{\mu}_f}{\delta t} = \underline{\mu}_f \times (\gamma_f \underline{H}_0 + \underline{\omega}_0) \tag{16}$$

If we compare this result with Eq. (14), we can make several very interesting observations. First, note that if we define an effective field,

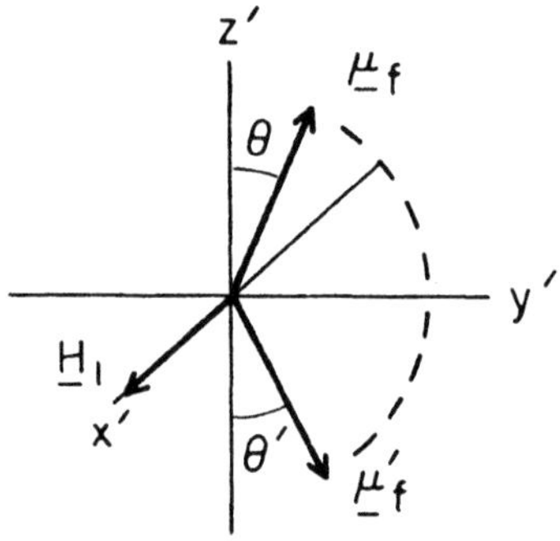

Fig. 3. Larmor precession in a rotating coordinate system, and the behavior of the magnetic moment with an oscillating field applied along the x' axis.

$$\underline{H}_{eff} = \underline{H}_0 + \frac{\omega_0}{\gamma_f} \tag{17}$$

the two equations, (14) and (16), are similar. That is, just as $\underline{\mu}_f$ precesses about $\underline{H}_0$ in a fixed coordinate system, $\underline{\mu}_f$ precesses about $\underline{H}_{eff}$ in the rotating coordinate system. Second, note that if $\gamma_f\underline{H}_0 = -\underline{\omega}_0$, $\underline{H}_{eff} = 0$ and $\delta\underline{\mu}_f/\delta t = 0$. In other words, $\underline{\mu}_f$ is fixed in the rotating coordinate system if that system rotates with an angular velocity $\underline{\omega}_0 = -\gamma_f H_0\underline{k}$. Finally, suppose we apply a second, oscillating magnetic field $\underline{H}_1(t)$ along the x axis of the laboratory coordinate system,

$$\underline{H}_1(t) = 2H_1^0 \cos(\omega t + \delta)\ \underline{i} \tag{18}$$

This may be expressed as the sum of two counter-rotating components, $\underline{H}_1^{\ell}$ and $\underline{H}_1^{r}$, one of which moves 'in-phase' with the rotating x' and y' axes. However, if ω is chosen to be equal in magnitude to ω_0, then either $\underline{H}_1^{\ell}$ or $\underline{H}_1^{r}$ will be fixed in the rotating coordinate system, say along x' (cf., Fig. 3). In that event, $\underline{\mu}_f$ will precess about $\underline{H}_{eff} = \underline{H}_1$, and move from its original position along the positive z' axis to a new position along the negative z' axis. Viewed from the perspective of the fixed coordinate system, $\underline{\mu}_f$ now precesses about the negative z axis with its characteristic Larmor frequency. Clearly, energy has been absorbed (or emitted) in the process, as the angle Θ is now different. This is the magnetic resonance experiment.

1.4. Thermal Equilibrium and Spin Relaxation.

Obviously, resonance absorption in a magnetic resonance experiment can be observed only if there is a population difference between the two spin levels connected by the $\underline{H_1}$ field. Suppose we have a macroscopic sample of N spins, distributed among the two energy levels of a spin 1/2 system with N_α in the lower level and N_β in the upper level. The relative population of the two levels is given by the Boltzmann factor,

$$\frac{N_\alpha}{N_\beta} = \exp\ (\gamma_f \hbar H_0/kT) \tag{19}$$

At ordinary temperatures, $\gamma_f \hbar H_0 << kT$, so

$$\frac{N_\alpha}{N_\beta} \cong 1 + \frac{\gamma_f \hbar H_0}{kT} \tag{20}$$

Since $N = N_\alpha + N_\beta$, we have

$$N_\alpha = \frac{N}{2}\ [1 + \gamma_f \hbar H_0/2kT] \tag{21a}$$

$$N_\beta = \frac{N}{2}\ [1 - \gamma_f \hbar H_0/2kT] \tag{21b}$$

If $\gamma_f \hbar H_0 \sim 10^{-3}$ cm^{-1}, then $N_\alpha - N_\beta \sim 10^{-5}$ at room temperature. Now, suppose we apply an oscillating magnetic field to the sample which satisfies the resonance condition $h\nu_0 = \gamma_f \hbar H_0$. Let the probability of a stimulated upward transition be $P_{\alpha\beta}$, and the probability of a stimulated downward transition be $P_{\beta\alpha}$. These transition probabilities can be calculated from the Golden Rule expressions

$$P_{\alpha\beta} = \frac{2\pi}{\hbar} |\langle\alpha|\hat{H}_1|\beta\rangle|^2\ \delta(E_\beta - E_\alpha - h\nu) \tag{22a}$$

$$P_{\beta\alpha} = \frac{2\pi}{\hbar} |\langle\beta|\hat{H}_1|\alpha\rangle|^2\ \delta(E_\alpha - E_\beta + h\nu) \tag{22b}$$

where $\hat{H}_1$ is the time-dependent perturbation

$$\hat{H}_1(t) = \gamma_f \hbar H_1^0 F_x \cos(\omega t + \delta) \tag{23}$$

It is relatively easy to show that $|\langle\beta|\hat{H}_1|\alpha\rangle|^2 = |\langle\alpha|\hat{H}_1|\beta\rangle|^2$ and therefore that $P_{\alpha\beta} = P_{\beta\alpha} = P$. Now, the rate of change of the population of the α spin state in the presence of the perturbation is

$$\frac{dN_\alpha}{dt} = N_\beta P_{\beta\alpha} - N_\alpha P_{\alpha\beta} = P(N_\beta - N_\alpha) \tag{24}$$

If we define $n = N_\alpha - N_\beta$, this can be written as

$$\frac{dN_\alpha}{dt} = -Pn \tag{25}$$

Similarly,

$$\frac{dN_\beta}{dt} = Pn \tag{26}$$

Therefore, the time dependence of the population difference can be expressed as

$$\frac{dn}{dt} = -2Pn \tag{27}$$

which has the solution

$$n = n_0 \, e^{-2Pt} \tag{28}$$

As the rate of absorption of energy from the field is proportional to n, we see that it decreases exponentially with the time. The explanation for this is, of course, that the perturbation tends to equalize the populations of the two connected levels. If this were to continue indefinitely, we would encounter a situation known as saturation, and the magnetic resonance signal would be very difficult to detect.

Fortunately, there exist other processes, known collectively as relaxation processes, which also influence the dynamic behavior of the spin system. We shall have much more to say about these later. But for the moment, we remark that there are two types of spin relaxation, longitudinal (T_1) and transverse (T_2) processes. T_1 processes affect the relative populations of the two spin levels via spin-lattice relaxation, an energy exchange mechanism involving the spins and the degrees of freedom of the surrounding 'lattice'. Suppose we examine the effect of spin-lattice relaxation on our two-level system, first in the absence of a perturbing H_1 field. If $W_{\beta\alpha}$ and $W_{\alpha\beta}$ are the corresponding relaxation probabilities, then at equilibrium

$$\frac{dN_\alpha}{dt} = N^o_\beta W_{\beta\alpha} - N^o_\alpha W_{\alpha\beta} = 0. \tag{29}$$

This gives

$$\frac{N^o_\beta}{N^o_\alpha} = \frac{W_{\alpha\beta}}{W_{\beta\alpha}} = \exp\,(-\gamma_f \hbar H_0/kT) \tag{30}$$

From this we see that $W_{\alpha\beta}$ and $W_{\beta\alpha}$ are not equal in the presence of a static field. If we write a relation similar to Eq. (29) for dN_β/dt, and express the result in terms of n and N, we find

$$\frac{dn}{dt} = \frac{-(n-n_0)}{T_1} \tag{31}$$

where

$$n_0 = N[(W_{\beta\alpha}-W_{\alpha\beta})/(W_{\beta\alpha}+W_{\alpha\beta})] \tag{32a}$$

$$T_1 = (W_{\alpha\beta}+W_{\beta\alpha})^{-1} \tag{32b}$$

Therefore, T_1 has the dimensions of time, and is a measure of the time taken for energy to be transferred to or from the lattice; i.e., for the spin system to approach thermal equilibrium. If we now include the effect of the perturbing field, we have

$$\frac{dn}{dt} = -2Pn - \frac{(n-n_0)}{T_1} \tag{33}$$

This yields, at equilibrium, the solution

$$n = \frac{n_0}{1+2PT_1} \tag{34}$$

Thus, saturation can be avoided if $2PT_1 << 1$. The transition probability P turns out to be proportional to $|\underline{H}_1|^2$, so the power of the $\underline{H}_1$ field must be kept low to allow the spins to relax.

1.5. The Bloch Equations.

A clearer picture of T_1 and T_2 can be obtained by taking a macroscopic view of the resonance experiment. We begin by considering the bulk magnetic moment $\underline{M}$ of a large assembly of spins at a certain temperature T. $\underline{M}$ is a vector quantity, so we can write

$$\underline{M} = M_x\underline{i} + M_y\underline{j} + M_z\underline{k} \tag{35}$$

Now, imagine that the spins have been aligned by placing them in a static magnetic field $\underline{H}_0$, so that N_α of them are in the α spin state and N_β of them are in the β spin state. The z component of $\underline{M}$ is

$$M_z = \gamma_f\hbar(N_\alpha-N_\beta) = \gamma_f\hbar n \tag{36}$$

If we then turn the field off, the population difference n decays exponentially in time owing to spin-lattice relaxation; the rate law is

$$\frac{dn}{dt} = -\frac{n}{T_1} \tag{37}$$

Since M_z is proportional to n, it obeys a similar equation

$$\frac{dM_z}{dt} = -\frac{M_z}{T_1} \tag{38}$$

approaching the value zero after several time intervals T_1. There is, of course, no physical distinction between the z axis and any other direction in the absence of a field, so M_x and M_y exhibit the same kinetic behavior.

The introduction of a static magnetic field $\underline{H}_0$ along the z axis alters the situation in several ways. First, M_z no longer vanishes at thermal equilibrium, but approaches a constant value M_0 which is proportional to the static magnetic susceptibility of all N spins. Thus, M_z obeys an equation analogous to Eq. (31),

$$\frac{dM_z}{dt} = -\frac{(M_z-M_0)}{T_1} \tag{39}$$

Further, although the transverse components of the magnetization, M_x and M_y, still decay to zero, in general the time constant is different from T_1. We introduce T_2 to describe these processes; thus,

$$\frac{dM_x}{dt} = -\frac{M_x}{T_2} \tag{40}$$

$$\frac{dM_y}{dt} = -\frac{M_y}{T_2} \tag{41}$$

T_1 and T_2 are, in general, not equal because they depend on different processes within the spin system. For example, dipolar coupling between two identical spins, one in the level E_α and the second in the level E_β, which results in an exchange of energy between the spins but no exchange of energy with the lattice, is a T_2 process but not a T_1 process. Thus, the second effect of a static field is to cause M_z and M_x,M_y to decay at different rates. Third, and finally, as we have already seen, the magnetic field exerts a torque on $\underline{M}$, which is described by the equation

$$\frac{d\underline{M}}{dt} = \gamma_f(\underline{M} \times \underline{H}_0) \tag{42}$$

This equation has the components

$$\frac{dM_x}{dt} = \omega_0 M_y \tag{43a}$$

$$\frac{dM_y}{dt} = -\omega_0 M_x \tag{43b}$$

$$\frac{dM_z}{dt} = 0 \tag{43c}$$

where $\omega_0 = \gamma_f H_0$, which express the fact that $\underline{M}$ precesses about $\underline{H}_0$ with a uniform angular velocity $\underline{\omega}_0$, tracing out a cone. Combining the effects of relaxation and precession, we have then the three equations

$$\frac{dM_x}{dt} = \omega_0 M_y - \frac{M_x}{T_2} \tag{44a}$$

$$\frac{dM_y}{dt} = -\omega_0 M_x - \frac{M_y}{T_2} \tag{44b}$$

$$\frac{dM_z}{dt} = -\left(\frac{M_z - M_0}{T_1}\right) \tag{44c}$$

Eqs. (44) were first written down by Felix Bloch (1) and have proven to be extremely useful for describing a wide variety of magnetic resonance phenomena. Of particular interest here are the solutions, found in standard texts (2-4), in the presence of an alternating field $\underline{H}_1$. We assume that $\underline{H}_1$ rotates in the xy plane, in the same sense as the Larmor precession, with a uniform angular velocity $\underline{\omega}$. As before, it is helpful to view the resulting motion of $\underline{M}$ in a coordinate system which rotates with $\underline{H}_1$, as shown in Fig. 4. We denote by u and v the transverse components of $\underline{M}$ along the x' and y' directions. The Bloch equations in the rotating frame are

$$\frac{du}{dt} = (\omega_0 - \omega)v - \frac{u}{T_2} \tag{45a}$$

$$\frac{dv}{dt} = -(\omega_0 - \omega)u + \gamma_f H_1 M_z - \frac{v}{T_2} \tag{45b}$$

$$\frac{dM_z}{dt} = -\gamma_f H_1 v - \frac{(M_z - M_0)}{T_1} \tag{45c}$$

Solutions to Eq. (45) are easily found, particularly in the steady-state approximation at low $\underline{H}_1$. In that event,

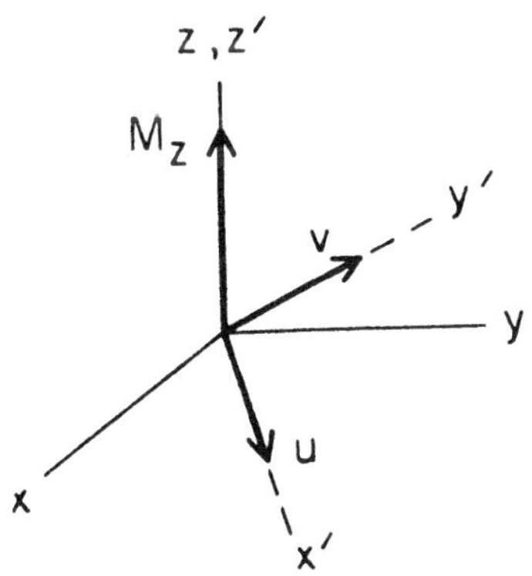

Fig. 4. Behavior of the components of $\underline{M}$ in the two coordinate systems.

$$u = \frac{\gamma_f H_1 T_2^2(\omega_0-\omega)}{1+T_2^2(\omega_0-\omega)^2+\gamma_f^2 H_1^2 T_1 T_2} M_0 \tag{46a}$$

$$v = \frac{\gamma_f H_1 T_2}{1+T_2^2(\omega_0-\omega)^2+\gamma_f^2 H_1^2 T_1 T_2} M_0 \tag{46b}$$

$$M_z = \frac{1+T_2^2(\omega_0-\omega)^2}{1+T_2^2(\omega_0-\omega)^2+\gamma_f^2 H_1^2 T_1 T_2} M_0 \tag{46c}$$

Equations (46) show that the magnetization $|\underline{M}|$ is a constant in the rotating reference frame, and therefore is rotating at frequency ω in the laboratory.

In a typical experimental arrangement, we observe the magnetization by studying the emf it induces in a fixed coil in the laboratory. If the coil is oriented along the x axis, we can calculate the emf from knowledge of the time-dependent component M_x. Since

$$M_x + iM_y = (u + iv)e^{-i\omega t} \tag{47}$$

we have

$$M_x = u \cos \omega t + v \sin \omega t \tag{48}$$

Bloch introduced a complex susceptibility $\chi(\omega)$ to describe these phase relations. The x components of the oscillating $\underline{H_1}$ field and the magnetization $\underline{M}$ are written as complex numbers $2H_1e^{-i\omega t}$ and $2H_1\chi(\omega)e^{-i\omega t}$, where the real and imaginary parts of χ are

$$\chi(\omega) = \chi'(\omega) + i\chi''(\omega) \tag{49}$$

The true magnetic moment is then the real part of this expression

$$M_x = H_1[\chi(\omega)e^{-i\omega t} + \chi^*(\omega)e^{i\omega t}] \tag{50}$$

$$= 2H_1\chi'(\omega) \cos \omega t + 2H_1\chi''(\omega) \sin \omega t$$

u and v are therefore proportional to $\chi'(\omega)$ and $\chi''(\omega)$, while a comparison between Eqs. (46), (48), and (50) shows that the spin susceptibilities are

$$\chi'(\omega) = \frac{1}{2}\chi_0\omega_0 \frac{T_2^2(\omega_0-\omega)}{1+T_2^2(\omega_0-\omega)^2 + \gamma_f^2 H_1^2 T_1 T_2} \tag{51a}$$

$$\chi''(\omega) = \frac{1}{2}\chi_0\omega_0 \frac{T_2}{1+T_2^2(\omega_0-\omega)^2 + \gamma_f^2 H_1^2 T_1 T_2} \tag{51b}$$

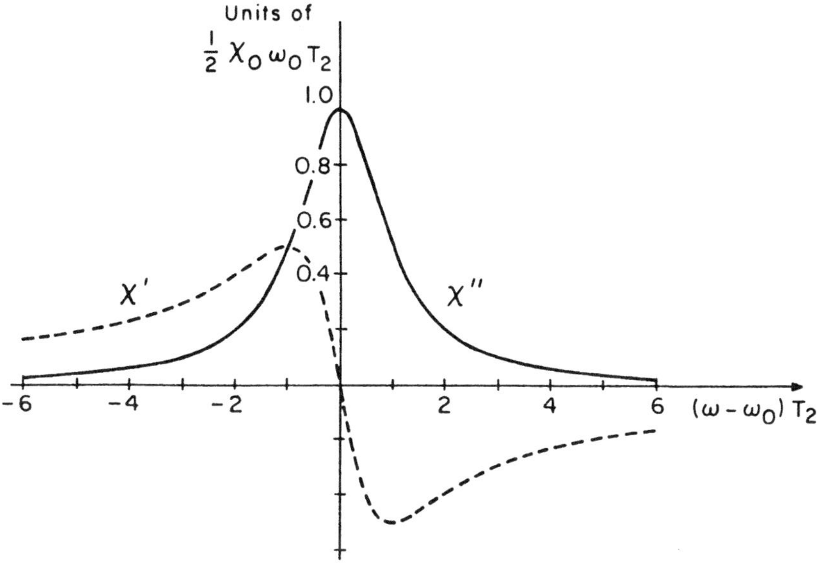

Fig. 5. Dispersion and absorption lineshapes derived from the Bloch equations (4).

Fig. 5 shows how these susceptibilities behave as ω passes through resonance. $\chi'(\omega)$ changes sign so that $\underline{M}$ is 180° out-of-phase with the perturbing field at high frequencies. On the other hand, χ'' becomes very large near resonance, and this causes a strong absorption of energy. The average power absorbed is

$$\frac{dE}{dt} = \frac{H_1{}^2\chi_0\omega\omega_0 T_2}{1+T_2{}^2(\omega_0-\omega)^2 + \gamma_f{}^2H_1{}^2T_1T_2} \tag{52}$$

When the resonance is unsaturated and $\gamma_f{}^2H_1{}^2T_1T_2$ is small, the absorption has the familiar Lorentzian shape

$$g(\omega) = \frac{T_2}{\pi}\;\frac{1}{1+T_2{}^2(\omega-\omega_0)^2} \tag{53}$$

with a FWHM of $(2/T_2)$. In strong H_1 fields, Eq. (52) has the form

$$\frac{dE}{dt} = n_0\Delta E\;\frac{P(\omega)}{1+2P(\omega)T_1} \tag{54}$$

where $P(\omega) = 1/2\pi\gamma_f{}^2H_1{}^2g(\omega)$. Thus, we see that the two relaxation times play quite different roles in the magnetic resonance experiment; T_1 determines the degree of saturation and T_2 determines the unsaturated linewidth. Both kinds of relaxation are caused by time-dependent magnetic or electric fields at the nucleus or the electron, and these fields in turn come from the random thermal motions which are present in any form of matter. A nuclear spin of 1/2, for example, may experience local magnetic fields from the spins of other nuclei, from unpaired electrons, or from spin-rotational interactions in which the molecular rotation itself generates magnetic fields at the nucleus. There may also be changes in the chemical shielding of the nucleus as a molecule rotates, and these modulate the total effective field which acts on the spin. Nuclei with electric quadrupole moments are further affected by changes in the local electrostatic field gradient when the molecule rotates or vibrates. Electron spins, too, may find themselves in a variety of fluctuating magnetic fields from other electrons or nuclei. To cite one example, radicals with an anistropic $\underline{g}$ tensor experience fluctuating Zeeman interactions with the external magnetic field when these radicals tumble in solution.

2. NUCLEAR MAGNETIC RESONANCE.

2.1. The Spin Hamiltonian.

An NMR spectrum measures the energy absorbed during the stimulated inversion of a nuclear spin in a magnetic field. It is therefore natural, as we turn to a detailed consideration of this experiment, to begin with a discussion of the interactions which influence the energies of the levels involved. These energies are the eigenvalues of a Hamiltonian involving spin operators, which we refer to as the spin Hamiltonian. A more subtle definition will become apparent as we proceed. It is important to note here, however, that operators which influence the energy but which do not involve the spin do not appear explicitly in this expression. This is tantamount to assuming that other degrees of freedom, such as the electronic, vibrational, and rotational motions of a polyatomic molecule, do not affect the magnetic resonance spectrum. In fact, they often do, but in such cases the consequences of these additional interactions are usually dealt with in the framework of perturbation theory. Thus, one often speaks of an effective spin Hamiltonian in magnetic resonance problems.

Let us write the spin Hamiltonian, $\hat{H}_I$, for a nucleus possessing spin angular momentum $\underline{I}$ in an external magnetic field as

$$\hat{H}_I = \hat{H}_{NZ} + \hat{H}_{DD} + \hat{H}_{HF} + \hat{H}_Q \qquad (55)$$

where

$$\hat{H}_{NZ} = -g_n\beta_n \underline{H} \cdot (\underset{\sim}{1} - \underset{\sim}{\sigma}) \cdot \underline{I} \qquad (56)$$

$$\hat{H}_{DD} = g_n^2 \beta_n^2 \underline{I}_1 \cdot \underset{\sim}{D} \cdot \underline{I}_2 \tag{57}$$

$$\hat{H}_{HF} = \underline{I} \cdot \underset{\sim}{T} \cdot \underline{S} \tag{58}$$

$$\hat{H}_{Q} = \underline{I} \cdot \underset{\sim}{Q} \cdot \underline{I} \tag{59}$$

Here, $\hat{H}_{NZ}$ is a term representing the Zeeman interaction of the nuclear magnetic moment with the external magnetic field, $\hat{H}_{DD}$ is a term representing the sum of scalar and dipolar interactions between the nucleus being studied (I_1) and a second nucleus (I_2) in its vicinity, $\hat{H}_{HF}$ is a term representing the sum of scalar and dipolar interactions between the nucleus and unpaired electrons in its vicinity, and $\hat{H}_Q$ is a term representing the quadrupole coupling, another type of electron-nucleus interaction. We shall have much more to say about each of these terms shortly. But, for the moment, we note only that all are written in their most general form, in tensor notation. We shall be concerned here only with second-rank tensors, which are linear vector operators used to represent the scalar product of two vectors. The simplest example of such a tensor is $\underset{\sim}{1}$, the unit tensor. It may be defined as the operator which, acting on any vector, yields that vector itself. Thus,

$$\underset{\sim}{1} \cdot \underline{A} = \underline{i}A_x + \underline{j}A_y + \underline{k}A_z = \underline{A} \tag{60}$$

More generally, a second-rank tensor has 9 components which can be arranged to form a 3 x 3 matrix. For example,

$$\underset{\sim}{\sigma} = \begin{pmatrix} \sigma_{xx} & \sigma_{xy} & \sigma_{xz} \\ \sigma_{yx} & \sigma_{yy} & \sigma_{yz} \\ \sigma_{zx} & \sigma_{zy} & \sigma_{zz} \end{pmatrix} \tag{61}$$

By multiplying the matrix of $\underset{\sim}{\sigma}$ on the left with a row vector, say $\underline{H}$, one obtains a new row vector $\underline{H}\cdot\underset{\sim}{\sigma}$. Similarly, matrix multiplication on the right by a column vector, say $\underline{I}$, yields a new column vector $\underset{\sim}{\sigma}\cdot\underline{I}$. Finally, it is possible to form a scalar $\underline{H}\cdot\underset{\sim}{\sigma}\cdot\underline{I}$ by the matrix multiplication

$$(H_x, H_y, H_z) \begin{pmatrix} \sigma_{xx} & \sigma_{xy} & \sigma_{xz} \\ \sigma_{yx} & \sigma_{yy} & \sigma_{yz} \\ \sigma_{zx} & \sigma_{zy} & \sigma_{zz} \end{pmatrix} \begin{pmatrix} I_x \\ I_y \\ I_z \end{pmatrix} \tag{62}$$

Note that $\underline{H}\cdot\underset{\sim}{\sigma}\cdot\underline{I}$ may also be regarded as the scalar product of two vectors; either $(\underline{H}\cdot\underset{\sim}{\sigma})$ with $\underline{I}$, or equally $\underline{H}$ with $(\underset{\sim}{\sigma}\cdot\underline{I})$. In any case, the product written in full is

$$H_x\sigma_{xx}I_x + H_x\sigma_{xy}I_y + \cdots + H_y\sigma_{yz}I_z + H_z\sigma_{zz}I_z = \sum_{i,j} H_i\sigma_{ij}I_j \quad (i,j = x,y,z) \tag{63}$$

where $H_x, H_y, \ldots, I_y, I_z$ are the components of $\underline{H}$ and $\underline{I}$ defined in terms of some coordinate system. Thus, the components of the tensor $\underset{\sim}{\sigma}$ must also be defined in that same coordinate system. Most tensors with which we will deal are symmetric, i.e., $\sigma_{xy} = \sigma_{yx}$, etc. In that event, they can be diagonalized by means of an appropriate matrix transformation

$$\underline{L} \quad \underset{\sim}{\sigma}\, \underline{L}^{-1} = \underset{\sim}{\sigma} \text{ ("diagonal")} \tag{64}$$

The transformation corresponds to a reorientation of the coordinate system in which $\underset{\sim}{\sigma}$ is defined, and the transformation matrix defines the orientation of the new "principal" axes with respect to the old. If such a coordinate system can be found, then the interaction term itself is also "diagonal", e.g.

$$\hat{H}_{NZ} \text{ ("diagonal")} = g_n\beta_n[H_x(1-\sigma_{xx})I_x + H_y(1-\sigma_{yy})I_y + H_z(1-\sigma_{zz})I_z] \tag{65}$$

The appearance of diagonal matrices in quantum mechanics can be connected in a simple way with the solution of the Schrödinger equation (5). Thus, in the interpretation of a magnetic resonance spectrum, we seek to determine the principal values and the principal axes of each of the relevant second-rank tensors in the problem. If successful, the energy eigenvalues and eigenfunctions are obtained in the process, and these may be used to develop a microscopic picture of the species under study.

2.2. The Chemical Shift.

Let us try to bring some of these ideas into a little better focus by considering a specific example, the helium atom in its electronic ground state, $(1s)^2$. The pair of electron spins are coupled into a singlet electronic state, so only the nucleus need be considered. The normal ^{4}He nucleus has no spin but there is a rare isotope ^{3}He which has $I = 1/2$. The spin has two states $|\alpha_n\rangle$ and $|\beta_n\rangle$ which split in a magnetic field; thus, according to Eq. (10), we can expect a single line in the NMR spectrum at the frequency $g_n\beta_nH/h$ $(= \gamma_nH/2\pi)$. Experimentally, it is found that the observed frequency in the He atom differs slightly from the value for the bare nucleus. This is called the screening effect or the "chemical shift" and it arises from the magnetic effect of induced electronic currents set up in the atom by the external magnetic field. The currents produce an additional, internal, field at the nucleus which opposes $\underline{H}$ and has a strength proportional to $\underline{H}$. To account for this, we write the nuclear Zeeman Hamiltonian as

$$\hat{H}_{NZ} = -g_n\beta_n\,(1-\sigma)\,HI_z \tag{66}$$

where σ is a small number of order 10^{-6}, called the screening constant. Magnetic resonance transitions in the atom at a fixed frequency ν_n therefore appear at a higher field than they do for the free nucleus.

The theory of the chemical shift in a closed shell atom is very simple. In an external magnetic field, the entire spherical electron cloud precesses about the field with angular velocity $\underline{\omega} = e\underline{H}/2mc$, as shown in Fig. 6. Since an electron at a distance $\underline{r}$ from the nucleus moves with velocity $\underline{v} = (\underline{\omega} \times \underline{r})$, it produces a secondary magnetic field at the nucleus which is given by

$$\underline{H}_\ell = -\frac{e}{c}\left(\frac{\underline{r} \times \underline{v}}{r^3}\right) = \frac{-e^2}{2mc^2}\left(\frac{\underline{r} \times (\underline{H} \times \underline{r})}{r^3}\right) \tag{67}$$

On the average, the electrons are distributed in space with a probability density $\rho(\underline{r})$; the resulting secondary field is

$$\underline{H}_\ell = -\frac{e^2}{2mc^2}\left(\frac{\underline{r} \times (\underline{H} \times \underline{r})}{r^3}\right)\rho(r)\,dr \tag{68}$$

The x and y components of $\underline{H}_\ell$ vanish in the spherically symmetric He atom, leaving a screening field $-\sigma H$ directed along the z axis.

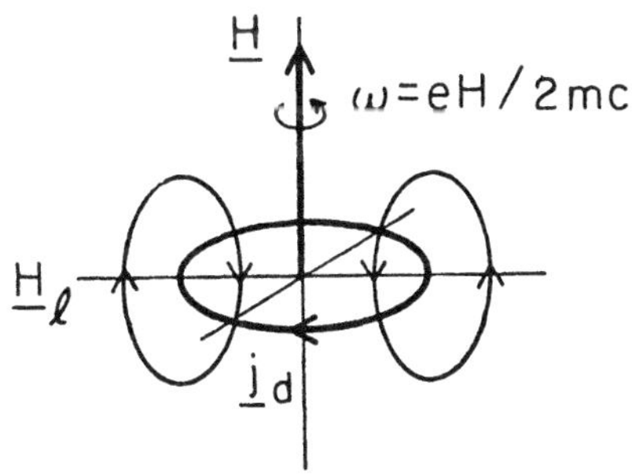

Fig. 6. Diamagnetic current flow in a spherically symmetric atom, and the magnetic fields produced by this current (3).

We evaluate the vector product in Eq. (68) and obtain

$$\sigma = \frac{e^2}{2mc^2}\int \frac{x^2+y^2}{r^3}\,\rho(\underline{r})dr = \frac{e^2}{3mc^2}\int \frac{\rho(\underline{r})}{r}\,dr = \frac{e^2}{3mc^2}\left\langle\frac{1}{r}\right\rangle \quad (69)$$

This important relation is called the Lamb formula, and we see that σ depends on the mean value of $1/r$ for all the electrons. The calculated screening constant for He is about 60×10^{-6}, but for larger atoms the shifts are larger.

The situation is both more complicated and more interesting in molecules (6). As in the atomic case, chemical shifts arise because of the simultaneous interaction of the electrons with the applied field and with the nucleus. Thus, we can expect contributions to the chemical shift similar to those observed in the He atom; these are called diamagnetic contributions because the secondary field opposes the static field. But in molecules there are also paramagnetic shielding terms. These have their origin in circulating electronic currents, as before, but these currents involve p electrons, or other electrons in non-spherically symmetric orbitals. A simple example will serve to illustrate this effect. Let us consider an atom with one electron outside closed shells in a p state. If the atom is free, the three p wavefunctions are degenerate and may be written as

$$p_{+1} = \frac{1}{\sqrt{2}}(x + iy)\,f(r) \quad (70a)$$

$$p_0 = z\,f(r) \quad (70b)$$

$$p_{-1} = \frac{1}{\sqrt{2}}(x - iy)\,f(r) \quad (70c)$$

where $f(r)$ is a spherically symmetric function. The states $m = +1$ and $m = -1$ correspond to electron circulation in opposite directions about the z axis. Thus, if the two states remain degenerate, there is zero net circulation of charge, and no net current. Of course, the application of a magnetic field will change things. We can see intuitively that a magnetic field in the z direction will cause one sense of circulation to be favored over the other, and that the magnitude of this effect will be proportional to the field, as before. But the state which is stabilized by the field is the $m = -1$ state; this corresponds to current flowing in the counterclockwise direction, as shown in Fig. 7. Thus, the secondary field is pointing in the same direction as the applied field, and is paramagnetic.

Now, in real molecules the degeneracy of the three p orbitals is lifted at least partially by the formation of chemical bonds. In that event, the orbital angular momentun which is the source of the electronic currents in an atom is quenched, and the appropriate p wavefunctions are the real wavefunctions $xf(r)$,

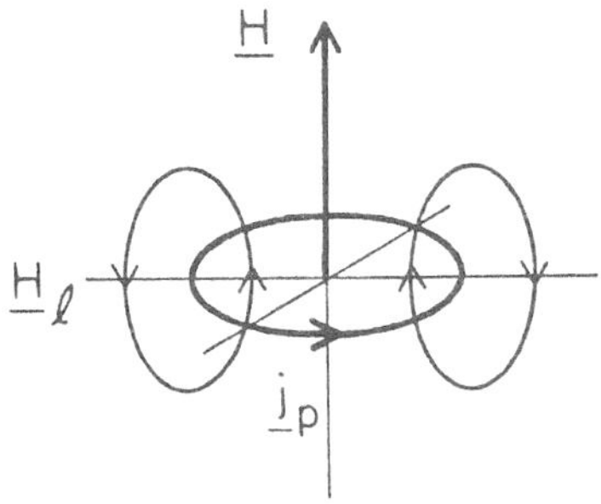

Fig. 7. Paramagnetic current flow in a p state, and the magnetic fields produced by this current (3).

yf(r), and zf(r), where

$$xf(r) = \frac{1}{\sqrt{2}} (p_{+1} + p_{-1}) \qquad (71a)$$

$$yf(r) = \frac{1}{\sqrt{2}} (p_{+1} - p_{-1}) \qquad (71b)$$

Suppose the ground state is xf(r), and the two excited states lie at an energy Δ above the ground state, as shown in Fig. 8. The resulting p electron distribution is cylindrically symmetric about x. If the field is applied along this axis, no current will flow, but if the field is along either y or z, electrons can circulate from xf(r) to yf(r) or from xf(r) to zf(r). This change results from the admixture of a small amount of the state yf(r) [or zf(r)] into the ground state xf(r); the amount which is mixed (and,

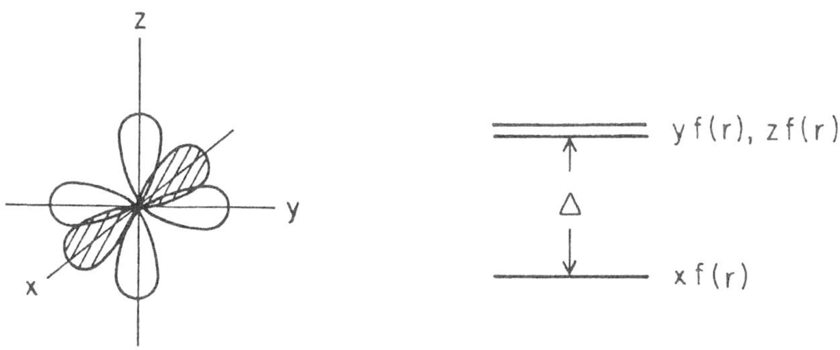

Fig. 8. "Crystal-field" energy level diagram for the three 2p orbitals showing the effect of bond formation along the x axis on their relative energies.

therefore, the current) is proportional to H/Δ. Furthermore, it is clear that the magnitude of the paramagnetic contribution to the chemical shift will depend, in general, on the orientation of the field relative to the molecule, and this is anisotropic. This is why the chemical shift term in the nuclear Zeeman Hamiltonian [Eq. (56)] is written in tensor notation.

A general theory of the chemical shift has been given by Ramsey and is discussed in the standard texts (2-6). Classically, the field produced at the nucleus by an electron at a position $\underline{r}$ with velocity $\underline{v}$ can be written as

$$\underline{H}_{\ell} = -\left(\frac{e\hbar}{2mc}\right)\frac{2\underline{L}}{r^3} - \left(\frac{e^2}{2mc^2}\right)\frac{\underline{r} \times (\underline{H} \times \underline{r})}{r^3} \qquad (72)$$

which reduces to Eq. (67) if $\langle \underline{L} \rangle = 0$. The quantum mechanical analog of Eq. (72) is, with $\underline{H}$ parallel to z,

$$H_{z\ell} = -\sigma_{zz}H = -\frac{e\hbar}{2mc}\left\langle\frac{2L_z}{r^3}\right\rangle - \left(\frac{e^2}{2mc^2}\right)H\left\langle\frac{x^2+y^2}{r^3}\right\rangle \qquad (73)$$

Substitution of the (perturbed) wavefunction yields, to first order in H,

$$\sigma_{zz} = \frac{e^2}{2mc^2}\langle 0|\frac{x^2+y^2}{r^3}|0\rangle - \left(\frac{e\hbar}{2mc}\right)^2 \sum_n \left\{\frac{\langle 0|L_z|n\rangle\langle n|\frac{2L_z}{r^3}|0\rangle}{E_n-E_0} + \frac{\langle 0|\frac{2L_z}{r^3}|n\rangle\langle n|L_z|0\rangle}{E_n-E_0}\right\} \qquad (74)$$

Eq. (74) is called Ramsey's shielding formula and applies to any closed shell molecule. The other components, σ_{xx} and σ_{yy}, are given by similar equations, and one may calculate from these three quantities the average screening constant σ, the trace of $\underline{\sigma}$. This is the parameter measured in the liquid or gas phase.

The diamagnetic part of σ, σ_d, is fairly easy to estimate theoretically because, as we have seen, it depends only on the electronic distribution in the ground state. However, the paramagnetic contribution, σ_p, also depends on the excited states. Thus to evaluate this term precisely, it is necessary to have good wavefunctions and energies of all the excited electronic states, including those in the continuum. Alternatively, one can invoke the average energy approximation; i.e., replace the sum over all of the excitation energies E_n-E_0 by an average value ΔE. If this is done, one finds

$$\sigma_p = -\frac{2}{3}\left(\frac{e\hbar}{mc}\right)^2 \frac{1}{\Delta}\left\langle\frac{1}{r^3}\right\rangle \qquad (75)$$

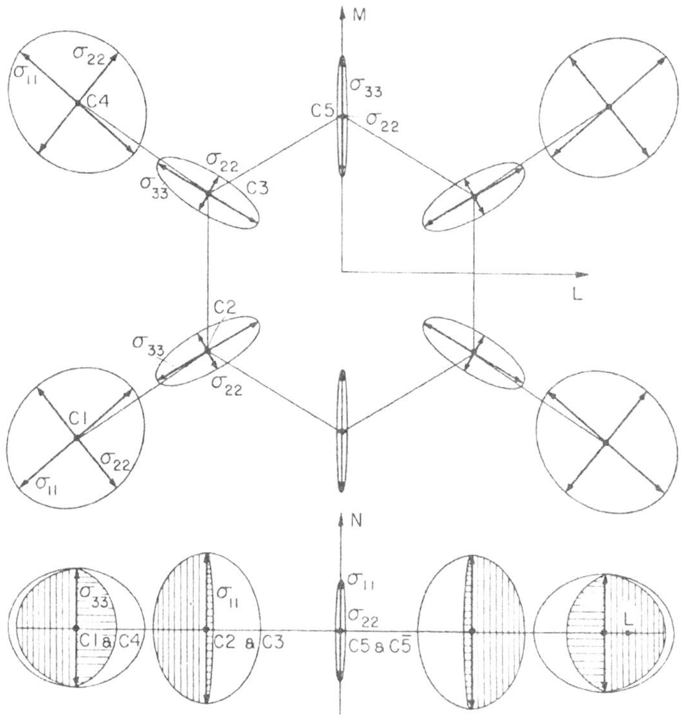

Fig. 9. Representation of the ^{13}C shielding tensors of durene. The shielding ellipsoids, shown in two projections, are such that the diameter in any direction is linear in the chemical shift for the applied field H in that direction (7).

Taking Δ = 4.3 eV and $\langle 1/r^3 \rangle = 8.89/a_0^3$, which are values appropriate to the 2p electrons of fluorine, one finds $\sigma_p = -20 \times 10^{-4}$, a value much larger than σ_d. Thus, paramagnetic contributions to σ, when present, usually dominate the diamagnetic ones. From this it is also clear why the range of fluorine (and other second-row element) chemical shifts is much larger than that of protons.

To be complete, it should also be noted that other nonlocal contributions to σ may be important as well. These may be either diamagnetic or paramagnetic depending on the nature of the atom or group of atoms which are the source of these local fields.

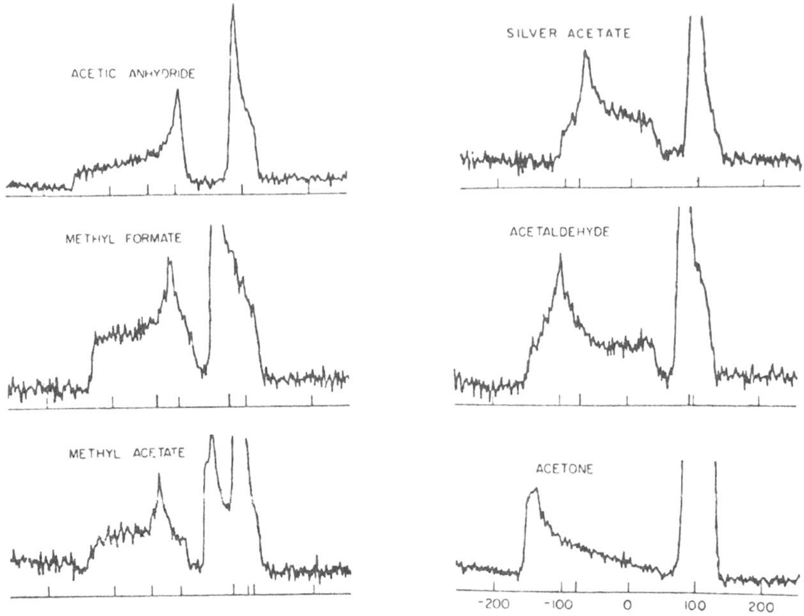

Fig. 10. Proton-enhanced ^{13}C spectra of polycrystalline compounds containing carbonyl groups at ~ 77K. The low-field peaks are from the carbonyl and the high-field ones from methyl groups. Note the sensitivity of σ_{22} to substituents. The horizontal axis is calibrated in ppm relative to an external reference of liquid benzene at 296K(8).

The best known example is the interatomic (or ring) current in aromatic molecules like benzene. We shall not dwell on these at length here. But, before closing this section, two examples of the work of Waugh and collaborators will be cited which demonstrate quite beautifully the anisotropic nature of the chemical shift. The first is a series of experiments on durene (1,2,4,5-tetramethylbenzene) (7). Here, proton-enhanced ^{13}C NMR techniques were employed on single crystal samples to obtain both the principal values and axes of the chemical shift tensors of each of the carbons in the molecule. Figure 9, taken from Ref. 7, summarizes the results obtained, which speak (elegantly) for themselves. The second example shows that at least the principal values are also obtainable from powder samples (8), as shown in Fig. 10. Clearly, much more information about molecular structure is obtained by studying the anisotropy of the shift, rather than its isotropic value. The type of spectra shown here may also be very useful for studying molecular motions.

2.3. Dipolar Coupling.

Normally, the kind of spectral detail revealed in the above experiments on solid samples is obscured by the contribution of the magnetic dipole coupling between the various nuclei to the width of the resonance line. A rough estimate of the magnitude of this effect is easily made. If typical neighboring nuclei are a distance r apart and have magnetic moment μ_n, they produce a magnetic field H_ℓ of the order $H_\ell = \mu_n/r^3$. With r = 2Å and $\mu_n = 10^{-23}$ erg/G, we find $H_\ell \cong 1$ G. Since this field may either aid or oppose the static field H, a spread in the resonance condition results, with significant absorption occurring over a range of H ~ 1 G. For typical laboratory fields of 10^4 G, this corresponds to a spread in frequency of 100 ppm, exceeding the anisotropy of many chemical shifts. So, on the one hand, clever experimental techniques must be used to extract this information from the typical solid state sample. On the other hand, it is also clear that the effect is not so large that one cannot extract useful information from a dipolar broadened line.

What kind of information can one obtain? To answer this question, let us consider the classical interaction energy E between two identical protons, with magnetic moments $\underline{\mu}_1$ and $\underline{\mu}_2$, which are separated by a distance $\underline{r}$ (Fig. 11):

$$E = \frac{\underline{\mu}_1 \cdot \underline{\mu}_2}{r^3} - \frac{3(\underline{\mu}_1 \cdot \underline{r})(\underline{\mu}_2 \cdot \underline{r})}{r^5} \qquad (76)$$

The quantum mechanical equivalent of this expression may be obtained by replacing $\underline{\mu}_1$ and $\underline{\mu}_2$ by their operator equivalents, yielding ($\underline{\mu}_n = g_n \beta_n \underline{I}$)

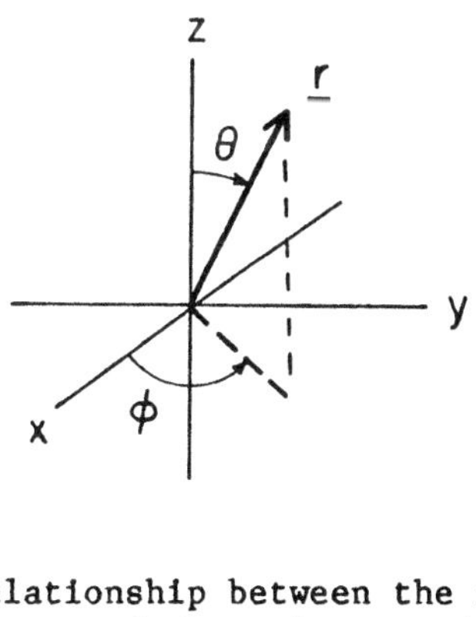

Fig. 11. Relationship between the rectangular coordinates x,y,z and the polar coordinates r,θ,φ.

$$\hat{H}_{DD} = \frac{g_n{}^2\beta_n{}^2}{r^5}\,[r^2(\underline{I}_1\cdot\underline{I}_2) - 3(\underline{I}_1\cdot\underline{r})(\underline{I}_2\cdot\underline{r})] \tag{77}$$

which is the dipolar, or fine-structure, Hamiltonian. Expanding each of the scalar products yields

$$\underline{I}_1\cdot\underline{I}_2 = I_{1x}I_{2x} + I_{1y}I_{2y} + I_{1z}I_{2z} \tag{78a}$$

$$\underline{I}_1\cdot\underline{r} = I_{1x}x + I_{1y}y + I_{1z}z \tag{78b}$$

$$\underline{I}_2\cdot\underline{r} = I_{2x}x + I_{2y}y + I_{2z}z \tag{78c}$$

Thus, if we insert Eq. (78) into Eq. (77), use the fact that $r = (x^2+y^2+z)^{1/2}$, and collect terms, we obtain

$$\begin{aligned}\hat{H}_{DD} = \frac{g_n{}^2\beta_n{}^2}{r^5}\,[&I_{1x}I_{2x}\,(r^2-3x^2) + I_{1y}I_{2y}\,(r^2-3y^2) \\ &+ I_{1z}I_{2z}\,(r^2-3z^2) - (I_{1x}I_{2y} + I_{1y}I_{2x})\,3xy \\ &-(I_{1y}I_{2z} + I_{1z}I_{2y})\,3yz - (I_{1z}I_{2x} + I_{1x}I_{2z})\,3xz]\end{aligned} \tag{79}$$

Clearly, this result can be written in tensor form

$$\hat{H}_{DD} = g_n{}^2\beta_n{}^2\,\underline{I}_1\cdot\underset{\sim}{D}\cdot\underline{I}_2 \tag{80}$$

[cf., Eq. (57)], where

$$\underset{\sim}{D} = \left(\frac{1}{r^5}\right)\begin{pmatrix} r^2-3x^2 & -3xy & -3xz \\ -3xy & r^2-3y^2 & -3yz \\ -3xz & -3yz & r^2-3z^2 \end{pmatrix} \tag{81}$$

Since this tensor is real and symmetric, like $\underset{\sim}{g}$, it can be diagonalized. The transformation required is one which makes xy = yz = xz = 0; i.e., one which brings $\underline{r}$ to a position where it points along one of the coordinate axes, say z'. In that event, x' = y' = 0, z' = $|\underline{r}|$ and the $\underset{\sim}{D}'$ tensor is cylindrically symmetric, viz.

$$D_{x'x'} = \left\langle \frac{(r')^2-3(x')^2}{(r')^5} \right\rangle = \left\langle \frac{1}{(r')^3} \right\rangle \tag{82a}$$

$$D_{y'y'} = \left\langle \frac{(r')^2-3(y')^2}{(r')^5} \right\rangle = \left\langle \frac{1}{(r')^3} \right\rangle \tag{82b}$$

$$D_{z'z'} = \left\langle \frac{(r')^2-3(z')^2}{(r')^5} \right\rangle = \left\langle \frac{-2}{(r')^3} \right\rangle \tag{82c}$$

Thus, one can determine the expectation value $\langle 1/r^3 \rangle$ by measuring the principal values of $\underset{\sim}{D}'$. Further, since the trace of $\underset{\sim}{D}'$ (and $\underset{\sim}{D}$) is zero, we also see that the dipole-dipole interaction can be made to vanish when the field spends, on the average, an equal time along each of the principal axes of $\underset{\sim}{D}'$, as in the case of a rapidly rotating molecule in solution or in certain multiple pulse schemes.

A very nice example of some of these effects can be found in the pioneering work of Pake (9) on a single crystal of gypsum, $CaSO_4 \cdot 2H_2O$. Gypsum contains no paramagnetic ions and is a relatively simple hydrate in that the only nuclei with non-vanishing spin, and therefore with a magnetic moment, are the protons in the water molecules. Large single crystals exist in nature and the structure is known from X-ray crystallography. Each unit cell contains eight water molecules but an examination of the structure (together with some reasonable assumptions) shows that there are only two permitted proton-proton (p-p) directions in the crystal, four of the eight water molecules in the unit cell belonging to each direction. Thus, if each nucleus of a given pair finds itself predominantly in the local field of its partner, one expects the proton NMR line to be split into a doublet since the local field may either add to or subtract from the static field. This is precisely what was found in Pake's experiment; measurements (at ~30 MHz) on a single crystal of gypsum showed a splitting of the line into four components corresponding to the two possible p-p directions in the unit cell. Moreover, as shown in Fig. 12, it was also found that the maximum separation of each pair of lines varied from 0-22 G as the orientation of the crystal relative to the magnetic field was changed.

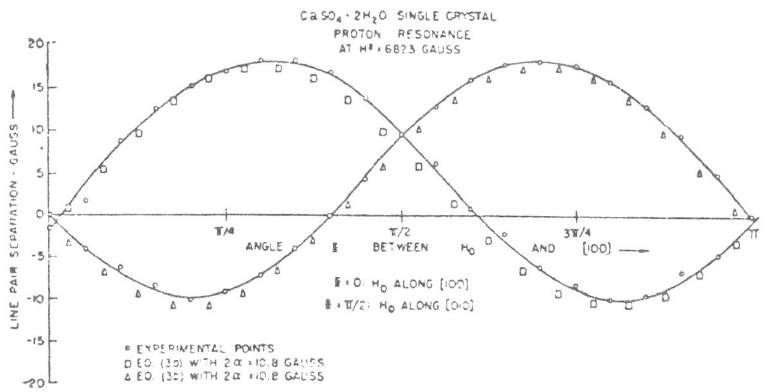

Fig. 12. Line pair separation as a function of the angle Φ between H_0 and [100]. Because two directions exist for proton-proton lines in a gypsum single crystal, there are two similar curves differing in phase (9).

A simple quantitative treatment of these results can be given. If we express the operators I_{1x}, I_{1y}, etc. in terms of the raising and lowering operators

$$I^+ = I_x + iI_y \tag{83}$$

$$I^- = I_x - iI_y \tag{84}$$

and the rectangular coordinates x,y,z in terms of the spherical polar coordinates r,θ,ϕ, then the dipolar Hamiltonian [Eq. (79)] may be written as

$$\hat{H}_{DD} = \frac{g_n{}^2\beta_n{}^2}{r^3}[A + B + C + D + E + F] \tag{85}$$

where

$$A = (1 - 3\cos^2\theta)I_{1z}I_{2z} \tag{86a}$$

$$B = -\frac{1}{4}(1-3\cos^2\theta)(I_1{}^+I_2{}^- + I_1{}^-I_2{}^+) \tag{86b}$$

$$C = -\frac{3}{2}\sin\theta\cos\theta\, e^{-i\phi}(I_{1z}I_2{}^+ + I_1{}^+I_{2z}) \tag{86c}$$

$$D = -\frac{3}{2}\sin\theta\cos\theta\, e^{i\phi}(I_{1z}I_2{}^- + I_1{}^-I_{2z}) \tag{86d}$$

$$E = -\frac{3}{4}\sin^2\theta\, e^{-2i\phi}\, I_1{}^+I_2{}^+ \tag{86e}$$

$$F = -\frac{3}{4}\sin^2\theta\, e^{2i\phi}\, I_1{}^-I_2{}^- \tag{86f}$$

This form is particularly convenient for computing matrix elements of the total Hamiltonian,

$$\hat{H}_I = \hat{H}_{NZ} + \hat{H}_{DD} \tag{87}$$

Suppose we assume the basis set

$$|t_1\rangle = |\alpha_1\alpha_2\rangle \tag{88a}$$

$$|t_0\rangle = \frac{1}{\sqrt{2}}|\alpha_1\beta_2 + \beta_1\alpha_2\rangle \tag{88b}$$

$$|t_{-1}\rangle = |\beta_1\beta_2\rangle \tag{88c}$$

$$|s\rangle = \frac{1}{\sqrt{2}}|\alpha_1\beta_2 - \beta_1\alpha_2\rangle \tag{88d}$$

In that event, it is clear that only the A and B terms of $\hat{H}_{DD}$ will contribute in first order; the other four terms (C-F) have only off-diagonal elements in this representation and can be (initially) neglected. The effective spin Hamiltonian is, then (setting $g = 0$)

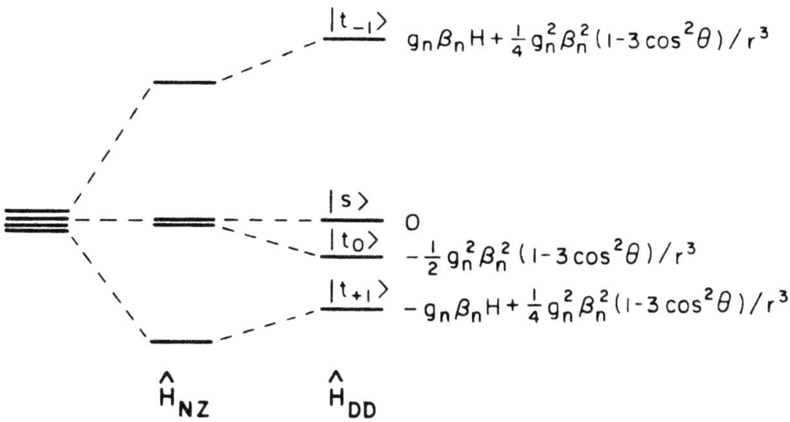

Fig. 13. Energy level diagram for the two-spin system $I_1 = I_2 = 1/2$, showing the effect of the perturbing dipole-dipole interaction.

$$\hat{H}_I = -g_n\beta_n H(I_{1z}+I_{2z}) + g_n^2\beta_n^2\left(\frac{1-3\cos^2\theta}{r^3}\right)\left[I_{1z}I_{2z} - \frac{1}{4}(I_1^+I_2^- + I_1^-I_2^+)\right] \quad (89)$$

This gives the energy level diagram shown in Fig. 13. The allowed transitions are $|t_{\pm1}\rangle \longleftrightarrow |t_0\rangle$ with energies

$$h\nu = g_n\beta_n H + \frac{3}{4} g_n^2\beta_n^2(1-3\cos^2\theta)/r^3 \quad (|t_0\rangle \rightarrow |t_{-1}\rangle) \quad (90a)$$

$$h\nu = g_n\beta_n H - \frac{3}{4} g_n^2\beta_n^2(1-3\cos^2\theta)/r^3 \quad (|t_{+1}\rangle \rightarrow |t_0\rangle) \quad (90b)$$

If we work at a fixed frequency ν for which the resonance of a single proton would occur at the field H^*, then the observed spectrum of a single pair of protons will consist of two lines at the field values

$$H = H^* \pm \frac{3}{4} g_n^2\beta_n^2 (1-3\cos^2\theta)/r^3 \quad (91)$$

The separation between the lines is, therefore, $\frac{3}{2}g_n^2\beta_n^2(1-3\cos^2\theta)/r^3$ gauss where θ, we recall, is the angle between the p-p vector and the magnetic field. From studies of the angular variation of the splitting it is thus possible to measure $1/r^3$ and determine the magnitude and direction of $\underline{r}$. In gypsum, $\frac{3}{2}g_n^2\beta_n^2/r^2 = 10.8$ G, giving a distance of 1.58 Å between protons in each water molecule. This distance can also be determined from the NMR spectrum of the powder if the dipolar splitting can be resolved.

Some additional remarks are in order here. Note from Fig. 12 that there are some orientations at which the dipolar splitting vanishes. Clearly, this is when $3\cos^2\theta = 1$, or $\theta = 54° 44'$. This angle is known as the "magic angle"; rapid rotation of any sample about this angle significantly narrows a dipolar-broadened resonance line, as was first demonstrated many years ago by Lowe and by Andrew. Complete narrowing by this technique is, of course, not possible because of the contribution of the heretofore neglected C-F terms in Eq. (85).

These terms are also an important source of spin relaxation in dipolar-coupled systems. Suppose we consider, for example, a water molecule in solution. In that event, the dipolar Hamiltonian is made time dependent by the tumbling motion. As A and B have diagonal matrix elements, they modulate the energy levels and contribute to T_2. But C,D,E, and F have off-diagonal elements between different energy levels; hence, they contribute to T_1. The corresponding relaxation probabilities, W_1 (for the one-photon process) and W_2 (for the two-photon process), are respectively proportional to the mean square fluctuations of the matrix element $\langle t_0|\hat{H}_{DD}|t_1\rangle$ at frequency ω_0 and $\langle t_{-1}|\hat{H}_{DD}|t_1\rangle$ at $2\omega_0$. Time-dependent perturbation theory gives (4)

$$W_1 = \frac{1}{\hbar^2} |\langle t_0|\hat{H}_{DD}|t_1\rangle|^2 \frac{2\tau_C}{1+\omega_0^2\tau_C^2} \tag{92a}$$

$$W_2 = \frac{1}{\hbar^2} |\langle t_{-1}|\hat{H}_{DD}|t_1\rangle|^2 \frac{2\tau_C}{1+4\omega_0^2\tau_C^2} \tag{92b}$$

where τ_C is the rotational correlation time. T_1 is, in turn, given by $T_1^{-1} = (W_1 + 2W_2)$; a straightforward calculation of the mean square values of the matrix elements gives (4)

$$\frac{1}{T_1} = \frac{3}{20} \left(\frac{g_n^4\beta_n^4}{\hbar^2 r^6}\right) \left\{\frac{2\tau_C}{1+\omega_0^2\tau_C^2} + \frac{8\tau_C}{1+4\omega_0^2\tau_C^2}\right\} \tag{93}$$

A similar calculation of T_2, somewhat more tedious, yields

$$\frac{1}{T_2} = \frac{3}{40} \left(\frac{g_n^4\beta_n^4}{\hbar^2 r^6}\right) \left\{6\tau_C + \frac{10\tau_C}{1+\omega_0^2\tau_C^2} + \frac{4\tau_C}{1+4\omega_0^2\tau_C^2}\right\} \tag{94}$$

A typical proton resonance experiment at 60 MHz has $\omega_0 = 1.9 \times 10^8$, whereas the correlation time for water at room temperature is 2.7×10^{-12} sec. Thus, $\omega_0 t_C \ll 1$, and we predict $T_1 = T_2$, with

$$\frac{1}{T_1} = \frac{3}{2} \left(\frac{g_n^4\beta_n^4}{\hbar^2 r^6}\right) \tau_C \tag{95}$$

Taking $r = 1.58$ Å and the above value of τ_C, we estimate $T_1 = 6.7$ sec. The observed relaxation times are indeed equal, and in reasonable agreement with this numerical value if a correction is made for diffusion. Of course, at lower temperatures (or higher viscosities), the approximation $\omega_0\tau_C << 1$ is no longer valid, and the values of T_1 and T_2 are no longer equal. This was first shown in the classic work of Bloembergen, et al. (10) about which much more will be said at this Conference.

Only in relatively rare cases does the crystal structure permit complete resolution of the dipolar splitting. One usually observes, even in single crystal samples, a very broad line with little or no resolved structure owing to the large number of intermolecular interactions with both near and distant nuclei. Despite these difficulties, valuable structural information can sometimes be obtained by other methods. One approach utilizes liquid crystals or other two-dimensional structures to orient guest molecules in relatively unique ways (11). Another, often employed, involves the measurement of the moments of the absorption line. Suppose, for example, a spectrum is obtained by varying the magnetic field at a fixed frequency. The effect of the dipolar coupling on the average field of absorption H_{av} is to produce a broadened line of Gaussian shape

$$f(H) = \frac{1}{\Delta H\sqrt{2\pi}} \exp\{-[(H-H_{av})^2/2(\Delta H)^2]\} \tag{96}$$

The second moment of the line is the mean square width $(\Delta H)^2$ measured from the center of the line

$$(\Delta H)^2 = \int_0^\infty (H-H_{av})^2 f(H)dH \tag{97}$$

which can be determined graphically. This quantity can also be calculated theoretically, since ΔH is "simply" the root mean square of the local magnetic fields produced at the nucleus N by all other nuclei in the sample. The calculation is, however, far from simple, since we must average each term of the dipolar Hamiltonian over all the possible spin states of the other nuclei, including spins which are not in resonance. Nonetheless, van Vleck has given a closed form expression for this quantity; it is

$$\begin{aligned}(\Delta H)^2 = {} & \frac{3}{5} g_n^2\beta_n^2 \left(\frac{1}{n}\right) I(I+1) \sum_{j,k} \frac{1}{r_{jk}^6} \\ & + \frac{4}{15} g_n'^2\beta_n^2 \left(\frac{1}{n'}\right) I'(I'+1) \sum_{j,j'} \frac{1}{r_{jj'}^6}\end{aligned} \tag{98}$$

where r_{jk} ($r_{jj'}$) is the distance between two nuclei j and k (j and j'), n and n' are the number of equivalent resonant and nonresonant nuclei, and I and I' are their corresponding spin quantum numbers. The sums run over each nucleus j in the unit cell and all its neighbors k (j') within the crystal. This expression, which is derived in Slichter (3), has found application in a number of problems. A particularly interesting example is provided by the work of Andrew and Eades (12) on solid benzene. By studying both protonated and deuterated samples, they were able to separate the second moment into its intra and intermolecular contributions, measure the characteristic proton-proton distances in the benzene molecule, and show that at temperatures above 90K, the molecule is relatively free to reorient about its six-fold axis in a solid, polycrystalline, sample.

2.4. The Effect of Unpaired Electrons.

We have already seen that an NMR spectrum can be influenced by the magnetic coupling of electrons to the nucleus. In diamagnetic materials, this coupling arises from magnetic fields generated by the motion of electrical charges, resulting in chemical shifts. But, in paramagnetic materials, there are also fields which originate from the magnetic moment associated with electron spin. These moments are much larger than those associated with nuclear spin. Thus, the presence of unpaired electrons can have profound effects on the appearance of an NMR spectrum.

Let us consider the simple example of a hydrogen atom in a magnetic field to illustrate some of these effects. The total spin Hamiltonian may be written as a sum of three terms,

$$\hat{H}_{S,I} = \hat{H}_{EZ} + \hat{H}_{NZ} + \hat{H}_{HF} \tag{99}$$

where $\hat{H}_{EZ}$ is the electron Zeeman coupling, $\hat{H}_{NZ}$ is the nuclear Zeeman coupling, and $\hat{H}_{HF}$ represents the interaction between the nuclear and electron spins, the so-called hyperfine Hamiltonian. This interaction, first introduced by Fermi to account for hyperfine structure in atomic spectra, can be written in general as

$$\hat{H}_{HF} = \underline{I} \cdot \underset{\sim}{T} \cdot \underline{S} \tag{100}$$

where $\underset{\sim}{T}$ is a second-rank tensor. This tensor may be decomposed into the sum of two terms,

$$\hat{H}_{HF} = a\,\underline{I} \cdot \underline{S} + \underline{I} \cdot \underset{\sim}{T}' \cdot \underline{S} \tag{101}$$

where $a\underline{I}\cdot\underline{S}$ is a term representing the isotropic, contact interaction between the nucleus and unpaired electrons having s character and $\underline{I}\cdot\underset{\sim}{T}'\cdot\underline{S}$ is a term representing the anisotropic, dipolar coupling between the nucleus and unpaired electrons in non-spherically symmetric orbitals. We shall have more to say about these couplings later on. But, for the moment, it is only necessary to note that the important hyperfine term in the case of the H atom in its electronic ground state is the isotropic

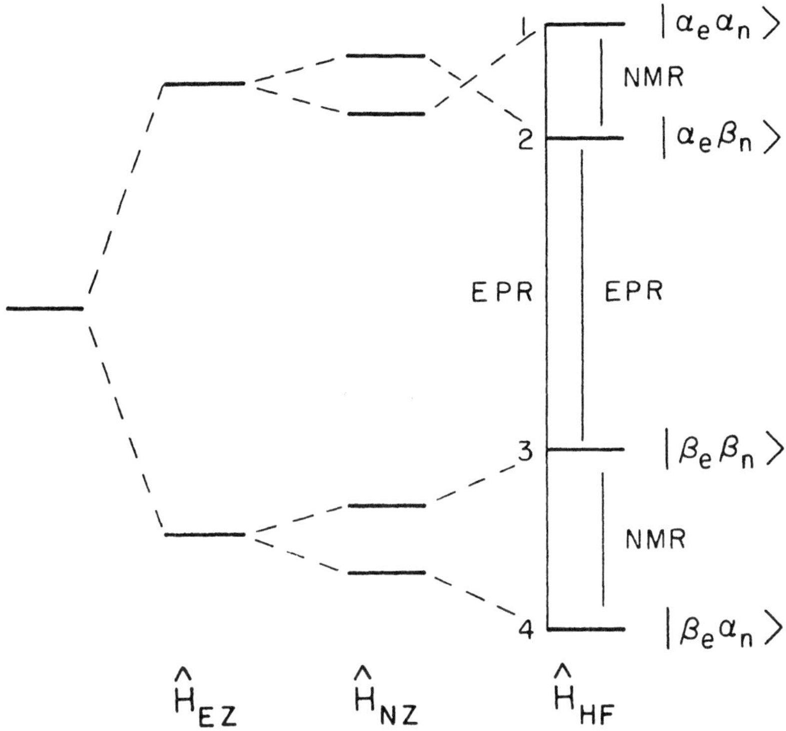

Fig. 14. Energy level diagram for the two-spin system I = S = 1/2, showing the effect of the perturbing hyperfine interaction.

one, and that we may further neglect terms like aI_xS_x and aI_yS_y at fields larger than ~ 1 kG. Thus, we have

$$\hat{H}_{S,I} = g_e\beta_eHS_z - g_n\beta_nHI_z + aI_zS_z \tag{102}$$

which is diagonal in the representation of the two-spin functions $|\alpha_e\alpha_n\rangle$, $|\alpha_e\beta_n\rangle$, $|\beta_e\alpha_n\rangle$, and $|\beta_e\beta_n\rangle$. This gives rise to the energy level diagram shown in Fig. 14, with eigenvalues

$$E_1 = \frac{1}{2}(g_e\beta_e - g_n\beta_n)H + \frac{a}{4} \tag{103a}$$

$$E_2 = \frac{1}{2}(g_e\beta_e + g_n\beta_n)H - \frac{a}{4} \tag{103b}$$

$$E_3 = -\frac{1}{2}(g_e\beta_e - g_n\beta_n)H + \frac{a}{4} \tag{103c}$$

$$E_4 = -\frac{1}{2}(g_e\beta_e + g_n\beta_n)H - \frac{a}{4} \tag{103d}$$

Now, physically the nuclear spin is precessing about the static field with an angular frequency $\omega = 2\pi g_n\beta_n H/h$; this is ~ $2\pi(42.58 \times 10^6)$ radian/sec at a field of 10 kG. But the electron spin is also precessing, at a much higher frequency. If the electron spin retains its orientation with respect to the period of the nuclear precession, then the nucleus will experience a local field owing to the presence of the electron spin. This field may either be aligned with or opposed to the static field. Therefore, we expect the single proton line of the H atom to be split into two lines, separated by twice the local, or hyperfine field. A simple calculation shows that this field is very large. Thus, from Eq. (103), we see that this line separation is just a, the hyperfine coupling constant, which is ~ 1400 MHz for the hydrogen atom. This is equivalent to a local field of ~ 1/2 (1400/42) (10^4) G, or ~ 160 kG, far exceeding the magnitude of the static field!

Now, in fact this model is totally unrealistic, not because of an error in the math, but because the electron spin does not retain its orientation for periods of 10^{-8} sec or so. Electron T_1's in the gas phase are on the order of 10^{-10} sec at room temperature for reasons that will be discussed. Therefore, during a single precession of the nucleus, it sees a time-averaged local field proportional to the mean value of $\langle S_z\rangle$. If the α and β electron spin states were equally populated, then $\langle S_z\rangle = 0$, and the magnetic field experienced by the nucleus would be just the static field. But the static field also polarizes the electron spin states, so that $\langle S_z\rangle \neq 0$. To calculate the magnitude of the local field under these conditions, we return to Eq. (102). That part of the energy which depends on the nuclear spin orientation can be written as

$$\hat{H}_I = -g_n\beta_n I_z (H + H_\ell) \tag{104}$$

where $H_\ell = -aS_z/g_n\beta_n$ is the local field produced by the unpaired electron at the nucleus. This field opposes the static field (for a > 0); hence, the NMR line will be shifted to high field by an amount

$$\Delta H = a\langle S_z\rangle/g_n\beta_n \tag{105}$$

The time-averaged value of S_z can be calculated from standard susceptibility formulae, yielding a temperature-dependent shift, in dimensionless units, of

$$\frac{\Delta H}{H} = -\frac{g_e\beta_e}{g_n\beta_n}\ \frac{aS(S+1)}{3kT} \tag{106}$$

Clearly, such shifts can be quite large, even for small hyperfine couplings, because of the amplification provided by the ratio of

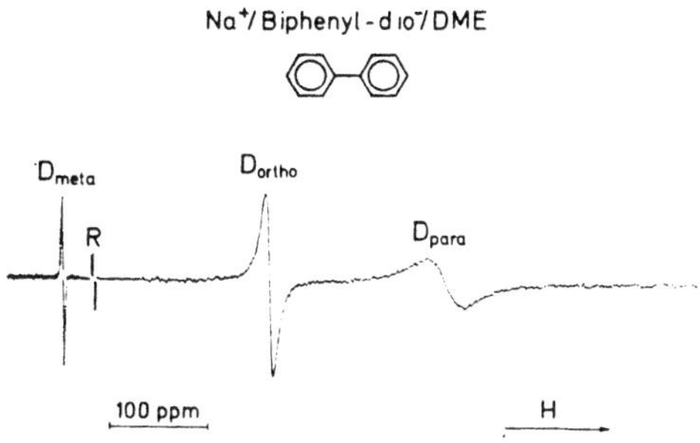

Fig. 15. ^2H NMR spectrum of a completely reduced solution of NaBp-$\underline{d}_{10}$ in 1,2-dimethoxyethane(DME) at 303K (13).

the nuclear and electronic masses.

Shifts of this type were first observed by Knight, who found that the NMR frequency of ^{63}Cu in metallic copper is 0.23% higher than in diamagnetic CuCl owing to the paramagnetism of the conduction electrons. This fractional shift is an order of magnitude larger than the chemical shifts among different diamagnetic compounds. Knight shifts also occur in solutions containing paramagnetic ions, in free radical solids, and in individual paramagnetic molecules, and are often employed for resolution enhancement in NMR spectroscopy, although the interactions responsible for these shifts are sometimes different. A nice example of Knight shifts in solution is provided by the ^2H NMR spectrum of the biphenyl-$\underline{d}_{10}$ radical anion in 1,2-dimethoxyethane, shown in Fig. 15 (13). In this case, the sample was completely reduced, so that the only lines appearing in the spectrum are the reference peak (R) and the peaks belonging to the ortho, meta, and para deuterons of the radical anion. We note immediately that the shifts are large, and that the o- and p-peaks are shifted upfield whereas the m-peak is shifted downfield. Studies of the temperature dependence of these shifts show conclusively that the isotropic coupling constants responsible for them are opposite in sign, thereby confirming the assumptions leading to Eq. (106). This method of measuring unpaired spin densities in paramagnetic molecules is very powerful.

There is a second, important, feature evident in Fig. 15. Note that the lines associated with deuterons having larger hyperfine couplings are broader than those associated with deuterons having smaller couplings. This is a consequence of the short electron spin relaxation times of the ion, which cause the strong local hyperfine field to fluctuate rapidly and induce rapid relaxation between the nuclear spin states. Brownian motion will also modulate anisotropic magnetic interactions in the usual way and contribute to the relaxation, whether the nuclei are attached to the ions themselves, or to other nuclei in solution. The discovery that even small traces of paramagnetic ions could dramatically increase the rate of nuclear spin relaxation was one of the landmarks in the history of magnetic resonance and has led to a much sounder understanding of relaxation processes in solids and liquids. The detailed interpretation of these effects involves many factors, and it is usually necessary to study both T_1 and T_2 as a function of several parameters before reaching any firm conclusions about the mechanism. In the case of the biphenyl-$\underline{d}_{10}$ radical anion, the principal mechanism is scalar relaxation of the "first kind" (2), for which

$$\frac{1}{T_{2n}} = \frac{a^2}{3\hbar^2} S(S+1)\, T_{1e} \tag{107}$$

thereby explaining differences in the linewidths in Fig. 15. Eq. (107) also suggests that the NMR line of a paramagnetic species may be so broad so as to be unobservable if a is large or T_{1e} is long, which is often the case. The effect of paramagnetic impurities on nuclear relaxation times has found important applications in biology, NMR imaging in medicine being a recent example (14).

Other, closely related, phenomena should be mentioned here. One is the effect of "cross-relaxation". In the above example, we have seen that the relaxation behavior of the electron spins affects the width of the NMR line, and therefore affects the relaxation behavior of the nuclear spins. Clearly, this is because of the coupling of the two spin systems _via_ the hyperfine interaction. But, in general, there exist a variety of spin-spin coupling mechanisms; one is the dipolar coupling of nuclear spins, already mentioned, and a second is the indirect scalar coupling of nuclear spins, a second-order effect discovered independently by Hahn and Maxwell, and Gutowsky and McCall. Space does not permit a detailed description of this coupling mechanism which is of great importance in chemistry; suffice it to say here that it is based on the polarization of paired electron spins in the vicinity of magnetic nuclei. In any event, if two spins, say A and B, are coupled together, then when cross relaxation occurs one can expect that any change in the populations of the spin states of A will cause secondary changes in the populations for spin B. Double resonance or "dynamic polarization" experiments are designed to take advantage of these changes. In a typical experiment two resonance lines are chosen. The first transition is saturated, thereby equalizing the populations of two spin states, and then one looks for changes in the intensity of resonance absorption at the second line. A famous example is the Overhauser effect. It occurs if one saturates the EPR line of a suitable paramagnetic sample while monitoring an NMR line. Enhancements in the NMR signal of several hundredfold are sometimes observed because, in effect, the relative populations of the nuclear spin states are governed by Boltzmann factors involving the electron spins. But there are many other examples of these double resonance schemes, some of which will be described at this Conference.

2.5. Quadrupole Coupling.

So far we have considered only the magnetic interactions of the nucleus with its surroundings. To be complete, we should also mention briefly one important electrical effect on the energy required to reorient the nucleus. This is called quadrupole coupling. Nonspherical nuclei (those with $I > 1$), if they experience nonspherical charge distributions from surrounding electrons, exhibit a kind of nuclear "Stark effect" which influences the relative energies of magnetic sublevels with different $|m_I|$. This is expressed in the quadrupole Hamiltonian (15),

$$\hat{H}_Q = \underline{I} \cdot Q \cdot \underline{I} \tag{108}$$

where Q, like earlier examples, is a second-rank tensor; thus Eq. (108) applies for an arbitrary orientation of the rectangular coordinates x,y,z. The tensor coupling can be simplified by choice of a set of principal axes x',y',z', giving

$$\begin{aligned}\hat{H}_Q &= \frac{eQ}{6I(2I-1)}[Q_{x'x'}(3I_{x'}{}^2-I^2)\\ &\quad + Q_{y'y'}(3I_{y'}{}^2-I^2) + Q_{z'z'}(3I_{z'}{}^2-I^2)]\end{aligned} \tag{109}$$

where eQ is the nuclear electric quadrupole moment. This expression can be rewritten, using LaPlace's equation, to give

$$\begin{aligned}\hat{H}_Q &= \frac{eQ}{4I(2I-1)}[Q_{z'z'}(3I_{z'}{}^2-I^2)\\ &\quad + (Q_{x'x'}-Q_{y'y'})(I_{x'}{}^2-I_{y'}{}^2)]\end{aligned} \tag{110}$$

Eq. (110) shows that only two parameters are needed to characterize the $Q_{i'j'}$ (= $\partial^2V/\partial_{i'}\partial_{j'}$); these are the field gradient eq and the asymmetry parameter η;

$$eq = Q_{z'z'} = \partial^2V/\partial z'\partial z' \tag{111a}$$

$$\eta = \frac{Q_{x'x'}-Q_{y'y'}}{Q_{z'z'}} \tag{111b}$$

The nuclear quadrupole coupling constant, e^2qQ/h, can either be very small (e.g., 0.225 MHz for the deuterons in D_2) or very large (e.g., 2420 MHz for ^{127}I in ICN), and may therefore be either negligible or enormous compared with the nuclear Zeeman energy. Numerous examples of such effects may be found in the literature.

3. ELECTRON PARAMAGNETIC RESONANCE.

Many of the basic principles which apply to the NMR experiment also apply to the EPR, or ESR (electron spin resonance), experiment. But there are some important differences. Perhaps the major one is the fact that nuclear properties such as spin, magnetic moment, and quadrupole moment are to a very good approximation unaffected by the surroundings, whereas the same is not true for electrons. Clearly, this is because of the larger spatial extent of the electronic wavefunction and the much smaller energies separating the ground state from electronically excited states. The sensitivity of the EPR experiment to the surroundings has several interesting consequences. For example, the requirements for spectral resolution in EPR are significantly reduced compared to the NMR experiment; important effects are often revealed with MHz (or even GHz) resolution, rather than the few Hz resolution often required in NMR. Second, owing to their spatial extent, unpaired electrons usually have much shorter spin relaxation times than nuclei, yielding much broader lines. For the same reason, the time-dependent perturbations which influence the dynamic behavior of electron spins occur at much higher frequency, thereby increasing greatly the types of motion which can be studied using magnetic resonance techniques. And, finally, the concept of the effective spin Hamiltonian itself is much more powerful (and necessary) in EPR because of the wider variety of interactions which can contribute to a typical spectrum. In this brief section, we will discuss some of the interactions that are important in EPR but not encountered in NMR, and illustrate these with a few simple examples.

3.1. The Spin Hamiltonian.

We begin, as before, by writing down the general form of the spin Hamiltonian, $\hat{H}_S$, for an electron possessing spin angular momentum $\underline{S}$ in an external magnetic field. This is

$$\hat{H}_S = \hat{H}_{EZ} + \hat{H}_{FS} + \hat{H}_{EX} + \hat{H}_I \qquad (112)$$

where $\hat{H}_{EZ}$ is the electron Zeeman interaction,

$$\hat{H}_{EZ} = \beta_e \, \underline{H} \cdot \underset{\sim}{g} \cdot \underline{S}, \qquad (113)$$

$\hat{H}_{FS}$ is the electron spin dipolar, or fine-structure, interaction

$$\hat{H}_{FS} = g_e^2 \beta_e^2 \; \underline{S}_1 \cdot \underset{\sim}{D} \cdot \underline{S}_2, \tag{114}$$

$\hat{H}_{EX}$ is the electron spin exchange interaction,

$$\hat{H}_{EX} = \underline{S}_1 \cdot \underset{\sim}{J} \cdot \underline{S}_2, \tag{115}$$

and $\hat{H}_I$ is the nuclear spin Hamiltonian, given in Eq. (55). Terms in $\hat{H}_I$ are necessary because, as we have seen, there is often strong magnetic coupling of the electron spin to one or more nuclei, thereby influencing the effective field experienced by the electron. Note that each of the terms in Eq. (112) is, again, written in tensor form.

3.2. The $\underset{\sim}{g}$ Tensor.

The g tensor is one of the most poorly understood, and yet most important, parameters in the EPR experiment. The lack of understanding derives, in part, from the fact that g is a tensor, and in part from the fact that the electronic Zeeman interaction is represented by the effective Hamiltonian $\hat{H}_{EZ}$ [Eq. (113)]. Let us deal with the second problem first by writing, again

$$\hat{H}_{EZ} = \beta_e \, \underline{H} \cdot \underset{\sim}{g} \cdot \underline{S} \tag{113}$$

When the electron possesses spin angular momentum only, the $\underset{\sim}{g}$ tensor is isotropic and has the free electron value $g_e = 2.00232$. But, as we have seen already in our discussion of the chemical shift, the electron may also possess orbital angular momentum $\underline{L}$, either by itself or via coupling with excited states. Thus, the true Hamiltonian for the magnetic interaction should be written as

$$\hat{H}_{EZ} = \beta_e \, \underline{H} \cdot \underline{L} + g_e \beta_e \, \underline{H} \cdot \underline{S} \tag{116a}$$

or

$$\hat{H}_{EZ} = \beta_e \, \underline{H} \cdot (\underline{L} + g_e \, \underline{S}) \tag{116b}$$

Now, at this point in the development, we have two choices. Following the example of NMR, we could write

$$\hat{H}_{EZ} = g_e \beta_e \, \underline{H} \cdot (\underset{\sim}{1} - \underset{\sim}{\sigma}') \cdot \underline{S} \tag{117}$$

where σ' is a kind of chemical shift parameter which accounts for the mixing in of orbital angular momentum. (Clearly, this mixing process will be anisotropic in molecules, so that σ' is a tensor, not a scalar). If this approach were taken, we would find tabulated chemical shifts for electron spins in particular environments, just as we do for nuclear spins in NMR spectroscopy. But the EPR spectroscopist takes an easier approach. Recognizing that the center of a spectrum can be defined by an effective g value, g_{eff}, where

$$g_{eff} = \frac{h\nu}{\beta H_0} \tag{118}$$

he replaces the quantity $(\underline{L} + g_e\underline{S})$ in Eq. (116) by the quantity $g_{eff}\,\hat{\underline{S}}$, where $\hat{\underline{S}}$ should now be regarded as the spin angular momentum of a fictitious spin of 1/2. In other words, the contributions of orbital angular momentum to the problem are taken up in g_{eff}, rather than explicitly written out. Then, anticipating that g_{eff} will be anisotropic, he writes

$$\hat{H}_{EZ} = \beta_e\,\underline{H} \cdot \underset{\sim}{g} \cdot \hat{\underline{S}} \tag{119}$$

which is the effective spin Hamiltonian for S = 1/2. Contributions from $\underline{L}$ can then be determined from the measured deviations of the principal values of $\underset{\sim}{g}$ from the free spin value of $g_e = 2.00232$.

At first sight, this does appear to be a bit complicated, but in reality it is not. We can prove this statement by first finding the conditions required to make the effective spin Hamiltonian of Eq. (119) equivalent to the "correct" Hamiltonian of Eq. (116). Following Carrington (9), let us consider a molecule having a single unpaired electron with spatial wavefunction $\psi_0(\underline{r})$ and spin wavefunction α or β. In the presence of a field along the z axis, the α spin state has energy

$$\begin{aligned} E_\alpha &= \beta_e H\langle\psi_0\alpha|L_z + g_e S_z|\psi_0\alpha\rangle \\ &= \beta_e H\langle\psi_0|L_z|\psi_0\rangle + g_e\beta_e H\langle\alpha|S_z|\alpha\rangle \\ &= \beta_e H\langle L_z\rangle + 1/2\ g_e\beta_e H \end{aligned} \tag{120}$$

However, the expectation value $\langle L_z\rangle$ vanishes in all molecules which do not have orbitally degenerate ground states. The only way the electron can acquire some orbital angular momentum is through the effect of spin-orbit coupling, which couples the

Fig. 16. Simple model for spin-orbit coupling in the hydrogen atom.

ground state (having $\langle L_z \rangle = 0$) with other electronic states having $\langle L_z \rangle \neq 0$. A simple physical picture of the nature of this coupling is possible. Suppose we consider, with the aid of Fig. 16, the Bohr model of the H atom in its ground state. Normally, we think of the electron in orbit about the nucleus, but it is more instructive in this case to consider the inverse; i.e., the apparent circulation of the nuclear charge about the electron. The magnetic field produced at the electron by this moving positive charge is

$$\underline{H}_\ell = -\frac{1}{r}\frac{dV}{dr}\frac{1}{mc}\ \underline{L} \tag{121}$$

where V is the electrostatic potential at the electron due to the nucleus. For a purely Coulombic attraction,

$$\frac{1}{r}\frac{dV}{dr} = -\frac{Ze}{r^3} \tag{122}$$

Thus, if the electron also has a magnetic moment $\underline{\mu}_e (= g_e \beta_e \underline{S})$ then there is an energy of interaction between this moment and the local field which is the spin-orbit energy,

$$\begin{aligned} E_{SO} &= -\underline{\mu}_e \cdot \underline{H}_\ell \\ &= -g_e\left(\frac{e}{2m^2c^2}\right)\left(\frac{Ze}{r^3}\right) \underline{L} \cdot \underline{S} \\ &= -\zeta\ \underline{L} \cdot \underline{S} \end{aligned} \tag{123}$$

where ζ is a constant, called the spin-orbit coupling constant. Note that ζ is proportional to the nuclear charge, and to $1/r^3$. Since this field may add to or subtract from the static field, depending on the relative orientation of $\underline{L}$ and $\underline{S}$, positive and negative g-shifts are possible.

To return to our calculation, the spin-orbit operator will mix the ground-state wavefunction with excited state functions. Treating this by first-order perturbation theory, the state $|\psi_0\alpha\rangle$ becomes

$$|+\rangle = |\psi_0\alpha\rangle - \sum_n \frac{\langle n|\zeta\underline{L}\cdot\underline{S}|\psi_0\alpha\rangle}{E_n - E_0}\ |n\rangle \tag{124}$$

Expanding $\zeta\ \underline{L}\cdot\underline{S}$, we have

$$\begin{aligned} |+\rangle = |\psi_0\alpha\rangle &- \frac{1}{2}\zeta \sum_n \frac{\langle n|L_z|0\rangle}{E_n - E_0}\ |n\alpha\rangle \\ &- \frac{1}{2}\zeta \sum_n \frac{\langle n|L^+|0\rangle}{E_n - E_0}\ |n\beta\rangle \end{aligned} \tag{125}$$

Spin-orbit coupling also converts the unperturbed state $|\psi_0\beta\rangle$ into a correspondingly mixed state

$$|-\rangle = |\psi_0\beta\rangle + \frac{1}{2}\zeta\sum_n \frac{\langle n|L_z|0\rangle}{E_n-E_0}|n\beta\rangle - \frac{1}{2}\zeta\sum_n \frac{\langle n|L^-|0\rangle}{E_n-E_0}|n\alpha\rangle \quad (126)$$

Clearly, the states $|+\rangle$ and $|-\rangle$ are no longer eigenstates of the true spin S_z. But if we define the fictitious spin $\underline{\hat{S}}$ by requiring that its components act on the states $|+\rangle$ and $|-\rangle$ in exactly the same way as the components of the true spin $\underline{S}$; i.e.

$$\hat{S}_z|+\rangle = \frac{1}{2}|+\rangle,\ \hat{S}_z|-\rangle = -\frac{1}{2}|-\rangle,$$

$$\hat{S}_x|+\rangle = \frac{1}{2}|-\rangle,\ \hat{S}_y|+\rangle = \frac{i}{2}|-\rangle \quad (127)$$

then the Zeeman Hamiltonian [Eq. (119)], with $\underline{H}\ ||\ z$, is

$$\begin{array}{cc} |+\rangle & |-\rangle \end{array}$$
$$\begin{pmatrix} \frac{1}{2}\beta_e H g_{zz} & \frac{1}{2}\beta_e H(g_{zx}-ig_{zy}) \\ \frac{1}{2}\beta_e H(g_{zx}+ig_{zy}) & -\frac{1}{2}\beta_e H g_{zz} \end{pmatrix} \quad (128)$$

On the other hand, the 'true' Zeeman Hamiltonian has matrix elements

$$\beta_e H \begin{pmatrix} \langle +|L_z+g_eS_z|+\rangle & \langle +|L_z+g_eS_z|-\rangle \\ \langle -|L_z+g_eS_z|+\rangle & \langle -|L_z+g_eS_z|-\rangle \end{pmatrix} \quad (129)$$

If the effective spin Hamiltonian has any meaning, it is clear that we must have

$$g_{zz} = 2\langle +|L_z+g_eS_z|+\rangle \quad (130a)$$

$$(g_{zx}+ig_{zy}) = 2\langle -|L_z+g_eS_z|+\rangle \quad (130b)$$

A straightforward calculation yields, using Eqs. (125), (126), and (130),

$$g_{zz} = g_e - 2\zeta\sum_n \frac{\langle 0|L_z|n\rangle\langle n|L_z|0\rangle}{E_n-E_0} \quad (131a)$$

$$g_{zx} = g_e - 2\zeta\sum_n \frac{\langle 0|L_z|n\rangle\langle n|L_x|0\rangle}{E_n-E_0} \quad (131b)$$

to first order in ζ. Note the similarity of these results to Ramsey's shielding formula, Eq. (74). Also note that the sign of $(E_n - E_0)$ changes as ψ_n changes from an occupied to an unoccupied orbital, so that filled and empty orbitals give contributions of opposite sign to the g value.

The spin Hamiltonian formalism, and the concept of fictitious spin, was first developed by Pryce and his co-workers at Oxford in order to interpret the early EPR results of Bleaney, Griffith, and others on transition metal ions in crystals. Their numerous publications contain a wealth of information and have been summarized in several places (16-18).

But we need not consider something so complicated as a transition metal ion in order to illustrate the simple ideas expressed here. Instead, we consider small triatomic radicals of the type AB_2, such as NO_2, NO_2^{2-}, and CO_2^-, with the aid of Figs. 17 and 18 (19). Figure 18 is a correlation diagram, known as a Walsh diagram, for the bending of a symmetrical AB_2 molecule based on the simple molecular orbitals illustrated in Fig. 17. Species with 16 or less valence electrons can accommodate all of the electrons in bonding orbitals; therefore, no strain need be relieved and the molecule is linear. However, when the next electron is introduced to form a 17-electron species, it will

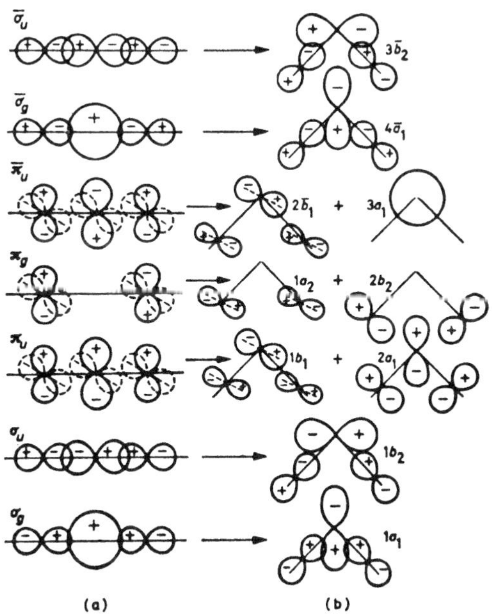

Fig. 17. Molecular orbitals in (a) linear and (b) 90° triatomic molecules (19).

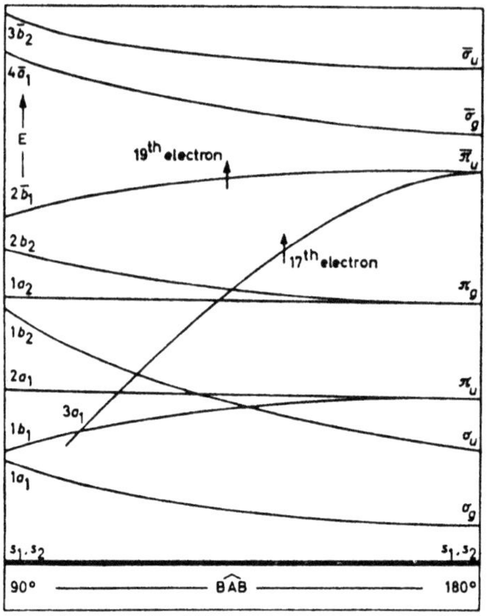

Fig. 18. Correlation diagram for AB_2 molecules (19).

enter the $\bar{\pi}_u$ antibonding orbital. As shown on the correlation diagram, if the molecule were now to bend, the "$\bar{\pi}_u$" orbital would become increasingly localized on A until, in the 90° molecule, it would occupy an s lone-pair orbital on A. The lowering in energy brought about by the inclusion of the s orbital favors a bent molecule. Now, irradiation of a single crystal of sodium formate, $NaHCO_2$, with γ rays results in loss of a hydrogen atom to form oriented CO_2^- radicals (20). The g tensor was found to be anisotropic with the principal values $g_{xx} = 2.0032$, $g_{yy} = 1.9975$, and $g_{zz} = 2.0014$. Since CO_2^- has 17 electrons, it is expected to have a bent structure with C_{2v} symmetry. Does the EPR data support this view? Yes! Note, first, that the g tensor is rhombic, with three different principal values. The deviations of the principal values from $g_e = 2.0023$ are

$$\Delta g_{xx} = +0.0009, \ \Delta g_{yy} = -0.0048, \text{ and } \Delta g_{zz} = -0.0009 \qquad (132)$$

Thus, by far the largest departure from the free spin value is observed when the magnetic field is in the y direction (the in-plane axis perpendicular to z, the C_2 axis). Now, the angular momentum operator L_y acts on the p orbitals to give

$$L_y|p_y\rangle = 0, \ L_y|p_z\rangle = i|p_x\rangle, \text{ and } L_y|p_x\rangle = -i|p_z\rangle \qquad (133)$$

so that it mixes the modified p_z orbital, $\psi(3a_1)$, which is singly occupied, with the lowest excited p_x orbital, $\psi(2b_1)$. Since this orbital lies to higher energy, a negative g-shift is observed, as expected from Eq. (131a) (with L_y substituted for L_z). A more quantitative treatment of the problem, using optical data to estimate the excitation energy for the transition $3a_1 \rightarrow 2b_1$, permits the determination of the coefficients in the LCAO MO wavefunction for the $3a_1$ electron. NO_2 has similar g values (g_{xx} = 2.006, g_{yy} = 1.991, g_{zz} = 2.002) which is expected, since it is isoelectronic with CO_2^-. But NO_2^{2-}, with two more electrons, has g_{xx} = 2.0038, g_{yy} = 2.009, and g_{zz} = 2.0070; the fact that g_{yy} is now larger than g_e provides further confirmation that our understanding of the origin of the g shift in molecules is qualitatively correct. Thus, the careful measurement and proper interpretation of g values provides valuable information about molecular structure, not only of the ground state, but also of excited states.

3.3. Hyperfine Coupling.

Complementary information is provided by detailed studies of the hyperfine interaction. It is the next most important term in the EPR Hamiltonian for dilute systems containing single unpaired spins since the dipolar coupling and electron exchange terms require the presence of two or more spins in close proximity. The hyperfine Hamiltonian has been considered before in our treatment of the hydrogen atom problem [cf., Eq. (101)]. But there we considered mainly the isotropic term whereas, in general, we can expect the hyperfine interaction to be anisotropic in nature. Thus, in the study of the EPR spectrum of CO_2^-, anisotropic hyperfine structure from ^{13}C was also observed, with principal values

$$T_{xx} = 436, \; T_{yy} = 422, \text{ and } T_{zz} = 546 \text{ MHz} \tag{134}$$

How can we interpret these results?

The isotropic hyperfine interaction has its origin in the finite electron density at the nucleus which is characterstic of s-type orbitals. This is immediately apparent on examination of the radial dependence of hydrogenic-type wavefunctions. Fermi has shown that for systems with one electron the isotropic interaction energy is given approximately by

$$a = \frac{8\pi}{3} g_e \beta_e g_n \beta_n \; |\psi(0)|^2 \tag{135}$$

where $|\psi(0)|^2$ is the squared amplitude of the electronic wavefunction at the nucleus. But other, non-spherically symmetric orbitals (e.g., p, d, f, etc.) having unpaired spin also can produce hyperfine splitting; in this case, the coupling between electron and nuclear spins is dipolar in character. It is precisely this interaction which gives rise to the anisotropic component of the hyperfine interaction. The classical expression for the dipolar interaction energy between a fixed electron and a nucleus separated by a distance $\underline{r}$ is exactly analogous to Eq. (76). Thus,

$$E = \frac{\underline{\mu}_e \cdot \underline{\mu}_n}{r^3} - \frac{3(\underline{\mu}_e \cdot \underline{r})(\underline{\mu}_n \cdot \underline{r})}{r^5} \tag{136}$$

Clearly, then, the quantum mechanical analog of this expression is the tensor interaction, $\underline{I} \cdot \underset{\sim}{T}' \cdot \underline{S}$. The principal values of $\underset{\sim}{T}'$ are usually denoted by t_1', t_2', and t_3'. The full hyperfine Hamiltonian requires the addition of the isotropic term, yielding

$$\hat{H}_{HF} = \underline{I} \cdot (a\underset{\sim}{1} + \underset{\sim}{T}') \cdot \underline{S} = \underline{I} \cdot \underset{\sim}{T} \cdot \underline{S} \tag{137}$$

Measurements of the principal values of $\underset{\sim}{T}$ therefore provide the values of a for s-type orbitals, the values of $\langle 1/r^3 \rangle$ for non s-type orbitals, and, if the principal axes can be determined as well, the orientations of these orbitals with respect to crystallographic or molecular axes. The experimental procedures for doing this are summarized nicely in the text by Wertz and Bolton (21). Parenthetically, we remark here that a full treatment of the problem requires inclusion of the nuclear Zeeman term, and the determination of the absolute signs of the hyperfine elements. This is often not possible, but ENDOR (electron-nuclear double resonance) methods (22) are sometimes helpful in this regard.

Let us return, briefly, to the example of CO_2^-. Summing the principal values of $\underset{\sim}{T}$ [Eq. (134)] yields the value $a = 1/3 \text{ Tr } \underset{\sim}{T}$ = 468 MHz which is due to unpaired spin density in the carbon 2s orbital. Since it is known that the isotropic splitting for a Hartree 2s atomic orbital on carbon is 3110 MHz (21), we can estimate that the s orbital spin density on the carbon atom in CO_2^- is $\rho^C = 468/3110 = 0.15$. The balance of the unpaired spin resides in 2p orbitals, mainly on the carbon atom. The principal values of $\underset{\sim}{T}'$ are

$$\underset{\sim}{T}' = \begin{pmatrix} 436-468 & & \\ & 422-468 & \\ & & 546-468 \end{pmatrix} = \begin{pmatrix} -32 & & \\ & -46 & \\ & & 78 \end{pmatrix} \text{ MHz} \tag{138}$$

Note that T' is traceless, and nearly cylindrically symmetric, with $t_3' \simeq -2t_1' \simeq -2t_2'$. By analogy with our discussion of the dipolar coupling problem in NMR, we can expect that

$$t_1' = -g_e\beta_e g_n\beta_n \left\langle \frac{r^2-3x^2}{r^5} \right\rangle \tag{139a}$$

$$t_2' = -g_e\beta_e g_n\beta_n \left\langle \frac{r^2-3y^2}{r^5} \right\rangle \tag{139b}$$

$$t_3' = -g_e\beta_e g_n\beta_n \left\langle \frac{r^2-3z^2}{r^5} \right\rangle \tag{139c}$$

where the brackets denote averages over the spatial distribution of the unpaired electron spin. Expressed in terms of spherical polar coordinates, these averages vary as the quantity $\langle(3\cos^2\Theta-1)/r^3\rangle$, where Θ is the angle between x, y, or z and $\underline{r}$. Now, consider Fig. 19, which is a schematic representation of the dipolar hyperfine coupling between a nucleus and a $2p_z$ electron on the same atom. As the electron is concentrated in the region where $(3\cos^2\Theta -1) > 0$, it is clear that the resulting contribution to T' would be axially symmetric, that the principal values would be $-(t/2)$, $-(t/2)$, and t, and that t would be positive. To be more precise, if the $2p_z$ wavefunction takes the standard form

$$\psi = \frac{3}{4\pi} \cos\theta \; \frac{f(r)}{r} \tag{140}$$

then

$$t = g_e\beta_e g_n\beta_n \frac{3}{4\pi} \int_0^{\pi}\int_0^{2\pi} (3\cos^2\Theta -1) \cos^2\Theta \sin\theta \, d\theta \, d\phi$$

$$\times \int_0^{\infty} \frac{1}{r^3} f^2(r) \, dr \tag{141a}$$

$$t = \frac{4}{5} g_e\beta_e g_n\beta_n \left\langle\frac{1}{r^3}\right\rangle \tag{141b}$$

In CO_2^-, there is also a minor contribution to $\underline{T}'$ from unpaired spin in the carbon $2p_x$ orbital. We can reconstruct the tensor components (−32, −46, 78) MHz by superposing two cylindrical tensors (−42, −41, +83) MHz and (+10, −5, −5) MHz oriented along the z and x axes, respectively. The value of t for a single electron in a pure carbon p orbital is ~182 MHz (21); thus, the $2p_z$ and $2p_x$ orbital spin densities are approximately 0.46 and 0.05. The rest of the unpaired spin is in oxygen $2p_z$ orbitals which make very little contribution to $\underline{T}'$ because of the $\langle 1/r^3\rangle$ dependence of the

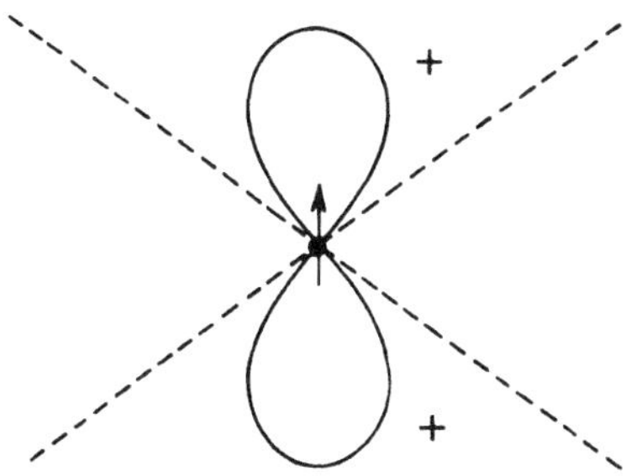

Fig. 19. Dipolar hyperfine coupling between a nucleus and a $2p_z$ electron on the same atom (4).

dipole-dipole coupling. We therefore conclude that the unpaired electron is mainly localized in a carbon sp^2 hybrid orbital which points along the (negative), in-plane z axis of the CO_2^- radical. The OCO bond angle is, in fact, somewhat less than 120° because the s/p ratio is less than 0.5 (19).

The hyperfine interaction has been extensively studied in organic, inorganic, and biological systems. Partly, this is because it often gives rise to incredibly rich and beautiful EPR spectra, such as is shown in Fig. 20. Here, the sample is a dilute solution of naphthalene in dimethoxyethane which has been electrolytically reduced at the surface of a Pt electrode. The spectrum is that of the naphthalene radical anion and shows hyperfine spliting due to the 4 equivalent α and β protons. Partly, the interest in the hyperfine interaction is more fundamental. For example, in $C_{10}H_8^-$, it was not clear for some time how the unpaired spin, residing in a π* orbital with a nodal plane at each C-H bond, could produce an isotropic coupling with proton spins lying in that plane. (We now know from the work of McConnell and Bersohn that exchange forces between the π* electron and the σ electrons in the C-H bonds polarize these bonding electrons, leading to a slight excess of negative spin density in each hydrogen 1s orbital. A similar mechanism is responsible for spin-spin coupling in NMR.) But, mostly, the interest in the hyperfine interaction stems from the fact that it is unique; no other interaction provides as much direct information about the nature of the molecular electronic wavefunction and the structure of paramagnetic species. Unfortunately, there is neither sufficient time nor space to provide more than a cursory treatment

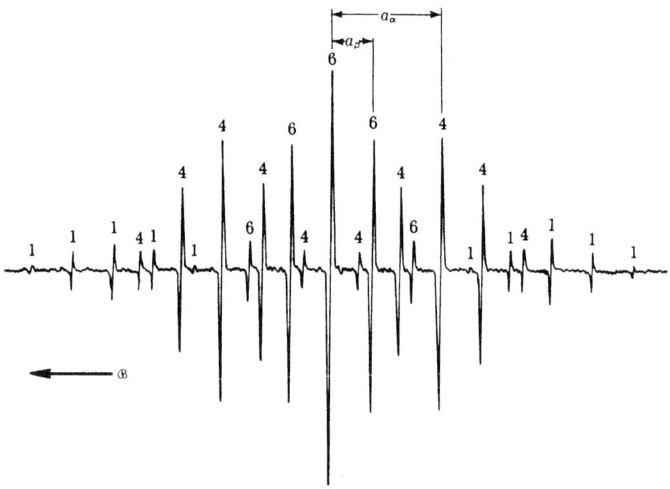

Fig. 20. EPR spectrum of the naphthalene negative ion.

of the subject here. But we can at least hope that most of the physical bases for understanding this effect are clear.

3.4. Interactions Between Unpaired Electrons.

Not all paramagnetic species have only a single unpaired electron. And not all paramagnetic samples are magnetically dilute. Therefore, we should also consider the influence of electron spin-electron spin interactions on EPR spectra. One effect, that of magnetic dipolar coupling between two (or more) electron spins, is exactly analogous to the problem of a dipolar coupling in NMR spectroscopy (and electron-nuclear dipolar contributions to the hyperfine interaction). This effect is expressed in the form of the fine-structure Hamiltonian, $\hat{H}_{FS}$, in Eq. (114). It includes the first-order contributions of the spin-spin interaction, and the second-order contributions of spin-orbit coupling, to the energies of the magnetic sublevels of triplet, quartet, and other higher multiplicity electronic states. Therefore, it is important in widely-diverse systems, from organic triplet states to paramagnetic rare earth ions. The reader is referred to a recent review on optically-detected magnetic resonance for a discussion of some of these effects (23).

But there is another phenomenon which is particularly important in concentrated paramagnetic samples. This is the effect of magnetic exchange. We can illustrate this effect by considering a biradical ($\dot{A}$-$\dot{B}$) in which each molecular fragment contains one magnetic nucleus of spin I which gives rise to a hyperfine splitting. The spin Hamiltonian appropriate to this system is

$$\hat{H}_S = g_e\beta_e(S_{1z}+S_{2z}) + a(I_{1z}S_{1z}+I_{2z}S_{2z}) + J\ \underline{S_1}\cdot\underline{S_2} \qquad (142)$$

Here, all interactions are assumed isotropic, $\underline{S_1}$ and $\underline{S_2}$ are the spin operators for electrons 1 and 2, and J is the exchange integral defined by

$$J = \langle\phi_A(1)\phi_B(2)|\frac{e^2}{r_{12}}|\phi_B(1)\phi_A(2)\rangle \qquad (143)$$

where ϕ_A is a one-electron orbital on fragment A and ϕ_B is an orbital on fragment B. r_{12} is the separation of the two unpaired electrons. (For a biradical in a solid, one should also include the term $\underline{S_1}\cdot\underline{D}\cdot S_2$, but this averages to zero in solution.) If $|J|>>|a|$, the zero-order Hamiltonian is

$$\hat{H}_S{}^\circ = g_e\beta_eH(S_{1z}+S_{2z}) + J\ \underline{S_1}\cdot\underline{S_2} \qquad (144)$$

If we expand $\underline{S_1}\cdot\underline{S_2}$ in terms of its components, and express the results in terms of the total spin $\underline{S}^2$ and its component S_z, then Eq. (144) has the eigenfunctions $|T_1\rangle = |1,1\rangle$, $|T_0\rangle = |1,0\rangle$,

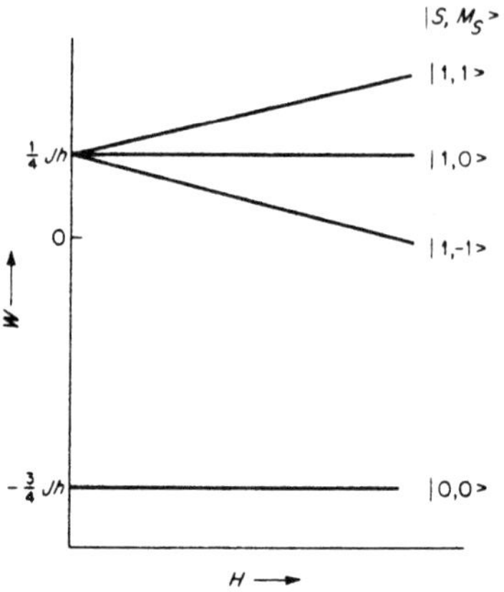

Fig. 21. The state energies of a system of two interacting electrons. When J is positive, the singlet state lies lower in energy. The states are labeled by their S, M_S values at high fields (21).

$|T_{-1}\rangle = |1,-1\rangle$, $|S_0\rangle = |0,0\rangle$ and eigenvalues

$$E_{1,1} = g_e\beta_e H + J/4 \; (+ \; am_I/2) \tag{145a}$$

$$E_{1,0} = J/4 \tag{145b}$$

$$E_{1,-1} = -g_e\beta_e H + J/4 \; (- \; am_I/2) \tag{145c}$$

$$E_{0,0} = -3J/4 \tag{145d}$$

These energies are plotted in Fig. 21 (21), which shows that the singlet state, $|S_0\rangle$, lies at lower energy if $J > 0$. Including the hyperfine term as a perturbation gives the first-order correction terms, enclosed in parentheses, in Eq. (145). Here, $m_I = m_I(1) + m_I(2)$. Thus, in the case of $I_1 = I_2 = 1$, the EPR spectrum would consist of five lines separated by a/2, with intensity ratios 1:2:3:2:1. This should be contrasted with the case of a monoradical in which one nucleus has I=1. In the latter case, pairs of lines would be separated by a. An example of these two limiting cases is shown in Fig. 22 (21). From this simple example, we learn two important things. The first is that, in a magnetically concentrated system, exchange coupling allows the electrons on neighboring molecules on sites to interact, giving rise to much structure in the spectrum which becomes increasingly difficult to resolve as the number of coupled spins increases. The second is that, if the coupling is antiferromagnetic in nature (with $J > 0$) and $|J|$ is very large, no EPR spectrum may be observed at all since the two spins are exchange coupled into a singlet ground

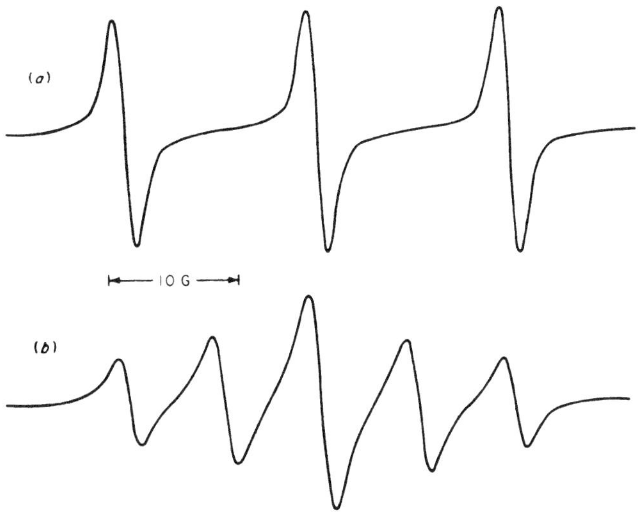

Fig. 22. (a) EPR spectrum of a typical monoradical with I = 1. (b) EPR spectrum of a typical biradical showing interaction with two I = 1 nuclei (21).

state. A number of transition metal ions behave in this way, including the famous example of copper acetate.

3.5. Electron Spin Relaxation.

This brings us, finally, to the subject of relaxation in EPR spectroscopy. From what has been said so far, it should be obvious that there are a number of processes which can contribute to the measured values of T_1 and T_2 in systems containing unpaired electron spins. Many of these are completely analogous to those occurring in NMR. These include the modulation of anisotropies in the g, T, and D tensors by molecular motion. In such cases, broad lines are often observed at low temperatures, due to incomplete averaging of these anisotropies, but "motional narrowing" effects are observed at high temperatures. Relations very similar to those considered earlier [e.g., Eqs. (93) and (94)] apply here as well, with one important modification. This is that the limit $\omega_0{}^2\tau_C{}^2 \ll 1$ is reached only at very small values of τ_C because of the much larger values of ω_0 in the typical EPR experiment. Thus, the spectrum of molecular motions which influence an EPR experiment is centered at much higher frequencies than in the case of NMR. The same comment applies to the effects of chemical exchange, an intriguing subject not considered here. Another effect, unique to EPR, also has important experimental consequences. For ions with strong spin-orbit coupling, the spin system is strongly coupled to the lattice

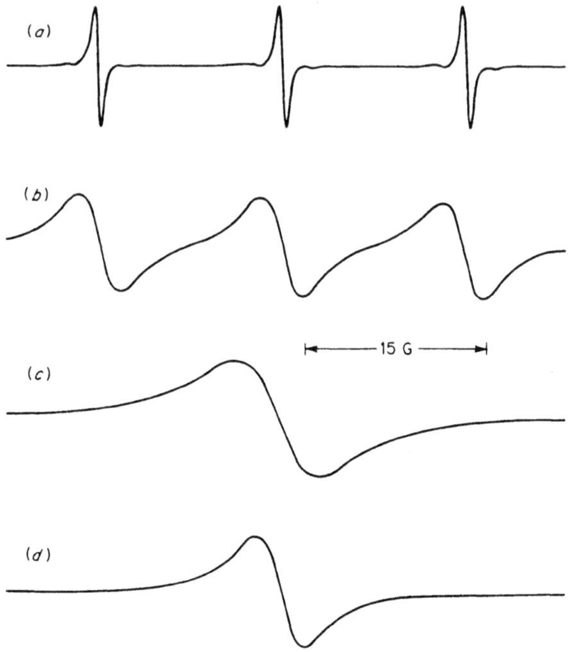

Fig. 23. First-derivative spectra of the di-t-butyl nitroxide radical in ethanol at room temperature at various radical concentrations. (a) 10^{-4}M. (b) 10^{-2}M. (c) 10^{-1}M. (d) Pure liquid nitroxide (21).

vibrations and the relaxation times are very short. This may yield EPR lines which, by simple uncertainty principle arguments, are too broad to be detectable. In that event, it is necessary to quench the lattice vibrations by working at very low temperatures. Contrasting examples are $Mn(H_2O)_6{}^{2+}$, which has relatively sharp lines at 300K, and $Mn(CN)_6{}^{4-}$, which can be observed only at 20K or below.

In the previous section, we considered the effect of magnetic exchange on the EPR spectrum of a system of coupled spins. There we saw that when the number of interacting species is small, the absorption curve may show some structure, but that a curve with no structure is expected when the number of coupled spins is large. Clearly, this effect is very similar to that of dipolar coupling in NMR spectroscopy, and can be treated by the method of moments. But the exchange interaction has an extra feature associated with it. One can show that the second moment of an absorption line is unaltered by an isotropic exchange term. However, van Vleck has found that the fourth moment is larger than it would be in the absence of exchange. This information is obviously insufficient to determine the complete lineshape, but it is equally obvious that increasing the fourth moment while keeping the second moment constant requires that the line be narrowed in the center and broadened in the wings. This is the phenomenon of "exchange narrowing". A nice example of this effect is shown in Fig. 23 (21). At low concentrations of the di-t-butylnitroxide radical in ethanol, the spectrum consists of three sharp lines owing to the hyperfine interaction with the ^{14}N (I=1) nucleus. At higher concentrations, the lines broaden, owing to the bimolecular exchange of the spin states of two free radicals. But, as the concentration is increased still further, the three lines coalesce into a single line, which becomes narrower at even higher concentrations. Under these conditions, one can say that the electron spins are exchanging so fast that the time-averaged value of the hyperfine field is close to zero. A similar exchange-narrowed spectrum is observed for most pure solid free radicals, such as are often found in coal. Here, as we have seen, the strong exchange arises from the overlap of molecular wavefunctions, making the identification of the individual paramagnetic species very difficult.

4. SPIN DYNAMICS.

The features of conventional NMR and EPR spectroscopy described in the preceding sections make it clear that a detailed knowledge of the dynamics of the process by which spin systems interact with radiofrequency or microwave fields is not necessary to interpret a typical spectrum. All one need do, given the results of a particular experiment, is adopt a "quantum-jump" picture of the absorption or emission process, assume a suitable spin Hamiltonian and basis set, and determine the set of eigenfunctions and eigenvalues which are required to fit the data. To be sure, this is sometimes a tedious process, but tedium has its rewards. Often it leads to important new discoveries, discoveries which would not have been made without sufficient attention to detail. But there is another dimension to the magnetic resonance experiment which is equally important. This is that the oscillating magnetic fields produced by typical radiofrequency or microwave sources are phase coherent, and affect each and every spin in the sample in exactly the same way. This fact is essential to the success of the experiment. For example, the ordinary incoherent emission from magnetized nuclei is far too weak to be detected by a crossed-coil NMR spectrometer. But, more importantly, the phase correlations induced by coherent sources greatly enhance the power of the magnetic resonance experiment. Studies of this coherence, and the mechanisms by which it decays, can provide a wealth of new information about the system under investigation. There is also a great deal of interesting physics contained in the description of coherent processes in atoms and molecules. Understanding this physics requires some knowledge of the properties of the density matrix, so we discuss it briefly before describing some of the simple, but elegant, consequences of coherence in magnetic resonance.

4.1. The Density Matrix (24).

We begin by considering a canonical ensemble of N identical quantum systems, each with its own set of internal coordinates q_n and wavefunction ψ_n. Each ψ_n is, of course, time dependent, but at any particular time it can be represented as a linear combination of functions $\phi_j(q_n)$ which are solutions of the time-independent Schrödinger equation. Thus,

$$\psi_n(q_n,t) = \sum_j c_{nj}(t)\ \phi_j(q_n) \tag{146}$$

If ψ_n varies in time, so must the c_{nj}'s. Now, suppose one wishes to calculate the expectation value of some physically observable property of the system, such as the x component of the magnetization, $\underline{M}$. We have, then,

$$\langle M_x\rangle_n = \int \psi_n^*\ \hat{M}_x\ \psi_n\ dq_n\ dt \tag{147a}$$

$$= \sum_{j,k} c_{nj}^* c_{nk}\ \langle j|\hat{M}_x|k\rangle \tag{147b}$$

If we perturb the system, $\langle M_x\rangle_n$ will change because the coefficients c_{nj} will change, but the matrix elements $\langle j|\hat{M}_x|k\rangle$ will remain the same. Therefore, to compute any observable, we need to specify the c_{nj}'s or the products $c_{nj}^*c_{nk}$. But since the products can be arranged to form a matrix they are more convenient. It may be imagined that the matrix $c_{nj}^*c_{nk}$ also represents some operator $\hat{P}_n$, the operator being defined by its matrix elements

$$\langle k|\hat{P}_n|j\rangle = c_{nk}c_{nj}^* \tag{148}$$

Thus, we can write

$$\langle M_x\rangle_n = \sum_{j,k} \langle k|\hat{P}_n|j\rangle\langle j|\hat{M}_x|k\rangle \tag{149a}$$

$$= \sum_k \langle k|\hat{P}_n\hat{M}_x|k\rangle \tag{149b}$$

$$= \mathrm{Tr}(\hat{P}_n\hat{M}_x) = \mathrm{Tr}(\hat{M}_x\hat{P}_n) \tag{149c}$$

since the ϕ_j form a complete set. In spite of its simplicity, this is a very important result which provides the "recipe" for finding the time-dependent expectation value of any quantum mechanical property of the system. But, before writing down this recipe, let us see how to compute the ensemble average of the property, since that is usually what one measures experimentally. If the N systems making up that ensemble are independent, we can write

$$\langle\overline{M_x}\rangle = \frac{1}{N}\sum_{n=1}^{N}\langle M_x\rangle_n \tag{150}$$

The products $c_{nj}c_{nk}^*$ may, of course, vary from system to system, but since the systems are identical, the matrix elements $\langle j|\hat{M}_x|k\rangle$ remain the same. Thus, Eq. (149c) may be used on the right-hand side of Eq. (150) to yield

$$\langle\overline{M_x}\rangle = \frac{1}{N}\sum_{n=1}^{N}\mathrm{Tr}(\hat{P}_n\hat{M}_x) \tag{151a}$$

$$= \mathrm{Tr}[(\frac{1}{N}\sum_{n=1}^{N}\hat{P}_n)\hat{M}_x] \tag{151b}$$

If we then define

$$\hat{\rho} \equiv \frac{1}{N}\sum_{n=1}^{N}\hat{P}_n \tag{152}$$

we have

$$\langle\overline{M_x}\rangle = \mathrm{Tr}(\hat{\rho}\hat{M}_x) = \mathrm{Tr}(\hat{M}_x\hat{\rho}) \tag{153}$$

From this result we see that the ensemble-averaged value of M_x is the trace of the product of the matrices of $\hat{M}_x$ and $\hat{\rho}$. The matrix of $\hat{\rho}$ is, in turn, the linear average of the $\hat{P}_n$ matrices representing individual systems of the ensemble. The time-dependent operator $\hat{\rho}$ is called the density operator, and its matrix is called the density matrix. The density matrix obeys the equation of motion

$$\frac{d\hat{\rho}}{dt} = \frac{i}{\hbar}[\hat{\rho},\hat{H}_n] \tag{154}$$

where $\hat{H}_n$ is the Hamiltonian of the nth system. Eq. (154) is the quantum-mechanical analog of the Liouville equation which occurs in classical statistical mechanics.

The procedure, then, for calculating the expectation value of any property at a particular time t, correctly averaged over all systems of the ensemble, is (1) choose any complete orthonormal set of time-independent wavefunctions ϕ_j and (2) form the matrix of the time-dependent operator $\hat{P}_n$ in that basis. Then, (3) choose an initial state for the ensemble by defining the j^2 elements of the density matrix $\hat{\rho}$ at time t = 0, using Eq. (152), and (4) solve Eq. (154) for the values of the matrix elements of $\hat{\rho}$ at the time of measurement. Next, (5) form the time-independent matrix of the operator representing the property of interest (e.g., $\hat{M}_x$) in the same basis. Finally, (6) multiply the two matrices together and sum the diagonal elements, as indicated in Eq. (153).

4.2. Properties of the Density Matrix. Coherence (25).

Clearly, this is a very powerful procedure. Before applying it to the magnetic resonance experiment, let us examine some of the properties of the density matrix. First, note that since the trace of each of the $\hat{P}_n$ matrices is 1, the trace of $\hat{\rho}$ will also be 1 for all time. The individual elements of $\hat{P}_n$ and $\hat{\rho}$ are, however, time dependent. This time dependence has its origin in the Hamiltonian $\hat{H}_n$ which drives the density matrix in accord with Eq. (154). Typically, $\hat{H}_n$ contains terms of three different kinds:

$$\hat{H}_n = \hat{H}_0 + \hat{H}_1 + \hat{H}_R \tag{155}$$

One group of terms, called $\hat{H}_0$, is time independent and has relatively large matrix elements. It establishes the stationary states of the system and is therefore the object of greatest attention is conventional spectroscopy. Another group of terms, called $\hat{H}_1$, is time dependent and provides the mechanisms for transitions to occur between the eigenstates of $\hat{H}_0$. The third group of terms, called $\hat{H}_R$, represents the relaxation processes; these provide the mechanisms for the ensemble of systems to relax back to an equilibrium distribution among the states of $\hat{H}_0$, whenever it is displaced from equilibrium by $\hat{H}_1$.

Suppose we choose as a basis the set of functions ϕ_j which are eigenfunctions of the time-independent operator $\hat{H}_0$. We can then express the time dependence of a typical element of $\hat{P}_n$ by writing

$$P_{jk}^{(n)} = |c_{nj}||c_{nk}| \exp[i\Omega_{jk}^{(n)}] \tag{156a}$$

$$= |P_{jk}^{(n)}| \exp[i\Omega_{jk}^{(n)}] \tag{156b}$$

where $\Omega_{jk}^{(n)}$ is a phase factor

$$\Omega_{jk}^{(n)} = \S_{jk}^{(n)} + (E_j - E_k)t/\hbar \tag{157}$$

The corresponding element of $\hat{\rho}$ can be written

$$\rho_{jk} = \frac{1}{N}\sum_{n=1}^{N} |P_{jk}^{(n)}| \exp[i\Omega_{jk}^{(n)}] \tag{158}$$

Consider, now, a diagonal element of $\hat{\rho}$,

$$\rho_{jj} = \frac{1}{N}\sum_{n=1}^{N} |P_{jj}^{(n)}| = \frac{1}{N}\sum_{n=1}^{N} |c_{nj}|^2 \tag{159}$$

Clearly, this represents the fraction of the systems in the ensemble which have the energy E_j; thus, at equilibrium,

$$\rho_{jj} = \frac{1}{Q} \exp(-E_j/kT) \tag{160}$$

where Q is the partition function. Next, consider a typical off-diagonal element

$$\rho_{jk} = \frac{1}{N}\sum_{n=1}^{N} |P_{jk}^{(n)}| \{\cos[\Omega_{jk}^{(n)}] + i \sin[\Omega_{jk}^{(n)}]\} \tag{161}$$

It is customary in statistical mechanics to assume that all phases $\Omega_{jk}^{(n)}$ are independent of the amplitudes $P_{jk}^{(n)}$, and that all phases are distributed at random throughout the ensemble. This hypothesis, called the "hypothesis of random phases", causes all of the off-diagonal elements to vanish at equilibrium. But suppose that, after equilibrium is established, we "turn on" $\hat{H}_1$, and that this driving field is coherent. Phase correlations are then induced among the matter waves of the systems in the ensemble. In other words, the matter waves begin to oscillate in step with one another. As the phases $\Omega_{jk}^{(n)}$ of the systems within the ensemble become more and more nearly the same, Eq. (161) shows that the density matrix approaches more and more closely the typical $\hat{P}_n$ matrix in <u>all</u> its elements. We then say that the ensemble has been excited coherently. With perfect coherence among the matter waves of the systems, the entire ensemble will mimic the behavior of each of its component systems. In this way, the microscopic behavior of the individual atoms, ions, or molecules will be "seen" in the macroscopic world. This is a remarkable result, and has several manifestations.

4.3. The Density Matrix for Magnetic Resonance (25).

Some of the most interesting manifestations of coherence in

quantum mechanical systems can be found in magnetic resonance. To keep things simple, let us consider a two-level system (TLS) in which there are only two stationary states, 1 and 2. In that event, the wavefunction for the nth system is, at any time t,

$$\psi_n(q_n,t) = c_{n1}(t)\ \phi_1(q_n) + c_{n2}\ \phi_2(q_n) \tag{162}$$

The density matrix at equilibrium is, then

$$\hat{\rho}_0 = \begin{pmatrix} \rho^0_{11} & \rho^0_{12} \\ \rho^0_{21} & \rho^0_{22} \end{pmatrix} = \frac{1}{Q}\begin{pmatrix} \exp(-E_1/kT) & 0 \\ 0 & \exp(-E_2/kT) \end{pmatrix} \tag{163}$$

Now, let us assume that the perturbation represented by $\hat{H}_1$ commences, driving the ensemble away from equilibrium. If, at some later time, $\hat{H}_1$ is turned off, each of the elements of $\hat{\rho}$ begins to decay back to its equilibrium value due to the effect of $\hat{H}_R$. If all decays are first order, we can write

$$\frac{d\rho_{jk}}{dt} = -\frac{(\rho_{jk}-\rho^0_{jk})}{T_{jk}} \tag{164}$$

where the (T_{jk}^{-1})'s are rate constants. Not all the T_{jk}'s are independent. Thus, in a TLS, since $\rho_{11} + \rho_{22} = 1$, the relaxation times for the diagonal elements must be equal. Conventionally, this time is called the longitudinal relaxation time (T_1). T_{12} must also equal T_{21}, because ρ_{12} is just the complex conjugate of ρ_{21} and must therefore decay to equilibrium at the same rate ($\hat{\rho}$ is Hermitian). Thus, there is only one possible additional relaxation time in a TLS; this is called the transverse relaxation time (T_2). The elements of the density matrix must therefore obey the following four differential equations:

$$\frac{d\rho_{11}}{dt} = \frac{[(\hat{H}_0+\hat{H}_1),\hat{\rho}]_{11}}{i\hbar} - \frac{\rho_{11}-\rho^0_{11}}{T_1} \tag{165a}$$

$$\frac{d\rho_{12}}{dt} = \frac{[(\hat{H}_0+\hat{H}_1),\hat{\rho}]_{12}}{i\hbar} - \frac{\rho_{12}}{T_2} \tag{165b}$$

$$\frac{d\rho_{21}}{dt} = \frac{[(\hat{H}_0+\hat{H}_1),\hat{\rho}]_{21}}{i\hbar} - \frac{\rho_{21}}{T_2} \tag{165c}$$

$$\frac{d\rho_{22}}{dt} = \frac{[(\hat{H}_0+\hat{H}_1),\hat{\rho}]_{22}}{i\hbar} - \frac{\rho_{22}-\rho^0_{22}}{T_1} \tag{165d}$$

The alert reader will immediately recognize the similarity between these relations and the Bloch equations, Eq. (44).

We can make this analogy a little stronger by requiring that $\hat{H}_1$ have only off-diagonal matrix elements. If $\hat{H}_1$, as originally defined, has diagonal elements, these are included in $\hat{H}_0$. In that event, Eq. (165) becomes

$$\frac{d\rho_{11}}{dt} = \frac{2\mathrm{Im}[(\hat{H}_1)_{12}\rho_{21}]}{\hbar} - \frac{\rho_{11}-\rho_{11}^0}{T_1} \tag{166a}$$

$$\frac{d\rho_{12}}{dt} = i\omega_0\rho_{12} + \frac{(\hat{H}_1)_{12}(\rho_{22}-\rho_{11})}{i\hbar} - \frac{\rho_{12}}{T_2} \tag{166b}$$

$$\frac{d\rho_{21}}{dt} = -i\omega_0\rho_{21} - \frac{(\hat{H}_1)_{12}^*(\rho_{22}-\rho_{11})}{i\hbar} - \frac{\rho_{21}}{T_2} \tag{166c}$$

$$\frac{d\rho_{22}}{dt} = -\frac{2\mathrm{Im}[(\hat{H}_1)_{12}\rho_{21}]}{\hbar} - \frac{\rho_{22}-\rho_{22}^0}{T_1} \tag{166d}$$

where $\omega_0 = (E_2-E_1)/\hbar$. If the total magnetic field experienced by the sample is

$$\underline{H}(t) = H_0\,\underline{k} + H_1(t)\,\underline{i} \tag{167}$$

then the Hamiltonian is

$$\hat{H}_n = \hat{H}_0 + \hat{H}_1 = -\gamma_f[H_0F_z + H_1(t)F_x] \tag{168}$$

where $H_1(t) = 2H_1^0 \cos \omega t$. In the basis of the two spin functions $|\alpha\rangle$ and $|\beta\rangle$ for F = 1/2,

$$F_x = \frac{\hbar}{2}\begin{pmatrix} 0 & 1 \\ 1 & 0 \end{pmatrix} \tag{169a}$$

$$F_y = \frac{\hbar}{2}\begin{pmatrix} 0 & -i \\ i & 0 \end{pmatrix} \tag{169b}$$

$$F_z = \frac{\hbar}{2}\begin{pmatrix} 1 & 0 \\ 0 & -1 \end{pmatrix} \tag{169c}$$

Therefore,

$$(\hat{H}_1)_{12} = (\hat{H}_1)_{21} = -\frac{1}{2}\gamma_f\hbar H_1(t) \tag{170a}$$

$$(\hat{H}_1)_{11} = (\hat{H}_1)_{22} = 0 \tag{170b}$$

and

$$\frac{d\rho_{11}}{dt} = -\gamma_f H_1(t)\mathrm{Im}(\rho_{21}) - \frac{\rho_{11}-\rho_{11}^0}{T_1} \tag{171a}$$

$$\frac{d\rho_{12}}{dt} = i\omega_0\rho_{12} + \frac{i\gamma_f H_1(t)(\rho_{22}-\rho_{11})}{2} - \frac{\rho_{12}}{T_2} \tag{171b}$$

$$\frac{d\rho_{21}}{dt} = -i\omega_0\rho_{21} - \frac{i\gamma_f H_1(t)(\rho_{22}-\rho_{11})}{2} - \frac{\rho_{21}}{T_2} \tag{171c}$$

$$\frac{d\rho_{22}}{dt} = \gamma_f H_1(t)\mathrm{Im}(\rho_{21}) - \frac{\rho_{22}-\rho_{22}^0}{T_1} \tag{171d}$$

Now, we already know that all magnetic moments in the sample are precessing in cones about the z axis at the Larmor frequency, ω_0. This fast motion in the laboratory coordinate system interferes with the visualization of the problem and may be eliminated by transforming everything into a rotating coordinate system. This has no effect on the diagonal elements of $\hat{\rho}$, but the off-diagonal elements become

$$\rho_{12}' = \rho_{12}\exp(-i\omega_0 t) \tag{172a}$$

$$\rho_{21}' = \rho_{21}\exp(+i\omega_0 t) \tag{172b}$$

If Eq. (172) are substituted into Eq. (171), and one makes the identifications

$$u \equiv \frac{1}{2} N'\gamma_f\hbar(\rho_{12}' + \rho_{21}') = N'\gamma_f\hbar\, \mathrm{Re}(\rho_{12}') \tag{173a}$$

$$v \equiv -\frac{1}{2i} N'\gamma_f\hbar(\rho_{12}'-\rho_{21}') = -N'\gamma_f\hbar\, \mathrm{Im}(\rho_{12}') \tag{173b}$$

$$M_z \equiv \frac{1}{2} N'\gamma_f\hbar(\rho_{11}-\rho_{22}) \tag{173c}$$

$$M_0 \equiv \frac{1}{2} N'\gamma_f\hbar(\rho_{11}^0-\rho_{22}^0) \tag{173d}$$

where N' is the number of spins per unit volume, one finds, after some algebra, that

$$\frac{du}{dt} = (\omega_0-\omega)v - \frac{u}{T_2} \tag{174a}$$

$$\frac{dv}{dt} = -(\omega_0-\omega)u + \omega_1 M_z - \frac{v}{T_2} \tag{174b}$$

$$\frac{dM_z}{dt} = -\omega_1 v - \left(\frac{M_z-M_0}{T_1}\right) \tag{174c}$$

where $\omega_1 = \gamma_f H_1^0$. These are the Bloch equations in the rotating frame, derived earlier [cf., Eq. (45)] based on a phenomenological approach to the magnetic resonance experiment. But now we have a much clearer understanding of the problem from a fundamental point of view. Thus, Eq. (173) show that u, the component of magnetization in phase with the driving field, is proportional to the sum of the off-diagonal elements of $\hat{\rho}$ in the rotating frame; that v, the out-of-phase component, is proportional to the difference of the off-diagonal elements; and that M_z is proportional to the population difference, $\rho_{11}-\rho_{22}$.

4.4. Applications. Magnetic Resonance in the Time Domain.

Equations (174) contain the answers to all questions that could possibly be asked about an ensemble of spin 1/2 particles. For example, we can easily calculate the ensemble-averaged value of the magnetization in the laboratory frame,

$$\langle\overline{M}\rangle = N'(\langle\overline{M}_x\rangle\,\underline{i} + \langle\overline{M}_y\rangle\,\underline{j} + \langle\overline{M}_z\rangle\,\underline{k}) \tag{175}$$

using the density matrix approach. The x component is given by [*cf.*, Eq. (153)]

$$\langle\overline{M}_x\rangle = \mathrm{Tr}\left\{\begin{pmatrix}\rho_{11} & \rho_{12}\\ \rho_{21} & \rho_{22}\end{pmatrix}\begin{pmatrix}(M_x)_{11} & (M_x)_{12}\\ (M_x)_{21} & (M_x)_{22}\end{pmatrix}\right\} \tag{176a}$$

$$= \rho_{11}(M_x)_{11} + \rho_{12}(M_x)_{21} + \rho_{21}(M_x)_{12} + \rho_{22}(M_x)_{22} \tag{176b}$$

Since $M_x = \gamma_f F_x$, we have, from Eq. (169a)

$$\langle\overline{M}_x\rangle = \frac{1}{2}\gamma_f\hbar(\rho_{12}+\rho_{21}) \tag{177}$$

From Eq. (172),

$$\langle\overline{M}_x\rangle = \frac{1}{2}\gamma_f\hbar[\rho'_{12}\exp(i\omega_0 t) + \rho'_{21}\exp(-i\omega_0 t)] \tag{178}$$

and from Eq. (173),

$$\langle\overline{M}_x\rangle = v\sin\omega t + u\cos\omega t \tag{179}$$

Similarly,

$$\langle\overline{M}_y\rangle = v\cos\omega t - u\sin\omega t \tag{180}$$

Thus,

$$\langle\overline{M}\rangle = (v\sin\omega t + u\cos\omega t)\,\underline{i} + (v\cos\omega t - u\sin\omega t)\,\underline{j} + M_z\,\underline{k} \tag{181}$$

We can also obtain the steady-state solutions to the Bloch equations under conditions which are appropriate for the conventional magnetic resonance experiment. But we have done this before, so it is much more interesting to solve these equations under different conditions.

Suppose we start with the ensemble of spins at equilibrium with a static field $\underline{H}_0$ along the z axis. As we have seen, this means that

$$u = 0,\quad v = 0,\ \text{and}\ M_z = M_0 \tag{182}$$

Now, let us turn on the $\underline{H_1}$ field suddenly at $t = 0$. If the field satisfies the condition $\omega = \omega_0$, and we neglect relaxation, the Bloch equations become

$$\frac{du}{dt} = 0 \tag{183a}$$

$$\frac{dv}{dt} = \omega_1 M_z \tag{183b}$$

$$\frac{dM_z}{dt} = -\omega_1 v \tag{183c}$$

These equations are readily solved; the solutions are

$$u = 0 \tag{184a}$$

$$v = M_0 \sin \omega_1 t \tag{184b}$$

$$M_z = M_0 \cos \omega_1 t \tag{184c}$$

A simple physical interpretation of these results is possible (cf. Fig. 24). When $\underline{H_1}(t) = 2H_1^0 \cos\omega_0 t$ is turned on, half of it appears as a vector of fixed orientation along the x' axis. Since this is the only magnetic field present in the rotating frame at exact resonance, the magnetization, of magnitude M_0 and initially aligned with the z' axis by relaxation processes, begins to precess around it. This precession occurs in a clockwise direction (for $\gamma_f > 0$) about the $\underline{x}'$ axis with a frequency of $\omega_1 = \gamma_f H_1^0$. Since H_1^0 varies from about 0.1 mG for high resolution NMR experiments to about 100 G for some pulse studies, the precession frequency for protons is $3\text{-}3\times10^6$ rad sec^{-1}. In a time t'(sec), the angle Θ through which $\underline{M}$ precesses is $\Theta = \omega_1 t'$(rad). Viewed from the laboratory frame, the motion is a precession about $\underline{H_0}$ at the high frequency $\omega_0 = \gamma_f H_0$, accompanied by a slow nutation, or change in the angle between $\underline{M}$ and $\underline{H_0}$, at the frequency ω_1. What is seen in the spectrometer is an absorption signal at ω_0 with a superimposed oscillation due to the nutation. The maxima in the oscillatory signal, corresponding to absorption of energy from the radiation field, occur as $\underline{M}$ passes the y' axis; the minima, corresponding to stimulated emission of the sample, occur as $\underline{M}$ passes the -y' axis. Transient nutations were first observed by Torrey for nuclear spins.

Suppose, now, that at some time t' during the course of a Torrey-type experiment, which is short compared to T_1 and T_2, $H_1(t)$ is turned off as suddenly as it was turned on. The Bloch equations which apply during the subsequent time interval are

$$\frac{du}{dt} = -\frac{u}{T_2} \tag{185a}$$

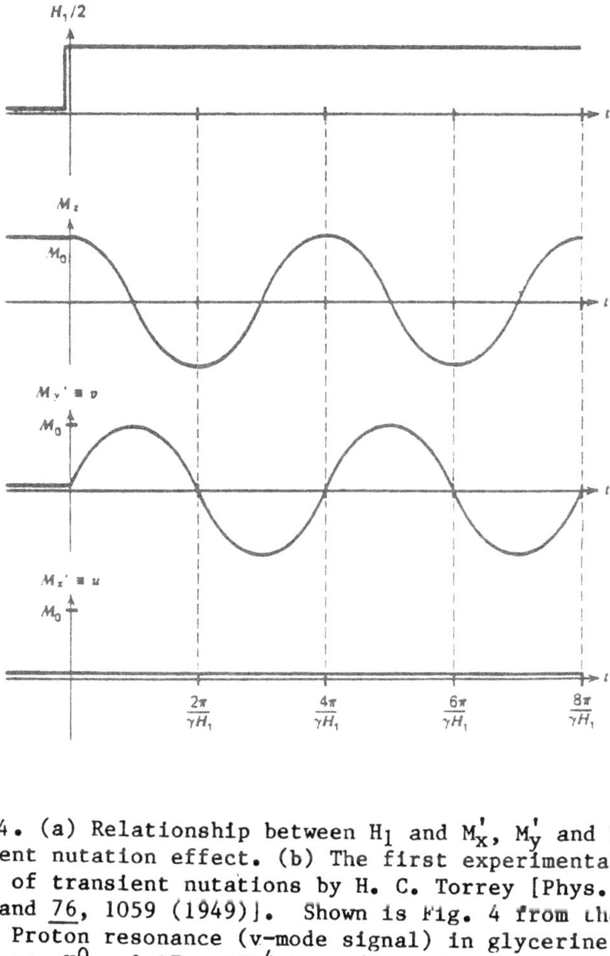

Fig. 24. (a) Relationship between H_1 and M_x', M_y' and M_z in the transient nutation effect. (b) The first experimental observation of transient nutations by H. C. Torrey [Phys. Rev. 75, 1326, and 76, 1059 (1949)]. Shown is Fig. 4 from the latter paper: Proton resonance (v-mode signal) in glycerine at exact resonance, $H_1^0 = 0.17 \times 10^{-4}$ T, $\omega_0/2\pi = 9.0$ MHz. The horizontal axis runs from 0 to 10 μs. Note the damping of the oscillations due to relaxation, corresponding to $T_1 \sim T_2 = 11$ μs (25).

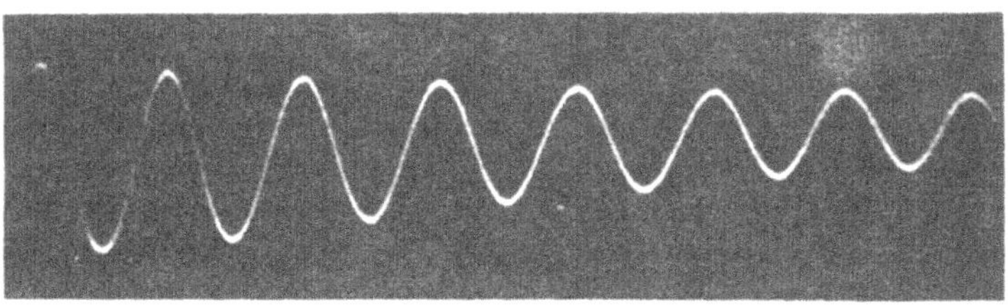

$$\frac{dv}{dt} = -\frac{v}{T_2} \tag{185b}$$

$$\frac{dM_z}{dt} = -\frac{(M_z - M_0)}{T_1} \tag{185c}$$

which have the solutions

$$u(t) = 0 \tag{186a}$$

$$v(t) = M_0 \sin\omega_1 t' \exp[-(t-t')/T_2] \tag{186b}$$

$$M_z(t) = M_0\{1 + [\cos\omega_1 t' - 1]\exp[-(t-t')/T_1]\} \tag{186c}$$

By properly choosing the amplitude H_1^0 so that

$$\omega_1 t' = \pi/2 \tag{187}$$

one sees from Eq. (186b) that this has the effect of bringing the entire magnetization to a position along the y' axis. The systems of the ensemble are then in superposition states halfway "between" α and β [i.e., $\rho_{11} - \rho_{22} = 0$]. The resulting magnetization, computed by substituting Eq. (186) into Eq. (181), is

$$\begin{aligned}\langle\overline{M}\rangle = M_0 \{ & \sin\omega_0 t \exp[-(t-t')/T_2]\, \underline{i} \\ & + \cos\omega_0 t \exp[-(t-t')/T_2]\, \underline{j} \\ & + [1 - \exp[-(t-t')/T_1]]\, \underline{k} \}\end{aligned} \tag{188}$$

Clearly, then, the x and y components of $\langle\overline{M}\rangle$ decay exponentially with a characteristic time T_2 whereas M_z grows exponentially back to its equilibrium value with a characteristic time T_1. Because these processes occur in the absence of an H_1 field, the behavior of a sample after a "$\pi/2$" pulse is called free-induction decay (FID). It was first observed by E. L. Hahn in an ensemble of nuclear spins.

The FID experiment is more than just an idle curiosity. This is because an exponential in the time domain and a Lorentzian in the frequency domain are Fourier transforms of each other. Mathematically, we can write

$$T_2/[1 + T_2^2(\omega-\omega_0)^2] = \int_0^\infty \exp(-t/T_2) \cos(\omega-\omega_0)t \, dt \tag{189}$$

The Lorentzian function on the left-hand side of this relation is the expression for the absorption mode that comes from the steady-state solutions of the Bloch equations [cf. Eq. (51b)], whereas the expression on the right is the cosine transform of the FID.

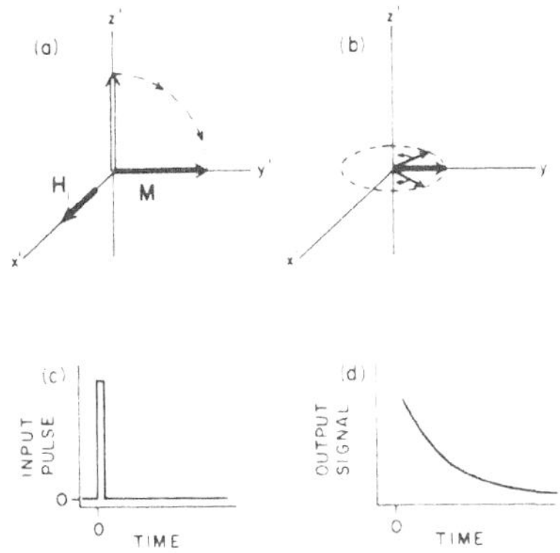

Fig. 25. (a) A 90° pulse along x' rotates M from the equilibrium position to the y' axis. (b) M decreases as magnetic moments dephase. (c) Input signal, a 90° pulse, corresponding to (a). (d) Exponential free induction decay, corresponding to (b) (27).

Thus, the Fourier transform of the FID following a 90° pulse gives the magnetic resonance spectrum in the frequency domain, provided certain conditions are met (27). This is important, because the experiment in the time domain provides spectral information in a much more efficient manner than does a conventional experiment in the frequency domain. Essentially, this is because the short pulse of RF or microwave energy contains component frequencies which can span the entire range of the spectrum; thus, the time required to record a spectrum is independent of its width. If 90° pulses are applied at frequent intervals and the resulting FID's are signal averaged in a digital computer, one can Fourier transform the accumulated signal and realize a significant improvement in signal-to-noise.

4.5. Some Typical Pulse Experiments.

In addition to providing significant advantages for the recording of spectra, relaxation times are also most conveniently measured using pulse methods. Let us consider a few of these here, beginning with the FID experiment (cf., Fig. 25). Suppose, again, that a resonant 90° pulse is applied along the x' axis in the rotating frame, bringing the entire M to the y' axis. Because of the processes, already discussed, that cause spins to exchange energy with each other, different spins begin to precess at different rates and the individual moments begin to spread out in the x'y' plane. Clearly, this corresponds to a dephasing of

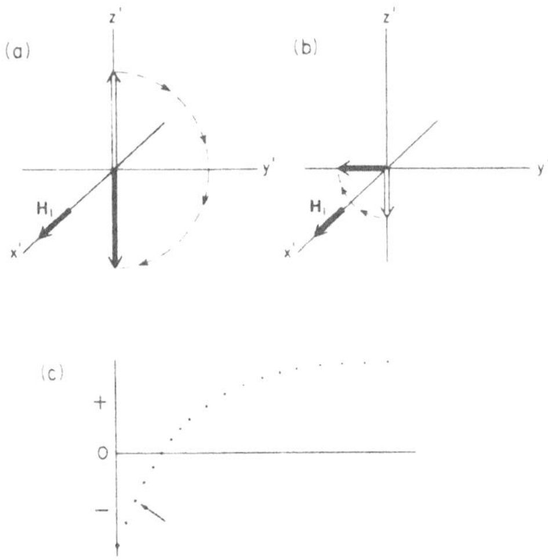

Fig. 26. Determination of T_1 by 180°, τ, 90 sequences. (a) $\underline{M}$ is inverted by a 180° pulse at time 0. (b) After a time τ a 90° pulse rotates $\underline{M}$ to the y' (or −y') axis. (c) The initial amplitude of the FID after the 90° pulse, which is proportional to the value of $\underline{M}$ at time τ, is plotted as a function of τ. Note that each point results from a separate 180°, τ, 90° sequence. The point corresponding to (b) is indicated by the arrow (27).

the coherent superposition state, with a time constant of T_2 that can be measured directly. Spin-lattice relaxation also occurs as the moments lose energy to their surroundings and thus relax, or tip back, toward the z' axis with a time constant of T_1. Subsequent repeated rapid scans of the spectrum with a weak frequency-swept field will reveal a gradual return of the signal to its initial strength, because the M_x component is proportional to M_z. In this way, one can determine T_1.

One can also determine T_1 by applying a π pulse (Fig. 26). Before applying the pulse, the sample is in equilibrium and therefore has a magnetization $M_0\underline{k}$. After the π pulse, the magnetization will be $-M_0\underline{k}$. (All pulses are presumed to be very brief in comparison with both T_1 and T_2). As the sample sits in the heat bath, it will again relax, but this time the relaxation occurs <u>along</u> the z' axis. $\underline{M}$ will shrink along the negative branch of that axis, pass through zero, and then regain its equilibrium value exponentially in a time T_1. How far this relaxation has proceeded after a given time interval τ can be

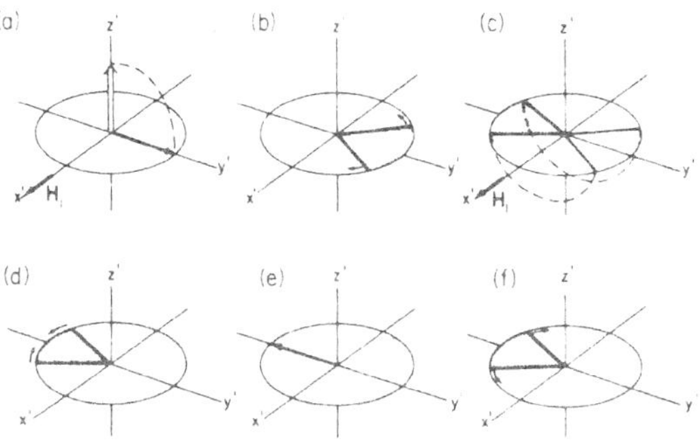

Fig. 27. The Hahn spin-echo experiment. (a) A 90° pulse applied along x' at time 0 causes M to tip to the positive y' axis. (b) The macroscopic magnetizations, m_i, of nuclei in different parts of the sample dephase as a result of the inhomogeneity in H_0. Those nuclei precessing faster than the average (the rotation rate of the frame) appear in the rotating frame to move toward the observer. Looking down from the positive z' axis they appear to move clockwise, while those slower than average move counterclockwise. (c) A 180° pulse along x' at time τ causes all m_i to rotate 180° about the x' axis. (d) The faster nuclei, still moving clockwise in the rotating frame, now go away from the observer, while the slower nuclei move counterclockwise toward the observer. (e) At time 2τ the m_i rephase along the -y' axis. (f) At time $>2\tau$ the m_i again dephase (27).

determined by measuring the length of the magnetization vector at that time. This can be done by applying a $\pi/2$ pulse; whatever M_z happens to be, it will be tipped into the xy plane and will generate an FID signal. By plotting the initial signal amplitudes versus pulse separation τ, the entire decay process can be mapped out and T_1 measured.

The most dramatic transient coherence effect in all magnetic resonance spectroscopy is the phenomenon of spin echoes, also discovered by Hahn. The method consists of the application of a 90°, τ, 180° pulse sequence and the observation at a time 2τ of a free induction "echo". The rationale of the method is shown in Fig. 27, which depicts the behavior of the magnetization in the rotating frame. In (a), M is shown being tipped through 90° by application of H_1 along the +x' axis. The total magnetization can be thought of as the vector sum of individual macroscopic magnetizations m_i arising from spins in different parts of the

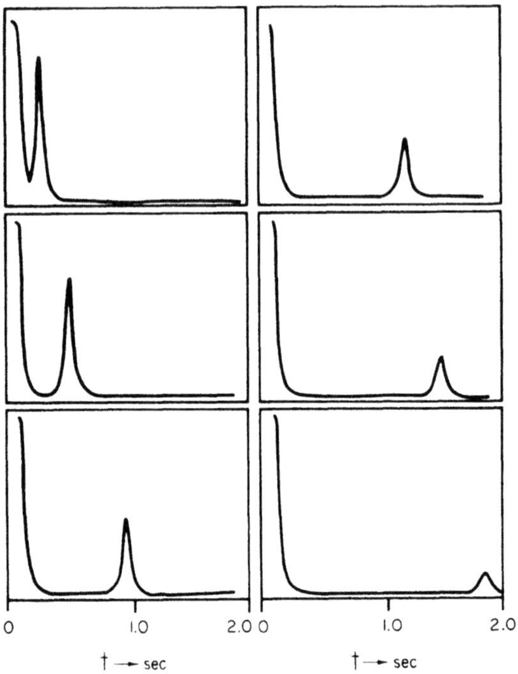

Fig. 28. A typical Hahn spin-echo experiment, consisting of six 90°, τ, 180° sequences, with τ varying from 0.1-1 sec. Note that the amplitude of the echo decreases as τ increases. In this experiment the echoes are positive, rather than negative as expected, since a diode detector was used, which measures only amplitude, not phase of the signal (27).

sample and hence experiencing slightly different values of the applied field, which is never perfectly homogeneous. There is thus a range of precession frequencies centered at ω_0. In (b), the $\underline{m}_i$ begin to fan out, as some spins precess faster and some slower than ω_0. At a time τ after the 90° pulse, a 180° pulse is applied, also along the +x' axis, as shown in (c). The effect of this pulse is to rotate each $\underline{m}_i$ by 180° about the x' axis. Those $\underline{m}_i$ that are moving faster than the frame naturally continue to move faster, but in (d) their motion is now away from the observer, rather than toward him. At time 2τ all $\underline{m}_i$ come into phase along the -y' axis, as shown in (e). This rephasing causes an FID signal to build to maximum at 2τ, but the signal is, of course, negative relative to the initial FID. Each $\underline{m}_i$ decreases in magnitude as τ is increased because of transverse relaxation. Thus, the echo amplitude depends on T_2, and this quantity may be determined from a plot of echo amplitude as a function of τ. A typical Hahn spin-echo experiment is shown in Fig. 28.

Spin echoes have been widely employed in NMR spectroscopy. More recently, they have found application in EPR as well. Many of these applications have been discussed by Mims (28), as well as by others at this Conference. We should also add that we have barely scratched the surface of possible pulse methods. There are, for example, a number of rotating frame experiments which have already proven very useful for obtaining, among other things, high resolution spectra in solids and information about relaxation processes occurring at much lower frequencies than those encountered in ordinary pulse or CW methods. But, as we have already exceeded the space and time allotted by the Editors of these Proceedings, I will leave these subjects for others to present. Suffice it to say that magnetic resonance is a rich and powerful technique which offers much to one who is fascinated with the microscopic, and macroscopic, world.

Acknowledgements.

Many persons have contributed to this article, directly and indirectly. They include the authors of the cited references, from which I have rather liberally taken material, and my students, from whom I have learned a great deal in the course of attempting to provide a coherent introduction to the subject. I am grateful to all of them. I also thank Ms. P. Napolitano, who expertly typed the manuscript with good cheer in spite of the constraints of time; Dr. L. Petrakis, for encouraging me to undertake this task; and the sponsors of this Conference for financial support. My own research in the field has been supported by the Department of Energy, the National Science Foundation, the Research Corporation, the Gulf Research Foundation, the Merck Foundation, the North Atlantic Treaty Organiztion, and the University of Pittsburgh.

5. REFERENCES.

(1). F. Bloch, Phys. 70, 460 (1946).

(2). A. Abragam, The Principles of Nuclear Magnetism, Oxford University Press, London, 1961.

(3). C. P. Slichter, Principles of Magnetic Resonance, Harper and Row, New York, 1963.

(4). A. Carrington and A. D. McLachlan, Introduction to Magnetic Resonance, Harper and Row, New York, 1967.

(5). L. I. Schiff, Quantum Mechanics, McGraw-Hill Book Co., New York, 1968.

(6). J. A. Pople, W. G. Schneider, and H. J. Bernstein, High-resolution Nuclear Magnetic Resonance, McGraw-Hill Book Co., New York, 1959.

(7). S. Pansak, A. Pines, and J. S. Waugh, J. Chem. Phys. 59, 591 (1973).

(8). A. Pines, M. G. Gibby, and J. S. Waugh, J. Chem. Phys. 59, 569 (1973).

(9). G. E. Pake, J. Chem. Phys. 16, 327 (1948).

(10). N. Bloembergen, E. M. Purcell, and R. V. Pound, Phys. Rev. 73, 679 (1948).

(11). For a review, see S. Meiboom and L. C. Snyder, Accounts Chem. Res. 4, 81 (1971).

(12). E. R. Andrew and R. G. Eades, Proc. Roy. Soc. (London) A218, 537 (1953).

(13). G. W. Canters, B. M. P. Hendriks, and E. de Boer, J. Chem. Phys. 53, 445 (1970).

(14). I. L. Pykett, Sci. Amer. 246, 78 (1982).

(15). T. P. Das and E. L. Hahn, Nuclear Quadrupole Resonance Spectroscopy, Academic Press, New york, 1958.

(16). W. Low, Paramagnetic Resonance in Solids, Academic Press, New York, 1960.

(17). A. Abragam and B. Bleaney, Electron Paramagnetic Resonance of Transition Ions, Clarendon Press, Oxford, 1970.

(18). G. E. Pake and T. L. Estle, The Physical Principles of Electron Paramagnetic Resonance, W. A. Benjamin, New York, 1973.

(19). P. W. Atkins and M. C. R. Symons, The Structure of Inorganic Radicals, Elsevier, Amsterdam, 1967.

(20). D. W. Ovenall and D. H. Whiffen, Mol. Phys. 4, 135 (1961).

(21). J. E. Wertz and J. R. Bolton, Electron Spin Resonance, McGraw-Hill Book Co., New York, 1972.

(22). L. Kevan and L. D. Kispert, Electron Spin Double Resonance Spectroscopy, J. Wiley & Sons, New York, 1976.

(23). D. W. Pratt, in Excited States, Vol. 4 (ed. E. C. Lim), Academic Press, New York, 1979, pp. 137-236.

(24). A nice discussion of the properties of the density matrix can be found in R. C. Tolman, The Principles of Statistical Mechanics, Oxford University Press, London, 1938. See also Refs. 3, 25 and 26.

(25). J. D. Macomber, The Dynamics of Spectroscopic Transitions, J. Wiley & Sons, New York, 1976.

(26). K. Blum, Density Matrix. Theory and Applications, Plenum Press, New York, 1981.

(27). T. C. Farrar and E. D. Becker, Pulse and Fourier Transform NMR, Academic Press, New York, 1971.

(28). W. B. Mims, in Electron Paramagnetic Resonance, (ed. S. Geschwind), Plenum Press, New York, 1972, pp. 263-351.

LINE-BROADENING INFLUENCES AND LINE-NARROWING TECHNIQUES IN SOLID STATE NMR

Gary E. Maciel

Department of Chemistry
Colorado State University
Ft. Collins, Colorado 80523 U.S.A.

ABSTRACT

The main spin interactions in NMR--chemical shift, magnetic dipole-dipole, nuclear electric quadrupole and indirect nuclear spin-spin (J) coupling--are introduced in terms of their effects on line widths in NMR. The relationships between NMR in liquids and solids are discussed. The techniques of decoupling and magic-angle spinning for line-narrowing in solids are described, with major emphasis of the latter. The use of line-narrowing techniques in solid-state NMR is examined.

I. INTRODUCTION.

A. High Resolution NMR.

Nuclear magnetic resonance (NMR) has achieved an almost unparallelled popularity for providing details of molecular structure and dynamics in chemistry and chemically-related areas of research and development. This popularity is largely based on the high-resolution character of most NMR experiments. For present purposes, we can define a "high-resolution" spectrum to be one in which individual details of molecular structure are manifested in specific regions and/or by specific patterns in the spectrum, regions that are distinguishable within the available characteristics of spectral dispersion and line widths. In NMR studies of liquid samples, this typically means that the dispersion of chemical shifts and the magnitudes of pertinent, indirect (J) spin-spin couplings are considerably larger than the natural widths of the resonances; otherwise, overlaps will preclude the

L. Petrakis and J. P. Fraissard (eds.), Magnetic Resonance. Introduction, Advanced Topics and Applications to Fossil Energy, 71–110.

distinctions of individual resonance lines on which structural interpretations of high-resolution spectra depend. For solid samples, the line widths that are normally attainable with available techniques usually obscure the small spin-spin couplings that are often so useful in liquid-state nmr; hence, the usual criterion for "high-resolution" in solid samples is the ability to distinguish resonances with reasonably different chemical shifts (e.g., the different ^{13}C chemical shifts in a substituted benzene).

A major technical goal of NMR has been to develop techniques for narrowing the resonance lines without sacrificing too much information on chemical structure or dynamics in the process. As we shall see below, this task is typically much more difficult for solid samples than for liquids. This fact is the reason why high-resolution NMR has been essentially a liquid-state technique until recently, and why liquid-sample NMR remains by far the most common type for high-resolution NMR applications.

B. Solids *vs.* Liquids in NMR.

As we shall see in greater detail below, the basic reason why line-broadening effects constitute more difficult problems in solid-state NMR than for liquid samples is that the rapid, random motions characteristic of liquids bring about a natural averaging of these line-broadening influences. Because of the more constrained motions present in solids, this kind of averaging must be brought about "artifically" by specific line-narrowing techniques; these techniques constitute the basis of modern high-resolution NMR approaches for solid samples.

In spite of the relative ease of obtaining high-resolution NMR spectra of liquid samples and the relative difficulty of employing line-narrowing techniques for solids, there are often compelling reasons for carrying out NMR experiments on solid samples. Of course, in some cases it is simply not possible to dissolve a solid sample of interest under conditions that would retain the structural integrity of the sample; examples are heterogeneous catalysts, coal and oil shale. In other cases it may be possible to dissolve a solid sample, so liquid-state experiments can be carried out, but it might be the properties of the solid that are primarily of interest; one expects this to be the case frequently with synthetic polymers. Also, the availability of high-resolution NMR capabilities for solids permits the use of NMR as a structure-determination bridge between the solid and liquid states of matter. This can prove useful in analytical work, where it is often pertinent to examine both a solid residue and an extract in an analytical procedure, and is highly relevant in fundamental structural chemistry. The most common source of detailed, accurate information on chemical structure is X-ray diffraction data on crystalline solids; yet chemical reactivity (and, hence, structure) is usually of more interest in the liquid-solution state. High-resolution NMR now has the capability of bridging this gap, in a sense extrapolating into the liquid state information that has been determined on crystals by X-ray diffraction.

Not only do the rapid, random motions in liquids greatly attenuate major line-broadening influences by motional averaging, these same motions also give rise to relaxation mechanisms. Relaxation is the process by which a nuclear spin set, having been perturbed from a condition of thermal equilibrium at the thermodynamic temperature of the entire sample, returns to the equilibrium state. The rate of this process is typically characterized by first-order kinetics, with a rate constant equal to the inverse of a relaxation time (T_p, where p specifies a particular type of relaxation process). In most cases of interest, the relaxation rate constant can be expressed as the sum of terms of the form, $Ck_p^2J(\omega_p)$. In this expression C is a collection of fundamental constants; k_p is a measure of the strength of the interaction giving rise to the relaxation; and $J(\omega_p)$ is the spectral density function characterizing the time dependence of the underlying interaction, evaluated at some frequency determined by the pertinent Larmor frequencies in the spin system. Hence, the contribution that a particular interaction (e.g., C-H dipole-dipole interaction) makes to a specific relaxation time is determined by the magnitude of the interaction and the components of the time dependence of this interaction at pertinent frequencies.

One important type of relaxation time is the <u>transverse</u> or spin-spin relaxation time, T_2, which is a measure of the rate at which equilibrium can be achieved in the components of nuclear spin magnetization in a plane transverse to the direction of the static magnetic field, H_0. An important component of this relaxation process proceeds by means of mutual spin-spin flip-flops. The pertinent rate constant ($1/T_2$) is directly related to the natural resonance line width (line width at half height, $\Delta\nu_{\frac{1}{2}}$) in the absence of inhomogeneous influences (e.g., inhomogeneous magnet, slightly different chemical shifts).

$$\Delta\nu_{\frac{1}{2}} = \frac{1}{\pi T_2} \tag{1}$$

These same fundamental spin interactions, the detailed time dependences of which determine the values of relaxation times, also give rise to both homogeneous and inhomogeneous contributions to NMR line widths in solids. In the next two sections we examine these line-broadening interactions in some detail and the means commonly employed to overcome their line-broadening effects on the NMR spectra of solids.

II. LINE BROADENING SPIN INTERACTIONS IN SOLIDS.

A. Types of Spin Interactions.

For spin-1/2 nuclei there are usually three types of nuclear spin interactions to consider in explaining the NMR spectrum of a diamagnetic solid. These are: 1) the chemical shift, which is due to the shielding of the applied field by electronic circulations, 2) the magnetic dipole-dipole interactions between nuclear magnetic moments and 3) the indirect (electronic-mediated) nuclear spin-spin coupling (J coupling). For nuclei with $I > 1/2$ there is also the nuclear electric quadrupole interaction, which is due to the electrical interaction between a non-spherically-symmetric nuclear electric charge distribution and the local electrical environment of the nucleus, as characterized by electric field gradients. In general all of these interactions depend on the orientation of the local molecular (or crystal) environment relative to the direction of $\underline{H}_0$, although the orientation dependence of J coupling is typically small. In fact, the magnitudes of J couplings are usually so much smaller than the intensities of competing spin interactions that contributions that J coupling can make to the line widths in solid-state NMR spectra are usually negligible. Another nuclear spin interaction that is important in NMR in special cases is the spin-rotation interaction, which involves a coupling between the nuclear spin angular momentum and angular momentum due to rapid rotations in molecules. However, this interaction is seldom of any importance in fossil fuel problems, and is not addressed further in this chapter.

B. Shielding. The Chemical Shift.

The spin interactions introduced above, including their orientation dependence, can be described quantum mechanically in terms of the Hamiltonian operator, which can be expressed in terms of tensors. A good example is the chemical shift or shielding interaction, brought about by the circulations of electrons in a sample in the presence of a magnetic field, thereby altering the field experienced by a nucleus. For a nucleus with spin angular momentum $\underline{I}$, the shielding-corrected Zeeman interaction of a given nucleus can be expressed (1,2) as

$$H_Z = -\gamma\hbar\underline{H}_0\cdot(\underset{\sim}{1} - \underset{\sim}{\sigma})\cdot\underline{I} = -\gamma\hbar\underline{H}_0\cdot\underline{I} + \gamma\hbar\underline{H}_0\cdot\underset{\sim}{\sigma}\cdot\underline{I} \qquad (2)$$

$$= -\gamma\hbar\,\underline{H}_0\cdot\underline{I} + H_{CS}$$

where $\underline{H}_0$ and $\underline{I}$ are vector quantities, and the shielding is represented by a second-rank tensor, $\underset{\sim}{\sigma}$. γ is the magnetogyric ratio of the nucleus and $\hbar = h/2\pi$, where h is Planck's constant. The chemical shift Hamiltonian, H_{cs}, can be expressed in terms of matrices.

$$H_{cs} = \gamma\hbar(H_{ox}\ H_{oy}\ H_{oz}) \begin{pmatrix} \sigma_{xx} & \sigma_{xy} & \sigma_{xz} \\ \sigma_{yx} & \sigma_{yy} & \sigma_{yz} \\ \sigma_{zx} & \sigma_{zy} & \sigma_{zz} \end{pmatrix} \begin{pmatrix} I_x \\ I_y \\ I_z \end{pmatrix} \quad (3)$$

The elements of $\underset{\sim}{\sigma}$ depend on the orientation of the molecular (or crystal) axes relative to $\underline{H}_0$ and the choice of an axis system. For one particular choice of axes called the principal axis system (PAS) of the chemical shift tensor, $\underset{\sim}{\sigma}$ is diagonal.

$$\underset{\sim}{\sigma} = \begin{pmatrix} \sigma_{11} & 0 & 0 \\ 0 & \sigma_{22} & 0 \\ 0 & 0 & \sigma_{33} \end{pmatrix} \quad (4)$$

The most important properties of $\underset{\sim}{\sigma}$ can be defined in terms of these principal elements as follows: If one adopts the convention, $\sigma_{11} < \sigma_{22} < \sigma_{33}$, then the anisotropy parameters δ and μ can be defined (3) as

$$\delta = \sigma_{33} - \frac{1}{3}\mathrm{Tr}\underset{\sim}{\sigma} \quad (5)$$

and

$$\mu = \sigma_{33} - \sigma_{11} \quad (6)$$

and the asymmetry parameter, ρ, is defined by

$$\rho = \frac{\sigma_{11} + \sigma_{33} - 2\sigma_{22}}{\sigma_{33} - \sigma_{11}} \quad (7)$$

If one chooses $\underline{H}_0$ to be directed along the z axis, then it can be shown that the shielding tensor element of interest is σ_{zz}; i.e., the "secular" chemical shift Hamiltonian (essentially that portion not averaged by Larmor precession), is given by:

$$H'_{cs} = \gamma\hbar H_0 \sigma_{zz} I_z \quad (8)$$

Hence, it is the value of σ_{zz} on which the chemical-shift-dependent resonance frequency is based, and σ_{zz} depends on the orientation of the principal axis system relative to the laboratory axis system.

For a sample consisting of a single crystal, σ_{zz} for a particular chemical environment (e.g., ^{13}C in H_2CO) has only a single value, determined by the values of σ_{11}, σ_{22} and σ_{33} and the orientation of the PAS relative to $\underline{H}_0$. In this case, considering only the chemical shift interaction, the spectrum consists of only a single line determined by the value of σ_{zz}, which depends on the orientation of the crystal. If, instead of a single crystal, one has a powder consisting of randomly oriented crystallites or an amorphous sample, then all possible orientations of the PAS relative to $\underline{H}_0$ exist in the sample, corresponding to all possible values of σ_{zz} between the limits, σ_{11} and σ_{33}. The resonance frequencies in such a sample will also range between values determined by ω_1 and ω_3, giving rise to what is referred to as a chemical shift anisotropy (CSA) powder pattern. This situation is depicted in Figure 1.

In Figure 1, σ_j represents the value of σ_{zz} for some specific, arbitrary orientation of the H_2CO crystal (and hence the orientation of the chemical shift PAS) relative to $\underline{H}_0$; ω_j is the corresponding resonance frequency (in radians/sec). ω_1, ω_2 and ω_3 are the resonance frequencies that arise from crystallite (H_2CO) orientations corresponding to values of σ_{11}, σ_{22} and σ_{33}, respectively, for σ_{zz}. ω_{iso} is the resonance frequency corresponding to the isotropic average of $\underset{\sim}{\sigma}$,

$$\sigma_{iso} = (1/3)\mathrm{Tr}\,\underset{\sim}{\sigma} = (1/3)(\sigma_{11} + \sigma_{22} + \sigma_{33}) \qquad (9)$$

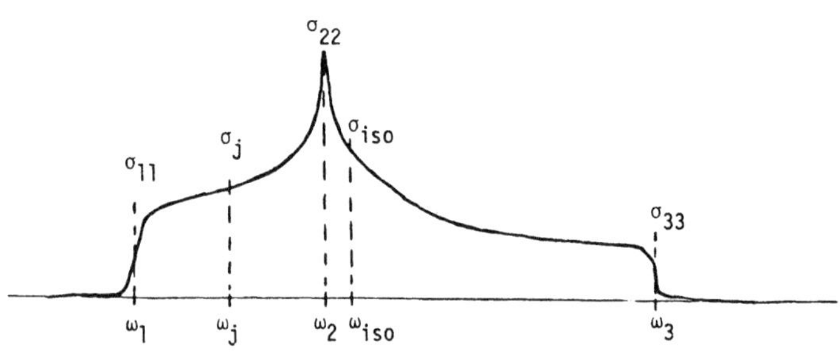

Figure 1. Powder pattern representing the ^{13}C chemical shift anisotropy of H_2CO. ω_1, ω_2 and ω_3 are resonance frequencies corresponding to σ_{zz} values of σ_{11}, σ_{22} and σ_{33}. ω_j is the resonance frequency corresponding to σ_j, which results from the jth orientation of a crystal.

The relative intensities represented in the CSA powder pattern shown in Figure 1 are determined by the relative probabilities of occurence of each orientation of the chemical shift PAS in a random collection of crystallites of a powder or in an amorphous sample. These can readily be computed analytically. Figure 1 resembles an experimental powder pattern (vide infra) in that sharp discontinuities at the extremes are absent because of the natural linewidth of the resonance corresponding to each orientation.

The CSA powder pattern shown in Figure 1 corresponds to a case for which $\sigma_{11} \neq \sigma_{22} \neq \sigma_{33}$. For some systems axial structural symmetry dictates an axially symmetric sheilding tensor; i.e., $\sigma_{11} = \sigma_{22}$ or $\sigma_{22} = \sigma_{33}$. In this type of system the expressions relating σ_{zz} or H'_{cs} to the principal shielding tensor elements take on particularly simple forms (1), e.g.,

$$H'_{cs} = \gamma\hbar H_0 \sigma_{11}(1 - 3\cos^2\beta)I_z - \hbar H_0(\sigma_{11} + \sigma_{22} + \sigma_{33})\cos^2\beta I_z \quad (10)$$

where $\sigma_{11} = \sigma_{22} < \sigma_{33}$ and β is the angle between $\underline{H}_0$ and the unique principal axis of the axially symmetric shielding tensor. An example of this kind of system is the ^{13}C chemical shift tensor of acetylene, as depicted in Figure 2. For this kind of chemical shift tensor, chemical shift powder patterns of the type shown in Figure 3 are observed.

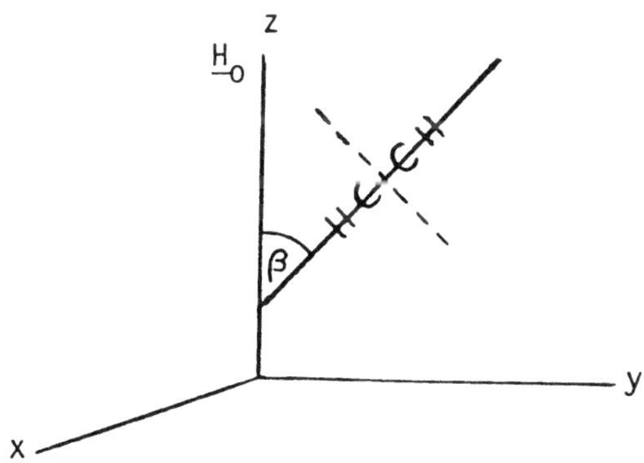

Figure 2. Orientation dependence of the ^{13}C shielding tensor of acetylene, showing the angle β between the direction of $\underline{H}_0$ (z axis) and the unique principal axis (bold line) of the shielding tensor, with the principal axis corresponding to $\sigma_{11} = \sigma_{22}$ shown as a dashed line.

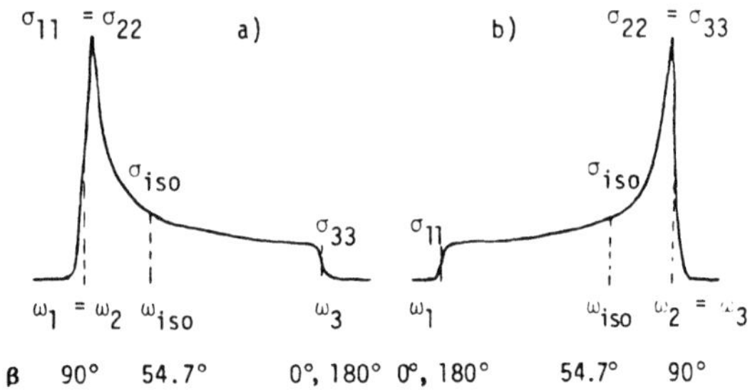

Figure 3. Chemical shift anisotropy powder patterns for axially symmetric shielding tensors, a) $\sigma_{11} = \sigma_{22} < \sigma_{33}$. b) $\sigma_{11} < \sigma_{22} = \sigma_{33}$. Note that in this and other theoretical plots in this chapter the resonance frequency increases from left to right (opposite to the usual convention for fixed-field experiments).

The widths of CSA powder patterns of the types shown in Figures 1 and 3 range from 5-to-20 ppm for protons to hundreds of ppm for nuclides such as ^{13}C or ^{15}N and even to thousands of ppm for some heavier nuclides. This can constitute a major or even dominant factor determining the overall character of the NMR spectrum of a solid. In the case of a very simple system, like solid CO, one could determine directly the resonance frequencies (ω_1, ω_2, ω_3) corresponding to σ_{11}, σ_{22}, and σ_{33} from the inflection points in the powder pattern. However, this analysis does not provide information on the orientations of the principal axes of the chemical shift tensor relative to crystallographic axes. That information comes most directly from experiments in which the resonance frequency of a single crystal is measured as a function of the systematic variation of the orientation of the crystal relative to the $\underline{H}_0$ axis (4).

If one investigates the detailed trigonometric relationships between σ_{zz} and σ_{11}, σ_{22} and σ_{33} in terms of the Euler angles (α, β,γ) that relate the chemical shift PAS to $\underline{H}_0$, one finds that random motional averaging yields an average shielding, $\bar{\sigma}$, which equals σ_{iso}. For the axially-symmetric shielding case, this result can be seen readily from Equation 10, the average value of $\cos^2\beta$ under isotropic rotational averaging is 1/3. Hence, the first term in Equation 10 vanishes, and the second term gives

$$\gamma\hbar H_0(\sigma_{11} + \sigma_{22} + \sigma_{33})(1/3)I_z = \gamma\hbar H_0\sigma_{iso}I_z \qquad (11)$$

Therefore, the result of the motion in typical liquids is that only an isotropic average resonance frequency, $\omega_{iso} = (1/3)(\omega_{11} + \omega_{22} + \omega_{33})$, is observed, rather than a powder pattern. This kind of motional averaging is the essence of motional line narrowing for liquids.

B. Magnetic Dipole-dipole Interactions.

A very important nuclear spin interaction, often dominating the lineshape in solid-state NMR and relaxation in liquid-state NMR, is the interaction between two nuclear magnetic dipoles, $\underline{\mu}_I = \gamma_I \hbar \underline{I}$ and $\underline{\mu}_S = \gamma_S \hbar \underline{S}$. In general this interaction can be represented by the dipolar Hamiltonian,

$$H_D = \underline{I} \cdot \underset{\sim}{D} \cdot \underline{S} \tag{12}$$

In this equation D is the dipole-dipole tensor, the elements of which depend on the orientation of the internuclear vector, $\underline{r}$, relative to $\underline{H}_o$, as seen schematically in Figure 4. The form of the tensor, $\underset{\sim}{D}$, is given by the following equation for its elements:

$$D_{ij} = \frac{\gamma_I \gamma_S \hbar^2}{r^5}(r^2 \delta_{ij} - 3r_i r_j) \tag{13}$$

where δ_{ij} is the Kronecker delta and $\underline{r} = r_x \underline{e}_x + r_y \underline{e}_y + r_z \underline{e}_z$ and with the $\underline{e}_i$ standing for unit vectors in the chosen coordinate system. For the heteronuclear case $(\gamma_I \neq \gamma_S)$ the secular dipolar Hamiltonian is found to have the form

$$H'_{IS} = \frac{\gamma_I \gamma_S \hbar^2}{r^3}(1 - 3\cos^2\beta) I_z S_z \tag{14}$$

If we view this Hamiltonian in terms of the Zeeman interaction of the S spin (say ^{13}C) as a perturbation due to a dipolar interaction with the I spin (say ^1H), then two cases can be considered, corresponding to two I spin states, with m = 1/2 and m = -1/2.

$$H_{IS}^{(1/2)} = \frac{\gamma_I \gamma_S \hbar^2}{2r^3}(1 - 3\cos^2\beta) S_z = (1/2) D\hbar (1 - 3\cos^2\beta) S_z \tag{15a}$$

$$H_{IS}^{(-1/2)} = -\frac{\gamma_I \gamma_S \hbar^2}{2r^3}(1 - 3\cos^2\beta) S_z = -(1/2) D\hbar (1 - 3\cos^2\beta) S_z \tag{15b}$$

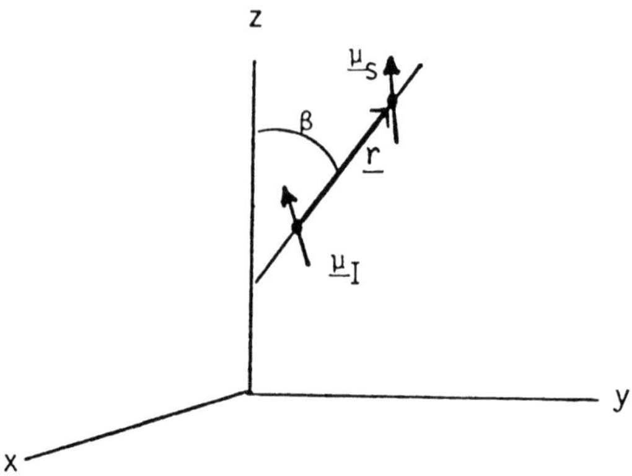

Figure 4. Schematic diagram showing the arrangement of two nuclear magnetic moments between which there is an internuclear vector, $\underline{r}$. $\underline{H}_0 = H_0\underline{e}_z$, and a dipole-dipole interaction.

The fact that Equation 13 shows that the dipolar tensor is symmetric ($D_{ij} = D_{ji}$) is consistent with the similarity between Equations 15 and the first term in the corresponding equation for an axially symmetric chemical shift tensor (Equation 10). Also, the fact that Equation 13 shows the dipolar tensor to be traceless corresponds to the fact that Equations 15 have no terms comparable to the second term in Equation 10. Hence, the effect of isotropic rotational averaging is to average the dipole-dipole interaction to zero (whereas $\underset{\sim}{\sigma}$ is averaged to $\sigma_{iso} \neq 0$). For this reason the resonance frequencies in liquid-state NMR show no evidence of dipole-dipole interactions, which are therefore manifested only as relaxation processes in liquids. This is another major line-narrowing consequence of molecular motion in liquids.

However, for powdered or amorphous solids, the random orientation of the internuclear vector ($\underline{r}$) relative to $\underline{H}_0$ -- i.e., the random variation of β -- can be expected to yield a range of resonance frequencies, just as is found to be the case for chemical shift anisotropy. Indeed, experimentally a dipolar powder pattern is found which has features quite similar to the CSA powder patterns for axially-symmetric systems (Figure 3). The form of the heteronuclear dipolar powder pattern is shown in Figure 5. In this figure the patterns due to $m_I = 1/2$ and $m_I = -1/2$ are shown separately with dashed lines, as predicted from Equations 15a and 15b, respectively, and the resultant of the two is shown above them.

The dipolar pattern represented in Figure 5, which has some broadening included in order to simulate the presence of a natural linewidth for individual values of β, shows a width of

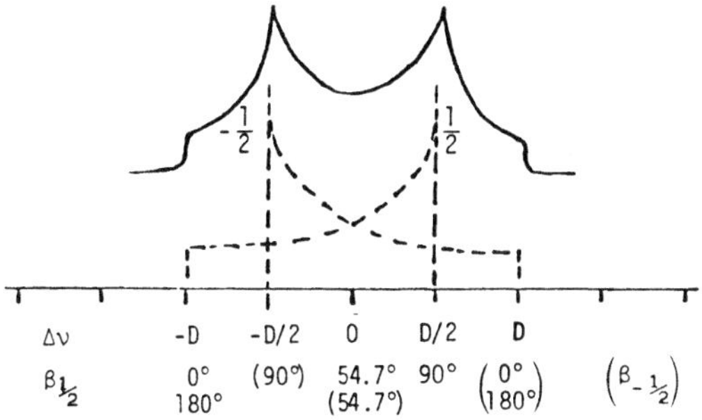

Figure 5. Powder pattern for the S resonance due to the heteronuclear dipolar interaction with a spin-1/2 I nucleus. Dashed lines show the contributions from the $m_I = 1/2$ and $m_I = -1/2$ states of I. Δν is the frequency shift due to the dipolar interaction. $\beta_{1/2}$ is the value of β for the $m_I = 1/2$ case. Corresponding numbers in parentheses show β values for $m_I = -1/2$.

roughly 2D. For a directly-bonded C-H pair, 2D corresponds to about 30 kHz; hence, the ^{13}C-^{1}H dipolar interaction can constitute a dominating line-broadening influence in the ^{13}C NMR spectrum of a solid, unless a suitable technique is used to eliminate it.

The development given so far for the heteronuclear dipolar interaction involves only two spins. In many solids a given nucleus experiences strong dipolar interactions with more than one other nucleus; and in most solids there is a manifold of weaker interactions with many nuclei. The net result is a complicated set of dipolar splittings. This is often exacerbated by a smearing effect due to spin-spin interactions (*vide infra* flip flops). The net result is usually a broad, featureless resonance band, unless a line-narowing influence, like molecular motion in liquids, is present to attenuate the dipolar effect on line width.

The form of Equation 14 shows that the heteronuclear dipolar interaction of the S spin with the I spin depends upon I_z. Hence, if rapid transitions of the I spins can be brought about, then a "spin averaging" of H_{IS} can be accomplished which has an effect analogous to that of motional averaging. This is the basis of heteronuclear decoupling, in which the application of rf energy at the I resonance frequency causes rapid I spin flips, so that the average of $\langle I_z \rangle$ is $\langle I_z \rangle = 0$. This is completely analogous to the ^{1}H spin decoupling used so commonly in obtaining ^{13}C spectra of liquid samples. The main difference is the magnitude of the interaction. For ^1H decoupling in a ^{13}C NMR experiment on a liquid, couplings of up to only about 160 Hz need be overcome because the ^{13}C-^1H dipolar interaction is eliminated by motional averaging and only the J coupling remains. For ^1H decoupling in the ^{13}C NMR spectrum of a solid, ^{13}C-^1H dipolar interactions of up to about 30 KHz must be averaged; hence, very strong ^1H rf fields (H_{1H}) must be employed ($\gamma_H H_{1H}$ > 30 KHz). Therefore, the rf technology used in the NMR probe is far more demanding for solid-state NMR than for liquids.

The case of homonuclear dipole-dipole interactions between nuclear spins is very similar in many respects to the heteronuclear case described above. Of course, the most important example of this case for organic solids is ^1H-^1H interactions, since ^1H is so highly abundant in nature. In this case, with $\gamma_I = \gamma_S$ and denoting $\underset{\sim}{I}$ as $\underset{\sim}{I}_i$ and $\underset{\sim}{S}$ as $\underline{I}_j$, a secular Hamiltonian analogous to that given in Equation 14 can be derived from Equation 12 and 13, and has the form,

$$H_{II} = \frac{\gamma_I^2 \hbar^2}{r^3}(1/2)(1-3\cos^2\beta)(3I_{z_i}I_{z_j} - \underline{I}_i \cdot \underline{I}_j) \tag{16}$$

This interaction leads to powder patterns analogous to those obtained for the heteronuclear case.

By analogy with Equations 10 and 14, the presence of the factor $1 - 3\cos^2\beta$ in Equation 16 ensures that a rapid, random, motion will average H_{II} to zero. Hence, the motion characteristic of the liquid state renders the homonuclear dipolar interaction inoperative in determining peak positions or contribution to inhomogenous relaxation in liquids. However, the modulation of H_{II} by motion in liquids is often the primary homogeneous source of relaxation, e.g., in proton spectra.

For solid samples, especially in ^1H or ^{19}F NMR, the homonuclear dipole-dipole interaction is often dominant, producing dipolar powder patterns that are so broad (several KHz) that they obscure any high resolution NMR information that one might hope to observe. In solids strong ^1H-^1H dipolar interactions are also responsible for the rapid spin-spin flip-flop transitions that give rise to efficient spin diffusion (1,2) and large natural (homogeneous) linewidths (1,2). These flip-flop transitions are represented schematically in Figure 6, and are made

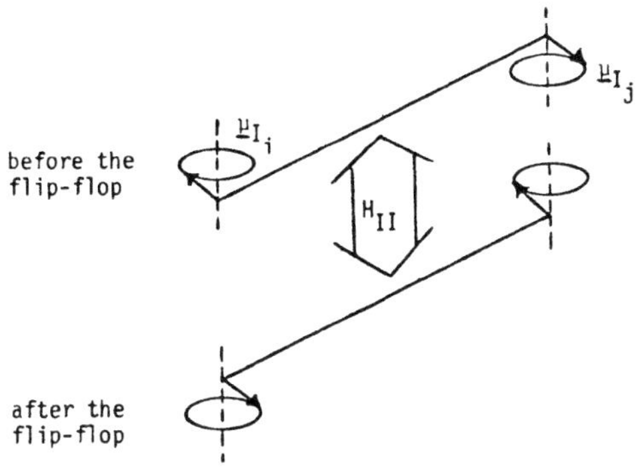

Figure 6. Schematic diagram of spin-spin flip-flops between two nuclei.

possible by the essentially equal Larmor frequency of two nuclei of the same species, which gives rise to a static component in the dipole-dipole interaction between them.

Elimination of homonuclear dipolar broadening in solid-state NMR spectroscopy, while still retaining high-resolution NMR information, has been a major challenge in NMR research. Clearly, one cannot simply use high-power decoupling, as one is trying to observe the same nuclei that one is trying to decouple. One approach that has shown considerable promise is the use of sophisticated multiple-pulse sequences, in which the spins represented in Equation 16 are manipulated to give a zero average of H_{II}, rather than relying on motional averaging of the trigonometric factor representing the orientation dependence (1,2,5-7). This multiple-pulse approach requires the generation of short, intense, precisely-defined pulses and rapid probe response, and has proved to be a difficult experiment to implement in several laboratories. For this reason there has been little ^{1}H data of consequence generated by this approach on fossil fuel samples. Considerably more promising is the use of ^{1}H multipulse methods as part of a solid-state ^{13}C NMR experiment to suppress the ^{1}H-^{1}H flip-flops which otherwise tend to undermine ^{13}C NMR experiments designed to distinguish among carbon sites of the -CH_3, $\geq CH_2$, $\geq$C-H and $\geq$C- (primary, secondary, tertiary and quarternary) types. Preliminary indications of this promise have already appeared in the literature (8-10), and one can expect rapid progress in this direction during the next few years.

Related to the nuclear-dipole/nuclear-dipole interactions described above is the case of magnetic dipolar interactions between paramagnetic centers (due to unpaired electron spins) and nuclear spins. Such interactions can be described in terms of a Hamiltonian very similar to that given in Equation 14. However, in the electron spin case what corresponds to γ_I in

Equation 14 is much larger than for a nuclear spin. Consequently, electron-spin/nuclear-spin interactions can be very strong. The rapid, random motion of a liquid would average the electron-spin/nuclear-spin interactions. However, the coupling of electron spin angular momentum with electronic orbital angular momentum, as characterized by the g tensor (11), is in general not isotropic, so the liquid-state motional average of the electron-spin/nuclear-spin dipolar interaction is not necessarily zero.

If it were not for the fact that electron spin relaxation is very rapid, one might therefore expect a splitting of the liquid-state NMR signal due to the motionally-averaged electron-spin/nuclear-spin dipolar interaction. However, the effect of electron spin relaxation is to reduce the splitting to a single peak at a position determined by the population-weighted average of the α and β spin states of the unpaired electron (assuming one unpaired electron); the resulting peak position determines what is called the pseudo contact shift in liquids. For solids, the motional averaging is absent, but averaging due to rapid electron spin relaxation is still important; the result of this average is, for each orientation of the vector between the nuclear and electron spins, a resonance frequency that is the weighted average of resonance frequencies corresponding to the interaction between the nucleus and the electron spin in either its α or β spin states. This situation is depicted pictorially in Figure 7, with the weighted average somewhat exagerated from ω_0. It is seen that a powder pattern is obtained which is substantially attenuated in width relative to what it would be in the absence of electron spin relaxation.

One should note that the α and β powder patterns in Figure 7a depend on the orientation dependence of the electron-spin/nuclear-spin dipolar Hamiltonian, which depends partly on the orientation dependence of the electron spin g tensor. Hence, the detailed shape of the relaxation-averaged powder pattern of Figure 7b depends upon the orientation dependence of the g tensor. In general, the powder pattern of Figure 7b would not be averaged to zero by random liquid-state motion. This corresponds to the existence of the pseudo-contact shift for liquids mentioned above.

C. Indirect (J) Coupling.

An interaction that is superficially similar to the nuclear-spin/nuclear-spin dipolar interaction is the indirect interaction between nuclear spins. This interaction can be described by a Hamiltonian (H_J) similar to the dipolar Hamiltonian given

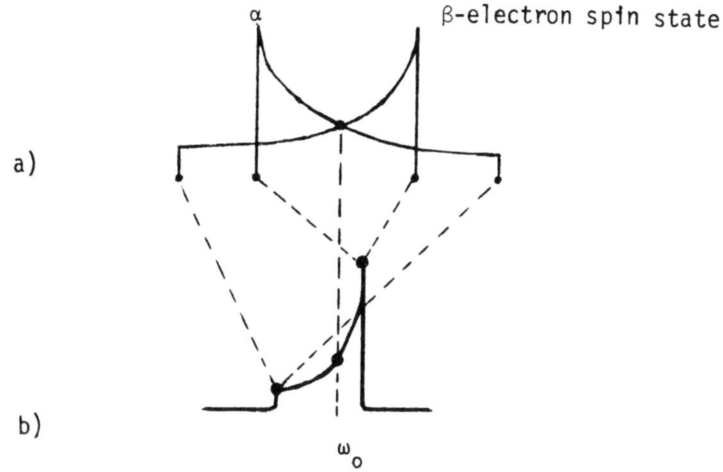

Figure 7. Hypothetical NMR powder pattern due to electron-spin/nuclear-spin dipole-dipole interaction, showing a) the individual powder patterns that would exist for α and β electron spin states, with the weighted-average resonance frequencies (•) derived from specific frequencies (·) in the α and β powder patterns; and b) the resulting weighted-average powder pattern.

$$H_J = \underline{I} \cdot \underset{\sim}{J} \cdot \underline{S} \qquad (17)$$

in Equation 12. However, the route of this J interaction is through the electronic network of the system and involves, therefore, interactions between the nuclei and bonding electrons in a diamagnetic system. There are three types of indirect (J) coupling mechanisms, named according to the nature of the electron-nuclear interaction involved. These are: a) the relativistic mechanism referred to as the Fermi contact mechanism, which involves the electron densities at both nuclei, b) the dipolar interactions between electron spins and both nuclear spins; and c) the interactions between both nuclear spins and electron orbital angular momentum. In principle, the J coupling is anisotropic, although most current evidence seems to suggest that, at least for H-H, C-H or N-H couplings of the type that would be important in fossil fuels, the anisotropy is very small, perhaps negligible. Hence, for the cases of primary interest in the present context, we write Equation 17 in the following approximate form:

$$H_J \approx J\underline{I}\cdot\underline{S} \qquad (18)$$

where J is the "scalar" coupling constant. Clearly, a Hamiltonian of the form shown on the right side of Equation 18 is independent of the orientation of the I-S internuclear vector. Hence, for a powdered or amorphous solid one would expect a powder pattern consisting of two sharp peaks, as shown for a hypothetical case in Figure 8. In this figure an arbitrary line shape has been included for each peak in the doublet as a reminder that there might be some anisotropy in the indirect coupling.

As the $J\underline{I}\cdot\underline{S}$ expression is independent of the orientation of the I-S internuclear vector, the random molecular motion of a liquid will not alter the splitting from what is seen in Figure 8; however, if there is a small anisotropic component to $\underset{\sim}{J}$, it will be eliminated by such motion, sharpening these peaks. For the heteronucler case $(\gamma_I \neq \gamma_S)$, because of greatly differing precessional frequencies $(\omega_{oI} \neq \omega_{oS})$, Equation 18 reduces to $H_J \approx JI_zS_z$. Hence, irradiation of the I spins at ω_{oI} averages the I-S coupling to zero, so that only a single peak appears in

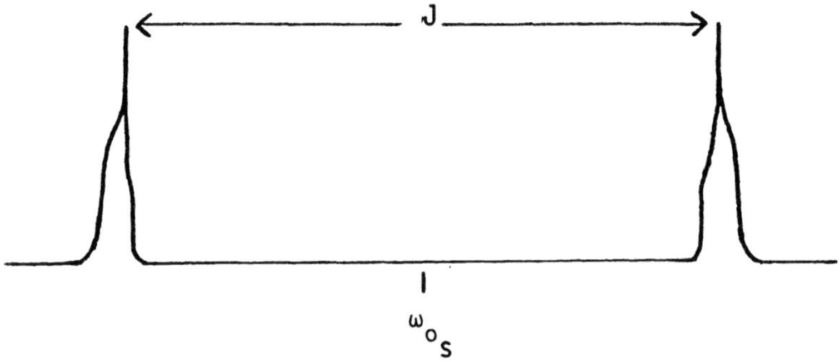

Figure 8. Resonance pattern expected for a powder in which there is nearly isotropic J coupling two spin-1/2 nuclei ($\underline{I}$ and $\underline{S}$).

the S spectrum. This is the essence of the ^1H decoupling of the J interaction in the spectra of ^{13}C in both solids and liquids.

Related to the Fermi contact I-S nuclear-spin/nuclear-spin coupling interaction in diamagnetic systems described above is

We have discussed above the effect on the S resonance of irradiation (decoupling) of the I resonance, where _I_ and _S_ both represent nuclear spins. It is pertinent to examine here the effect on the I nuclear resonance of irradiation of the electron spin resonance of S in a paramagnetic system. Rapid electron spin relaxation already brings about a collapse of the dipolar or contact splitting patterns shown in Figures 7 and 9, but application of the electron spin (S) resonance radiation changes the nature of the averaging by changing the weighting factors. If enough power at ω_{0S} could be supplied to overcome the competing influence of S spin lattice relaxation, then the populations of the S spin states would be equalized, and the averaging would result in NMR resonances at ω_{0I}; ordinarily sufficient power at ω_{0S} cannot be supplied to accomplish this, so a residual shift of the I resonance frequency away from ω_{0I} remains. Furthermore, the change in populations of the electron spin states, together with I-S relaxation processes, gives rise to very large changes in the NMR signal intensity. Depending on the details of the I-S relaxation processes involved, these effects are referred to as the electron-nuclear _Overhauser effect_, the _solid-state effect_ or the _thermal effect_ (12-14).

D. Nuclear Electric Quadrupole Interactions.

The nuclear electric quadrupole interaction is an electrostatic interaction between the electric field environment of a nucleus and the non-spherical charge distribution in the nucleus. Only nuclei with I > 1/2 can have this kind of electric charge distribution. The quadrupolar Hamiltonian (H_Q) can be defined in terms of elements of the electric field gradient tensor--second partial derivatives of the electrical potential, V, with respect to the principal (primed) axis system of that tensor, evaluated at the nucleus:

$$\text{e.g.,} \qquad V'_{xx} = \frac{\partial^2 V}{\partial x'^2} \; ; \; V'_{yz} = \frac{\partial^2 V}{\partial y' \partial z'}$$

In these terms, H_Q can be expressed as

$$H_Q = \frac{h\nu_Q}{9}\left\{3I_{z'}^2 - I(I+1) + \eta(I_{x'}^2 - I_{y'}^2)\right\} \qquad (19)$$

where $\nu_Q = \dfrac{3e^2qQ}{2I(I-1)h}$ is called the quadrupolar coupling constant and the spin operators are defined with respect to the principal (primed) axis system of the quadrupolar tensor. Q is the nuclear electric quadrupole moment; the quadrupole asymmetry parameter (η) and q are defined as follows:

$$eq = V'_{zz} \text{ and } \eta = \frac{V'_{xx} - V'_{yy}}{V'_{zz}}$$

the corresponding Fermi contact interaction between unpaired electrons and nuclei. This interaction can be represented by the equation

$$H_c = \underline{I} \cdot \underset{\sim}{a} \cdot \underline{S}$$

where S represents an electron spin and $\underset{\sim}{a}$ is the hyperfine tensor. If we again assume that this interaction is nearly isotropic, we obtain an approximate equation analogous to Equation 18, and we anticipate a powder pattern analogous to that shown in Figure 8. However, recalling that electron spin relaxation is rapid, we expect this kind of pattern to reduce to the weighted-average powder pattern shown symbolically in Figure 9b. The resulting weighted-average powder pattern has a width that depends on the anisotropy of $\underset{\sim}{a}$ (and on homogeneous relaxation rates) and a position that depends on the magnitude of $\underset{\sim}{a}$ and the electron spin populations that govern the weighted averaging.

The effect that the random motions of the liquid state would have on the pattern shown in Figure 9b would be an averaging of the anisotropic character of $\underset{\sim}{a}$ and hence a narrowing of the peak. The position of the resulting narrowed peak determines what is called the contact shift.

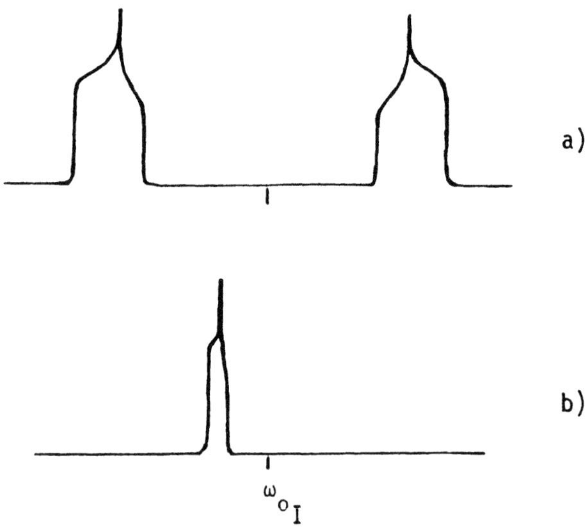

Figure 9. Effect of the electron-spin/nuclear-spin contact interaction on the resonance pattern of the I nuclear spin. a) The splitting due to the contact interaction with an electron. b) The weighted-average powder pattern that results from rapid electron spin relaxation.

with $|V'_{zz}| \geq |V'_{yy}| \geq |V'_{xx}|$

For the so-called high-field case, in which the magnitude of the nuclear Zeeman interaction is much larger than that of the nuclear electric quadrupole interaction, perturbation theory yields expressions for the quadrupolar corrections to the total energy of the nuclear spin. The first-order correction, $E_m^{(1)}$, is given(15,16) by the equation,

$$E_m^{(1)} = \frac{h\nu_Q}{4}[3\cos^2\beta - 1 - \eta\cos^2\alpha(1-\cos^2\beta)][m^2 - \frac{I(I+1)}{3}] \qquad (20)$$

In this equation, $m\hbar$ is the component of I spin angular momentum along the Zeeman (z) axis. α and β are angles relating the quadrupolar (primed) principal axis system to the laboratory axis system; specifically β is the angle between z' and z. The expressions for higher-order energy corrections are analogous to what is given in Equation 20, but more complicated(15,16). The expression for $E^{(2)}$ includes a factor ν_Q^2/ν_L, where ν_L is the Larmor frequency; hence, the second-order quadrupolar terms are less important at higher field.

There are two especially important features of the $E^{(1)}$ expression given above. First, this expression takes the following simple and trigonometrically familiar form, if the electric field gradient is axially symmetric, i.e., $\eta = 0$:

$$E_m^{(1)} = \frac{h\nu_Q}{4}[3\cos^2\beta - 1][m^2 - \frac{I(I+1)}{3}] \qquad (21)$$

Second, the dependence of $E_m^{(1)}$ on the quantum number, m, is quadratic, so it is the same for $m = 1/2$ and $m = -1/2$.

Figure 10 represents the effect that the quadrupolar interaction has on the energy levels and transitions ($\Delta m = \pm 1$) in the high-field case for $I = 1$ (e.g., ^{14}N) and for $I = 3/2$ (e.g., ^{23}Na); these two cases are representative of the situations in which I is an even multiple of 1/2 and an odd multiple of 1/2.

The energy levels and transitions (double-headed arrows) and transition frequencies shown in Figure 10 are for a single arbitrary choice of ν_Q and for a particular orientation of the principal axes of the electric field gradient tensor relative to the laboratory axis system. For the case with $I = 1$, it is seen that to first-order or second-order, the degenerate Zeeman transition energies and frequency become two separate peaks. The splitting between these two peaks depends in the general case on ν_Q, α and β; the size of the splitting is on the order of ν_Q,

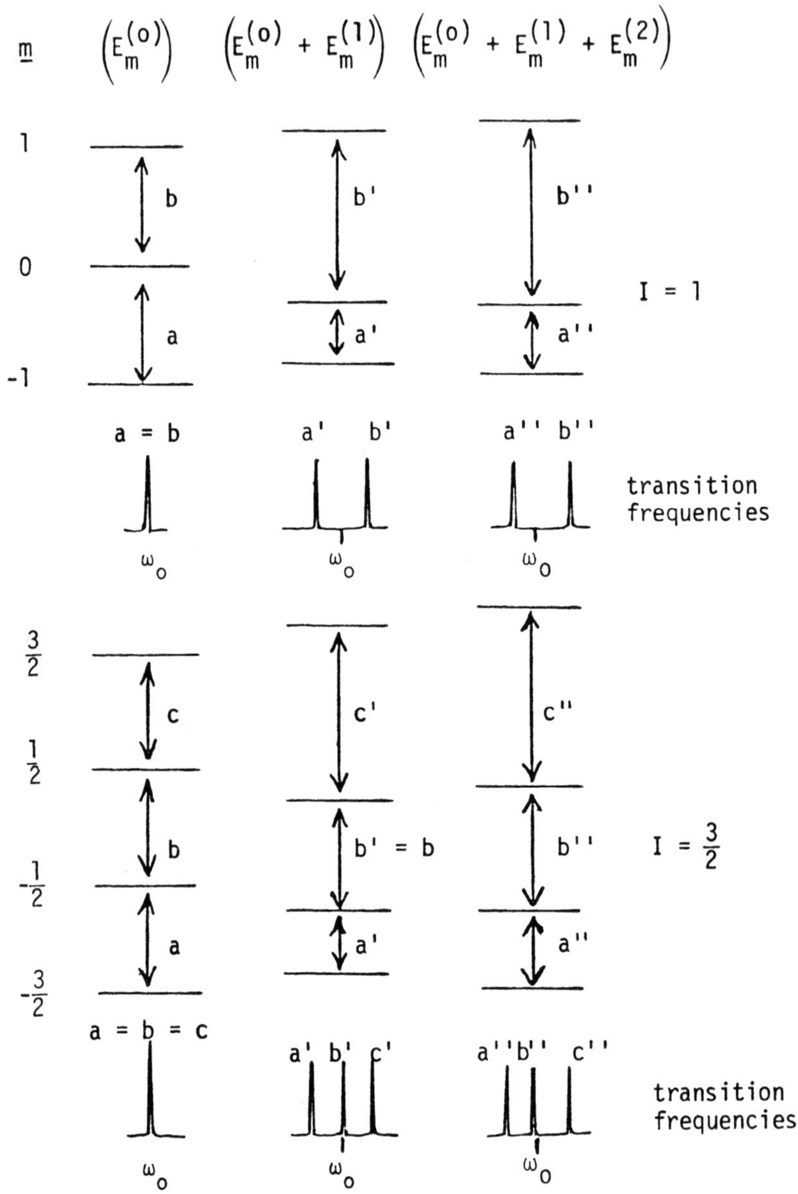

Figure 10. Diagram of energy levels, transitions (double-headed arrows) and transition frequencies at three levels of perturbation theory for the Zeeman (E_m) and quadrupolar effects. ω_o is the Larmor frequency in the absence of a quadrupolar interaction.

which can be very large. Indeed, ν_Q can be hundreds of KHz or even several MHz; it is small only for a few quadrupolar nuclides with very small values of Q or if the nuclear environment is very nearly cubic in symmetry.

For discussing the resonance patterns one expects from quadrupolar nuclei it is most convenient to consider the special case of axial symmetry, as represented to first order in Equation 21. For this case, one has two equations for the two transitions, m: $-1\leftrightarrow 0$ and m: $0\leftrightarrow 1$, These two equations correspond to Equations 15a and 15b for the heteronuclear dipolar interaction. The two new equations, expressed in terms of first-order corrections to the transition frequencies, are:

$$\nu^{(1)}_{-1\leftrightarrow 0} = -\frac{\nu_Q}{4}(3\cos^2\beta - 1) \tag{22a}$$

$$\nu^{(1)}_{0\leftrightarrow 1} = \frac{\nu_Q}{4}(3\cos^2\beta - 1) \tag{22b}$$

These two equations lead to quadrupolar powder patterns (Figure 11) analogous to what were seen in Figure 5 for the dipolar powder pattern.

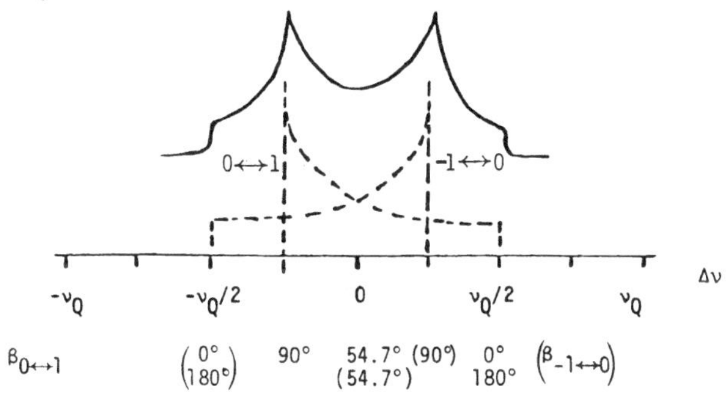

Figure 11. Powder pattern due to the quadrupolar interaction for I = 1 and η = 0. Dashed lines show the contributions due to the indicated m: $-1\leftrightarrow 0$ and m: $0\leftrightarrow 1$ transitions. $\Delta\nu = \Delta\omega/2\pi$.

The splitting between the two maxima in this powder pattern is $\nu_Q/2$, and the total powder-pattern width is ν_Q.

For the case, I = 3/2, there are three pertinent transitions, m :$-3/2\leftrightarrow -1/2$, $-1/2\leftrightarrow 1/2$, $1/2\leftrightarrow 3/2$. For the case η = 0, the first-order corrections to the transition frequencies are obtained from Equation 21, yielding

$$\nu^{(1)}_{-\frac{3}{2}\leftrightarrow -\frac{1}{2}} = -\frac{\nu_Q}{2}(3\cos^2\beta - 1) \quad (23a)$$

$$\nu^{(1)}_{-\frac{1}{2}\leftrightarrow \frac{1}{2}} = 0 \quad (23b)$$

$$\nu^{(1)}_{\frac{1}{2}\leftrightarrow \frac{3}{2}} = \frac{\nu_Q}{2}(3\cos^2\beta - 1) \quad (23c)$$

Clearly, Equations 23a and 23c predict powder patterns similar to what is shown in Figure 11, except that for the case I = 3/2 the splitting between the maxima is ν_Q and the total powder pattern width is $2\nu_Q$. For the transition, m:-1/2 ↔ 1/2, referred to as the central resonance, there is no first-order quadrupolar contribution; hence, to first-order the resonance frequency of the central transition is simply ν_o, the Zeeman transition frequency. The overall result is that one expects for the case I = 3/2 a powder pattern of the form seen in Figure 12.

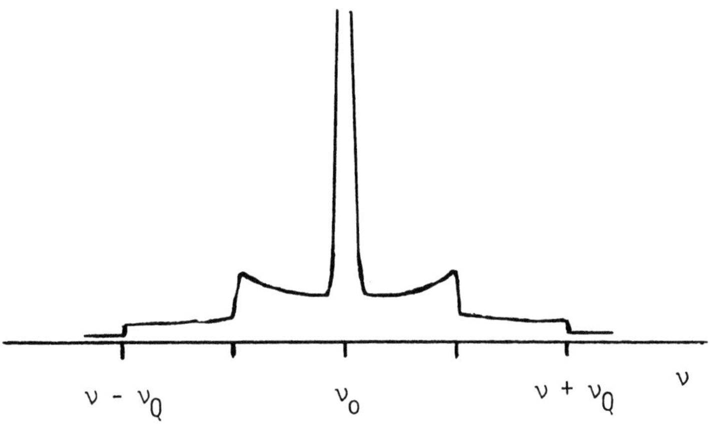

Figure 12. Quadrupolar powder pattern for the case I = 3/2, showing the central (m: -1/2 ↔ 1/2) transition superimposed on the broad powder pattern for the other two transitions with m: -3/2 ↔ -1/2 and 1/2 ↔ 3/2. With broadening added.

The powder patterns shown in Figures 11 and 12 are based on the results of a perturbation theory treatment carried only to first order. The second-order corrections make significant, but not dominant, contributions to the powder patterns. Perhaps the most significant effect that the second-order corrections have is on the central transition for nuclides with I equal to odd multiples of 1/2 (e.g., $I = 3/2$). In this case, as first-order quadrupolar contributions are absent, the second-order contributions usually assume a major importance (17-19).

Because of the trigonometric factors present in the expressions for the quadrupolar contributions to energy levels and transition frequencies, it can be seen that the rapid motions of liquids average the quadrupolar powder patterns to single resonance lines. For this reason, the nuclear electric quadrupole effect is not manifested in peak positions in the NMR spectra of liquids and shows up only as a relaxation mechanism. Indeed, as the magnitude of the effect can be so large (e.g., hundreds of KHz or even several MHz), this is usually the dominant relaxation mechanism in the NMR of diamagnetic samples for $I > 1/2$.

III. LINE NARROWING TECHNIQUES FOR SOLIDS

A. Overview.

In the preceding section we have seen that fundamental spin interactions lead to broad powder patterns in the NMR spectra of powdered or amorphous solids obtained by the classical, straightforward methods of NMR detection. We have also seen that there are essentially three types of approaches for avoiding these broad patterns. These approaches are: 1) converting the sample into a liquid, so that the random motions of the liquid state bring about rotational averaging of the orientation-dependent interactions; 2) working with a single crystal, so there is only one set of relevant orientations (relevant angles, α, β, etc), and sharp lines are obtained rather than an envelope of such lines (powder pattern); and 3) manipulation of the nucler spins themselves to average out the interactions, e.g., in heteronuclear decoupling or homonuclear multiple-pulse experiments. The first two approaches do not permit one to study powdered or amorphous solids directly, and the third approach has so far been employed successfully only to average dipolar interactions. Hence, another method is needed.

B. Magic-Angle Spinning.

A powerful method of averaging spin interactions of the types described in Section II is rapid sample spinning about a specific axis. This method was originally introduced by Andrew and coworkers (20) and Lowe and Norberg (21) for addressing dipolar interactions. The essence of this approach can be seen by considering the effect of mechanically rotating the entire sample rapidly about an axis that makes some arbitrary angle with respect to the direction of $\underline{H_0}$ (z axis), as shown in Figure

13 for the heteronuclear dipolar interaction between a specific I-S pair. In this figure the bold line shows the the instantaneous orientation of the internuclear vector at some arbitrary origin of time (t = 0); the Hamiltonian shown in Equation 14 depends on this orientation, through the angle (β) that this vector makes with the z axis. The dashed line in Figure 13 represents the axis of rotation, and the circles drawn in perspective show the locus of points followed by the I and S nuclei during rotation. Derivation of the rotational average of H_{IS}' given in Equation 14 yields an expression of the form,

$$\overline{H'_{IS}} = \frac{\gamma_I \gamma_S \hbar^2}{r^3}(3\cos^2\Theta - 1)f(\Theta,\beta)I_z S_z \tag{24}$$

where the function f relates the orientation of the sample-spinning axis (Θ) to the orientation of the internuclear vector

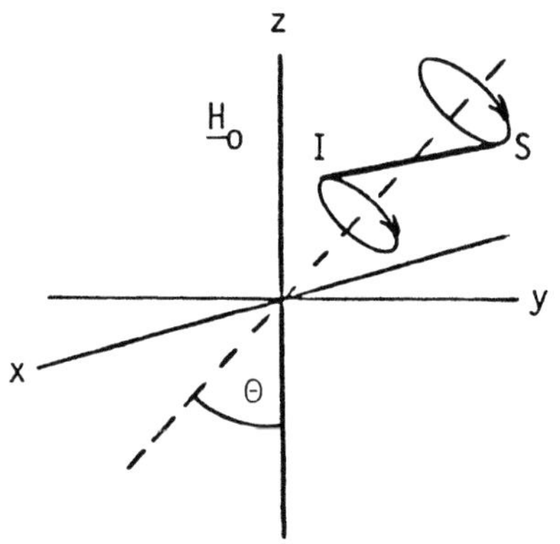

Figure 13. Effect of rapid sample spinning on the heteronuclear dipolar interaction. The spinning axis (dashed line) makes an angle Θ with the direction of the static magnetic field (z axis).

(α, β) at t = 0. For an amorphous or powdered solid, f varies from one site (or molecule) to another throughout the sample; and a range of resonance frequencies would be the result. A "sharp" averaging of H_{IS} and sharp resonances are not obtained except for a sample rotation angle for which $3\cos\Theta^2 - 1 = 0$. This is the angle, ψ, that is called the magic angle.

$$\psi \equiv \cos^{-1}\sqrt{1/3} \equiv \text{magic angle}. \tag{25}$$

From Equation 24 we can see that sample spinning at any angle other than ψ gives a residual I-S dipolar effect. Only with magic-angle spinning (MAS) is the I-S dipolar effect removed from the spectrum for all heteronuclear I-S pairs, irrespective of their orientations within the sample at t = 0.*

The net result of rapid spinning of the sample about the magic-angle axis is to eliminate the dipolar coupling by mechanical motional averaging--magic-angle spinning. As will be shown later in this section, in order to accomplish truly effective spectral simplification by this method, it is necessary to spin the sample rapidly (in Hz) relative to the size (in Hz) of the interaction to be overcome, in this case a heteronuclear dipolar interaction. For directly-bonded C-H pairs, the size of the interaction is sufficiently large to have precluded so far the use of magic-angle spinning to eliminate the effects of this interaction in the ^{13}C spectra of solids. Of course, this is not a major limitation, as ^{1}H decoupling can be used to remove ^{1}H-^{13}C dipolar splittings from the ^{13}C NMR spectra of solid samples.

In order for MAS to be effective in averaging a dipolar interaction to zero, it is necessary that the spins not change states (between m = 1/2 and m = -1/2) during a rotation. ^{1}H-^{1}H flip-flops of the type depicted in Figure 6 can be a problem in this regard. However, if the spins change states sufficiently rapidly, self decoupling will occur, and MAS is not necessary. In any case, for some combinations of nuclei MAS can be an effective approach for eliminating the effects of I-S dipolar interaction from the spectrum.

Similar conclusions can be drawn for the effect of magic angle spinning on the narrowing (averaging) of powder patterns due to chemical shift anisotropy and the nuclear electric quadrupolar interaction. For example, Figure 14 shows schematically

---

*Superficial inspection of Figure 13 might lead one to conclude directly that $H_{IS} = 0$ for $\Theta = \psi$, because the average orientation of the internuclear vector lies along the spinning axis, so it would seem intuitive that the average of $1 - 3\cos^2\beta = 0$ for $\beta = \psi$. While this superficial reasoning leads to the correct conclusion in this specific instance, not recognizing that in Equation 14 it is $\cos^2\beta$ (not β itself) that must be averaged directly over sample spinning can lead to erroneous conclusions on related matters, e.g., the residual dipolar broadening for other spinning angles. Similar comments apply to Figures 14 and 15.

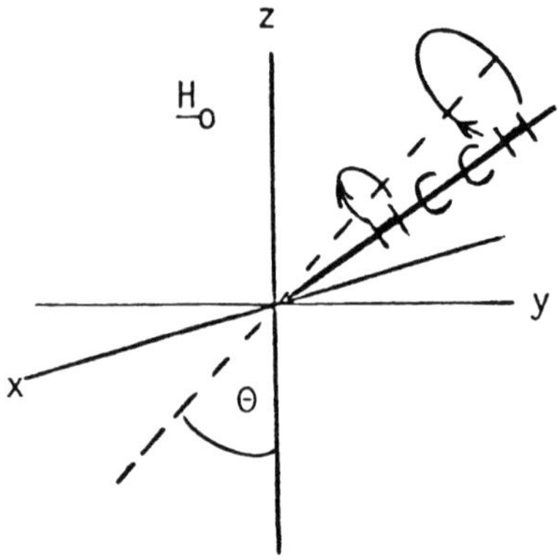

Figure 14. Effect of rapid sample rotation on the chemical shift anisotropy of acetylene. The spinning axis (dashed line) makes an angle Θ with the direction of $\underline{H}_o$.

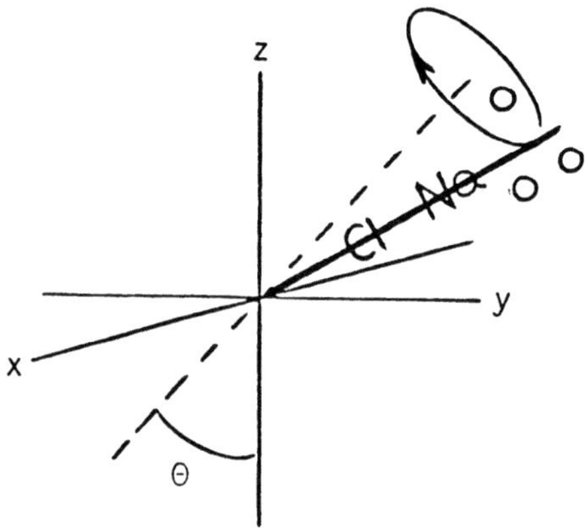

Figure 15. Effect of rapid sample rotation on the nuclear electric quadrupolar effect of ^{23}Na in a hypothetical axially-symmetric environment of one Cl^- ion and three O^- ions.

the spinning of a sample of solid acetylene, showing one specific acetylene molecule in an amorphous or powdered sample; and Figure 15 shows the rotation of a sample of a powdered or amorphous hypothetical sodium compound in which the ^{23}Na nucleus is in an axially symmetric environment (with three oxygen ions located about the Na-Cl axis in such a manner that yields C_{3v} symmetry). In both figures the dashed line represents the axis of sample rotation and the bold line represents the unique principal axis of the appropriate tensor (shielding tensor for Figure 14 and electric field gradient tensor for Figure 15).

If one averages the chemical shift Hamiltonian for an axially symmetric case (such as that given in Equation 10) over one rotation of the sample,one obtains the following result:

$$H'_{CS} = \frac{\gamma h H_0}{2} Tr\underset{\sim}{\sigma} \sin^2\Theta + \frac{\gamma h H_0}{2}(3\cos^2\Theta - 1)\sigma_{rr} \qquad (26)$$

where σ_{rr} is the component (at t = 0) of the shielding tensor with respect to the rotation axis. Clearly, σ_{rr} depends on the orientation of the acetylene molecules relative to the sample spinning axis, and consequently will cover the entire range of values throughout the sample. Hence, the only way a single resonance frequency (narrow line) can result is if $3\cos^2\Theta - 1 = 0$, i.e., if $\Theta = \psi$, the magic angle. In this case,

$$\overline{H}'_{CS} = \frac{\gamma \hbar H_0}{2}(Tr\underset{\sim}{\sigma})\sin^2\psi = \frac{\gamma \hbar H_0 Tr\underset{\sim}{\sigma}}{3} = \gamma \hbar H_0 \sigma_{iso} \qquad (27)$$

Thus, magic-angle spinning yields the same chemical shift result mechanically on a solid sample as does the random motion in liquids. The idealized result of magic-angle spinning for removing broadening due to chemical shift anisotropy is, then, what is shown in Figure 16. This is to date the most important application of magic-angle spinning.

In Equations 21, 22 and 23, for the first-order contributions of the nuclear electric quadrupolar interaction in the case of axially symmetric electric field gradient, we see the familiar trigonometric factor, $1 - 3\cos^2\beta$. For the central transition of a nuclide with I = 3/2, 5/2, 7/2, ..., there is no first-order quadrupole correction; the first-order transition frequencies of other transitions can in principle be averaged in the same way that we have seen for the other second rank tensors discussed above, shielding and dipolar interactions. Again,

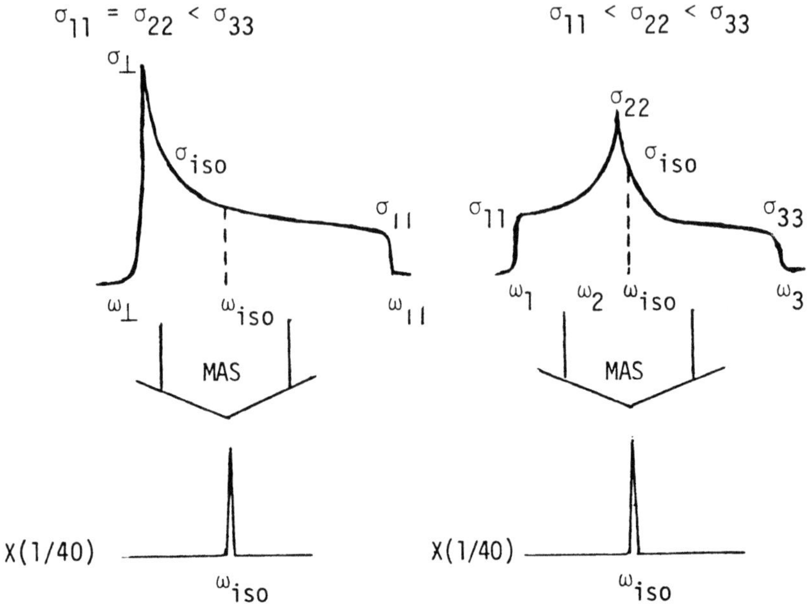

Figure 16. Theoretical effect of magic-angle spinning on the powder patterns due to chemical shift anisotropy. In the lower spectra the total intensities have been scaled down by a factor of about 40.

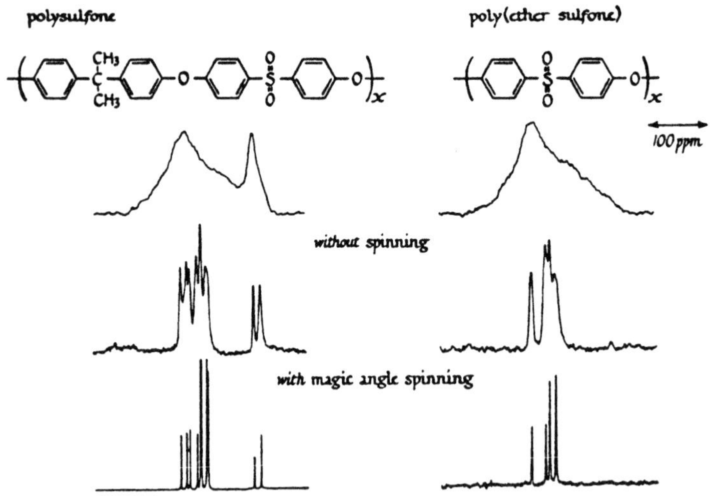

Figure 17. Experimental effect of magic-angle spinning on ^{13}C powder spectra. Schaefer and Stejskal (22).

spinning at the magic angle would seem to promise a major line-narrowing effect. However, there are three reasons why magic-angle spinning is usually not as effective in removing the quadrupolar interaction as one might hope. First, the quadrupolar coupling constant is often orders of magnitude larger than the speeds with which one can spin a sample, so extensive arrays of sidebands are produced (vide infra). Second, because of the very large magnitude of the interaction and corresponding width of the powder pattern, the sample rotation axis must be maintained extremely close to the magic angle to accomplish the desired line narrowing. Third, the second-order quadrupole contributions to transition frequencies are not averaged to zero at any angle, even for rapid spinning at the magic angle. Nevertheless, as we shall see below, very useful line-narrowing can often be achieved with quadrupolar nuclei, especially when it is the central ($-1/2 \leftrightarrow 1/2$) transition that is observed, so there is no quadrupolar broadening to first order, and chemical shift anisotropy and/or dipolar broadening can be eliminated by MAS.

C. Characteristics of Magic Angle Spinning (MAS).

Magic-angle spinning has become such a widespread and well-developed art, that a complete discussion would be beyond the scope of this chapter. Nevertheless, there are a few characteristics that are sufficiently important to warrant discussion here.

One important matter concerns the importance of having the spinning device adjusted with its spinning axis very near the magic angle. From equations like Equation 26 for the chemical shift one can determine from the second term how much of a "residual powder pattern" remains if Θ varies from ψ by a given amount. (Recall that the factor σ_{rr} in the second term of Equation 26 spans the entire range of the shielding tensor.) Figure 18 shows the experimental variation of the line shapes in a ^{13}C NMR spectrm of solid poly(methylmethacrylate); it is very clear that the sharpest lines are obtained at the magic angle (54.7°), and that severe line broadening occurs for other angles. A detailed analyses shows that spinning at Θ = 0° (β = 0° in Figure 18) yields the same spectrum as if there were no sample rotation at all (as shown by Equation 26).

Figure 19 displays a very important characteristic of spectra obtained by magic-angle spinning -- spinning sidebands. These sidebands appear as satellites about each peak, with separations between the main peaks and their sidebands equal to the spinning frequency. From this figure one sees that the

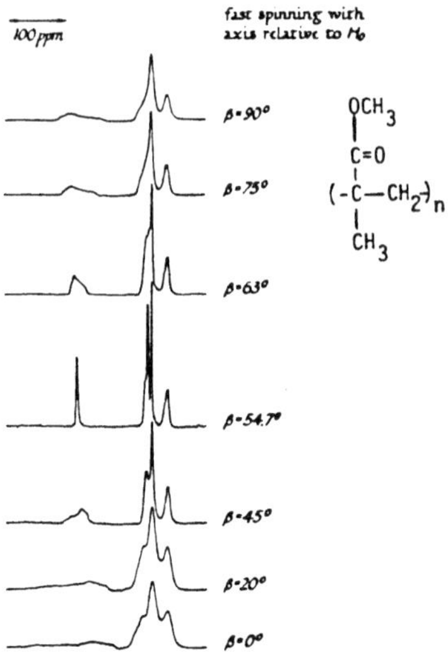

Figure 18. ^{13}C NMR spectra of solid poly(methylmethacrylate) obtained as a function of spinning angle (β) and with ^{1}H decoupling. From Stejskal, Schaefer and McKay (23).

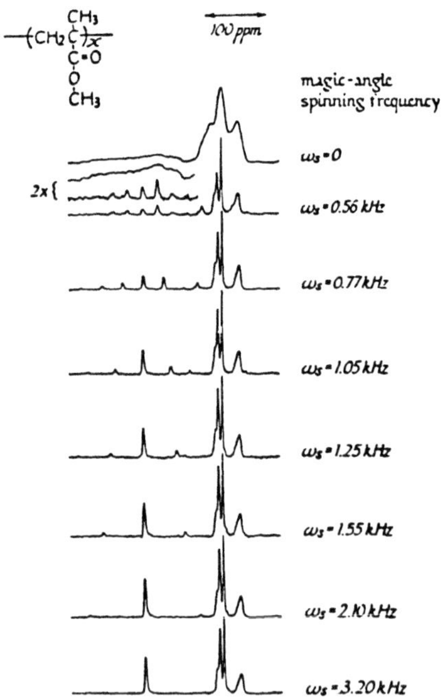

Figure 19. The effect of spinning speed on the ^{13}C MAS spectra of poly(methylmethacrylate) obtained with ^{1}H decoupling at 15 MHz. From Stejskal, Schaefer and McKay (23).

intensities of spinning sidebands decrease with increasing spinning speed. For low spinning speeds the sideband patterns become very extensive for the carbonyl carbon resonance. Close inspection shows that the envelope of the sideband intensity pattern of a given resonance (e.g., the carbonyl resonance) maps out the chemical shift anisotropy pattern of that resonance and that resonances with small chemical shift anisotropies (sp^3 carbons) show no direct evidence of spinning sidebands in these spectra.

The kind of spinning sideband behavior represented in Figure 19 can be understood conveniently if one views sample rotation as a means of modulating spin interactions, with sidebands manifesting this modulation. Maricq and Waugh (24,25) have looked at this problem in detail and have developed the concept of the "rotational echo". This concept can be seen in Figure 20. The top part of this figure shows the CSA powder

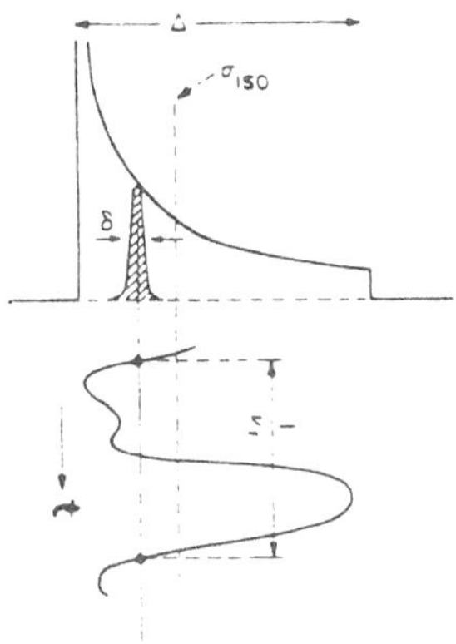

Figure 20. The effect of sample rotation on the resonance frequency for the case of an axially symmetric shielding tensor. Top, the CSA pattern, showing the resonance (shaded) for an instantaneous orientation. Bottom, the instantaneous resonance frequency as the sample is rotated.

pattern of an axially symmetric system, indicating the resonance position for a particular molecular orientation as a shaded peak. As the sample is rotated, the resonance frequency of this nucleus sweeps out the pattern shown in the lower part of the figure. Each crystallite in a powdered system or each molecule in an amorphous system has its own rotationally-dependent orientation; so each sweeps out its own particular pattern analogous to that shown in the bottom portion of Figure 20, constrained by the fact that the average resonance frequency over one rotation must be ω_{iso} for each crystallite. Hence, as the rotation period proceeds, the corresponding magnetization vectors get out of phase with each other, and the net transverse (xy) magnetization decays to zero. However, after one complete revolution these individual magnetization vectors return to their original phases and therefore reinforce each other (assuming they were in phase at t = 0, e.g., at the end of a 90° pulse or spin lock condition). The resulting refocussed magnetization is referred to as a "rotational echo". As the sample rotation continues, the transverse magnetization again decays, and then refocusses after each period. Hence, a train of rotational echoes, as seen in Figure 21, is produced. It is this train of rotational echoes that is acquired in memory during data acquisition, and Fourier transformed later to provide the final spectrum in the frequency domain.

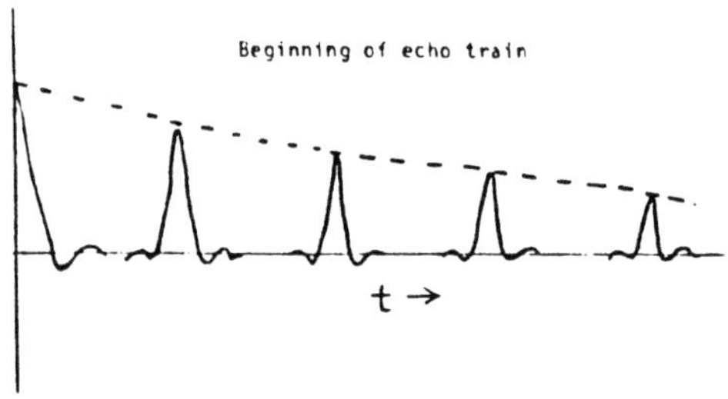

Figure 21. A train of rotational echoes appearing at periods of the sample rotation, showing an envelope decay dependent on ω_{iso}.

Fourier transformation of the echo train transforms the modulation in the time domain at the rotor period ν_r^{-1}, into sidebands in the frequency domain, with spacings equal to ν_r. The overall decay envelope of the echo maxima carries the information on the desired frequency spectrum, including linewidths, without sidebands. The spinning sideband intensities in the Fourier transformed spectrum depend on how rapidly each rotational echo decays within a rotor period, which in turn depends on the size of the chemical shift anisotropy relative to the rotor period. For nuclei with large chemical shift anisotropies relative to the rotor frequency, the decay of individual echoes is pronounced, and an extensive spinning sideband array is produced in the frequency spectrum. For nuclei with very small chemical shift anisotropies relative to ν_r, the decay of individual echoes is much less pronounced, and little or no spinning sideband intensity is detected in the Fourier transformed spectrum. For ^{13}C NMR, sp^2- and sp-hybridized carbon sites provide nuclei of the former type, sp^3-hybridized carbons the latter type.

These considerations lead to the conclusion that the CSA spinning sideband problem in solid-state NMR with magic-angle spinning increases as the field strength, H_0, is increased and decreases as the sample rotation speed increases. Hence, the optimum H_0 value for a type of sample with small chemical shift anisotropies (say ^{13}C in polyethylenes) will typically be much higher than for a sample with large chemical shift anisotropies (say, ^{13}C resonances of aromatic carbons in coal). There have been some very elegant tricks developed for minimizing the spinning sideband problem; and some approaches, especially the approach developed by Dixon (26-28), are very promising. Nevertheless, each method currently has certain disadvantages. Hence, to date the most reliable way to obtain spectra that are plagued by neither sidebands nor by distortions introduced because of attempts to eliminate sidebands is to use a spectrometer with a low enough H_0 value so that spinning sidebands are manageable with available spinning speeds. With current upper limits of about 4 to 5 kHz for routine spinning of samples of reasonable size, this limits routine work with coal and oil shale to spectrometers with field strengths of about 2.1 to 2.3 T. Figure 22 shows the situation in a comparison of the ^{13}C MAS spectra of a coal sample obtained at three field strengths. Aside from the spinning sideband problem, one usually wishes to operate at as high a magnetic field strength as possible, because of gains in sensitivity and possible gains in spectral dispersion (often disappointing in solid-state ^{13}C NMR).

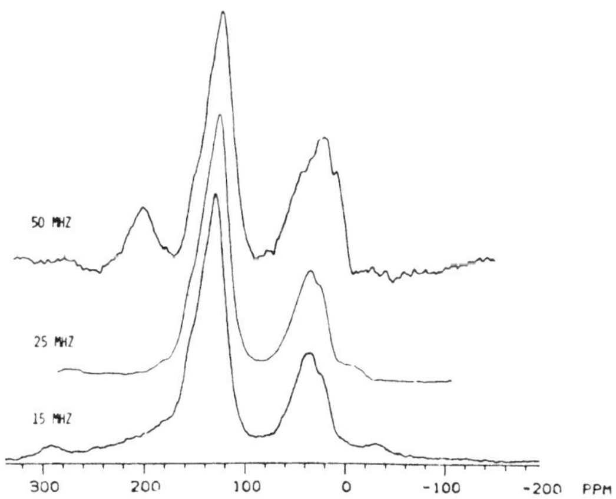

Figure 22. ^{13}C MAS NMR spectra of Powhatan No. 5 coal, obtained at 1.4 T, 2.35 T and 4.7 T. From Sullivan and Maciel (29).

The concepts of modulation and rotational echoes have been applied above explicitly only to the chemical shift anisotropy. However, these ideas apply also to the other anisotropic spin interactions that can be modulated by sample rotation, i.e., dipole-dipole interactions and the nuclear electric quadrupole interactions.

Some interesting examples of compromises and choices that have to be made for the optimum magnetic field strength in magic-angle spinning NMR experiments on solids occur with quadrupolar nuclei. As indicated above, the magnitudes of quadrupolar coupling constants are often so large that for non-central transitions the requirements on sample spinning speeds and setting the angle Θ to ψ exceed practicable laboratory capabilities. However, we have seen above that the central ($-1/2 \leftrightarrow 1/2$) transition for nuclei with I equal to an odd multiple of 1/2 is not broadened to first order by the quadrupolar interaction; hence, in such cases one has to worry only about the effect of magic-angle spinning on the second-order quadrupolar effect and on other anisotropic interactions---dipole-dipole and chemical shift anisotropy. Modulation of any or all three of these effects can give rise to spinning sidebands. As indicated in the discussion of quadrupolar powder patterns, the second-order quadrupolar contribution is inversely proportional to the Larmor frequency (and, hence, H_0). From this point of view, as well as gains in sensitivity and (in some cases) spectral dispersion, it is advantageous to obtain spectra at the highest field available. However, since the width of the chemical shift anisotropy powder pattern in Hz scales linearly with H_0, the spinning sideband problem associated with chemical shift anisotropy increases with increasing H_0. Hence, there is a compromise to be addressed, and the optimum field depends on the relative sizes of the powder patterns due to chemical shift anisotropy and the quadrupolar interaction.

Figure 23 shows ^{27}Al spectra (I = 5/2) obtained on γ-alumina. The top spectrum was obtained on a static sample. The bottom spectrum, obtained with spinning at the magic angle, displays both a sideband pattern and chemical shift dispersion (octahedral and tetrahedral aluminum sites) for the central transition. Figure 24 shows ^{27}Al spectra obtained on $KAl(SO_4)_2 \cdot 12H_2O$. The top spectrum was obtained on a static sample, and shows a powder pattern that is analogous to that sketched in Figure 12; the middle spectrum was obtained with spinning near the magic angle. In the bottom spectrum (Figure 24c), obtained with spinning at the magic angle, an expansive set of sidebands is observed; these sidebands result from modulation of the noncentral transitions, and appear only when the spinning axis is set to ψ with very close tolerances.

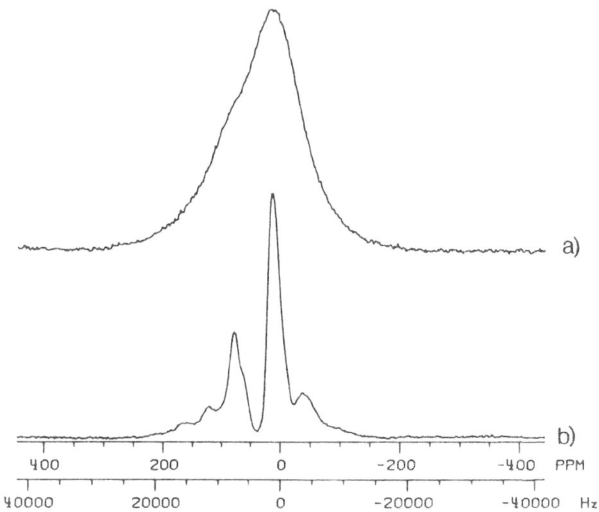

Figure 23. 93-MHz ^{27}Al NMR spectra of γ-alumina a) Static sample. b) MAS spectrum. From M. Seger (19).

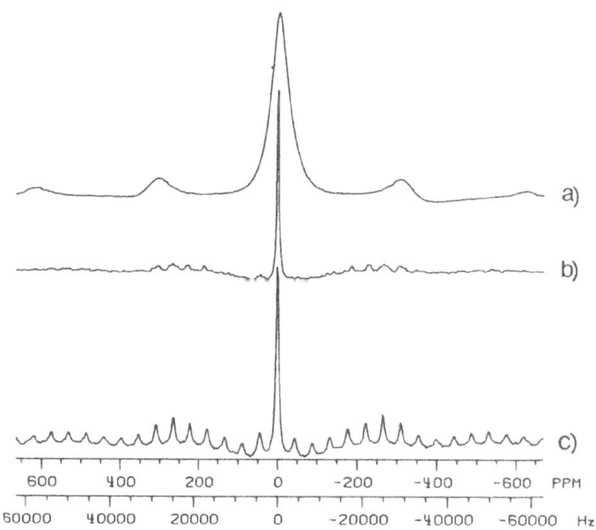

Figure 24. 93-MHz ^{27}Al NMR spectra obtained on $KAl(SO_4)_2 \cdot 12 H_2O$. a) Static sample, b) spinning at 4.0 kHz near the magic angle, c) spinning at 4.0 kHz at the magic angle. From M. Seger (19).

Another manifestation of the quadrupolar interaction in magic-angle spinning can be seen in the spectra of ^{13}C nuclei that are directly bonded to ^{14}N. One often finds that ^{13}C resonances of such nuclei are much sharper in higher-field experiments than in lower-field experiments. A dramatic example of this phenomenon is seen in Figure 25, which shows the ^{13}C magic-angle spinning NMR spectra of a urea-formaldehyde resin obtained at two fields. The reason for the dramatic improvement in the spectrum at higher field is the improved efficiency of magic-angle spinning in removing ^{13}C-^{14}N dipolar broadening at higher field. At lower field, the ^{14}N quadrupolar interaction is sufficiently competitive with the ^{14}N Zeeman interaction that the axis of ^{14}N spin quantization deviates substantially from the direction of $\underline{H}_o$; hence, Equations 14 and 15 are no longer strictly valid, and magic angle spinning is not effective in eliminating the ^{13}C-^{14}N dipolar interaction. At higher field the ^{14}N Zeeman interaction dominates over the ^{14}N quadrupolar interaction, so the assumptions of the earlier treatment are more closely approached.

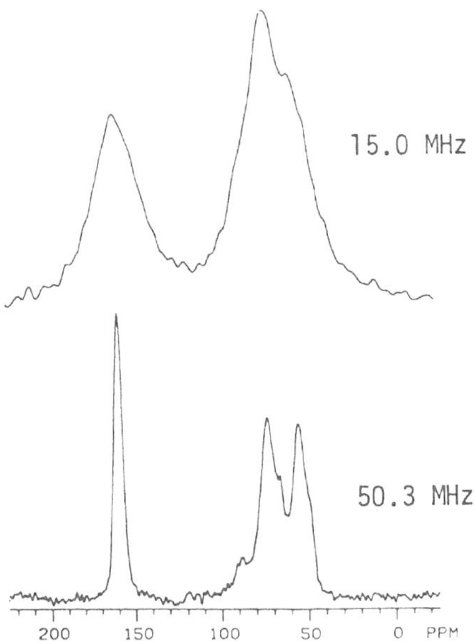

Figure 25. ^{13}C magic-angle spinning spectrum of a urea-formaldehyde spectrum otained at 15.0 MHz and 50.3 MHz. From Maciel <u>et al</u> (30).

Another point to be made regarding magic-angle spinning is that the sharpness of the resulting spectra depend dramatically on the chemical nature of the sample, even for a particular nuclide. Figure 26 demonstrates this fact for the case of ^{13}C.

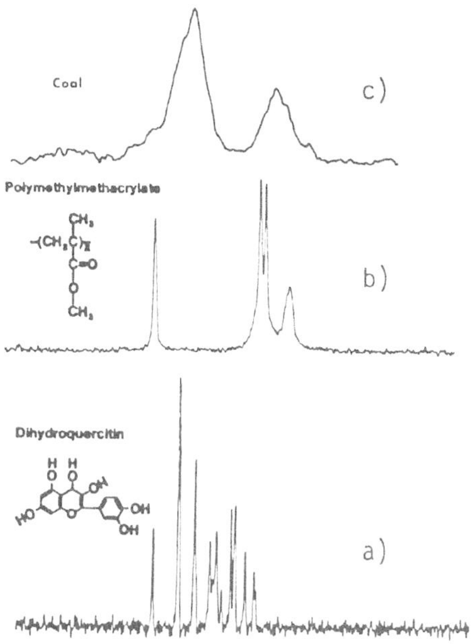

Figure 26. 15.1 MHz ^{13}C magic-angle spinning NMR spectra of a) dehydroquercitin, b) poly(methylmethacrylate) and c) coal.

This figure shows magic-angle spinning ^{13}C spectra of a) a crystalline organic compound (dihydroquercitin), b) a synthetic organic polymer (polymethylmethacrylate), and c) a complex macromolecular organic (coal). For the crystalline material the resolution is essentially comparable to that of a liquid-state spectrum. This reflects the fact that in a crystalline material there are no structural variations; each unit cell is a replicate of any other. For the synthetic polymer there are conformational variations that give rise to discrete ranges of isotropic chemical shifts, and these manifest themselves as broadened envelopes. Coal represents a complex mixture of organic patterns. For most of the resonances in coal the natural widths of individual lines are only two or three ppm, but the envelope of such lines is very broad. In fact, the spectrum consists mainly of a broad band for sp^2 carbons and a broad band for sp^3 carbons. The ^{13}C NMR resonance bands in coal and oil shale spectra are so broad and the chemical shift anisotropies of aromatic carbons so large, that without magic-angle spinning it would be difficult to even to identify the presence of aromatic carbons in some shale samples. An example of this is seen in Figure 27.

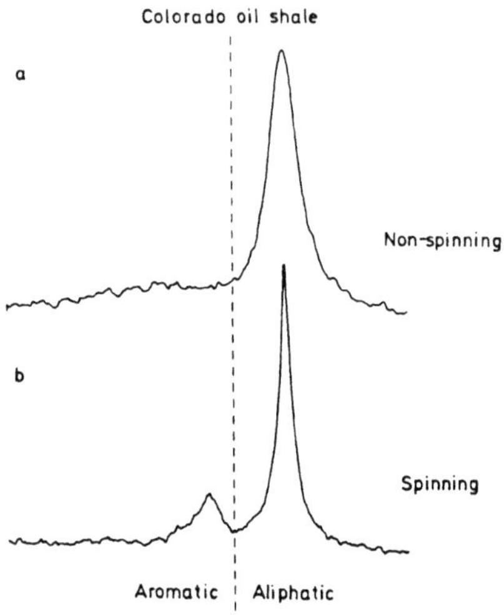

Figure 27. 15.1 MHz ^{13}C NMR spectra of Colorado oil shale a) without spinning, b) with magic-angle spinning.

References

1. Mehring, M., "High Resolution NMR Spectroscopy in Solids," Second Edition, Springer-Verlag, New York, 1983.
2. Haeberlen, U., "High Resolution NMR in Solids. Selective Averaging." Academic Press, New York, 1976.
3a. Herzfeld, J. and Berger, A., 1980, J. Chem. Phys., 73, 6021.
3b. Waugh, J.S., Maricq, M.M. and Cantor, R., 1978, J. Magn. Reson., 29, 183.
4. Pausak, S., Pines, A, and Waugh, J.S., 1973, J. Chem. Phys., 59, 591.
5a. Waugh, J.S., Huber, L.M. and Haeberlen, U. 1968, Phys. Rev. Letters, 20, 180.
5b. Mansfield, P., Orchard, M.J., Stalker, D.C. and Richards, K.H.B., 1973, Phys. Rev. B, 7, 90.
6a. Rhim, W.K., Elleman, D.P. and Vaughan, R.W., 1973, J. Chem. Phys., 59, 3740.

6b. Rhim, W.-K., Ellemann, D.D., Schreiber, L.B. and Vaughan, R.W., 1974, J. Chem. Phys., 60, 4595.
6c. Burum, D.P. and Rhim, W.K., 1979, J. Chem. Phys., 71, 944.
6d. Burum, D.P., Linder, M. and Ernst, R.R., 1981, J. Magn. Reson., 44, 173.
7. Rosenberger, H., 1983, Z. Chem., 23, 34.
8a. Terao, T., Miura, H. and Saika, A., 1981, J. Chem. Phys., 75, 1573.
8b. Terao, T., Miura, H. and Saika, A., 1982, J. Amer. Chem. Soc., 104, 5228.
9. Zilm, K.W. and Grant, D.M., 1982, J.Magn. Reson., 48, 524.
10. Caravatti, P. Bodenhausen, G. and Ernst, R.R., 1982, Chem. Phys. Lett., 89, 363.
11. Slichter, C.P., "Principles of Magnetic Resonance," Second Edition, Springer-Verlag, New York, 1978.
12a. Wind, R.A., Trommel, J. and Smidt, J., 1979, Fuel, 58, 900.
12b. Wind, R.A., Trommel, J. and Smidt, J., 1982, Fuel, 61, 398.
13. Wind, R.A., Anthonio, F.E., Duijvestijn, M.J., Smidt, J., Trommel, J. and de Vette, G.M.C., 1983, J. Magn. Reson., 52, 424.
14. Wind, R.A. , this volume, p. 461.
15. Cohen, M.H. and Reif, F., 1957, Solid State Physics, 5, 321.
16. Samoson, A., Kundla, E. and Lippmaa, E., 1982, J. Magn. Reson., 49, 350.
17. Muller, D., Gessner, W., Behrens, H.-J. and Scheller, G., 1981, Chem. Phys. Lett., 79, 59.
18. Ganapathy, S., Schramm, S. and Oldfield, E., 1982, J. Chem. Phys., 77, 4360.
19. Seger, M., Ph.D. Dissertation, Colorado State University, 1983.
20. Andrew, E.R., 1971, Prog. Nucl. Magn. Reson. Spectrosc., 8, 1.
21a. Low, I.J., 1959, Phys. Rev. Letters, 2, 285.
21b. Kessemeier, H.; Norberg, R.E., 1967, Phys. Rev., 155, 321.
22. Schaefer, J. and Stejskal, E.O., "Topics in Carbon-13 NMR Specroscpy," G.C. Levy, ed., vol. 3, Wiley-Interscience, New York, 1979.
23. Stejskal, E.O., Schaeger, J. and McKay, R.A., 1977, J. Magn. Reson., 25, 569.
24. Maricq, M.M. and Waugh, J.S., 1977, Chem. Phys. Lett., 47, 327.
25. Maricq, M.M. and Waugh, J.S., 1979, J. Chem. Phys., 70, 3300.
26. Dixon, W.T., 1981, J. Magn. Reson., 44, 220.
27a. Dixon, W.T., Schaefer, J., Sefcik, M.D., Stejskal, E.O. and McKay, R.A., 1981, J. Magn. Reson., 45, 173.
27b. Dixon, W.T., Schaefer, J., Sefcik, M.D., Stejskal, E.O. and

McKay, R.A., 1982, J. Magn. Reson., 49, 341.
28. Dixon, W.T., 1982, J. Chem. Phys., 77, 1800.
29. Sullivan, M.J. and Maciel, G.E., 1982, Anal. Chem., 54,1606.
30. Maciel, G.E., Szeverenyi, N.M., Early, T.E. and Myers, G.E., 1983, Macromol., 16, 598.

MATHEMATICAL TECHNIQUES IMPORTANT IN MAGNETIC RESONANCE

J. S. Waugh

M.I.T., Cambridge, MA 02139

The useful applications of NMR depend on the local interactions between nuclear spins and their surroundings (chemical shifts, quadrupole splittings, ...) or other nuclear species (dipolar or J couplings). The spectra to which these interactions give rise are modified if the spins at the same time experience much stronger disturbances from externally applied fields. These include the static field always used to define the Larmor frequency as well as radiofrequency fields, either steady or modulated in frequency, phase, and/or amplitude. The technique of the average hamiltonian is designed to analyze these modifications, and provides a framework within which one can tailor external fields to control these modifications — in particular to find modifications which eliminate one or another of the natural spin interactions.

The average hamiltonian theory begins by expressing the interactions in a frame of reference (interaction representation) which suppresses the external disturbance at the cost of making the natural spin interactions $\mathcal{H}$ appear time-dependent, and in particular periodic with period t_c called the cycle time: $\mathcal{H} \to \mathcal{H}(t)$. $\mathcal{H}(t)$ can then be replaced by an <u>average hamiltonian</u> $\bar{\mathcal{H}} = \bar{\mathcal{H}}^{(0)} + \bar{\mathcal{H}}^{(1)} + \cdots$, with

$$\bar{\mathcal{H}}^{(0)} = \frac{1}{t_c} \int_0^{t_c} \mathcal{H}(t')dt'$$

$$\bar{\mathcal{H}}^{(1)} = \frac{-i}{2t_c} \int_0^{t_c} \int_0^{t'} [\mathcal{H}(t'), \mathcal{H}(t'')]dt''dt'$$

L. Petrakis and J. P. Fraissard (eds.), Magnetic Resonance. Introduction, Advanced Topics and Applications to Fossil Energy, 111–135.

Often $\bar{\mathcal{H}}^{(0)}$ by itself is an adequate approximation.

The form of $\mathcal{H}(t)$ and hence of $\bar{\mathcal{H}}$ is controlled by the externally applied fields. Some examples and their consequences are

External Field	Effect
1) static field H_0	"truncation" of spin interactions.
2) steady RF field at resonance	spin-locking, heteronuclear decoupling.
3) steady RF field off resonance	Lee-Goldburg dipolar narrowing.
4) train of 180° pulses	spin echoes-suppression of field inhomogeneity and chemical shifts.
5) trains of 90° pulses of four phases	"multiple-pulse" line narrowing.
6) RF fields at two frequencies	cross-polarization.

Periodically time-dependent interaction can also be produced by mechanical rotation of the sample, leading *inter alia* to magic-angle line narrowing.

The exploitation of the average hamiltonian method requires a familiarity with the properties of angular momentum operators.

A. QUANTUM MECHANICAL PRELIMINARIES

Functions of Operators. Rotations.

A function f(Q) of an operator Q is defined by its power series expansion

$$f(Q) = \sum_n a_n Q^n \tag{A1}$$

An important example is the complex exponential function whose argument contains a component of angular momentum, say I_y.

$$R_y(\Theta) = e^{-i\Theta I_y} = \sum_{n=0}^{\infty} \frac{(-i\Theta)^n}{n!} I_y^n$$

This kind of unitary operator (e^{iA} is unitary if A is hermitian) generates a useful set of relations when it is applied to the components of $\tilde{I}$ itself:

$\tilde{I}_x = e^{i\Theta I_z} I_x e^{-i\Theta I_z}$ etc. Such expressions are easily evaluated. One way would be by expanding the exponentials, repeated use of the commutation rules for angular momentum, and regrouping terms. A more powerful and generally useful method is by <u>parameter differentiation</u>. For example, differentiate $\tilde{I}_x$ with respect to Θ directly: by the chain rule

$$\frac{\partial}{\partial\Theta}\tilde{I}_x = ie^{i\Theta I_z}[I_z, I_x]e^{-i\Theta I_z}$$

$$= e^{-i\Theta I_z} I_y e^{-i\Theta I_z}$$

$$\frac{\partial^2}{\partial\Theta^2}\tilde{I}_x = -ie^{i\Theta I_z}[I_z, I_y]e^{-i\Theta I_z}$$

$$= -e^{i\Theta I_z} I_x e^{i\Theta I_z} = \tilde{I}_x.$$

The differential equation $\frac{d^2 \tilde{I}_x}{d\Theta^2} = \tilde{I}_x$ has the general solution

$$\tilde{I}_x = A\cos\Theta + B\sin\Theta.$$

The operators A and B can be evaluated from the boundary condition $\Theta = 0$. From the definitions and solutions for $\tilde{I}_z$ we have $A = I_z$. Similarly from the expression for the first derivative of $\tilde{I}_z$ we have $B = -I_y$. Thus

$$\tilde{I}_x = I_x\cos\Theta - I_y\sin\Theta.$$

All relations of this type are summarized as follows:

$$\tilde{I}_\mu = e^{+i\Theta I_\nu} I_\mu e^{-i\Theta I_\nu}$$

μ \ ν	x	y	z
x	I_x	$I_x\cos\Theta + I_z\sin\Theta$	$I_x\cos\Theta - I_y\sin\Theta$
y	$I_y\cos\Theta - I_z\sin\Theta$	I_y	$I_y\cos\Theta + I_x\sin\Theta$
z	$I_z\cos\Theta + I_y\sin\Theta$	$I_z\cos\Theta - I_x\sin\Theta$	I_z

From these relations we also have $e^{i\Theta I_z} I_\pm e^{-i\Theta I_z} = I_\pm e^{\pm i\Theta}$. These express the result of negative (left-handed) rotations of the axis with respect to a fixed vector $\underset{\sim}{I}$, or of positive rotations of $\underset{\sim}{I}$ with respect to fixed axes. In such a situation confusion of signs is probably inevitable and a trial-and-error approach is recommended.

The unitary transformation of a function of an operator has an important property which can be seen from (A1) by repeated insertion of the identity $U^{-1}U = 1$:

$$Uf(Q)U^{-1} = \sum_n a_n UQ^nU^{-1}$$

$$\begin{aligned} UQ^nU^{-1} &= UQQQ\ldots U^{-1} \\ &= UQU^{-1}UQU^{-1}\ldots U^{-1} \\ &= (UQU^{-1})^n \end{aligned}$$

Thus

$$Uf(Q)U^{-1} = f(UQU^{-1}) \tag{A3}$$

An example is provided by the rotation of an angular momentum vector about an axis other than one of the cartesian axes, e.g., about an arbitrary axis in the xy plane. The operator is

$$\begin{aligned} R(\Theta) &= e^{-i\Theta\left(I_x \cos\Phi + I_y \sin\Phi\right)} \\ &= \exp\left(-i\Theta\, e^{-i\Phi I_z} I_x e^{i\Phi I_z}\right) \end{aligned}$$

$$e^{-i\Phi I_z} e^{-i\Theta I_x} e^{i\Phi I_z}$$

which expresses the result as equivalent to three successive rotations about cartesian axes. (Note that the three exponents do not commute.) Similarly, it is possible to express a general rotation of the coordinate system in three dimensions through arbitrary Euler angles α, β, γ. A rotation of axes through α is first made about the z axis, followed by a rotation through β about the new y axis, y' and finally by a rotation through γ about the resulting z axis, z". The overall rotation thus defined is

$$D(\alpha\beta\gamma) = e^{-i\gamma I_z''} e^{-i\beta I_y'} e^{-i\alpha I_z}$$

This can be re-expressed, using (A3) repeatedly, in terms of successive rotations about the original axes in the reverse order.

$$D(\alpha\beta\gamma) = e^{-i\alpha I_z} e^{-i\beta I_y} e^{-i\gamma I_z} \tag{A4}$$

For systems with fixed angular momentum quantum number I, D can be expressed in the basis where I_z is diagonal with magnetic quantum number m in terms of the Wigner rotation matrices

$$\langle m|D(\alpha\beta\gamma)|m'\rangle = D^I_{mm'}(\alpha\beta\gamma) = e^{i(\alpha m-\gamma m')} d^I_{mm'}(\beta) \tag{A5}$$

$$d^I_{mm'}(\beta) = \langle m|e^{-i\beta I_y}|m'\rangle \tag{A6}$$

Sylvester's theorem shows how to construct any function of an N x N matrix such as I_y as a linear combination of the first N powers of the matrix. For $I = \frac{1}{2}$ we have

$$d^{\frac{1}{2}}_{mm'}(\beta) = \begin{pmatrix} \cos\frac{\beta}{2} & \sin\frac{\beta}{2} \\ -\sin\frac{\beta}{2} & \cos\frac{\beta}{2} \end{pmatrix} \tag{A7}$$

and for I = 1.

$$d^1_{mm'}(\beta) = \begin{pmatrix} \frac{1}{2}(1+\cos\beta) & -\frac{1}{\sqrt{2}}\sin\beta & \frac{1}{2}(1-\cos\beta) \\ \frac{1}{\sqrt{2}}\sin\beta & \cos\beta & -\frac{1}{\sqrt{2}}\sin\beta \\ \frac{1}{2}(1-\cos\beta) & \frac{1}{\sqrt{2}}\sin\beta & \frac{1}{2}(1+\cos\beta) \end{pmatrix}$$

Further details can be found in the book by Edmonds.

Representations and Frames of Reference

Any unitary transformation U has the property that when it is applied to all the state vectors of a system and to all the operators representing observables, $\tilde{Q} = UQU^{-1}$, none of the physically measurable quantities (eigenvalues, expectation values, expansion coefficients) is altered: that is, the system has not been changed but its representation has. A familiar example arises in solving stationary eigenvalue problems, in which we wish to find the representation $|n\rangle$ in which some operator, e.g., the Hamiltonian, is diagonal with eigenvalues

$$\langle n'|\mathcal{H}|n\rangle = E_n\delta_{n'n}.$$

In an arbitrary representation $|k\rangle$, in which $|n\rangle$ has the expansion.

$$|n\rangle = \sum_k C_{nk}\,|k\rangle = \sum_k |\rangle\langle|n\rangle, \tag{A9}$$

This is

$$\sum_{k'}\sum_{k} \langle n'|k'\rangle\langle'|\mathcal{H}\;|\rangle\langle|n\rangle = E_n\delta_{nn'}$$

$$= \sum_{k'}\sum_{k} U_{n'k'}\;\mathcal{H}_{k'k}U^{-1}_{kn} = (U\mathcal{H}U^{-1})_{n'n}$$

The problem is to find the unitary operator U which reduces the nondiagonal $\mathcal{H}_{k'k}$ to diagonal form. Such a change of basis corresponds to a symbolic rotation of the set of basis vectors $|k\rangle$ into the set $|n\rangle$ in a multidimensional vector space.

A unitary transformation may correspond to a physical rotation in ordinary space as well: viz., the rotation operators of the preceding section, in which case the term "frame of reference" or simply "frame" is used in place of representations. Often the physical rotations needed to simplify a problem are more obvious than the purely symbolic rotations just mentioned: A spin $\underset{\sim}{I}$ in a magnetic field in the xz plane can be described by the Hamiltonian operator

$$= -\gamma\underset{\sim}{H}\cdot\underset{\sim}{I} = -\gamma J\,(I_z\cos\alpha + I_x\sin\alpha).$$

whose eigenvalues are not obvious. But applying $U = e^{+i\Theta I_y}$ we have

$$\tilde{\mathcal{H}} = -\gamma H\left[\left(I_z \cos\Theta - I_x \sin\Theta\right)\cos\alpha + \left(I_x \cos\Theta + I_z \sin\Theta\right)\sin\alpha\right]$$

which, for $\alpha = \Theta$, is just

$$\tilde{\mathcal{H}} = -\gamma H I_z$$

with eigenvalues $-\gamma Hm$.

Note that the rotation operator acts only on the part of a system whose angular momentum it contains. In a system of two spins, I and S, for example

$$e^{-i\frac{\pi}{2}I_y} I_z S_z e^{i\frac{\pi}{2}I_y} = I_x S_z .$$

Moreover, in a system of nuclear spins I plus lattice, operators of the above type do not affect the lattice variables. It pays to be careful, in speaking of frames of reference, to remember that they may not represent a change in the viewpoint of the observer toward the whole physical world.

Time-Dependent Transformations

A transformation may also contain an explicit time dependence, and this leads to a new feature: if $U = U(t)$ the states and operators are transformed as always, and all expectation values are computed in the usual way, including the energy $E = \langle\tilde{a}|\tilde{\mathcal{H}}|\tilde{a}\rangle$. However the motion of the system with respect to the moving frame of reference is modified. In the untransformed frame the system obeys the time-dependent Schrodinger equation.

$$\frac{\partial}{\partial t}|a\rangle = -i\mathcal{H}|a\rangle \qquad \text{(A10)}$$

Substituting $|a\rangle = U^{-1}|\tilde{a}\rangle$ and then applying U in the left, we have

$$\frac{\partial}{\partial t}\,|\,\tilde{a}\rangle = iU \quad U^{-1}\,|\,\tilde{a}\rangle - VU\frac{\partial}{\partial t}U^{-1}\,|\tilde{a}\rangle$$

That is, the system obeys a Schrödinger equation of the form (A10) only if the Hamiltonian is replaced by an <u>effective</u> Hamiltonian

$$\mathcal{H}_{eff} = U\mathcal{H}U^{-1} - iU\frac{\partial}{\partial t}U^{-1} = U(\mathcal{H} - i\frac{\partial}{\partial t})U^{-1}. \qquad \text{(A11)}$$

The extra term corresponds to fictitious forces which one expects to find in an accelerated frame of reference.

The "rotating frame" is extremely widely used in magnetic resonance. It is obtained by a rotation at uniform angular velocity ω about the direction of some constant magnetic field, conventionally defined as the z-axis: $U(t) = e^{-i\omega t I_z}$. The extra term in $\mathcal{H}_{eff}$ according to (A11) is then ωI_z. A spin Hamiltonian containing a Zeeman effect $-\gamma H_0 I_z$ plus some other terms $\mathcal{H}'$ then becomes

$${}_{eff} = -\gamma(H_0 - \frac{\omega}{\gamma})I_z + \tilde{\mathcal{H}}' \qquad \text{(A12)}$$

which corresponds to a change in the apparent magnetic field, in particular its reduction to a small value if $\omega \sim \gamma H_0$.

Time-dependent transformations of course need not correspond to simple rotations. For any system with a time-independent Hamiltonian $\mathcal{H}$, the Heisenberg representation results from $U(t) = e^{i\mathcal{H}t}$ which gives $\mathcal{H}_{eff} = 0$. The state vectors in this representation therefore are stationary. This is also evident from the fact that in the original (Schrödinger) representation, the state vector develops according to the Schrödinger equation (A10), whose integrated form is

$$|a(t)\rangle = e^{-i\mathcal{H}t}|a(0)\rangle \qquad \text{(A13)}$$

so

$$|\tilde{a}(t)\rangle = e^{i\mathcal{H}t}|a(t)\rangle = |a(0)\rangle$$

However the operators become time dependent Heisenberg operators

$$Q(t0 = e^{i\mathcal{H}t}Qe^{-i\mathcal{H}t} \tag{A14}$$

When the Hamiltonian consiste of a large simple part $\mathcal{H}_0$ and a small perturbation V, the interaction representation induced by $U(t) = e^{+i\mathcal{H}_0 t}$ is useful, in that

$$\mathcal{H}_{eff} = e^{i\mathcal{H}_0 t}\, Ve^{-i\mathcal{H}_0 t} \tag{A15}$$

is now small (through rapidly "oscillating") and appropriate grist for the mill of time-dependent perturbation theory.

Time-Ordered Operators

The time dependence referred to above are all of the simple type e^{iAt} with A independent of time. The action of such an operator over successive time intervals t_1 and t_2 is then e^{iAt_2} $e^{iAt_1} = e^{iA(t_1+t_2)}$. However, A may be time dependent in such a way that its values in the two intervals do not commute, and the two exponentials cannot be combined into one. An obvious example is provided by successive rotations about two different axes (cf the discussion following (A3).) One can approximate a general time variation of A(t) by a succession of intervals δt in each of which A is constant, the transformations during the first such interval after t = 0 being $U(\delta t) = e^{iA(0)\delta t}$. Over a longer interval, then

$$U(t) = U(N\delta t) = e^{-i\delta tA[(N-1)\delta t]}\ldots e^{-i\delta tA(\delta t)}\, e^{-i\delta tA(0)} \tag{A16}$$

A common notation for the limit of this transformation as $N \to \infty$, $t \to 0$ with $N\delta t$ constant is

$$U(t) = Te^{-i\int_0^t A(t')dt'} \tag{A17}$$

where the "time-ordering operator" T is a reminder that (A17) is merely a shorthand for (A16), whose factors must be kept in the proper order.

A Factoring Theorem

Time-ordered operators appear when one wishes to enter an interaction representation: the time-development operator generated by (A13) is a simple case in point. That result can be obtained from a useful factoring theorem, which for the case of two time-independent operators A and B can be expressed

$$e^{-i(A+B)t} = e^{-iAt} T e^{-i\int_0^t \tilde{B}(t')} = T e^{-i\int_0^t \tilde{B}(-t')dt'} e^{-iAt} \tag{A18}$$

$$\tilde{B}(t') = e^{iAt} B e^{-iAt} \tag{A19}$$

The proof is as follows: let

$$S = e^{iAt} e^{-i(A+B)t}$$

Then

$$\frac{\partial}{\partial t} S(t) = iAS - ie^{iAt}(A+B)e^{-i(A+B)t}$$

$$= -ie^{iAt} B e^{-i(A+B)t}$$

$$= -ie^{iAt} e^{-iAt} \tilde{B}(t) e^{iAt} e^{-i(A+B)t}$$

$$= -i\tilde{B}(t)S(t)$$

whence

$$S(t) = Tw^{-i\int_0^t \tilde{B}(t')dt'}$$

as desired. The theorem is also adaptable to cases in which A or B or both depend explicitly on time: then

$$Te^{-i\int_0^t [A(t') + B(t')]dt'} = Te^{-i\int_0^t A(t')} Te^{-i\int_0^t \tilde{B}(t')dt'} \tag{A20}$$

$$\tilde{B}(t') = T^{-1} e^{i\int_0^t A(t'')dt''} B(t')\, T e^{-i\int_0^t A(t'')dt''} \qquad (A21)$$

Here T^{-1} indicates an inverse time ordering, i.e., in a defining expression such as (A16) time increases from left to right. The proof follows the same lines as that given above.

Average Hamiltonians

Time-dependent hamiltonians arise whenever one enters an interaction representation or (which is the same thing) employs the factoring theorem of the previous section. Sometimes it is useful to replace such a time dependent hamiltonian with a time-independent one which generates the same propagator over some chosen time t_c. This can always be done formally: the "true" propagator is a sequence of rotations in Hilbert space which, because of the group property, is equivalent to the single rotation generated by some fictitious fixed hamiltonian $\bar{\mathcal{H}}$. It is generally neither easy nor useful to construct $\bar{\mathcal{H}}$ explicitly except when $\mathcal{H}$ is "small" and/or t_c is short, i.e., the propagator $U(T_c,0) \approx 1$.

When this is true one can treat the problem perturbatively: conventionally one iterates

$$U(t,0) = 1 + \int_0^t \dot{U}(t')dt' \qquad (A22)$$

$$\dot{U}(t) = -i\mathcal{H}(t)\, U(t) \qquad (A23)$$

to obtain the usual time-dependent perturbation expansion

$$U(t_c,0)=1 -i\int_0^{t_c} \mathcal{H}(t')dt' - \int_0^{t_c} dt' \int_0^{t'} dt''\mathcal{H}(t')\mathcal{H}(t'') + \ldots$$

(A24)

As we shall see, this result suffers from the defect, for certain applications, that $U(t_c,0)$ is not unitary when truncated at any finite order.

An alternative form of perturbation theory avoids this difficulty by writing

$$U(t_c,0) = e^{-i\bar{\mathcal{H}}t_c} \tag{A25}$$

$$\bar{\mathcal{H}} = \sum_{n=0}^{\infty} \bar{\mathcal{H}}^{(n)} \tag{A26}$$

with every $\bar{\mathcal{H}}^{(n)}$ hermitian. (A26) is called the Magnus expansion. Its terms can be found by expanding (A32) in powers of $\hbar^{-1}$ and comparing term by term with (A24). One gets

$$\bar{\mathcal{H}}^{(0)} = \frac{1}{t_c} \int_0^{t_c} dt\, \mathcal{H}(t)$$

$$\bar{\mathcal{H}}^{(1)} = \frac{-i}{2t_c} \int_0^{t_c} dt \int_0^{t} dt' \left[\mathcal{H}(t), \mathcal{H}(t')\right]$$

$$\bar{\mathcal{H}}^{(2)} = \frac{1}{6t_c} \int_0^{t_c} dt \int_0^{t} dt' \int_0^{t'} dt'' \{[\mathcal{H}(t),[\mathcal{H}(t'),\mathcal{H}(t'')]] + [\mathcal{H}(t''), (\mathcal{H}(t'),\mathcal{H}(t)]] \tag{A27}$$

This form is especially useful when $\mathcal{H}(t)$ is <u>periodic</u> modulo the upper limit t_c of the integrals: then since the history of $\mathcal{H}(t)$ is that same in each such interval

$$U(Nt_c,0) = (e^{-i\bar{\mathcal{H}}t_c})^N = e^{-i\bar{\mathcal{H}}(Nt_c)} \tag{A28}$$

That is, even though $\bar{\mathcal{H}}$ had to be evaluated perturbatively over a short time t_c, the result describes the behavior of the system to arbitrarily long times, as long as one is content with "stroboscopic" observation of the properties of the system at the instants Nt_c.

B. TRACES AND THE DENSITY MATRIX

Expectation Values as Traces

The expectation value of an observable Q for a system in the quantum state $|a\rangle$ can be written, using the closure relation (A9) with respect to some basis $|n\rangle$ as

$$\begin{aligned} \langle Q\rangle &= \langle a|Q|a\rangle = \sum_n \langle a|n\rangle\langle n|Q|a\rangle = \sum_n \langle n|Q|a\rangle\langle a|n\rangle \\ &= \sum_n (Q|a\rangle\langle a|)_{nn} = \mathrm{Tr}(Q|a\rangle\langle a|). \end{aligned} \tag{B1}$$

that is as the sum of the diagonal elements, or trace of a matrix which is the product of Q with an operator $|a\rangle\langle a|$. . The chief convenience of this relation lies in the fact that the result is independent of the basis chosen which need not, in particular, be the eigenbasis of the Hamiltonian but can be chosen for convenience. Of course this does not constitute a magical short cut to the solution of quantum mechanical problems: one still needs to know the state $|a\rangle$. In fact, the matrix of $|a\rangle\langle a|$ in the representation $|n\rangle$ is just

$$\langle n'|a\rangle\langle a|n\rangle = C_{an'}C^*_{an} \tag{B2}$$

i.e., is composed of the expansion coefficients of $|a\rangle$ in the chosen basis. It is only in problems having a statistical aspect that this matrix assumes real importance (Vide infra).

Anticipating that, we summarize several useful properties of traces, all of which follow from the first:

$$\begin{aligned} &\mathrm{Tr}(AB) = \sum_k (\textstyle\sum_\ell A_{k\ell}B_{\ell k}) = \sum_k (\sum_\ell B_{\ell k}A_{k}) = \mathrm{Tr}\ (BA) \\ &\mathrm{Tr}(UAU^{-1}) = \mathrm{Tr}A \text{ for any unitary } U \end{aligned} \tag{B3}$$

$$\mathrm{Tr}[A,B] = 0 \tag{B4}$$

$$\mathrm{Tr}\{[A,B]C\} = \mathrm{Tr}\{[C,A]B\} = \mathrm{Tr}\{[B,C]A\} \tag{B5}$$

$$\mathrm{Tr}\{[A,[B,C]D\} = \mathrm{Tr}\{[D,A][B,C]\} = \mathrm{Tr}\{[D,[C,B]A\}$$
$$= \mathrm{Tr}\{[C,[D,A]B\} = \mathrm{Tr}\{[B,[A,D]C\} \tag{B6}$$

Example of a Trace Calculation

Suppose, for a system composed of N spins I, we desire the value of TrI_x^2 where $I_x = \sum_{i=1}^{N} I_{xi}$, that is

$$\mathrm{TrI}_x^2 = \sum_n \langle n| \sum_i \sum_j I_{xi} I_{xj} |n\rangle \; *$$

where the set of functions n has $(2I + 1)^N$ members. Remembering the invariance of the trace to unitary transformation we choose to deal instead with the operator

$$e^{-i\frac{\pi}{2}I_y} I_x^2 \, e^{i\frac{\pi}{2}I_y} = \left(e^{-i\frac{\pi}{2}I_y} I_x e^{i\frac{\pi}{2}I_y}\right)^2 = I_z^2$$

The operator I_z^2 is convenient because it is diagonal in the familiar basis labelled by the magnetic quantum numbers of the individual spins:

$$|n\rangle = |m_1 m_2 \ldots\ldots m_N\rangle$$

Thus

$$\mathrm{TrI}_z^2 = \sum_i \sum_j \sum_{m_1} \sum_{m_2} \ldots\ldots \sum_{m_N} m_i m_j$$

In every term with $i \neq j$, the sum of m_i or m_j is $-I + (-I + 1) + \ldots\ldots + I = 0$, so only the terms with $i = j$ can contribute and we drop the sum on j. Also, all spins i contribute in the same way, so the result must be N times the value for any particular i, say i = 1:

$$\mathrm{Tr}I_z^{\,2} = N \sum_{m_1} \sum_{m_2} \cdots \sum_{m_N} m_1^{\,2}$$

$$= N \sum_{m_1} m_1^{\,2} \cdot \sum_{m_2} 1 \cdot \ldots \cdot \sum_{m_N}$$

where in the last line we have factored $m_1^{\,2}$ out of all the sums which do not depend on it, leaving unity as the other factor. Now $\sum 1 = 1 + 1 + 1 \ldots$, $2I + 1$ times, or $2I + 1$. The sums over $m_2, \ldots, m_N$ thus give a factor $(2I+1)^{N+1}$.

$$\mathrm{Tr}I_z^2 = N(2I + 1)^{N-1} \sum_{m_1=-I}^{I} m_1^{\,2}$$

Sums of the type which remains can be calculated (e.g., by using the Euler-Maclaurin formula to relate them to integrals of the type $\int^{I} m^2 dm$). The value of the present sum is 1/3 $I(I+1)(2I+1)$. Thus

$$\mathrm{Tr}I_x^{\,2} = \mathrm{Tr}I_z^{\,2} = 1/3\ NI\,(I+1)(2I+1)^N$$

The traces of other combinations of I_x, I_y, I_z, $I_\pm$ to various powers have been tabulated (Am 62, Am 62a).

A Useful Trick for Recognizing Traces Which Vanish

Traces of very complicated expressions frequently occur in calculations, and it is most annoying to go to the trouble of evaluating such a trace only to find at the end that it is zero. A very valuable means for recognizing when this will be so is based on (B3), the invariance of the trace to unitary transformation. Suppose that a transformation can be found which inverts the sign of the operator whose trace is needed, and therefore of the trace itself. Since the trace cannot change, it can only have been zero.

Of course the success of this procedure depends on one's prescience in trying the right unitary transformations. In magnetic resonance calculations this is not so problematical as it might seem, because of the fact that nearly all operators are written as sums and products of angular momentum operators for the spins. The transformations (A2), often with $\Theta = \pi$, acting on one or all of the spins, are especially valuable, as can be seen from the following examples of operators and transformations which change their signs

Operator	Transformation
I_z	$e^{\pm i\pi I_x}$ or $e^{\pm i\pi I_y}$
$I_{zi}I_{zj}$, $i \neq j$	$e^{i\pi I_{xi}}$ (but not $e^{i\pi(I_{xi}+I_{xj})}$)

Operators which can be written as a sum of terms may allow the terms to be treated separately using different transformations. The truncated dipolar hamiltonian between spins i and j

$$\mathcal{H}_d^0(ij) = b_{ij}(\underset{\sim}{I}_i \cdot \underset{\sim}{I}_j - 3I_{zi}I_{zj})$$

$$= b_{ij}I_{xi}I_{xj} + b_{ij}I_{yi}I_{yj} - 2b_{ij}I_{zi}I_{zj}$$

has the property that the first and second terms can be inverted by $e^{i\pi I_{zi}}$ and the third by $e^{i\pi I_{xi}}$, for example, so that the trace of each vanishes.

Greater advantage could in fact have been taken of unitary transformations in the preceding calculation of TrI_x^2: By use of $\pi/2$ rotations we see immediately $TrI_x^2 = TrI_y^2 = TrI_z^2$. But $I_x^2 + I_y^2 = \underset{\sim}{I}\cdot\underset{\sim}{I}$, which has eigenvalues $I(I+1)$ in the m representation. Thus

$$TrI_x^2 = (1/3)\, I(I+1)Tr1$$

and Tr1 over N spins is trivially found to be $N(2I+1)^N$.

The Density Matrix

The precise quantum state $|a\rangle$ of a complicated system (such as a sample in an NMR experiment) is pratically never known: in general we know only that it is one of the many states which is consistent with the values of certain macroscopic properties, such as the total magnetic moment, with which it was prepared. We then consider an ensemble of systems which are alike in these measured properties but among which all the consistent states are represented, and we wish to calculate other properties as statistical averages over the ensemble.* The "average state" has no clear meaning, but the average of the complex number $C_{an'}C_{an}{}^*$ does. Accordingly the density matrix is defined as

$$\sigma_{n'n} = \overline{\langle n' | a\rangle\langle a | n\rangle} \tag{B7}$$

the bar denoting the statistical average. Similarly we can define the density operator as

$$\sigma = \overline{|a\rangle\langle a|} \tag{B8}$$

since the states $|n\rangle$, $|n'\rangle$ available to all members of the ensemble are the same.

The average expectation value of an observable Q over the ensemble is then

$$\langle\overline{Q}\rangle = \mathrm{Tr}(Q\sigma) \tag{B9}$$

In most cases we will drop the bar for typographical convenience where there is no likelihood of confusing $\langle\overline{Q}\rangle$ with the purely quantum mechanical quantity $\langle Q\rangle$. The diagonal elements σ_{nn} are the ensemble average probabilities P_n for finding the system in $|n\rangle$, and the off diagonal elements express any coherence these states which persist in the ensemble.

---

* For many purposes the experimental sample itself may be through of as the ensemble, the systems composing it being regions of the sample which are small compared ot the total but yet large enough to be "representative." From a purist's viewpoint this is quite wrong, but does no harm in cases where we can convince ourselves that the result of an experiment would be unaffected by inserting a large number of rigid, thermally insulting partitions into the sample.

It is easily verified from the definition that:

a) σ is hermitian
b) $\mathrm{Tr}\,\sigma = 1$
c) If all members of the ensemble are in the same state

$$|a\rangle, \text{ then } \sigma^2 = \sigma: \ \overline{|a\rangle\langle a|}\ \overline{|a\rangle\langle a|} = |a\rangle\langle a|a\rangle\langle a| = |a\rangle\langle a|$$

Equation of Motion.

Each member of the ensemble obeys the time-dependent Schrödinger equation (A10). Accordingly

$$\frac{\partial}{\partial t} |a\rangle\langle a| = \frac{\partial |a\rangle}{\partial t} \langle a| + |a\rangle \frac{\partial |a\rangle}{\partial t} = -i[\mathcal{H}, |a\rangle\langle a|]$$

and, averaging over the ensemble (whose members have the same Hamiltonian) one obtains the Liouville equation (sometimes called the von Neumann equation)

$$\frac{\partial \sigma}{\partial t} = -i\ [\mathcal{H}, \sigma] \qquad \text{(B10)}$$

whose integrated form for $\mathcal{H}$ = const is

$$\sigma(t) = e^{-i\mathcal{H}t}\, \sigma(0) e^{i\mathcal{H}t} \qquad \text{(B11)}$$

Equation (B10) is sometimes written in terms of the Liouville operator

$$L = -[\mathcal{H},\quad] \qquad \text{(B12)}$$

which is interpreted as having the operand inserted in the bracket. In this notation (B10) is

$$\frac{\partial \sigma}{\partial t} = iL\sigma \qquad \text{(B13)}$$

and (B11) is

$$\sigma(t) = e^{iLt}\,\sigma(0) \tag{B14}$$

as can be verified by expanding the exponentials in (B11), collecting terms of equal powers of t, and using (B12). L is of course a sort of "superoperator" which, if σ is represented by an N x N matrix, must be represented by an N x N x N x N supermatrix.

$$\frac{\partial \sigma_{k\ell}}{\partial t} = i \sum_{m,n} L_{k\ell mn}\,\sigma_{mn}$$

Alternatively the elements of can be regrouped by some chosen convention into a column vector of N^2 elements, and L becomes a square matrix of dimension N^2.

Thermodynamic Equilibrium.

The macroscopic properties of a complicated system, left to itself, tend to evolve in time toward limiting values which thereafter do not change: one says that the system has reached equilibrium. What actually occurs in an isolated system ($\mathcal{H}$ constant) is readily seen by writing (B11) in the energy representation, where $\mathcal{H}$ is diagonal:

$$\begin{aligned}\sigma_{nn'}(t) &= \langle n|e^{-i\mathcal{H}t}\,\sigma(0)\,e^{i\mathcal{H}t}|n'\rangle \\ &= e^{i(E_{n'}-E_n)t}\,\sigma_{nn'}(0)\end{aligned} \tag{B15}$$

That is, the diagonal elements remain fixed and the off-diagonal elements oscillate at a variety of frequencies determined by the eigenvalue spectrum of $\mathcal{H}$. An arbitrary observable Q displays the ensemble average behavior

$$\langle Q(t)\rangle = \mathrm{Tr}\left[\sigma(t)Q\right] = \sum_n \sum_{n'} e^{i(E_{n'}-E_n)t}\,\sigma_{nn'}(0)Q_{n-n}. \tag{B16}$$

Decompose the matrix of Q into its diagonal and off-diagonal parts $Q_d + Q_n$ in the energy representation. From (B16) it is clear that, as far as the desired diagonal elements of $\sigma(t)Q$ are concerned, Q_d picks out only the diagonal elements $\sigma_{nn}(0) = P_n(0)$ from the density matrix, for which there are no oscillations, so that $\langle Q_d(t)\rangle$ is constant in time. In the same sense Q_n picks out only the purely oscillatory off diagonal terms. The approach to a time-independent $\langle Q\rangle$ for large t is to be understood on the basis that there are many such oscillatory terms which frequencies which are not rationally related; their contributions to $\langle Q_d(t)\rangle$ interfere destructively in such a way that $\langle Q_n(t)\rangle\to)$ in a time which is characteristic of the spread in energy-level spacings of the system.

If we are satisfied that all observables Q of interest will behave in this manner, the final equilibrium values $\langle Q(\infty)\rangle = \langle Q_d\rangle$ can be calculated in an alternative way: simply replace the actual σ, whose structure is complicated, by a fictitous density matrix σ_e in which all off diagonal elements have been set equal to zero.

$$(\sigma_e)_{nn'} = P_n\delta_{nn'} \tag{B17}$$

then

$$\langle Q(\infty)\rangle = \sum_n P_n Q_{nn} = \langle Q_d\rangle$$

as desired.

It is to be emphasized that this replacement is a fiction, inasmuch as the off diagonal elements of σ for an isolated system really continue to oscillate indefinately according to (B15): it however gives the correct results for the values of observables at equilibrium once it is presupposed that this equilibrium is reached. The great value of the substitution hinges on the replacement of an enormous N x N array of numbers, where knowledge would require a solution of all the dynamical equations, by the much smaller set of N values of the P_n.

The <u>equilibrium ensemble</u> whose density matrix is σ_e need not be one in which the system is known to be in an eigenstate of $\mathcal{H}$. It is merely necessary that arbitrary states $|a\rangle$, for which $\langle n|a\rangle\langle a|n'\rangle$ may have off diagonal parts, be represented in the ensemble with a variety of phases-- $|a_0\rangle e^{i\Phi}$ --in such a way

~~that~~ the off diagonal elements of the ensemble average $|a\rangle\langle a|$ vanish (random phase assumption).

This, of course, still leaves open the question of how the values of the P_n are to be determined for the equilibrium ensemble, and indeed whether they can be determined with sufficient uniqueness on the basis of the extremely limited information we usually have about the state of the system of interest. The theory of statistical mechanics shows that they can be: For a system of specified Hamiltonian, otherwise known only to be at some temperature T, the appropriate choice is the canonical ensemble, in which

$$\sum \rho_n = 1; \qquad \sum \rho_n E_n = \overline{E} \tag{B18}$$

where $\overline{E}$ is a fixed average energy appropriate to T, and is to be identified with the thermodynamic internal energy E. The P_n are chosen to maximize the uncertainty H in the distribution of dynamical states in the ensemble subject to the constraints (B18). Shannon has shown that this uncertainty has the natural definition

$$H = -K\sum p_n \ell n p_n \tag{B19}$$

(which is later shown to be the same as the entropy S if K is taken to be the Boltzmann constant k.) The maximization of S leads directly to the Boltzmann distribution

$$\begin{aligned} p_n &= \frac{1}{z} e^{-\hbar E_n/kT} \\ Z &= \sum_n e^{-\hbar E_n/kT} \end{aligned} \tag{B20}$$

Because of the important invariance of traces to unitary transformation, it is useful to be able to write the $p_n = \sigma_{nn}$ defined by (B20) in a form which invites such transformation. Clearly

$$\sigma_{nn} = \frac{1}{Z} \langle n|e^{-h \quad /kT}|n\rangle$$

so we can write σ_e in operator form as follows:

$$\sigma_{nn} = \frac{1}{Z} e^{-\hbar\mathcal{H}/kT}; \quad Z = \mathrm{Tr}\, e^{-\hbar\mathcal{H}/kT} \tag{B21}$$

Similarly (B19) can be written

$$S = -k\ \mathrm{Tr}\left(\sigma_e \ln \sigma_e\right) \tag{B22}$$

Given σ_e, all the other equilibrium thermodynamic quantities can be calculated, for example from (B21) and the Helmholtz free energy

$$F \ -kT \ln Z \tag{B23}$$

The other thermodynamic functions can be constructed in the usual way. In this connection it is to be noted that we will be interested in magnetic systems where pressure and volume have essentially fixed values. Their roles are assumed by the magnetization $\underset{\sim}{M}$ and the external field $\underset{\sim}{H}$, respectively,* so that

$$dE = TdS - \underset{\sim}{M} \cdot d\underset{\sim}{H}$$

$$dF = -Sdt - \underset{\sim}{M} \cdot d\underset{\sim}{H}$$

etc. For example (for $\underset{\sim}{M}$ parallel to $\underset{\sim}{H}$)

$$M = \left(\frac{\partial F}{\partial H}\right) = \frac{kT}{Z}\left(\frac{\partial Z}{\partial H}\right)_T = \frac{kT}{z}\frac{\partial}{\partial H}\mathrm{Tr}\, e^{-\hbar\mathcal{H}/kT}$$

$$= \frac{h}{z}\mathrm{Tr}\left\{\left(-\frac{\partial}{\partial H}\right) e^{-\hbar\mathcal{H}/kT}\right\} = -\hbar\left\langle\frac{\partial\mathcal{H}}{\partial H}\right\rangle. \tag{B24}$$

* Why PdV should be replaced by $\underset{\sim}{M} \cdot d\underset{\sim}{H}$ rather than $\underset{\sim}{H} \cdot d\underset{\sim}{M}$ is a confusing question which depends on the way the thermodynamic system and its surroundings are defined. Either procedure is satisfactory if consistently applied: the one given is conventional in the literature of magnetic resonance.

Extensive discussions of these matter are given by Gorter and by Guggenheim.

Explicit Representations

An important feature of (B21) is that all NMR experiments begin (or are imagined to begin) with a sample at equilibrium at some temperature T. Moreover, σ_e is expressed as an explicit operator function whose variations during an experiment can be treated with all the machinery of quantum mechanical operator calculus. The adoption of a specific set of basis functions, with all the bookkeeping of indices that entails, can be put off to a late stage or even to the end of a calculation, where a trace may perhaps have to be evaluated to obtain a numerical result.

In the case of arbitrary temperature this convenience is partly illusory, because of the practical difficulties of manipulating the exponential function of a Hamiltonian operator which may be complicated. A rare example in which the full form (B15) can be easily exploited is provided by a system consisting of a single spin-1/2 particle subject to a magnetic field H in the direction of a unit vector $\underset{\sim}{n}$.

$$= -\gamma H \underset{\sim}{n} \cdot \underset{\sim}{I} = -\frac{\gamma H}{2} \underset{\sim}{n} \cdot \underset{\sim}{\tau}$$

where the components of τ are the Pauli matrices.

$$\sigma = \frac{1}{Z} e^{\alpha \underset{\sim}{n} \cdot \underset{\sim}{\tau}}, \qquad \alpha = \frac{\gamma H}{2kT}$$

Expanding and using the special properties $\tau_x \tau_y = i\tau_z$, etc.,

$$Z\sigma = \frac{1}{2} \operatorname{sech} \alpha \left[\cosh \alpha + \underset{\sim}{n} \cdot \underset{\sim}{\tau} \sinh \alpha\right] = \frac{1}{2}\left[1 + n \cdot \sigma \tanh \alpha\right]$$

The equilibrium magnetic moment is then the expectation value of the operator $\frac{1}{2} \gamma h \underset{\sim}{\tau}$:

$$\langle \underset{\sim}{\mu} \rangle = \frac{1}{2} \gamma h \tanh \frac{\gamma h H}{2kT} \underset{\sim}{n},$$

the famous Brillouin function, which saturates at $\langle \underset{\sim}{\mu} \rangle = \frac{1}{2}\gamma h$ for $\frac{H}{T} \to \infty$, when the spin is perfectly polarized.

The High Temperature Approximation.

For all but the most recherche´ NMR experiments, the interactions of spins with external fields and with their surroundings are associated with energies very much smaller than kT. The value of $\gamma h H_0/kT$ for a proton in 10kG at room temperature is $<10^{-5}$. Accordingly the equilibrium density matrix can be expanded to extremely high accuracy as follows:*

$$\sigma_e = \frac{1}{Z}\left(1 - \frac{h\omega_0}{kT}\right); \qquad Z = \mathrm{Tr}\ 1 \qquad \text{(B25)}$$

In this high temperature approximation (HTA), Tr 1 is equal to the total number of quantum states accessible to the system: for N spins I it is $(2I+1)^N$. The approximation will be adequate for all our purposes. [But note that z must be taken to the next order if thermodynamic quantities are to be obtained directly from its derivatives in the manner of (B23).]

The Reduced Density Matrix

It is desirable in extended calculations based on the high temperature approximation to simplify the algebra by dispensing with a part of (B24). We define a reduced density matrix at equilibrium, ρ_e, by

$$\rho_e = \frac{\hbar \mathcal{H}}{kT}; \qquad \sigma_e = \frac{1}{\mathrm{Tr}\ 1}(1 + \rho_e). \qquad \text{(B26)}$$

the system begins at equilibrium and then develops under some set of internal or imposed forces represented by a Hamiltonian $\mathcal{H}$ (t), the density matrix at a later time is

$$\sigma(t) = U\sigma_e U^{-1}\ ; \quad U = Te^{-i\int_0^t (t)'dt'} \qquad \text{(B27)}$$

---

* It might seem that the Hamiltonian for a system of many spins, having some eigenvalues >> kT, would invalidate this expansion. However, what is important is the mean energy per spin, as is clearly explained in an appendix of Slichter's book (Sℓ 78). It is clear on physical grounds that the high temperature of the system cannot be changed by increasing the size of the sample.

$$\begin{aligned}\sigma(t) &= \frac{1}{\text{Tr } 1}\left(1 + U\rho_e U^{-1}\right) \\ &= \frac{1}{\text{Tr } 1}\left[\left(1 + \rho(t)\right]\right.\end{aligned} \tag{B28}$$

That is, all of the time dependences reside in ρ, which itself obeys the Liouville equation. Moreover the operators whose expectation values we wish to evaluate usually have (or can be defined to have) zero trace themselves. Then

$$\langle Q\rangle = \text{Tr}(Q\sigma) = \frac{1}{\text{Tr } 1}\,\text{Tr}(Q\rho) \tag{B29}$$

That is, ρ is the only part of σ which has any physical interest. Note $\text{Tr}\rho = 0$.

The reduced density matrix has a physical interpretation within the high temperature approximation: the entropy is

$$S = -k\text{Tr}(\sigma\ell n\sigma) = -\text{Tr}\left\{\sigma\left(\rho - \ell n \text{ Tr } 1\right)\right\}$$

Now $k\ell n$ Tr 1 is S_B, the entropy for a completely disordered system (infinite temperature), so

$$S_\infty - S = k\text{Tr}\left(\sigma\rho\right) = k\langle\rho\rangle \tag{B30}$$

i.e., ρ is an "order" operator.

TWO-DIMENSIONAL NMR SPECTROSCOPY

Ad Bax, Bruce L. Hawkins, Gary E. Maciel and
Nikolaus M. Szeverenyi

Department of Chemistry
Colorado State University
Fort Collins, Colorado 80523

Abstract

Two-dimensional (2D) NMR spectroscopy is a powerful indirect method of studying spin systems. It allows the measurement of NMR parameters that cannot be determined with conventional one-dimensional spectroscopic techniques. Examples of applications of 2D experiments to the exact measurement of coupling constants, to sensitivity enhanced detection of ^{15}N and to removal of chemical shift anisotropy in solids are discussed.

I INTRODUCTION

The concept of two-dimensional (2D) Fourier transformation in NMR was first introduced by Jeener (1) in 1971. Since that time, several hundred different 2D experiments have been proposed. The selection of experiments described here is based on the insight they provide into the basis and generality of two-dimensional NMR.

In general, four different time intervals in a 2D NMR experiment can be distinguished: a preparation period, an evolution period, a mixing period and a detection period (Fig.1a). As an example of a 2D pulse scheme, the oldest 2D experiment (1) is illustrated in Fig.1b. The operation of this pulse sequence will be briefly discussed for the simple case of a set of isolated, non-coupled spins. At the end of the preparation period, which has a duration

L. Petrakis and J. P. Fraissard (eds.), Magnetic Resonance. Introduction, Advanced Topics and Applications to Fossil Energy, 137–145.

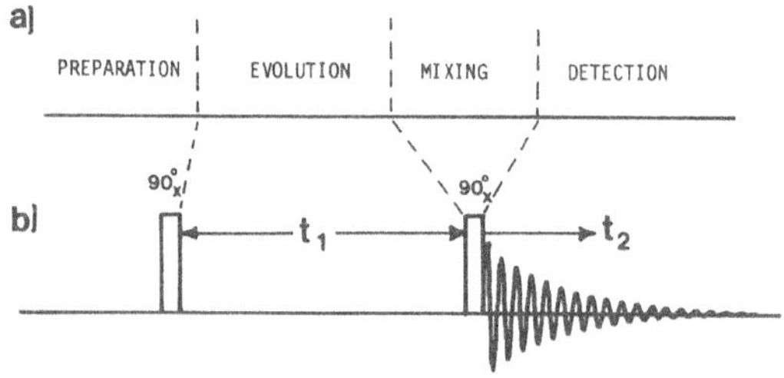

Fig.1 (a) General subdivision of the time axis in a 2D experiment. (b) Schematic representation of the oldest 2D NMR experiment. The sequence is repeated for many different durations of the evolution period, yielding a 2D time signal $s(t_1,t_2)$.

of several times the longitudinal relaxation time T_1, a 90°_x pulse rotates the thermal equilibrium longitudinal magnetization into a position parallel to the y axis of the rotating frame (Fig.2a). The magnetization then rotates for a time t_1, with angular chemical shift frequency Ω_1. A second 90°_x pulse rotates this transverse magnetization into the xz plane, leaving a transverse component with amplitude proportional to $\sin(\Omega_1 t_1)$ parallel to the x axis (Fig.2c). This transverse component is detected during the time t_2, and gives, neglecting relaxation effects, a signal

$$s(t_1,t_2) = C \sin(\Omega_1 t_1)\cos(\Omega_2 t_2) \qquad [1]$$

The experiment is repeated for a large number of different values for the length of the evolution period, t_1, and the acquired free induction decay (FID) is stored separately for each value of t_1. In Fig.3a the Fourier transform of a set of FID's, obtained with different lengths of the evolution period is displayed. As expected, the amplitude of the resonance is modulated as a function of t_1, with frequency Ω_1. If one now takes slices parallel to the

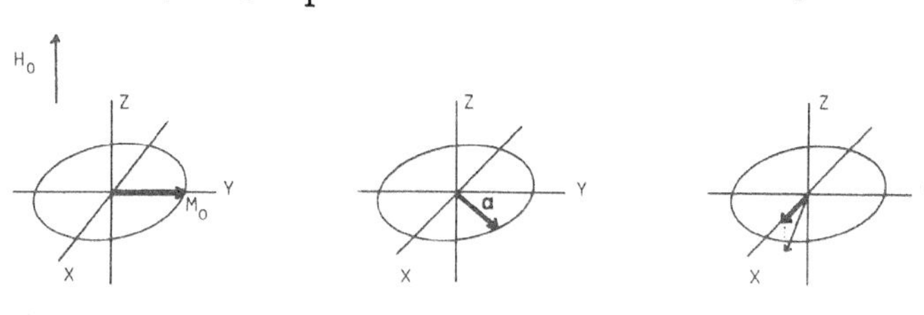

Fig.2 Evolution of the magnetization of a set of isolated spins during the sequence of Fig.1b. (A) Just after the first 90° pulse, (B) just before the second 90° pulse and (C) just after the second 90° pulse.

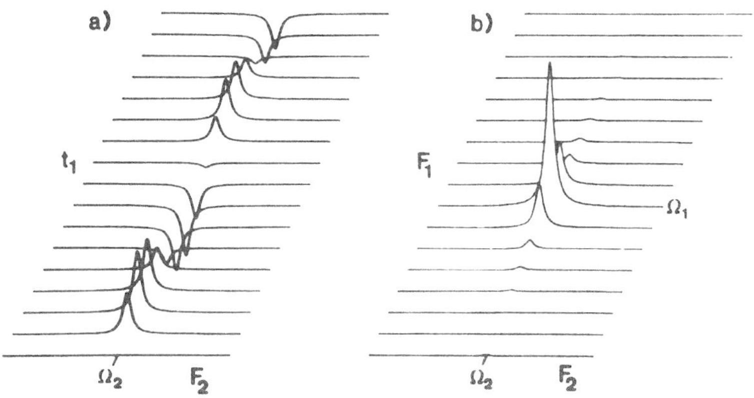

Fig.3. (a) The absorptive part of a set of spectra obtained with the sequence of Fig.1b and modulated in amplitude as a function of t_1. (b) A 2D spectrum obtained from (a) by Fourier transformation of cross-sections parallel to the t_1 axis.

t_1 axis through the data matrix of Fig.3a, all slices except the one taken at $F_2=\Omega_2$ will show zero intensity. The t_1-slice, taken at $F_2=\Omega_2$ shows intensity modulated with frequency Ω_1. Fourier transformation with respect to t_1 of all those slices will then give a resonance in the final 2D spectrum at $(F_1,F_2)=(\Omega_1,\Omega_2)$ (Fig.3b). In this simple case, Ω_1 and Ω_2 are identical, and the resonance will be on the diagonal of the 2D spectrum. For a system with homonuclear spin-spin coupling, a theoretical treatment becomes more complicated (1,2), as magnetization transfer from one spin to another is introduced by the second pulse in the scheme of Fig.1b, yielding important connectivity information in the final 2D spectrum (1-6). The basic idea however, is clear from the simple example given above: the rotation frequency of the magnetization during the evolution period is measured indirectly by studying the effect of that rotation on the signals observed during the detection period, t_2. Hence, in a 2D spectrum, the F_2 coordinate of a resonance equals the precession frequency of the transverse magnetization during the detection period. The F_1 coordinate of a resonance equals the <u>average</u> precession frequency during the evolution period.

The major power of 2D NMR is three-fold:

1. The study of how magnetization is transferred from one spin to another during the mixing period, yielding important conformational and structural information.
2. The study of non-equilibrium phase coherences between spin states which are not directly observable because they do not correspond to transverse magnetization (multiple quantum coherences).

3. The study of the average precession frequency, which allows great simplification over the study of real-time frequencies that are measured in most direct-detection experiments.
Examples of various experiments demonstrating those features will be discussed.

II DETECTION OF PROTON-CARBON-13 LONG-RANGE COUPLINGS

Consider a ^{13}C multiplet component in a proton-coupled ^{13}C spectrum at frequency $\nu_C+J/2$, where J is the coupling to an arbitrary proton, P, in the m=1/2 spin state. ν_C is determined by the chemical shift frequency of the ^{13}C nucleus considered, and by couplings to protons other than P. If the pulse sequence of Fig.4 is applied to this spin system, during the first half of the evolution period the transverse magnetization of the multiplet component considered precesses with frequency $\nu_C+J/2$, and evolves through an angle α, given by

$$\alpha = 2\pi(\nu_C+J/2)t_1/2. \quad [2]$$

This angle is inverted by the ^{13}C 180^o_y pulse, and the spin state of proton P is inverted by the selective proton 180° pulse, to m=-1/2. The new precession frequency of the component considered is therefore equal to $\nu_C-J/2$. At the end of the evolution period, t_1, the total angle precessed through is then given by $-\pi Jt_1$. The average precession frequency during the evolution period is therefore determined by the coupling to proton P, and is independent of the chemical shift frequency and of couplings to protons other than P. In a 2D spectrum we will then get two resonances in the F_1 dimension, at $F_1= \pm J/2$. This can give a dramatic simplification (7) over studying conventional proton-coupled ^{13}C spectra. Fig.5 gives an example of sections parallel to the F_1 axis of the 2D spectrum taken at the F_2 resonance frequencies of the aromatic carbons in 2-acetonaphtalene, where proton H1 was selectively inverted at the center of the evolution period.

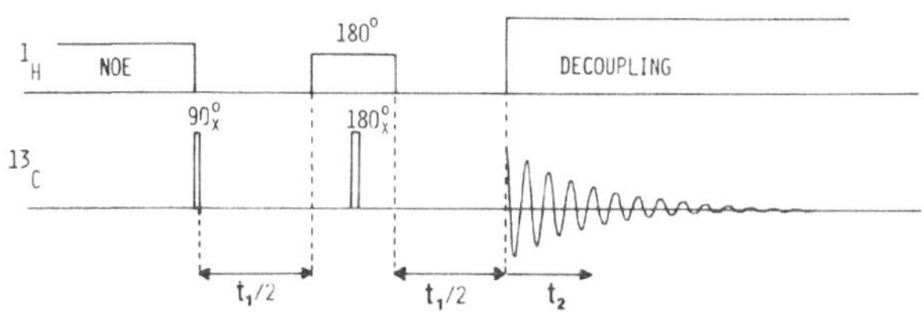

Fig.4 Pulse scheme of the selective proton-flip experiment. The selective proton pulse rotates the magnetization of a preselected proton by 180°. Only couplings to this proton will produce a splitting in the 2D ^{13}C spectrum.

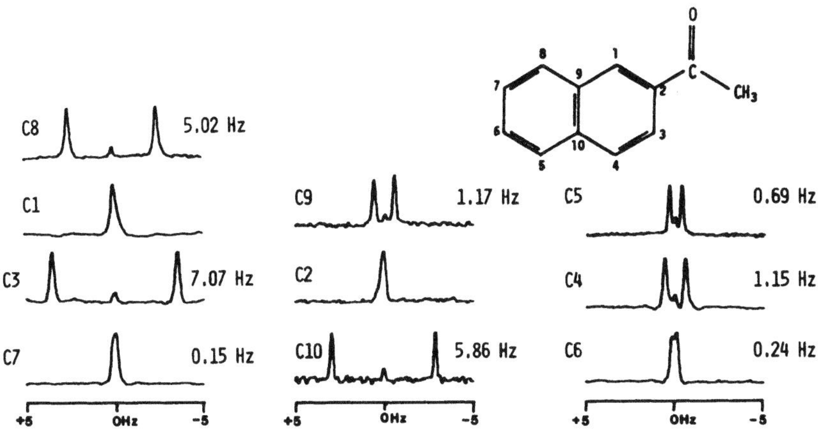

Fig.5 Cross-sections parallel to the F_1 axis through the aromatic resonances of the selective 2D J-spectrum of 2-acetonaphtalene. Proton H1 was selectively inverted at the center of the evolution period, and only couplings to this proton give rise to a splitting in the F_1 dimension.

III INDIRECT DETECTION OF ^{15}N CHEMICAL SHIFTS

Detection of ^{15}N chemical shifts is limited by its low NMR sensitivity, due to its low magnetogyric ratio, and by its low natural abundance (0.37%). A new approach (8-10) to improve ^{15}N sensitivity utilizes the concept of multiple quantum coherence (11-14).

Consider an isolated ^{15}N-^{1}H spin system. The appropriate energy level diagram is sketched in Fig.6. The proton resonance frequencies equal $\delta_H \pm J/2$, where δ_H is the proton chemical shift frequency and J is the ^{15}N-^{1}H scalar coupling constant. The ^{15}N resonance frequencies equal $\delta_N \pm J/2$. It follows immediately from Fig.6, that the double quantum coherence between levels 1 and 4 equals $\delta_H - \delta_N$. If this frequency can be measured indirectly, in a 2D experiment, the value of δ_N can be calculated if δ_H is known. As NMR sensitivity is several orders of magnitude better for protons than for ^{15}N, and the multiple quantum coherences are detected via the protons, a significant shortening of measuring time can be expected.

One of the most suitable sequences for creation and detection of ^{1}H-^{15}N multiple quantum coherences is shown in Fig.7. It has been calculated (15), that a $90^o(^1H)$-$1/(2J)$-$90^o(^{15}N)$ sequence converts <u>all</u> longitudinal proton magnetization into equal amounts of zero and double quantum coherence. Those coherences then evolve during the evolution period, t_1, and are converted back into transverse proton magnetization by a final $90^o(^{15}N)$ pulse. The detected proton magnetization is then modulated as a function of t_1 with the

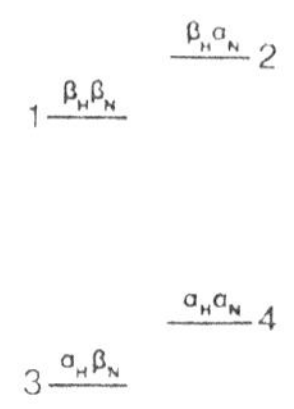

Fig.6 Energy level diagram and wave functions for an isolated ^{15}N-^{1}H spin pair.

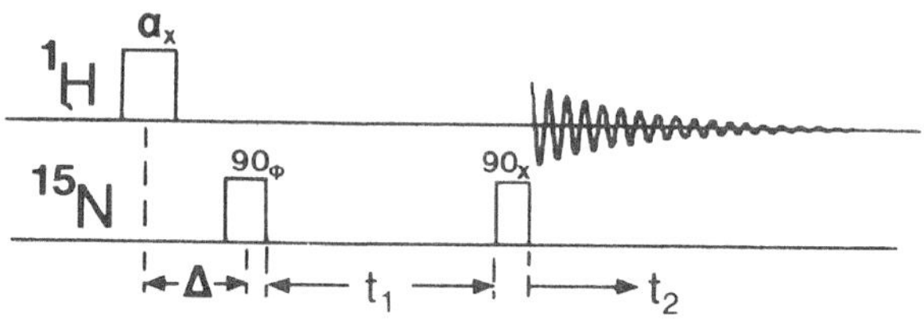

Fig.7 Pulse sequence for the indirect observation of ^{15}N via zero and double quantum coherence. The phase ϕ is cycled in the four steps of the experiment.

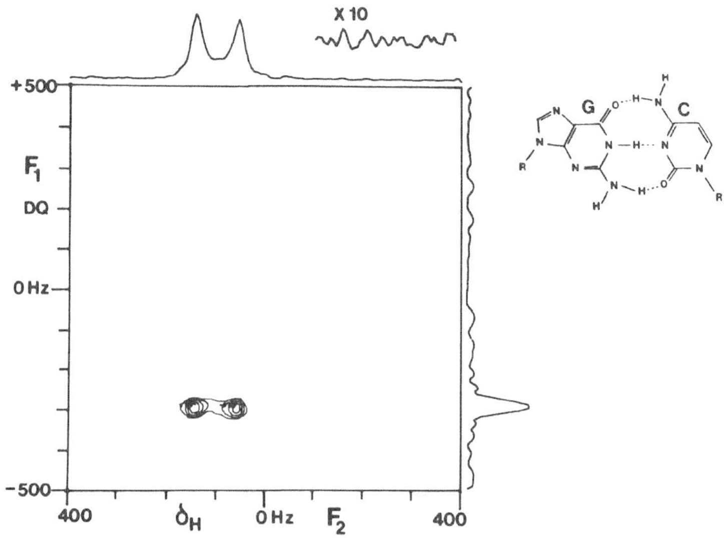

Fig.8 Two-dimensional heteronuclear ^{1}H-^{15}N zero quantum spectrum of the molecule shown in the inset. A selective proton pulse (flip angle of 45° on resonance) was used to minimize excitation of resonances of protons not coupled to the nitrogen, avoiding dynamic range problems. A 32x256 data matrix was acquired, and 800 transients were recorded for each t_1 value. The total measuring time was 2 hrs.

frequencies that existed during t_1, i.e. the zero and double quantum frequencies. Suitable phase cycling of the first $90^o(^{15}N)$ pulse allows the selection of either the zero or the double quantum component (8,10). A 2D Fourier transformation will then show that the two detected proton doublet components are both modulated by the zero or double quantum frequency.

An example of a double quantum spectrum obtained for a 0.25 M sample solution with natural abundance ^{15}N concentration in a 5 mm sample tube, is shown in Fig.8. The total measuring time was 2 hrs. With conventional ^{15}N detection, similar sensitivity would be obtained in a measuring time on the order of 2 years.

IV MAGIC-ANGLE HOPPING

Unlike in liquid samples, in solids the chemical shift anisotropy (CSA) does not average to zero. Therefore, in solid-state NMR, the resonance frequency of a nucleus depends on the orientation of the crystallite with respect to the static magnetic field. In a polycrystalline sample, all possible orientations of crystallites contribute to the spectrum, giving rise to so-called anisotropy powder patterns. Usually, the width of these powder patterns is larger than the difference in isotropic resonance frequencies of the different sites, and overlap will occur. A way around this overlap problem is the use of magic-angle spinning (16-18). One of the problems with this approach is that very high spinning speeds are required for many practical applications, which leads to mechanical difficulties.

A two-dimensional solution to remove the CSA is based on the same principle as magic angle spinning; the resonance frequency of, for example, a ^{13}C nucleus in a certain crystallite is given by (19)

$$\Omega(\beta) = \Omega_I + C_1\cos(\beta+\phi_1) + C_2\cos(2\beta+\phi_2) \qquad [3]$$

where C_1, C_2, ϕ_1 and ψ_2 are constants, and β is the angle through which the sample has been rotated about the magic angle axis. Ω_I is the isotropic chemical shift frequency. It follows immediately from Eq.[3], that if the magnetization evolves during three consecutive intervals, each of length $t_1/3$, with angles $\beta = 0^o$, 120° and 240° respectively, that the <u>average</u> precession frequency during t_1 equals Ω_I. A two-dimensional experiment which is based on this principle is described in reference 20; the detected (t_2) static powder pattern is modulated as a function of t_1 with the isotropic chemical shift frequencies. A projection onto the F_1 axis of such a 2D ^{13}C spectrum, obtained for para-diethoxybenzene, is shown in Figure 9. Also shown are the cross-sections through the 2D spectrum, taken parallel to the F_2 axis, showing the CSA powder

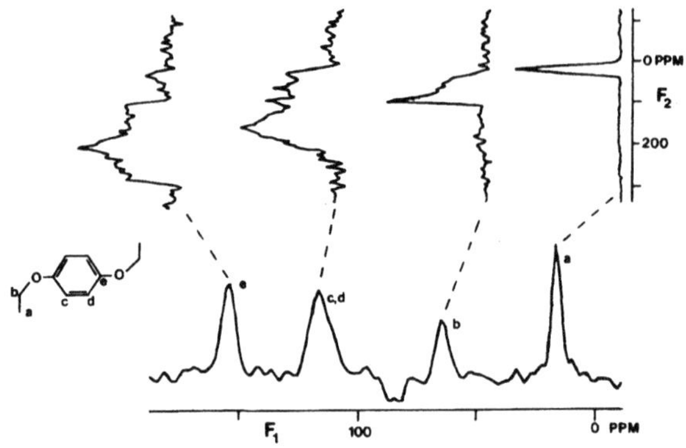

Fig.9 Projection of a 2D magic-angle hopping ^{13}C spectrum of paradiethoxybenzene onto the F_1 axis, and cross-sections parallel to the F_2 axis taken at the isotropic shift frequencies in the F_1 dimension, displaying the CSA powder patterns for the various sites.

patterns for the different sites. This latter information is not directly available from a magic-angle spinning spectrum, nor from a non-spinning spectrum because overlap of the various powder patterns often makes those spectra intractible. Resolution in the isotropic shift dimension (F_1) in Figure 9 is mainly limited by probe design and natural abundance $^{13}C-^{13}C$ coupling.

V CONCLUSIONS

Although only a very small number out of the large collection of two-dimensional NMR experiments have been discussed, it is clear that the 2D approach extends the capabilities of NMR enormously. Many applications of 2D NMR in organic chemistry and biochemistry to structural and conformational problems have appeared in the recent literature.

The practical implementation of the various 2D experiments is still rather complicated, but as almost all experiments can be completely automated by proper spectrometer design and support software, it is to be expected that in the future the ease of operation will become much better. The only price one then will have to pay is a large amount of data storage space and calculation time to store and compute the 2D spectrum. However, in view of current developments in computer technology, this probably will not impede a wide-spread use of two-dimensional NMR.

ACKNOWLEDGMENTS

The authors gratefully acknowledge partial support of this work by the U.S. Department of Energy (Laramie Energy Technology Center) and use of the Colorado State University Regional NMR Center, supported by National Science Foundation Grant CHE-8208821.

References

1. J. Jeener, Ampere International Summer School, Basko Polje, Yugoslavia, 1971.
2. W.P. Aue, E. Bartholdi and R.R. Ernst, J. Chem. Phys. 64, 2229 (1976).
3. A. Bax and R. Freeman, J. Magn. Reson. 44, 542 (1981).
4. K. Nagayama, A. Kumar, K. Wuthrich and R.R. Ernst, J. Magn. Reson. 40, 321 (1980).
5. K. Nagayama, K. Wuthrich and R.R. Ernst, Biochem. Biophys. Res. Commun., 90, 305 (1979).
6. A. Bax, "Two-dimensional Nuclear Magnetic Resonance in Liquids", Ch.3, Reidel, Boston/Dordrecht/London, 1982.
7. A. Bax and R. Freeman, J. Am. Chem. Soc. 104, 1099 (1982).
8. A. Bax, R.H. Griffey and B.L. Hawkins, J. Am. Chem. Soc., in press.
9. R.H. Griffey, C.D. Poulter, A. Bax, B.L. Hawkins, Z. Yamaizumi and S. Nishimura, Proc. Nat. Acad. Sci., in press.
10. A. Bax, R.H. Griffey and B.L. Hawkins, J. Magn. Reson., submitted for publication.
11. A. Minoretti, W.P. Aue, M. Reinhold and R.R. Ernst, J. Magn. Reson. 40, 175 (1980).
12. L. Muller, J. Am. Chem. Soc. 101, 4486 (1979).
13. G. Bodenhausen, Progr. in NMR Spectrosc. 14, 137 (1981).
14. P.H. Bolton, J. Magn. Reson. 52, 326 (1983).
15. A. Bax, J. Magn. Reson. 52, 76 (1983).
16. E.R. Andrew, A. Bradbury and R.G. Eades, Nature (London), 182, 1659 (1958).
17. E.R. Andrew, Phil. Trans. R. Soc. London. Ser. A. 299, 505 (1981).
18. I.J. Lowe, Phys. Rev. Lett. 2, 285 (1959).
19. M.M. Maricq and J.S. Waugh, J. Chem. Phys. 70, 3300 (1979).
20. A. Bax, N.M. Szeverenyi and G.E. Maciel, J. Magn. Reson. 52, 147 (1983).

ELECTRON SPIN ECHO SPECTROSCOPY AND THE STUDY OF HYPERFINE INTERACTIONS IN SOLIDS

H.Barkhuijsen, R. de Beer and D. van Ormondt

Laboratorium voor Technische Natuurkunde, Technische Hogeschool Delft, P.O. Box 5046, 2600 GA Delft, The Netherlands

Time domain (pulsed) electron paramagnetic resonance (TD EPR) is well suited for studying nonequilibrium properties of electron spin systems. Also the determination of hyperfine interactions in solids is well feasible with this method. In this paper some aspects of TD EPR instrumentation and theory will be discussed. Moreover some examples of applications of the method will be given.

1. INTRODUCTION

In the past decade time domain (pulsed) electron paramagnetic resonance (TD EPR), especially in the form of electron spin echoes, has proven to be complementary to continuous wave (CW) EPR in various cases [1,2]. Particularly, the method is very well suited for studying nonequilibrium properties of electron spin systems, such as relaxation phenomena [3-6], paramagnetic properties of short-lived excited states [7-9] and chemical kinetics of unstable radicals [10,11]. Also the measurement of hyperfine interactions in solids, especially of nuclei with a weak magnetic moment, is well feasible in electron spin echo (ESE) spectroscopy [12,13].

In an ESE experiment the paramagnetic sample is irradiated by two or three short, intense, resonant microwave pulses, mostly in the X-band region (9 GHz). If two pulses are applied at a distance τ an echo signal appears (in the form of a build-up of microwaves in the cavity) at a time $t=2\tau$. The echo is maximal for a so-called $90^0-\tau-180^0$ pulse sequence,

L. Petrakis and J. P. Fraissard (eds.), Magnetic Resonance. Introduction, Advanced Topics and Applications to Fossil Energy, 147–163.

where 90^0 and 180^0 refer to the rotation angle of the sample magnetization in the rotating frame during the first and second pulse respectively. In a three-pulse scheme with time intervals τ and T (see figure 1) a second echo (the so-called stimulated echo) occurs at $t=2\tau+T$. This echo reaches its maximum for a 90^0-τ-90^0-T-90^0 sequence. Typically, τ ranges from 0.1 to 10 and T from 1 to 100 μsec. Mostly, the pulse widths (10 to 100 nsec.) are short compared to the time intervals. In this paper we also shall use the symbol $\tau'=\tau+T$, which denotes the time interval between the first and third microwave pulse. When increasing the time interval τ the two- or three-pulse echo decays more or less exponentially with a characteristic time (the phase memory time), that typically is of the order of a few microseconds. In a three-pulse experiment also the second time interval T can be increased, which leads to a decay of the stimulated echo with generally a characteristic time of the order of the spin lattice relaxation time T_1.This is because during the time T the sample magnetization is stored along the z-axis, i.e. the direction of the external static magnetic field. Hence often the stimulated echo can be observed for a much longer time than the two-pulse echo, which in several ESE applications can be an important advantage [13,14]. Moreover, the fact that two time paramaters can be varied offers the unique opportunity of performing two-dimensional EPR in the time domain [12].

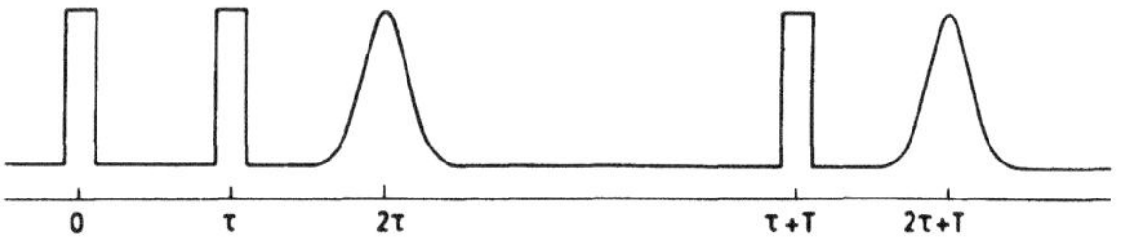

Figure 1. Pulse sequence of a three-pulse (stimulated) electron spin echo scheme. The time interval between the first- and third pulse is called $\tau'=\tau+T$.

Generally, in pulsed EPR the spectral range of the microwave pulses is of the order of the EPR linewidths (or less). This means, that in contrast to pulsed NMR only a single EPR line (or a part) can be covered by the pulse and consequently the sensitivity of pulsed EPR is of the same order of magnitude as of CW EPR [1]. The advantage of pulsed EPR over CW EPR, in fact, does not concern the question of sensitivity but rather the ability of performing EPR in the time domain. As a result of this various time-domain parameters of electron spin systems can be measured very accurately.

In sections 2 and 3 of this paper certain aspects of ESE instrumentation and theory are described respectively. Section 4 includes a brief example of measuring nonequilibrium properties of electron spin systems. Finally, section 5 concerns with the major topic of this paper, namely the determination of hyperfine interactions in solids via the so-called nuclear modulation effect of electron spin echoes, also often referred to as ESE envelope modulation [15].

2. ESE INSTRUMENTATION

Over the years several papers have been published concerning ESE instrumentation [1,2]. In this section we only briefly describe a typical form of an ESE spectrometer. The basic sections of such a spectrometer are diagrammed in figure 2. One can distinguish: (a) a microwave circuit for providing the transmitter pulses at the sample, (b) a programmable pulse generator for producing the various control pulses of an ESE experiment, and (c) a data station for acquiring, analysing and presenting the data.

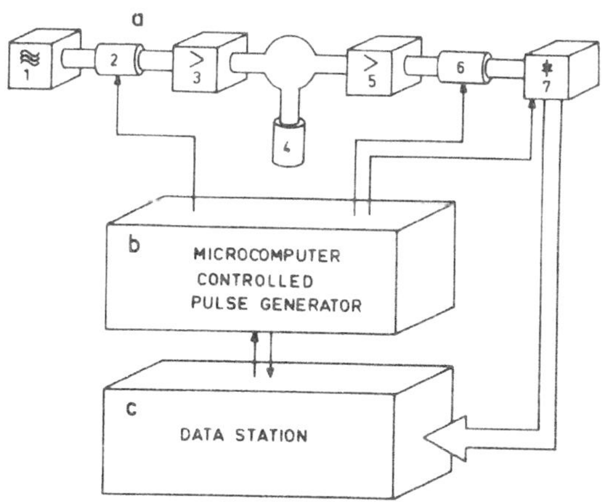

Figure 2. Basic sections of an ESE spectrometer. The numbering of the components of the microwave circuit agrees with that of the text.

Generally, an ESE microwave circuit contains the following essential parts (the numbering agrees with that of figure 2): (1) a low-power CW microwave source, (2) a fast (PIN-diode) microwave switch (isolation 80 db and switching time 10 nsec. typically) for producing pulses of the correct length, (3) a pulsed TWT amplifier (typical power 100 W) for increasing the amplitude of the microwave pulses, (4) a reflection-type microwave cavity with a relatively low Q-value (in order to decrease cavity ringing), (5) a high-gain low-noise microwave amplifier (TWT or GaAsFET) for increasing the echo signal to noise ratio [16], (6) a second microwave switch for protecting the spectrometer receiver system from damaging or overloading and, finally, (7) the spectrometer receiver system consisting of a microwave detector, a broad-banded preamplifier and a gated integrator.

In our spectrometer the programmable pulse generator is controlled by a separate microcomputer [13]. In the current form the repetition rate of, for instance, a three-pulse ESE envelope modulation experiment (in which one of the time intervals is swept digitally) is about 30 kHz. This number depends, among other things, on the final length of the swept time interval. The relatively high repetition rate is achieved partly by dedicated microcomputer programming and partly by averaging the echo data in the time domain on hardware basis.

On-line time-domain data acquisition and further avaraging (on software basis) is accomplished by the data station, that in our case is a minicomputer system. The same system is also used for interactive (graphical) data processing and analysis. Since we are particularly dealing with measuring hyperfine interactions in paramagnetic solids via the ESE envelope modulation method software packages have been written for specific purposes such as digital filtering and windowing of discrete time series [17] and spectrum analysis of discrete time series. Concerning the latter, the traditional direct method via the FFT operation can be employed as well as the so-called autoregressive (AR) model [18,19] (also often referred to as the maximum entropy method (MEM)). It has been found in several cases [13,18] that AR modelling is especially well suited for analysing short time series with only a limited number of sinusoids.

3. FORMULAE OF ESE ENVELOPE MODULATION

In this section we discuss some formulae of the ESE theory, namely those that are related to the ESE phenomenon known as envelope modulation (also the nuclear modulation effect). It will be assumed that the reader is familiar with the basic theory of electron spin echoes [1]. In essence the nuclear modulation effect amounts to the following. In an ESE experiment one observes the echo resulting from two or three successive microwave pulses. The hyperfine interaction between the paramagnetic centre under investigation and the nuclei that surround it influences the ESE signal height, the strength of the effect depending in a sinusoidal manner on the time intervals. Thus, when varying the time intervals, the envelope of ensuing echoes will show a modulated structure. It has been shown theoretically [1,15] and observed in several ESE envelope modulation experiments [13,20] that the frequencies of the sinusoids contained in the envelope modulation pattern are equal to the resonance frequencies of the nuclei involved (and, under certain conditions, combinations thereof). Hence, with pulsed EPR one can do an ENDOR-like investigation [13] when using the envelope modulation phenomenon.

The theory concerning ESE envelope modulation shows [15] that the phenomenon essentially is based on intermixing of the nuclear states. This intermixing can be caused by anisotropic hyperfine interaction or quadrupole interaction. Using the density matrix approach the following expressing can be derived for the nuclear modulation function of the three-pulse (stimulated) echo [13]

$$E_{\mathrm{mod}}(\tau,\tau') = \frac{1}{2(2I+1)}$$
$$\times \mathrm{Re}\,\mathrm{Tr}[e^{iW^{\beta}\tau}U^{\dagger}e^{iW^{\alpha}\tau'}Ue^{-iW^{\beta}\tau}U^{\dagger}e^{-iW^{\alpha}\tau'}U + e^{iW^{\beta}\tau'}U^{\dagger}e^{iW^{\alpha}\tau}Ue^{-iW^{\beta}\tau'}U^{\dagger}e^{-iW^{\alpha}\tau}U], \quad [1]$$

in which τ and τ' are the time intervals in figure 1, W^{α} and W^{β} are $(2I+1)$-dimensional matrices with elements $W^{\alpha}_{ij}=\lambda^{\alpha}_{i}\delta_{ij}$ and $W^{\beta}_{ij}=\lambda^{\beta}_{i}\delta_{ij}$, U is a unitary matrix relating the nuclear spin wave functions for the electron spin up (α) or down (β), and U^{+} is the inverse of U. Furthermore $\hbar\lambda^{\alpha}_{i}$ and $\hbar\lambda^{\beta}_{i}$ denote the nuclear levels for the electron spin up or down. In absence of anisotropic hyperfine interaction or quadrupole interaction U becomes diagonal and the sinusoidal dependence on τ or τ' can then be seen to vanish. Equation [1] is valid in the high-temperature approximation for a paramagnetic centre with spin S=1/2 and an interacting nucleus with arbitrary spin I. Also it should be noted that the effects of relaxation have been ignored. If in equation [1] matrix multiplications involving diagonal matrices are worked out first an expressing is obtained that can be programmed very easily for computer calculations

$$E_{\mathrm{mod}}(\tau,\tau') = 1/2[V(\tau,\tau') + V(\tau',\tau)], \quad [2]$$

with

$$V(\tau,\tau') = \frac{1}{2I+1} \times \mathrm{Re}[\sum_{j,k,l,m} e^{-i(\lambda^{\beta}_{j}-\lambda^{\beta}_{k})\tau}e^{-i(\lambda^{\alpha}_{l}-\lambda^{\alpha}_{m})\tau'}U^{*}_{lj}U^{*}_{mk}U_{mj}U_{lk}]. \quad [3]$$

If N nuclei are involved the individual contributions are to be multilplied as

$$E_{\mathrm{mod}}(\tau, \tau') = 1/2[\prod_{n=1}^{N} V^n(\tau, \tau') + \prod_{n=1}^{N} V^n(\tau', \tau)], \qquad [4]$$

From equation [4] it can be concluded that the stimulated echo modulation function is completely symmetric. with respect to τ and τ'. In section 5 an application is given of this fact. Furthermore, equation [4] shows that Fourier transformation of a modulation pattern in the τ- or τ'-dimension (or both) yields a one-dimensional (1D) spectrum (or a 2D-spectrum).

4. SPIN-LATTICE RELAXATION MEASUREMENTS

As was mentioned already in the introduction (section 1) a particularly useful application of ESE spectroscopy is the measurement of nonequilibrium properties of electron spin systems. A striking example of this is the determination of relaxation times, such as the phase memory time or the spin-lattice relaxation time [3-6]. Other examples are the study of the time dependent behaviour of short-lived photoexcited triplet-state molecules [7,8] or the study of the chemical kinetics of unstable radicals [10,11].

In this section the measurement of the spin-lattice relaxation time T_1 will be illustrated. This is done on the hand of the results that we have obtained for the free radical $CH_3\dot{C}(COOH)_2$ in an X-ray irradiated single crystal of methyl malonic acid [21] (the crystal was kindly provided by Dr.R.A.Wind of the University of Technology Delft). The

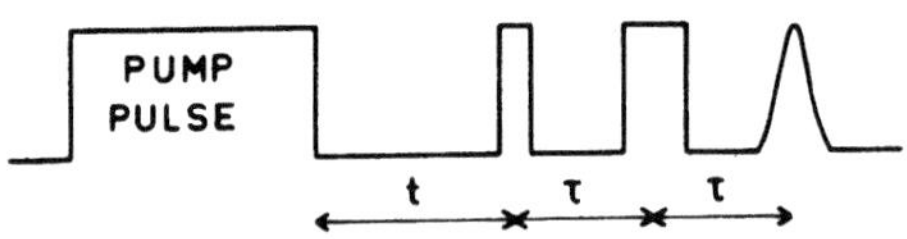

Figure 3. Pulse sequence for the measurement of spin-lattice relaxation times.

purpose of the investigation was to establish whether or not the temperature dependence of the spin-lattice relaxation process (in the temperature region of 1.4 to 4.2 K) is determined by the hindered rotation of the methyl group [22,23]. We have chosen to measure T_1 via the ESE pulse sequence proposed by Mims [3] (see figure 3). In this pulse sequence the Boltzmann distribution of the electron spin system first is disturbed by a high-power resonant microwave pulse (pump pulse) of relatively long duration (typically milliseconds) and then the recovery of the magnetization is monitored after a variable time interval by means of the two-pulse spin echo signal. The advantage of this method is that effects of cross-relaxation may be studied by varying the duration of the pump pulse. For instance, if a (mostly) fast recovery of the sample magnetization due to cross-relaxation is superimposed on a slower spin-lattice relaxation recovery the former may be suppressed by choosing a pump pulse length longer than the cross-relaxtion time [3]. However, if the two relaxation times have close values it often is difficult to resolve the two effects. In that case the characteristic time of the (slower) spin-lattice relaxation process should be determined from the last part of the recovery curve, i.e. it is then important to measure the asymptotic behaviour of the recovery curve with a good signal to noise ratio. Also, the value at infinity (in our case the signal strength of the two-pulse echo without having applied a pump pulse first) should be known then accurately. In our experiments the

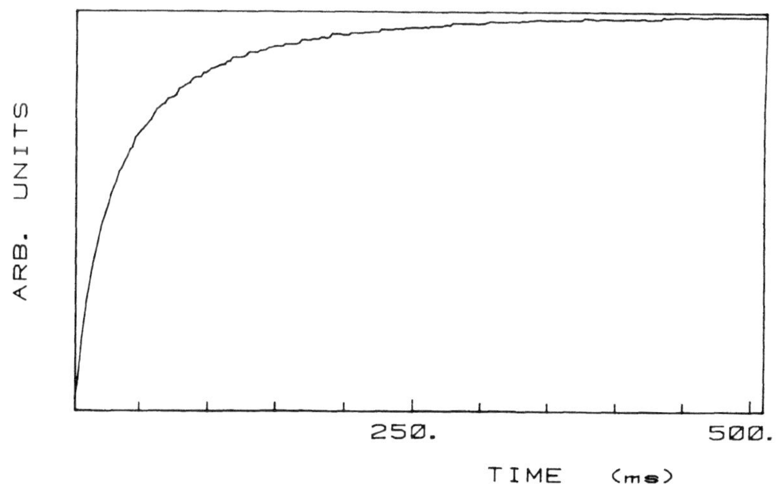

Figure 4. Typical relaxation curve of the $CH_3\dot{C}(COOH)_2$ radical in irradiated methyl malonic acid.

requirement of having a good signal to noise ratio was more or less fulfilled by means of signal averaging. Furthermore the question of determining the value at infinity of the recovery curve was solved in the following way. Alternately an ESE pulse sequence was applied with and without a pump pulse. The resulting two monitor echoes were fed into the two channels of a dual-channel boxcar integrator and then subtracted. This process was repeated for successive digitally increased values of the time interval between the pump pulse and the monitor echo sequence. In this way drift effects during lengthy relaxation experiments could be eliminated [21].

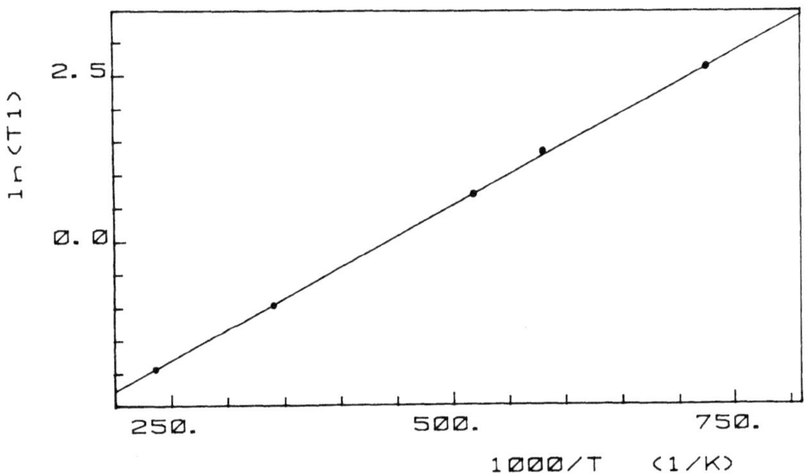

Figure 5. Temperature dependence of the spin-lattice relaxation of the EPR transition 4,5 of irradiated methyl malonic acid.

Figure 4 shows a typical relaxation (recovery) curve of the $CH_3\dot{C}(COOH)_2$ radical. The result was obtained after 10 scans of 256 data points. As a rule the repetition time of the pulse sequence was chosen to be approximately three times the spin-lattice relaxation time. It could be established that the relaxation curve has a fast and a slow component. The fast component could be fitted to a characterictic time of 35 msec. This value appeared to be independent of the temperature (in the chosen temperature region) which is typical of cross-relaxation processes [3]. The slow component of the relaxation curve was temperature dependent and therefore it was assumed that this component is due to spin-lattice relaxation. In figure 5 the temperature dependence of the relaxation behaviour of one of the EPR lines (transition 4,5; see ref.[23]) is illustrated. The curve (and also those of other lines of the EPR spectrum) could be fitted to the expression

$$\ln T_1 = -4.4 + \frac{10}{T} .$$

If hindered rotation of the methyl group is responsible for this temperature dependence it would mean that the separation between the lowest and first excited rotational level is of the order of 10 K. This spacing of the lowest two levels, in turn, can only be explained theoretically by assuming that the potential barrier at the methyl group has dominantly a six-fold character [24]. However, a six-fold potential barrier contradicts the observed temperature independence of the tunneling frequency in methyl malonic acid [25]. Hence the final conclusion should be that in the temperature region of 1.4 to 4.2 K the spin-lattice relaxation of the $CH_3\dot{C}(COOH)_2$ radical is not determined by hindered rotation of the methyl group.

5. APPLICATIONS OF ESE ENVELOPE MODULATION

The phenomenon of ESE envelope modulation has been exploited now for numerous paramagnetic systems in order to obtain ENDOR-like information about nearby nuclei. Among the various applications we like to mention: (a) the local mapping of equivalent nuclei involved in the modulation effect from an absolute analysis of the ESE modulation signal [26-28] (with conventional CW ENDOR this is very difficult), (b) the study of hyperfine interactions in short-lived paramagnetic systems [8,9], and (c) the study of hyperfine interactions of nuclei with a very weak magnetic moment [12,13]. In this section we only illustrate an example of case (b) and (c).

5.1. Hyperfine interactions of triplet excitons in 1,2,4,5-tetrachlorobenzene.

A few years ago it has been established that ESE spectroscopy combined with pulsed laser excitation is very attractive for studying the dynamics of triplet excitons in single crystals. Triplet excitons can be described by collective states of the whole crystal. They may arise as a result of interactions between the individual triplet-state molecules [29]. In the case of triplet excitons in 1,2,4,5-tetrachlorobenzene it was found [7] that the laser flash causes a selective population distribution over the exciton states that corresponds to a strong peak in the ESE detected lineshape. This strong non-Boltzmann peak decays to the thermal equilibrium value in about 15 μsec. Recently ESE envelope modulations have been observed of the same system [30]. The data processing and spectrum analysis of these measurements was performed in our laboratory. Figure 6 shows a typical example of a recorded modulation pattern. The result was obtained at T=4.2 K for $\bar{H}$//Y (low-field EPR line; see ref.[29] for a definition of the reference frame).

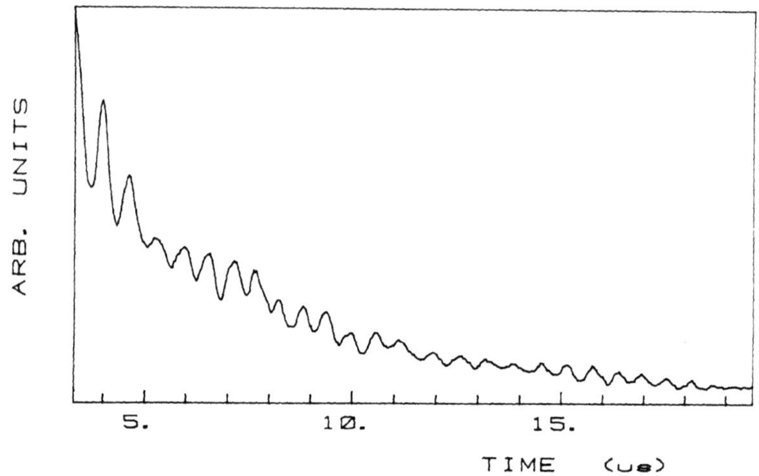

Figure 6. Typical modulation pattern of triplet excitons in 1,2,4,5-tetrachlorobenzene as recorded at T=4.2 K for $\bar{H}$//Y (low-field EPR line).

Before actually applying spectrum analysis of a discrete time series it often is recommendable to perform some form of data preprocessing. The various steps of such a preprocessing mostly depend on the method of spectrum analysis that is applied afterwards. Figure 7a illustrates the data preprocessing that we have performed before using the FFT algorithm. The following steps could be distinguished. First the background (relaxation) decay was removed by means of a first-order recursive lowpass filter [31]. Then the resulting pattern was multiplied by a Gaussian time window in order to suppress the appearance of side-lobes (spectral leakage) [13,17]. Figure 7b shows as an alternative the preprocessing that was performed before applying the AR model. In this case the background decay was removed by using a finite impulse response bandpass filter [32]. Furthermore the resulting time-domain signal was made stationary by multiplying it with

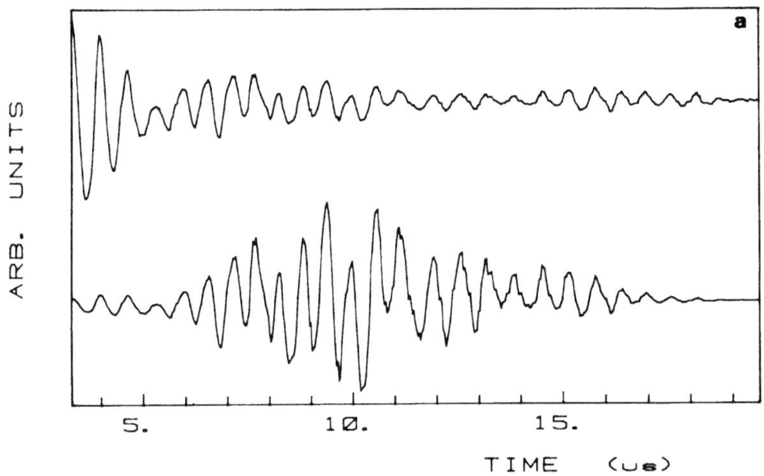

(a) Data preprocessing before applying FFT.

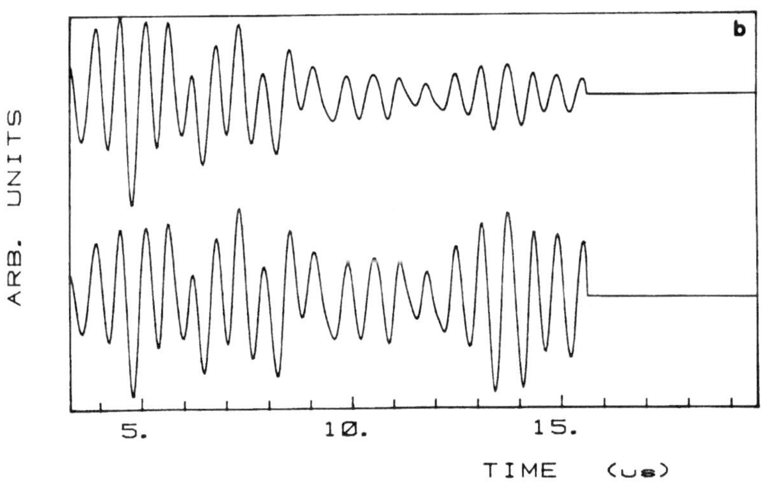

(b) Data preprocessing before applying AR modelling.

Figure 7. Various steps of data preprocessing. The raw time series is shown in figure 6.

an increasing exponential function. The latter is preferable because stationarity is one of the assumptions of the AR model [18,19].

The results for the two forms of spectrum analysis can be compared in figure 8. It is clear that for time series as short as 20 μsec. (see figure 6) the resolution of the AR model is superior to that of FFT [18]. Since the measurements described above were preliminary there is no complete interpretation yet of the observed frequencies. Nevertheless it can be concluded that the ESE envelope modulation method is capable of determining hyperfine interactions in such short-lived systems.

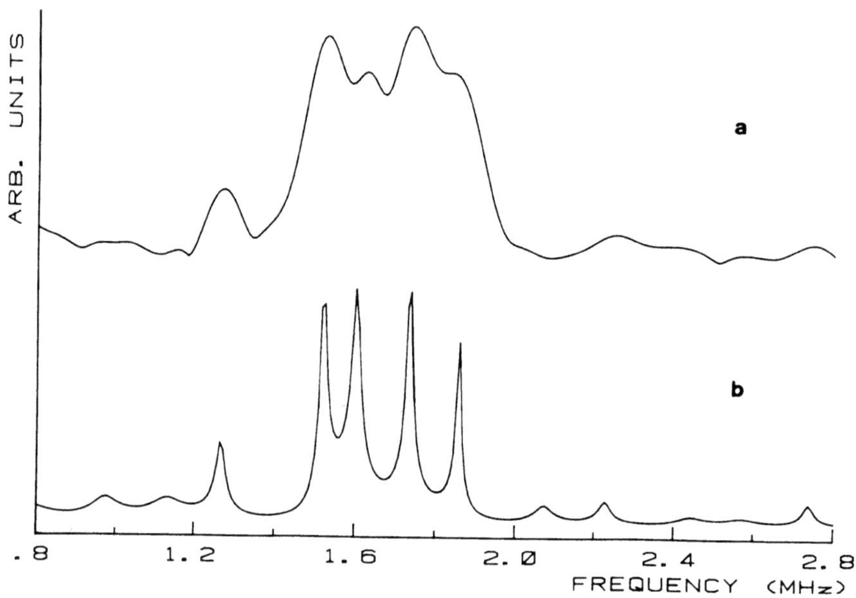

Figure 8. Spectrum analysis of the modulation pattern shown in figure 6. Part (a) corresponds to the FFT operation and part (b) to AR modelling.

5.2. Case study of ESE envelope modulation of F centres in KCl.

An important aspect of the ESE technique is that the amplitude of the modulation is independent of the magnetic moment of the nuclei involved (ignoring quadrupole interaction) [1,33]. Consequently, signals from nuclei with weak magnetic moments such as ^{14}N, $^{35,37}Cl$ or $^{39,41}K$ do not yield to those from stronger moments as is the case with CW ENDOR. Hence ESE envelope modulation is very well suited for detecting hyperfine interaction frequencies in the very low frequency region (< 1 MHz) [12,13].

In order to test the application mentioned above we have performed a case study of F centres in additively coloured single crystals of KCl [13]. This system was chosen by us because it has been intensively studied by CW ENDOR [34,35]. Signals from nuclei in shells as distant as the ninth could be observed by that technique. As shown below ESE is also capable of this. In addition, signals from ^{41}K (7% natural abundance) in the third and fifth shell could be detected.

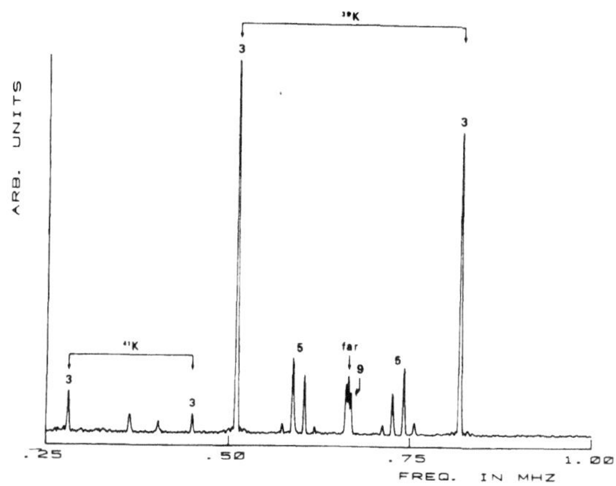

Figure 9. 1D-ESE spectrum (based on FFT) of F-centres in KCl as measured at 4.2 K for $\bar{H}$//[001].

The ESE measurements on KCl were performed at a temperature of 4.2 K (however, modulations could also be observed at room temperature). The modulation patterns were sampled as a function of τ' (1D-measurement) or as a function of both τ and τ' (2D-measurement; see figure 1 and section 3). A typical 1D-ESE spectrum (based on FFT) is shown in figure 9. The frequency peaks could be assigned to the potassium shells 3,5 and 9. Also peaks due to ^{41}K and the matrix peak at the ^{39}K free precession frequency were observed. In order to test the accuracy of the determination of spin-Hamiltonian parameters with the ESE method, the angular dependence of the 1D-modulation due to nuclei in the fifth shell has been measured. Figure 10 shows a detail of the angular plot of the corresponding FFT spectrum. By fitting the spin-Hamiltonian paramaters to the experimental points it could be established that the ESE method is as accurate as CW ENDOR [13].

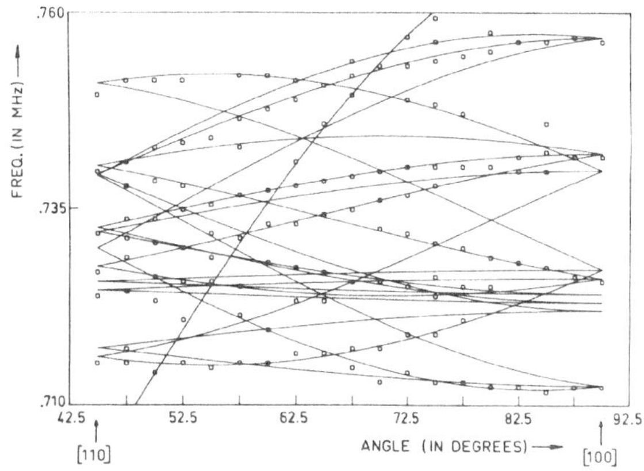

Figure 10. Detail of the angular dependence of the 1D-ESE spectrum (based on FFT) due to nuclei in the fifth shell of additively coloured KCl.

Finally, figure 11 serves to illustrate the potential use of 2D-ESE spectroscopy. It shows a contour plot of the 2D-spectrum corresponding to the orientation $\bar{H}$//[110]. The observed 2D-peaks are due to $^{35,37}Cl$ in shell 4. The 2D-spectrum was obtained by employing FFT along the τ'-dimension and (high-resolution) AR modelling along the τ-dimension (the so-called method of separable spectrum estimators [36]). As predicted by theory (see section 3) the 2D-spectrum shows reflection symmetry relative to the line $\nu=\nu'$. The two peaks forming a symmetry pair are due to equivalent nuclei, but with the nearby paramagnetic centre in a different state (M_S or M_S-1, of the transition to which the microwave frequency is tuned). This reflection symmetry property of the 2D-spectrum thus offers additional information about the assignment of the hyperfine interaction frequencies.

Concluding this section the following can be said: (a) with ESE the accuracy of the measurement of hyperfine interactions can be as high as with CW ENDOR. (b) the resolution of ESE in the frequency domain matches that of CW ENDOR, (c) substantial reduction of the linewidth in the frequency domain can be obtained by applying AR spectrum analysis rather than FFT, and (d) 2D-ESE spectroscopy yields additional hyperfine interaction information.

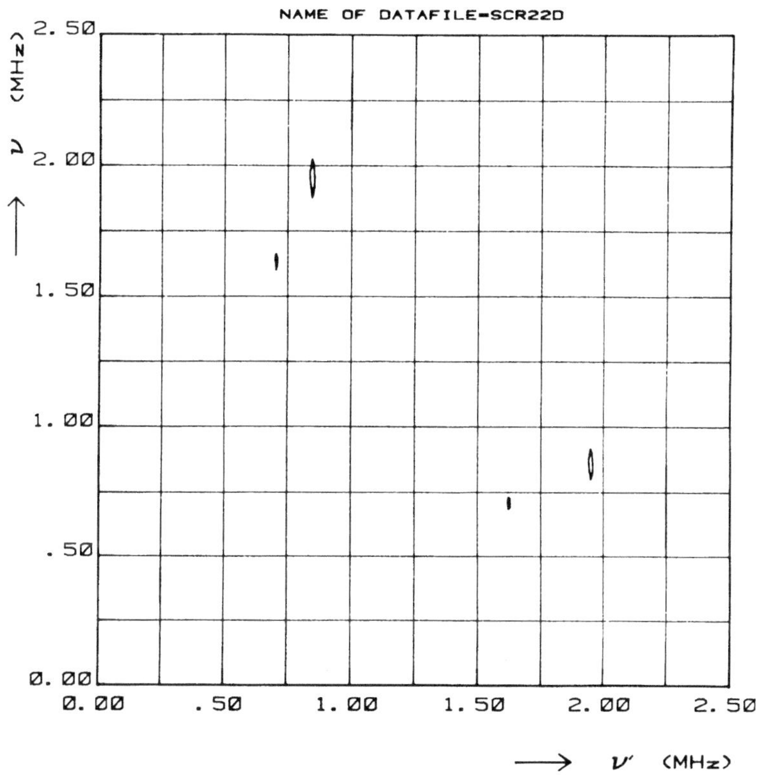

Figure 11. Contour plot of the 2D-ESE spectrum of F-centres in KCl corresponding to the orientation $\bar{H}$//[110]. The observed 2D-peaks are due to chlorine nuclei in shell 4. In the ν'-direction FFT and in the ν-direction AR modelling has been employed.

ACKNOWLEDGEMENTS

This project is supported by the "Stichting voor Fundamenteel Onderzoek der Materie" (Grant 80.82.201). The authors thank Prof.J.M.Spaeth (University of Paderborn) and Dr.R.A.Wind (University of Technology Delft) for providing single crystals of KCl and methyl malonic acid respectively.

REFERENCES

[1] Mims,W.B.,"Electron Spin Echoes" in Electron Paramagnetic Resonance, editor S.Geschwind (Plenum, New York 1972) chap.4,pp.263-351.
[2] Norris,J.R.,Thurnauer,M.C. and Bowman,M.K.,"Electron Spin Echo Spectroscopy and the Study of Biological Structure and Function" in Advances in Biological and Medical Physics,Vol.17,editor J.H.Lawrence,J.W.Gofman and T.L.Hayes (Academic Press, New York 1980) pp.365-416.
[3] Mims,W.B. and McGee,J.D. 1960,Phys.Rev.119,pp.1233-1237.
[4] Kiel,A. and Mims,W.B. 1967,Phys.Rev.161,pp.386-397.
[5] Völkel,G.,Brunner,W.,Windsch,W. and Steudel,D. 1979,Phys. Stat.Sol.95, pp.99-105.
[6] Stillman,A.E. and Schwartz,R.N. 1981,J.Phys.Chem.85, pp.3031-3040.
[7] Strien,A.J.van,Kooten,J.F.C.van and Schmidt,J. 1980,Chem. Phys.Letters 76,pp.7-12.
[8] Sloop,D.J.,Hsiang-Lin Yu,Tien-Sung Lin and Weissman,S.I. 1981,J.Chem. Phys.75,pp.3746-3757.
[9] Hsiang-Lin Yu,Sloop,D.J.,Weissman,S.I.,Tien-Sung Lin,Norris,J.R. and Bowman,M.K. 1982,J.Phys.Chem.86,pp. 4287-4290.
[10]Trifunac,A.D. and Norris,J.R. 1978,Chem.Phys.Letters 59,pp.140-142.
[11]Narayana,P.A.,Li,A.S.W. and Kevan,L. 1982, J.Am.Chem. Soc.104,pp.6502-6505.
[12]Merks,R.P.J. and Beer,R.de 1979,J.Phys.Chem.83,pp.3319-3322.
[13]Barkhuijsen,H.,Beer,R.de,Wild,E.L.de and Ormondt,D.van 1982,J.Magn. Reson.50,pp.299-315.
[14]Mims,W.B. 1965,Proc.Royal Soc.283,pp.452.
[15]Mims,W.B. 1972,Phys.Rev.B5,pp.2409-2419.
[16]Hoentzsch,Ch.,Niklas,J.R. and Spaeth,J.M. 1978, Rev.Sci.Instrum.49, pp.1100-1102.
[17]Harris,F.J. 1978,Proc.IEEE 66,pp.51-83.
[18]Kay,S.M. and Marple,S.L. 1981,Proc.IEEE 69,pp.1380-1419.
[19]Ormondt,D.van and Nederveen,K. 1981,Chem.Phys.Letters 82,pp.443-446.
[20]Dikanov,S.A.,Tsvetkov,Yu.D.,Bowman,M.K. and Astashkin, A.V. 1982, Chem.Phys.Letters 90,pp.149-153.

[21]Banen,R.A.,MSc. thesis,Delft 1980.
[22]Gamble,W.L.,Miyagawa,I. and Hartman,R.L. 1968, Phys.Rev.Letters 20, pp.415-418.
[23]Clough,S.,Hill,J.R. and Poldy,F. 1972,J.Phys.C:Solid State Phys.5,pp.518-528.
[24]Ligthelm,D.J.,private communication.
[25]Clough,S. and Hill,J.R. 1974,J.Phys.C:Solid State Phys.7,pp.L20-L21.
[26]Kevan,L.,Bowman,M.K.,Narayana,P.A.,Boeckman,R.K., Yudanov,V.F. and Tsvetkov,Yu.D. 1975,J.Chem.Phys.63,pp. 409-416.
[27]Mims,W.B.,Peisach,J. and Davis,J.L. 1977,J.Chem.Phys. 66,pp.5536-5550.
[28]Ichikawa,T.,Kevan,L.,Bowman,M.K.,Dikanov,S.A. and Tsvetkov,Yu.D. 1979,J.Chem.Phys.71,pp.1167-1174.
[29]Strien,A.J.van,"Triplet Exiton Dynamics",thesis, University of Leiden 1981.
[30]Strien,A.J.van,Kooten,J.F.C.van and Schmidt,J.,private communication.
[31]Merks,R.P.J. and Beer,R.de 1980,J.Magn.Reson.37,pp.305-319.
[32]Rabiner,L.B. and Gold,B.,"Theory and Application of Digital Signal Processing" (Prentice-Hall,Englewood Cliffs,N.J. 1975).
[33]Dikanov,S.A.,Shubin,A.A. and Parmon,V.N. 1981,J.Magn. Reson.42,pp.474-487.
[34]Kersten,R. 1968,Phys.Stat.Sol.29,pp.575-586.
[35]Duvarney,R.C. and Spaeth,J.M. 1979,Sol.Stat.Commun. 32,pp.1237-1240.
[36]McClellan,J.H. 1982,Proc.IEEE 70,pp.1029-1039.

ELECTRON-NUCLEAR DOUBLE RESONANCE OF CHARCOALS

Jean Uebersfeld

Département d'électronique
Université Pierre et Marie Curie (Paris VI) - France

I. INTRODUCTION

ENDOR (Electron-Nuclear Double Resonance), which was proposed and demonstrated by G. Feher in 1956[1] is a modification of ESR (Electron Spin Resonance) involving the interaction between nuclear and electron spins in paramagnetic materials. ENDOR belongs to the class of methods called "Double Resonance" in which one transition between atomic or molecular levels belonging to one physical system is observed, while one other transition belonging to the same physical system or to an other physical system is saturated.

It is customary to call "signal transition" the observed transition and "pump transition" the saturated transition.

The change in the signal transition when the pump transition is saturated gives physical information on the structure of the physical system (if the two transitions belong to the same physical system) or on the interactions between the two physical systems (if the signal and pump transitions belong to different physical systems).

In case of ENDOR the signal transition is the ESR transition belonging to the electron spin system, and the pump transition is an NMR (Nuclear Magnetic Resonance) transition belonging to the nuclear spin system . Basically ENDOR is an ESR experiment in which the electron spin system is used as a sunsitive monitor of paramagnetic structures (free radicals, or paramagnetic ions), molecular motions (in the solid or liquid which contains the paramagnetic structure) and electron-nuclear spin interactions[2].

L. Petrakis and J. P. Fraissard (eds.), Magnetic Resonance. Introduction, Advanced Topics and Applications to Fossil Energy, 165–175.

The ENDOR apparatus is therefore essentially an ESR spectrometer with an additional source of radiofrequency (rf) magnetic field (rf H_1 field). As we will see, the ENDOR technique enables to resolve the hyperfine structure in a broad unresolved ESR line as well to determine the hyperfine constants in a very complex hyperfine structure.

II.- THEORY OF THE ENDOR

Let us consider an electron spin $\vec{S}$ (S = 1/2) and a nuclear spin $\vec{I}$ (I = 1/2) in a strong magnetic field H. $m_S = \pm 1/2$ and $m_I = \pm 1/2$ are the magnetic quantum numbers for the electron and the nuclear spin respectively.

g and g_n are respectively the "g factor" of the electron and nuclear magnetic moments.

β and β_n are the electron and nuclear Bohr magnetons.

The electron-nuclear spin coupling constant A (hyperfine coupling) is expressed by hA (h Planck constant) ; therefore A is expressed in frequency units. A is assumed to be isotropic.

The spin Hamiltonian $\mathcal{H}$ of this system is of the form :

$$\mathcal{H} = g\beta\vec{H}.\vec{S} - g_n\beta_n\vec{H}.\vec{I} + hA\vec{S}.\vec{I}$$

The gyromagnetic ratio of the nucleus is assumed to be positive and in the Hamiltonian g, g_n, β and β_n are positive quantities.

The 4 magnetic energy levels of the system may then be written as

$$E(m_s,m_e) = g\beta Hm_S - g_n\beta_n Hm_I + hAm_S m_I$$

If A = 0 the ESR and NMR frequencies ν_e and ν_n are given by

$$\nu_e = \frac{g\beta H}{h} \qquad \nu_n = \frac{g_n\beta_n H}{h}$$

In such a case the 4 magnetic levels are the levels 1, 2, 3, 4 of the figure 1.

In the case where A is negative and $|A| < 2\nu_n$, the four magnetic levels are the levels 1', 2', 3', 4' of the figure 1.

The frequency of the ESR transition (defined by the selection rules $\Delta m_S = \pm 1$ and $\Delta m_I = 0$) becomes :

$$\nu'_e = \nu_e \pm \frac{|A|}{2}$$

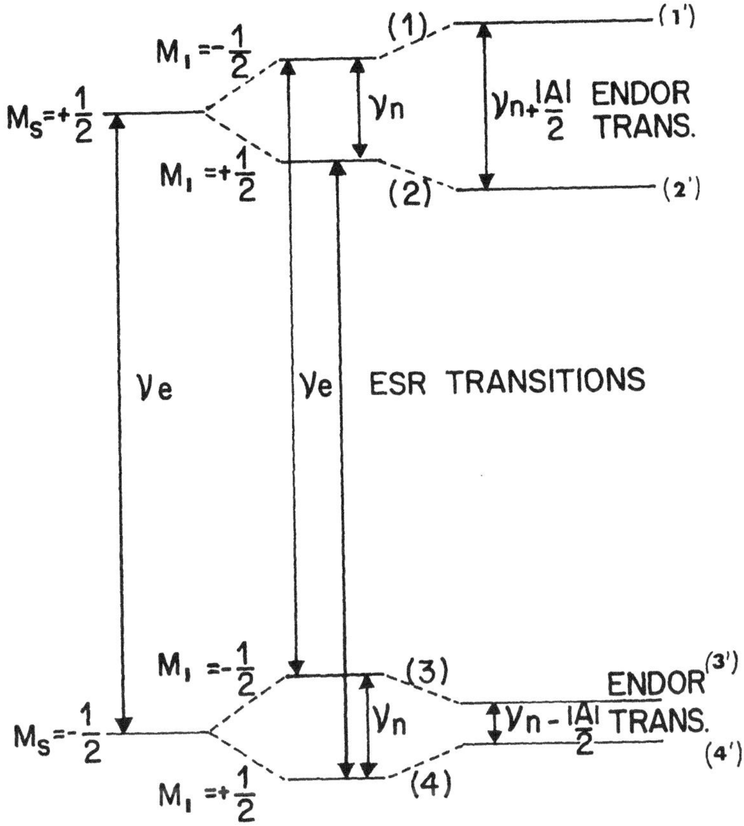

Fig. 1.- Magnetic sublevels of an electron spin $S = 1/2$ and nuclear spin $I = 1/2$ in contact interaction ($A < 0$ and $|A|/2 < \nu_n$).

The spectrum consists in 2 lines, the splitting being equal to A. As it is well known, these two lines represent the hyperfine structure of the E.S.R. line due to the COUPLING of the electron spin with one nucleus of spin 1/2.

If the number of nuclei which are coupled with the electron spin is large, and if there are various coupling constants, the hyperfine structure becomes extremely complex. An exemple is given in figure 2 with the hyperfine structure of the perinaphtyl radical as obtained by L.S. Singer[3]. The determination of the various coupling constants can be performed only with a computer program.

If the number of nuclei or of coupling constants increases, the hyperfine structure of ESR line is no more resolved and one gets only a broad line called "inhomogeneously broadened" line. It is clear that in absence of hyperfine structure the unambiguous determination of the free radical structure is impossible. The ENDOR technique enables us to determine the hyperfine coupling constants. The frequency of the nuclear transition (defined by the selection rules $\Delta m_I = \pm 1$ and $\Delta m_S = 0$) in case where $A \neq 0$ becomes

$$\nu'_n = \nu_n \pm \frac{A}{2}$$

By measuring ν'_n it is possible to determine A.

The ENDOR provides a tool to measure ν'_n. Let us assume that the microwave field H_1 -which produces the ESR in the static magnetic field H- is strong enough to partially saturate the ESR line. The degree of saturation of the ESR line depends on two factors affecting the saturation on opposite directions.

The first factor is the microwave field H_1 which tends to equalize the populations of the levels (1') and (3') (if its frequency corresponds to that transition).

The second factor is the electron spin-lattice relaxation mechanism (spin-lattice relaxation time T_1) which tends to maintain the thermal equilibrium between levels, this relaxation mechanism being active between the levels (1') and (3') and the levels (2') and (4') (relaxation pathways). Therefore at given level of microwave field, any change in the relaxation pathways will change the saturation level.

By saturating the (1') <--> (2') transition at frequency ν'_n, the pathway (2') <--> (4') will become effective for the level (1') ; therefore saturation of the (1') <--> (2') transition will "desaturate" the E.S.R. transition. The details of the quantitative effects depend strongly on the relaxation mechanism and will

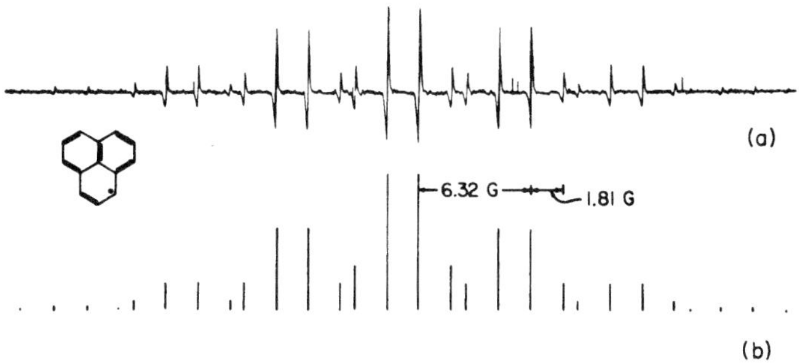

Fig. 2.- ESR spectrum of perinaphtyl and its theoretical interpretation. REF 3

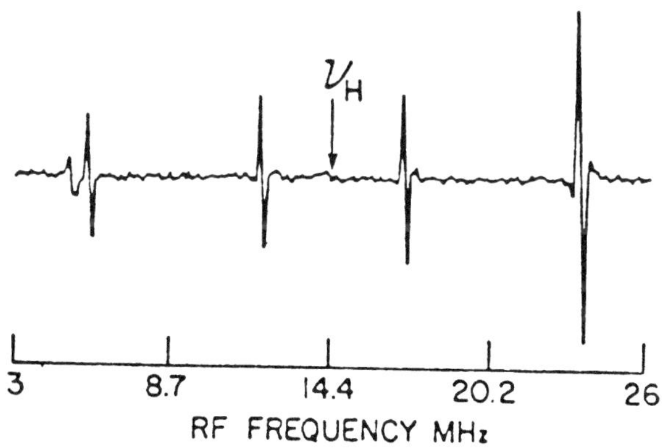

Fig. 3.- ENDOR spectrum of perinaphtyl. REF 4

not be discussed here.

Therefore by partially saturing the ESR of an inhomogeneously broadened line and by applying a strong rf magnetic field of variable frequency, one will get an increase in the E.S.R. intensity when $\nu = \nu'_n$.

The ENDOR spectrum is plotted as a function of the frequency of the rf H_1 field (usually called "rf frequency"). Figure 3 shows the ENDOR spectrum of the perinaphtyl free radical [4].

This spectrum is very simple as compared to the ESR spectrum of figure 2. In particular the 2 different values of the hyperfine coupling A, which were obtained from the ESR spectrum after difficult computer aided analysis [3], appear clearly and directly from the ENDOR spectrum.

The ENDOR spectrum, as contemplated, is centered around the frequency ν_n, which is in this case of the order of 14 MHz (protons in a constant applied magnetic field H = 0,3 T).

III.- MATRIX OR DISTANT ENDOR

The ENDOR described in the preceding paragraph is related to the system of an electron in contact interaction with nuclei in the vicinity of the electron. It is sometimes called for this reason "LOCAL ENDOR".

On the other hand in the ENDOR spectrum of a solid containing dilute paramagnetic centers surrounded by nuclei in dipole-dipole interaction with the unpaired electron spin, one usually observes an ENDOR line at the magnetic resonance frequency ν_n of the nuclei in the applied magnetic field. For example, the figure 4 shows the ENDOR spectrum given by a charcoal obtained by heat treating saccharose sugar at 400° C and in vacuum [5].

As it is well known this charcoal (as most of them) contains unpaired electron spins and the main nucleus which have a non vanishing nuclear spin is the proton. The ENDOR spectrum is recorded in a static magnetic field of 0,3 T and ν_n is closed to 14 MHz.

The ENDOR line at frequency ν_n produced by nuclei having a negligible contact interaction with the electron is called "Distant ENDOR" or "Matrix ENDOR" as it is produced by the long range dipole-dipole interaction. The first observation of a Matrix ENDOR line is due to Lambe and al. [6] and can be attributed to the two following mechanisms :

- the spins of the distant nuclei are polarized (dynamic nuclear polarization by solid state effect) by the microwave field

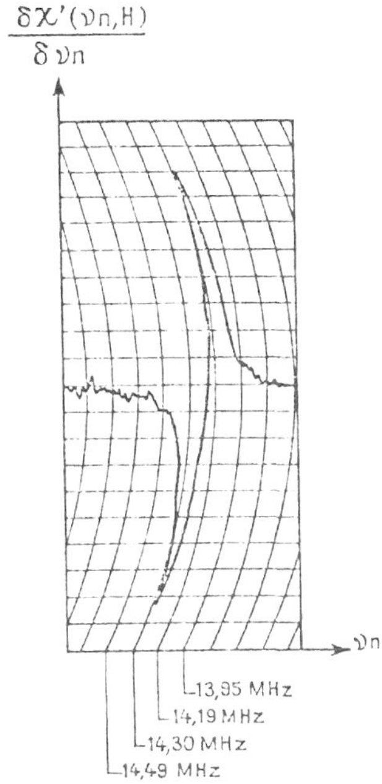

Fig. 4.- Distant ENDOR spectrum of a saccharose char observed on the dispersion mode of the ESR. The constant magnetic field value is 0,3 T.

partially saturating the ESR line (inhomogeneously broadened) ;

- saturation of the NMR transition will destroy the dynamic polarization and at same time reduce the electron spin-lattice relaxation time and finally desaturate the ESR line.

The Matrix ENDOR signal is maximum when the dipole-dipole interaction can be considered as static, i.e. when the dynamic polarization by solid state effect is maximum.

When a molecular motion appears, the Matrix ENDOR is affected and the line shape may show a significant change [6] [7]. When the molecular motion becomes important, the Matrix ENDOR signal disappears. Also when the electron spin is highly delocalized the Matrix ENDOR disappears.

The sunsitivity of the matrix (or distant) ENDOR to motion, environment and electron spin delocalization makes this technique very valuable to probe those parameters in great detail.

IV.- INSTRUMENTATION

As pointed out in the introduction one needs for an ENDOR experiment, to have a very stable ESR spectrometer capable of saturating and at same time monitoring a given point of a selected ESR line for a rather long period of time without drift and in presence of a large H_1 rf field. The very important point of a useful ENDOR spectrometer is the cavity which :

- must operate at various temperatures (change in molecular motion and in spin-lattice relaxation time) ;
- must have a very efficient temperature stabilization system in presence of large rf power (up to 100 W) ;
- must let the rf H_1 field penetrate inside the cavity.

The helix, which is small and can be a resonant structure both at microwave and rf frequencies, is an intensity tool to meet this 3 requirements [9].

Commercial ENDOR spectrometers are available. Usually the rf frequency is modulated so that the derivative of the ENDOR line is plotted.

V.- APPLICATION TO COALS AND CHARCOALS (CARBONACEOUS MATERIALS)

Since the discovery in 1954 of ESR in natural coals [10] [11] it is well known that practically all coals and charcoals (organic chemical compounds, including oil, heat treated at temperature below 1000° C) contains highly stable free radicals.

The structure of those free radicals is believed to be associated with the odd-alternative aromatic hydrocarbons like perinaphtyl, benzanthranyl[3]. The presence of hydrogen atoms in those carbonaceous substances makes possible to study ENDOR.

The first observation of ENDOR in chars has been performed in 1967 in a 400° C heat treated sugar char[5]. Matrix ENDOR is observed and the ENDOR line is shown in figure 4. The same matrix ENDOR line has been observed later on coals[12]. More recent work by Retcofsky and al.[13] shows that some evacuated coal sample shows a local ENDOR signal which not yet interpreted.

For coal in air as well as for asphaltenes, the same authors observe the distant (matrix) ENDOR line. But for oil produced from coal by hydrogenation, two local ENDOR lines (giving an hyperfine coupling of 3,6 MHz) are observed. These lines decrease as the temperature is lowered, while at same time the matrix peak increases. At -40° C, the local ENDOR lines have disappeared.

The most recent work on ENDOR of carbonaceous materials is the work of Lewis and Singer[14] on petroleum pitch. Figure 5 shows the ENDOR spectra of a petroleum mesophase pitch at two different temperatures. At 250° C, above the so called "glass transition temperature", the matrix peak, strongly reduced, coexists with a doublet which can be attributed to a superposition of local ENDOR lines. The authors have been able to follow the decrease of the matrix line as a function of temperature and correlate it with the glass transition temperature. This is a remarkable application of the ENDOR technique to the study of internal motions of solid.

CONCLUSION

Although up to now only few ENDOR research have been devoted to carbonaceous substances, remarkable results have been already obtained in the study of large aromatic free radicals which appear in the carbonization process, as well as in the study of molecular motions.

ENDOR appears definitively as a powerfull tool to study the carbonaceous substances. But, of course, it is highly desirable to use ENDOR in conjonction with other magnetic resonance techniques like ESR, NMR and dynamic nuclear polarization.

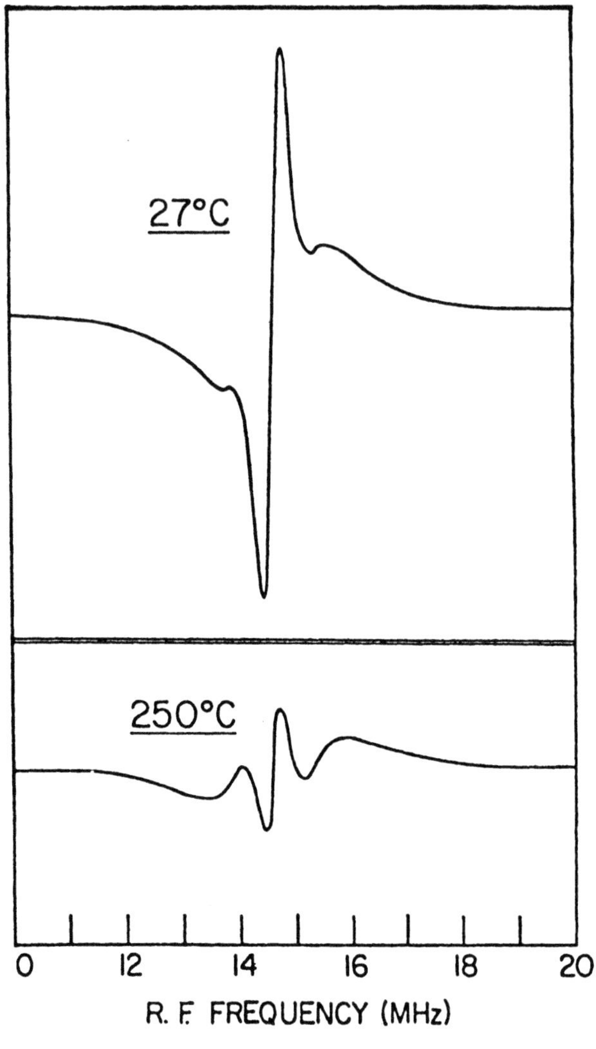

Fig. 5.- ENDOR spectra of a Petroleum Mesophase Pitch at 27° C and 250° C. REF 14

REFERENCES

1. Feher, G. 1956, Phys. Rev., 103, 834.

2. Kevan, L. and Kispert, L.D. 1976, "Electron Spin Double Resonance Spectroscopy", John Wiley & Sons.

3. Lewis, I.C. and Singer L.S. 1981, "Electron Spin Resonance and the Mechanism of Carbonization", in Chemistry and Physics of Carbon, edited by P.L. Walker Jr, and P.A. Thrower, Vol. 17, pp. 1-88, Marcel Dekker 1981.

4. Lewis, I.C. and Singer, L.S., 16th Bienn. Conf. on Carbon, San Diego, July 1983.

5. Decaillot, M. and Uebersfeld, J., Compt. Rend., 265, Series B 155, 1967.

6. Lambe, J., Laurance, N., Mc Irvine, E.C. and Terhune, R.W., Phys. Rev., 122, 1161, 1961.

7. Hyde, J.S., Rist, G.H. and Eriksson, L.E.G., J. Chem. Phys. 72, 4269, 1968.

8. Leniard, D.S., Hyde, J.S. and Vedrine, J.C., J. Chem. Phys. 72, 2079, 1972.

9. Fourrier-Lamer, A., J. Phys. E. Sc. Instrum., 12, 430, 1979 and Thèse de Doctorat, Paris VI, 1981.

10 Uebersfeld, J., Etienne, A. and Combrisson, J., Nature, 174, 614, 1954.

11 Ingram, D.J.E., Tapley, J.G., Jackson, R., Bond, R.L. and Murnahgan, A.R., Nature, 174, 797, 1954.

12 Schlick, S., Narayana, P.A. and Kevan, L., JACS, 100, 3322, 1978.

13 Retcofsky, H.L., Hough, M.R., Maguire, M.M. and Clarkson, R.B., "Nature of the free radicals in coals, pyrolyzed coals, solvent-refined coal, and coal liquefaction products" in Coal Structure edited by Martin L. Gorbaty and K. Ouchi, pp. 37-58, Amer. Chem. Soc., 1981.

14. Lewis, I.C. and Singer, L.S., 16th Bienn. Conf. on Carbon, San Diego, July 1983.

MÖSSBAUER SPECTROSCOPY AS AN ANALYTICAL TOOL IN COAL UTILIZATION

P.A. Montano

Department of Physics, West Virginia University
Morgantown, WV 26506

ABSTRACT

Moessbauer spectroscopy has been successfully applied to the study of iron bearing minerals in coals. Recently this technique was also used to study the coordination of ^{119}Sn derivatives in coal. The position of oxygen in the coal structure was inferred from these measurements. The deep penetration of the γ-rays allows in situ studies of the transformation of the minerals during coal liquefaction and/or gasification. This spectroscopy was applied with great success in the study of iron sulfide stoichiometries and their role in direct coal liquefaction. A review of the applications of Moessbauer spectroscopy is presented relative to the areas of mineral characterizations to synthetic fuels.

1. INTRODUCTION

In recent years resonance techniques have become powerful tools in the study of different components of coal (1,2,3). The inorganic components of coal are of central importance from the point of view of Moessbauer spectroscopy. This spectroscopy has been successfully used in the characterization of iron bearing minerals in coal (4,5,6,7). In general X-ray diffraction methods represent a competing technique to Moessbauer spectroscopy. However, the bulk mineralogy of the ashes in particular clays appear to be of low crystallinity, indicated by the poor diffraction patterns of the coal ashes as compared to identically-prepared mineral mixtures. By contrast, Moessbauer spectroscopy can be successfully applied to materials with poor

L. Petrakis and J. P. Fraissard (eds.), Magnetic Resonance. Introduction, Advanced Topics and Applications to Fossil Energy, 177–202.

crystallinity. Although the major emphasis in the application of Moessbauer spectroscopy to coal research has been on the characterization of the iron bearing minerals, this technique was recently used to obtain information about the heteroatom population near the hydroxyl groups (8). The Moessbauer effect of ^{119}Sn was used as a probe. The Moessbauer effect has been applied to the study of the transformation of coal minerals during utilization. This technique is used to characterize liquefaction residues, chars, coke, ashes, etc. (9,10,11,12). An area of considerable importance to coal conversion is the characterization in situ of catalysts for direct and indirect coal liquefaction (13,14,15). This technique can be applied to an increasingly large number of interesting problems in coal chemistry and physics. In this paper we will attempt to offer an overall review of the Moessbauer effect and its applications to coal research.

2. MOESSBAUER PARAMETERS

The parameter which gives information about the vibration of the Moessbauer atom in the lattice is the Debye-Waller factor (16). The second-order Doppler effect (17) is also related to the lattice dynamic behavior of the Moessbauer atom. The parameters which describe the local environment (valence, symmetry, magnetic moment) of the Moessbauer atom or ion are the isomer shift (IS), quadrupole splitting (QS), and magnetic hyperfine splitting (MHS). A brief description of these parameters will be given in the following paragraphs.

2.1. Moessbauer Spectral Area

The Moessbauer spectral area has been calculated by several authors using series expansions on $t=nf\sigma_o$ (where n is the number of Moessbauer atoms per cm^2, σ_o is the absorption cross section at resonance and f is the Debye-Waller factor) or Bessel functions of imaginary argument (18,19). The last method will be used to describe the spectral area; for a single absorption line one obtains:

$$A = \pi \frac{\Gamma_o}{2} f_s t_a \exp\{-\tfrac{1}{2}t_a\}[I_o(\tfrac{1}{2}it_a) + I_1(\tfrac{1}{2}it_a)] \quad (1)$$

where Γ_o is the natural linewidth, f_s is the Debye-Waller factor of the source, I_o is a zero order Bessel function of imaginary argument and I_1 the first order Bessel function of imaginary argument. If the Moessbauer spectrum consists of a doublet, the spectral area is given by:

$$A= \frac{\Gamma_o}{2} \pi f_s \{K(t_{1a})+K(t_{2a}) - \frac{K(t_{1a})K(t_{2a})X(t_{1a},t_{2a})}{[X(t_{1a},t_{2a})]^2+\Delta^2}\}$$

$$K(\lambda) \equiv \lambda e^{-\lambda/2} [I_o(\frac{\lambda}{2}) + I_1(\frac{\lambda}{2})] \qquad (2)$$

$$X(t_{1a},t_{2a}) = \frac{K(t_{1a})+K(t_{2a})}{K(t_{1a})+K_2(t_{2a})-K(t_{1a}+t_{2a})}$$

where $t_{ia} = \beta_i t_a$ and β_i is the relative intensity of the ith component. Δ is the energy difference between the two lines in units of $\Gamma_{o/2}$. In the case of single crystal absorbers, the dependence of the cross section on the polarization has to be considered (20). When obtaining quantitative information from the Moessbauer spectral area, one has to be aware of the effect of background radiation. One should record the total gamma-ray spectrum going through the sample and accurately determining the amount of non-resonant radiation (21) in order to obtain reliable quantitative information.

2.2. Debye-Waller Factor

The Debye-Waller factor, also called Moessbauer fraction, is temperature-dependent and is given in the harmonic approximation by

$$f = \exp[-k^2(x^2)_T] \qquad (3)$$

where k is the wave vector of the gamma-ray and $(x^2)_T$ is the thermal average of the mean square displacement. For temperatures where $h\omega_{max} \leq 2\pi k_B T$ (ω_{max}, maximum vibration frequency, h Planck's constant, k_B Boltzmann's constant) one can use the Thirring expansion to describe the Debye-Waller factor (22) for the isotropic case:

$$-\ell n\ f(T) = \frac{E_\gamma}{mC^2} < (h\omega)^{-2} > T + \frac{1}{12}(\frac{h}{k_B})^2 T^{-1} - \frac{1}{720}(\frac{h}{K_B})^4 .. \qquad (4)$$

where E_γ is the Moessbauer transition energy, and $<\omega^n>$ are the moments of the frequency spectrum given by

$$<\omega^n> = \int_o^\infty g(\omega)\omega^n d\omega \qquad (5)$$

Where $g(\omega)$ is the density of the vibrational modes.

$$\Theta_D(n) = \frac{h}{k_B} [\frac{1}{3} (n+3) (\omega^n)]^{1/n} \tag{6}$$

Each term in Equation (4) can be described by an effective weighed Debye temperature (22). One can clearly observe from Equation (4) that the Debye-Waller factor is very sensitive to low frequency modes of the Moessbauer lattice. If the particle size is very small, considerable variation in the effective Debye temperature will be observed as compared to the bulk.

2.3. Second Order Doppler Shift (SODS)

In addition to the IS there is a shift of the Moessbauer centroid due to the second order Doppler effect (17). This shift is given by the following relation

$$\delta E_o = \frac{-E_\gamma \langle V^2 \rangle_T}{2C^2} \tag{7}$$

where $\langle V^2 \rangle_T$ is the thermal average of the velocity of the Moessbauer nucleus in the lattice. The SODS is sensitive to the high frequency modes in the phonon spectrum. The effective Debye temperature obtained from the SODS will always be larger than that obtained from the Debye-Waller factor.

2.4. Isomer Shift (IS)

The IS results from the electrostatic interaction of the nuclear and electron charge distributions inside the nuclear region. The observed shift in the Moessbauer spectrum is given by the difference between the shifts in the source and absorber (Figure 1). The isomer shift is given by the following relation (23):

$$IS = \frac{2\pi}{5} Ze^2R^2(\frac{\delta R}{R})S(Z) \{|\psi_s(0)|_a^2 - |\psi_s(0)|_s^2 \}$$

$$\delta R = R_{ex} - R_g \tag{8}$$

$$R = \frac{R_{ex} + R_g}{2}$$

where $|\psi_s(0)|_a^2$, $|\psi_s(0)|_s^2$ are the electron densities at the source and absorber respectively; Z is the atomic number; R_{ex}, R_g are the radii of the excited and ground state of the Moessbauer nucleus; and S(Z) takes relativistic effects into consideration

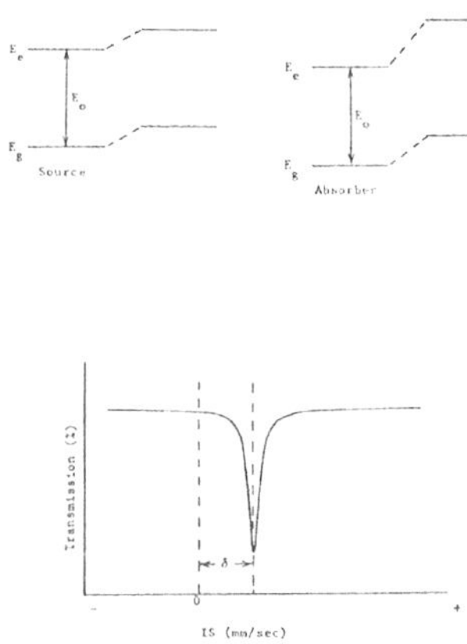

FIGURE 1 Energy shifts of the ground and excited states of the Moessbauer atom produces the IS.

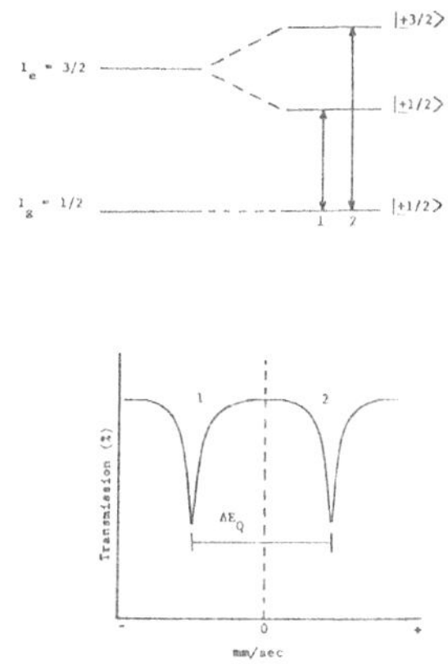

FIGURE 2 Quadrupole splitting for $I_e = 3/2$ to $I_g = 1/2$ transition.

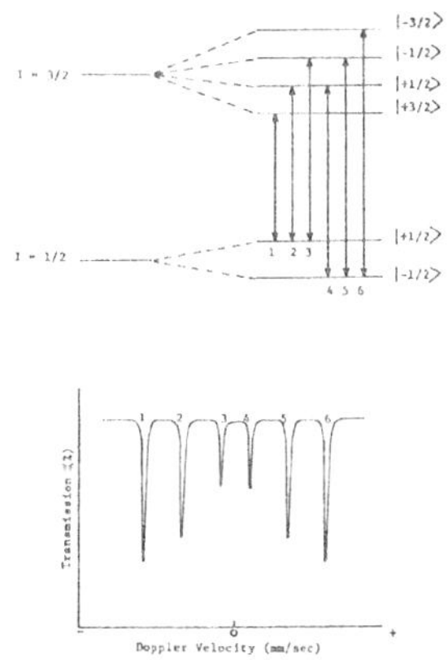

FIGURE 3 MHS for ^{57}Fe and the resulting Moessbauer spectrum.

(24). The IS gives unique information about the valence state of the Moessbauer atom or ion, e.g. for ^{57}Fe compounds, the range of values covers between -0.78mm/sec to +1.9mm/sec (all the IS given in reference to α-Fe, 1mm/sec=0.48x10-7 eV). The IS values reported in the literature always include the SODS. One can distinguish the SODS contribution by making use of its strong temperature dependence.

2.5. Quadrupole Splitting

This interaction exists when the Moessbauer atom owns electrons or the neighboring atoms (or ions) produce an inhomogeneous electric field at the nucleus and when the nucleus possesses a quadrupole moment Q. The splitting of the Moessbauer levels is for a transition with nuclear spin I=3/2 to I=1/2 by (25)

$$\Delta E_q = (\tfrac{1}{2})\, e^2 qQ(1 + \eta^2/3)^{\frac{1}{2}} \tag{9}$$

where eq = V_{ZZ} is the principal axis of the electric field gradient (EFG) and $\eta = V_{XX}-V_{YY} / V_{ZZ}$ is the asymmetry parameter.

In general, the crystal field affects the electrons of the Moessbauer atom, even those possessing spherical symmetry. These electrons are distorted, producing an EFG at the nucleus which can be larger than that produced directly by the crystal field. The electrons that do not possess spherical symmetry also produce a distortion of the closed electronic shells and an additional EFG. The effective EFG is given by

$$q = (1-R)q_{valence} + (1-\gamma_\infty)q_{lattice} \tag{10}$$

The term (1-R) is the Sternheimer factor (26) that corrects for the polarization of the core by the valence electrons, and $(1-\gamma_\infty)$ is the correction which takes into consideration the polarization of the core electrons by the crystal field. Figure 2 illustrates the effect of a non zero EFG on the Moessbauer spectrum.

Both q and η depend on $(1/r_3)$, and it is expected that the valence contribution will be dominant unless the ion has high symmetry in which case the lattice contribution will dominate. The valence contribution can show pronounced temperature dependence, e.g. Fe^{2+} high spin compounds.

2.6. Magnetic Hyperfine Splitting

The MHS is produced by the interaction of the nuclear magnetic moment μ with a magnetic field at the nucleus (Zeeman effect). This interaction is given by (25)

$$H_I = -g\mu_n \vec{I}\cdot\vec{B} \tag{11}$$

where g is the nuclear gyromagnetic factor, μ_n the nuclear Bohr magneton, I the nuclear spin, and B the magnetic field at the nucleus. The latter field can originate either within the atom itself, within the crystal via exchange interaction, or as a result of applying an external field. The Hamiltonian given above has the following eigenvalues:

$$\begin{aligned} E_m &= -g\mu_n Bm_I \\ m_I &= I, I-1, \ldots, -I \end{aligned} \tag{12}$$

where m_I is the magnetic quantum number. Figure 3 shows the MHS for ^{57}Fe, the allowed transitions are $\Delta m=0, \pm 1$. The internal field at the nucleus is given by

$$B = B_S + B_L + B_D \tag{13}$$

where B_S is the Fermi contact interaction, B_L is the orbital contribution, and B_D is the dipolar term. The MHS is very sensitive to the local environment of the Moessbauer atom; consequently, it is very useful in the identification of iron compounds in coal (21). This parameter has been shown to be extremely important in the determination of the pyrrhotite stoichiometries (9).

If an EFG and a magnetic field are present at the nucleus, then the position of the sublevels will depend on the ratio of the magnetic to the electric interactions, on the symmetry of the EFG and on the angle between the principal axis of the EFG and B. It is possible to write the Hamiltonian for this general case but no general solution is possible. Approximations exist for some special cases (16).

3. EXPERIMENTAL TECHNIQUES

There are different ways of recording a Moessbauer spectrum. The most common technique involves the determination of the counting rate of an appropriate Moessbauer gamma radiation passing through an absorber (transmission mode). The energy of the source is varied by imparting a Doppler velocity to the Moessbauer source. One can use two types of drives:

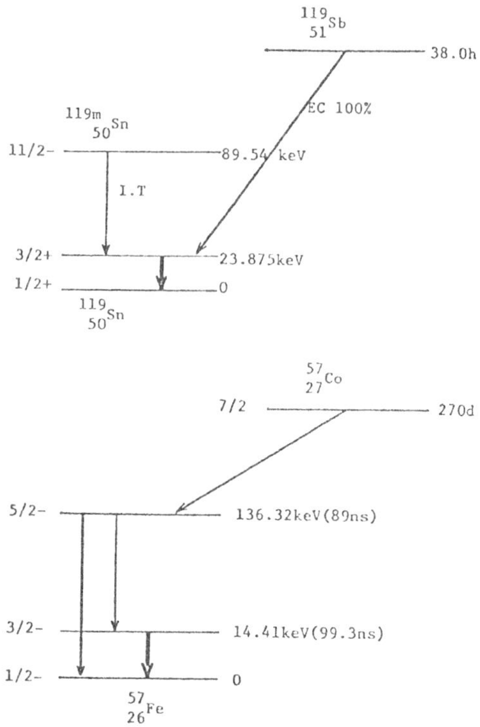

FIGURE 4 Decay schemes of ^{57}Co and ^{119}Sn, two commonly used Moessbauer isotopes.

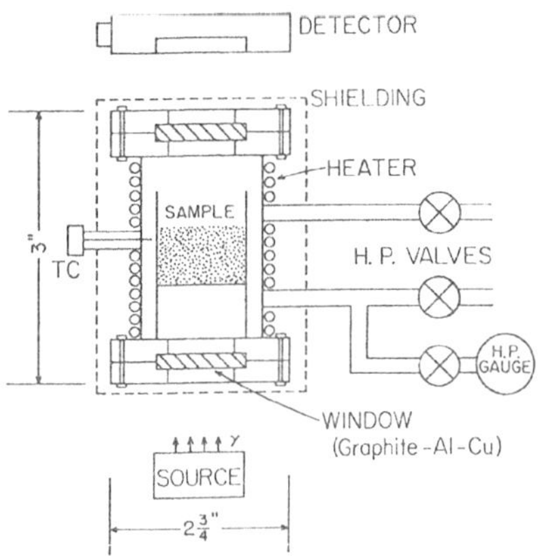

FIGURE 5 Diagram of the reactor used for *in situ* Moessbauer measurements at high pressures.

constant velocity and constant acceleration. In the former mode the spectrum is scanned point by point, a convenient technique when a limited number of peaks need to be detected. However, when a full spectrum is recorded, the constant acceleration mode seems to be more reliable due to its insensitivity to small electronic drifts in the system. The Moessbauer spectrum can also be recorded in the emission mode, either gamma or x-ray fluorescence, or conversion and Auger electron emission. The latter method is very sensitive to surface composition especially when it is combined with energy analysis.

The Moessbauer sources more commonly used in coal research are ^{57}Co and ^{119}Sn. In the present work we used ^{57}Co:Rh (220mCi) and $Ca^{119}SnO_3$ (10 mCi). The detectors used were a proportional counter (Kr 3%CH) and NaI scintillator for the ^{119}Sn experiments. The decay schemes for ^{57}Co and ^{119}Sn are shown in Figure 4.

The coal conversion studies were performed in two different types of reactors. The high pressure reactor, shown in Figure 5, is described in detail in Reference 13. The reactor shown in Figure 6 was used for the Fischer-Tropsch synthesis work. The latter reactor design allows the simultaneous recording of mass spectra of the products (15).

4. CHARACTERIZATION OF COAL COMPONENTS

Moessbauer spectroscopy finds its wider application in the identification of iron bearing minerals (4,5,6,7). The identification of such minerals requires a very careful analysis of the spectrum. A first condition is that the sample under study is representative of the coal seam. Several samples from the same source should be analyzed in order to check for consistency of the results.

The major minerals in coal are the clays which represent a large percentage of the inorganic matter in coal. The clay minerals present a great diversity of structures (27). The major clay minerals in coal are kaolinite, illite, chlorite, mixed layer clays, etc. The most frequently observed in USA coals are the illites. Illite is a term denoting the mica in soils, consolidated sediments, and sedimentary rocks. In general, the total amount of iron in the illites is small. The Moessbauer spectra of the illites show a great diversity (27). The other major minerals in coal are the sulfides (pyrite, marcasite, sphalerite, galena, etc.), carbonates (calcite, ankerite, siderite, dolomite), quartz, and other minerals in lesser amounts such as anatase, rutile, hematite, feldspar, etc. The presence of iron sulfates is invariably an indication of

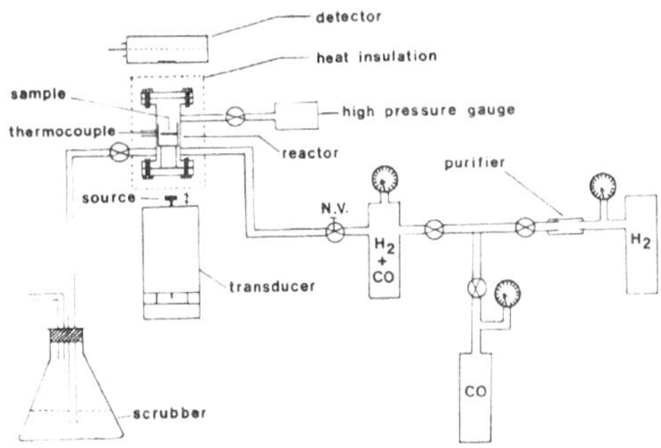

FIGURE 6 Diagram of the system used for _in situ_ measurements of Fischer-Tropsch catalysts.

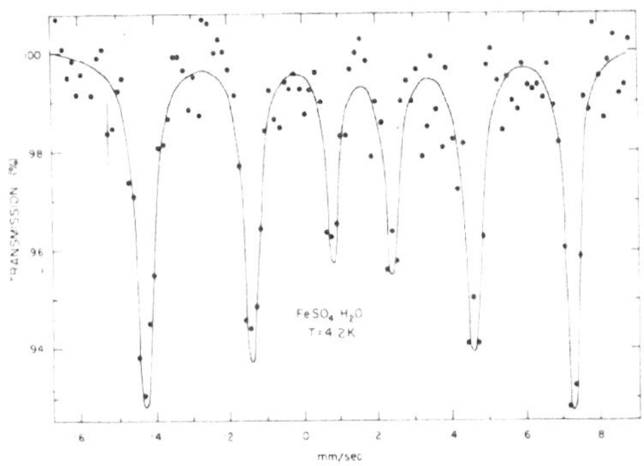

FIGURE 7 Moessbauer spectrum of szomolnokite at 4.2 K; clear evidence of magnetic ordering is observed (Ref 28).

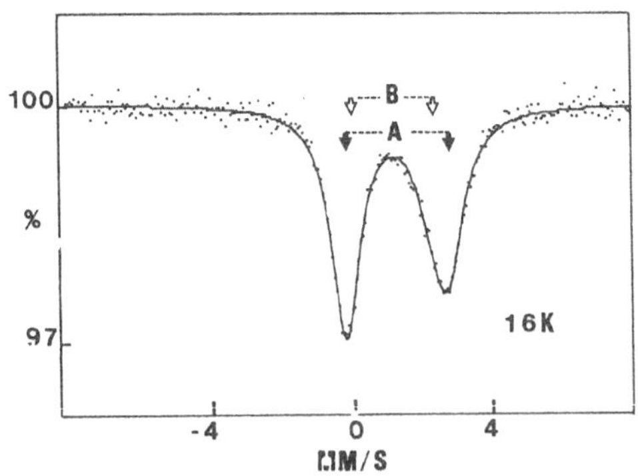

FIGURE 8 ^{119}Sn Moessbauer spectrum at 16 K of the tri-n-butyl derivative of Rawhide coal (Ref 8).

weathering of the coal. The most frequent sulfates observed in coal are szomolnokite, coquimbite, and jarosites. Gypsum is rarely observed in fresh coals. The most important minerals from the point of view of Moessbauer spectroscopy are the iron sulfides which are also of considerable technical importance in coal utilization. For complicated Moessbauer spectra, one needs to introduce constraints in order to reduce the number of free parameters in the fitting. To identify the minerals, one has to compare the parameters obtained from the fit, with those of standard minerals. A serious problem is the overlap of the Moessbauer spectral lines for different compounds. Several clay minerals are difficult to distinguish from each other; they also have overlapping lines with some of the iron sulfates. Trivalent iron is particularly annoying since it overlaps strongly with the pyrite lines. In many instances we find it necessary to chemically treat the sample (e.g. with dilute HCl) to eliminate the sulfates. Another powerful technique is to carry out low temperature measurements (in some cases in the presence of an external magnetic field (21,28)). In Figure 7, the Moessbauer spectrum of szomolnokite is shown at 4.2 K; a clear MHS allows uniquivocal identification of the compound (28). Another common technique is to analyze the low temperature ashes. In this case the spectra are usually easier to characterize. The major handicap of the latter technique is the formation of new species that are originally absent in the coal (7).

Recently Moessbauer spectroscopy has been successfully applied to the understanding of the spatial distribution of oxygen in coal. In Illinois #6 and Rawhide, Larsen and coworkers (8) have observed that the oxygen atoms are not randomly distributed and that most hydroxyl groups have neighboring heteroatoms capable of accepting a hydrogen bond. The hydroxyl groups in the coals were converted to $-O-SnBN_3$ derivatives by reacting with $(n-BN_3Sn)_2O$ in toluene. The Moessbauer spectrum shows that almost all the tin is trigonal bipyramidal (Figure 8). This recent experiment is clear evidence that Moessbauer spectroscopy in coal research can be extended beyond iron and inorganic components.

Several reviews have been written on the use of Moessbauer spectroscopy for pyritic sulfur analysis. In a recent publication we presented a critical review of this subject; we refer the reader to that paper for additional information (21,29).

5. COAL UTILIZATION

5.1. Combustion

The most common use of coal involves its combustion to produce steam in order to generate electricity. One of the major problems associated with this use of coal is related to the presence of sulfur compounds in the coal. Pyrite is the most deleterious of all the coal minerals and major source of SO_2 pollution of the atmosphere. Several methods are available for coal cleaning (30). Moessbauer spectroscopy has been successfully applied to follow the transformation of pyrite during coal cleaning (30).

Moessbauer spectroscopy can be used to observe the transformation of iron-bearing minerals during oxidation. Such a technique was used to study the oxidation of pyrite in an ILL#6 coal (31). Figure 9 shows the different phases observed during oxidation. The oxidation occurs in three steps: (1) to iron sulfates between 25 to 310° C, (2) to γ-Fe_2O_3 between 310 and 325° C, (3) to α-Fe_2O_3 above 325° C. The oxidation of pyrite is also particle size dependent (31). The Moessbauer effect has also been used by Huffman and Huggin to study the residual ashes in coal (12). They observed the transformation of pyrite to hematite, of siderite to hematite, and of ankerite and Fe^{2+} in clays to ferric ions in glassy materials. The composition of the ashes was strongly dependent on the original minerals and the cooling conditions.

5.2. Coke

Coke, one of the most fundamental components in the iron and steel industries, is produced by exposing a properly chosen coal or blend of coals to high temperatures in an air deficient environment. Heavily weathered coals cannot be used in the coking industry. Moessbauer spectroscopy was used to identify the iron compounds present in coke (12); troilite, pyrrhotite, hematite masnetite, metallic iron and ferrous ions in glasses were detected. This spectroscopy has also been used to screen the coal before its utilization in the coking industry (32). This work dealt mainly with the detection of oxidation products prior to its use.

5.3. Acid Mine Drainage

Coal oxidizes when exposed to the atmosphere for a period of time. The degree of oxidation depends on the intrinsic characteristics of the coal. This oxidation also has a detrimental effect on the environment. One of the major problems associated with coal mining is the pollution of the

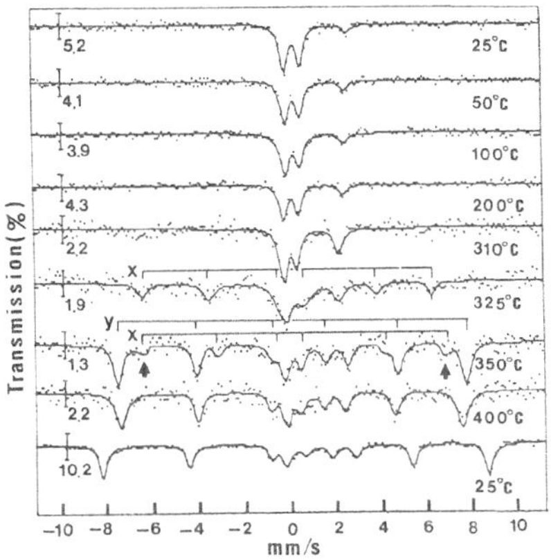

FIGURE 9 *In situ* ^{57}Fe Moessbauer spectra of an ILL#6 coal at various temperatures in the presence of oxygen (Ref 31).

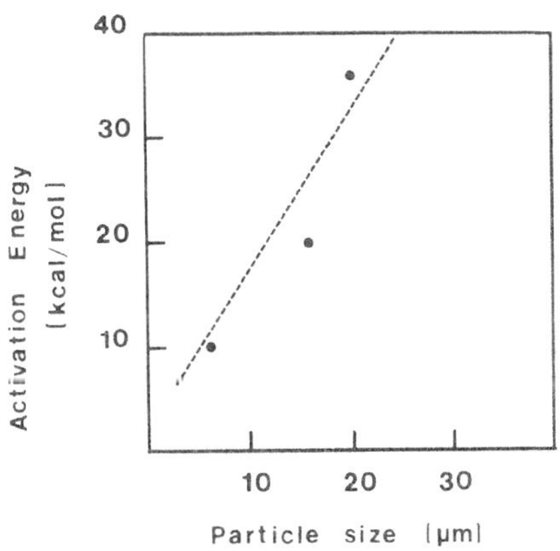

FIGURE 10 Particle size dependence of the activation energy for pyrite decomposition (Ref 37).

surrounding streams due to acid mine drainage. It is generally accepted that pyrite is a precursor for acid formation (33). When coal is exposed to the atmosphere, pyrite is oxidized to a series of hydrous iron sulfates which subsequently dissolve in water to form the so-called acid mine drainage. There is a lack of correlation between the total amount of pyritic sulfur and the amount of acid produced. The particle size seems to be more important than the total amount of pyrite. The Moessbauer effect has been used with great success to identify potential acid mine drainage-producing coal seams (34). Because of its resonance character, the identification of small amounts of iron compounds become feasible. The use of Moessbauer spectroscopy to screen coals for further utilization represents one of the most promising industrial applications of this technique.

6. SYNTHETIC FUELS

6.1. Direct Coal Liquefaction

Direct coal liquefaction is a process that has been known for a number of years. The liquefaction of coal is a complex process involving the interaction between coal, hydrogen donor solvent, and catalysts. In recent years it has been realized that the mineral matter plays a significant role in the conversion of coal to liquid products (35,36). Addition of pyrite, pyrrhotite and liquefaction residues to coal has been shown to affect the coal conversion yields and the viscosity of the products. Moessbauer spectroscopy has found extensive application in the characterization of the residues and in monitoring the transformation of the iron sulfides during coal liquefaction (9,13).

Under liquefaction conditions, FeS_2 is transformed into a non-stoichiometric iron sulfide, $Fe_{1-x}S$. In a recent Moessbauer and SEM study, it was shown that the activation energy for pyrite reduction is strongly particle size-dependent (37, see Figure 10). It was noticed that the transformation of pyrite to pyrrhotite commences around 300° C. The decomposition of pyrite seems to be controlled by the availability of anions on the surface and by the rate of diffusion from the bulk (38). This model is in good agreement with the experimental observations.

Moessbauer spectroscopy has been developed to a standard analytical tool for the determination of pyrrhotite stoichiometry. This method is based on the observation that MHS in the pyrrhotites is strongly dependent on the number of vacancies around the iron ions (39). The MHS is reduced as the number of nearest neighbor vacancies increases. This feature is used to calculate the stoichiometry of the pyrrhotites. The

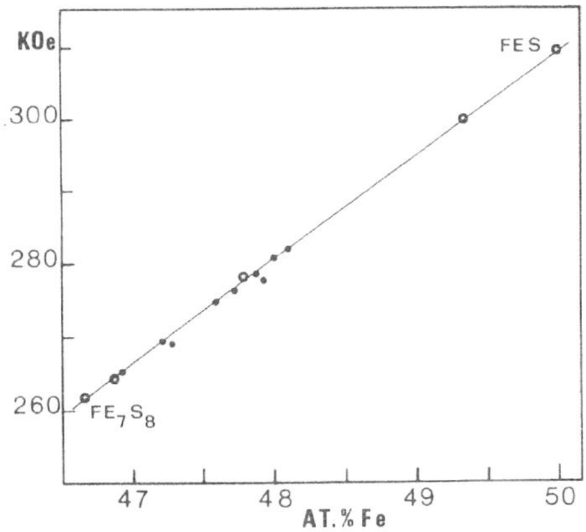

FIGURE 11 Relation between average magnetic hyperfine field and atomic %iron in $Fe_{1-x}S$.

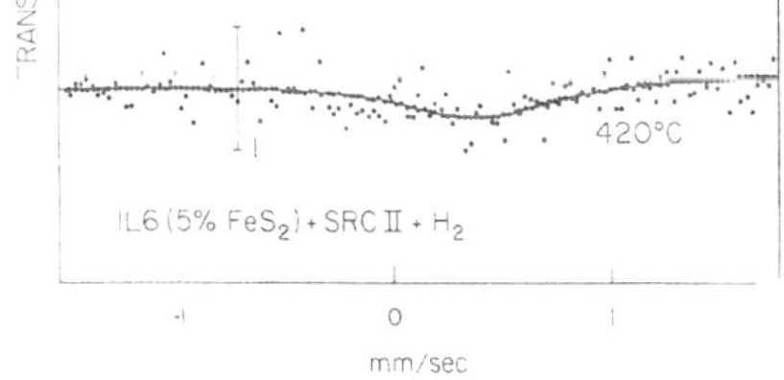

FIGURE 12 In situ Moessbauer spectra of an ILL#6 coal with 5% FeS_2 added by weight. The measurements were performed in a solvent and at 1.4 MPa of hydrogen (Ref 13).

atomic percentage of iron is calculated according to (9):

$$\text{atomic \% Fe} = 100\text{x}[0.6836\text{x}10^{-3}\text{x}\bar{B}_{MHS} + 0.2881]$$

$$\bar{B}_{MHS} = \frac{1}{A_{TOTAL}} \Sigma A_i B_i$$

$\bar{B}_{MHS}$: average magnetic hyperfine field

A_{TOTAL}: total Moessbauer spectral area

A_i: spectral area under i-th site

B_i: magnetic hyperfine field on i-th site

In Figure 11 we show our calibration results using standard samples (stars) and liquefaction residues. The identification of troilite is very easy due to its simple six line Moessbauer spectrum.

A very important use of Moessbauer spectroscopy is to study in situ reactions. We have employed this technique using the reactors described earlier to follow the transformation of pyrite under coal liquefaction conditions (13,14). These experiments were carried out with four well-characterized coals. In these experiments, a considerable decrease in the IS and increase in linewidth was observed at high temperatures (> 400 ^{o}C). Two factors contribute to the line broadening: diffusional broadening due to vacancies motion, and motional broadening due to the coal-derived liquids where the pyrrhotites are in suspension. The IS is significantly more negative than for runs carried with pure pyrite (TABLE I). The time scale for a process at the Moessbauer atom is determined by the lifetime of the Moessbauer transition (10-7 sec). Any process that takes place in a longer time scale will appear as static from the point of view of Moessbauer spectroscopy. That such effects are observable are indicative of large surface areas at the reaction temperatures (above 350° C). Figure 12 shows a series of in situ Moessbauer spectra for an ILL#6 coal with 5% pyrite (solvent to coal ratio = 2, H_2 pressure 4.2 MPa). A factor that can contribute to the line broadening is a distribution of IS, due to reactions with multiple components of the coal. The observed interaction at high temperatures occurs even in the absence of solvent or hydrogen, however the effect is more pronounced in their presence (14).

The effects of additives in coal liquefaction were studied by Stephens et al (40). They carried out a series of experiments with various additives on an ILL#6 coal and observed an enhancement in conversion to benzene solubles with the

TABLE I

ISOMER SHIFT AT 420° C

Runs at 420° C in the pressence of solvent (SRC-II Heavy distillates) and H_2 (4MPa)

Sample	I.S. ($mmsec^{-1}$) (at 420° C)
IL6	0.18 ± 0.05
KY 9/14	0.34 ± 0.05
Powhatan 5	0.35 ± 0.05
Blacksville #2	0.21 ± 0.05
Fe_7S_8	0.50 ± 0.03
FeS	0.51 ± 0.05

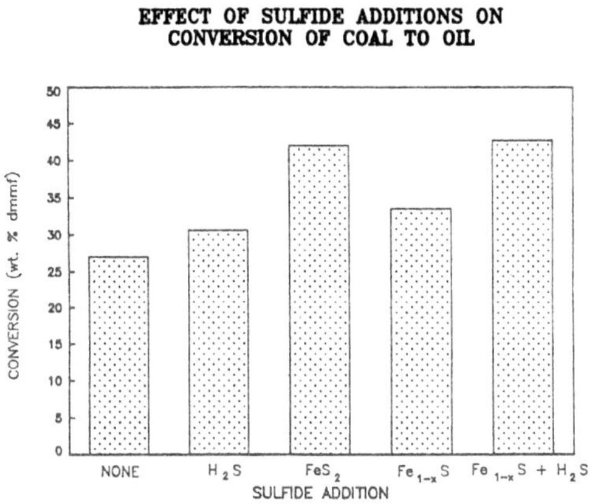

FIGURE 13 Effect of sulfide addition on conversion of coal to oil (Ref 40).

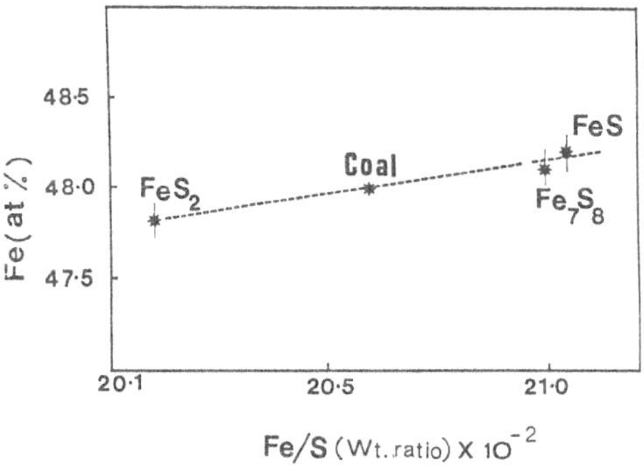

FIGURE 14 Effect of iron sulfides additives to a Powhatan #5 coal on the stoichiometry of $Fe_{1-x}S$. The atomic % iron is plotted VS the weight ratio of Fe/S in the coal.

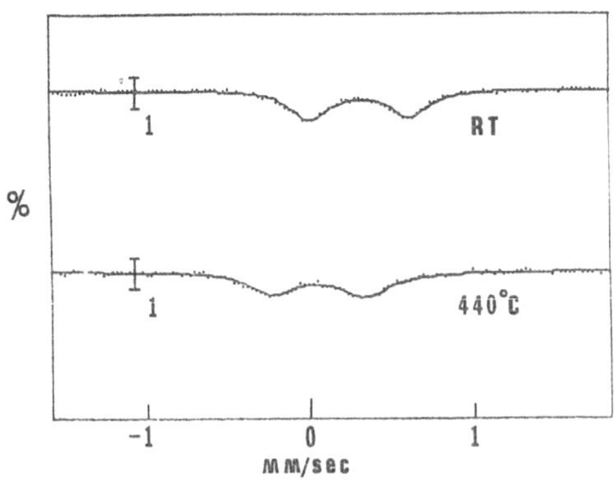

FIGURE 15 Moessbauer spectra *in situ* of a mixture of FeS_2 and dibenzothiophene (Ref. 46).

addition of pyrrhotite and H_2S. A similar result was obtained with pyrite alone (see Figure 13). These results strongly suggest that the pyrrhotites and H_2S play a catalytic role in coal conversion. It appears that H_2S controls the stoichiometry and surface composition of the pyrrhotites. At a given temperature, and above 300° C, the atomic % of iron in $Fe_{1-x}S$ will decrease as the partial pressure of H_2S increases. It is expected that the total amount of sulfur in coal will be a crucial factor in determining the stoichiometry of $Fe_{1-x}S$. The Moessbauer effect has been used to study, under liquefaction conditions, the effect of additives on the transformation of pyrite to pyrrhotite (14,41). In Figure 14 the variation of the atomic % of iron in $Fe_{1-x}S$ is plotted versus the Fe/S ratio by weight for a Powhatan #5 coal. From the figure it is clear that the addition of pyrite produces a more iron deficient pyrrhotite.

In a complex material such as coal, it is very difficult to identify the roles of vastly differing organic functional groups and their interactions with pyrrhotite or H_2S. A simpler approach involves the study of model compound reactions; a limited number of such studies has appeared in the literature (42,43,44,45). We have used Moessbauer spectroscopy to study the reactions of FeS_2 and Fe_7S_8 with several model compounds. There was strong evidence of interaction with a number of model compounds (46). Figure 15 shows the Moessbauer spectra <u>in situ</u> of dibenzothiophene and 10% pyrite (by weight). Full conversion of pyrite to pyrrhotite occurs in one hour, even in an inert atmosphere. Reactions also occur with phenanthrene, pyrene and naphthoquinone; in the latter compound, there is clear evidence of the formation of an intermediate oxide. In the case of Fe_7S_8 evidence also exists of high temperature reactions with the compounds. However, in several instances a transformation of Fe_7S_8 to troilite takes place during the reactions (Figure 16). This difference in behavior is attributed to the low partial pressure of H_2S in the system. A recent Moessbauer study using Fe_7S_8 as a catalyst and diphenylmethane (47) as the model compound strongly points out the synergistic effect between $Fe_{1-x}S$ and H_2S. The experimental evidence indicates hydrocracking activity for $Fe_{1-x}S$ and H_2S. We also suggest that one of the mechanisms by which pyrrhotite enhances conversion is by interaction with oxygen in the coal (46). These observations are in good agreement with the earlier experiments that found a correlation between the extent of conversion of coal to benzene or THF solubles, and the iron deficiency in the pyrrhotites present in the residues. This indicates that the iron vacancies on the surface are the catalytic sites where H_2S can attach and dissociate leaving highly reactive sulfur. In such a process hydrogen will be available to the solvent and the organic functional groups in the coal (47).

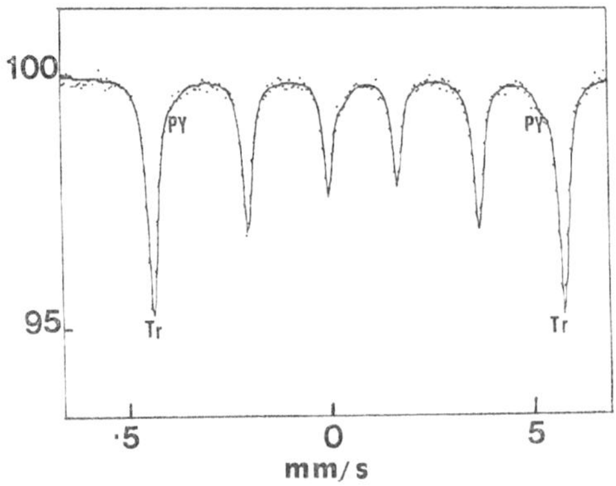

FIGURE 16 Moessbauer spectrum of the residue of the reaction of Fe_7S_8 and pyrene. Troilite (Tr), pyrrhotite (Py). (Ref 46).

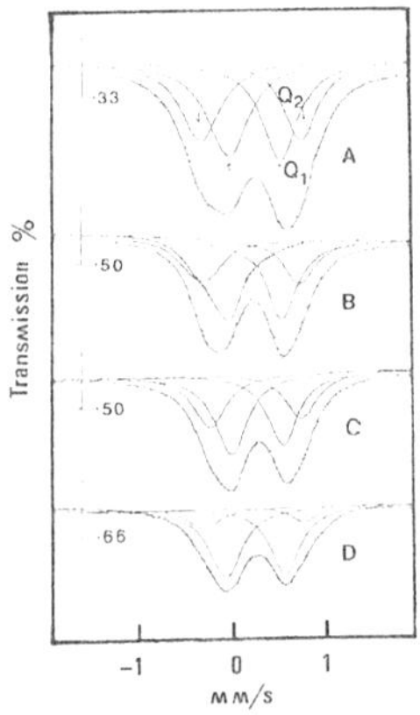

FIGURE 17 Room-temperature ^{57}Fe Moessbauer spectra of sulfided Fe-Mo/Al_2O_3 catalysts. A) R=0.05 B) R=0.25 C) R=0.50 and D) R=0.75 (Ref 48).

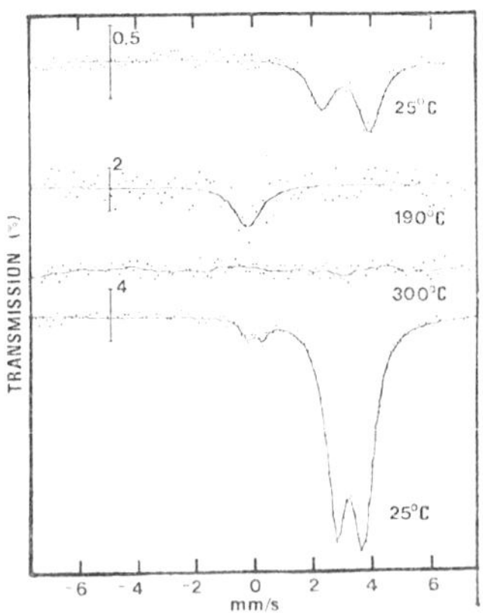

FIGURE 18 In situ ^{119}Sn Moessbauer spectra of ILL#6 coal dry-mixed with a tin chloride catalyst, SRC-II solvent in the presence of hydrogen. a) 25°C b) 190°C c) 300°C and d) 25°C (Ref 49).

The use of Moessbauer spectroscopy is not limited to the pure iron sulfides but can also be used to study other catalytic systems for direct coal liquefaction. Such a system is the sulfided catalyst Fe-Mo supported on Al_2O_3 (48). It was observed that the activity of the Mo/Al_2O_3 catalyst is highly influenced by the presence of Fe in the sample. Figure 17 illustrates the variation in the Moessbauer spectra as a function of the ratio R = Mo/Mo+Fe. The two doublets were identified as iron oxides on alumina (Q_2) and small superparamagnetic particles of iron sulfides (Q_1). The addition of iron produces a decrease in the conversion of coal to THF or pentane soluble. We attribute this to blocking by iron oxides and sulfides of the most active Mo sulfides.

A very important coal liquefaction catalyst is $SnCl_2$. This compound contains a Moessbauer isotope which gives a reasonably large effect at room temperature. In situ Moessbauer spectroscopy was used to study the catalytic role of $SnCl_2$ and its interaction with an ILL#6 coal during hydrogenation (with and without solvent).

The study showed that both tin and chlorine can enter the coal structure at an early stage of hydrogenation (49). The presence of chlorine can cause the cleavage of ether-oxygen linkages in the coal. Tin oxides (or hydroxides) are formed as intermediates and finally tin (II) sulfide is formed as the predominant form of the catalyst (see Figure 18). This study has also shown the formation of $FeCl_2 \cdot XH_2O$ as a result of the reaction between pyrrhotites and hydrochloric acid.

6.2. Indirect Coal Liquefaction

One of the most promising methods of producing liquid fuels from coal is through indirect liquefaction where the coal is gasified to obtain synthesis gas (CO and H_2). The major application of Moessbauer spectroscopy is related to the use of iron catalysts in the production of fuels and chemicals from synthesis gas (15). The synthesis gas reacts over an iron catalyst to produce a mixture of hydrocarbons and oxygenates. Two examples will be given of such an application of Moessbauer spectroscopy.

In situ ^{57}Fe Moessbauer spectroscopy was used to study the process of reduction and Fischer-Tropsch synthesis on $FeCl_3$ intercalated in graphite (15,50). First, higher-stage iron chloride, and mixed ferric chloride-potassium intercalated graphite catalysts were investigated under reduction (H_2 flow) and Fischer-Tropsch reaction conditions. A simultaneous measurement of the product distribution was performed using this

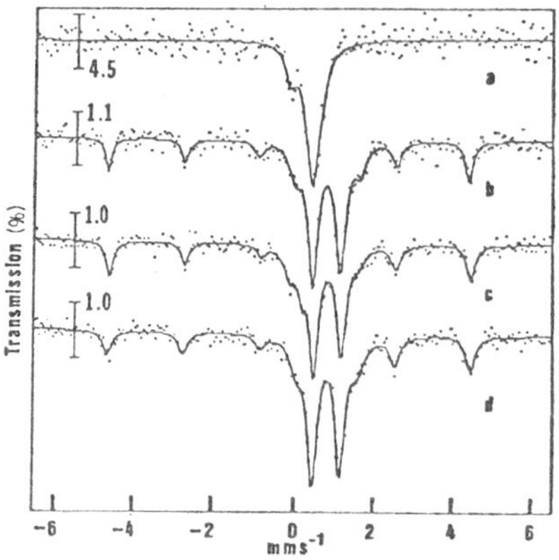

FIGURE 19 Room temperature ^{57}Fe Moessbauer spectra for a ferric chloride-potassium chloride intercalated graphite. a) fresh catalyst, b) after 21 hrs reduction in H_2 at 400°C. c) after 48 hrs reduction in H_2 at 400°C, d) after Fischer-Tropsch reaction (Ref 50).

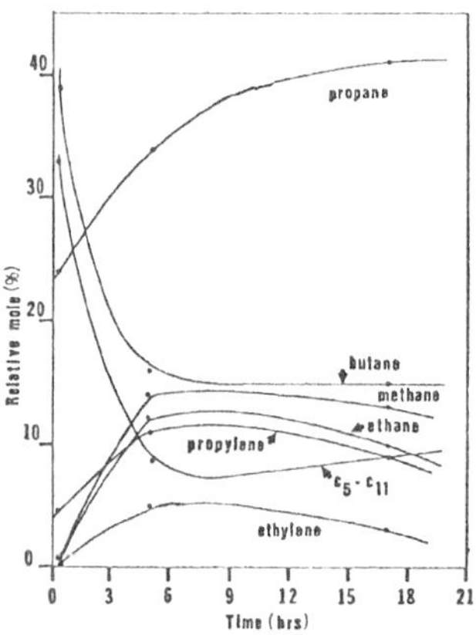

FIGURE 20 Analysis of the product distribution in the effluent gases during Fischer-Tropsch synthesis reactions for a ferric chloride-potassium chloride-graphite catalyst (Ref. 50).

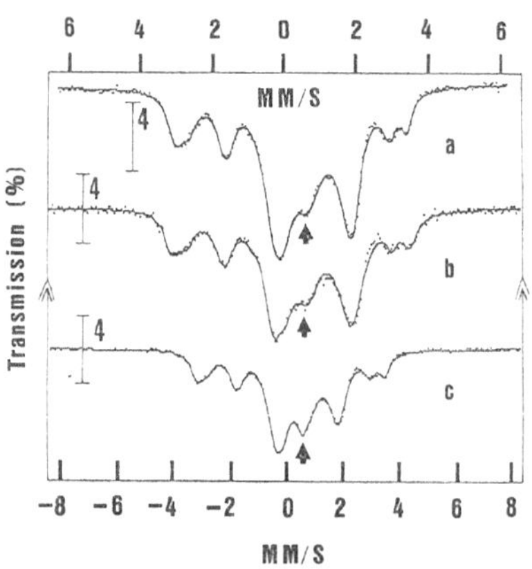

FIGURE 21 Moessbauer spectra of an Fe/Al_2O_3 catalyst after a) 8 hrs of carburization, b) 42 hrs carburization, and c) 7 hrs of Fischer-Tropsch reactions.

technique. The first two catalysts predominantly possess a higher selectivity for methane, whereas the third catalyst has higher selectivity for the formation of propane (50). The differences in selectivity were attributed to geometrical effects due to the various degrees of intercalation. Figure 19 shows the Moessbauer spectra for a ferric chloride-potassium chloride intercalated in graphite, the top spectrum corresponds to the fresh catalyst. During reduction α-Fe is formed as well as ferrous chloride. There is less reduction of the ferric chloride to α-Fe for this catalyst than for all the others. This is attributed to a strong association of the potassium with the ferric chloride. A great selectivity for the formation of propane was observed for this catalyst (Figure 20).

Another example of the use of Moessbauer spectroscopy in the characterization of catalysts is given in Figure 21. The figure shows the Moessbauer spectra of Fe on Al_2O_3 after (a) 18 hrs of carburization, (b) 42 hrs carburization and (c) 7 hrs of Fischer-Tropsch reaction (51). The spectra show evidence of metal support interaction through the formation of $FeAl_2O_4$ and a carbide with a composition close to that of Hagg's carbide. It is clear that this spectroscopic technique is very sensitive to small changes in the catalyst composition.

We have tried to give a general review of the importance of Moessbauer spectroscopy in coal sciences. However, such a task becomes extremely difficult and space limitation forces us to omit several contributions in this area of research. The author wishes to acknowledge the fruitful collaborations of P.P. Vaishnava, A.S. Bommannavar, V. Shah, H.J. Shyu, D. Walker and P. Russell. The financial support of the U.S. DOE and the West Virginia University Energy Research Center is greatly appreciated.

REFERENCES

1. Retcofsky, H.L. 1977, Appl. Spectr. 31, p. 116.

2. Gerstein, B.C., Ryan, L.M., and Murphy, P.D. 1981, "Coal Structure," Eds. M.L. Gorbaty and K. Ouchi, ACS 192, Washington, D.C.

3. Bartuska, V.S., Maciel, G.E., Schaefer, J., and Stejskal, E.O. 1977, FUEL 56, p. 354.

4. Lefelhocz, J.F., Friedel, R.A., and Kohman, T.P. 1967, Geochim. Cosmochim. Acta. 31, p. 2261.

5. Montano, P.A. 1977, FUEL 56, p. 397.

6. Levinson, L.M. and Jacobs, I.S. 1977, FUEL 56, p. 453.

7. Huffman, G.P. and Huggins, F.E. 1978, FUEL 57, p. 597.

8. Larsen, J.W., Nadar, P.A., Mohammadi, M., and Montano, P.A. 1982, FUEL 61, p. 889.

9. Montano, P.A. and Granoff, B. 1980, FUEL 59, p. 214.

10. Montano, P.A., Shah, V., Reddy, S., and Bommannavar, A.S. 1981, Ch. 21, "New Approaches in Coal Chemistry," eds. B.D. Blaustein, B.C. Bockrath, and S. Friedman, ACS Symposium Series No. 169.

11. Huggins, F.E. and Huffman, G.P. 1980, "Analytical Methods for Coal and Coal Products," 13, Academic Press.

12. Huffman, G.P. and Huggins, F.E. 1980, in "Recent Chemical Applications of Moessbauer Spectroscopy," eds. J. Stevens and G.K. Shenoy, ACS, Advances in Chemistry, p. 194.

13. Montano, P.A., Bommannavar, A.S., and Shah, V. 1981, FUEL 60, p. 703.

14. Bommannavar, A.S. and Montano, P.A. 1982, FUEL 61, p. 523.

15. Vaishnava, P.P. and Montano, P.A. 1982, J. Phys. Chem. Solids 43, p. 809.

16. Werteheim, G.K. 1969, "Moessbauer Effect: Principles and Applications," a good introduction to this technique, Academic Press, N.Y.

17. Pound, R.V., Rebka, G.A. Jr. 1960, Phys. Rev. Lett. 4, p. 274; Josephson, B.D. 1960, Phys. Rev. Lett. 4, p. 342.

18. Lang, G. 1963, Nucl. Instr. Method 24, p. 425.

19. Bykov, G.A. and Pham Xuy Xien, 1963, "Proceedings of the Dubna Conference on the Moessbauer Effect," pp. 2-27, Consultants Bureau Enterprises, Inc., NY.

20. Housley, R.M., Grant, R.W. and Gonser, U. 1969, Phys. Rev. 178, p. 514.

21. Montano, P.A. 1981, Ch. 7 in "Moessbauer Spectroscopy and Its Chemical Applications," Eds. J.G. Stevens and G.K. Shenoy, Advances in Chemistry Series 194, Washington, D.C.

22. Housley, R.M. and Hess, F. 1966, Phys. Rev. 146, p. 517.

23. Kistner, O.C., Sunyar, A.W. 1960, Phys. Rev. Lett. 4, p. 412.

24. Wegener, H. 1965, "Der Moessbauer Effect," Bibliographisches Inst. Mannheim.

25. Abragam, A. 1964, "L'Effect Moessbauer et ses applications à l'étude des Champs Internes," Gordon and Brach, Paris.

26. Sternheimer, R.L. 1950, Phys. Rev. 80, p. 102.

27. Coey, J.M.D. 1975, Int. Conf. on Moessbauer Spectroscopy V2, Cracow.

28. Russell, P. and Montano, P.A. 1978, J. Appl. Phys. 49(3), p. 1573.

29. Montano, P.A. 1982, Nuclear Magnetic Resonance Review Vol.7, N2, p. 175.

30. Levinson, L.M. 1979, "Research Project 267-2 Final Report," EPRI FP-1228; Jacobs, I.S., Levinson, L.M., and Hart, H.R. Jr., 1978, J. Appl. Phys. 49(3), p. 1775.

31. Shyu, H.J., Vaishnava, P.P., and Montano, P.A. 1981, FUEL 60, p. 1023.

32. Huggins, F.E., Huffman, G.P., Kosmack, D.A., and Lowenhaupt, D.E. 1980, Int. Journal of Coal Geology 1, p. 75.

33. Barnes, H.L. and Romberger, S.R. 1968, J. Water Pollut. Contr. Fed. 40, p. 371.

34. Stiller, H.H., Renton, J.J., Montano, P.A., and Russell, P.E. 1978, FUEL 57, p. 447.

35. Davidson, R.M. 1983, "Minerals Effects in Coal Conversion," Report No. ICTIS/TR22, IEA Coal Research, London, An excellent review of the subject.

36. Granoff, G., and Thomas, M.G. 1977, ACS Div. Fuel. Chem. Preprints 22, p. 183.

37. Montano, P.A., Vaishnava, P.P., King, J.A., and Eisentrout, E.N. 1981, FUEL 60, p. 713.

38. Attar, A. 1978, FUEL 57, p. 201.

39. Ovanesyan, N.S., Trukhtanov, V.A., Odinets, G. Yu. and Novikov, G.V. 1971, G.V. Soviet Physics JETP 33, p. 1193.

40. Stephens, H.P., Stohl, F.V., and Padrick, T.D. 1981, Proceedings of the International Conference on Coal Science, p. 368, Düsseldorf Verlag Glück Auf, GMBH, Essen.

41. Bommannavar, A. and Montano, P.A. 1983, FUEL.

42. Guin, J.A., Lee, J.M., Fan, C.W., Curtis, C.W., Lloyd, J.L., Tanner, A.R. 1980, Ind. Eng. Chem. Process 19, p. 440.

43. Jolly, J.P. and Ollis, D. 1979, J. Catal. 60, p. 216.

44. Derbyshire, F.J., Varghese, P., and Whitehurst, D.D. 1981, Proceedings of the International Conference on Coal Science, Düsseldorf, Verlag Glück Auf, GMBH Essen.

45. Bockrath, B.C. and Schroeder, K.T. 1981, "New Approaches in Coal Chemistry," Eds. D.D. Blaustein, B.C. Bockrath and S. Friedman, ACS Symposium, 169, Ch. 11.

46. Montano, P.A. and Bommannavar, A.S. 1983, J. Molecular Catalysis.

47. Stenberg, V., Ogawa, O. and Montano, P.A., to be published.

48, Vaishnava, P.P., Montano, P.A., Tischer, R.E., and Pollack, S.S. 1982, J. Catalysis 78, p. 454.

49. Vaishnava, P.P., Shyu, H.J., and Montano, P.A. 1981, FUEL 60, p. 625.

50. Vaishnava, P.P. and Montano, P.A. 1983, Proceedings of the Materials Research Society 20, p. 397.

51. Vaishnava, P.P., Ktorides, P., Montano, P.A., Mbcadam, K., and Melson, G., to be published.

CHALLENGES IN FOSSIL ENERGY CHEMISTRY

Martin L. Gorbaty

Exxon Research and Engineering Company
P. O. Box 45
Linden, New Jersey 07036

ABSTRACT

More fundamental knowledge of fossil fuels, particularly coal and shale in terms of structure and reactivity is essential for generating the new technologies necessary for their efficient use in the future. This paper is intended to review the present state of the art and to suggest specific basic research opportunities and challenges in the areas of coal characterization, gasification, and liquefaction. Many of the same opportunities apply to shale. Assessments of the impact such research programs could have are offered.

A major fraction of the energy used today comes from petroleum and natural gas. The inexorable decline of these resources will increase the need for reliance on less desirable, more refractory materials such as coal and oil shale for liquid fuels. While the timeframe of impact for clean liquid and gaseous fuels from coal and shale is uncertain, the question is one of "when" rather than "if." Technologies are available today for the conversion of coal and shale but generally involve severe processing conditions leading to conversions, thermal inefficiencies and hydrogen utilizations well below those theoretically possible. Major intellectually stimulating and useful opportunities exist to provide new knowledge about the chemical structures and reactivities of coal and oil shale. This new knowledge is critically needed in order to find significantly better methods for utilizing these resources.

L. Petrakis and J. P. Fraissard (eds.), Magnetic Resonance. Introduction, Advanced Topics and Applications to Fossil Energy, 203–218.

What follows is an assessment of critical science needs which could provide the base for new technologies for improving current conversion processes and for entirely new and better processes. The main emphasis will be on coal and coal products but many of the needs described apply to oil shale as well.

CHALLENGES IN COAL CHEMISTRY

At the outset, it should be mentioned that today sophisticated and elegant engineering work is being carried out to develop processes to convert coal into liquid and gaseous products (1). For example such processes as EDS, SRC II and H-coal for liquefaction, Lurgi and Koppers-Totzek and Exxon's CCG (2) for gasification represent the state of the art today. However, the chemistry upon which these processes are based is quite old, and in some cases, not well understood. The chemistries are also severe and harsh, requiring high temperatures and pressures. Because of the thermal and/or hydrogen utilization inefficiencies associated with these chemistries, the economics for producing synthetic fuels from coal and shale are unfavorable today. Newer, milder and more selective chemistries are needed to help overcome current limitations. These will arise from a better understanding of the chemical and physical structures of coal and of the reactivities of its various constituents.

The state of coal science today is comparable to where biochemistry was about 50-75 years ago. At that time key questions had been asked and the tools for getting answers were becoming available. So too with coal science today, significant advances in spectroscopic and other analytical instrumentation are providing the tools to obtain answers to critical structure and reactivity questions, and opportunities for basic chemical research on coal and shale are numerous.

CHARACTERIZATION

Coal is a complex heterogeneous organic rock, made up of fossilized remnants of primordial plant matter and incorporated inorganic materials (3). As such, it has an organic and an inorganic structure. It is also a porous rock and has an associated physical structure. Unraveling the basic chemistry of coal requires a more detailed knowledge of these structures. However, because of coal's inherent heterogeneity, systematic studies of these structures have been limited. With today's modern characterization tools and techniques, a clearer, more precise, and in-depth understanding of each of these structures, the extent to which they interact and the ways they affect reactivity becomes possible.

ORGANIC STRUCTURE

Many average molecular structures representative of the organic material in coal have been and continue to be proposed (4-6) and as expected, become more refined as new information is obtained (Fig. 1). An excellent review of this area is available (7). These models are useful to guide thinking, but it must be remembered they are average structures meant to represent functional group distributions and may not be absolutely accurate. Since coals are very heterogeneous and vary from millimeter to millimeter in the same seam, and considerably more between various ranks of coals formed from different plant matter subjected to varying degrees of temperature and pressure, it may never be possible to know the "real" structure. Nevertheless it is possible to refine these structures by more accurate determinations of key structural features which affect utility and reactivity.

After carbon and hydrogen, oxygen is the most abundant element in coals, and it is present both in the organic functional groups and inorganics. Techniques to differentiate and quantify these functional groups are becoming available (8). This knowledge makes possible the determination of the role of oxygen functionalities in coal structure and reactivity. For example the effect of hydroxyl group hydrogen bonding on caking properties of bituminous coals has been demonstrated (9).

While adequate methods are available to determine total sulfur and inorganic sulfur, organic sulfur is determined by difference. An independent method for determining organic sulfur directly is essential. The forms in which the organic sulfur exists in coal have been proposed based on analyses of coal liquids (3). Methods to qualitatively and quantitatively characterize forms of organic sulfur as they exist in the native structure are required. Ultimately, knowledge of the forms of organic sulfur will allow us to develop the proper organic chemistry required to remove it from a coal or char.

Little is known with certainty about the forms and amounts of each form of organic nitrogen in coal, though assumptions have been made that they exist as basic pyridines or quinolines or weakly acidic or neutral pyrrole, indole or carbazole derivatives (3). It is necessary to know how nitrogen is incorporated in the organic matrix, since it is becoming apparent that nitrogen compounds play a key role in coal asphaltene behavior (10), and it is likely this heteroatom plays a role in coal conversion chemistry (11).

Magnetic resonance studies continue to provide valuable information about coal structure (12). In particular solid state

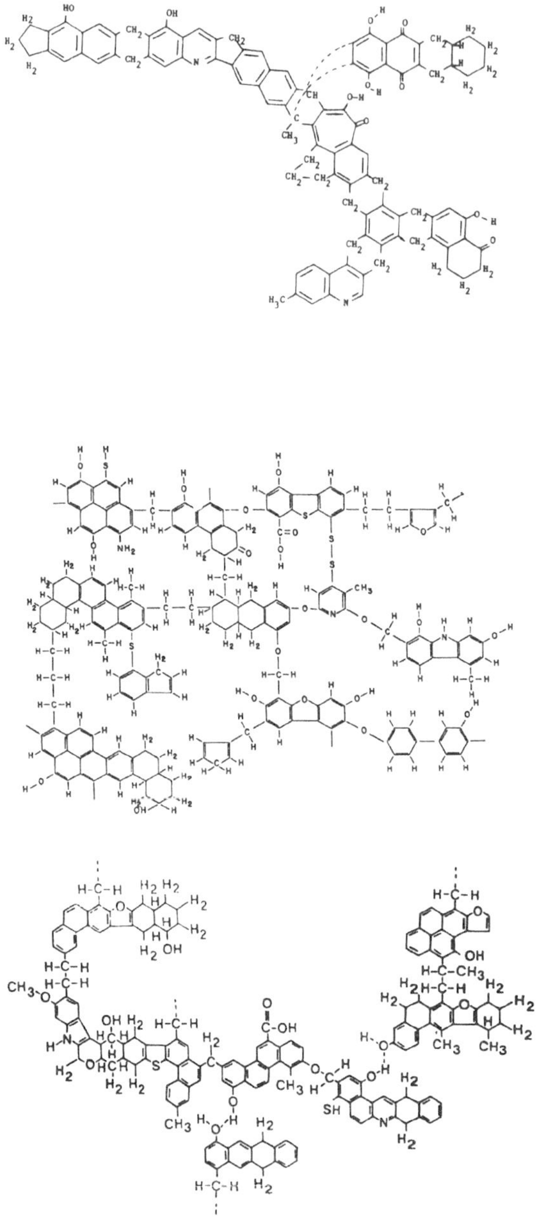

(a)

(b)

(c)

Figure 1. Representative Structures for Coal. (a) Reference 4, (b) Reference 5, (c) Reference 6.

^{13}C nuclear magnetic resonance allows determination of how carbon is distributed in coals (13). With the increasing resolution available today through the coupling of ultra-high fields using super conducting magnets with pulse sequencing and "magic angle" spinning (14), it may be possible to define more precisely how the carbons are bound to one another and to the various oxygen, sulfur and nitrogen heteroatoms. Some high temperature ESR work has been reported on model (15) and real coal systems (16). Similar experiments using ^{13}C nmr must be carried out. They could provide critical information needed to determine the mechanisms of many coal conversion reactions.

Given the detailed knowledge of the functional groups and their distributions, appropriate model systems can be chosen to study the chemistry of converting coal to liquids either by pyrolysis or hydroliquefaction. In particular, it would be important to determine which bonds break first under a given set of conditions and to quantify the kinetics of these reactions.

INORGANIC STRUCTURE

Although the amounts of mineral matter in coals vary considerably most contain about 10% (17). Mineral matter is basically inorganic material deposited before, during or after coalification of partially decayed biomass. This mineral matter is not necessarily inert during coal conversion processes, and has been reported to act beneficially in some cases as a catalyst for gasification (18) and liquefaction (19). Therefore it is important to know the structure of the minerals and their effects on reactivity.

Methods are available to remove carbon (20) (Low Temperature Plasma Asher) and to study the residue. Techniques more sensitive than X-ray diffraction and Fourier-Transform Infrared Spectroscopy are required to characterize this mixture.

When coal is burned, the mixture of remaining inorganic materials is called ash. Techniques to characterize the chemical compounds in ash, many of which are non-stoichiometric, are required. Sensitive techniques to determine the compounds in which trace elements exist in native coals and in ash are needed.

This information can impact on all forms of coal utilization including combustion, gasification and liquefaction. For example, if mineral matter is catalytic in liquefaction, identification of the chemical species responsible could lead to ways of enhancing its catalytic behavior. Determination of compounds that make up the slag found in furnaces might lead to ways of controlling the slagging phenomenon.

PHYSICAL STRUCTURE

The physical structure of coal has been described as a porous macromolecular gel (21). Methods for determining surface areas and pore volumes are needed. Depending on the coal, the adsorbing gas, the equation used for the calculation and the procedures, values of from 2-200 m^2/g have been reported for the same sample (22,23). The difficulty in making these measurements is compounded because pore structures collapse as some coals, particularly subbituminous coals and coals of lower rank are dried (24). Knowledge of surface areas and pore volumes are needed in order to help establish correlations between surface areas and chemical reactivity, particularly regarding access of gases and liquids to the interior.

Opportunities exist for characterizing coal physical structure as a polymer (25) (Figure 2). Ways to determine swelling and modulus properties could lead to definition of coal structure in terms of $\overline{M_c}$, average molecular weight between crosslinks. This vital parameter could be used to define how many of which bonds need to be cleaved to reduce the "molecular weight" of coal from a solid macromolecular network to liquids.

COAL LIQUIDS

Coal liquids are chemically different from petroleum. They must be upgraded, generally by catalytic hydrogenation to remove the nitrogen, sulfur and oxygen heteroatoms and to increase the atomic hydrogen to carbon ratio in order for coal liquids to be used as suitable substitutes for conventional petroleum products (26). Fundamental research on the chemistry and mechanisms for these catalytic reactions and on the catalytic materials themselves is needed to improve current upgrading processes and to provide the scientific basis for new ones.

Characterization techniques are needed in several key areas. Number-average molecular weight ($\overline{M_n}$) is usually obtained by vapor pressure osmometry, but no satisfactory method is available for determining weight-average molecular weight ($\overline{M_w}$), which is useful in predicting rheological properties. Gel permeation chromatography (GPC) is capable of providing this information if a set of satisfactory calibration standards of known molecular weight ranges and structures (both chemical and physical) similar to those materials being analyzed were available. Without these standards GPC can only be used on a

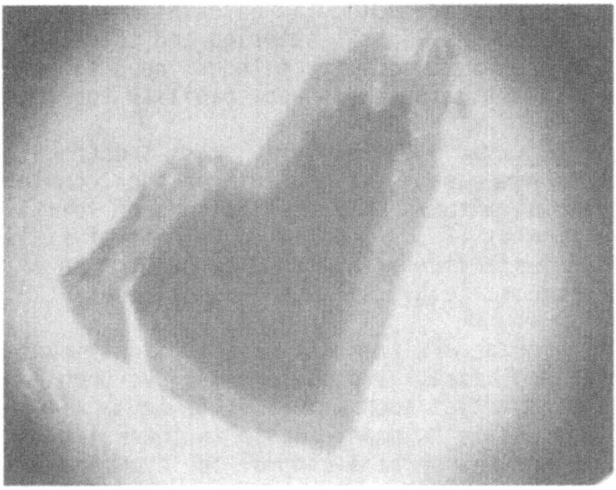

Figure 2. Superposition of pyridine-swollen and untreated coal thin sections illustrate coals can swell isotropically like elastomers. (Reference 25)

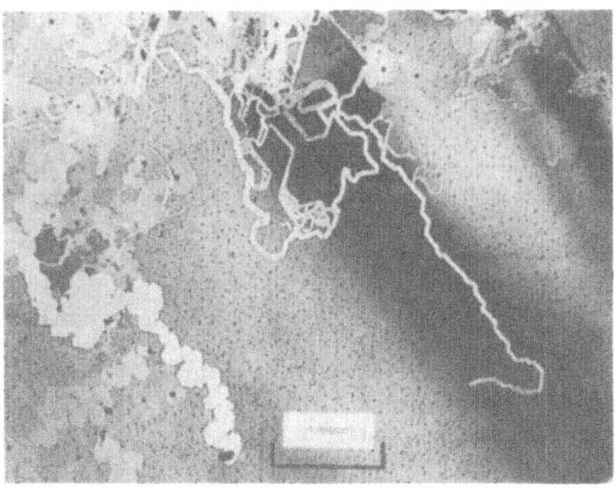

Figure 3. Controlled Atmosphere Electron Microscope allows observation in real time of catalytic channeling of Pt through graphite in oxygen at 700°C. (Reference 36)

relative basis. Another drawback which must be overcome is to account for possible molecular associations in solution. Techniques such as light scattering and ultracentrifugation, which are considered primary methods, need to be critically assessed to establish their applicability for determining $\overline{M_W}$.

Sophisticated high resolution mass spectrometry (MS) techniques are capable of providing rather complete molecular weight and structural characterizations of volatile components of coal liquids (27). The new techniques of field desorption and field ionization are being extended to non-volatile coal liquid products (e.g., SRC, coal asphaltenes) (28).

Coal liquids are known to contain significant quantities of tar acids and bases, components which hydrogen bond and affect physical properties such as viscosity (29). Techniques which can help identify monomeric units in these liquids would allow us to understand and perhaps control the chemical and physical properties of coal products.

Finally, a conceivable limitation to utilization of coal liquids is toxicology. While significant research is ongoing in this area, more studies to pinpoint toxicological and other environmental problems associated with transportation and utilization of coal liquids are needed.

COAL UTILIZATION

Coal can be utilized as an energy source by burning it directly to generate heat, or by converting it into gaseous or liquid products which can be burned (30). Research opportunities in combustion of coal include understanding the complex chemistry of mineral matter conversion to ash, combustion of polynuclear aromatic hydrocarbons and control of nitric oxide formation and emission. These have been described elsewhere (31). Research opportunities in gasification and liquefaction of coal are discussed below.

GASIFICATION

ALthough the reaction of carbon with steam to produce synthesis gas has been known for at least 200 years (32):

$$C + H_2O \longrightarrow CO + H_2 \qquad (1)$$

and the reaction of carbon monoxide and hydrogen to make methane for about 80 years (33):

$$CO + 3H_2 \xrightarrow{\text{Catalyst}} CH_4 + H_2O \qquad (2)$$

fundamental knowledge of the mechanisms of these reactions is lacking. An example of the impact of new knowledge on the gasification area is illustrated by the discovery of Exxon's Catalytic Coal Gasification (CCG) Process (2).

Reaction (1) above is endothermic by about 32 kcal/mole while reaction (2) is exothermic by about 49 kcal/mole. Conceptually it should be possible to gasify carbon thermoneutrally as follows:

		ΔH_{298K}
$2C + 2H_2O \longrightarrow$	$2CO + 2H_2$	+62.8 kcal/mole
$CO + H_2O \longrightarrow$	$CO_2 + H_2$	- 9.8
$CO + 3H_2 \longrightarrow$	$CH_4 + H_2O$	-49.3
$2C + 2H_2O \longrightarrow$	$CH_4 + CO_2$	+3.7

However, reaction (1) is endothermic at relatively high temperatures (∿900°C) while reaction (2) is exothermic at relatively low temperature (∿300°C), and Carnot's principle prevents input of low temperature heat into a high temperature reaction. Consequently, gasification of carbon (coal) to methane has always involved at least two separate steps leading to high thermal inefficiencies.

It has been known for over a century that alkali and alkaline earth metals catalyze the rate of the carbon-steam reaction (34). However, several years ago it was discovered that potassium not only catalyzed the rate of reaction (1) but also brought the gas phase into the equilibrium state for CO, CO_2, H_2, H_2O and methane. This discovery led to the CCG process which involves brining the gas phase into methane equilibrium, separating the methane, and recycling back the other products. The net result is a process which operates with reactions 1 and 2 taking place in the same vessel at the same temperature, thus at higher thermal efficiencies.

Major opportunities still exist today for applying modern surface characterization techniques to gain a better understanding of the mechanism or mechanisms involved in gasification. Studies undertaken to elucidate the mechanism of gasification reactions invariably stress the controlling role exercised by surface oxide complexes which are formed as soon as a fresh carbon surface comes into contact with oxygen. It has been shown that at least five different oxides of varying stability can thus be formed.[35] There is a need to know more about the nature of these oxides and their chemistry, particularly under reaction conditions, i.e., at high temperatures and pressures. In this regard, with controlled atmosphere electron microscopy it is possible to observe gas-solid reactions, such as carbon gasification, as the reactions take place, down to almost the molecular level [36] (Figure 3). Thus, the modern tools of surface physics should be particularly useful and find a fertile field of application here. The understanding gained would be useful for devising ways to improve current processes and to devise newer more efficient ones. For example, if a catalyst could be found which would do what potassium does, but at about 400°C instead of about 650°C, the gas phase equilibrium level of methane would be sufficiently high that separation and recycle would no longer be necessary. This would translate into even higher thermal efficiencies.

LIQUEFACTION

Coal has an atomic hydrogen to carbon ratio less than unity. To make desirable liquids, with atomic H/C ratios of 1.5-2, one must either remove carbon or add hydrogen. The former method of producing liquids is pyrolysis, the latter hydroliquefaction. Coal pyrolysis is an exceptionally old technology, which probably was started on a commercial scale concommitant with the industrial revolution. In 1867 Berthelot reported treating coal with hydrogen to make liquids.[37] In the 1910-1920 period, Bergius developed a commercial coal hydroliquefaction process.[38]

In many respects liquefaction is closely related to pyrolysis. They share an identical step, the thermal generation of radicals from the coal via homolytic bond scission. In pyrolysis these radicals are either capped by an internally transferred hydrogen atom or combined with a carbon site to form heavier molecular weight material (char) (Figure 4). In liquefaction these two events still occur along with transfer of hydrogen to the radicals from a hydrogen source[39] (Figure 5). The net effect is that hydroliquefaction will produce greater amounts of liquid and gaseous products compared with conventional pyrolysis but at the expense of additional hydrogen consumption. Liquids from hydroliquefaction are substantially depleted of heteroatoms as compared with either the parent coal or pyrolysis liquids.

A wide range of different techniques is used to make liquids from coal,[1] even though they all share the thermal conversion step. These differ in whether the hydrogen is provided from an organic donor or from molecular hydrogen either catalytically or noncatalytically. They also differ in whether and what kind of solvent is used. Thus, study of the physical solution

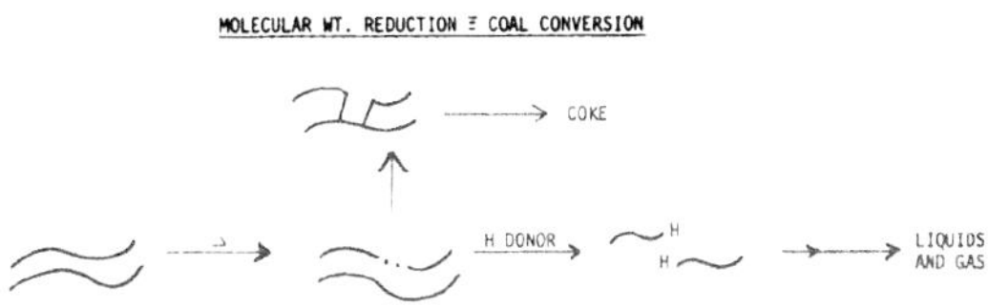

Figure 4: Coal conversion to liquids begins with a thermolysis. Capping radicals with hydrogen leads to liquids; recombination leads to coke.

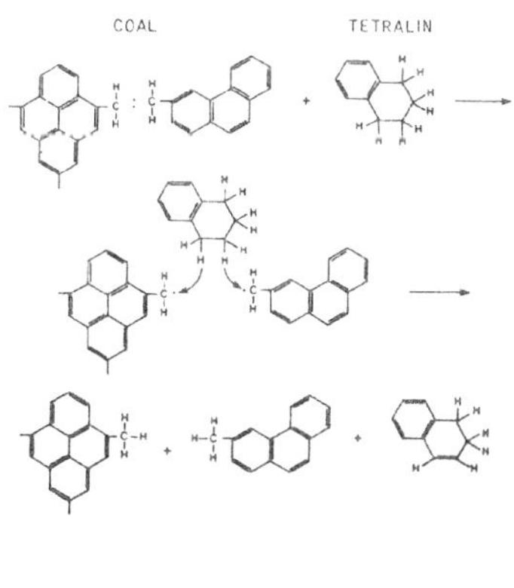

Figure 5: Proposed chemistry of hydrogen donation.

properties of macromolecules from coal in various solvents as well as the colloidal nature of the solutions would be helpful. The phase behavior at high temperatures and under high hydrogen pressures should also be determined to help understand the liquefaction process.

A better understanding of the kinetics and mechanism of the first step - thermal cleavage of bonds - is critical to improving liquefaction processes. It is known that these initial reactions are rapid, as coals can be rendered soluble in a few minutes at temperatures of 400°C.[39] Questions such as which bonds break first and how can these reactions be controlled need to be answered.

Catalysis of hydroliquefaction has received much attention. More research is needed to develop catalysts which will positively affect the initial coal conversion. It may be relatively easy to affect the course of reactions after the primary products are out of the coal particle. However, by this time the product distribution may already be determined. If a catalyst could be found which could influence the product distribution of the primary products as they are formed, entirely different types and quantities of products might be formed.

The question of catalysis by inherent mineral matter is still open. There have been reports of catalysis by the sulfur containing minerals pyrite and pyrrhotite.[40] However, there is other evidence[41,42] to indicate that H_2S derived from these minerals is acting as the catalyst. Obvious processing consequences could result by resolution of this question.

HYDROCARBON SYNTHESIS

Another route for making clean fuels and chemicals from coal involves gasification with steam to make carbon monoxide and hydrogen (synthesis gas) followed by catalytic conversion of the synthesis gas to methane (synthetic natural gas, SNG) or higher hydrocarbons or oxygenates. This sequence can be viewed as breaking down the coal structure to its simplest carbon and hydrogen components and then rebuilding or reassembling these elements into useful hydrocarbons. Although not fossil fuels in the strict sense, they are made from fossil fuels. Research opportunities in this area are numerous.

Hydrocarbon synthesis from CO and H_2 is practiced today on a relatively small commercial scale in South Africa.[43] The chemistry upon which these processes are based was discovered by Fischer and Tropsch in 1925.[44] The key issue is hydrocarbon product selectivity. Traditional catalysts appear to be able to make products containing one carbon selectively (e.g. methane, or methanol) but above one carbon, a broad distribution of products results with poor selectivity to any given compound. The challenge is to discover and develop heterogeneous or homogeneous catalysts that will selectively convert synthesis gas to desired chemical or fuel products.

The elegant studies of Pettit et al. (45) have shown that for the catalysts and systems they studied, the mechanism of hydrocarbon synthesis involves polymerization of $-CH_2-$ groups on the surface and ruled out carbon monoxide insertion reactions. However, much more work is needed to define the surface properties of catalysts which allow this behavior to take place. The modern tools of surface science should be applied to gain an understanding of surface properties of conventional and new catalysts. This understanding could provide the ability to tailor catalysts for preparation of specific molecules or product distributions.

CHALLENGES IN SHALE

Oil shale, like coal, is a rock containing organic material which is potentially convertible into usable liquid products (46). The oil shale rock contains in general from about 5-30% organic material (kerogen) compared with coal which is, in general 90% organic. The organic matter in oil shale has a much higher hydrogen to carbon atomic ratio, that is, it is richer in hydrogen, than is coal. Consequently, liquid products derived from oil shale would be expected to be more like petroleum, and would be more easily refined and utilized in today's society. Conventional processes recover oil from oil shale by retorting, i.e. by pyrolysis of the rock to depolymerize the organic material into volatile liquids. However, retorting has significant drawbacks including thermal inefficiency, incomplete organic conversion to liquids and large solids handling devices.

Fundamental understanding of chemical and physical structures of oil shale and their interactions will be crucial to finding better ways to utilize this resource. Many of the same opportunities described above for coal also apply to oil shale. In many ways the science of coal is more advanced than that for shale, and one obvious research direction would be to extend the same techniques and approaches now being used to study coal to shale. Functional group distribution of the kerogen and its chemical and physical properties need to be identified, and since shale is mostly inorganic, the chemistry of the rock and its constituents and how the organic and inorganic portions interact need to be defined. This information will provide the critical foundation needed to improve current approaches to oil shale conversion and to devise better ones.

By their nature, fossil fuels are very complex materials and many different approaches are required to gain the scientific understanding needed. In general, these approaches will require researchers with multiple disciplines, including organic, inorganic, physical and analytical chemistry, solid state/condensed matter physics, geology and geochemistry and chemical and mechanical engineering. As with recent advances in other fields involving complex materials, the greatest chance of success will come as a result of interactions among people in these disciplines. Like the solid state chemistry of semiconductors in the 1950's and genetic biochemistry of the 1970's, the time for fossil fuels chemistry at the molecular level has arrived and awaits the contributions of the chemical community.

REFERENCES

1. Summaries of many of these coal conversion processes may be found in "Coal Handbook" (R. A. Meyers, Ed.) 1981, Marcel Dekker: New York.
2. Hirsch, R. L., Gallagher, J. E., Lessard, R. R., and Wesselhoft, R. D. 1982, Science 215, p. 121.
3. van Krevelen, D. W. 1961, "Coal," Elsevier: Amsterdam.
4. Given, P. H. 1960, Fuel 39, pp. 147-53.
5. Wiser, W. H. 1973, "Proceedings of the Electric Power Research Institute Conference on Coal Catalysis," Santa Monica, CA.
6. Solomon, P. R. 1981, "Am. Chem. Soc. Symposium Series" 10º, pp. 61-71; Spiro, C. L. and Kosky, P. G., 1982, Fuel 61, pp. 1080-1084.
7. Davidson, R. M. 1982, "Coal Science and Technology Vol. I" (M. L. Gorbaty, J. W. Larsen, I. Wender, Eds.) Academic: New York, pp. 83-160.
8. Liotta, R. 1981, J. Am. Chem. Soc. 103, pp. 1735-42.
9. Liotta, R. 1981, Fuel 60, pp. 453-454.
10. Sternberg, H. W., Raymond, R., and Schweighardt, F. K. 1975, Science 188, p. 49.
11. Atherton, L. F. and Kulik, C. J., EPRI Journal 1982 (May), pp. 31-34.
12. Retkofsky, H. L. 1982, in "Coal Science and Technology, Vol. I" (M. L. Gorbaty, J. W. Larsen, I. Wender, eds.) Academic: New York, pp. 43-82; Silbernagel, B. G., Gebhard, L. A., Dyrkacz, G. R., Am. Chem. Soc. Div. Fuel Chem. Prepr. 28, pp. 115-124.
13. Bartuska, V. J., Maciel, G. E., Schaefer, J., and Stejshal, E. O. 1977, Fuel 56, pp. 354-58.
14. Gerstein, B. C., Chow, C., Pembleton, R. G., and Wilson, R. C. 1977, J. Phys. Chem. 81, pp. 565-70; Stejskal, E. O., Schaefer, J., and Waugh, J. S. 1977, J. Mag. Resonance 28, pp. 105-112.

15. Livingston, R. and Zeldes, H. 1981, Rev. Sci. Instrum. 52, pp. 1352-7.
16. Petrakis, L., Grandy, D. W. and Jones, G. L. 1983, Fuel, 62, 665-689 and references cited therein.
17. O'Gorman, J. V. and Walker, P. L. 1971, Fuel 50, pp. 135-51.
18. Johnson, J. L. 1975, Am. Chem. Soc. Div. Fuel Chem. Prepr. 20, p. 85.
19. Granoff, B., Thomas, M. G., Baca, P. M., and Noles, G. T. 1978, Am. Chem. Soc. Div. Fuel Chem. Prepr. 23, pp. 23-29.
20. Gluskoter, H. J. 1965, Fuel 44, pp. 285-91.
21. Silbernagel, B. G., Ebert, L. B., Schlosberg, R. H. and Long, R. B. 1981, "Am. Chem. Soc. Adv. in Chem. Series v. 192," (M. L. Gorbaty and K. Ouchi, Eds.) pp. 23-35.
22. Gan, H., Nandi, S. P., and Walker, P. L. 1972, Fuel 51, p. 272.
23. Grimes, W. 1982, in "Coal Science and Technology, Vol. I" (M. L. Gorbaty, J. W. Larsen, I. Wender, Eds.) Academic: New York, pp. 21-42.
24. Gorbaty, M. L. 1978, Fuel 57, 796-797.
25. Brenner, D. 1982, Am. Chem. Soc. Div. Fuel Chem. Prepr. 27, pp. 244-253.
26. "Am. Chem. Soc. Adv. in Chem., V 179, 1979 (M. L. Gorbaty, B. M. Harney, Eds.).
27. Aczel, T., Williams, R. B., Brown, R. A., and Pancirov, R. J. 1978, in "Analytical Methods for Coal and Coal Products, Vol. I" (C. Karr, Ed.) Academic: New York, pp. 499-540.
28. Scheppele, S. E. 1978, "Characterization of Coal Derived Liquids and Other Fossil Fuel Related Materials Employing Mass Spectroscopy" Department of Energy: Washington, Report No. FE-2537-7; Anbar, M., Kempster, E., Scolnick, M. E., and St. John, G. 1976, in Proceedings of the 24th Annual Conference on Mass Spectroscopy and Allied Topics 1976 (American Society of Mass Spectrometry, San Diego, CA) pp. 467-69.
29. Gould, K. A., Gorbaty, M. L., Miller, J. 1978, Fuel 57, p. 510.
30. Excellent reviews of the early history of coal utilization may be found in "Chemistry of Coal Utilization" (H. H. Lowry, Ed.) Wiley: New York 1945, and Elliott, M. A. 1974, Am. Chem. Soc. Div. Fuel Chem. Prepn. 19, p. 140.
31. Gorbaty, M. L., Wright, F. J., Lyon, R. K., Long, R. B., Schlosberg, R. H., Baset, Z., Liotta, R., Silbernagel, B. G., and Neskora, D. R. 1979, Science, 206, pp. 1029-34; "The Department of Energy: Some Aspects of Basic Research in the Chemical Sciences" 1979, National Academy of Sciences, pp. 11-31.
32. Fontana, F. 1777, Mem. Mat. F. S. Soc. Ital. 1, p. 679.
33. Sabatier, P. and Senderens, J. B. 1905, Ann. Chim. Phys. 4, p. 319.

34. Tessi du Montay, C. M. and Marechal, C. R., British Patent 2548 1867; Taylor, H. S. and Neville, H. A., 1921, J. Am. Chem. Soc. 43, p. 2055.
35. Smith, R. N. 1959, Q. Review 13, p. 287.
36. Baker, R. T. K. 1979, Catal. Rev. Sci. Eng. 19, pp. 161-209.
37. Berthelot, M. P. E. 1869, Bull. Chim. Soc. Fr. 11, p. 278.
38. Bergius, F. 1913, Deutsches-Reichs-Patent 201231; 1934 Proc. World Pet. Conf. F, 282.
39. Neavel, R. C. 1976, Fuel 55, p. 237.
40. Montano, P. A. and Bommannavar, A. S. 1982, Fuel 61, p. 523.
41. Lambert, J. M. 1982, Fuel 61, pp. 777-8.
42. Baldwin, R. M. and Vinciguerra, S., 1983, Fuel 62, pp. 498-501.
43. Hoogendoorn, J. C. 1978, in "Symposium Papers, Clean Fuels from Coal, "Institute of Gas Technology: Chicago, pp. 353-66; Dry, M. E. and Hoogendoorn, J. C., 1981, Catal. Rev. Sci. Eng. 23, pp. 265-278.
44. Fischer, F., and Tropsch, H. 1925, German Patent 484,337; Storch, H. H., Golumbic, N. and Anderson, R. B. 1951, The Fischer Tropsch and Related Synthesis Wiley: New York.
45. Brady, R. C. and Pettit, R. 1980, J. Am. Chem. Soc. 102, p. 6181; 1981 J. Am. Chem. Soc. 103, pp. 1287-1289.
46. "Oil Shale," (Yen, T. F. and Chilingarian, G. V., Eds.) 1976, Elsevier: Amsterdam.

OIL SHALES
What they are, and what are their resources

R. Pelet

Institut Français du Pétrole

1. Oil shale is not a concept of geology, but of economics. In their natural state, they are often not shales and they contain very little oil. They are rocks which, upon retorting, generate sufficient quantities of an oil of acceptable quality. "Sufficient" and "acceptable" depend entirely on the price and availability of petroleum on a given market at a given moment. Due to these limitations, oil shale exploitations flourished until the years'50, when they steadily decreased ; it does not seem that the raise in actual value of petroleum prices since 1973 reversed the tendency. Oil shale must contain high quantities of organic matter, itself yielding high quantities of good quality oil. It results from those considerations that algal material (kerogen type I) generated in lacustrine environments makes up the bulk of oil shales.

2. Due to geological considerations, all the stratigraphic periods are not equally favorable to the deposition of oil shales. The principal industrial accumulations are to be found :
- in Estonia (USSR) for the Devonian kukkersite
- in Canada (New Brunswick) for the mississipian Albertite
- in Brazil (Southern states) for the permian Irati Shales
- in USA (Central Rockies) for the eocene Green River shales.
Other accumulations, more important from an historical than an industrial standpoint in the present day world are also described.

3. Coorongite and balkhashite, which forms today in arid lakes or lagoons, are exact modern counterparts of some ancient oil shales.

L. Petrakis and J. P. Fraissard (eds.), Magnetic Resonance. Introduction, Advanced Topics and Applications to Fossil Energy, 219–233.

1. WHAT ARE OIL SHALES

The term oil shale is quite unadequate, since the rocks so called are generally not shales and do not contain any oil. "Oil shales" are any type of rocks which, upon retorting, yield sufficient quantities of a product resembling petroleum. The threshold which quantifies the term "sufficient" is economy dependent, and consequently vary with economic conditions. At present time, this threshold is of the order of 100 l/t (100 liters of oil per ton of rock) but, even at a given moment, other factors can play a more determinant role in the total economics of the recovery process, and influence the value of the threshold. Shale oil, compared to petroleum, is always for the refiner a material of poor quality ; the worst of them can be, at some periods, simply refused by the refineries. The recovery by mining and the disposal of spent shale can create acute environmental problems. Finally, retorting consumes water which must consequently be available in great quantities. All these problems, and others, will be dealt with in the "Recovery" section.

Although oil shales show very diverse mineralogies, they have however some features in common, due to the requirements on the high quantity of high quality products they must yield. First of all, they must be organic matter rich. Second, their OM must belong to type I or to the best portion of type II kerogens in order to yield high quantities of hydrocarbons. Lithologically, they are generally fine grain laminated rocks, since there is an undirect relationship between organic richness,small size of mineral particles and lamination for type I and II kerogens. A favorite, although not exclusive, environment for oil shale deposition is lakes. When a sufficient input, or initial stock, of nutrients (essentially nitrates and phosphates) exists in a lake, a high development of algal plankton takes place in the water surface. Due to the small size and the shallow depth, relative to the sea, of a lake, a thermal stratification can be easily maintained ; the high surface productivity results in a high OM sedimentation rate which, joined to the stratification, obtains rapidly the total anoxy of the bottom of the lake ; benthic life is then completely destroyed ; even the swimming forms (nekton) are intoxicated by toxic substances (e.g. H_2S) diffusing from the anoxic sink, resulting in a drastic decrease, if not a complete disappearance, of animal life, which itself decreases the fraction of algal material biologically recycled, etc... This process, called eutrophication, is limited only by the rate of turnover of the plankton, resulting in a preservation rate of OM in the sedimentary record exceptionnally high. This sketch explains why lakes can be the source of sediments exceptionnally organic rich ; it does not explain why, furthermore, the OM is of high hydrocarbons content (type I) ; this is only a fact of observation : fresh or brackish water algae are,

TABLE 1

OIL SHALES MINERALOGY

	SILICEOUS MINERALS (Opal, Quartz, Feldspars)	CLAY MINERALS	CARBONATES (Calcite, dolomite)	OTHERS
TASMANITE (Tasmania)	62%	24%	–	14%
TORBANITE (South Africa)	55%	30%	2%	13%
CANNEL COAL (Scotland)	28%	46%	3%	23%
KUKERSITE (Estonia)	15%	14%	65%	6%
GREEN RIVER SHALES (in part)	15%	36%	48%	1%
GREEN RIVER SHALES (Highly Saline member)		35% Dawsonite	65% Nahcolite	

TABLE 2

ELEMENTAL COMPOSITION OF THE KEROGEN FROM VARIOUS OIL SHALES

		C	H	O	N	S
PICEANCE (COLORADO)	EOCENE	76.5	10.0	10.3	0.6	2.6
UINTA (UTAH)	EOCENE	82.7	4.1	8.3	1.7	3.2
UINTA (UTAH)	EOCENE	81	9	6	2	2
DOUALA	L-CRETACEOUS	72.7	6.0	19.0	2.3	–
ENGLAND	KIMMERIDGE	68.5	7.3	14.5	1.2	8.5
PARIS BASIN	L-TOARCIAN	85.4	7.1	5.0	2.3	0.2
NEW SOUTH WALES	PERMIAN	84	10.5	3.5	1.0	1.0
TASMANIA	PERMIAN	72.5	9.1	10.5	1.1	3.0
S-AFRICA	KARROO	80	8	9	1.8	1.2
CANNEL COAL	CARBONIFEROUS	74.6	4.5	13.2	4.5	1.2
ESTONIA	ORDOVICIAN	66.4	7.1	18.0	0.7	1.1
ESTONIA	ORDOVICIAN	76	9	13	0.2	1.8

TABLE 3

SOME INTERESTING PROPERTIES OF VARIOUS OIL SHALES

	OIL CONTENT (l/t)	SPECIFIC GRAVITY (SHALE)	HEATING VALUE (MJ/kg)	ORGANIC CARBON wt%	ASH wt%
AUSTRALIA	346	1.60	19.0	40	52
THAILAND	298	1.61	15.5	31	56
NEW ZEALAND	276	1.46	21.4	46	33
NOVA SCOTIA	214		12.7	26	62
SPAIN	196	1.80	12.6	26	63
S-AFRICA	190	1.58	19.3	44	43
BRAZIL	119	1.70	8.2	17	77
FRANCE	108	2.03	8.9	19	71
COLORADO	102	2.23	5.1	11	67
SCOTLAND	93	2.22	5.9	12	78
SWEDEN	58	2.09	9.1	19	72
MANDCHURIA	32	2.29	3.4	8	83

on the whole, richer in hydrocarbons structures than their marine counterparts. This coincidence explains why - when coals are omitted - there is a direct relationship between the OM content of a rock and the H/C value of its kerogen.

The preceding considerations explain also the peculiar mineralogical variety of oil shales. The necessity of high supplies of nutritive salts implies an erosional stage advanced at least to the beginning of "chemical erosion", i.e. when topography is eroded to the point when water can percolate slowly along the profiles, dissolving high quantities of ions, but carrying a limited quantity of solids in the colloidal range. Hence the clayey to carbonate to other chemicals dominated mineralogy. Examples are given in table 1. From an essentially siliceous composition, at a rather early stage of erosion, such as the south-african sample one comes to the purely evaporitic composition, characteristic of the chemical stage, of the Green River sample in the dawsonite-nahcolite zone. Note that the economics of Green River shales exploitation took into consideration, for some designs, the recovery of these chemicals (dawsonite is an aluminium ore, less easy to process than bauxite, but much easier than the clay minerals).

Table 2 gives some examples of chemical composition of the kerogen of various oil shales. As, once again, the definition is only a matter of economics, some easily exploitable and processible shales may have a rather low H/C, or OM content (but not both). Despite high OM content, coal can hardly be used as an oil shale, due to its low yield of liquid products, reflected by a low H/C. However, this is true only of usual, humic coals ; "cannel coals", which are an accumulation of higher vegetals spores, are historically among the first "oil shales" exploited. Incidentally, higher vegetals spores are functionnaly and, as a consequence, chemically, similar to unicellular algae, whose only role is to propagate the species. In usual coals, spores play only a subordinate role.

Table 3 gives some interesting parameters relative to "whole" oil shales. The relationship between oil yield, specific gravity, heating value and organic carbon is clear, but far from a single curve, due to influence of the chemical composition of OM. However, for a given formation, this composition is either constant or varies accordingly to another parameter, e.g. organic carbon content, resulting in relationships which can be perfectly used for quantitative predictions. The measurement of specific gravity, which is easy to perform even on the mining sites, is a very convenient substitute for the OM content.

From what precedes,it seems that no particular period of the Earth history appears as especially suitable for oil shale generation. However, the type of aquatic environment - shallow lakes or lagoons of restricted area - favorable to oil shales deposition is more frequently found in those times when sea expands largely on continental shelves. As an example, the retreat of ice after the gondwanan permocarboniferous glaciation led to the invasion by water (fresh or marine, such as in present day Scandinavia) of the depressed (by the weight of the icecaps) continents, resulting in environments depositing Irati shales in Brazil, Ermelo shales in South Africa, and others of lesser importance else here in Gondwana . Incidentally, in other parts of Gondwana , oil shales are replaced by coals : the same geometric patterns of small shallow continental aquatoria is equally favourable to the deposition of coal. The pertinent factor is the availability of lignin-rich vegetation (i.e. other than algae). In that case, humic compounds of phenolic origin appear in the water, which are toxic for almost all the living beings, prevent algal development in the water body, thus permitting coal sedimentation. Once again, different and independent factors can obtain similar results : the process described above is triggered by an abundant supply of vegetal debris in the water body. During times of very restricted erosion, only some leaves, spores and pollens will be delivered to the lake, thus not preventing algal development on the other hand enhanced by the chemicals in solution. On the contrary, in times of moderate erosion, the quantity of organic lignin -rich debris which can reach the lake and accumulate may be sufficient to initiate the process described. If erosion is too important, the vegetal biomass available decreases due to the temporal instability of soils.

All those considerations apply even at a very local scale ; in the Coal Measures of Scotland, for instance, the most frequent facies is usual, humic coal ; but all the intermediates exist with cannel coal we described earlier and with bogheads, which are algae dominated ; these two latter fuels have been exploited per se in the past.

2. WHERE OIL SHALES ARE FOUND

In the present day world, the exploitation of oil shales has greatly decreased compared to what occurred at the beginning of the century. The continuous decrease of the petroleum absolute price during more than 100 years (up to 1973) led to the progressive abandonment of almost all of the exploitations in developed countries (the main exception being the estonian kukersite). The sawtooth movement of oil prices since 1973 led to some reevaluation of the economics ; as a result, small scale pilot plants opened in the Green River area in USA and a big scale pilot plant operated in Irati shales in Brazil. But nowadays, USA announced

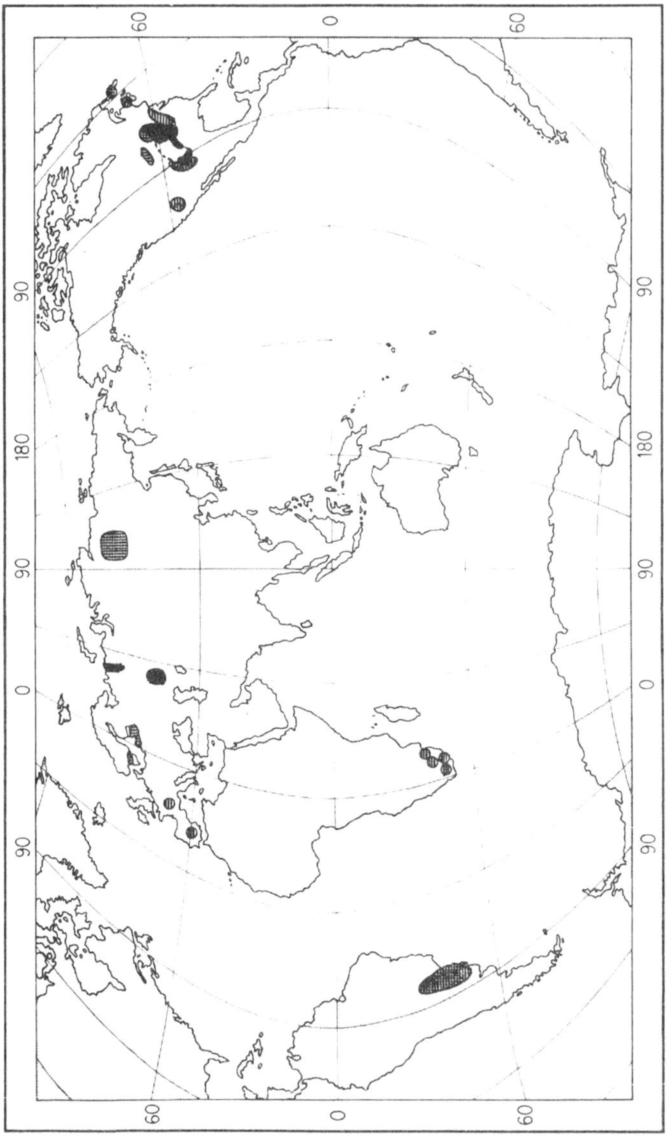
OIL SHALES WORLD RESOURCES
PALAEOZOIC
60
0
60
90
180
90
0
90
LOWER
MIDDLE
UPPER

Figure 1.

the abandonment of any industrial exploitation in the near future. The following considerations are consequently half-historical and half-prospective. They are half-historical, because they will describe some occurrences which are now uneconomical, and shall certainly remain such in the future, but were important in the past. They are half-prospective, because the other occurrences described are those with easily exploitable reserves above 10^9t of shale or above 10^8 t of producible oil. We shall follow the stratigraphic scale.

TABLE 4

PRINCIPAL OIL SHALES OCCURRENCES IN U.S.A.

SYSTEM	FORMATION	OCCURRENCE
MIOCENE	MONTEREY	CALIFORNIA
EOCENE	GREEN RIVER	COLORADO, UTAH, WYOMING
JURASSIC		ALASKA
TRIASSIC		ALASKA
PERMIAN	PHOSPHORIA	MONTANA
PENNSYLVANIAN	CANNEL COAL	EASTERN STATES
MISSISSIPPIAN	CHATTANOOGA	CENTRAL AND EASTERN STATES
DEVONIAN	CHATTANOOGA	CENTRAL AND EASTERN STATES
ORDOVICIAN	MAQUOKETA	APPALACHIAN BASIN
	UTICA	APPALACHIAN BASIN
PRECAMBRIAN	NONESUCH	MICHIGAN, WISCONSIN

2.1. Palaeozoic

In lower Palaeozoic, noticeable resources are situated in USSR (fig. 1). In Siberia, the Olenek basin contains shale layers up to 15 m of total thickness, on 150 000 km^2 in the lower to middle Cambrian . They are low grade shales with reserves estimated at 2.10^9 t of shale.

In Ordovician of the Baltic shield , lower ordovician Dictyonema shales are poor graded and not exploited ; on the contrary, middle ordovician Kukkersite is the most important industrial exploitation in the world at present day. In Estonia (fig. 2) up to 30 layers may represent a total thickness of 15 m. The exploitable area extends on some 3500 km^2, with reserves of 3,5.10^9 t of oil ; the shales are high-grade, from 150 to 280 l/t. The rock is largely carbonated, the spent shale is used to manufacture rather low quality concrete blocks, which is anyway a very smart use of what is usually an annoying refuse.

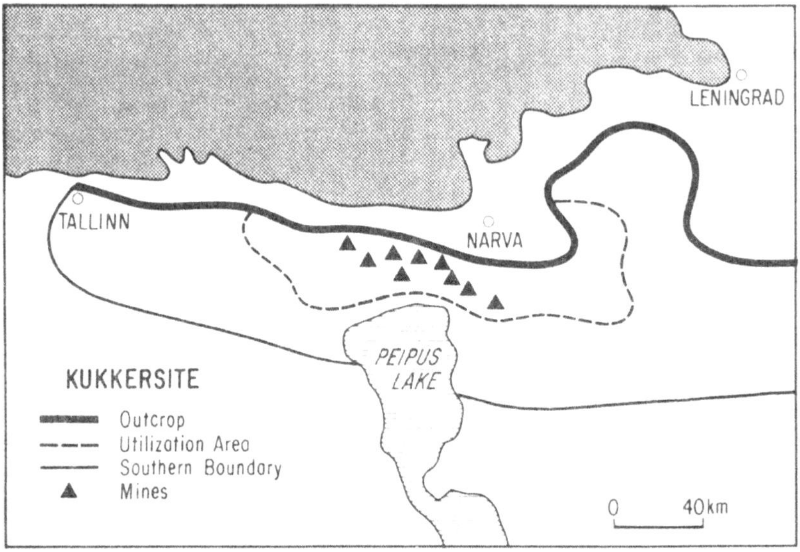

Figure 2

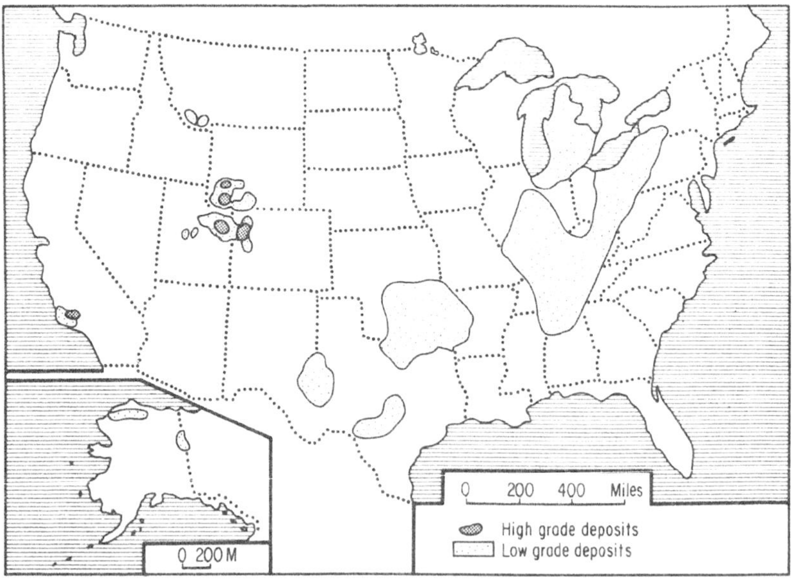

Figure 3

An occurrence of historical interest is represented by the swedish alum shales. These cambrian shales, which contain alunite (a complex Al, K and Na Sulfate) are low-grade (40 to 80 l/t) with reserves estimated at 4.10^8 t of oil and were mined until 1965.

The main resources in middle palaeozoic (fig. 1) are in USA and Canada. A lot of separate formations, from the devonian to the mississipian extends on the appalachian trend from Canadian border and great lakes to Texas, on 650 000 km^2 (Chattanooga, New Albany, Antrim , Ohio, Sunbury, Marcellus shales, and others, fig. 3). They are low grade, with estimated reserves about 8.10^9 t of oil, but not easily recoverable. There is no planned attempt to exploit them to date, they must be considered as a possibility in remote future. In Canada, lower Mississipian "albertite shales" of New Brunswick are known since 1815. They are medium grade, with reserves about 5.10^9 t of oil.

In Scotland, lower carboniferous of oil shale group formation was exploited as soon as 1851. These medium to high grade (65 to 195 l/t) shales which underlie coal measures were among the first to be exploited in Europe, and were the subject of the first scientific consideration on oil shales.

In upper palaeozoic (fig. 1) the most important accumulation is in Brazil. The permian Irati shales (fig. 4) are low to medium grade (40 to 125 l/t), with thickness up to 12 m in two layers, extending on 25 000 km^2. An important pilot plant is operated at San Mateos do Sul, but recently the forecast extension to the industrial scale has been postponed, if not cancelled. Reserves are tremendous, up to 120.10^9 t of oil - but the postponed commercial exploitation was intended to recover only 0,1% of this figure.

The permian shales near Autun in France were exploited even before the Scottish carboniferous shales ; these low to medium grade (50 to 100 l/t) shales, intensively mined from 1838 to the end of the last century, were intermittently exploited till 1953, when they were abandoned, the possible reserves being almost completely exhausted. As the scottish shales, they stimulated the first scientific studies about oil shales geology, and especially how and why they differ from coal.

Australian torbanites from New South Wales belong to the same sedimentary system as Irati shales. These extremely high grade (up to 500 l/t) shales, known since 1802 were exported to Europe and USA until 1911. Oil resources amount to 3.10^9 t. Belonging also to the same sedimentary system are the tasmanites from Tasmania, with reserves up to 4.10^9 t of oil and the Ermelo shales from South Africa.

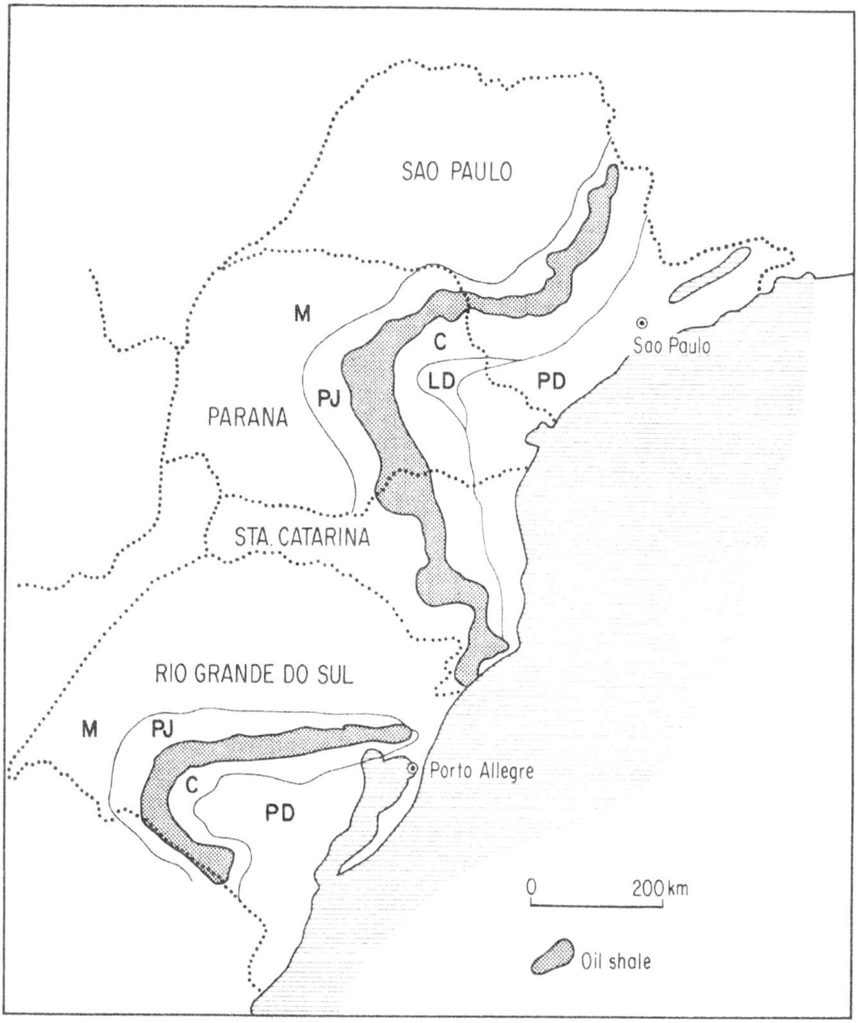
SAO PAULO
M
C
Sao Paulo
LD
PD
PJ
PARANA
STA. CATARINA
RIO GRANDE DO SUL
M
PJ
C
PD
Porto Allegre
0 200 km
Oil shale

Figure 4

2.2. Mesozoic

The triassic and cretaceous shales of Zaire are the most prominent in the mesozoic realm (fig.5). They are medium grade (70 to 150 l/t) and reserves are estimated to 16.10^9t of oil.

Other mesozoic occurences are of lesser industrial interest. In Europe, lower toarcian shales outcrop in many countries, but especially in France, Austria, West Germany and Luxembourg... They are low grade, below 60 l/t, and the extent of the best grade reserves are modest, certainly below 10^8t. They have a great historical interest, since they have been mined intermittently in the four countries cited above during the past century, and since, due to the energy hunger and search for self-sufficiency of European countries, there is always at work some planning, or intention of planning, for their exploitation in an ever remote future.

In Alaska, triassic to cretaceous shales outcrop in the Brooks Range on some 10 000 km^2. Details are scarce, but resources seem as considerable as remoteness of the site. Richnesses up to 520 l/t have been reported. In Canada, Cretaceous shales in the Saskatchewan, close to u.s. border are evaluated at 3.10^9t of oil "in place".

In USSR, low to high grades Jurassic and Cretaceous shales occur in various parts of the country : Volga-Ural basin, Kuibyshev region, Timan-Pechora basin. In Siberia, the famous Bazhenovo formation occupies more than 1,2 10^6 km^2. Resources have been tentatively estimated at 12.10^9t of oil, but this figure is almost meaningless, due to our ignorance of the actual conditions of recovery.

2.3. Tertiary

Two sites dominate the picture for tertiary oil shales : the celebrated Green River region in the USA, and the Fushun district of Manchuria (fig. 5).

The Green River region shows a complex organization (fig.6) both areally and stratigraphically. The eocene shales occupy about 65 000 km^2, but the thickest and richest shales are concentrated in much smaller areas, with an oil content from 100 to 400 l/t. The beds are thin, from 10 cm to 1,3 m, but numerous (up to 63), totalling on the average about 15 m. These lacustrine sediments show a highly variable and sometimes unusual mineralogy. Apart from dawsonite and nahcolite already encountered, some beds of low organic content are composed of minerals of interest : rock salt (halite), trona and shortite, which are complex sodium carbonates, phosphates, and uranium minerals. The reserves are the

OIL SHALES WORLD RESOURCES
MESOZOIC & TERTIARY

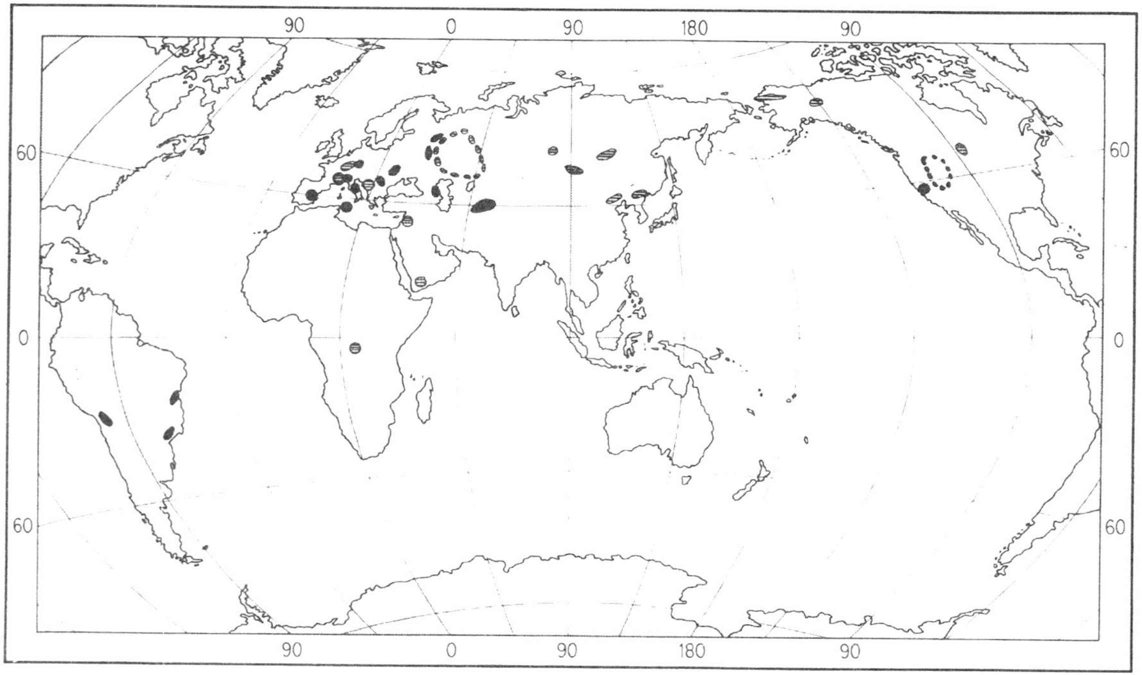

Figure 5. TERTIARY MESOZOIC

highest in the world since we are dealing with physically recoverable reserves, and not only with areal extension ; they amount to 300 10^9 t of oil. The richest basin is the Piceance Creek basin ; there, the Mahogany zone of the Parachute Creek member contains on 3000 km^2 25.10^9t of oil in shales yielding 100-125 l/t. The deepest cover of sterile shale (in fact, of poorer oil shale) is at most 70 m in the center of the basin.

The Fushun district of Manchuria - I have no map - is the best known of the Chinese occurrences of oil shale. These oligocene shales overlie coals, equally mined. Fushun shales are actively exploited since 1920. They are low grade (30 to 60 l/t) and the reserves are about 20.10^9 t of oil.

Other occurrences have chiefly an historical interest. In Europe, tertiary shales are found in Sicily (Miocene), Yugoslavia (Miocene), France (Miocene) and Germany. In this last country, the Eocene Messel shales, exploited from 1884 to 1962, are low to medium grade but with modest reserves. They are famous for the fine fossils found abundantly and preserved in excellent condition ; Propalaeotherium messelensis is an essential link in the evolutionnary chain which leads to horses. Another unusual occurence is near Ulm, where the crater (20 km diameter, 5 km depth) caused by the impact of a giant meteorite at the end of oligocene was occupied by a lake which deposited oil shales. They are medium grades, with reserves about 10^8t of oil.

3. PRESENT DAY FORMATION OF OIL SHALES

In periodically flooded rainlakes, or lagoons along the Australian south-western coast, in a flat and arid country, thrive algae and microorganisms ; especially the algal genus Botryococcus, which is a fresh-brackish water form, with a high content of lipids which can, in some instances not entirely elucidated, increase up to half the dry weight of the cells. These communities are for the most part killed during periods of dryness and fossilized ; thus the bottom of some of these lakes is covered by a material named coorongite, after the Coorong, which is one of this lagoons. Coorongite can be considered as a modern equivalent of oil shales. A very similar material, but less well known, named balkhashite, is found in some parts of the half-fresh, half-saline lake Balkhash in the USSR. Coorongites and balkhashites have properties very similar to torbanites or permian Autun shales ; this is no wonder since it has been proved that essentially the same environments and even the same genus of algae, Botryococcus, prevailed in the formation of these materials.

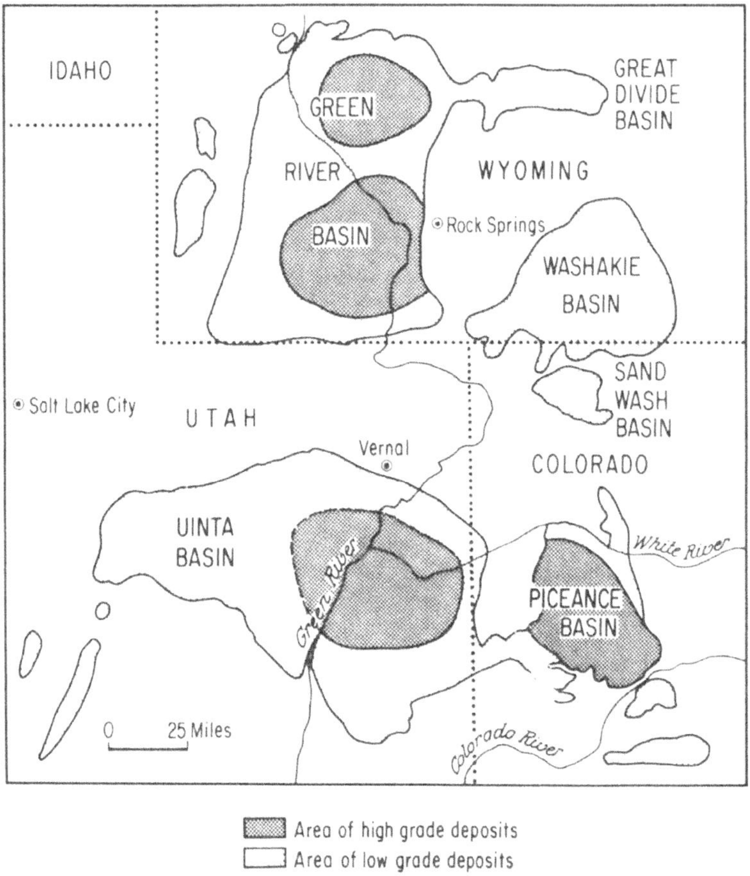
IDAHO
GREEN
RIVER
BASIN
GREAT
DIVIDE
BASIN
WYOMING
Rock Springs
WASHAKIE
BASIN
SAND
WASH
BASIN
Salt Lake City
UTAH
Vernal
COLORADO
UINTA
BASIN
Green River
White River
PICEANCE
BASIN
0 25 Miles
Colorado River
Area of high grade deposits
Area of low grade deposits

Figure 6

SUGGESTED READINGS

The literature on oil shales is very sparse ; nevertheless, the following readings can be suggested :

Duncan D.C., Swanson V.E., 1965
Organic rich shales of the United States and world land areas
U.S. Geol. Survey, circular n° 523.

Yen T.F., Chilingarian G.V., 1976
Oil shale
Elsevier publishing house, Amsterdam.

Rühl W., 1982
Tar (Extra Heavy Oil) sands and oil shales
Ferdinand Enke Publishers, Stuttgart.

and for people fluent in russian language :

Proceedings of the UNO symposium on the development and utilization of oil shale resources
Valgus publishing house, Tallinn 1970, E.F. Petukhov editor.

KEROGEN

A brief description of sedimentary organic matter and of its study by analytical methods other than NMR

R. Pelet and B. Durand

Institut Français du Pétrole

1. Kerogen is that fraction of the sedimentary organic matter (SOM) which is insoluble in organic solvents. It represents - except in petroleum reservoirs - from 70% up to 99% of the total SOM. Although its average content in sedimentary rocks is low - about 1% weight - it is a normal and universal component of the sedimentary record. It originates almost totally from the fossilization of former living beings, which are essentially plants, with the shares of marine planktonic algae and of continental higher vegetals of the same order of magnitude.

2. Except for coals,which are an extreme type of kerogen, kerogen cannot be recovered pure by physical means only. The chemical destruction of the mineral fraction of the rocks is an aggressive operation which modifies kerogen. However, except from recent and young sediments, this modification can be generally neglected.

3. Elemental analysis, when plotted on Van Krevelen diagram (atomic H/C versus atomic O/C) enables to recognize three types of compositional regularities :

- samples at the same geological evolutionary stage (e.g. recent sediments) occupy definite areas on the diagram, depending on the origin of the SOM. Three main types (I, II, III) are recognized

- for one kerogen type, points representative of increasing geological evolution (increasing age, increasing depth, i.e. temperature) follow a regular curve (evolution path)

- samples submitted to subaerial weathering follow a definite alteration path.

L. Petrakis and J. P. Fraissard (eds.), Magnetic Resonance. Introduction, Advanced Topics and Applications to Fossil Energy, 235–286.

4. Infrared spectrometry, electronic diffraction are powerful means to precise the chemical structure of kerogens. Optical examination enables to determine the origin of SOM (type of kerogen). Pyrolysis comes in two versions : one is a convenient, rapid and cheap method to characterize the degree of evolution and the type of the kerogen ; the other is a sophisticated technique which permits the elucidation of structural details. EPR and chemical methods are only of secondary utility. All these methods do not give a direct estimation of aromaticity - there enters NMR.

1. WHAT IS KEROGEN

Organic matter (OM), i.e. covalent compounds of reduced carbon, is a normal constituent of sedimentary rocks. It can be considered, almost in totality, as arising from the geological evolution of former living matter. Its composition is doubly complex, since it is a complex mixture of molecules of complex structure. Being a mixture, the first analytical operations attempted at are fractionations. Thus, geochemists now agree to call kerogen sensu stricto the fraction of sedimentary OM which is insoluble in the usual organic solvents, as opposed to the soluble fraction called bitumen. However we do not limit our use of the word to organic matter disseminated in ancient sediments as is the case of current uses of the term, and we apply it to all sedimentary OM. Thus we include in sedimentary OM the humic coals of various ranks (peat, lignite, bituminous coal, anthracite), boghead coal, cannel coal, asphaltoid substances (asphalts, bitumens, tar in tar sands), OM in recent sediments and soils.

The use of a single term for designating the insoluble fraction of sedimentary OM in general is justified by the progressive convergence now observed between the methods and philosophies of the various disciplines dealing with sedimentary OM. To an increasing degree a general conception of sedimentary OM is being developed which we are presenting here.

According to this view, the organic materials found in soils, recent and ancient sediments, coals, etc... are considered as special cases. Formation and evolutionary processes account for the transitions from one special case to the next. Thus the types of OM generating hydrocarbons appear as less and less contrasted with those which are not capable of doing so. There does not seem to be any criterion whether of a genetic or physico-chemical kind which seems to be sufficient to establish such sharply defined categories in spite of technical developments, and there are transitions between the various evolutionary stages of the OM and between the types of OM (We will, however, see that some types are better represented, and this phenomenon is the basis for classifications which are too sharply defined).

With this definition, kerogen represents always the bulk of sedimentary OM, up to 99% by weight in recent sediments, less in ancient ones but with a high dispersion of values due to the mobile nature of the remainder of sedimentary OM (oil and gas).

Historically speaking, the term "kerogen" was coined by the british geologist Crum-Brown in the beginning of this century to name the organic matter from a scottish oil shale which produced, upon distillation, highly waxy oils, hence the name from the greek (which generates wax). The term came rapidly in current use for the organic matter of all oil shales, but it was only in the fifties-sixties that it was applied to every sedimentary OM, with some uncertainties about its definition by different authors. It is clear that the definition adopted here makes coal a special variety of kerogen, but leaves apart petroleum (oil and gas).

Organic carbon is not the only form of carbon to be found in sedimentary rocks ; on the contrary (table 1) it represents roughly 15% of this carbon, the remainder being accounted for by carbonates. These two reservoirs are not independent ; consumption of atmospheric or dissolved CO_2 by plants on the one hand, dissociation or precipitation of carbonates controlled by CO_2 concentrations on the other hand, show that there are constant exchanges. But the very mechanisms of this regulation are only roughly known, since complex kinetic factors play the main role in determining the level of each carbon species at a given moment. For instance, the massive combustion of fossil fuels in the last decades led to a steady increase of atmospheric CO_2 concentration : obviously this combustion takes place at a rate faster than those of the controlling processes.

If now we compare, as in fig. 1, the quantities of the different varieties of kerogen, and the quantities of petroleum, oil and gas as well as asphaltic products (heavy oils, tar sands), it is clear that disseminated kerogen makes up more than 99,9% of the total. Fossil fuels appear as highly exceptional variation of kerogen. But care must be taken that the definition of oil shales is entirely a matter of economics ; the definition of coal and petroleum which may seem at first sight more science oriented is in fact largely controlled by economics : a coal veneer 10 cm thick will be called "coal" in the description of the section by the geologist, but none of the certainly numerous coal veneers 10 cm thick present in the sedimentary record are taken into account in the figure of fig. 1, because they are definitely unexploitable. Thus, we shall forget in the following sections all these subtilities and treat kerogen as a whole. However, due to the special importance of coal, which will be the subject of a special talk, we shall not deal with method or considerations applicable only to this material.

Table 2 gives an idea of the distribution of organic matter in the various types of sediments. This compilation is now outdated for what concerns oceanic sediments, where OM contents are definitely lower. What it shows is that fine grained sediments, either clays and shales or fine grained carbonates, co ntain about 90% of the sedimentary OM. This is to be kept in mind.

The distribution of organic matter according to stratigraphic age is more difficult to establish, since there are few studies dealing with a set of data of sufficient representativity to mean something. Fig. 2 gives two of these compilations, together with the distribution of petroleum production and reserves. It appears a sort of general increasing trend with decreasing age, with some oscillations, and clearer for petroleum than for OM. Several reasons can be imagined to account for this tendency, but none is absolutely convincing. The first to come into consideration is the progressive development of life on Earth, beginning with a sterile protoplanet up to the present explosion of living forms. But at the time scale of the Earth development, Cambrian is close to present days, and if it almost certain that life was less diversified in Cambrian than nowadays, there is no convincing argument to demonstrate, or even to suggest, that the total biomass was lower. The problem remains unsolved.

2. HOW TO ISOLATE KEROGEN

The definition set up in the preceding section defines a concept. How make this concept to become a piece of matter ? Obviously, since kerogen is insoluble in any solvent, the only solution is to destroy the mineral fraction of the rock and recover the kerogen by, so to speak difference. Now, this destruction of minerals requires a specific sequence of acidic attacks :

- HCl attack, in order to dissolve carbonates
- HF attack, in order to dissolve silicates and silica
- HCl rinse , to destroy any artefact formed in the previous steps (essentially calcium fluosilicates or aluminates)

After these treatments, the mineral fraction is usually drastically reduced. But some minerals are not destroyed by acidic attack : pyrite and the different forms of TiO_2, zircon are the most frequent. They need for their destruction oxidative or reducing treatments.

All these attacks can equally modify the OM of the rock. It has been empirically determined that, provided that only acidic and not oxidative or reducing attacks are used, kerogen from old rocks remains unchanged. This is not the case for recent sediments, which release during kerogen isolation from 10% up to 80% of their

OM into solution, by hydrolysis. Young sediments lie in between. The loss can be monitored by a material balance of organic carbon before and after attack, but this has two detrimental effects. The first, which adds to the analytical burden , is the necessity of establishing carbon balances. The second, much more boring, is that it is extremely difficult, if not impossible, to separate and study the kerogen of recent sediments, and that some difficulty remains with young rocks.

As it is hardly possible to use oxidative or reducing treatments, it is extremely difficult to get rid of the residual minerals. The most unwelcome is pyrite, which causes errors in elemental analysis, either directly (sulfur) or indirectly (oxygen). It is always systematically attempted to separate pyrite by densimetry in liquids of appropriate density ($ZnCl_2$ aq, or $CH_2 Cl_2$) but the results are never warranted. It is a safe practice to discard kerogens with pyrite content higher than 30% by weight.

The complete operation is long, tedious and expensive. Attempts to improve convenience and efficiency have been done. In IFP, we use an apparatus as in fig. 3, which avoids intermediate manipulations of the rock under attack between the various steps, and which operates semi-automatically, permitting the operation of 12 to 24 cells by a single technician. One cycle lasts 72 h. After isolation, kerogens for optical examination must be kept in suspension in water to avoid physical irreversible transformations (shrinkage, blackening).

3. ELEMENTAL ANALYSIS OF KEROGEN

3.1. Introduction

Elemental analysis has been historically the first method used in kerogen analysis. More precisely, the first analyses were made on more or less pure OM, such as coals, peats, asphalts, extremely organic-rich oil shales etc..., to bypass the difficulties of kerogen isolation. Van Krevelen - a dutch coal scientist - proposed a representation known as Van Krevelen diagram, where the ordinates display the atomic ratio H/C and the abscissas the atomic ratio O/C. Sulfur, and less importantly nitrogen, are absent from that representation, which is however the most condensed and convenient of all the diagrams proposed before, and after it. Fig. 4 shows the main properties of Van Krevelen diagrams.

We shall comment only briefly on the analytical techniques. Suffice it to say that carbon and hydrogen are measured with good accuracy and suffer a minimum of interferences with the remaining minerals in the kerogen. Nitrogen, due to its low content and the stability of its compounds in kerogen, is not measured with good accuracy. The measurement of oxygen is, by itself, more delicate and is somewhat influenced by the presence of pyrite. The case is worsefor sulfur, whose determination becomes almost meaningless when pyrite is present. In that case, Fe is measured, and supposed to be present only as FeS_2 ; hence a correction on sulfur.But when corrections become much more important than corrected values, the results are never warranted. These considerations explain also why Van Krevelen diagram, which omits sulfur, is such a good representation.

Some people report oxygen values "by difference" ; i.e. they substract C, H, N, S, from 100%. When sulfur and nitrogen values are not close to 0, these oxygen values "by difference" are almost meaningless. They must be avoided in good scientific practice.

3.2. Regularities of kerogen elemental composition

A compilation of values taken from the literature, such as in fig. 5 is at first sight discouraging. Nearly all the field of physically possible values for H/C and O/C is covered. However, even at this stage, it appears that the density of coverage is not the same anywhere.

The fundamental double rationalization appears when are plotted separ ately, as in fig. 6, kerogens belonging to a geologically homogeneous facies : Green River shales for what is called type I, Western Europe lower Toarcian shales for what is called type II, humic coals and Cameroon upper Cretaceous for what is called type III. Not only the points cover three separate ribbons in the plane, but the order along each ribbon is not at random : from the right to the left, and from the top to the bottom, samples come from deeper and deeper wells for the sediments, or increase in age for the coals. We encounter here for the first time a fundamental process : the geological evolution.

Almost all of the components of a sediment freshly deposited are out of thermodynamic equilibrium. This is especially true for the organic compounds. However, it is only below the sedimentary column, in the zones of metamorphism, that a tendency towards equilibrium assemblages appear clearly. In the sedimentary column, minerals, but especially organics, undergo complex transformations controlled by kinetic factors. These factors are highly dependent on temperature ; as temperature in the subsurface increases regularly (35°C per 1000 m, on the average), it is equivalent to say that kinetics factors depend on depth. There appears the two main factors controlling geological evolution : time - i.e. stratigraphic age - and temperature - i.e. depth. These two factors do not play a symmetrical role : evolution is proportional to time, whereas temperature enters in exponential factors. The result is that the so-called "compensation" of time by temperature is of intrinsic complex nature ; as an example two similar formations of the same age and at same depths in present time may reveal quite different evolution degree depending on the particular geological history of each basin, i.e. the type of dependence of temperature upon time which is the consequence of this history.

Now the three kerogen types distinguished in fig. 6 were not chosen at random. Type I represents the most hydrogenated kerogens existing in the sedimentary record, type III representing on the contrary the less hydrogenated. Type II represents the commonest type found in typical marine shales, i.e. fine-grained silicate rocks deposited in oceanic basins, with an OM largely of marine planktonic origin. Type III is also a quite common type ; usual (humic) coals belong to this type, as well as the dispersed organic matter largely of continental, higher vegetals, origin. Type I is less frequent, and is frequently associated with algal material from lakes or lagoons.

A second type of orderly evolution obtains when are plotted points of increasing secondary surface alteration, as in fig. 7. The trend is now almost a reflexion on a vertical mirror of the preceding geological evolution trend : as alteration increases, points move from the left to the right and from the top to the bottom of the Van Krevelen diagram.

These considerations enable to interpret apparently ambiguous cases such as those of fig. 8. Points in region D must result from alteration, since they lie below the type III region, which is the limiting region for unaltered material. Points in region B are more difficult to interpret. Three different interpretations are a priori possible :

- we are dealing with a special type of OM
- there is a mixture of planktonic and higher plants OM
- this is altered type II OM

The geological context permits generally to conclude. In the case of fig. 8, the conclusion was we had a mixture of type II and III materials (the corresponding samples were from DSDP drillings in the North Atlantic).

3.3. Petroleum genesis

Finally, it is possible to synthetize all the preceding considerations into the semi-theoretical diagram of fig. 9, which shows evolutionary pathways for the three principal kerogen types. The graduation in Ro represents vitrinite reflectances, or equivalent vitrinite reflectances. This parameter, which will be examined in the following, is a measure of geological evolution originally used with coals and which, for this particular kerogen, proved quite successful.

We must now say something about petroleum (oil and gas) generation. Petroleum (together with CO_2 and water) can be considered as a by-product of the geological evolution of the kerogen. It is obvious from the Van Krevelen diagram that in the first steps of evolution, before the value Ro = 0.5, the by-products must be oxygen-rich. This stage is called diagenesis, and the main by products are CO_2 and water. Later, between Ro = 0.5 and Ro = 2.0, comes the time of catagenesis, where the principal by-products are hydrocarbons,chiefly oil before Ro = 1.0, chiefly gas after Ro = 1.0. Beyond Ro = 2.0 is the metagenesis stage, preceding metamorphism. In that stage, kerogen yields but small quantities of gas, chiefly methane. At that stage, incidentally, oil which has not been preserved of further evolution by migration to upwards reservoirs, is itself cracked to gas (and carbonaceous residue) such as de picted in fig. 10, where we can see on the left what is generally found in rocks at varying degrees of evolution, and on the right what would have been present if no migration had occurred.

Finally, we are able to reestablish the unicity of OM : the solvent extractable fraction, which was separated from kerogen for analytical reasons, appear, in its major part, as a by-product of the geological evolution of kerogen itself (a subordinate fraction may originate directly from formely living beings). Petroleum demands a moderate to strong evolution of sedimentary OM; coal quality will be the better the stronger the evolution (anthracite) ; on the contrary oil shale quality will be the better the weaker the evolution (high oil yield).

4. THE INFRARED SPECTROSCOPY

4.1. General

In its principle, IR spectroscopy is able to give a more detailed picture of the functional groups making up a kerogen molecule than elemental analysis. But experience shows that some limitations to the use of IR spectroscopy appear rapidly. Fig. 11 gives an example of a typical IR spectrum of kerogen. Apart from the rather good peaks in the region 2840-2960 cm-1, the rest of the spectrum shows more or less well defined bands, not to say humps. This poor definition is inherent to the kerogen itself, since it persists in spectra recorded at liquid nitrogen temperatures, or on interferometer instruments (FT/IR spectrometers). It is due to the fact that each functional group, which in a simple molecule would result in a sharp peak because of the unicity of its environment, is no more in that case ; due to the size and the complexity of the kerogen molecule, each identical group is situated in a particular environment, which causes a definite small shift of the vibration frequency relative to its standard central frequency. The result is a broadening of the peak, more important for groups capable of strong interactions (e.g. oxygen containing groups) than for groups with weak affinities (e.g. alkanes groups). Hence the appearance of the spectra, with hydrocarbon regions rather well defined, and on the contrary regions corresponding to reactive functional groups showing broad bands.

The principal attributions are as follows :

.Hydrocarbons (C-H) bands :
Saturates : several bands between 2840 - 2960 cm -1
one band about 1450 cm-1
one band about 1375 cm-1

Aromatics : one band about 3050 cm-1
several bands between 750 - 870 cm-1

.Carbon-oxygen bands :
C = 0 bands : ketones, acids, esters, one band from 1700 to 1750 cm-1
quinones one band about 1620 cm-1

C - 0 bands : several broad bands between 1050 - 1200 cm-1

.Hydroxyl bands : they show the maximum of complexity. OH functions occupy a domain between 2400 - 3600 cm-1 ; with bands largely overlapping. The lower the frequency, the stronger the hydroxyl association (OH from COOH absorbs around 2400 cm-1).

In immature kerogens, the C-O band from the glycosidic band of polysaccharides is often prominent ; also in immature kerogen often appear two bands characteristic of N-H bonds of proteinaceous (peptidic) material, at 1660 and 1450 cm-1.

If the kerogen has not been carefully isolated, some residual minerals can give trouble some interferences. Fortunately, pyrite has no really interfering bands ; but it increases the total absorbance of the sample ; thus reducing rapidly the spectrum quality. The other unavoidable residual minerals are usually not detrimental.

4.2. The correlation with elemental composition

As we said in the preceding paragraph, there is in the general case only loose relations between IR spectra and elemental composition. But due to the definite variations in this elemental composition described upper, either in the case of geological evolution or in the transition between various kerogen types, some empirical correlation appears. The most conspicuous (fig. 12) is between the hydrogen content and the area of the 2840-2960 cm-1 region (CH, CH_2, CH_3 from alkanes). It expresses simply that the other hydrogen-containing functions, such as OH or aromatic C-H either vary little or are in content so low as not to influence the hydrogen content. This latter interpretation is the correct one, and the greater dispersion for high hydrogen contents expresses the fact that, for these kerogens, the OH content is no more negligible. Weaker correlations - if related to all kerogens - but which remain important for a given type of kerogen exist between oxygen content and C = O representative band (C - O bands cannot be used for quantitative purposes, due to their very low resolution). These correlations lead to a very interesting conclusion : the information given by elemental analysis is richer than what could be forecast a priori. Reciprocally, it makes IR spectrometry a little less attractive when elemental analysis has been performed before. However, some structural details can be perceived by IR, which cannot be seen by elemental analysis alone. For instance (fig. 13), the increase, then decrease of aromatic H during kerogen evolution could not be suspected on elemental composition. It is also possible to see how the oxygen - containing functions disappear during evolution, the ester functions disappearing early and some OH functions being the last to disappear.

The principal reason, up to now, for an use no more than moderate of IR spectroscopy in kerogen studies is the difficulty to quantify the spectra. The present development of FT/IR spectrometers, which deliver spectra in digital form, and supply computers and softwares especially designed for IR interpretation will certainly cause rapid progress in IR quantification. It is my opinion that IR can be classified in the group of new old methods.

5. WHAT KEROGEN LOOKS LIKE

5.1. General

To the naked eye, kerogen is an apparently homogeneous brown powder ; but even with this rough examination, it is possible to have an idea of the kerogen composition : high H and O contents give a dull , light brown ; the tint darkens as H and O decrease. We will see in the next section the reasons for this correlation.

A simple lens reveals that this apparently homogeneous powder is in fact highly heterogeneous. Hence the interest of microscopic examination. Historically speaking, this examination was initiated by coal petrographers. Consequently, the whole rock, without any isolation treatment, was examined ; obviously, this method provides a better insight of organic structures than the examination of a ground , chemically processed material. But when the carbon content of the rock decreases, the recognizance of organic particles becomes cumbersome ; finally, the bulk, if not the totality, of OM becomes so finely dispersed that it is no more possible to see anything. For all these reasons, the examination of isolated kerogen is now a standard method (palynofacies).

Historically again, examination was made under natural or polarized transmitted light. Soon the use of reflected light was recognized as useful. Recently, a major development has been the examination of fluorescence excited usually by blue light illumination (390 or 360 nm). All these modalities will be now detailed.

5.2. The transmitted light

Fig. 14, 15 and 16 give an idea of what can be seen when examining a standard kerogen in transmitted light. The most fundamental distinction is between structured and structureless particles. The former can be recognized as fragments of organisms, either vegetal (wood debris, cuticles, pollens and spores) or animal (bones, claws, scales) ; the main drawback is that non-organic shells have been destroyed when isolating kerogen, resulting generally in the disparition of any recognizable structure in the organic remains, which cannot maintain their shape without a protective shell. The latter, improperly termed "amorphous particles", are classified according their general appearance : angular, fluffy, flaky etc... they are traditionnally often related to planktonic algal material, but this is not always the case ; humic gels, of perfectly continental higher plants origin, are also perfectly amorphous. Generally, this ambiguity cannot be solved on optical grounds only. The result of the transmitted light examination is usually given as percentages of more or less conventional categories : woody fragments, cuticles, spores and pollens, other fossils, "amorphous" material (itself sometimes divided in morphological categories).

The colour of these particles is as already said related to their H and O content ; consequently, in the same sample, particles of different origin have different colours. By "different origin" we mean either biological (different parts of different organisms) or temporal origin (old organic particles reworked or resedimented). These different colours merge macroscopically in an average tint which is therefore but loosely linked to the degree of geological evolution of the sample. Attempts have been made to use selected forms of the same species, or at least genus (pollens or spores) where geological evolution is the only variable causing the colour changes, and to derive a general correspondence scale between colour and degree of evolution. Due to numerous measurement difficulties, the practical use of this technique has not been very successful.

Finally, with the thickness of standard thin slides (50 m), OM becomes opaque at a medium stage of evolution (humid gas zone, beginning of thermal cracking of oil already formed). Further examination can be made only in reflected light.

5.3. The reflected light

Coal petrographers have classified the organic debris making up coal in categories called "macerals" (after "minerals") and whose diagnosis is to be made in reflected light (fig. 17). The result of this examination is a percentage "maceral composition" which is not exactly identical to the transmitted light examination, but which is not essentially different either. But the main interest of the examination in reflected light is not there. Coals petrographers have established that a certain maceral, called vitrinite, has reflective properties directly related to the degree of evolution or, to use coal terminology, coal rank (lignite is low, anthracite is high). They have consequently derived a rank scale based on the measurement of vitrinite reflectance in certain conditions. In the case of coals, vitrinite is generally readily recognized, and its reflectance can be measured satisfactorily. The result is that the vitrinite reflectance evolution scale is excellent for coals. In the late twenty years, this technique has been applied to disperse organic matter. But there the situation is not very satisfactory. First of all, vitrinite stricto sensu exists only in type III kerogens ; it is therefore, strictly speaking, impossible to extend this scale to type I, II and mixed situations. Second, even with sediments with type III kerogens, two main drawbacks appear ; the first is the difficulty to recognize vitrinite, difficulty more important than for coals ; the second, if the first difficulty is overcome, is the difficulty to have particles suitable for correct reflectance measurements (size, homogeneity, orientation). The result is that, despite the quasi-universal use of vitrinite reflectance in evolution studies of all-types kerogens (fig. 18), and the implicit or explicit claim of its success, it cannot be considered as a correct quantification of the degree of evolution, but at best as a rough estimate, fraught with major uncertainties.

5.4. Fluorescence

Reflectance and fluorescence are, for fundamental electronic reasons, opposite properties. Macerals always reflectant (fusinite, inertinite) are never fluorescent ; fluorescence decreases when evolution increases, and in deed disappears at an early stage (practically at the beginning oi oil generation). Fluorescence is in fact a property of free, small size aromatic systems. Macerals of the exinite class (spores, pollens, cuticles) are the most fluorescent ; liquid products (oil) are also fluorescent, but they can appear in a kerogen preparation only in two cases : either a bad quality preparation, or an in situ generation by accidental heating - such as when polishing a surface for reflected light examination. As a method to precise morphological examination, fluorescence is more useful with whole rocks than with kerogen preparation : fluorescent minerals are either rare or easy to recognize, thus permitting a more precise localization of OM than transmitted light (fig. 19, 20). But this use is not the reason for the present development of fluorescence techniques.

Even if all the limitations of the use of vitrinite reflectance as a quantitative measurement of the degree of evolution were overcome, it would remain that at the earlier stage (before the beginning of oil generation) reflectance is low, hence difficult to appreciate, and vary little. On the contrary, as we already said, at this stage kerogen fluorescence vary rapidly, and both in intensity (which decreases with evolution) and in colour (which goes from bright yellowish green to dull brown - liquid products remaining always green to blue green). This double variation, which permits a crosscheck of the results, is the base for the design of a fluorescence evolution scale, usable before the vitrinite reflectance scale, roughly in the same domain as the colour variation of spores and pollens - which proved unsuccessful. It is now too early to conclude about the use of this scale. Some successes have been reported, but the measurement itself is not easy, due particularly to the so-called "fading" : a preparation illuminated for some minutes does not maintain a constant fluorescent reemission : its intensity decreases, and its colour shifts to the red. More work is necessary.

6. THE DIFFRACTION TECHNIQUES

6.1. General

As we saw in section 2, the geological evolution of kerogen can be viewed as a progressive loss of heteroatoms from a matrix more and more carbonaceous. The final step thermodynamically expected would be graphite, hence the search for a pregraphitic structural organization in kerogens. X-ray as well as electron diffraction can equally be used for that purpose, but, due to the greater versatility of electronic optics, the latter technique gives the better results.

6.2. X-Ray diffraction

X-ray diffractograms remain almost structureless up to a high degree of evolution. At this stage, appears a broad band centered on 0.3 45 nm (graphite spacing 0.02 : 0.335 nm). This band tends to sharpen and shift to the graphite 0.02 spacing as evolution proceeds, but very slowly. This type of study gives but little information.

6.3. Electron diffraction

Electron diffractograms are no more informative than X-Ray diffractograms, but the existence of electronic lenses makes possible a much more interesting technique. Known as "dark field imagery" it consists to physically isolate by a diaphragm in the focal plane of the main lens the dot corresponding to the electronic rays making with the optical axis the Bragg angle corresponding to graphite 0.02 spacing (more precisely corresponding to the interval 0,335 - 0,345 nm, fig. 22-23). This results in a generally black image (because of the diaphragm) where appear as bright spots only those domains where exist fragments of quasi-crystalline lattices with the right spacing. Another technique, known as "lattice fringes" obtains, by a precise slight defocussing, a diffraction pattern in the focal plane which is equivalent to a projection on this plane of the crystalline stacking perpendicular to the electron beam - provided the crystal is thin enough and properly oriented : in our case, this is equivalent to an edge view of the aromatic sheets. Examples of the resulting imagery are given in fig. 23-24. Fig. 25 gives the dark field image of an immature kerogen : tiny medium-bright dots are interspersed in a black matrix, without any ordering. There is no lattice fringes. The interpretation is straightforward : aromatic sheets are very small, both on a-b dimensions (3-4 rings) and c dimension (2-3 sheets), they do not show any ordering one relative to the other, and they are separated by quantitatively important non-aromatic regions. Fig. 26-27-28 give the pictures for evolved kerogens of types I, II, III. Compared to fig. 25, in the three cases bright dots are brighter and show some grouping, which corresponds to an a-b ordering ; correlatively, lattice fringes appear which show also this ordering. But the intensity of the process depends strongly on the kerogen type : type I gives an a-b ordering extending on distances $\sim$50 nm and an apparition of a good c-ordering ; type III shows an a-b ordering for distances $\sim$10 nm and practically no c-ordering. Type II lies in between.

All these results can be summarized with a plate like fig. 29. Strictly speaking, this is an artist's view of immature type II kerogen. For type I, the relative content of aromatic units would have to be divided by 2-3, and the balance between saturated rings and chains shifted towards the latter ; this last operation would have to be done as well for type III kerogen, but now the relative aromatic units content would have to be multiplied, and not divided by 2-3. In the three cases, the path of geological evolution is the same : first are eliminated the O (and S) containing groups, taking away some carbon in the form of CO_2. Then, essentially saturated hydrocarbons are eliminated (petroleum genesis), drastically reducing the total quantity of kerogen, and relatively concentrating the polyaromatic units, and permitting their mutual spatial reorientation. This tendency to crystalline ordering will be the easier the greater the room polyaromatics can use to reorientate ; hence the better crystalline order obtained with type I kerogens. During these processes, new aromatic units can also be synthetized de novo, for example by dehydrogenation of saturated rings, but the increase in aromaticity of the kerogen is chiefly the result of the relative enrichment described upper.

The structure of the non-aromatic part of the kerogen molecule in fig. 29 is obviously not given by electron diffraction studies. It is interesting to notice that direct evidence for this structure is provided by IR and NMR spectroscopy, and pyrolysis ; petroleum composition gives indirect evidence.

Rather deceivingly, electron diffraction is not a sensitive means of estimating geological evolution. The pictures of fig. 26-27-28 appear only at high evolution stages ; for weaker evolution, observed patterns are very difficult to interpret.

7. CARBON ISOTOPES STUDY

Due to differences in photosynthetic pathways, and differences in isotopic composition of atmospheric CO_2 compared to dissolved CO_2, marine and continental plants have different carbon isotopic composition, terrestrial carbon being lighter than marine carbon (fig. 30). This difference, shifted to lighter values, persists in the kerogens, and even in the petroleum they generate (fig. 31). In principle, this would enable to assign a marine or continental origin to kerogens. But this technique is very global, and optical examination, joined to general geological consideration, is a much more powerful and precise tool to elucidate OM origin.

8. PYROLYSIS

Pyrolysis can be considered as a sort of simulation of natural evolution. This indicates that pyrolysis is a priori a powerful method of kerogen study. Pyrolysis is controlled heating : temperature programming as well as other experimental conditions must be controlled. The standard IFP method uses a device such as in fig. 32. Heating must be performed in an inert atmosphere (helium, nitrogen) ; the temperature cycle and the parameters measured are given in fig. 33. Four parameters are used : S1, which represents the simple vaporization of mobile (liquid) compounds eventually present in the sample ; S2, which represents the compounds generated by the complete thermal cracking of the kerogen, measured as hydrocarbons ; S3 which represents the low-temperature generated CO_2, during the same cracking ; T_M which is the temperature of peak generation of hydrocarbons (S_2 peak).

The significance of S1 is clear, but its interpretation is not straightforward, since S1 can be improverished by migration out of the sample, or enriched by migration into the sample (reservoir). S2 represents the maximum hydrocarbon yield possible for the sample ; it depends evidently on the evolution stage, and on the nature of the kerogen ; with equally evolved samples (in that case, immature) fig. 34 shows this difference depending on kerogen nature : type I kerogen yields almost three times more hydrocarbons than type III. S3 is related to the oxygen content of the kerogen, just as S2 is related to its hydrogen content ; a plot such as in fig. 35 gives a pattern somewhat reminiscent of the Van Krevelen diagram. In practical use, S3 is more delicate to use than the ratio O/C. T_M is a very interesting value, since it depends strongly on the degree of evolution, and only slightly on the nature of the kerogen. Fig. 36 gives the correlation between vitrinite reflectance (on coals !) and T_M : it is quite good, but we will notice that it is not linear (indeed there was no reason to expect linearity).

Fig. 37 shows a theoretical plot of all these values on samples taken during drilling of a borehole ; T_M increases with depth (= with geological evolution, time - age - and temperature both increasing generally with depth) and the interpretation of the remaining columns is self-explanatory. Fig. 38 shows an actual plot, in a real borehole, of the same parameters ; interpretation is the same.

Now, a further dimension can be given if S2 content is analyzed by detailed techniques, such as gas-phase chromatography. Fig.39 shows that, effectively, it is possible to discriminate between types of kerogen by this technique. Furthermore, this differentiation enables to have an insight into the structural units which are not of aromatic polycondensed type. This last type of analysis can be even refined by use of mass spectrometry, but there is an essential difference between these detailed analysis and the simple Rock Eval analysis described upper. Detailed analyses must be made on isolated kerogens, due to complex interactions between pyrolysis products and minerals in the sample, if any. Although these effects exist also in simple, global pyrolysis, they are of lesser extent, and can be corrected in some way. Then, the total time required for one operation on Rock Eval device is about 20 mn , including sample preparation ; a detailed analysis requires 2-3 hr of work, and a total time span of some 3 days. Consequently, the prices of these two types of analysis, or the total number of samples that can be reasonably processed, are not of the same order of magnitude.

9. ELECTRONIC PARAMAGNETIC RESONANCE

This method will be addressed at length by others in this meeting essentially for structural determinations. However, we shall speak here briefly of one special application. In the mid 70, the measurement of the number of paramagnetic centers per unit mass of kerogen (= intensity of EPR signal) was proposed as a reliable means of evaluation of geological maturation.

The main claim was that EPR intensity is a measurement of geological evolution not influenced by kerogen type. As can be seen from fig. 40 and 41, the latter affirmation is not true and, moreover, the EPR curve versus evolution is bell-shaped, with a maximum (fig. 42) ; consequently, a given value of EPR intensity can correspond to two different degrees of evolution. This was to be expected, since graphite is diamagnetic and consequently has an EPR intensity of 0. But this fact, joined to the uncertainty of the measurements themselves makes EPR intensity totally unsuitable for evolution estimation.

After an enthusiastic period, these observations led to an almost complete abandonment of EPR techniques in kerogen studies. Just as the first enthusiasm was unreasonable, it can be guessed that the present situation is not satisfactory as well.

10. OTHER METHODS

Speaking of geochemistry, there seems to be an unexpected absent: organic chemistry itself. But indeed, the very numerous attempts to use chemical methods have proved, if not entirely unsuccessful, difficult to interpret and, in some instances, contradictory with the results of physical methods. The reason for that has been already given : the infinite variety of chemical environments of the functional groups, which causes the broadening of IR bands - and, incidentally, has the same result with NMR bands. This variety is even more detrimental with chemical than with physical methods, since a given group can be sterically hindered to the point that it becomes unaccessible to the reactants. Unexpected side reactions can also occur. On the whole, functional analysis appears as unreliable, except in its most drastic versions - but in that case, it becomes equivalent to elemental analysis.

11. CONCLUSION

At the end of this review of the methods actually operative for kerogen studies, we can see that these methods enable a rather detailed insight in kerogen structure. However, some points remain a little obscure. The most important is aromaticity. Despite its elegance, electron diffraction cannot be considered as a method of practical use. It requires a lengthy and tedious preparation of the samples, and is itself lengthy ; as it operates on very small domains, it could be considered as quantitative only if a statistically significant number of particles could be analyzed, which is never the case for practical, and financial, reasons. IR is not a good method to investigate aromatic groups, since the specific absorbance of C-H and even more C-C bands is very weak compared to the saturated C-H absorbances. Finally, elemental analysis remains an indirect and, at best, gross method to study aromaticity. There can enter NMR.

SUGGESTED READINGS

1. A general textbook on organic geochemistry :
TISSOT B., WELTE D. (1978)
Petroleum formation and occurrence
Springer Verlag, Berlin, Heidelberg, New York.

2. A book especially devoted to kerogen :
DURAND B., editor (1980)
Kerogen
Editions Technip, Paris.

3. For pyrolysis, which is not covered by "Kerogen", the proceedings of the "International Symposium on Analytical Pyrolysis" are useful, especially the 6th one, presently in press, held in Vail, Colorado, in March 1983.

4. Finally, the biennial books "Advances in Organic Geochemistry" contain the totality of what is being done on kerogen geochemistry. Several publishers, depending on the year, the last one (Advances in organic geochemistry 1981, M. Bjoroy ed.) being published by John Wiley and Sons.

TABLE 1

QUANTITIES OF THE DIFFERENT FORMS OF CARBON IN THE EARTH'S CRUST

FORMS OF CARBON	QUANTITIES (t)
KEROGEN	1.10^{16}
CARBONATES	
IN SEDIMENTARY ROCKS	6.10^{16}
IN NON SEDIMENTARY ROCKS	1.10^{16}
DISSOLVED IN OCEANS	5.10^{13}
"ELEMENTAL" CARBON (MAINLY IN BASALTIC, GRANITIC AND METAMORPHIC ROCKS)	1.10^{16}
DISSOLVED IN OCEANS (CO_2 ORGANICS)	1.10^{12}
ATMOSPHERE (CO_2)	1.10^{12}
BIOMASS (LIVING ORGANISMS)	$0.3 - 3.10^{12}$

TABLE 2

ORGANIC CARBON IN SEDIMENTS

LOCATION	MEAN VALUES (wt%)	MASS (10^{16} t)
CONTINENTS, SHELF AND SLOPE		
CLAYS AND SHALES	0.99	0.82
CARBONATES	0.33	0.08
SANDS	0.28	0.09
OCEANIC		
CLAYS AND SHALES	0.22	0.07
CARBONATES	0.28	0.10
SILICEOUS	0.26	0.04

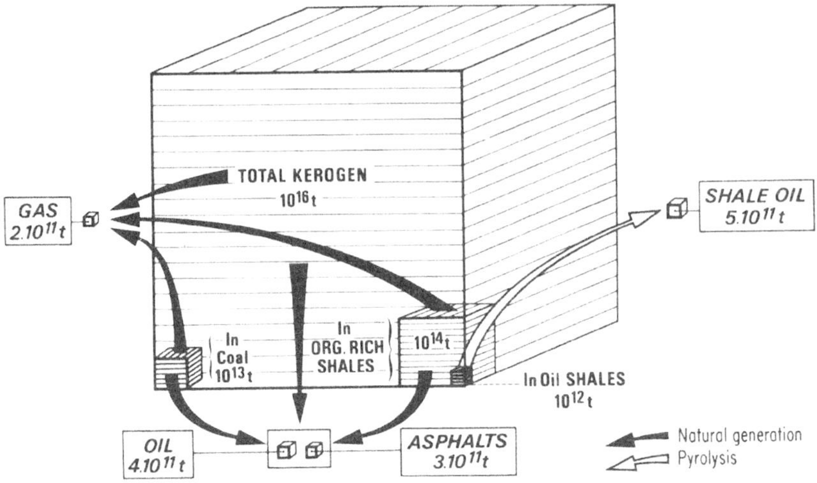

Figure 1.

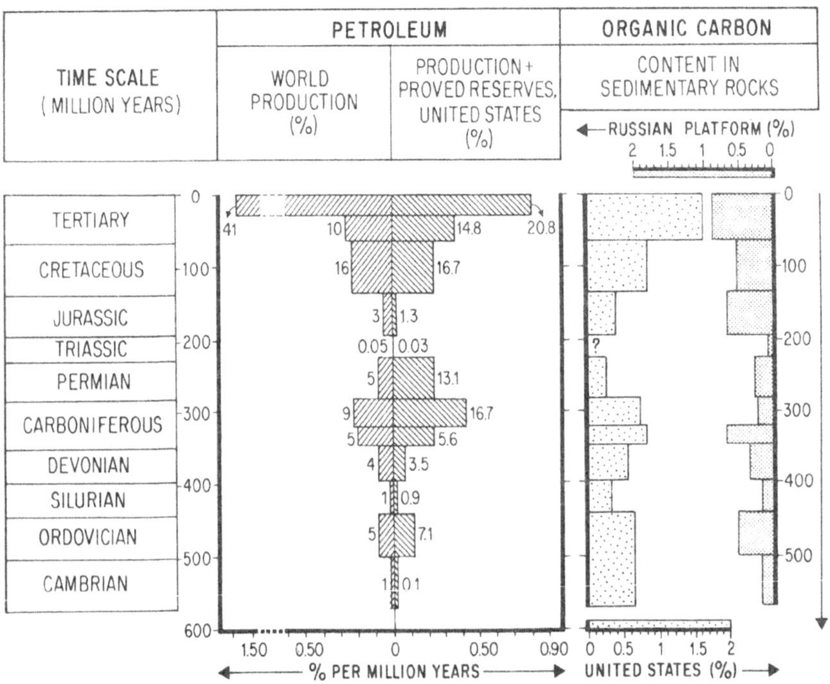

Figure 2.

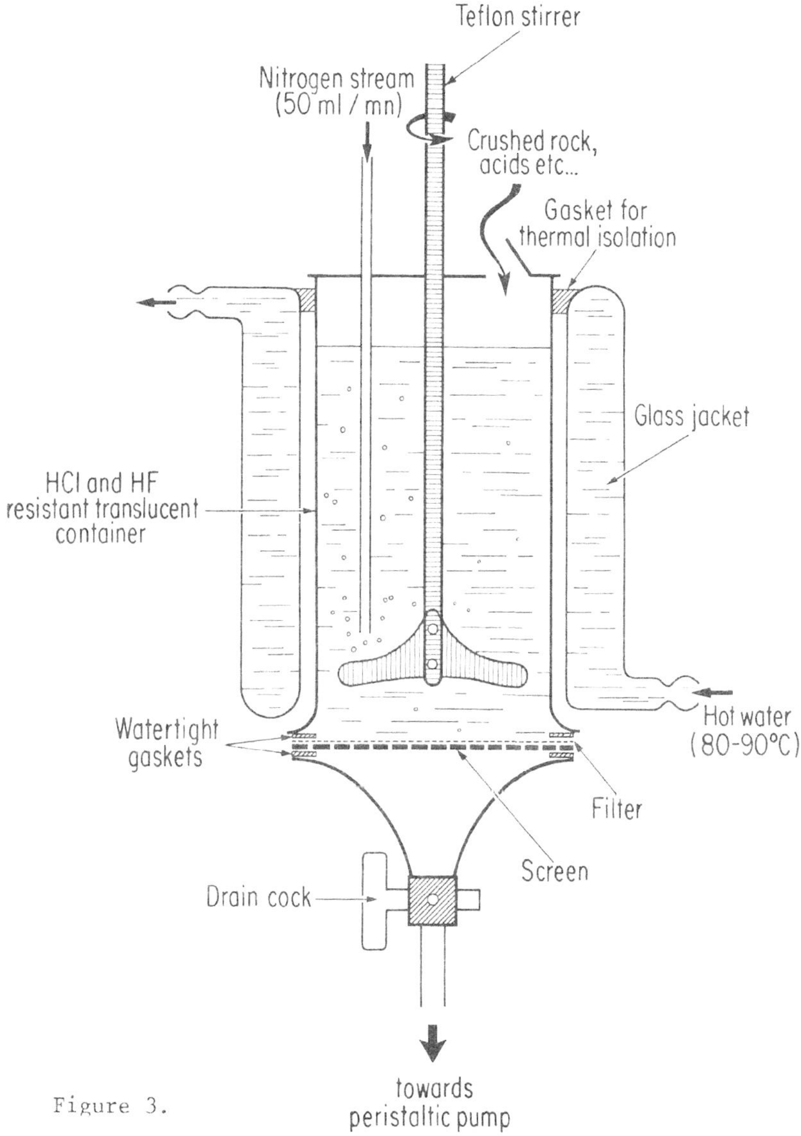

Teflon stirrer
Nitrogen stream
(50 ml / mn)
Crushed rock,
acids etc...
Gasket for
thermal isolation
Glass jacket
HCl and HF
resistant translucent
container
Watertight
gaskets
Hot water
(80-90°C)
Filter
Screen
Drain cock
towards
peristaltic pump

Figure 3.

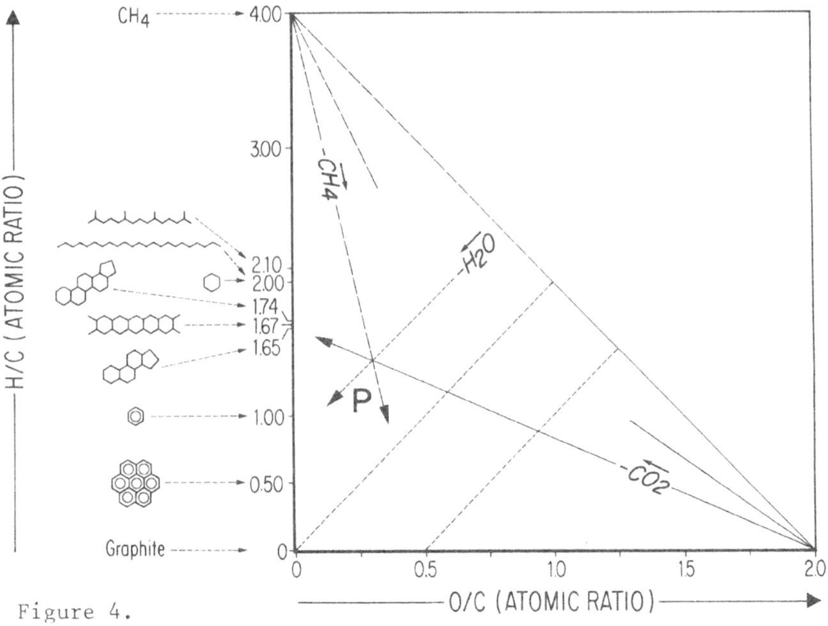

Figure 4.

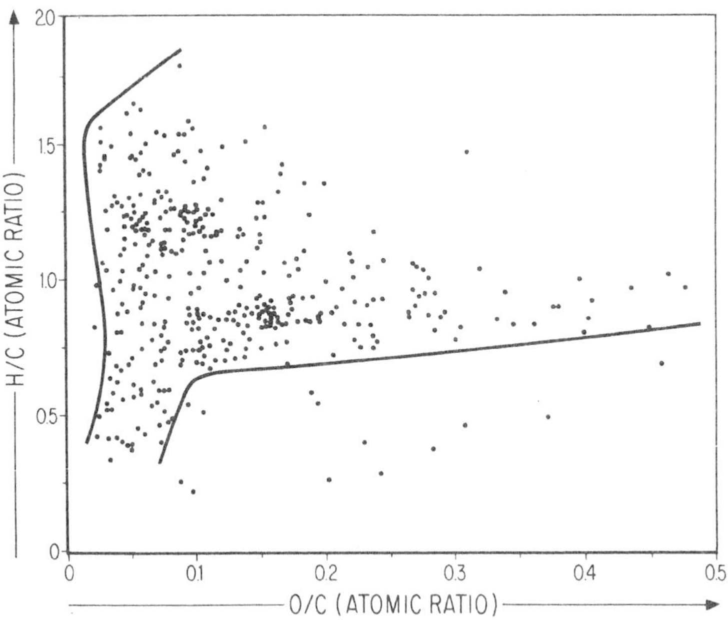

Figure 5.

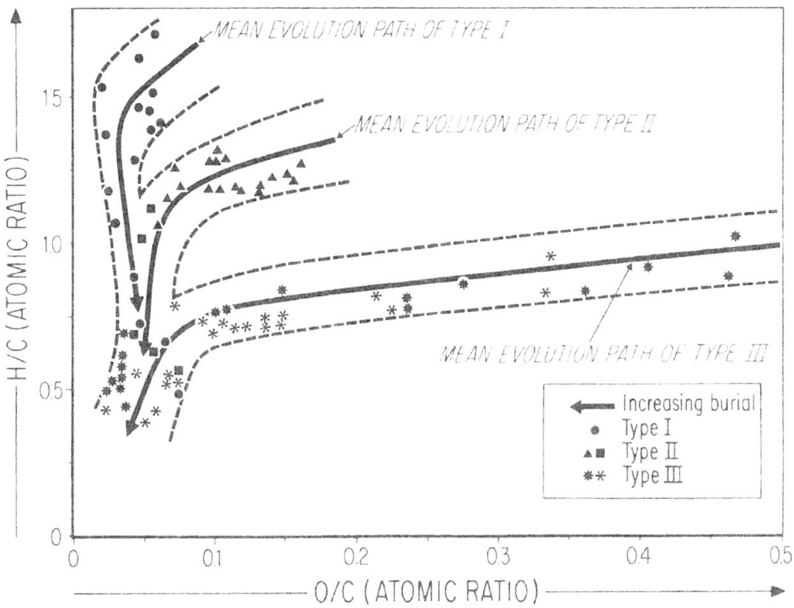

Figure 6.

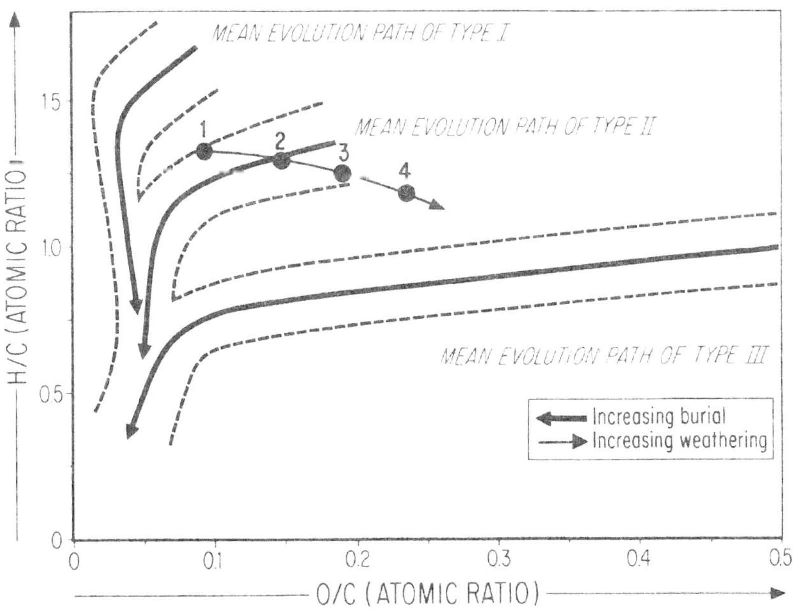

Figure 7.

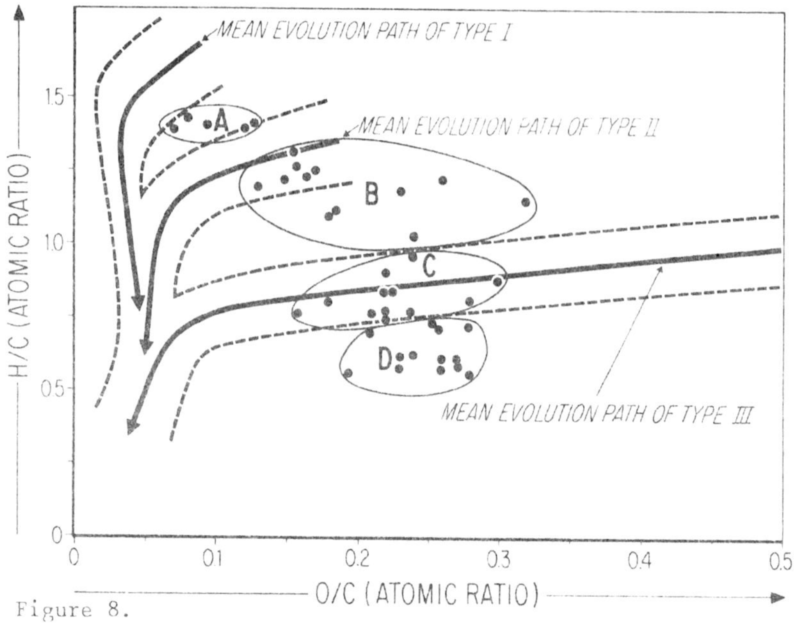

Figure 8.

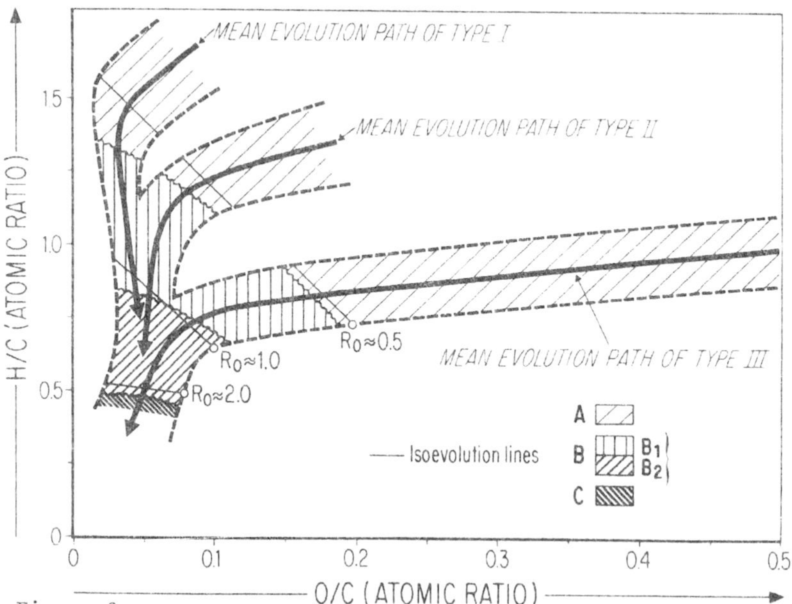

Figure 9.

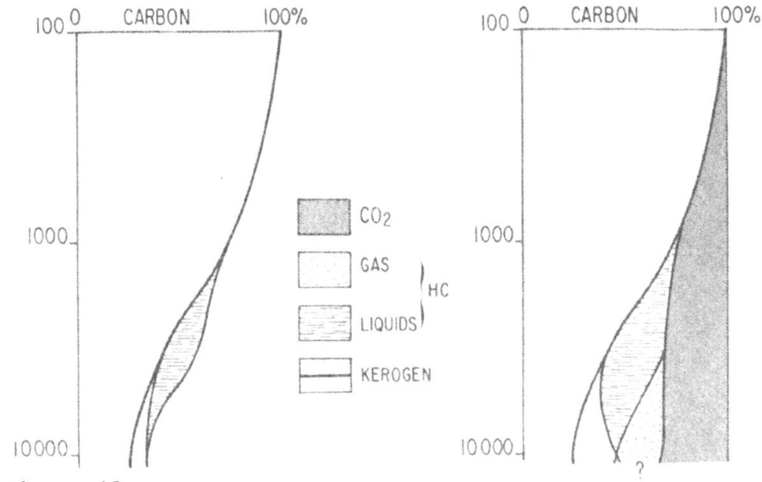

Figure 10.

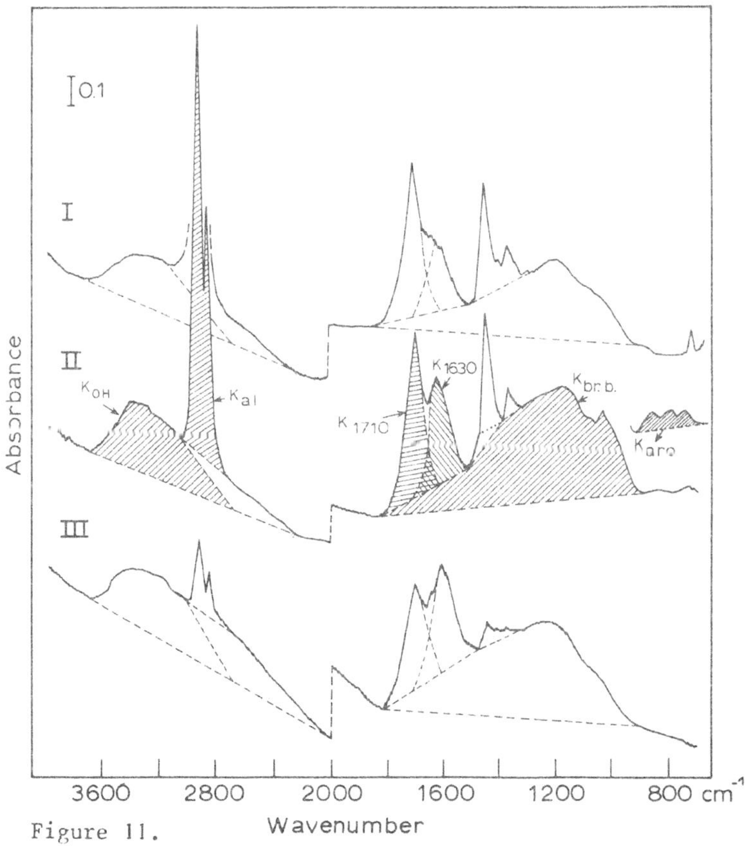

Figure 11.

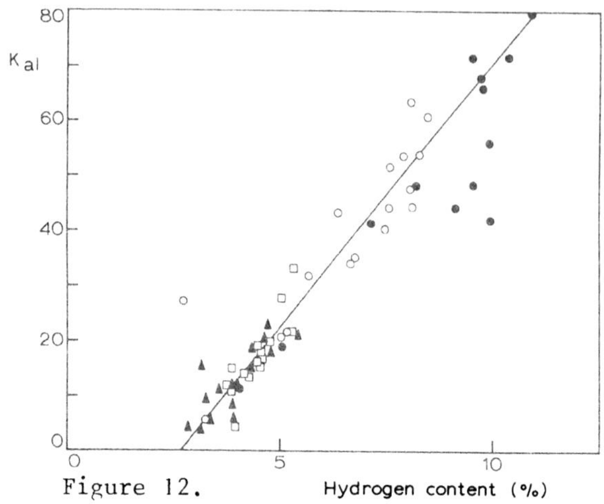

Figure 12.

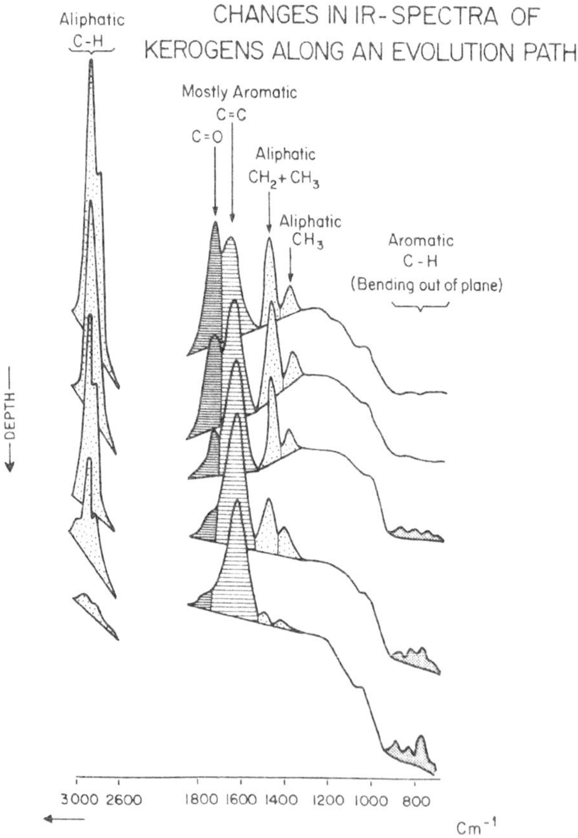

Figure 13.

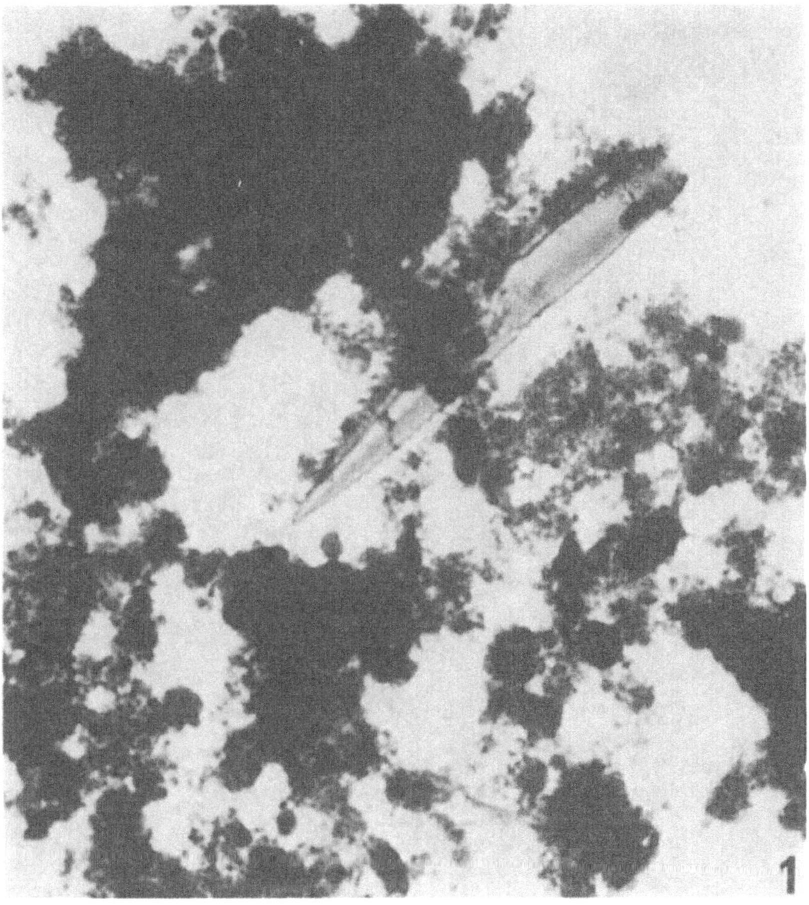

Figure 14.

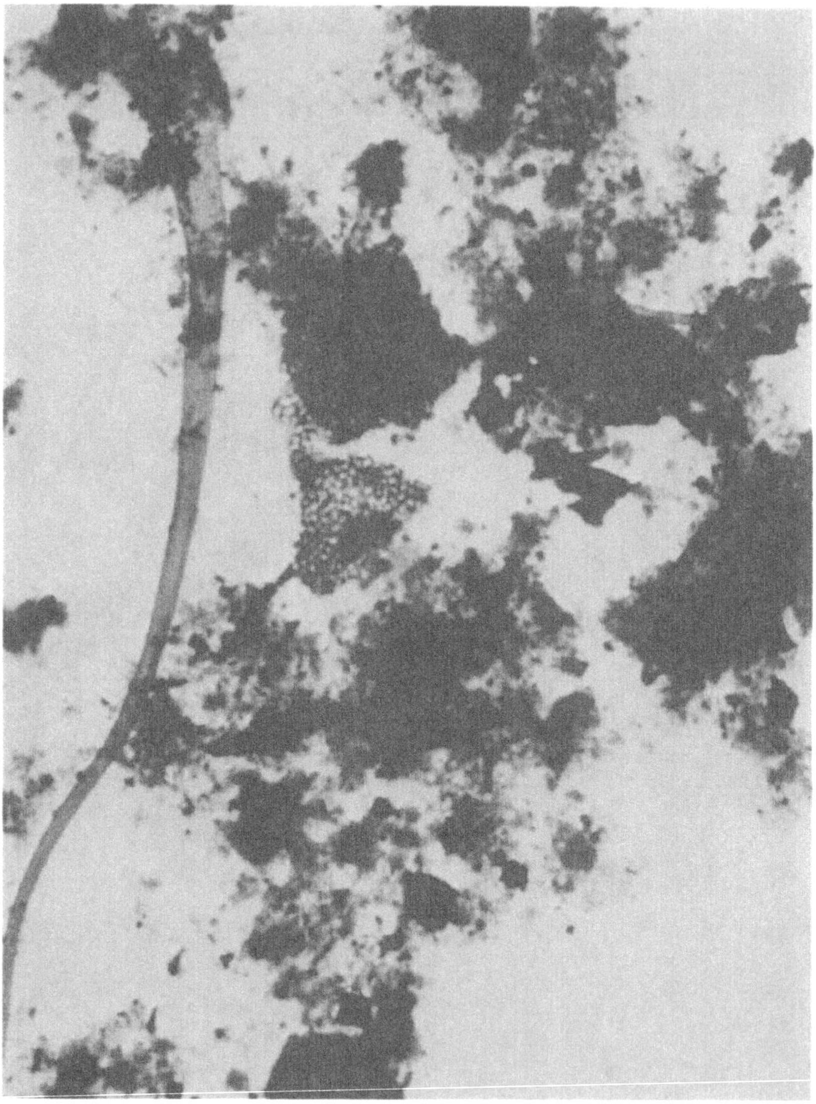

Figure 15.

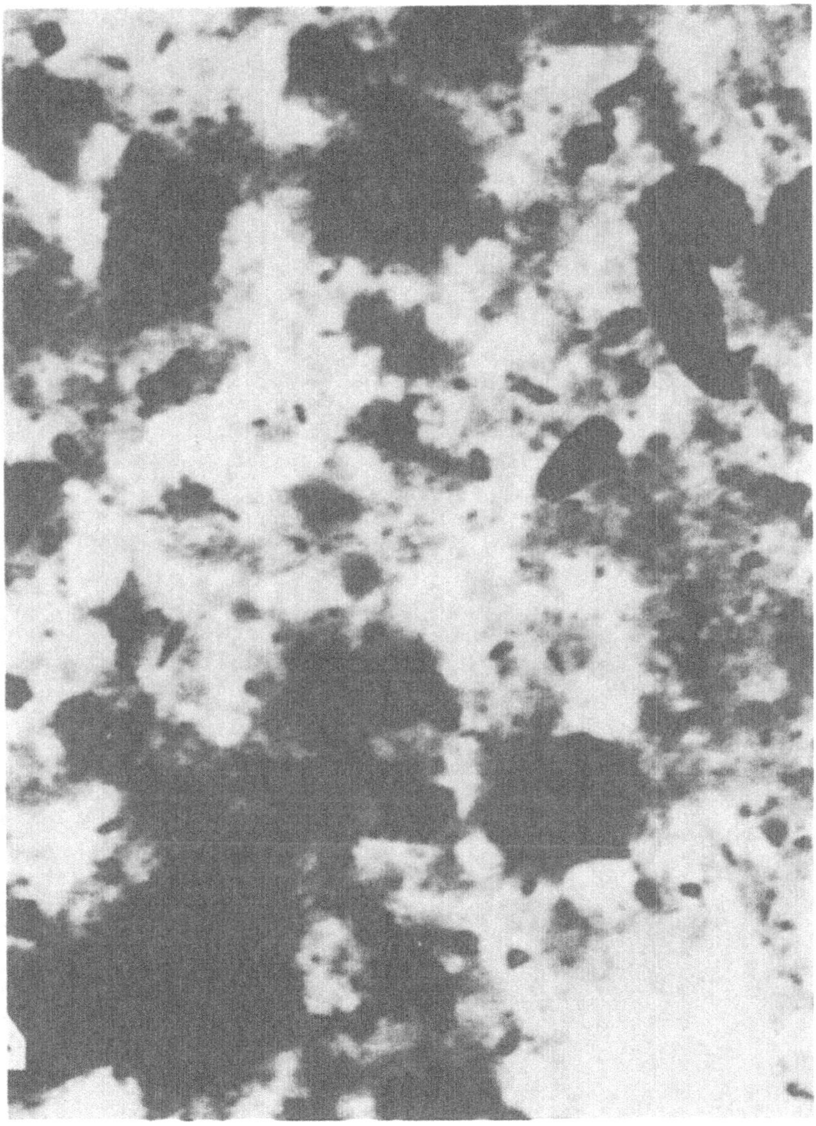

Figure 16.

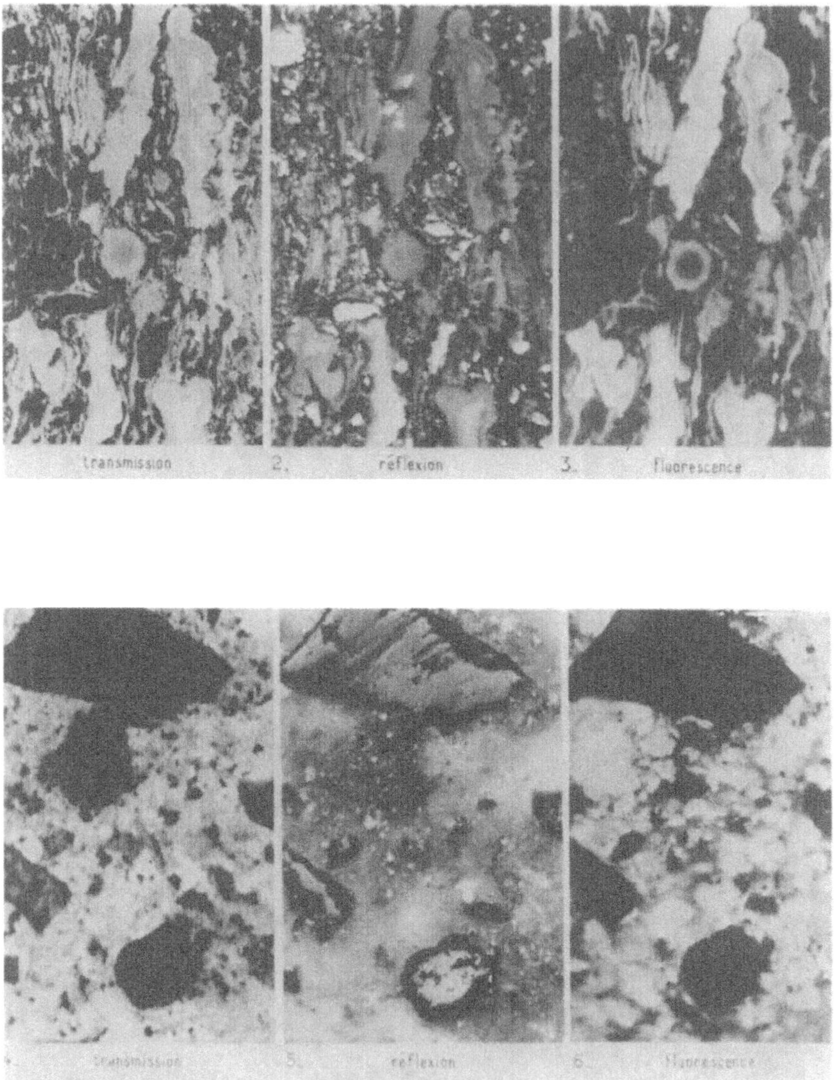

Figure 17.

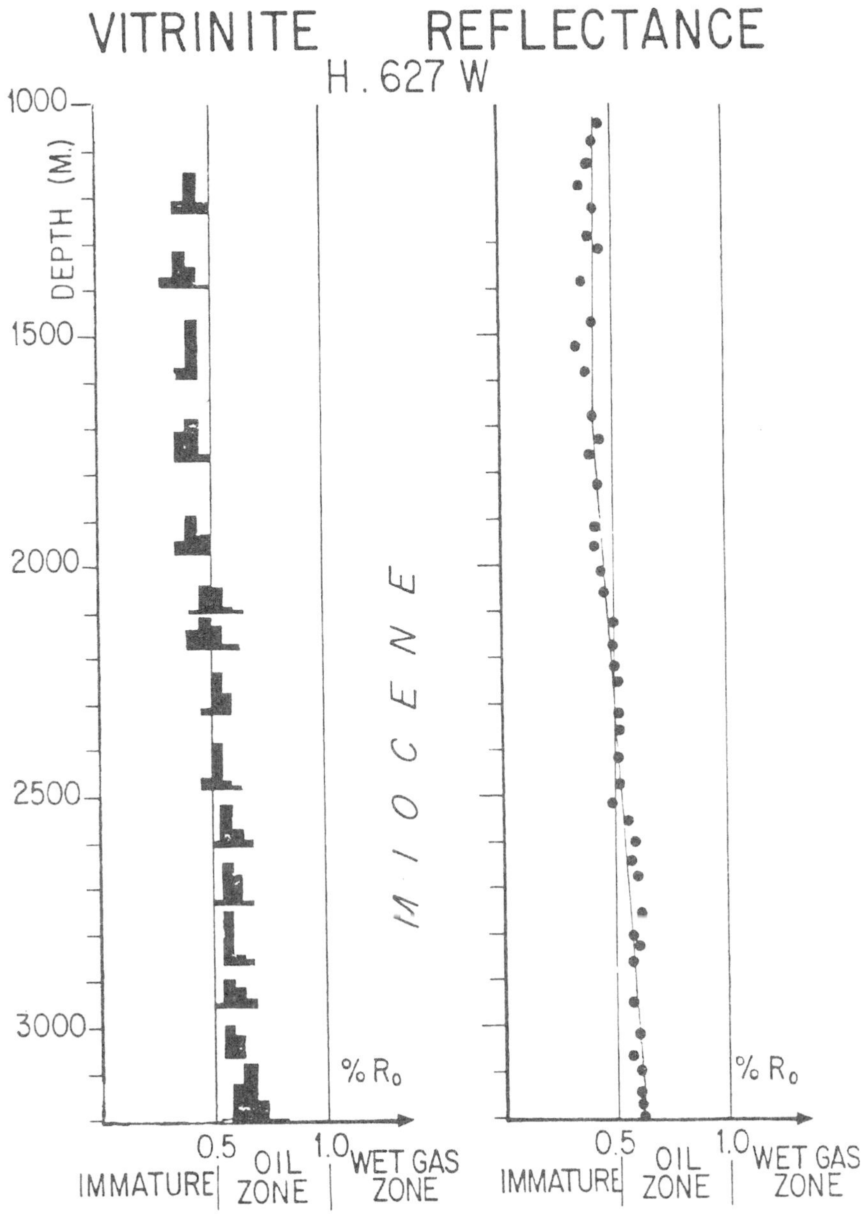
VITRINITE REFLECTANCE
H . 627 W
DEPTH (M.)
1000
1500
2000
2500
3000
M I O C E N E
% Ro
0.5
1.0
IMMATURE
OIL ZONE
WET GAS ZONE
% Ro
0.5
1.0
IMMATURE
OIL ZONE
WET GAS ZONE

Figure 18.

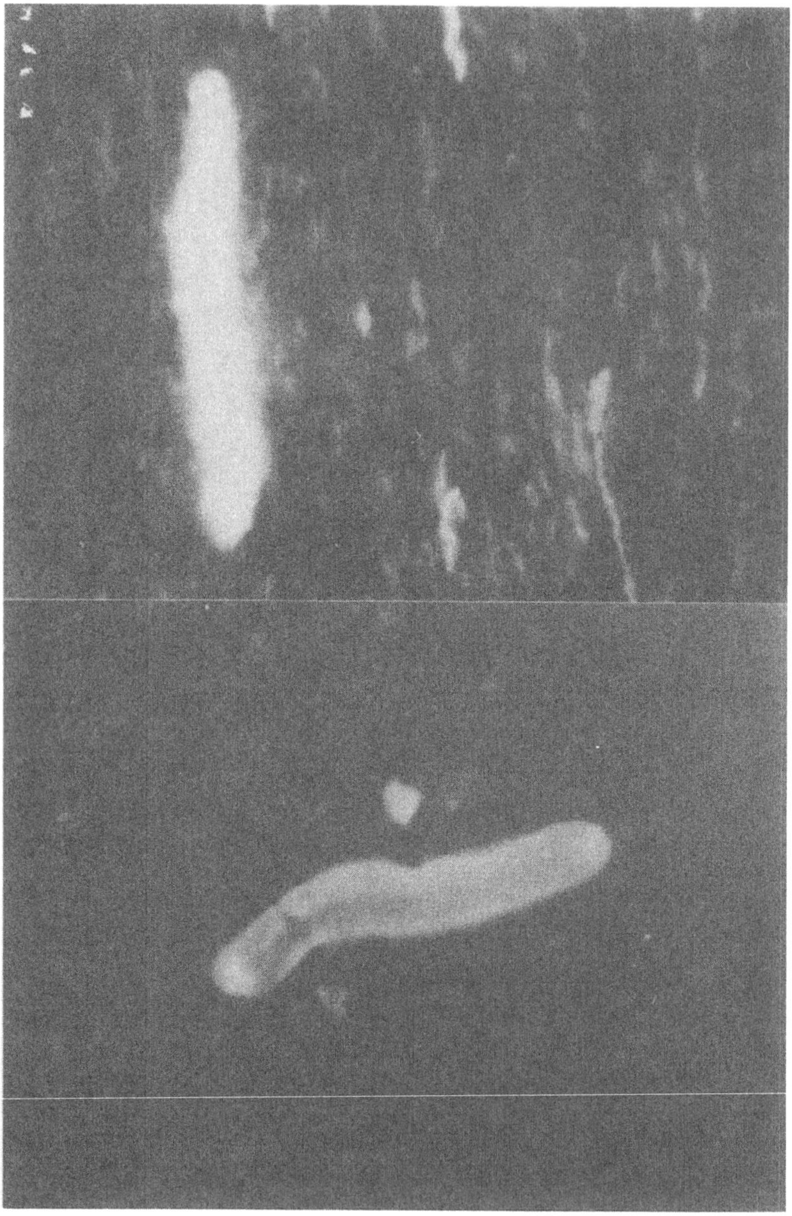

Figure 19.

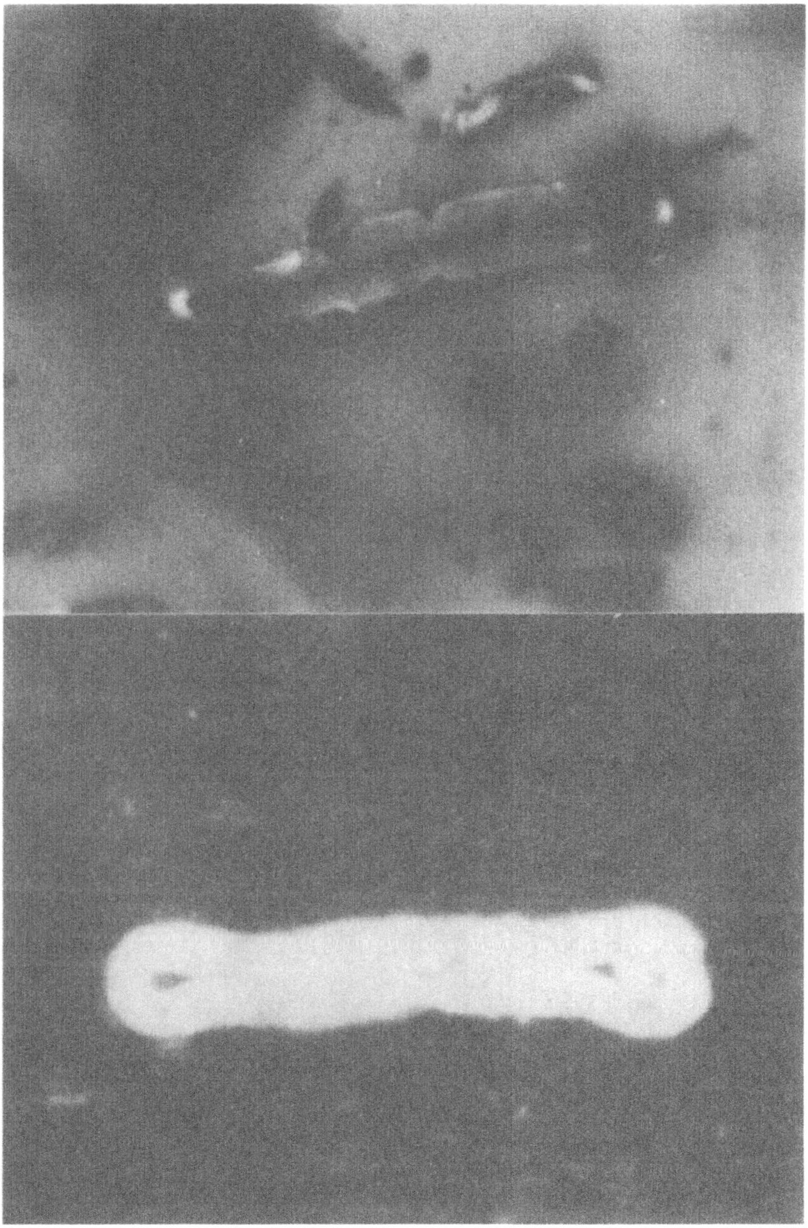

Figure 20.

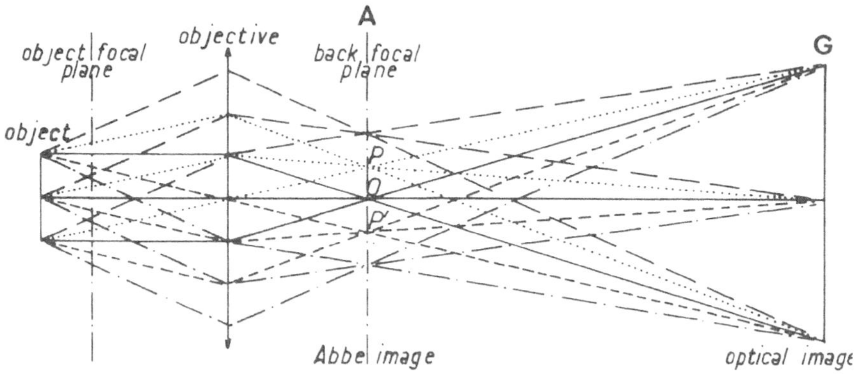

Figure 21

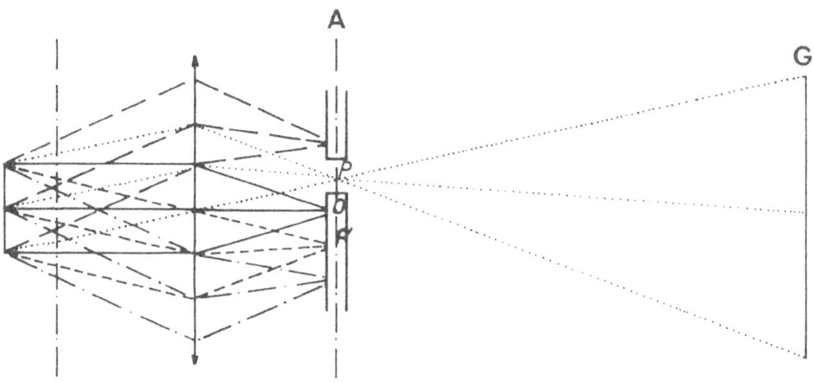

Figure 22

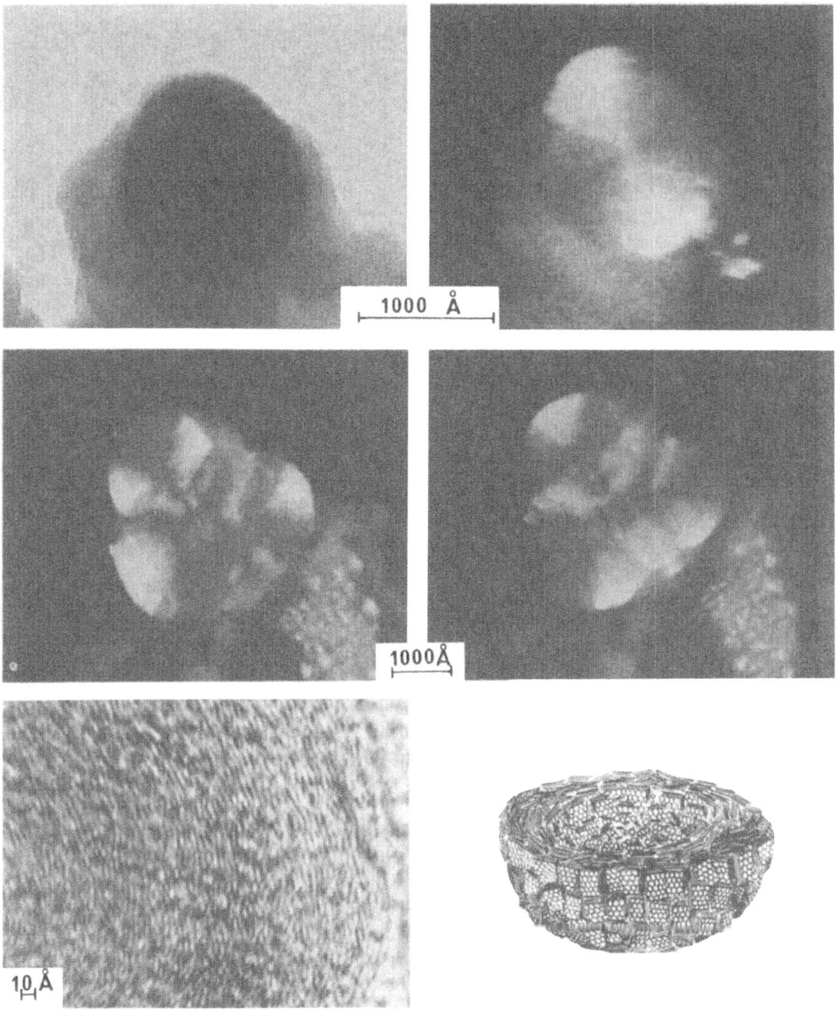

Figure 23.

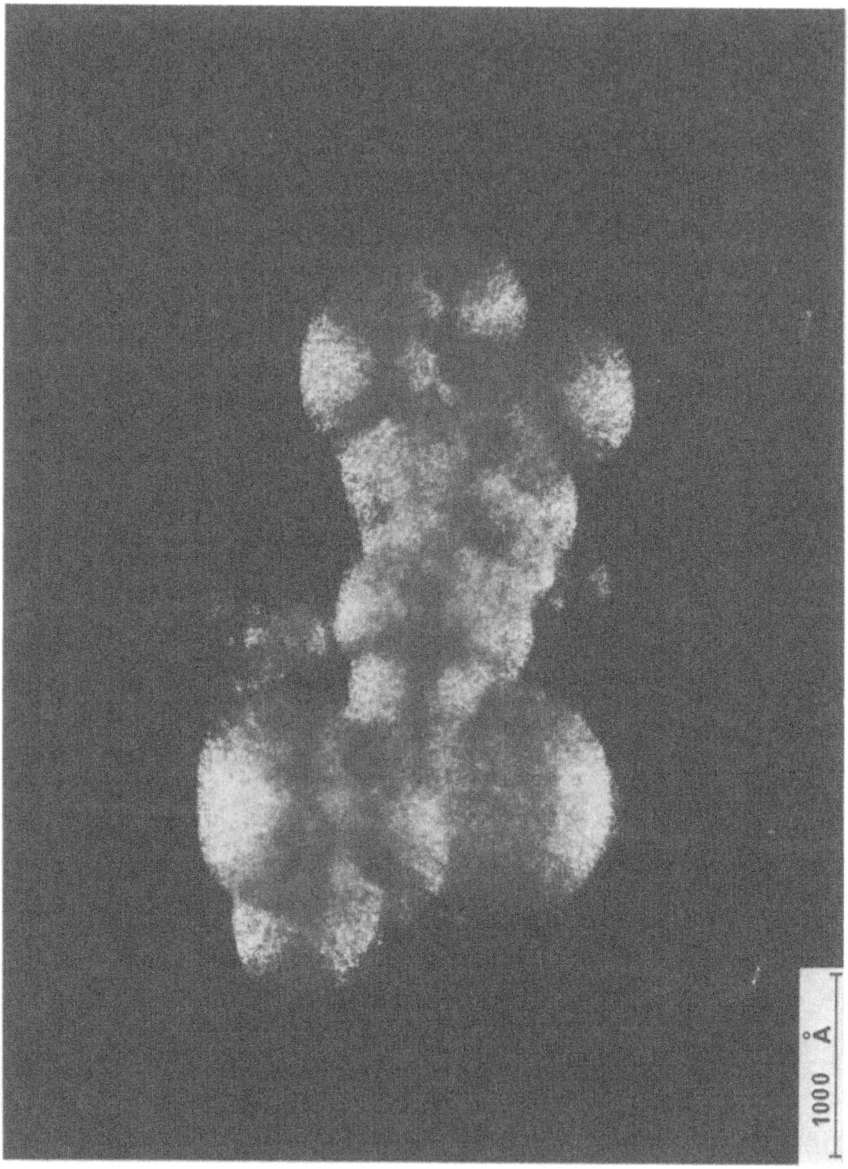

Figure 24.

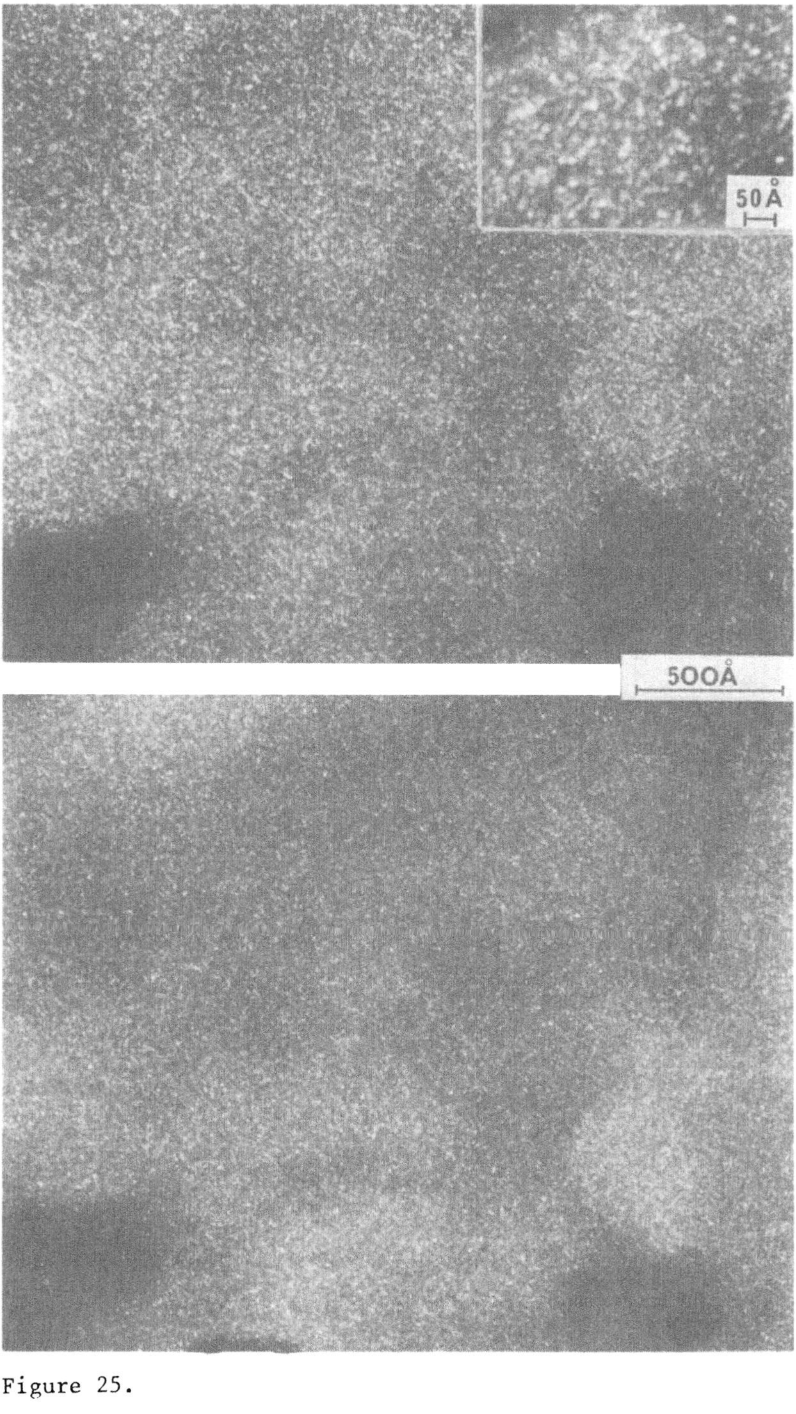

Figure 25.

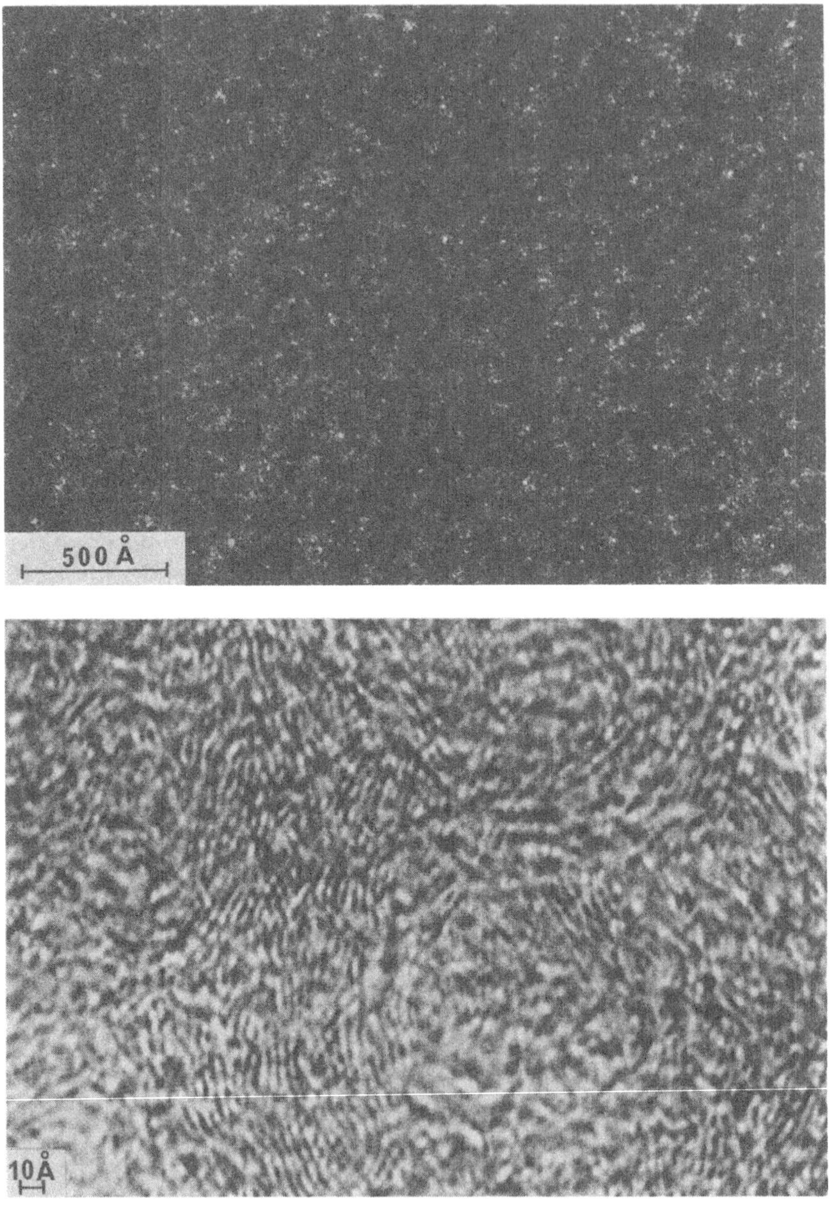

Figure 26.

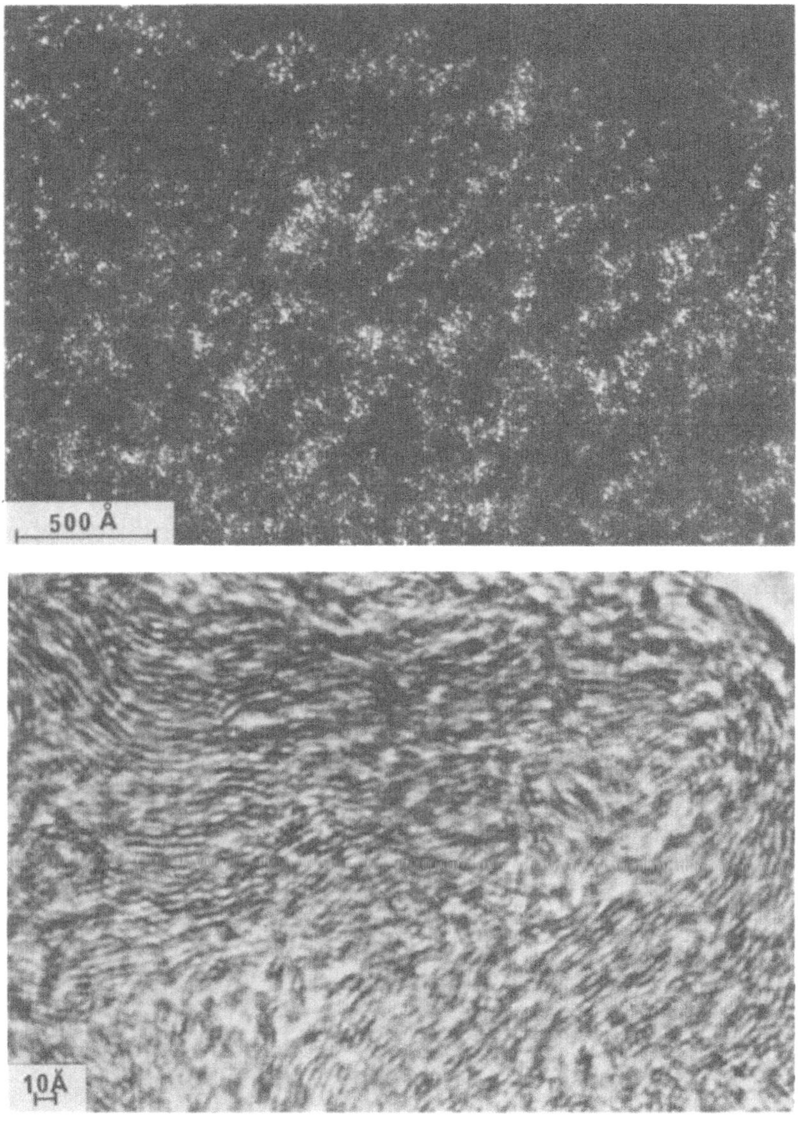

Figure 27.

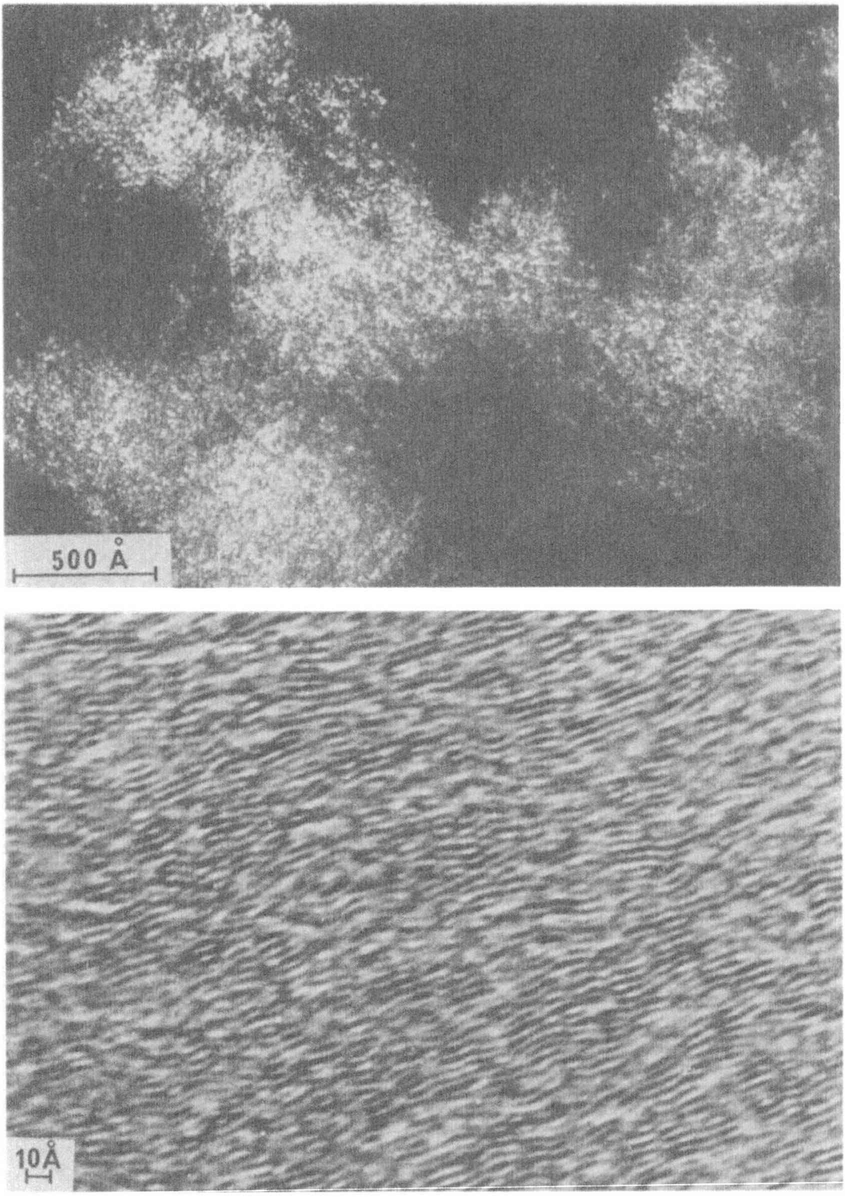

Figure 28.

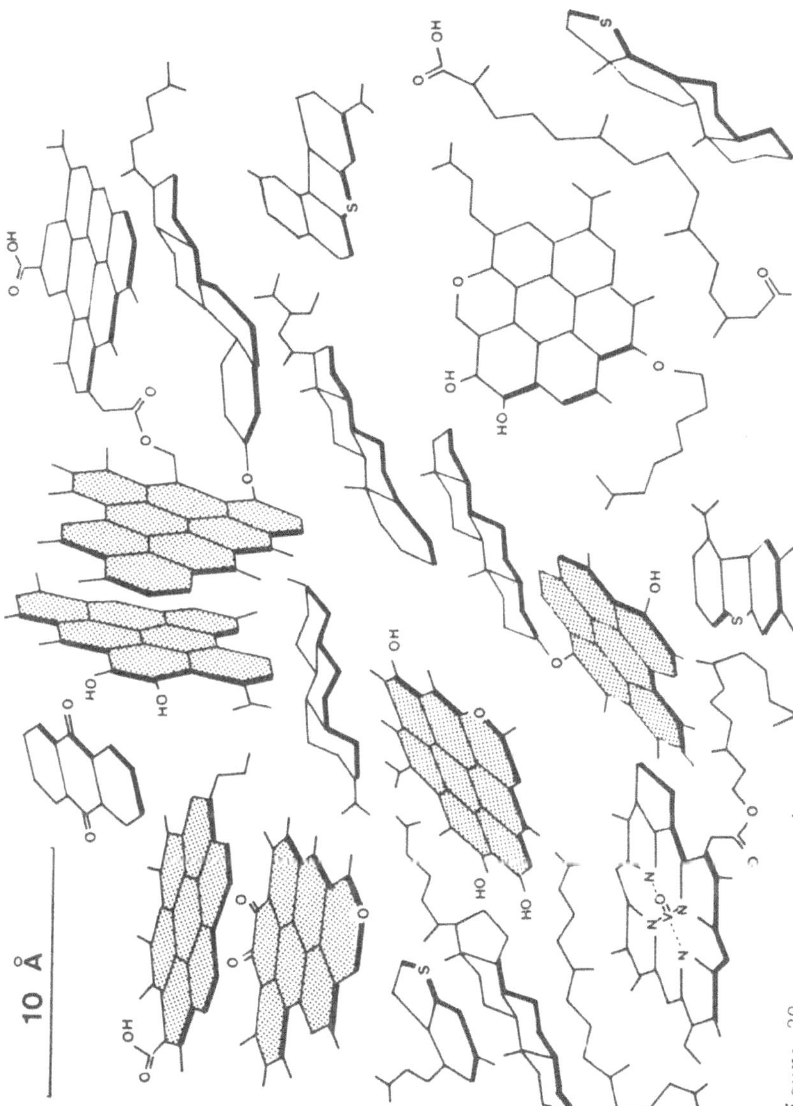

Figure 29.

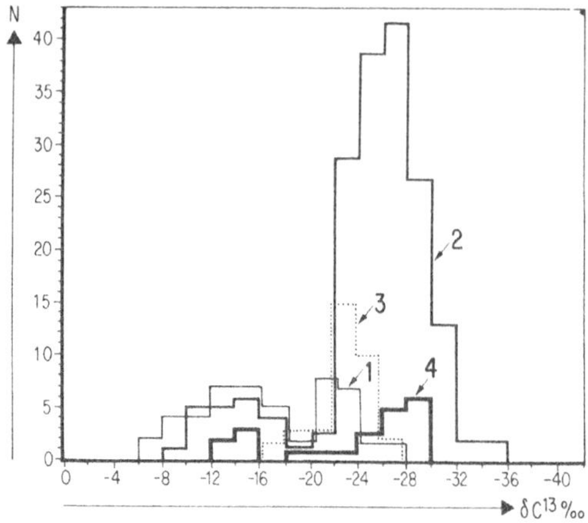

Figure 30.

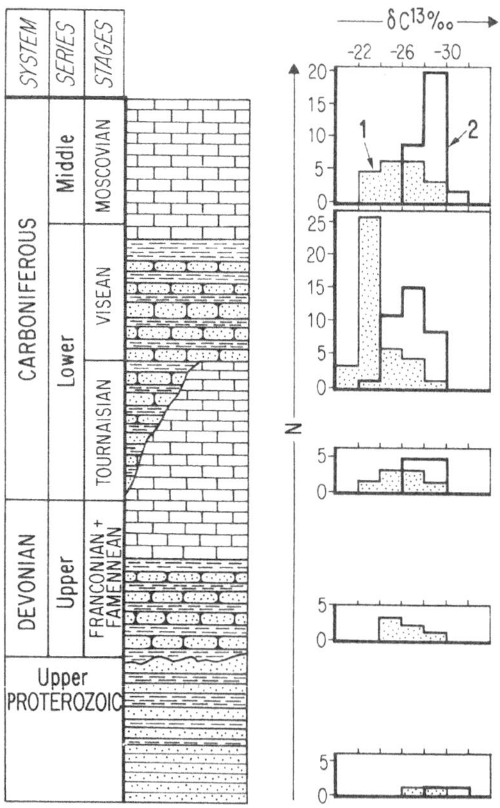

Figure 31.

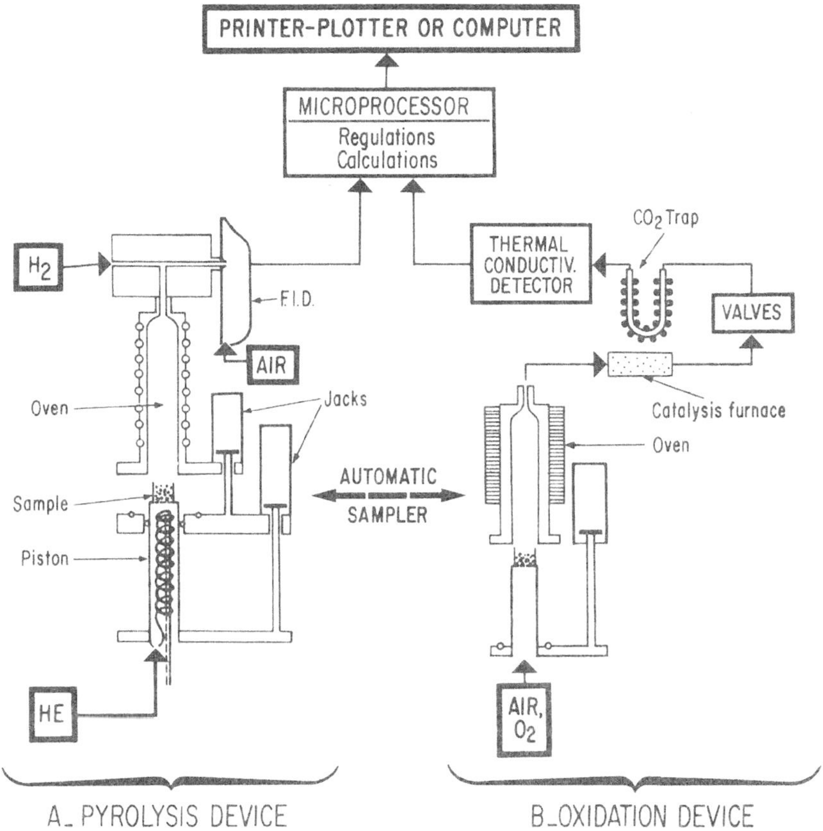
PRINTER-PLOTTER OR COMPUTER
MICROPROCESSOR
Regulations
Calculations
H_2
F.I.D.
AIR
Oven
Jacks
Sample
Piston
HE
THERMAL
CONDUCTIV.
DETECTOR
CO_2 Trap
VALVES
Catalysis furnace
Oven
AUTOMATIC
SAMPLER
AIR,
O_2
A_ PYROLYSIS DEVICE
B_OXIDATION DEVICE

Figure 32.

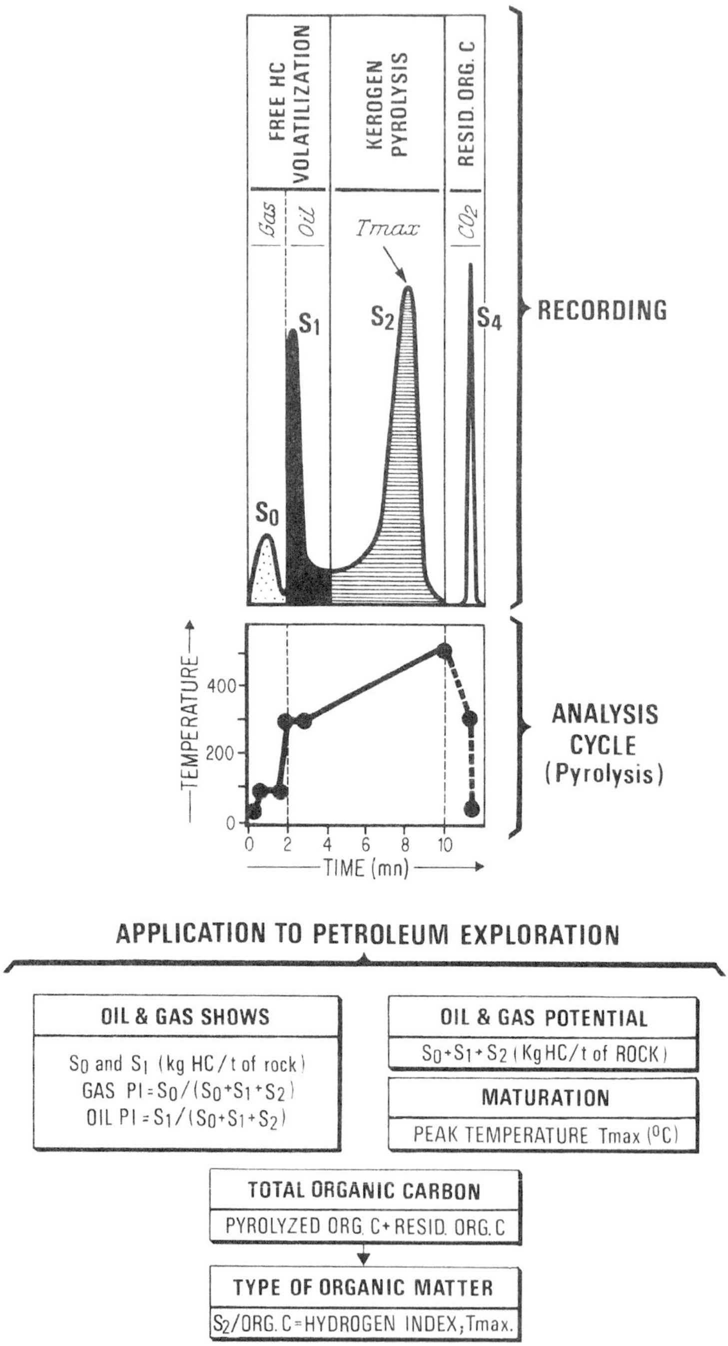
FREE HC VOLATILIZATION
KEROGEN PYROLYSIS
RESID. ORG. C
Gas
Oil
Tmax
CO2
S0
S1
S2
S4
RECORDING
TEMPERATURE
400
200
0
0 2 4 6 8 10
TIME (mn)
ANALYSIS CYCLE (Pyrolysis)
APPLICATION TO PETROLEUM EXPLORATION
OIL & GAS SHOWS
S0 and S1 (kg HC/t of rock)
GAS PI = S0/(S0+S1+S2)
OIL PI = S1/(S0+S1+S2)
OIL & GAS POTENTIAL
S0+S1+S2 (Kg HC/t of ROCK)
MATURATION
PEAK TEMPERATURE Tmax (°C)
TOTAL ORGANIC CARBON
PYROLYZED ORG. C + RESID. ORG. C
TYPE OF ORGANIC MATTER
S2/ORG. C = HYDROGEN INDEX; Tmax.

Figure 33.

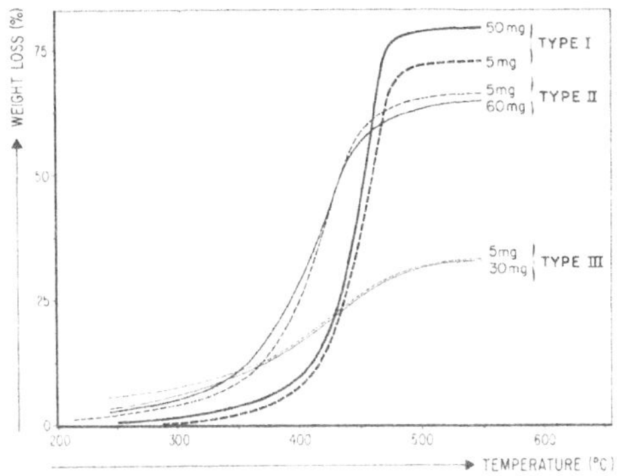

Figure 34.

CARACTÉRISATION RAPIDE
DES TYPES DE ROCHES-MÈRES

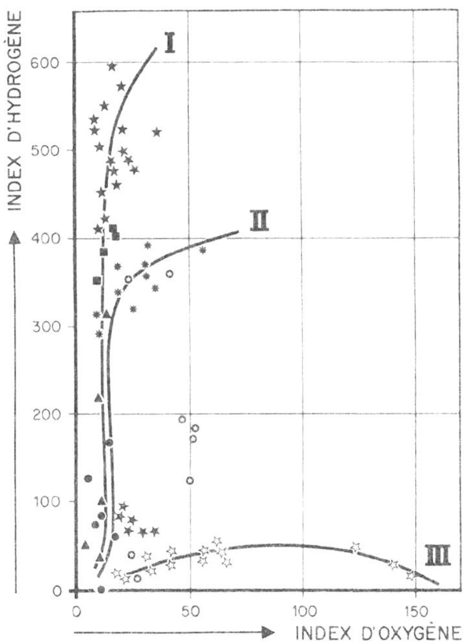

★ *GREEN RIVER SHALES*
✱ *TOARCIEN, BASSIN DE PARIS*
▲ *SILURO-DÉVONIEN, SAHARA_LIBYE*
● *PALÉOZOÏQUE SUP.-TRIAS, SPITZBERG*
☆ *CRÉTACÉ SUP., BASSIN DE DOUALA*
■ *CRÉTACÉ, GOLFE PERSIQUE (Calc. à Oligostégines)*
○ *JURASSIQUE SUP., AQUITAINE NORD*

Figure 35.

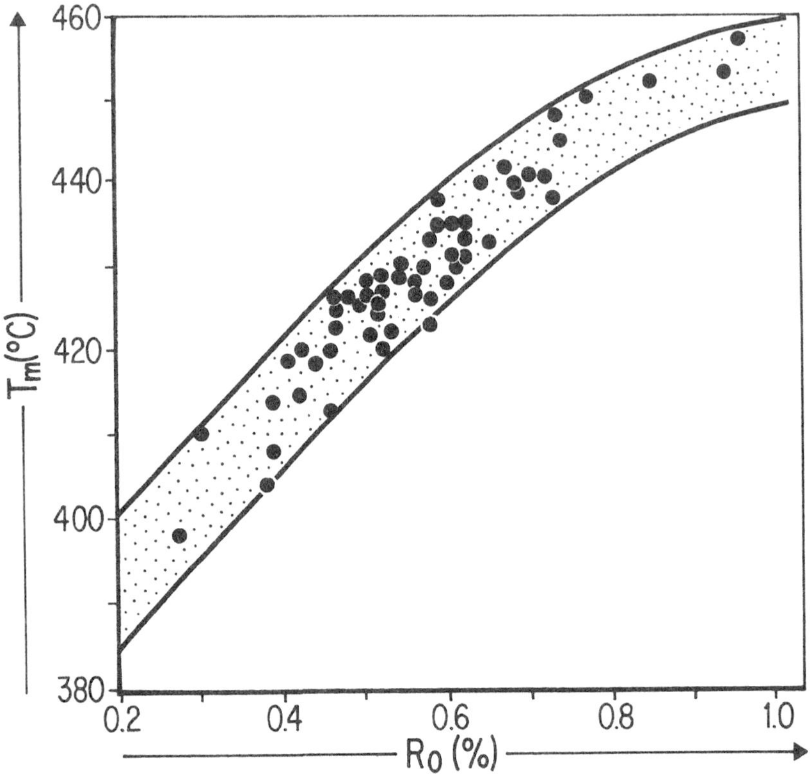

Figure 36.

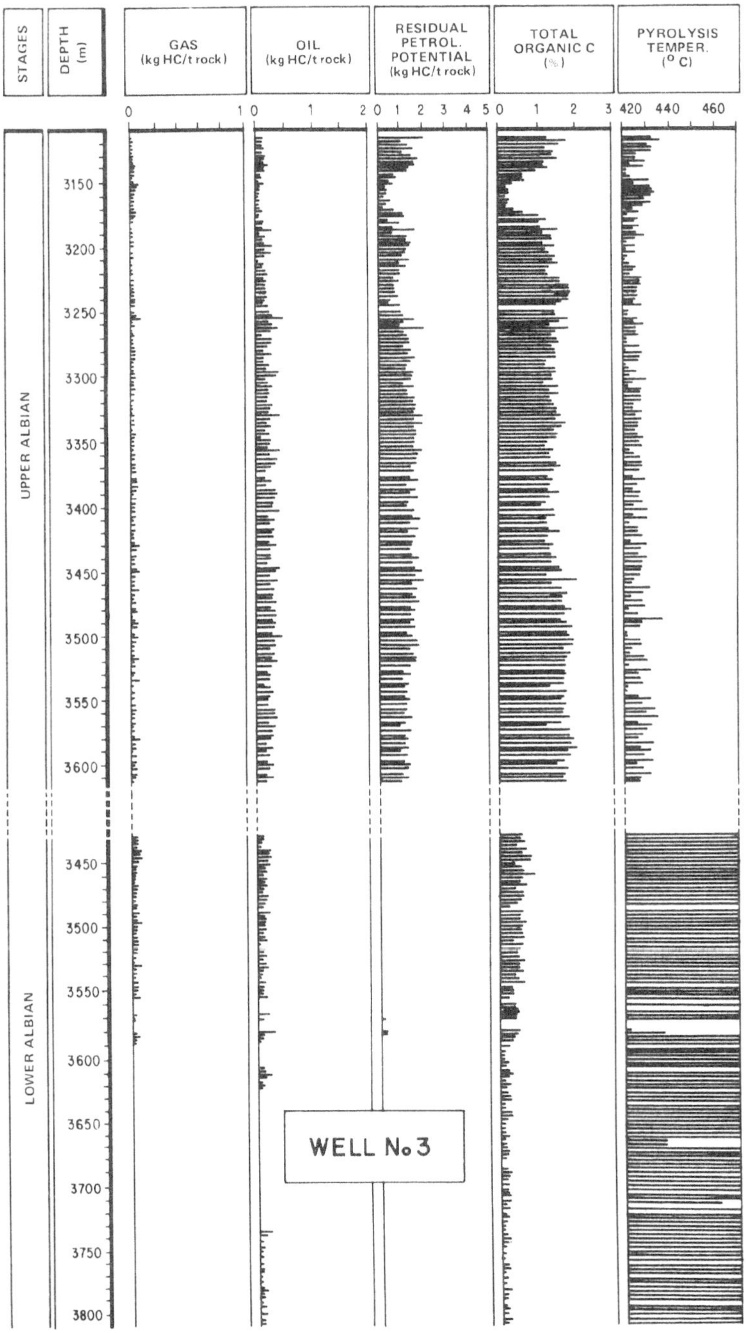
STAGES
DEPTH (m)
GAS (kg HC/t rock)
OIL (kg HC/t rock)
RESIDUAL PETROL. POTENTIAL (kg HC/t rock)
TOTAL ORGANIC C (%)
PYROLYSIS TEMPER. (°C)
UPPER ALBIAN
LOWER ALBIAN
WELL No 3

Figure 37.

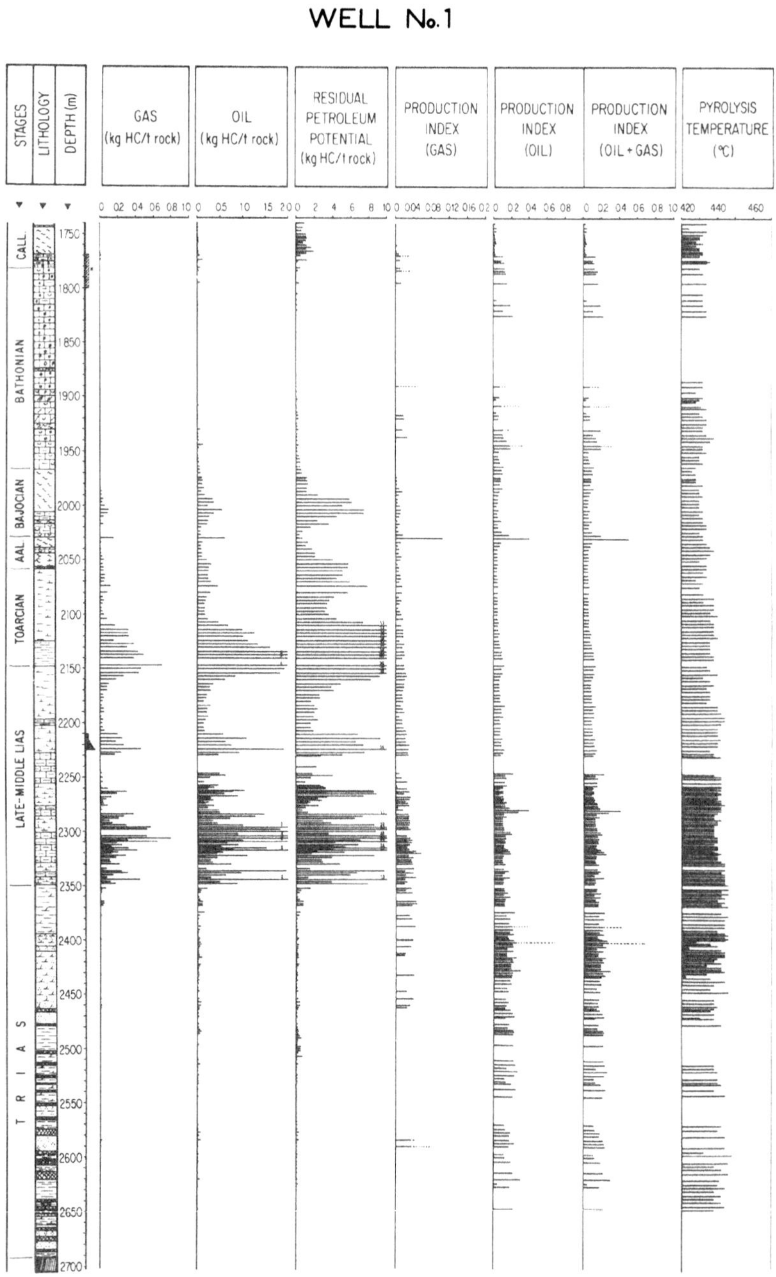
WELL No.1
STAGES
LITHOLOGY
DEPTH (m)
GAS (kg HC/t rock)
OIL (kg HC/t rock)
RESIDUAL PETROLEUM POTENTIAL (kg HC/t rock)
PRODUCTION INDEX (GAS)
PRODUCTION INDEX (OIL)
PRODUCTION INDEX (OIL + GAS)
PYROLYSIS TEMPERATURE (°C)
CALL.
BATHONIAN
BAJOCIAN
AAL.
TOARCIAN
LATE-MIDDLE LIAS
T R I A S

Figure 38.

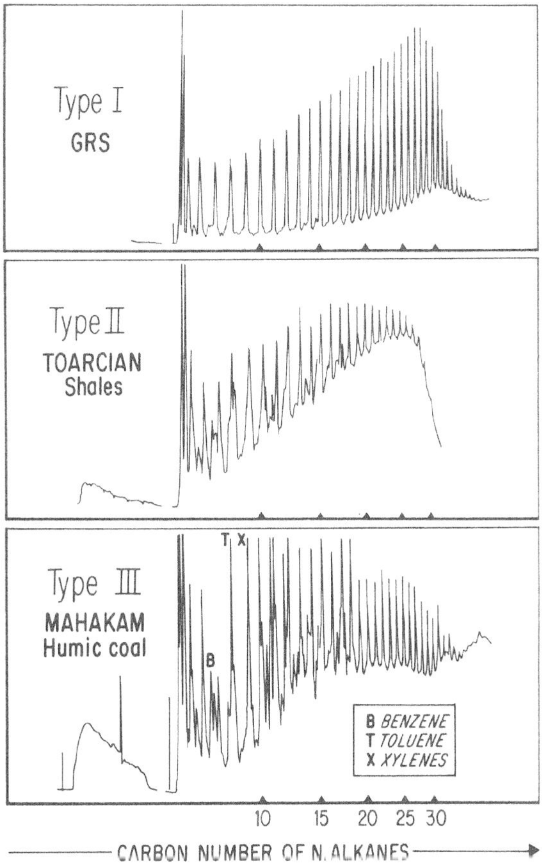
Type I
GRS
Type II
TOARCIAN
Shales
Type III
MAHAKAM
Humic coal
T
X
B
B BENZENE
T TOLUENE
X XYLENES
10
15
20
25
30
CARBON NUMBER OF N. ALKANES

Figure 39.

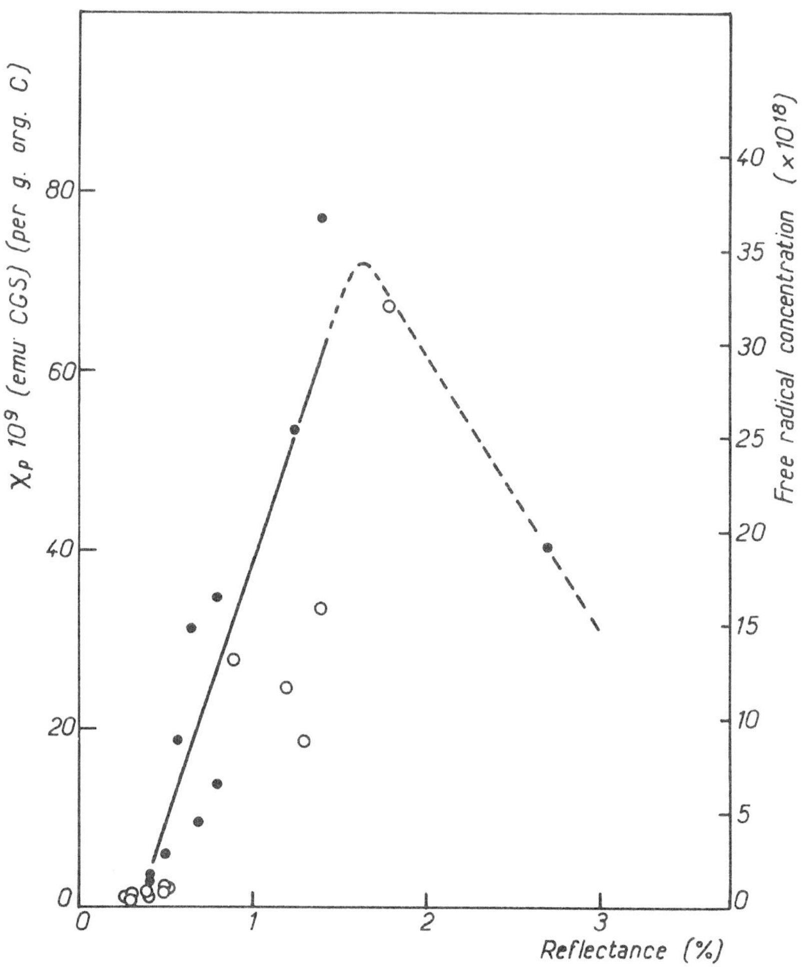
χ_p 10^9 (emu CGS) (per g. org. C)
80
60
40
20
0
Free radical concentration (x10^{18})
40
35
30
25
20
15
10
5
0
0
1
2
3
Reflectance (%)

Figure 40.

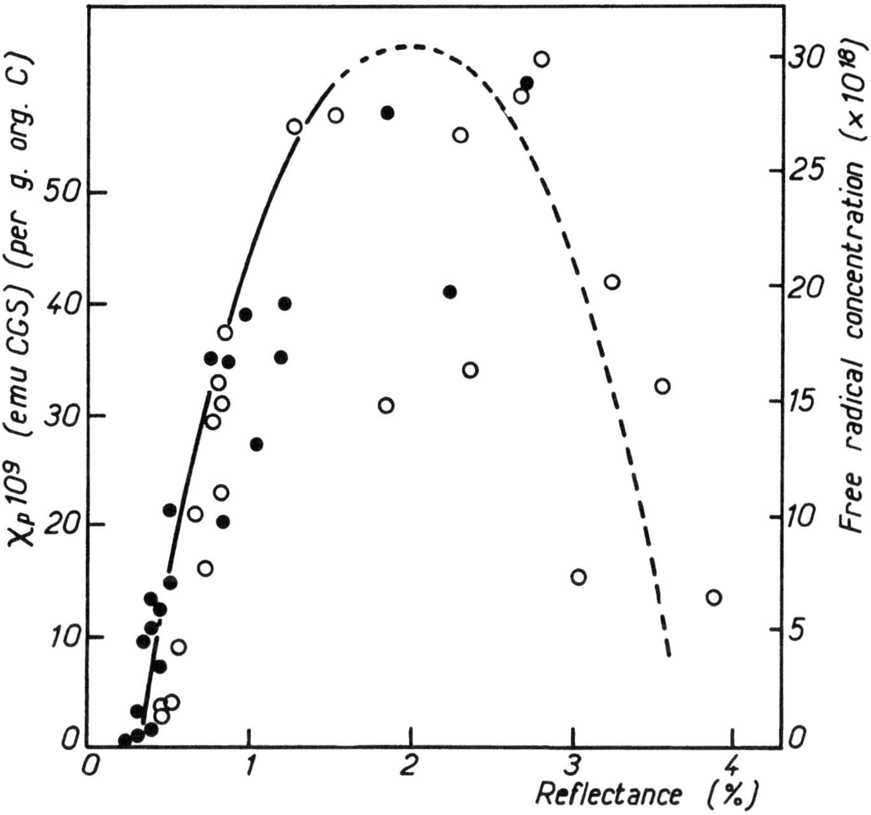
Xp 10⁹ (emu CGS) (per g. org. C)
0
10
20
30
40
50
Free radical concentration (x10¹⁸)
0
5
10
15
20
25
30
0
1
2
3
4
Reflectance (%)

Figure 41.

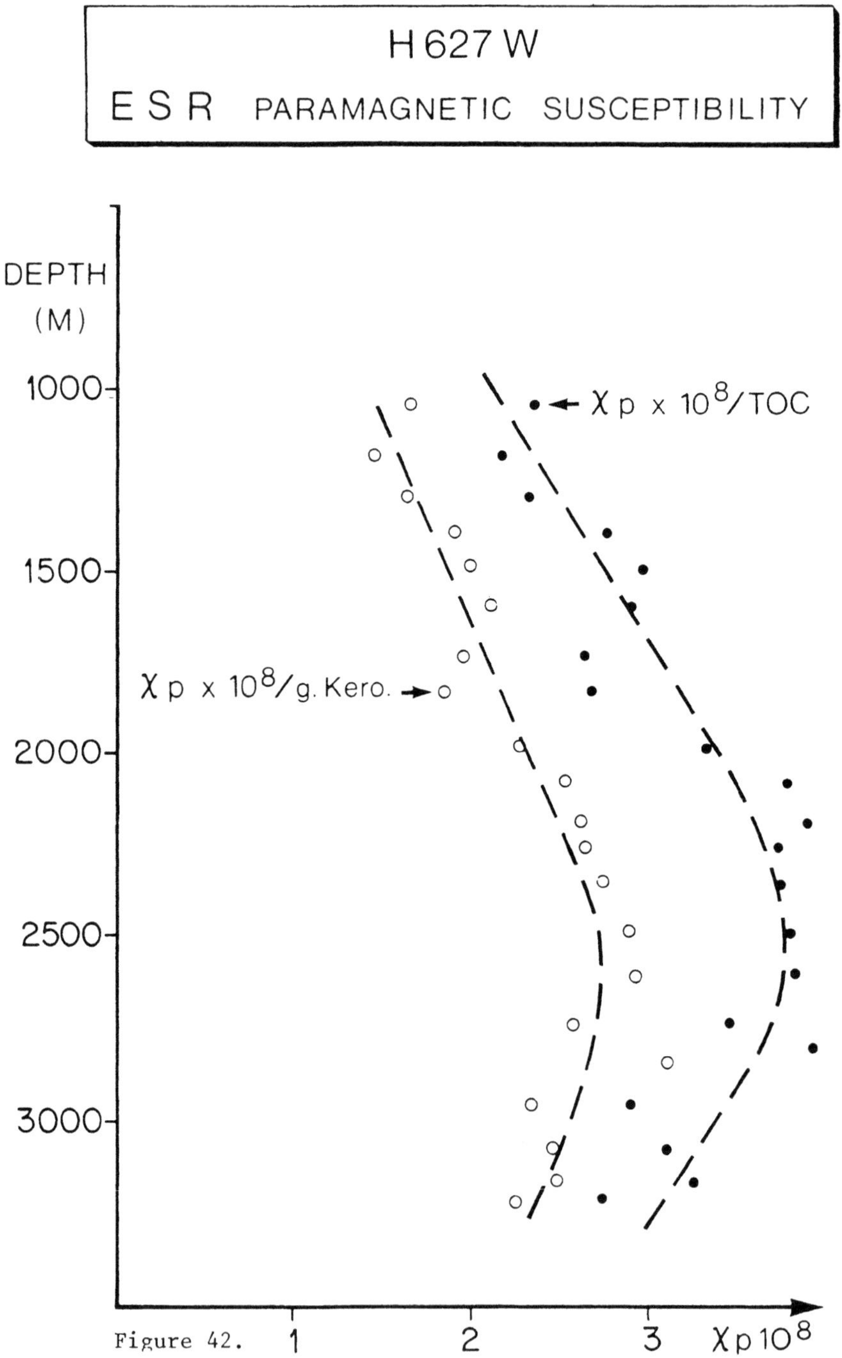
H 627 W
E S R PARAMAGNETIC SUSCEPTIBILITY
DEPTH
(M)
1000
1500
2000
2500
3000
χ p x 10⁸/TOC
χ p x 10⁸/g. Kero.
1
2
3
χ p 10⁸

Figure 42.

COAL PROPERTIES, RESOURCES AND UTILIZATION

P. CHICHE

Centre d'Etudes et Recherches
des Charbonnages de France
60550 Verneuil-en-Halatte, France

Coal is a heterogeneous mineral consisting essentially of carbon, hydrogen and oxygen, and resulting from the gradual transformation of plant debris into peat, lignite, bituminous coal and anthracite under the predominant effect of time and temperature. The petrographic constituents called macerals belong to the three main groups of vitrinite, exinite and inertinite and the degree of coalification can be determined by several rank parameters such as vitrinite reflectance. Mild degradation methods combined with powerful analytical techniques have shown that low and medium rank coals are composed of bi- to pentacyclic aromatic-hydroaromatic monomer units associated in a macromolecule. Coals and cokes possess a microporous structure characterized by the presence of constrictions which hinder the accessibility to gases and liquids. Traditional uses of coal are combustion and cokefaction, but conversion into gases and liquids will be achieved in the future to replace natural gas and petroleum feedstocks.

1. DEFINITION

Coal is a physically and chemically heterogeneous mineral or rock consisting essentially of carbon, hydrogen and oxygen, with lesser amounts of sulphur and nitrogen. Inorganic, ash-forming compounds are distributed as discrete particles of mineral matter throughout the coal substance. When heated, all coals give off tars and gases ; some of them melt and become plastic, leaving a residue of coke, while others simply leave a friable char residue.

L. Petrakis and J. P. Fraissard (eds.), Magnetic Resonance. Introduction, Advanced Topics and Applications to Fossil Energy, 287–323.

Coal may be burned to generate heat or steam, or carbonised to produce metallurgical coke. It may also be converted to liquid fuels by hydrogenation, or to synthesis gas or gaseous fuels by gasification with oxygen or air and steam.

2. SOME PROMINENT LANDMARKS IN COAL HISTORY

Around 1215 : first common use of coal as fuel
1587 : coal used to melt iron
1718 : coke used in blast furnace
1792 : coal gas produced for lighting
1851 : invention of horizontal-chamber coke oven
1913 : hydrogenation of coal (Bergius process)
1913 : Fischer-Tropsch synthesis
1934 : first work on underground gasification

However, at the beginning of the 20th century, the scientific study of coal was still in its infancy.

3. ORIGIN OF COAL

Coal results from the accumulation of plant debris that were later covered, compacted and gradually changed into an organic rock through the stages of peat, lignite (brown coal), bituminous coal and ultimately anthracite, with a corresponding increase in the percentage of carbon, a decrease in the "volatile matter content" and in the percentage of oxygen. This process, which is called "coalification" or "advance in rank", can be divided into two parts :

. diagenesis, or biochemical stage, in which biochemical factors have played an important part in the very first stage of the process, including in particular the whole of the peat-forming period, and which brought the carbon content from an initial value of 40-45 % to more that 60 %, with a corresponding increase in aromaticity.

. catagenesis, or metamorphosis, or geochemical stage, in which only chemical reactions could take place, the time temperature factors having been predominant to promote the chemical changes of coalification.

4. PETROGRAPHIC CONSTITUTION

Coal is an aggregate of microscopically distinguishable, physically distinct and chemically different substances called macerals (organic fraction) and minerals (inorganic fraction).

Coal's heterogeneity is inherited from the diversity of source materials (trees, ferns, algae, fungi), of their components (woody tissues, spores, waxes, resins ...) and of the particular conditions of accumulation and maceration as a function mainly of time and temperature, but also of pressure, acidity, and the presence or absence of oxygen (aerobic or anaerobic conditions).

As early as 1919, M.C. STOPES (1) distinguished four macroscopic components, or lithotypes (suffix -ain), occurring in coal (table 1) :

. vitrain, coherent and uniform, bright and vitreous in texture.

. clarain, which has a smooth and shiny surface when broken at right angles to the bedding plane.

. durain, hard, with a close and firm texture, rough and mat broken faces.

. fusain, powdery material with somewhat fibrous strands.

Lithotypes	Maceral group	Maceral	Origin	Chemistry
VITRAIN bright	VITRINITE	Telinite (cell walls) Collinite (cell fillings, structureless vitrinite)	Woody tissues, bark, leaves, etc.	Intermediate hydrogen content and volatiles
CLARAIN smooth and shiny	EXINITE (or LIPTINITE)	Sporinite Cutinite Resinite Alginite Suberinite Fluorinite Bituminite Exudatinite	Spore exines Cuticles Resins Algae Cork Oils and Fats	Higher hydrogen content and volatiles, more aliphatic
FUSAIN powdery or fibrous DURAIN hard and mat	INERTINITE	Fusinite Semifusinite Macrinite Micrinite Sclerotinite	Woody tissues, etc. Plant materials Fungal sclerotia, hyphae	Lower hydrogen content and volatiles, more aromatic

Microlithotypes		
Monomaceralic		Volume %
Vitrite	V	>95
Exite (liptite)	E	>95
Inertite	I	>95
Bimaceralic		
Clarite	V+E	>95
Vitrinertite	V+I	>95
Durite	E+I	>95
Trimaceralic		
Duroclarite	V+I+E	V>I+E
Clarodurite	"	I>V+E
Vitrinertoexite	"	E>V+I
Carbominerite		
	V+I+E +MM*	$20 < MM < 60$
*MM: mineral matter		

Table 1. Petrographic constituents of coal.

Later, the examination of polished coal faces by reflected light under the metallurgical microscope led to the distinction of microlithotypes (suffix -ite) and their more or less homogeneous petrographic constituents called macerals (suffix -inite) which are, for all practical purposes, classified into three maceral groups :

. vitrinite
. exinite (or liptinite)
. inertinite

4.1. Vitrinite group

Vitrinite is the predominant maceral group in most coal seams, commonly accounting for more than 70-80 % of any given coal ; thus the gross properties of a coal reflect essentielly the properties of its vitrinite, which may be differentiated essentielly into :

. telinite, where the cell walls are still apparent

. collinite, which does not display cell structure under normal microscopic observation.

4.2. Exinite (or liptinite) group

The exinite group macerals consist of the outer coats of plant spores and pollen grains (sporinite), leaf cuticles (cutinite), resinous material (resinite), algae (alginite, relatively rare), cork cell walls commonly impregnated with wax (suberinite), and various hydrocarbons of uncertain origin ("oily" macerals : fluorinite, bituminite, exudatinite). These macerals have a comparatively high hydrogen content and are believed to have the most aliphatic chemical structure. However, as the rank of coal increases, the properties of exinite approach those of vitrinite and it becomes difficult to distinguish between the two groups of macerals under the microscope.

4.3. Inertinite group

The inertinite group consists of macerals that are unreactive or only slightly reactive in coal carbonization and liquefaction processes :

. fusinite and semifusinite are derived from plant tissues similar to those which form vitrinite, the "fusinization" being the result either of charring by fires in the peat swamp (pyrofusinite) or of a severe biochemical degradation prior to burial (degradofusinite).

. macrinite and micrinite come from degradation products of plant matter during diagenesis and early stages of coalification.

. sclerotinite consists of coalified fungal remains.

Figure 1 illustrates some typical aspects of the different macerals.

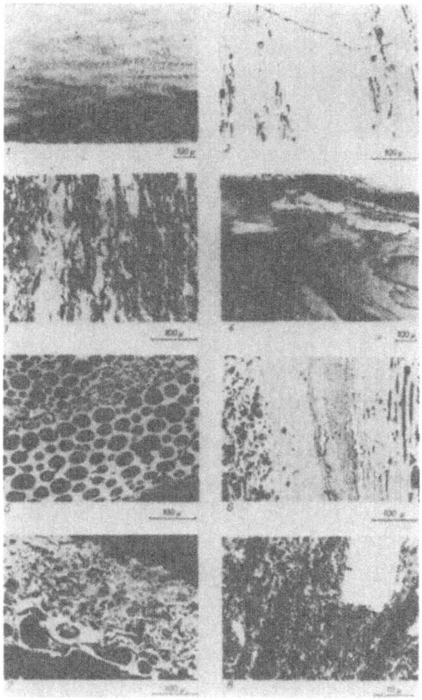

Figure 1. Some typical aspects of the different macerals.

4.4. Microlithotypes

It has been operationally useful to consider some associations of macerals that occur in particles of crushed coal. These associations are referred to as microlithotypes, the main ones appearing in the last column of table 1.

The maceral or petrographic analysis of a coal allows the coal type to be designated according to what maceral group is preponderant :

- gelified type : more than 66 % vitrinite
- fusinized type : more than 33 % inertinite
- sapropelic type : more than 33 % exinite

For the most part, coal found in the northern hemisphere is of the gelified type whereas the fusinized type is found principally in southern hemisphere regions once part of ancient continent Gondwana (South Africa, Australia, India). The relatively rare <u>sapropelic</u> type forms the transition between coals and oil shales ; cannel coals (composed of micrinite in which large quantities of microspores are dispersed) and boghead coals (which consist essentially of alginite with an admixture of micrinite) belong to this type.

Other than its type, the "nature" of coal is also determined by its degree of coalification, termed rank.

5. COAL RANK AND CLASSIFICATION

Coal is a metamorphic sedimentary rock which, in addition to its maceral composition, should be classified by its position along the chain of evolution ranging progressively from the initial plant matter remains to peat, lignite, bituminous coal, and ultimately anthracite.

If one plots the elemental composition of all coals using carbon, hydrogen and oxygen as coordinates (fig. 2), one finds that nearly all of the points fall within a narrow curved band which in essence represents the above-mentioned evolution. A rank criterion should be able to define the position of a coal along this curve.

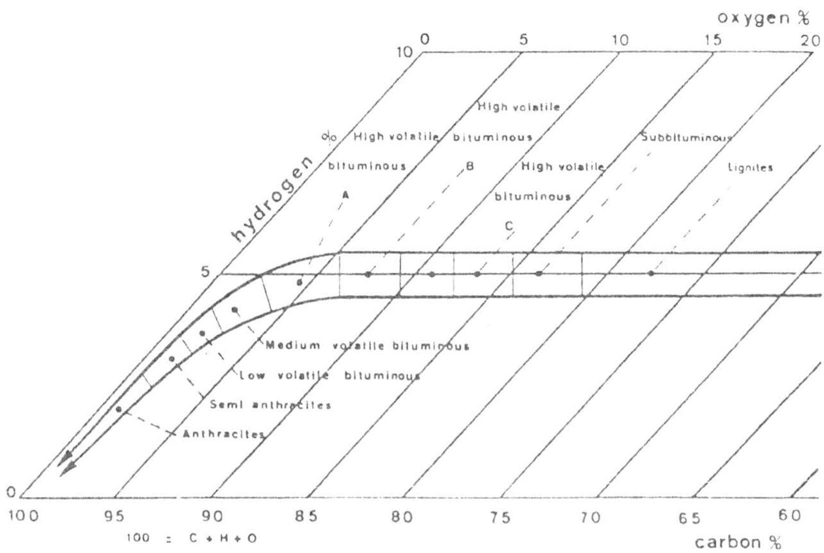

Figure 2. Seyler diagramm of coals.

Several parameters can be more or less satisfactorily adopted in this regard. The majority of them are global quantities, the parameter being determined on the whole coal without taking the maceral composition into account. Possible parameters to describe rank are :

- carbon, oxygen or hydrogen content
- volatile matter (VM) content
- calorific value or moisture

However, it is clearly more advisable to use the reflectance of visible light, which is a point function, measured on collinite, the maceral which appears to be the most homogeneous from the group vitrinite.

In addition to its point function character and rapid execution due to an automated apparatus, reflectance has the advantage over other coal rank parameters to be able to discern whether the coal is homogeneous or is a mixture of coals from diverse sources, in which case the reflectance can give their relative proportions with a good level of accuracy. For certain utilizations, coking for one, such a distinction is absolutely critical.

Figure 3 illustrates the variation of reflectance with rank.

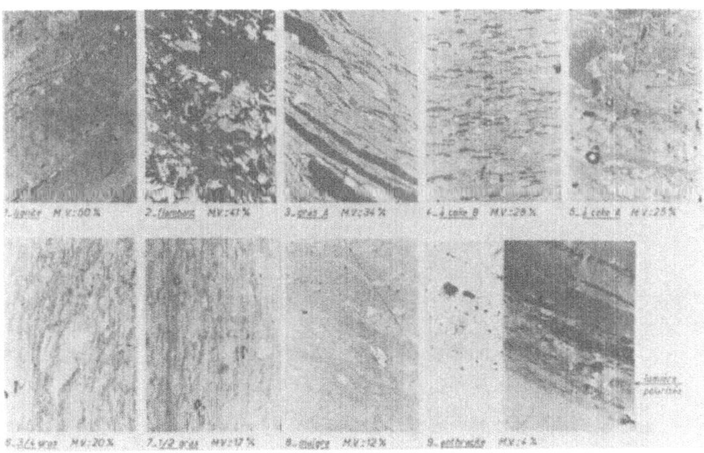

Figure 3. Dependence of reflectance on rank.

Figure 4 shows how various rank parameters (VM, O, H, and R) vary as a function of carbon content. Note that the hydrogen content does not drop until the beginning of the anthracitic stage and thus is not suitable as a rank criterion. At first glance, oxygen content seems more suitable, however its direct determination is difficult and most often it is determined by difference, accumulating the errors of all the other elemental analyses. Meanwhile, there is a good bit of scatter in the volatile matter content for low rank coals. Unfortunately, a universal classification of coals based on petrographic analysis and the reflectance of vitrinite does not yet exist. We are still using the international classification of coals published in 1956, dividing coal into 10 classes, 3 groups, and 5 sub-groups (table 2).

Two rank parameters are used to define the 10 classes (as a function of rank) : volatile matter content (dmmf) between 0 and 33 % (classes 0 to 5) and gross calorific value (with moisture, but mineral matter free) for VM higher than 33 %.

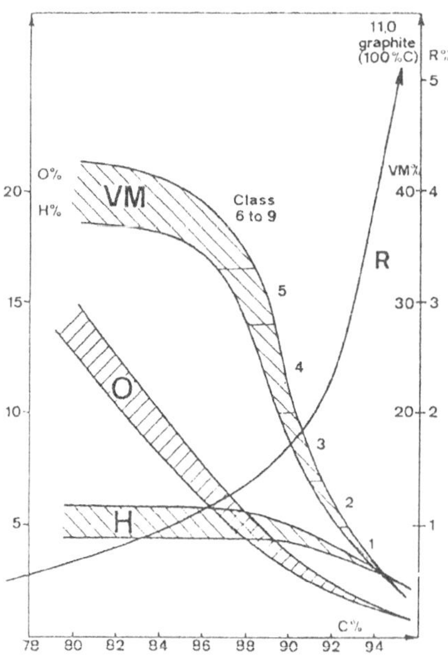

Figure 4. Principal rank parameters.

The 3 groups (as a function of caking power) are determined according to the free swelling index or Roga index.

The 5 sub-groups (as a function of coking power) are defined based on the dilatometer test or the Gray-King test.

Thusly, every coal is characterized by a three digit number representing in order the class, group, and sub-group. For each class of coal, table 3 shows the typical range of values for VM, C, H, O and R on a dmmf basis.

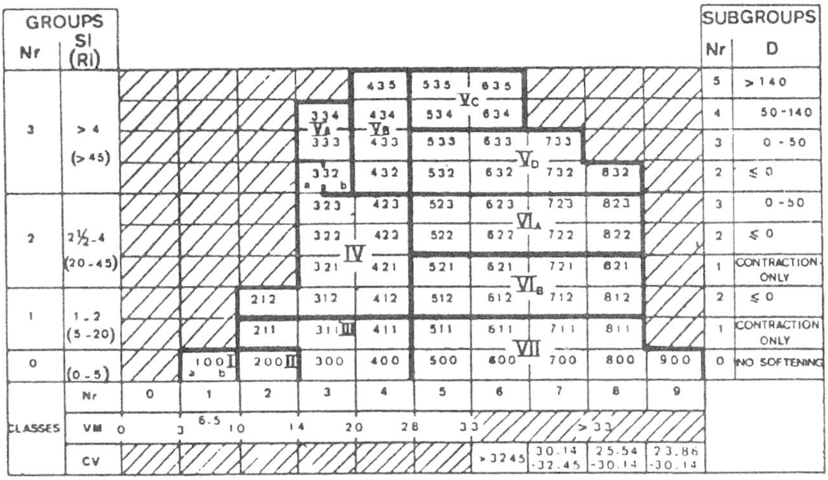

VM : volatile matter (dmmf)%

CV : gross calorific value MJ/Kg

SI : swelling index RI : Roga caking index

D : dilatation

Table 2. International classification of coal.

Class nr	1	2	3	4	5	6	7	8	9
VM % dmmf	3-10	10-14	14-20	20-28	28-33	33-41	33-44	35-50	42-50
C	91-94	90-92	89-91	86-91	84-86	81-84	80-84	74-83	74-80
H	3-4	4-4.5	4.5-5	4.8-5.2	5-5.4	5.2-5.7	5.2-5.7	5-6	5-6
O	1.3-2	2-3	2.5-4	3-6.5	5.5-8	6-11	7-12	10-18	12-20
CV MJ/kg dmmf	36.2-36.3	36.3-36.5	36.5	36.2-36.5	35.6-36.5	34.3-35.8	33-34.8	30.1-32.7	29.3-31
Reflectance vitrinite %	2-3.5	1.9-2.3	1.6-1.9	1.3-1.6	1.1-1.3	0.8-1.1	0.6-0.85	0.5-0.65	0.4-0.5

Table 3. Typical values of rank parameters.

6. CHEMICAL COMPOSITION

Any raw coal contains moisture, mineral matter (often improperly called ash), and organic matter.

6.1. Moisture

In its natural state, coal is generally saturated with water, the concentration of which varies with rank. For bituminous coals and anthracites, the concentration is between 1.5 % and 10 % while for lignite and peat, the moisture content is much greater.

6.2. Mineral matter

The mineral matter in coal comes from diverse origins :

- constitutive minerals found in the original plant tissue and intimately mixed with the organic matter in a proportion of less than 1 % (constitutive mineral matter).

- animal skeletons or sediments deposited during the coal forming process and thus dispersed throughout the organic matter (deposited mineral matter).

- intercalated rock beds or walls which are external to the organic matter.

The mineral matter content of a coal after dense medium washing is dependent on the distribution of mineral matter types and is rarely below 7-8 %.

The mineral matter itself is composed of complex minerals. In general, the primary constituents are hydrated aluminosilicates (clays such as illite, kaolinite ..., shales) with a concentration typically greater than 50-75 % of the mineral matter, followed by carbonates (calcite, dolomite, siderite). Smaller amounts of sulphur containing compounds (pyrite, marcasite), sulphates (gypsum) and silica (quartz) are often present.

6.3. Organic matter

The study of the chemical composition of the organic matter in coal is an extremely arduous task due to the organic matter complexity, macromolecular character and low solubility.

Elemental analysis gives the concentrations of the major constituents : carbon, hydrogen, oxygen (which vary, as we have seen, as a function of rank), sulphur and nitrogen (always present in variable quantities independent of rank).

The number of experimental techniques that can be used to study solid coal is relatively limited. Maceral analysis, reflectance, and pore structure measurements don't give information concerning chemical structure. At the present time, infrared spectroscopy, electron spin resonance, and above all solid state ^{1}H and ^{13}C nuclear magnetic resonance are about the only techniques that are capable of providing direct information about coal chemical structure.

One can obtain indirectly such information by mild degradation methods, where the coal is decomposed into smaller fragments with molecular weights small enough to be handled by standard analytical techniques, but still representative of the initial coal structure. It should be pointed out however that only low and medium rank coals are amenable to such an analysis ; high rank coals possess such a high degree of aromatic condensation that no chemical methods are capable of selectively degrading it.

6.4. Degradation methods

Solvent extraction. For simple extraction, one generally uses solvents such as chloroform, benzene, or pyridine, the latter providing the highest comparative yields as shown in table 4 for a series of vitrains of various rank (2) :

Carbon content %	Pyridine soluble %
81.7	28.3
82.8	30.1
84.3	22.3
84.7	31.4
87.1	37.1
89.0	37.9
90.9	0.6

Table 4. Pyridine soluble yield of a series of vitrains

One may question the representativeness of these results with respect to the entire coal sample. However, it has been shown, in certain cases, that the composition of the extracts does not vary significantly as a function of the level of extraction so that such an apprehension may be less founded than one thought.

The most soluble part of coal consists of straight chain paraffins from C_{13} to C_{34}, and also numerous isoprenic compounds derived from phytol of chlorophyll $C_{20}H_{39}OH$ and present in the constituents of the group exinite.

Acid catalyzed depolymerization. Coal may be "depolymerized" by using it as an alkylating agent at temperatures as low as 100°C. This method is based on the assumption that coal contains aromatic structures linked by aliphatic bridges such as methylene bridges - CH_2 - sufficiently reactive to participate in an acid catalyzed transaryl-alkylation reaction.

In certain cases, the extraction yield is quite high, as can be seen in table 5 (3).

Coal type	Carbon % dmmf	Solubility %
Lignite	70.6	75.2
Subbituminous	76.7	23.4
hva bitum.	82.4	47.4
hvb bitum.	85.1	28.8
hvc bitum.	86.8	25.0
lv bitum.	90.7	9.8

Table 5. Depolymerization of coals of different rank with phenol - BF_3 at 100°C.

However, the extracts contain quite a bit of high molecular weight colloidal products, contrary to what one would expect from the supposedly reactive bonds.

Alkylation and acylation. In the presence of a Friedel-Crafts ($AlCl_3$, BF_3) catalyst, the solubility increases only slightly with alkyl chlorides while remarkable results for medium rank coals are obtained with long chain acyl chlorides (4). Alkylation takes place even with olefinic chains in the presence of a catalytic mixture HF-BF_3 (table 6).

These data demonstrate clearly that such reactions are best applicable to bituminous coals of 85 to 90 % carbon content. This result can be interpreted as an indication that coals of this rank contain the largest number and the most easily accessible aromatic C-H groups.

Coal type	VM%	Pyridine soluble %		
		Before treatment	Acylation C_nH_{2n+1} COCl (max.eff.)	Alkylation $BF_3 + C_{16}H_{32}$
High volatile bitum.	34.4	18.5	28.1 (n=11)	26
Medium " "	24.2	27.2	85.3 (n=11)	68
Low " "	19.4	3.5	85.0 (n=15)	65
Low " "	14.4	2.0	17.2 (n=9)	8

Table 6. Solubility increase of coals by acylation and alkylation.

Flash pyrolysis, qualified by a heating rate of several hundreds to thousands of degrees per second, breaks the weakest bonds in the coal "macromolecule" without profoundly disturbing the structure of "monomer units" subsequently liberated. These monomer units can evaporate into the gas phase and thus furnish "primary" tar in very high yields on the condition that no secondary cracking takes place.

In practice, flash pyrolysis is routinely accomplished by heating the coal in a fluidized bed or by mixing with a solid heat-carrying medium, which corresponds to the first step in the combustion of pulverized coal. On the laboratory scale, several experimental devices can be used, such as a vertical electric oven in which the pulverized coal particles are allowed to fall freely, or a metallic mesh screen placed in an air-tight vessel, the coal put inside the mesh being heated by passing a controlled electrical current of perscribed duration through the mesh screen (5).

Flash pyrolysis can increase enormously the tar yields with respect to ordinary slow pyrolysis. For example, using the metallic mesh screen technique, a coal with 34 % volatile matter was heated at 1500°C/s and gave a 45 % tar yield while a yield of only 15 % was obtained when the coal was heated at 1 to 5°C/min. As the tar yield increases, the tar composition becomes more and more similar to that of the parent coal. Subsequent tar analysis reveals the presence of "monomer units" from two to four condensed aromatic rings in size.

Mild hydrogenolysis. Hydrogenolysis is a combination of pyrolysis and hydrogenation. Heated to just above the threshold of pyrolysis (420-430°C) under hydrogen pressure, coal decomposes into fragments that for the most part are prevented from recombining by the hydrogen, which quenches the free radicals. Generally, this is done in the presence of a hydrogen donor solvent so that it is almost impossible to distinguish between what product comes from the coal and from the solvent. For structural studies, it is preferable to work without solvent. Accordingly, it was possible to solubilize 80 to 90 % of the organic matter of a low rank coal, and in particular 98 % of its exinite, in normal hexane in the presence of the catalyst MoS_2 (6). In parallel, experiments using a series of model compounds representing different structural types or groups supposedly present in the coal have established that, in the corresponding operational conditions :

- ring cleavage is very rare, the condensed polycyclic clusters remain intact, except for a possible partial hydrogenation

- C - C bonds are broken easily but without leading to condensation reactions : no polycyclic compound has been detected in hydrogenation products from a phenol-formol resin

- aliphatic chains, even long ones at that, are released in their entirety with relatively little cleavage.

- for the most part, hydroxyl groups remain intact while carbonyl groups are eliminated

- only bi- to tetracyclic compounds, but never monocyclic, have been observed in the low rank coals studied

- the polar fractions are composed of compounds containing oxygen (30 to 40 %) and nitrogen (60 to 70 %) having only one heteroatom per molecule.

Oxidation. Coal oxidation is of utmost importance in many respects : combustion of course, but also spontaneous heating, more or less rapid even at temperatures lower than 100°C, which is capable of starting a fire in the mine or in stockpiles of coal, and also degrading the coking properties of coal, etc. But from the structural analysis point of view, the degree of oxidation has been in the past too severe to allow truly interesting conclusions to be made.

More recently though, new techniques have been reported in the literature which claim to be able to preferentially attack either the aliphatic part (using dichromate, permanganate, peracetic acid) (7) or the aromatic part (using trifluoroacetic acid with hydrogen peroxide) of coal (8).

6.5. Analytical methods

The above mentioned more or less controlled degradation methods lead to product mixtures that are quite often highly complex, whereby only the combined application of an arsenal of complementary techniques is capable, imperfectly at that, of separating identifiable constituents or groups of constituents. Let us cite rapidly :

. elemental analysis (above all C, H, O, N, S)

. different chromatographic techniques : thin layer chromatography, high performance liquid chromatography (HPLC), capillary column gas phase chromatography, which gives access to polycyclic hydrocarbons containing 24-26 carbon atoms. However, the problem with chromatography is one of identifying the separated components because the calibrated retention times and volumes obtained from pure compounds do not always suffice in making unequivocal assignments.

. coupled gas phase chromatography - mass spectrometry, which combines the resolving power of chromatography with identification by mass spectrometry.

. mass spectrometry, which can be used with different ionization energies and at different resolution levels, reaching masses of several thousands, corresponding to a boiling point on the order of 500°C under atmospheric pressure, that is to a polycondensed aromatic structure containing 34 carbon atoms.

. infrared spectroscopy, and especially its Fourier transform version (FTIR) which improves the sensitivity, speeds up the acquisition rate, and gives access to the far infrared region (10 to 200 cm^{-1}).

. ^{1}H and ^{13}C nuclear magnetic resonance (NMR), which has been used to characterize liquid degradation products and recently, thanks to cross polarization and magic angle spinning, to study solid coal.

. electron spin resonance, giving information about the number and nature of paramagnetic centers (free radicals).

. molecular weight determination by vapour phase osmometry or gel permeation chromatography, one area in which much progress can be made, the reliability of the present results being questionable.

6.6. Present conceptions

Except for high rank coals (more than about 95 % carbon), it is generally agreed today that coal is formed of relatively small condensed aromatic units containing 2 to 3, then 3 to 5 condensed rings. The non aromatic part essentially consists of hydroaromatic rings (i.e. completely hydrogenated), besides a few alkyl chains most often reduced to methyl groups - CH_3. These condensed clusters form the "monomer units" attached to each other by either direct C - C bonds or as in most cases by bridges (predominantly methylene - CH_2 - and ether - O - bridges).

The functional distribution of carbon varies with increasing rank (fig. 5) : the concentration of methyl groups diminishes, the concentration of non-hydrogenated carbon (condensed or substituted) increases while that of aromatic CH groups increases initially, then decreases when the polycyclic condensation prevails over the aromatization of naphthenic rings. This evolution is just as clear in a representation where the ratio aromatic carbon not bound to a hydrogen atom/aromatic carbon of CH groups $C_{ar}/C_{CH_{ar}}$ is plotted as a function of volatile matter content (fig. 6). This ratio is the smallest in the coking coal region (coals very rich in CH_{ar}) and increases on either side : on the low rank side because of increased substitution, on the high rank side due to extended aromatic condensation (9).

Only 1 to 2 % of coal is comprised of true paraffins, which can account for up to 15 % of exinite. Most of the aliphatic carbon is found in methyl groups CH_3 but also in long chain paraffins all the way up to C_{30}, with odd numbered chains and branched structures being more abundant.

The aromaticity, defined as the ratio aromatic carbon/ total carbon (C_{ar}/C), increases with rank, from 65 to 75 % for low rank coals (C 75-80 %) all the way up to 100 % for anthracites. In a given coal the aromaticity varies from one maceral group to another, as seen from the data presented below for a 85.8 % carbon coal (hva bituminous) (10) :

- vitrinite 85 %
- exinite 66 %
- micrinite 85 %
- fusinite 93-96 %

In general, exinite is less aromatic and inertinite more aromatic than the associated vitrinite.

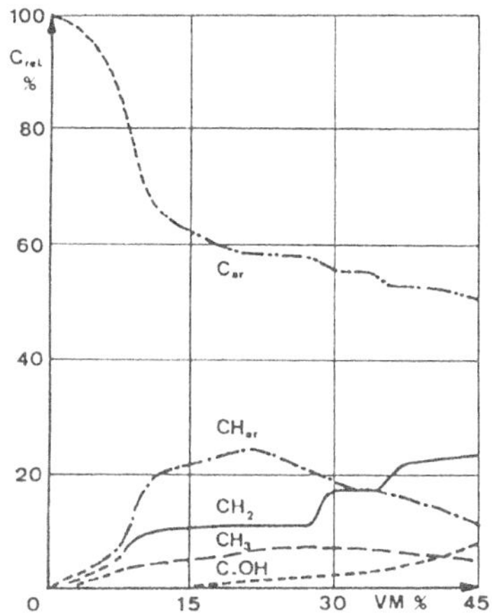

Figure 5. Distribution of carbon as a function of rank.

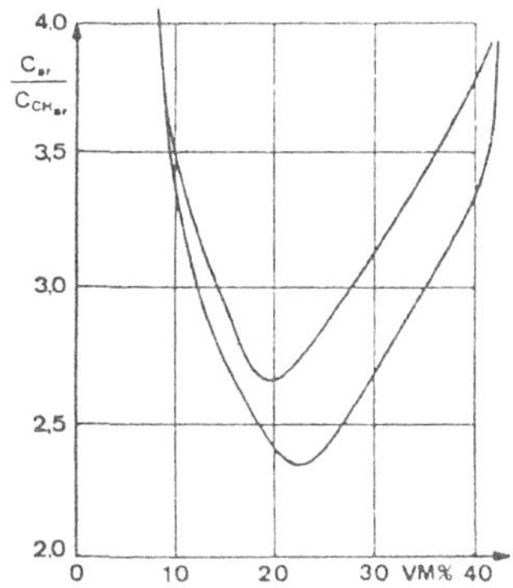

Figure 6. Dependence of aromatic carbon on rank.

The preceding consideration about the functional distribution of carbon has a direct bearing on that of hydrogen whose large majority is bound to carbon atoms in rings (aromatic or naphthenic hydrogen) or in CH_3 and CH_2 groups directly attached to rings. It can be pointed out that for low rank coals (80 % C), hydrogen is 2/3 aliphatic (including naphthenic) and 1/3 aromatic while for high rank coals (93 % C) the fractions are approximately reversed. Hydrogen can also be bound to atoms other than carbon such as sulphur, nitrogen and oxygen.

All coal contains some amount of oxygen, decreasing with increasing rank. A priori, many kinds of oxygen containing groups could exist in coal. However, only very low concentrations of alcoholic (CH_2OH or CHOH), carboxylic (COOH) and methoxylic (CH_3O) groups have been detected in bituminous coals even though these functional groups may have existed in the original organic matter and in lower rank coals (lignites and subbituminous coals). Essentially, oxygen is encountered in the form of hydroxylic, carbonylic (including quinones) and etheric (including five or six membered heterocycles) groups.

To give an idea of the oxygen distribution in coal, two examples are presented in table 7 (11).

Coal	Burning Star bituminous 71.5 %C	Belle Ayr subbituminous 69.3 %C
Oxygen content as:		
hydroxyl groups(-OH)	2.4	5.6
carboxyl .. (-COOH)	0.7	4.4
carbonyl .. (=CO)	0.4	1.0
ether .. (-O-)	2.8	0.9
Total	6.3	11.9

Table 7. Oxygen distribution in two coals.

It has been found that the proportion of phenolic oxygen in the organic matter in coal is a good rank parameter even though the proportion of phenolic oxygen with respect to the total oxygen content varies at random with rank.

Even though one can obtain precise, coherent data concerning the different forms of the main elements in coal, carbon, hydrogen and oxygen, whose concentrations evolve with rank, the same is not true for organic sulphur and nitrogen, whose concentrations are generally small and vary at random as a function of rank.

Organic sulphur probably results from sulphur-containing amino-acides in the organic plant matter, and is generally considered to be present in coal in the form of thioethers, both aliphatic (sulphur or disulphur bridges) and heterocyclic (highly stable).

There is also a lack of knowledge about the origin and chemical evolution of nitrogen, which is almost solely present in organic form. It is very probable that a contribution is made from alkaloids and related compounds where the nitrogen is in heterocycles sufficiently resistant to have survived the coalification process.

Many structural models for coal have been proposed. The most generally accepted ones are "aromatic-hydroaromatic" in character, such as the model proposed by Wiser (12) (fig. 7), where one also finds polycyclic monomer units and the different types of bonds interconnecting them to form the coal "macromolecule". Such a model cannot claim to represent any existing portion of an actual coal, but can aid in visualizing the different types of probable structures, explaining or predicting the changes in coal during coalification or carbonization processes.

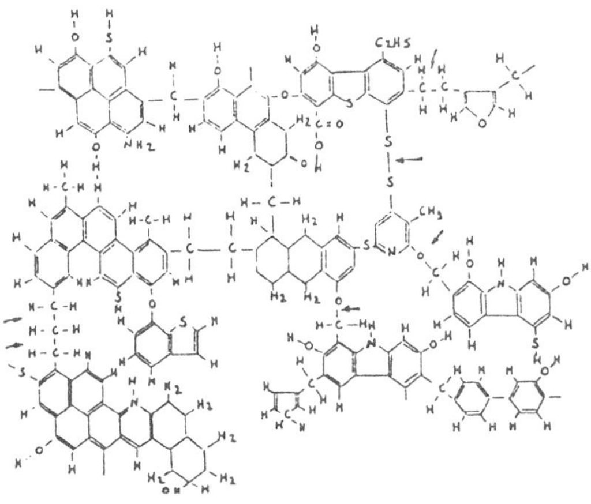

Figure 7. WISER model of coal chemical constitution.

7. PYROLYSIS AND CARBONIZATION

When a coal is heated in the absence of air at the rate of a few degrees per minute, one can distinguish four temperature zones where very different phenomena occur.

7.1. Below 350°C

At least for bituminous coals and anthracites, virtually the sole phenomenon that can be observed is the desorption of pre-existing gases from the interior of the pores : water, between 100 and 150°C, methane and carbon dioxide.

7.2. Between 350 and 500°C

In this temperature range, the most profound transformations take place accompanied by the release of most of the volatile matter and, for medium rank coals, by a softening followed by a resolidification to semi-coke.

The volatile matter essentially consists of tar. In the condensed phase, a competition exists between cracking reactions, that break carbon-carbon bonds to form less polymerized constituents than the parent coal, and aromatization and condensation reactions which lead, on the contrary, to an extension of the aromatic structure by both the dehydrogenation of saturated rings and the formation of new carbon-carbon bonds.

The relative importance between both types of reactions, the first of which produces mainly tars while the second leads to a solid residue, depends on the available hydrogen that can combine with oxygen and inhibit recondensation by saturating the unfilled valence shells. Also even a moderate hydrogenation greatly increases the yields in tar and benzol, while oxidation has just the opposite effect.

Coals can be classified into three groups, based on their behaviour in this temperature range, using two broad limits (38-40 % VM or 0.65-0.75 % R and 16-18 % VM or 1.5-1.7 % R) :

. low rank coals are non-softening during carbonization due to an elevated hydroxyl group concentration that favours cross-linking and condensation reactions at the expense of cracking ;

. medium rank coals possess coking properties, exhibiting thermoplastic behaviour above about 350°C followed by thermosetting behaviour above 470-500°C leading to a mechanically resistant coke residue ;

. high rank coals are once again non-softening due to their highly condensed aromatic structure.

In the case of softening coals, characterized by a softening or even a true melting over a temperature range of 100°C, the beginning of the process is a physically reversible and not a pyrolytic phenomenon, rapidly followed by cracking reactions. As the temperature increases, condensation reactions eventually take over, so that the plastic region is upperbound by the temperature of resolidification which is irreversible as soon as the residual volatile matter content falls below 10 to 15 % whatever the rank of the initial coal may be.

The phenomenon of structural reorganization preceding resolidification is of paramount importance for all of the physico-chemical properties of the resulting solid coke or char (porosity, reactivity, friability ...).

Medium rank coals exhibit a truly remarkable phenomenon, which is even more spectacular in high temperature pitch, the distillation residue of a coke oven tar (13, 14) : within an initially isotropic substance, tiny droplets of insoluble anisotropic liquid crystals - mesophase - appear, formed by stacks of approximately planar layers of carbon atoms, but deformed at the surface (fig. 8 and 9). The small liquid crystal spheres, initially well dispersed, grow with increasing temperature until they touch one another and eventually

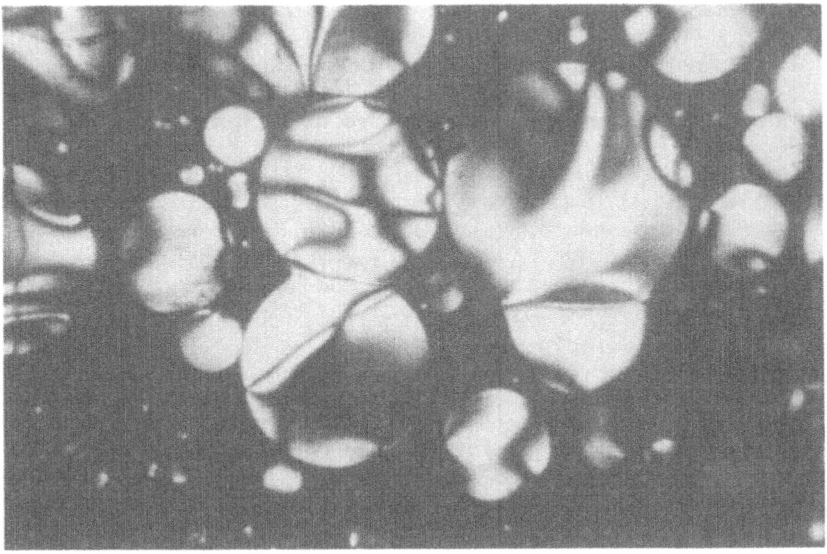

Figure 8. Mesophase in pitch.

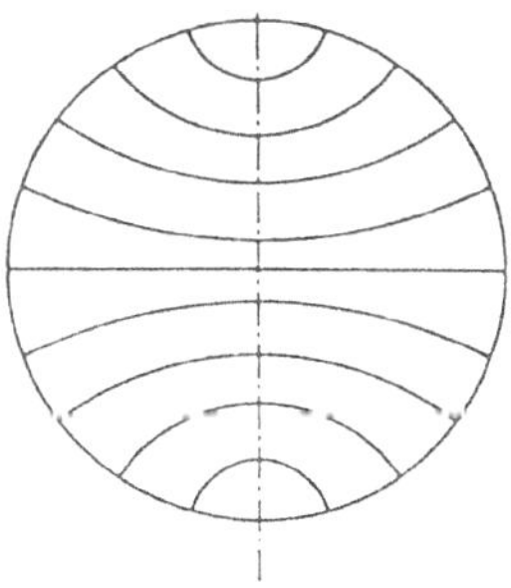

Figure 9. Schematic représentation of a droplet of mesophase.

coalesce, the isotropy progressively giving way to anisotropy all the way up to the resolidification point. According to the degree of extension of each crystal, one can end up with either a mosaic type texture or a fibrous texture, in certain cases, composed of bands characterized by long range order without the short range order being noticeably different. This type of long rang order is specific to graphitizable carbon substances, that is, substances that are capable of transforming into graphite at 2500°C. Medium rank coal is such a substance, producing a dense coke, low in reactivity and porosity.

On the other hand, carbonization of low rank coals leads to non graphitizable chars that are porous and reactive because cross-linking by oxygen does not allow the melting and mobility necessary for molecular reorientation.

High rank coals, although non softening, possess a pregraphitic structure, by virtue of its place in the coalification scheme, that is completely transformed at high temperatures.

7.3. Between 500 and 1000°C

From an irreversibly solid residue, methane is released in rapidly decreasing quantities accompanied by carbon monoxide (especially between 600 and 800°C) and hydrogen (with its characteristic maximum around 750°C).

7.4. Above 1000°C

Weight loss is almost nil. Only solid state reactions take place, such as reduction of some of the mineral matter by carbon, removal of organic sulphur and nitrogen around 1500°C and ultimately, in some cases, graphitization between 2500 and 3000°C.

8. PORE STRUCTURE OF COALS AND COKES

One of the fundamental characteristics of coals and cokes is the presence of a fully developed network of pores, a large portion of which consists of micropores with diameters less than 2 nm and constrictions which make it difficult or even impossible for gas and liquid molecules to pass through the pores. The pore structure plays an important role in most processes of coal and coke utilization : release of methane or carbon dioxide in the mines, water uptake in coal-water slurries, reactivity of coke during its gasification, production of very adsorbent active carbons by partial gasification of porous reactive chars, used in air or water purification (15).

Generally, pores are classified according to their diameter :

- macropores : diameter greater than 50 nm
- transitional pores or mesopores : diameter between 2 and 50 nm
- micropores : diameter less than 2 nm.

Several methods are capable of characterizing quantitatively the internal pore structure of coals and cokes. The measurement of the true density using helium coupled with apparent density measurements using different fluids gives the total pore volume and the accessible pore volume with respect to the different fluids based or their molecular dimensions. Mercury penetration under pressure gives the pore volume distribution as a function of pore diameter for macropores and the larger mesopores.

Adsorption of gases and vapours gives the total surface area of the pores and the pore volume distribution as a function of pore diameter for the mesopores and part of the micropores.

Small angle X-ray scattering allows one to measure the total surface area and porosity including all of the sealed off pores that are inaccessible to any fluid.

8.1. Density

Using helium as the displacement fluid to measure the true density of coal assumes that the helium does not adsorb appreciably at room temperature. Even though this assumption is not always valid, especially in the case of high surface area adsorbents such as active carbons, the helium densities found for raw coal are not too different from the results given by water and are considered to be close to the true densities (fig. 10).

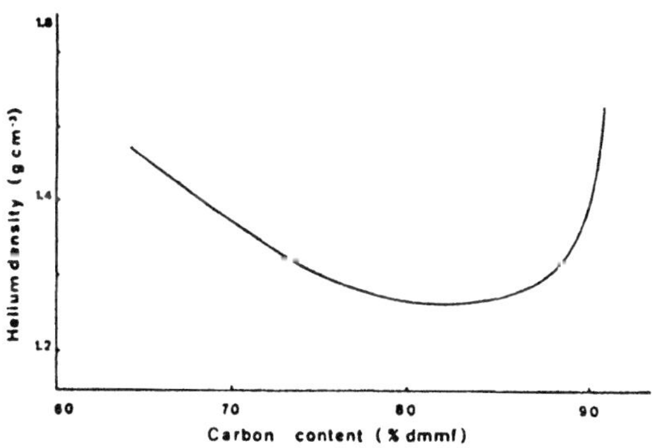

Figure 10. Helium densities of coals as a function of rank.

The true density of coal goes through a broad minimum as a function of carbon content at around 80 to 90 % C, increasing sharply above 90 % C and gently below 80 %.

Table 8 shows the densities of several coals measured with different fluids (16).

Coal rank	Helium	Mercury	Methanol	Benzene	Tetralin
anthracite	1.565	1.391	1.598	1.521	1.395
lv bitum.	1.326	1.256	1.402	1.340	1.295
mv "	1.294	1.233	1.326	1.286	1.264
hva "	1.282	1.255	1.348	1.283	1.262
hvb "	1.292	1.157	1.409	1.355	1.285
hvc "	1.323	1.213	1.465	1.372	1.274
Subbitum. a	1.340	1.206	1.497	1.392	1.312
" b	1.334	1.232	1.512	1.348	1.305
lignite	1.364	1.136	1.564	1.379	1.350

Table 8. Densities (in g/cm^3) of different coals at 298 K (dmmf).

8.2. Surface area

The conventional BET method to determine the surface area using nitrogen at 77 K measures only a small part of the total surface area in coal because of diffusion limitations in the micropores. Consequently, the most credible surface areas are measured with gases such as xenon or carbon dioxide at room temperature, as table 9 shows (17).

Coal rank C % dmmf	N_2 77 K	Kr 195K	CO_2 195 K	Xe 273K	CO_2 298K
95.2	34	176	246	226	224
90.0	0	96	146	141	146
86.2	0	34	107	109	125
83.6	0	20	80	62	104
79.2	11	17	92	84	132
72:7	22	84	198	149	139

Table 9. Surface areas of coals as measured by different adsorbates and at different temperatures (in m^2/g).

If one uses the BET method with carbon dioxide at room temperature, one must operate under pressure (vapour pressure of CO_2 at 298 K = 6.4 MPa). However, one can operate under atmospheric pressure by using the Polanyi-Dubinin equation :

$$\log V = \log V_o - \frac{0.434\ BT^2}{\beta 2} (\log p/p_o)^2$$

with : V = amount adsorbed at equilibrium relative pressure p/p_o

V_o = micropore capacity

B = a constant that is a measure of the micropore size

β = affinity coefficient of the adsorbate relative to nitrogen or benzene

V_o is obtained from the intercept of the plot of log V versus $(\log p/p_o)^2$, and converted to a micropore surface area if the area occupied by the adsorbate molecule is known.

Also, diffusion in the micropores is generally activated, with the diffusion coefficient given by the following equation

$$D = D_o \exp(-E/RT)$$

with : D_o = constant

E = activation energy
R = gas constant
T = absolute temperature

The rate of penetration or desorption varies dramatically as a function of temperature and above all as a function of the molecular diameter of the diffusing gas, as can be seen in table 10 for a coal with 19 % volatile matter.

	Molecular diameter nm	E kJ/mole	D/D_{CH_4}	D_{323}/D_{298}	$t(u/u_\infty = 0.9)$
Methane CH_4	0.33	28.5	1	2.4	504
Benzene C_6H_6	0.47	62.8	1×10^{-6}	6.8	6×10^{3} years
n-hexane C_6H_{12}	0.52	104.6	6×10^{-14}	24.5	6×10^{11} years

Table 10. Dependence of diffusion rate of different gases on molecular size and temperature (VM 19 %).

Note in particular, in the last column the calculated time needed to reach 90 % of the equilibrium adsorption which varies astonomically as a function of the molecular diameter.

Other than the rank of the original coal, the specific surface area depends on at least one other essential parameter : the highest temperature of treatment (HTT). But as the HTT increases above 1000°C, the pore network of all porous carbon solids becomes more and more inaccessible, at least at room temperature, complicating its characterization. Thus, at higher temperatures, adsorption techniques are totally unable to give useful results, and small angle X-ray scattering is the only technique that can lead to reliable data on the evolution of the pore structure between 1000 and 3000°C.

A systematic study using both types of techniques has been made (18) (fig. 11), showing that non graphitizable chars form low rank coals retain their low density and high porosity at all temperatures and that the total specific surface area, including that which becomes inaccessible, continues to develop all the way up to 1500°C. The same is true for high rank coals but graphitization commences starting at 2000°C and is accompanied by an increase in density and a reduction in specific surface area. The behaviour of cokes from medium rank

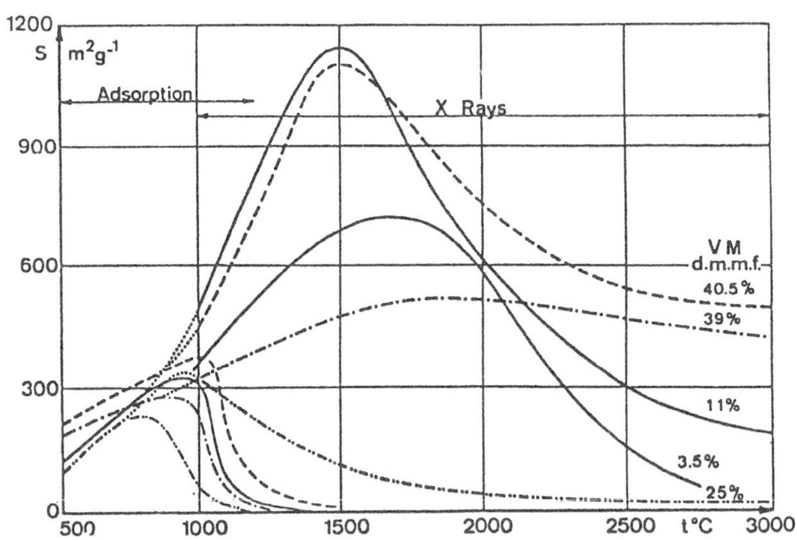

Figure 11. Surface area of coke as a function of carbonization temperature.

coals appears quite different, the structural reorganization that took place in the plastic phase as early as 400-450°C leads to a dense material, not very porous, with a specific surface area becoming very small above 1200-1500°C.

Not much is known about the possible accessibility of the closed porosity at high temperatures, even though it has been shown in some cases that there is a reversible opening and closing effect of the pores below the HTT. A more complete knowledge of such a phenomenon would be very precious in evaluating the reactive surface area of coke in the first steps of gasification. Indeed, beyond a weight loss of 15 à 20 %, the accessibility is recovered even at low temperature, so that the diffusion problems are much less crucial at this point.

Finally it should be pointed out that in certain cokes, their highly developed microporous structure gives them veritable molecular sieve properties that can be used to separate oxygen from nitrogen in air, for example.

9. RESOURCES, RESERVES AND PRODUCTION

Usually the distinction is made between total geological resources and proven recoverable reserves. Currently, the generally accepted values are (table 11) :

	Total resources	Proven recoverable reserves
USSR	3200	115
USA	1900	135
China	1000	70
Austria	450	25
Western Europa	300	65
Eastern Europa	100	30
South Africa	125	25
India	75	10
Canada	265	5
Latin America	30	5
Total	7445	485

Table 11. World coal resources and reserves (bituminous and anthracite) (G mt of oil equivalent*).

* 1 t of oil equivalent = 1,5 t of coal equivalent.

In table 12, the proven recoverable reserves of different fossil fuels are compared both in terms of tons and in years of consumption at the present rate.

	G mt of oil equivalent	Years of consumption
Coal	480	255
Oil	98	52
Gas	66	37

Table 12. Compared reserves and years of consumption of different fossil fuels.

Finally, tables 13 and 14 give 1981 coal production and trade.

USA	696.0
China	600.0
USSR	486.0
Poland	162.0
South Africa	130.0
Great-Britain	125.3
India	123.0
F.R.Germany	95.5
Australia	92.0
Canada	34.0
Other countries	194.2
TOTAL	2738.0

Table 13. 1981 coal production (M mt of coal).

Principal exporting countries		Principal importing countries	
USA	102.0	Japan	78.3
Australia	51.0	France	27.4
South Africa	29.9	Italy	18.4
USSR	22.0	Canada	14.8
F.R.Germany	17.2	Other countries	136.9
Canada	16.3		
Poland	15.0		
Other countries	22.4		
TOTAL	275.8	TOTAL	275.8

Table 14. 1981 coal trade (M mt of coal).

10. COAL UTILIZATION

Basically, there are four categories of coal utilization :

- combustion (thermal electric power plants, industrial and domestic heating)

- coking (blast furnace coke production)

- gasification (conversion into fuel gas or synthesis gas)

- liquefaction (conversion into liquid fuels or chemical products).

Combustion and cokefaction are ways in which coal has been used traditionaly and for a long time to come will continue to consume most of the coal produced.

Gasification and liquefaction are conversion processes, whose basic principles go back to the turn of the century. However, to be used on a large scale, these conversion processes still need technical and economic improvements.

10.1. Combustion

Combustion of coal will not be dealt with in great detail here, even though it is the oldest and most natural way of using coal and represents by far the largest tonnage of coal consumed. However, that is not to say that technical improvements cannot be made.

. The transfer of the technology used in coal fired power plants to medium sized furnaces to develop the use of coal in industry and for community heating. The essential problem is to obtain compact flames in a confined environment.

. The use of coal-oil or coal-water mixtures that, a priori, have the advantage of being pseudo-liquids. Even though results of coal-oil mixtures have been disappointing, much work is being done on the formulation, rheological behaviour and combustion of dense suspensions of coal in water.

. The use of micronized coal, that is, ground to just a few microns instead of the normal 80-100 µm. The rate of combustion is expected to be almost as rapid as that for a liquid fuel.

. Combustion in a fluidized bed, which has not yet found commercial success in spite of much work oven the past few decades, and a variant operating under pressure with coupling to a gas turbine. These techniques are particularly attractive for high sulphur coals due to the sulphur retaining properties of certain oxides (CaO, MgO) added to the reactor.

10.2. Coking

Blast furnace coke fulfils several functions at the same time : it acts both as a fuel and reducing agent, supports the blast furnace charge and regulates the liquid and gas flow. Above all, it should have a uniform granulometry (20 to 80 mm) and sufficient mechanical strength characterized by the shatter test and the abrasion index respectively. Blast furnace coke is made by the slow pyrolysis (16 to 18 hours) of a judicious blend of coals, in a coke oven, consisting of a battery chambers 45 cm wide, 4 to 7 m high and 12 to 15 m length.

A typical material balance for the carbonization of a coal blend having 30 % volatile matter is :

1000 kg dry charge = 750 kg dry coke + 35 kg coal tar
+ 10 kg benzol + 170 kg gas (350 m^3)
+ 75 kg water

The coke gas is rich in hydrogen (55 to 60 %) and methane (25 to 30 %) and has a high calorific value (about 20 MJ/m^3).

Coke oven tar, or high temperature coal tar, has served in the past as the base of the carbochemical industry. The most abundant constituents of coal tar are :

naphthalene	:	about 10 %
phenanthrene	:	5 %
fluoranthene	:	3.5 %
acenaphth-(yl)-ene	:	2.5 %

Tar distillation gives :

light oil (raw benzol)
phenol oil
naphthalene oil
acenaphthene oil
anthracene oil
chrysene oil

and a residue, coal tar pitch, used essentially as a binder (briquetting of coal fines, fabrication of electrometallurgical electrodes, in particular for aluminium).

Conventionally, coal is charged into the coke oven chamber by free fall through holes in the top of the oven. But the need to enlarge the range of coals that can be used to make coke has instigated diverse ameliorations :

. horizontal stamped charging notably increases the charge density.

. preheated charging (around 250°C) improves the mechanical quality of the coke and increases the oven productivity by some 30 %.

. formed coal charging, that is, prebriquetting or adding coking additives, hydrocarbon binders that increase the charge density.

. fabrication of formed coke, independent of a standard coke oven, obtained by carbonizing a briquetted coal in a special oven. After raising hopes, this new technique has not yet passed the industrial pilot plant stage contrary to the preheated or stamped charging techniques.

The coke industry consumes approximately 450 Mt of coal annually, about 17 % of the total world production.

It must be noted that several flash pyrolysis processes related to coking, or at least to carbonization, are under study at the pilot plant scale with the goal of obtaining higher yields in tar and gas and gasifying the residual char or using it as fuel.

10.3. Conversion processes : gasification and liquefaction

The conversion of coal into liquid and gaseous products which are much easier to handle and use than solid coal requires additional hydrogen as one can see easily by comparing the atomic ratio H/C of certain substances :

bituminous coals and anthracites	0.4-0.9
benzene	1.0
crudc oil	1.5-2.0
synthetic fuels	2.0
butane	2.5
ethane	3.0
methane	4.0
methanol	4.0

The necessary hydrogen can be furnished by several means (fig. 12).

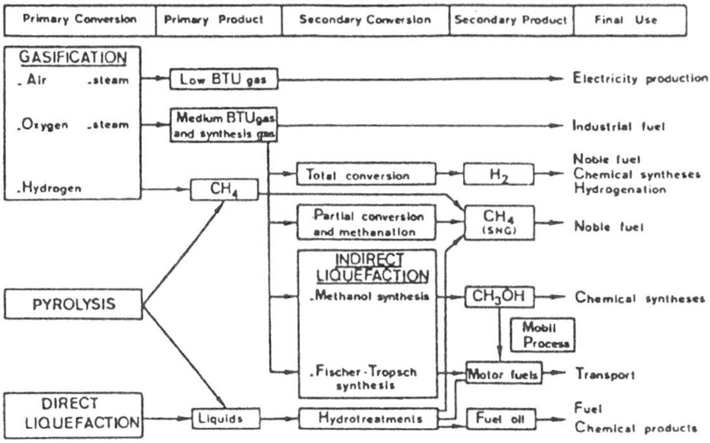

Figure 12. Coal conversion processes.

1°/ By direct hydrogenation of the coke : however the reaction rate is very low and only successful catalyst research currently under way can lead to an industrial scale process. Still, this method has the advantage of being exothermic and forms methane directly :

$$C + 2\,H_2 \rightarrow CH_4 + 74.9\ KJ$$

2°/ By steam gasification that produces a mixture of CO and H_2 (synthesis gas) which has many uses :

. fuel

. conversion to methane by the exothermic reaction :

$$CO + 3\,H_2 \rightarrow CH_4 + H_2O + 206.3\ KJ$$

. methanol synthesis by the exothermic reaction :

$$CO + 2\,H_2 \rightarrow CH_3OH + 128.1\ KJ$$

. Fischer-Tropsch synthesis (indirect liquefaction) leading to a variety of essentially aliphatic liquid products, such as alcohols, motor fuels, and various oils.

But the first step is the highly endothermic reaction :

$$C + H_2O \rightarrow CO + H_2 - 131.4 \text{ KJ}$$

The necessary heat often is provided by burning part of the coal by air or oxygen (air-steam or oxygen-steam gasification). Also research is under way on processes that furnich the heat by an external source : electricity, helium heated in a high temperature nuclear reactor ...

3°/ By a hydrogen donor solvent (direct liquefaction): operating at about 450°C under 15 to 30 MPa of hydrogen. The coal is pyrolyzed and the fragments are prevented from recombining to form heavier compounds or char by hydrogenation. This process produces 15 to 20 % gas, C_1 to C_4, and a syncrude that can be used as a desulphurized fuel or upgraded by complementary hydrotreatment to motors fuels or chemical feedstocks.

Unlike Fischer-Tropsch synthesis products, coal liquids are essentially aromatic in character.

Numerous coal conversion processes have been studied at the pilot plant level, but to this day the only industrial unit in operation is SASOL (South Africa) that shall soon process almost 30 Mt of coal per year by gasification in Lurgi gasifiers followed by Fischer-Tropsch synthesis to produce motor fuels.

Gasification. There are three types of industrial gasifiers :

- the Lurgi gasifier : mobile bed under pressure (2.5-3.0 MPa)
- the Winkler gasifier : Fluidized bed
- the Koppers-Totzek gasifier : entrained bed

Several so called "second generation" gasifiers have been developed recently, especially in the U.S., all operating in the fluidized bed mode under pressure hoping to increase the direct production of methane in a manner that reduces the overall endothermicity of the process (Bigas, Hygas, Synthane ...).

But the results have not been too encouraging and consequently current research is directed towards improving the conventional gasifiers :

- the Lurgi gasifier, by increasing the temperature (slagging gasifier) or the pressure (Lurgi Ruhr 100)

- by increasing the pressure in the Winkler gasifier (0.8 to 1.0 MPa) and the Koppers-Totzek gasifier (3 MPa).

To feed the solid coal into the gasifier under pressure is a difficult operation that in general requires the use of lock hoppers. However, another solution to this technical problem has been explored, namely the use of either a coal-oil slurry (Hygas process mentioned above) or a coal-water slurry (Texaco).

Finally, a few words should be said about underground coal gasification, or in situ gasification, that, in addition to the advantage of mining without miners, has the capability of reaching coal seams too deep to be mined by conventional means.

At shallow depths, underground gasification cannot be conducted under pressure, is difficult to control, faces the problem of underground water seepage and produces a low Btu gas.

On the other hand, at profound depths, one should be able to operate at elevated pressures under dry air tight conditions resembling a conventional gasifier. The main difficulty is in linking the injection borehole with the producing borehole about 60 to 80 m apart by hydraulic fracturation, electrocarbonization or explosion. Experiments are in progress in Belgium and France, but require several years before any conclusions can be made concerning this new technique that is e pected to produce gas of the same quality as a conventional gasifier.

Liquefaction. Even though all of the motor fuel produced in Germany during the Second World War came from direct or indirect coal liquefaction, the installations were not what we would call today industrial sized units. Also numerous projects have been proposed, especially after 1973 energy crisis, to improve the process technology and increase the yields in high value products. Although the basiç principles of liquefaction are well known, differences appear in the operating conditions (pressure, temperature, residence time ...), in the choice of the proper hydrogenation catalyst, if one is used at all, and at what point it is employed, in the solid residue separation process (distillation, filtration, precipitation ...), etc.

The largest coal liquefaction pilot plants are :

- SRC (solvent refined coal) USA 50 t/d

- H-Coal, USA 200 to 600 t/d depending on whether the process operates in the syncrude or fuel oil mode

- EDS (Exxon-donor-solvent) USA 200 t/d

- Ruhrkohle-Veba Oel : FRG 200 t/d

SRC and H-Coal scale-up projects (6000 t/d) have been postponed due to new governmental policies in the US and also to the falling cost of crude oil pushing the economic break-even point too far into the future. However, a change in context, either economic or political, could very well reverse the present policy.

11. CONCLUSION

It is of course quite impossible, within the scope of the present paper, to depict all scientific, technical and economic aspects of such a variable and complex substance as coal. A raw material of the past, but also of the future due to its extended resources and reserves, coal will continue to demand, for a long time to come, a great research effort : research aimed at a better understanding of coal's structure and properties, at improving the conventional means of coal utilization such as combustion and coking as well as the new conversion processes into liquids and gases that, sooner or later, will replace natural gas and petroleum feedstocks and give rise to the revival of carbochemistry.

REFERENCES

1. Stopes, M.C. 1919, Proc. Roy. Soc. London B 90, pp. 470-487.

2. Durie, R.A., Shewchyk, Y. and Sternhell, S. 1966, Fuel 45, pp. 99-113.

3. Heredy, L.A., Kostyo, A.C. and Neuworth, M.B. 1965, Fuel 44, pp. 125-133.

4. Hodek, W. and Kolling, G. 1973, Fuel 52, pp. 220-225.

5. Loison, R. and Chauvin, R. 1964, Chim. et Ind. 91, pp. 269-275.

Chauvin, R. and Deelder, R. 1969, Bull. Soc. chim. France, pp. 3916-3927.

6. Boyer, A.F. and Payen, P. 1965, Bull. Soc. chim. France, pp. 2765-2770.

Develotte, J., Mazza, M. and Payen, P. 1969, Bull. Soc. chim. France, pp. 341-348.

7. Hayatsu, R., Scott, R.G., Moore, L.P. and Studier, M.H. 1975, Nature (London) 257, pp. 378-380.

8. Deno, N.C., Greigger, B.A. and Stroud, S.G. 1978, Fuel 57, pp. 455-459.

9. Oelert, H.H. 1967, Brennstoff-Chem. 48, pp. 331-339.

Oelert, H.H. 1969, Brennstoff-Chem. 50, pp. 178-182.

10. Retcofsky, H.L. and Venderhardt, D.L. 1978, Fuel 57, pp. 421-423.

11. Ruberto, R.G. and Cronauer, D.C. 1978, in Organic Chemistry of Coal, ACS Symp. Series 71, pp. 50-70.

12. Wiser, W.H., in Sternberg, H.W. 1975, Research in coal technol. : the university's role, Conf. 741091 Buffalo, NY, pp. II-57-II-72.

13. Brooks, J.D. and Taylor, G.H. 1986, in Chemistry and Physics of Carbon 4 (Walker, P.L. Jr., Ed.), Marcel Dekker, New-York, pp. 243-286.

14. Ihnatowicz, M., Chiche, P., Deduit, J., Pregermain, S. and Tournant, R. 1966, Carbon 4, pp. 41-50.

15. Walker, P.L., Jr 1981, Phil. Trans. Roy. Soc. Lond. A 300, pp. 65-81.

16. Neslon, J.R. 1979, Ph. D. thesis, Pennsylvania State University.

17. Walker, P.L., Jr and Kini, K.A. 1965, Fuel 44, pp. 453-459.

18. Chiche, P., Durif, S. and Pregermain, S. 1965, Fuel 44, pp. 5-28.

GENERAL REFERENCES

Van Krevelen, D.W. 1961, Coal, Elsevier, Amsterdam.

Stach, E., Mackowsky, M. Th., Teichmüller, M. and R.,

Taylor, G.H. et Chandra, D. 1975, Stach's Textbook of Coal Petrology, Gebrüder Borntraeger, Berlin.

Davidson, R.M. 1980, Molecular structure of coal, IEA Coal research, Report nr. ICTIS/TR 08.

Elliott, M.A. 1981, Chemistry of Coal Utilization, 2nd supplementary volume, J. Wiley and sons, New-York.

CONVERSION OF BITUMINOUS COAL TO LIQUIDS AND GASES: CHEMISTRY AND REPRESENTATIVE PROCESSES

Wendell H. Wiser

Professor, Dept. of Fuels Engineering, University of Utah, Salt Lake City, Utah 84112, U.S.A.

Abstract

Bituminous coal consists principally of condensed ring structures containing aromatic and hydroaromatic rings. The condensed units are connected mainly by short aliphatic chains and ether linkages. Coal liquefaction reactions in the usual temperature range (400-500°C) proceed chiefly via a free radical mechanism, thermally produced radicals being stabilized mostly by hydrogen from available sources. Several approaches to achieve liquids production are illustrated with specific processes. Gasification of coal to yield a gas of low heating value and one of high heating value is discussed and the principles illustrated with specific processes.

Coal Liquefaction

I think it would be useful to begin this discussion by indicating what must be done to bituminous coal to convert it to the liquid state. The similarities and differences in advanced coal liquefaction technologies can then be discussed and the nature of the problems and challenges more readily understood.

Figure 1 contains a schematic representation of structural groups and connecting bridges in bituminous coal, drawn by the author. These structural units are ring-type structures varying in size from a single ring to perhaps six or seven condensed rings for bituminous coal. An average size unit is about three rings. When these units are joined together in an extended structure as illustrated in Figure 1, the vapor pressure is very low and the melting point high. Hence, coal is a solid at room temperature.

L. Petrakis and J. P. Fraissard (eds.), Magnetic Resonance. Introduction, Advanced Topics and Applications to Fossil Energy, 325–350.

FIGURE 1. SCHEMATIC REPRESENTATION OF STRUCTURAL GROUPS AND CONNECTING BRIDGES IN BITUMINOUS COAL

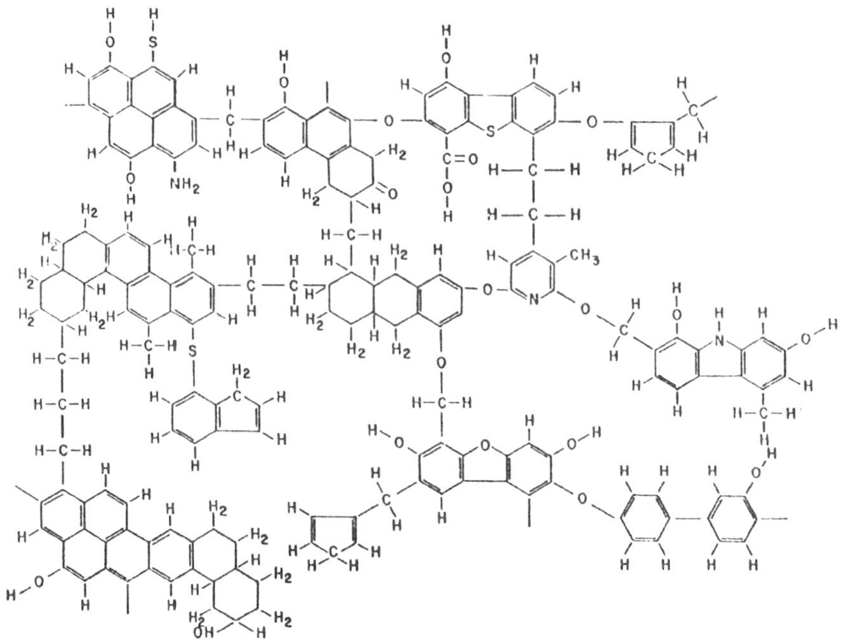

To convert bituminous coal to a liquid, it is necessary first to break these condensed structures apart so that the structural unit becomes a molecule. Thermal bond rupture between condensed units becomes appreciable at temperatures above 375°C and can be rather complete at temperatures up to 500°C. Coal liquefaction technologies, therefore, usually operate between 400° and 500°C.

When the bridges connecting the units are ruptured, free radicals are formed. By nature, free radicals have a very short life in this temperature range. They must either be capped by a small entity such as a hydrogen atom, or they will combine in condensation or polymerization reactions, thereby producing large structures which will again be solid at room temperature. Since the thermal bond rupture of these connecting bridges is common to all coal liquefaction processes in this temperature range, the different technologies represent in part different approaches to stabilizing these radicals to produce small, liquid-size molecules.

Compared with petroleum, coal contains large amounts of oxygen and appreciably higher amounts of sulfur and nitrogen. These heteroatoms must be removed during the liquefaction and upgrading processes if a product is to be obtained which can substitute for petroleum products. Coal also contains much more inorganic material than does petroleum. This inorganic material appears as solid residue following the liquefaction process. It must be completely removed from the liquid products. Separation of the solid residue from the liquids can be a major problem. With reference to Figure 1, a gasoline product cannot contain very much ring-type material larger than a single ring. One can see that to produce a gasoline product, the condensed configurations must also be broken up. This is much more difficult than rupturing the original connecting bridges and involves extensive catalytic upgrading with the addition of hydrogen.

There exist a great many processes for converting coal to a liquid product. These processes can be grouped generally into about five groups. The simplest of these approaches to liquefaction is coal pyrolysis. By definition, pyrolysis means heating the coal in the absence of any reacting atmosphere. In the temperature range of 400° to 500^oC the connecting bridges between condensed units are thermally ruptured. Hydrogen is abstracted from some of the structural units in the coal to stabilize these free radicals. Since the amount of hydrogen in the coal available for this purpose is limited, a limit is placed upon the amount of liquid-size molecules which can be formed before condensation and polymerization reactions become the dominant means of radical stabilization. The total yield of liquids and gases by pyrolysis is usually of the order of 40 percent by weight of the coal. This leaves something like 60 weight percent of the coal as a solid residue known as char. In order for a pyrolysis process to be economically attractive, there must be a market for this char. Since the char has lost nearly all of its volatile matter, its combustion characteristics have changed relative to coal. Also, the inorganic matter is concentrated in the char. This char is not particularly attractive to the electric utilities.

One of the most successful coal pyrolysis operations has been the COED process. Figure 2 contains a schematic representation of the COED process. It utilizes a multi-stage pyrolysis with the temperature in each succeeding stage being higher than the preceding stage. This scheme was used to combat the caking properties of a bituminous coal. A typical yield by this process is 40 weight percent of the coal as liquids and gases, with about one-half of this product being liquids. This liquid product has been upgraded and used to fuel a naval surface vessel in order to demonstrate the suitability of the liquid product as a fuel. However, there was not and is not now a market for the solid char. The char was then gasified in a process known as the CoGas Process. Both the COED and the CoGas Processes have operated at pilot plant. However, due to discontinuing of funding by the U.S. government for pilot plant coal processes, both the COED and CoGas Processes have been terminated.

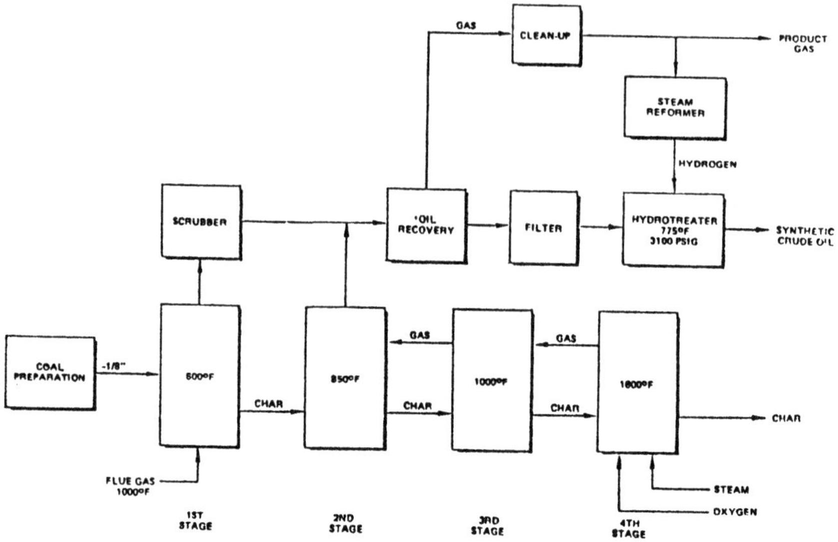

FIGURE 2. COED PROCESS

Figure 3 contains a fundamentally different approach to coal pyrolysis for the production of liquids. This process has been under development for several years by Occidental Petroleum Company. Inasmuch as organic materials are sensitive to high temperatures, and liquid-size molecules tend to crack to form gas-size molecules at temperatures around 500°C, the Occidental Process uses very short residence times in the high temperature zones. As a consequence, liquid products as high as 35 percent by weight of the coal have been reported by this process. The hydrocarbon gas yield is thus low, at about 7 percent by weight of the coal. Although a higher liquid yield is obtained, the total net conversion is still around 40 to 45 percent by weight of the coal, a limit imposed by the very structure of the coal itself. There is still a need for a market for the solid char. This process is presently also in a holding pattern awaiting more favorable economics or a more readily available market for the char which would improve the economics. There are many pyrolysis processes which have been patented, each with a slightly different approach somewhere in the process. However at this date, the overall economics, imposed partly by the low conversions, do not appear very attractive.

The second group of coal liquefaction processes can be identified as solvent extraction processes. By definition, these processes involve dissolving the coal in some type of solvent, usually a hydrogen donor solvent, in the absence of any

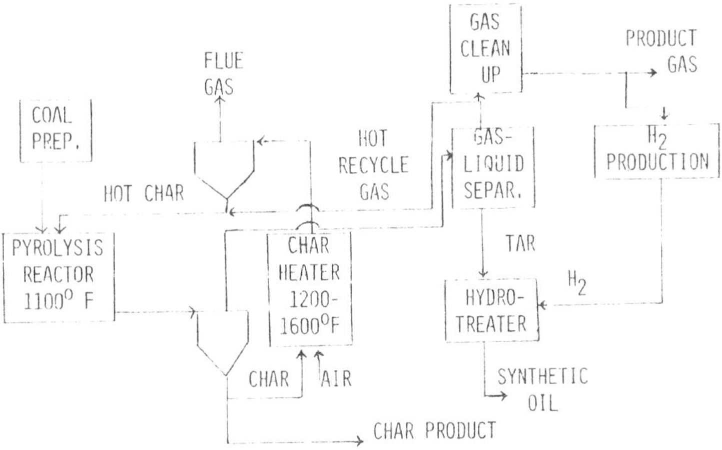

FIGURE 3. OCCIDENTAL PROCESS

externally applied catalyst. A catalyst usually appears later in the process, but is not introduced into the raw coal environment. By adding hydrogen to the coal from the solvent, to help stabilize the free radicals, in addition to the hydrogen already available in the coal, it is possible to convert the organic matter of the coal almost completely to liquid and gaseous materials. Such conversion usually involves the addition of three to four weight percent hydrogen to the coal in order for the product to be a liquid at room temperature. Lesser amounts of hydrogen usually result in a product which is solid at room temperature, such as the SRC I product. The addition of the solvent generally serves at least three purposes: (1) it facilitates pumping into the pressurized regions; (2) it aids in preventing agglomeration of the particles or sticking of the materials to the walls of the reactor; (3) it provides hydrogen to assist in stabilizing the thermally-produced radicals.

One of the more advanced processes utilizing this approach to liquefaction is the SRC II Process, now a part of the Gulf Oil Company program. A schematic diagram of this process is presented in Figure 4. A process-derived hydrogen donor solvent is added to the coal and the slurry pumped into a region at a temperature of about 430°C. Hydrogen gas is added to the system and the inorganic material of the coal, usually referred to as ash, is recycled for the catalytic benefit from this inorganic material. Residence time in the heated region is about one hour. The product is liquid at room temperature. A pilot plant operating in the SRC II mode to produce a liquid product has operated at Tacoma, Washington, U.S.A., with a capacity of 30 tons of coal per day. A conceptual design for a demonstration plant with a

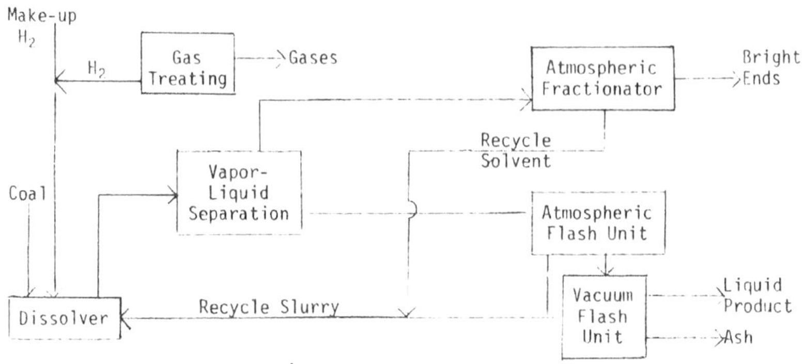

FIGURE 4. SRC-II PROCESS

capacity of 6,000 tons per day has been completed. A joint venture agreement between the United States, the Federal Republic of Germany and Japan was terminated by the U.S. Department of Energy July 2, 1981. The project is now in a holding pattern because of the absence of U.S. government participation in funding the program.

Another application of this general technology is the SRC-I pilot plant operating at 6 tons per day in Wilsonville, Alabama, U.S.A. However, since that process produces a product which is solid at room temperature, it will not be further discussed in this presentation.

Another advanced technology based upon the solvent extraction concept is the Exxon Donor Solvent Process, illustrated schematically in Figure 5. This process has operated at a pilot plant at Baytown, Texas, with a capacity of 250 tons of coal per day. The Process has operated very satisfactorily on bituminous coal, subbituminous coal and lignite. Future continued development of the process is clouded at this date because of the termination of U.S. government funding support to all large-scale coal conversion operations. There does not appear to be any serious technological or engineering problem associated with this process. If the process development does not continue, it will be principally because of financial considerations.

A third class of coal liquefaction processes is identified as catalytic hydrogenation with coal slurried in a solvent. By definition, an external catalyst is introduced into the coal-solvent system where the catalyst comes into contact with the raw coal. This presents a very difficult environment for most coal liquefaction catalysts. Radical stabilization by the molecular hydrogen through the medium of the catalyst is possible in this process in addition to the hydrogen in the donor solvent and from the coal itself. Hydrogen addition to the coal is of the order of four to five weight percent and the product is liquid at room temperature.

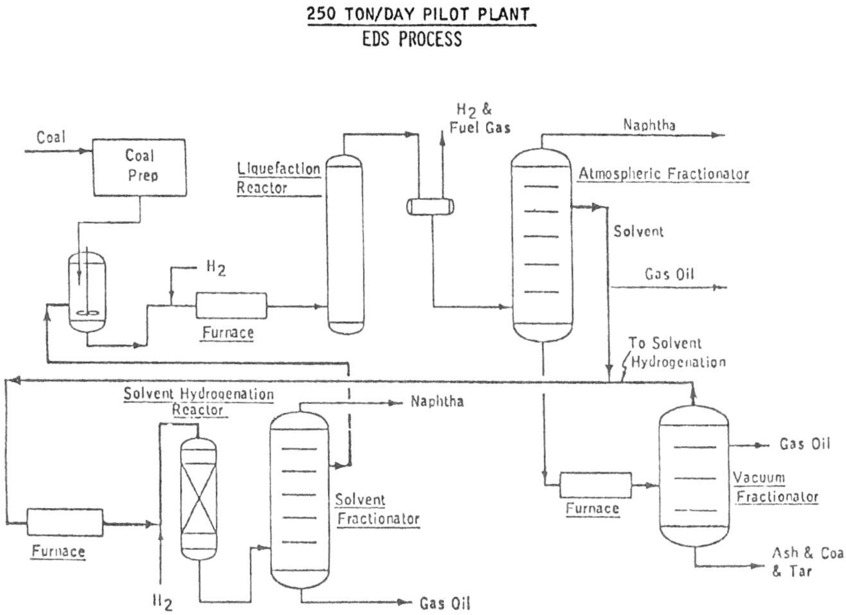

FIGURE 5

Perhaps the most successful process utilizing this concept is the H-Coal Process developed by Hydrocarbon Research, Inc. and schematically represented in Figure 6. A key feature of this process is the ebullated bed reactor which permits continual withdrawal of catalyst for regeneration. This is necessary because of the severe environment imposed by the raw coal. The H-Coal Process has operated in a pilot plant in Kentucky in the U.S. with a capacity of 600 tons of coal per day. Since this pilot plant has operated with funding support from the U.S. Department of Energy, its future operation is also clouded by the discontinuation of U.S. government funds for large coal conversion projects. This process has operated in a completely satisfactory manner at the large pilot plant level. Catalyst activity is maintained at a satisfactory level, and catalyst life for the cobalt-moly catalyst is also satisfactory. The principal problems associated with continuation of this process appear also to be financial rather than technological.

Another approach to coal liquefaction utilizes catalytic hydrogenation in the absence of a solvent. In this process, coal is fed dry to the reactor system. In the absence of a solvent in the feed, heat and mass transfer are greatly increased. The hot molecular hydrogen in turbulent flow is utilized as the heat transfer medium to bring the coal particles to temperature very rapidly. A catalyst exhibiting high mobility is utilized to

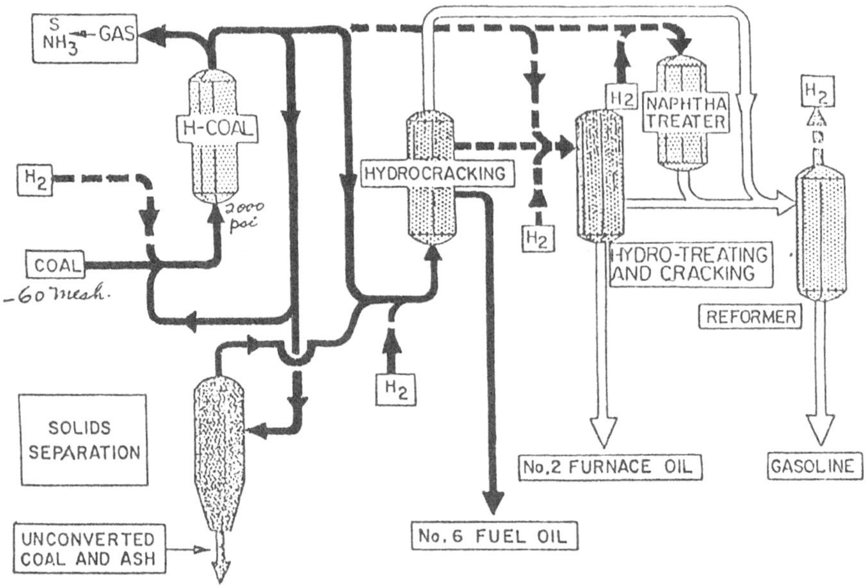

FIGURE 6
H-COAL REFINERY

promote the liquefaction reactions. Solids residence times in the heated reactor are of the order of a few seconds, the time depending upon the coal particle size (compared with 20 minutes to one hour for Donor Solvent Processes). Conversions of the order of 75 weight percent of the coal are obtained with high ratios of liquids to gases of the order of ten to one. The high liquid ratios are a consequence of the short time in the heated reactor. Low hydrogen consumption is a consequence of low hydrocarbon gas production. High space utilization results from the short residence time in the heated reactor. These factors indicate an economic advantage in this approach to coal liquefaction. A process utilizing this approach to coal liquefaction is under development at the University of Utah where one reactor configuration has operated in a PDU at a capacity of 0.6 tons per day. The process is illustrated schematically in Figure 7.

The only technology for production of liquids from coal operating at a commercial scale in the world utilizes the Fischer-Tropsch technology. The essential features of this approach are illustrated in Figure 8. This technology introduces molecular oxygen and steam into the coal and operates at gasifier temperatures generally above 950°C. Most of the carbon in the coal is oxidized to carbon monoxide. The carbon monoxide is then hydrogenated at temperatures of the order of 430°C, in

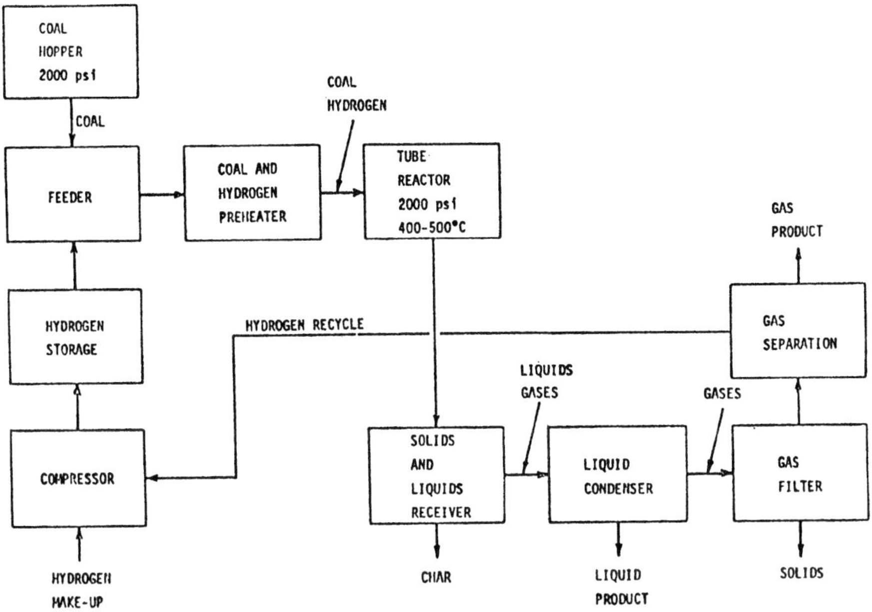

FIGURE 7. UNIVERSITY OF UTAH PROCESS FOR DIRECT HYDROGENATION OF COAL

the presence of an appropriate catalyst, to form liquid-size molecules. The technology is well developed and operates well. However, the thermal efficiencies associated with this approach are lower than those in the direct liquefaction approaches. Consequently, the cost per barrel of oil produced by this means is generally considerably higher than the anticipated cost by the direct technologies described above. This approach has not been under continual development in the United States, chiefly because of the low thermal efficiencies and high cost associated with this process. However, the synthetic liquid fuel plants operating commercially in South Africa are based upon this technology because it is a well-developed and reliable technology.

A more recent approach to the production of gasoline from coal related to the Fischer-Tropsch technology is the Mobil Process. This Process is illustrated schematically in Figure 9. Coal is gasified with oxygen and steam to produce a synthesis gas consisting principally of carbon monoxide and hydrogen. This synthesis gas is then catalytically converted to methanol by a reliable and straightforward technology. The methanol is then introduced into the Mobil ZSM-5 catalyst. This catalyst is sometimes referred to as a shape-selective catalyst because most of the molecules which exit from the pores of the catalyst are molecules in the gasoline range. This technology is considered by

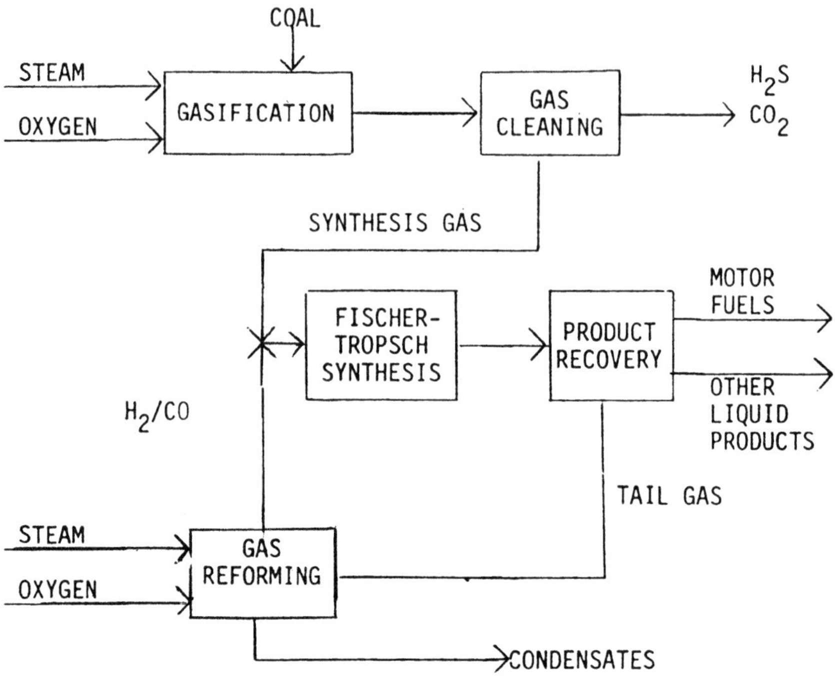

FIGURE 8. FISCHER-TROPSCH PROCESS

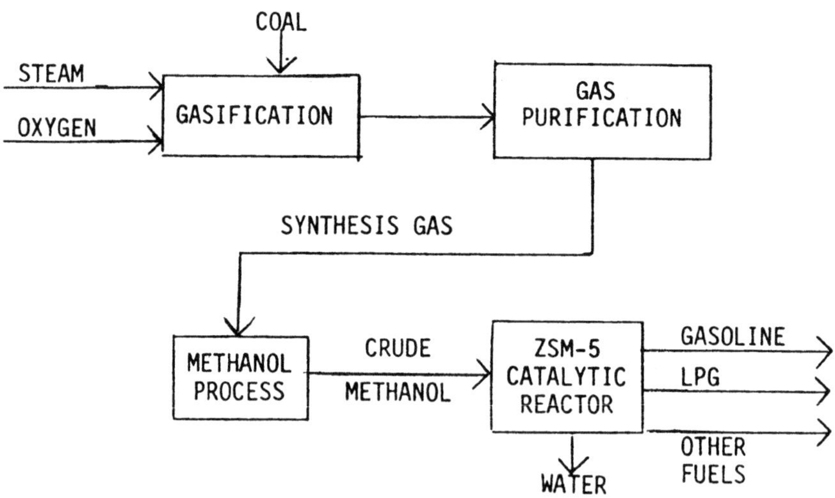

FIGURE 9. MOBIL PROCESS: COAL TO GASOLINE

Mobil to be very promising. A schematic flow diagram for the conversion of coal to methanol is contained in Figure 10.

A commercial plant for the conversion of methanol to gasoline utilizing the ZSM-5 catalyst in the Mobil Process is to be built in New Zealand. The methanol for this process will be produced from natural gas. The economics for this conversion from natural gas to gasoline via the Mobil Process appear to be good. Plans have been announced for a second plant to be built in Germany. This plant would utilize coal for the production of methanol to be followed by the conversion of the methanol to gasoline in the Mobil Process. A definite timetable for this plant construction has not been announced. The economics of this process are somewhat less attractive and less certain than the natural gas to gasoline plant announced for New Zealand.

This investigator is not aware of any announced, firm timetable for the construction of a commercial coal liquefaction plant in the United States. As indicated above, there are several technologies which are progressing well without any serious engineering or technological difficulties. The problems which may prevent construction of commercial plants utilizing any of these technologies are principally financial problems closely associated with political considerations.

It is estimated that the capital investment to build a commercial plant of 100,000 barrels per day capacity, based upon the most advanced technologies described above, but excluding Fischer-Tropsch, would be between \$4 and \$5 billion, in 1982 U.S. dollars. A synthetic crude oil by these technologies would have a price tag of the order of \$50 to \$60 per barrel. This compares with a

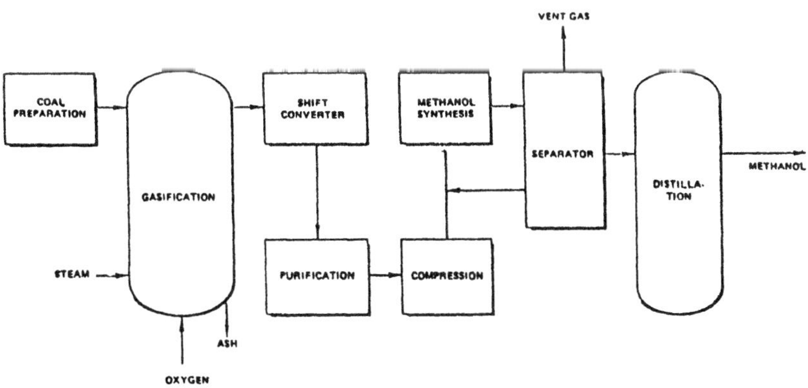

FIGURE 10. METHANOL SYNTHESIS

current world oil price of \$29 per barrel. There is no guaranteed market in the United States for a liquid product which would need to be priced appreciably above world oil market prices. At this date, there is no indication that a market in the U.S. will be guaranteed through the political process. Under present policies as announced by the U.S. Department of Energy, the continued development of these processes to commercial scale in the United States must be funded by private sources. The combination of multi-billion dollar investment coupled with a synthetic crude price appreciably above world oil prices, and the absence of a guaranteed market appears to represent an insurmountable barrier to commercialization of coal liquefaction in the United States.

Coal Gasification

The representation of coal structural groups contained in Figure 1 indicates the nature of the skeletal structure in bituminous coal, which is principally a carbon structure. The conversion of coal to gaseous fuels involves breaking down this carbon skeletal structure, usually to individual carbon atoms. This factor dictates the reactants and the conditions for converting coal to gaseous fuels.

Early technologies for coal gasification usually involved the reaction of coal with a limited supply of air. The combustible component in this product gas was mostly carbon monoxide.

$$C + \frac{1}{2} O_2 = CO \qquad (1)$$

The CO was diluted by the nitrogen from the incoming air resulting in a gaseous product with a heating value of the order of 4000 to 6000 kJ per standard cubic meter (SCM). The equipment for this purpose was called a gas producer and the product gas came to be known as producer gas.

Somewhat later in history it was observed that carbon in coal could react with steam to yield a gas with two combustible components, namely carbon monoxide and hydrogen.

$$C + H_2O = CO + H_2 \qquad (2)$$

Since the steam-carbon reaction is strongly endothermic, an energy supply to sustain the reaction was necessary. Early practice involved blowing air through a burning bed of coal or coke to achieve a high bed temperature. The air inlet was then terminated and steam was blown through the hot bed. The gases collected from the steam cycle contained principally carbon monoxide and hydrogen. This gas came to be known as water gas with a heating value of the order of 11,000 kJ/SCM.

Subsequently the technology developed whereby air and steam are injected simultaneously into the coal bed. The desired reaction for producing the fuel gas is the steam-carbon reaction. Sufficient air is injected to provide for the carbon-oxygen reaction, producing carbon monoxide, sufficient to supply the energy for the endothermic steam-carbon reaction and the energy to maintain the bed temperature at about 950°C or higher. The energy consumed for these two purposes represents about 22 to 33 percent of the heating value of the original coal. Because of the dilution effect of the nitrogen from the air, the heating value of this gas is about 5500 kJ/SCM. This gas has come to be known as low-calorie gas, or in some countries low-Btu gas.

With the development of improved and more economical technology for separating oxygen from nitrogen in air, molecular oxygen came to be used for the carbon-oxygen reaction to supply the energy for the steam-carbon reaction, instead of using air, thus avoiding the dilution effect of the nitrogen. Again the product gases consist principally of carbon monoxide and hydrogen with varying amounts of hydrocarbons, the amounts depending upon the operating temperature. This gas has a heating value of the order of 11,000-13,000 kJ/SCM and is often referred to as intermediate-Btu gas. Because carbon monoxide and hydrogen in this mixture often are used for the synthesis of other products, including a fuel gas of higher heating value, this gas has come to be known as synthesis gas. This is the basis of operation for many of the coal gasification technologies either available commercially or under development.

Because of the low heating value of this gas it is not practical to transmit it over very long distances. The gas has been used in some countries for residential and commercial heating under the name Town's gas. It is used close to the site where it is produced. The gas is also attractive for in-plant heating in the industrial sector and as a fuel gas for electric power generation.

The Lurgi gasifier operates commercially in several countries and was designed to gasify coal either with air and steam or with oxygen and steam. The Lurgi gasifier is illustrated in Figure 11. Crushed coal is fed through lock hoppers to the top of the gasifier. Air and steam or oxygen and steam are injected at the bottom. The gasifier normally operates at a pressure of 350-500 psia. The temperature at the bottom of the gasifier where molecular oxygen reacts with carbon is between 950° and 1150°C. Product gases exit from the top of the gasifier at a temperature in the range of 450° to 550°C.

The analysis of a typical product gas from the Lurgi gasifier is shown in Table I. The rather low exit temperature of this

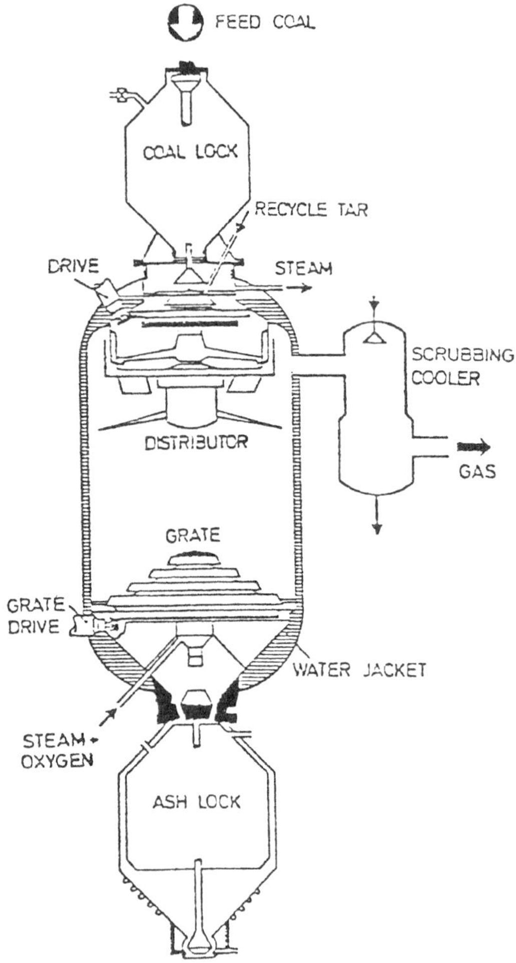

FIGURE 11. LURGI GASIFIER

gasifier permits the formation of a significant amount of methane, usually of the order of 10 percent by volume of the product gases. However, the balance of chemical reactions arising from this temperature range also results in a fairly high carbon dioxide content which tends to offset the methane content effect on heating value of the gas.

TABLE I

LURGI PROCESS

TYPICAL GASIFIER RAW GAS ANALYSIS
(DRY BASIS)

COMPONENT	VOLUME %
H_2	37.9
CO	20.2
CH_4	11.0
C_2H_6, C_2H_4	1.0
CO_2	28.6
H_2S, COS	0.5
Other	0.8
	100.0

H.V. = 295 BTU/SCF

In an attempt to increase the through-put in the gasifier by promoting faster reactions, as well as to enhance steam decomposition, some gasifiers operate at much higher temperatures. Illustrative of these gasifiers is the Koppers-Totzek gasifier represented in Figure 12. This gasifier operates as an entrained

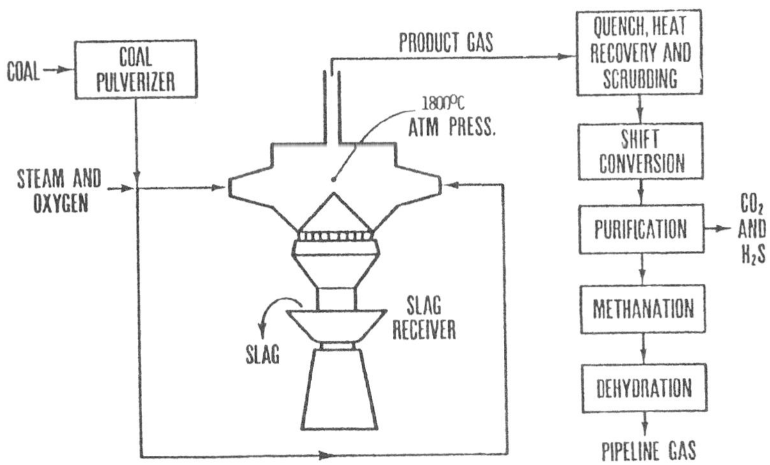

FIGURE 12. KOPPERS-TOTZEK

flow system with pulverized coal, steam and oxygen or air admitted concurrently. This gasifier operates at atmospheric pressure and a temperature of the order of 1800°C. Whereas the inorganic material (often referred to as ash) leaves the Lurgi gasifier as a fine solid, the inorganic material leaves the Koppers-Totzek process as a slag, being above the melting point. A typical analysis of the product gas from the Koppers-Totzek process is shown in Table II. Because of the high temperature of operation, methane formation is negligible. However, the CO_2 concentration of this gas is low, such that this product gas also has a heating value of the order of 11,000 kJ/SCM.

There are several coal gasifiers available today on the market all of which operate through the chemistry involving coal, steam and oxygen but each with some variation to promote a particular principle. In general there are advantages and disadvantages relative to each. The product gas, when operating with oxygen and steam, usually consists principally of carbon monoxide and hydrogen with a heating value of the order of 11,000 to 13,000 kJ/SCM.

There are a number of uses for fuel gas where the heating value must be much higher than 11,000 kJ/SCM. There are also markets for a gaseous fuel which are far from any practical location for operating a coal gasifier. For these purposes as well as others, a gas with a heating value near that of natural gas, with combustion characteristics very similar to those of natural gas, is often desireable. If a synthetic fuel gas is to be completely interchangeable with natural gas it must consist

TABLE II

KOPPERS - TOTZEK PROCESS

TYPICAL GASIFIER RAW GAS ANALYSIS
(DRY BASIS)

COMPONENT	VOLUME %
H_2	33
CO	60
CO_2	6
Other	1
	100

H.V. = 300 BTU/SCF

almost totally of methane. Such a gas is referred to as high-calorie or high-Btu gas or pipeline quality gas.

The atomic hydrogen to carbon ratio in coal is nearly always less than one. Since the atomic hydrogen to carbon ratio in methane is four, a large amount of hydrogen must be added to coal if the product is to be principally methane. Realistically speaking this hydrogen must ultimately come from water. Coal gasification technologies for producing a gas high in methane usually begin with direct interaction between coal and steam.

Carbon can react with steam by at least two pathways depending upon the temperature. One pathway was indicated in equation (2) producing carbon monoxide and hydrogen. That reaction proceeds as the principal route at higher temperatures, the equilibrium constant, K_p, being greater than 1 at temperatures higher than 674°C. At lower temperatures the reaction can become

$$2C + 2H_2O = CH_4 + CO_2 \qquad (3)$$

This reaction is thermodynamically favored at temperatures below 260°C. This chemical reaction is extremely slow at that temperature and a suitable catalyst to provide adequate reaction rates has not been found. Hence, in any practical coal gasification system, the steam-carbon reaction proceeds according to equation (2). The chemical reaction according to equation (2) is endothermic to the extent of about 32 kcal per gram mole, hence must be sustained by an energy input.

As indicated above, the initial gasification system nearly always introduces coal (carbon), steam and oxygen. At equilibrium the system then consists of six principal species, namely H_2, C, CO, CO_2, H_2O and CH_4. The relative quantities of these components at equilibrium may be adjusted by adjusting the temperature, pressure, the amount of molecular oxygen injected in relation to the other components, or by injecting molecular hydrogen instead of molecular oxygen as a reactant.

Figure 13 presents equilibrium yields of methane as mole percent in the product, as a function of temperature for three cases. The upper curve is for the case where molecular hydrogen is injected with steam at a molar ratio of 1/2 mole of hydrogen per mole of steam. The middle curve represents the steam-carbon reaction only. The lower curve represents the injection of 1/2 mole of oxygen per mole of steam. These curves are equilibrium values at a total pressure in the system of 1000 psia. As one would expect, a higher molar percent of methane can be achieved when hydrogen and steam are injected than when oxygen and steam are injected. A very important feature of these equilibria is the manner in which the methane yield decreases as the temperature increases. If one desires to obtain a high yield of methane

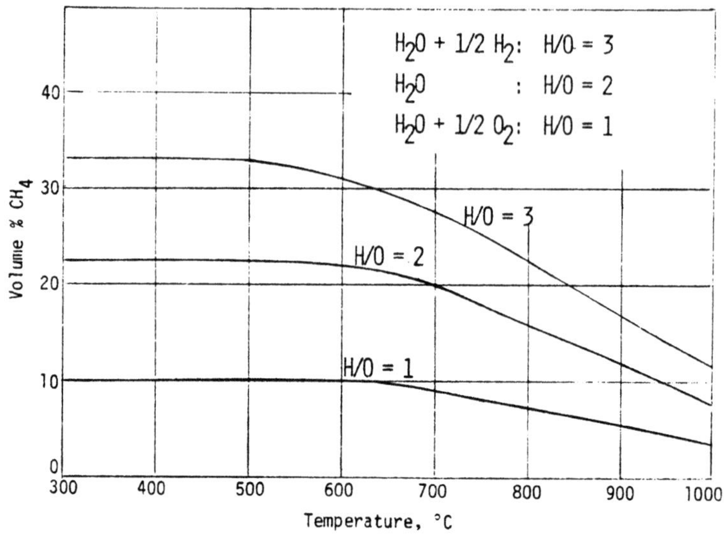

EQUILIBRIUM YIELDS IN SYSTEM OF H_2, C, CO, CO_2, H_2O AND CH_4; PRESSURE = 1000 PSIA.

FIGURE 13

in the gasifier where these species are all present, perhaps the most important parameter which can be adjusted is the temperature.

Figure 14 presents equilibrium molar concentrations of methane as a function of temperature at different pressures, each of these representations being for a system where 1/2 mole of hydrogen is introduced per mole of steam. One observes that if the pressure is increased it is possible to obtain a higher yield of methane in the product gases at equilibrium at any given temperature. It thus becomes possible to obtain reasonable methane yields in the product gas at satisfactory operating temperatures if the pressure is high. It may be stated that the benefits to be gained by increasing pressure are generally less than those from reducing temperature.

If a product gas is to be at least 95 percent methane, in order to perform like natural gas, the product gas from the gasifier must usually be processed further. For example, with a composition of the product gas from the Lurgi gasifier, indicated in Table I, considerably more methane must be produced. The methane can be produced by two principal reactions:

$$C + 2H_2 = CH_4 \quad (4)$$

$$CO + 3H_2 = CH_4 + H_2O \quad (5)$$

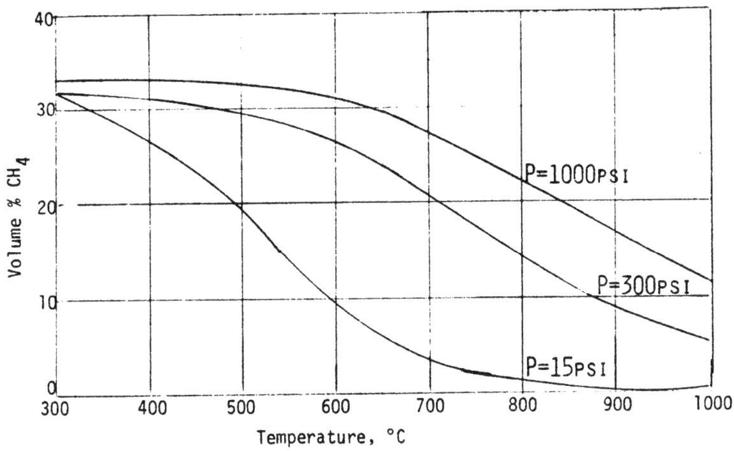

EQUILIBRIUM YIELDS IN SYSTEM OF H_2, C, CO, CO_2, H_2O AND CH_4; H/O RATIO = 3

FIGURE 14

Carbon may react directly with molecular hydrogen according to equation (4). This reaction is promoted by increasing the hydrogen partial pressure in the gas phase. At equilibrium, methane formation is favored by lower temperatures. Methane is also produced according to equation (5). It is noted that at least three moles of hydrogen are consumed per mole of carbon monoxide according to this reaction path. One therefore needs to increase the ratio of hydrogen to carbon monoxide in this gas to a value of 3.0 or greater. This can be achieved through the water-gas shift reaction according to the following equation:

$$CO + H_2O = H_2 + CO_2 \tag{6}$$

In order to achieve a molecular ratio of hydrogen to carbon monoxide of the order of three or greater, reaction (6) must be conducted at a temperature lower than 400°C. Since the reaction rate for this reaction is slow at this temperature, a catalyst must be used to increase the rate. The gas from the shift converter can then be passed through a methanator where molecular hydrogen will react with carbon monoxide to produce methane according to equation (5). However, the reaction of equation (5) tends to proceed to the right only at lower temperatures. This reaction is usually conducted in the temperature range of 430°C using an appropriate catalyst to improve the rate.

A large number of coal gasification processes have been developed for production of high-Btu gas. Most of these processes have tried to capitalize upon the adjustment of one or more parameters which affect the equilibrium yield of methane in the gasifier. Even after achieving the highest possible methane yield in the gasifier, a further shift reactor and methanator are usually required.

Figure 15 is a schematic drawing of the HYGAS process developed in the United States, which is representative of a large number of technologies under development. Examination of this schematic drawing reveals that the principal effort toward improving the methane yield in this gasifier is to increase the hydrogen content of the reactant stream to the bottom of the gasifier. The temperature of the gases leaving the gasifier is also held as low as possible, compatible with non-catalytic reaction rates. Pressure is also fairly high, 1500 psia, to gain benefit as indicated above. In order to achieve a high methane yield in the final product a shift converter and catalytic methanator follow the gasifier.

Table III presents a typical raw gas analysis from the HYGAS gasifier. This is the composition of the gas leaving the gasifier but prior to the shift converter and methanator. One notes the substantial quantity of methane in this gasifier product, 18.7 percent by volume, and the resulting heating value of the gasifier product.

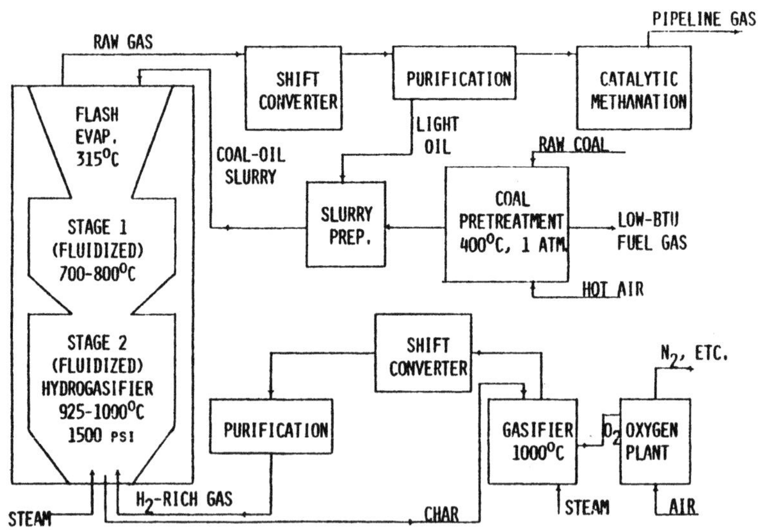

HYGAS PROCESS
FIGURE 15

TABLE III

HYGAS PROCESS (OXYGEN MODE)

TYPICAL GASIFIER RAW GAS ANALYSIS (DRY BASIS)

COMPONENT	VOLUME %
H_2	30.2
CO	23.8
CH_4	18.7
C_2H_6	0.7
CO_2	24.5
H_2S, COS	1.2
Other	0.9
	100.0

H.V. = 390 BTU/SCF

There are some significantly different approaches to the gasification of coal to produce a high Btu gas. One such process is under development by Exxon Oil Company and is illustrated in Figure 16. In this process the steam-carbon reaction, which normally requires a temperature in excess of 950°C for satisfactory rates if conducted non-catalytically, is catalyzed by the use of potassium carbonate. This catalyst, or other similar catalyst, can permit the steam-carbon reaction to be conducted at temperatures of the order of 700°C. In this temperature range, as indicated in Figure 13, equilibrium yields of methane can be between 30 and 40 volume percent. This product gas is then passed through a series of component-removal systems, to remove water vapor, H_2S and CO_2, leaving a stream consisting of CH_4, H_2 and CO. The methane is then removed in a cryogenic distillation system. The H_2 and CO are recycled to the process, methane being the sole fuel product.

Another substantially different gasification approach is illustrated in Figure 17 for the COGAS process. This process was designed to gasify the char residue from a multi-stage coal pyrolysis process, the char usually representing about 60 percent of the weight of the coal. The heart of this char gasification process is represented in Figure 18. This process operates at essentially atmospheric pressure. The char is gasified with steam but with the heat required to sustain the steam-carbon

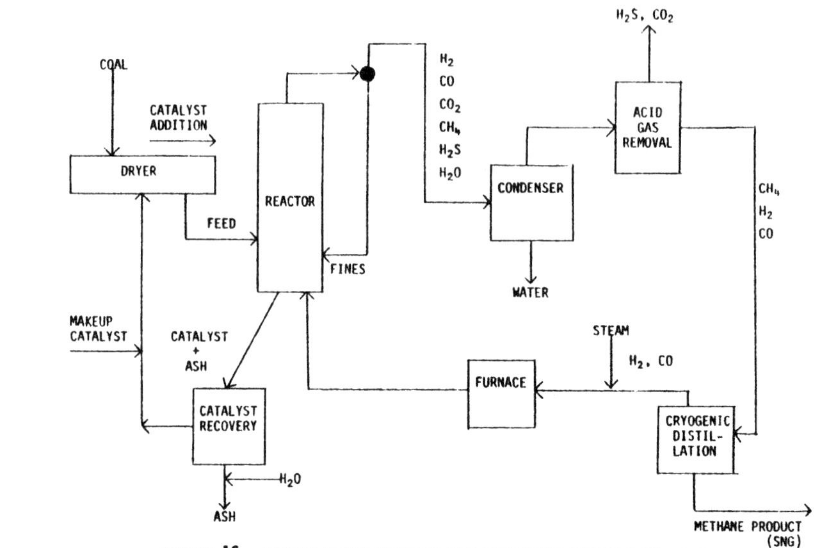

FIGURE 16. SINGLE-STAGE REACTOR AND CRYOGENIC METHANE RECOVERY: EXXON'S CATALYTIC GASIFICATION PROCESS

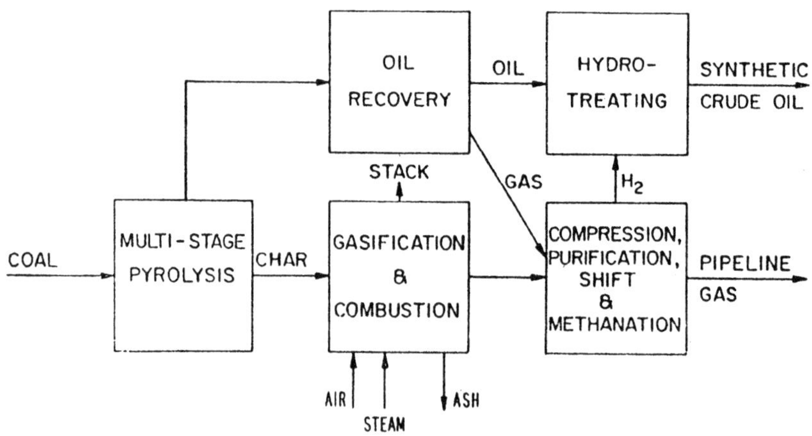

FIGURE 17. COGAS PROCESS CONCEPT

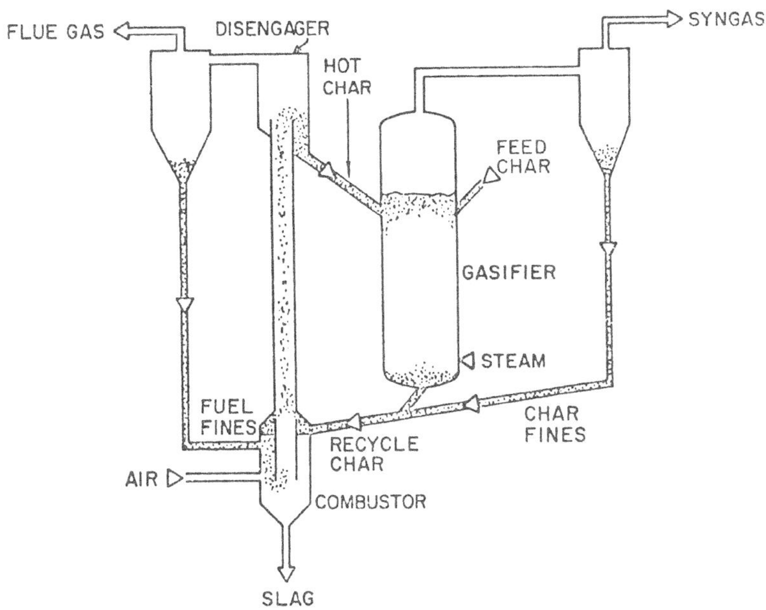

FIGURE 18. CHAR HEAT-CARRIER PROCESS
COGAS

reaction being supplied by recycle of hot char particles. Hot char is generated by burning part of the char with air. An alternative system recycles ceramic beads as the heat carrier instead of char particles. The technology has proven to work rather well. A typical analysis of raw product gas from the COGAS gasifier is given in Table IV. This gas is then fed to a shift converter and methanator for the production of a gaseous product which is at least 95 percent methane.

There are in the world very large reserves of low-rank coals known as lignite and as brown coals. These coals are characterized by high oxygen content. Much of the oxygen exists such that CO_2 in large quantity is produced when these coals are heated. A process, known as the CO_2-acceptor process, has been developed in the United States to apply particularly to these coals. Again the coal is gasified with steam. The heat required to sustain this steam-carbon reaction is supplied through the exothermic interaction of CO_2 from the low rank coal with calcined dolomite, namely calcium oxide and magnesium oxide, yielding the respective carbonates. These carbonates are then recycled through a recuperator where char from the gasifier is burned with air, supplying heat to drive off the CO_2, regenerating the oxides for recycle. This technology also operates well. A typical product

TABLE IV

COGAS PROCESS
(Char Feed)

TYPICAL GASIFIER RAW GAS ANALYSIS
(DRY BASIS)

COMPONENT	VOLUME %
H_2	53.4
CO	29.4
CH_4	1.4
CO_2	15.8
	100.0

H.V. = 290 BTU/SCF

gas from the CO_2-acceptor process is illustrated in Table V. This gasifier product gas can then be processed through a shift converter and methanator to produce a high-Btu gas as indicated previously. A schematic drawing for the CO_2-acceptor process is shown in Figure 19.

TABLE V

CO_2-ACCEPTOR PROCESS
(LIGNITE)

TYPICAL GASIFIER RAW GAS ANALYSIS
(DRY BASIS)

COMPONENT	VOLUME %
H_2	32.1
CO	29.3
CH_4	15.6
CO_2	21.5
H_2S	0.8
Other	0.7
	100.0

H.V. = 375 BTU/SCF

It should be mentioned that there are large reserves of coal in the world existing in thin seams or under very thick overburden or in formations of such nature that mining techniques, either surface mining or deep mining, do not appear to be practical. Some development work is underway in various parts of the world to gasify these coals in place. One receives mixed reviews concerning the potential of these various attempts. In situ coal gasification is not proposed as a competitor for above ground gasification for coals which can be readily mined. Rather, in situ gasification is expected to find application for coals which cannot realistically be recovered by other techniques. Gases produced from in situ gasification usually have heating values of the order of 4500-6000 kJ/SCM if gasified with air, or of the order of 11,000 kJ/SCM when gasified with oxygen. One often finds that the influx of ground water into the formation supplies water necessary for the steam-carbon reaction as indicated for above-ground gasification. Typical analyses for the product gas from in situ gasification for the air blown and the oxygen blown modes are given in Table VI.

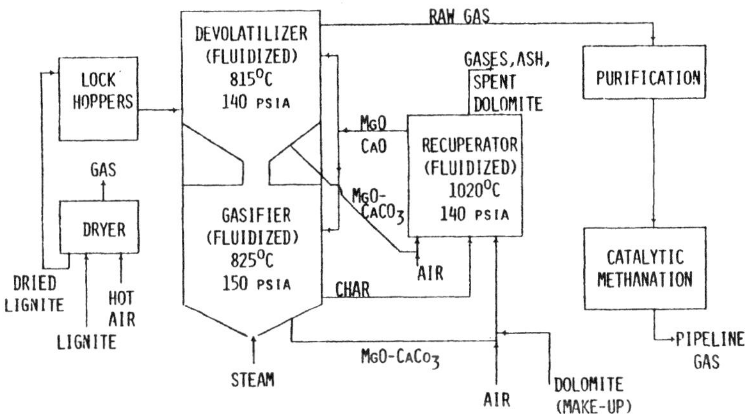

CO_2-ACCEPTOR PROCESS

FIGURE 19

TABLE VI

UNDERGROUND GASIFICATION (SUB-BITUMINOUS COAL)

COMPONENT	AIR BLOWN	STEAM/O_2 BLOWN
H_2	16.0	34.0
CO	6.3	13.4
CH_4	3.9	8.3
C_2H_6, C_3H_8	0.6	1.3
CO_2	19.2	40.8
N_2	53.2	0.3
Other	0.8	1.9
	100.0	100.0
H.V., BTU/SCF	124	274

QUANTITATIVE ^{13}C AND ^{1}H NMR OF CRUDE OIL PETROLEUM AND COAL LIQUEFACTION PRODUCTS

J.-J. Delpuech

Laboratoire de Chimie Physique Organique, ERA CNRS 222,
Université de Nancy I, France.

ABSTRACT

Optimum experimental conditions are given for a quantitative analysis of liquid fuels by both ^{1}H CW-NMR and ^{13}C FT-NMR. General procedures are outlined to determine the optimum instrumental parameters (acquisition time, flip angle, pulse delay, solvent and relaxation reagent) to be used in quantitative FT-NMR. Digital corrections of the signal are shown to be necessary to improve the expected accuracy up to *ca.* 1 %.

I - INTRODUCTION

The uses of high resolution nuclear magnetic resonance (NMR) spectrometry in fossil fuel chemistry have expanded considerably during the past few years (1-20). This is a consequence of a general need for detailed compositional characterization of petroleum products and of coal liquids. Compositional data are important both for evaluation of fossil fuels for their intended end use and as an aid in the upgrading of petroleum or coal residues.

The development of processes for the conversion of crude oil heavy ends or of coal to valuable light oils or chemicals in effect requires easy and reliable analytical techniques to follow changes in composition at the various stages of transformation and to control the quality of the products finally obtained.

The success of NMR spectroscopy in the analysis of liquid fuels relies on two properties : the existence of frequency ranges which can be used as fingerprints of characteristic structural

L. Petrakis and J. P. Fraissard (eds.), Magnetic Resonance. Introduction, Advanced Topics and Applications to Fossil Energy, 351–376.

units, and the possibility of accurate quantitative measurements, the final objective being to derive statistical average parameters of chemical significance for the potential users from the observed spectra.

Except in the case of light oil cuts such as gasoline or of light fractions from coal liquids, the recognition of individual components is impossible, and moreover unnecessary. Ignoring the presence of other elements than carbon and hydrogen as a first approximation, ^{1}H and ^{13}C spectroscopies are mainly used to assign structural units characteristic of hydrocarbons. ^{1}H NMR spectroscopy was first used for this purpose (1,2). A significant piece of information brought about by this spectroscopy is a clear distinction between the aromatic (HA) and saturated (HS) protons. The latter category may be further divided into three subgroups, HS_{α}, HS_{β} and HS_{γ} according to the position of the observed protons (α, β or γ, respectively) with respect to the aromatic ring to which they are (eventually) attached (see the nomenclature and the correlation chart in Table I). Carbon-13 NMR was introduced in oil analysis (4) primarily to separate aromatic carbons (CA) from saturated carbons (CS), since the aromatic domain (118-160 ppm) lies far apart from the saturated domain (0-70 ppm) (Table II). In most NMR investigations, the aromatic domain is further dissected (4, 8, 21-24) into the protonated (118-130 ppm) and substituted (130-160 ppm) subdomains (CAH and CAS, respectively). In fact this assignment proved to be partially erroneous owing to some overlap of the assumed shift ranges. An inspection of the available data from recent literature (25-40) and from our own observations led us to distinguish various types of aromatic carbons and to assign the shift ranges displayed in Table II to each of them (19).

TABLE I. Proton chemical shift correlation chart for hydrocarbons (19)

Shift range (ppm from TMS)	Assignment
9.30-6.30	HA : aromatic carbons
6.30-4.50	HO : olefinic carbons
4.50-1.85	HS_{α}: protons attached to a saturated α-carbon
1.85-1.00	HS_{β}: protons attached to a saturated β-carbon (CH_3) or to a β , γ, δ (or more) carbon (CH_2, CH)
1.00-0.50	HS_{γ}: methylic protons on a γ, δ, (or more) carbon

TABLE II. Carbon-13 chemical shift correlation chart for hydrocarbon (19)

Shift range (ppm from TMS)	Assignment	
0-70.0	CL :	aliphatic carbons
20.0-45.0	CN :	naphtenic carbons
118.0-130.5	CAH :	aromatic protonated carbons
123.5-126.0	CAAA :	carbons at the junction of three aromatic rings
128.5-136.0	CAA :	carbons at the junction of two aromatic rings
129.0-137.0	$CACH_3$:	methyl-substituted aromatic carbons
132.0-137.0	CAN :	carbons at the junction of an aromatic and a naphtenic ring
137.0-160.0	CAL :	alkyl-substituted (methyl group excluded) aromatic carbons

Great progress in line assignment problems is expected from the use of selective pulses which can differentiate between CH_3, CH_2, CH and quaternary carbons, obtaining separate subspectra for each type of carbon. The intensity of the resulting lines is however not faithfully reproduced, although the most recent procedures (DEPT sequence) claim for an accuracy of up to 5 % (41-45).

It is then possible to obtain important *average structure parameters* (ASP's), such as the aromatic factor FA = C(aromatic)/ C(total) and carbon to hydrogen (C/H) ratios (total, aliphatic or aromatic). If the average molecular weight is known from independent measurements (GPC or VPO), the above structural parameters can be converted into *average molecular parameters* (AMP's), e.g. the number of total, saturated, aromatic, etc... carbon atoms per average molecule. Finally these methods aim at giving molecular models for petroleum or coal liquids, as shown originally by Williams (1) from ^{1}H NMR and elemental analysis data, then by Knight (4), and Petrakis et al. (8), from a combined use of ^{1}H and ^{13}C NMR.

An important point in all these investigations is the accuracy of quantitative measurements, which in turn critically controls the reliability of average structural factors. Peak areas in the NMR spectra are assumed to be directly proportional to the number of contributing hydrogen and carbon atoms. However high resolution NMR, especially proton decoupled ^{13}C FT-NMR, is not quantitative normally under operating conditions used typically in routine qualitative analysis. This important problem will be examined with some detail in this paper which is mainly devoted to theoretical and practical aspects of quantitative analysis by NMR spectroscopy.

II - QUANTITATIVE ANALYSIS BY ^{1}H CW-NMR

Continuous wave (CW) NMR is often used for protons analysis in single-scan experiments using an electronic integrator. In this case the NMR signal arises from the transverse nuclear magnetization with components u and v in the xy rotating frame (dispersion and absorption components, respectively). The area below the absorption curve is computed as (46)

$$A = \int_{-\infty}^{+\infty} v(\omega)d\omega = \gamma B_1 M_o \pi / \sqrt{1+\gamma^2 B_1^2 T_1 T_2} \qquad (\underline{1})$$

where γ is the gyromagnetic ratio of ^{1}H nuclei, T_1 and T_2 their relaxation times, B_1 the RF field intensity, ω the RF frequency, and M_o the equilibrium nuclear magnetization, itself proportional to the number of resonant nuclei per unit volume.

Relaxation times T_1, T_2 are likely to differ from peak to peak. The first requirement to obtain reliable relative intensities is to use an RF power B_1^2 far from the so called <u>saturation level</u> :

$$\gamma^2 B_1^2 T_1 T_2 << 1 \qquad (\underline{2})$$

in which case, eqn. 1 reduces to

$$A = \gamma B_1 M_o \pi \qquad (\underline{3})$$

a value which is strictly proportional to the number of resonant nuclei in each line.

The point to which saturation is reached in fact depends upon *the sweep rate* and varies according to wheter the spectrum or its integral is desired. In the former case, the saturation point is determined by recording a narrow line at a slow sweep rate a, compatible with the required resolution : $a <\sim 2 Hz.s^{-1}$ for ^{1}H spectra, and B_1 is increased until a maximum in signal height is observed ($\gamma^2 B_1^2 T_1 T_2 = 1$). This is the RF level generally adopted for routine analysis. This is in fact a poor choice, too high for recording the spectrum (the resolution is degraded by 40 %) and too low for recording the integral. In the latter case, faster sweep rates are preferable, up to $a \simeq 10 Hz.s^{-1}$. Line distortion is still acceptable, apart from wiggles whose mean area is close to zero. This allows in turn to use high RF power, and then to achieve good accuracy of the integral. In practice, B_1 is increased continuously up to the point where the ratios between the various integrals within the spectrum begin to deviate from a constant value. Another major reason for high sweep rates is that the rms noise error increases proportionnaly to the square root of the time difference (t_2-t_1) betwwen the two measurements of the integral line - before and after passage through resonance - assuming "white" noise. This time difference in turn is inversely proportional to the sweep rate a.

Another problem is raised in the course of this discussion : *the choice of the two instants* t_1 *and* t_2, since in practice the integration is performed between finite boundaries. A compromise must be made in the choice of t_1, t_2. If they are chosen far apart, the neglected intensity contribution from the tails of the signal outside this interval is small (a 1 % error requires an integration over 63 times the full width at half height for a Lorenzian line), but the error caused by the random noise can be substantial. If the interval (t_2-t_1) is too small, the reverse statements are valid.

Another factor in this connection is the possibility of *overlap with neighbouring resonances*. If individual integrals are to be obtained for closely spaced lines, a curve-fitting procedure is a prerequisite to disentangle each line which can further be integrated through numerical procedures.

It may also happen that individual integrals are not necessary, and that integration is performed over certain frequency ranges of interest. This is the case for ^{1}H spectra of petroleum products, where integration is carried out over characteristic frequency ranges. Such complex spectra are in fact composed of a great number of small overlapping lines merged into broad envelopes which can hardly be distinguished from the baseline in some instances (see e.g. Fig. 4). In this case, integration is performed between two given boundaries, neglecting the very small contribution from the tails of the signal outside the interval. It should be noted that a sufficient S/N ratio over the integral can be obtained even in this case in spite of very poor S/N ratios for individual peaks, since the integral S/N ratio is then approximately increasing as the square root of the integrated length.

Observing the *absorption* curve free from any *dispersion* component, as it is assumed in eqns. 1 and 3, requires a correct adjustement of the *phase angle* $\emptyset$ of the phase-sensitive detector ($\emptyset = \pi/2$). If it is not so, the detected signal V_r is a mixture of the dispersion and absorption modes :

$$V_r \propto v \sin \emptyset + u \cos \emptyset \qquad (4)$$

The dispersion signal has a zero total area and an erroneous phase setting should be simply accompanied at first sight by a uniform reduction by a factor of $\sin \emptyset$ according to eqn. 4. In fact the area under each arch of opposite sign in function u is infinite, this results in continuous drifts of the line integral before and after passage through resonance, thus preventing reliable measurements.

The RF intensity of the transmitter, the sweep rate, and the phase angle of the receiver should be carefully adjusted in quantitative CW analysis. This requires observing one sharp line in the spectrum, generally that of the standard reference sample, e.g. dioxane (D) or hexamethyldisiloxane (HMDS) in the case of petroleum products.

III - OPTIMUM INSTRUMENTAL PARAMETERS FOR QUANTITATIVE ^{13}C FT-NMR

1 - Problems in quantitative FT-NMR (46,47)

Quantitative analysis is known to be difficult in the case of ^{13}C FT-NMR, because of both the problems attached to pulse Fourier Transform NMR and the particular characteristics of ^{13}C nuclei.

(a) *The RF pulse should equally rotate the equilibrium magnetization M_0 of each set of ^{13}C nuclei.* In fact, the flip angle θ depends on the resonance frequency. Deviations of θ from the nominal flip angle $\theta_0 = \gamma B_1 P_W$ (P_W is the pulse width) obtained at the exact frequency of the carrier wave are observed as a function of the frequency offset $\Delta\nu$. This results in turn in intensity and phase deviations depending upon the parameter $\xi = 4P_W\Delta\nu$. These deviations are at a maximum for the large θ_0 values ($\theta_0 = 90°$) which are necessary for quantitative NMR (see below). They should be kept to a minimum using pulse widths as small as possible. We use typically : $\theta_0 = 90°$; $P_W = 10\ \mu s$; $\Delta\nu_{(max)} = 12500$ Hz (^{13}C frequency range at 62.86 MHz), so that $\xi_{max} = 0.5$. Under these conditions it can be computed that the relative magnitude of signals are not altered (from 1 to 0.9986 as $\Delta\nu$ goes from 0 to 12500 Hz), but that their phase angle $\emptyset$ is changed from 90° ($\Delta\nu = 0$) to 61° ($\Delta\nu = 12500$ Hz). This is one cause among others, generally ignored, for frequency-dependent phase angle corrections (see below).

(b) *The receiver circuitry should amplify the signals equally all along the frequency window defined by the frequency filter bandwidth.* This is only approximately achieved since the frequency filter has not a sharp cut-off frequency. The intensity of the signals is reduced as the offset $\Delta\nu$ goes from 0 to $\Delta\nu_{(max)}$ (10). The frequency filter also introduces frequency-dependent phase angle variations.

(c) The sampling procedure introduces a third *phase error* due to the necessary *delay time* DE between the end of the RF pulse and the acquisition of the first point of the free induction decay (FID). All these phase errors are frequency-dependent and cannot be avoided by an instrumental phase setting prior to recording of the spectrum, as it is the case for CW-NMR. This correction is however of fundamental importance to obtain pure absorption signals and is performed after acquisition of data and Fourier Transformation. This operation carriers the signal I(t) from the time to the frequency domain, i.e. I(t) is transformed into

$$F(\omega) = \int_{-\infty}^{+\infty} \exp(-i\omega t) I(t) dt = F_C - iF_S \qquad (\underline{5})$$

The real and imaginary parts of $F(\omega)$, F_C and F_S (Fourier cosine and sine transforms, respectively), are a mixture of both absorption and dispersion modes

$$F_C \propto [v \cos \emptyset - u \sin \emptyset].\sin \theta \qquad (\underline{6})$$

$$F_S \propto [u \cos \emptyset + v \sin \emptyset].\sin \theta \qquad (\underline{7})$$

The pure absorption spectrum is then restored through the appropriate combination of F_C and F_S :

$$v \sin \theta \propto F_C \cos \emptyset + F_S \sin \emptyset \qquad (\underline{8})$$

This correction should be made point by point since the phase angle $\emptyset(\omega)$ is frequency-dependent.

(d) Finally, the *integral* of the absorption curve is obtained by *digital integration* over a discrete distribution of sampled points, thus possibly introducing truncation errors.

2 - Problems specific to ^{13}C NMR (47)

The low receptivity of this nucleus requires fast accumulation of a large number of scans n, so as to increase the initial value $(S/N)_1$ of the signal-to-noise ratio to a final value $(S/N)_f$:

$$(S/N)_f = (S/N)_1.\sqrt{n} \qquad (\underline{9})$$

Besides a high $(S/N)f$ ratio for a high level of accuracy, another requirement is a *uniform sensitivity* throughout the whole spectrum for reliable intensity comparisons. This is difficult owing to (i) the length of ^{13}C and ^{1}H relaxation times and (ii) to the existence of the nuclear Overhauser effect (NOE).

(a) The fact that $T_1 >> T_2$ for ^{13}C nuclei requires *waiting a given length of time* t_3 (*pulse delay*) after the acquisition time t_2. Ideally, an infinite length of time $(P_W+t_2+t_3) \simeq (t_2+t_3)$ is necessary for the longitudinal ^{13}C magnetization M_{zc} to recover its equilibrium value M_{oc}. In fact, the waiting time will be limited to a value $t_{3r} \simeq 4.55\ T_{1C}$ such that the fractional decrease z_f of M_{oc}

is lower than a given limit of 1 % :

$$z_f = |(M_{zc}-M_{oc})/M_{oc}| \leqslant 0.01 \qquad (\underline{10})$$

If several lines are simultaneously observed, the pulse delay will be chosen as the longest t_{3r} value (t_{3m}) which ensures condition (10) to the set of ^{13}C nuclei endowed with the longest relaxation time $T_{1C(max)}$. This condition is never respected in routine qualitative analysis, but is of fundamental importance in quantitative analysis. This condition is opposed to the requirement of a maximum $(S/N)_f$ ratio, and an *optimum* is to be found as shown in the next section.

(b) *The NOE effect* is defined by the fractional increase η of the ^{13}C nuclear magnetization in the presence of the broad-band decoupling of protons. The apparent equilibrium magnetization is raised to a value :

$$M'_{oc} = M_{oc}(1+\eta) \qquad (\underline{11})$$

which is proportionally reflected in the line intensities. However this enhancement is not the same for all lines : the NOE factor varies from 0 to 1.998, mainly depending on the number n_H of protons directly attached to the observed nucleus and eventually on the number of more remote neighbours and on the reorientation rate of molecules in solution. The NOE effect is suppressed by the gated decoupling procedure which consists in switching on the decoupler only during the pulse and acquisition times. In fact the NOE effect requires a finite length of time to build up (during the acquisition time) and to disappear (during the waiting period t_3). The time constant for these processes is approximately the relaxation time of the coupled protons T_{1H}. The gated decoupling experiment therefore needs much longer waiting times t_3 than stated above, except for systems with $\eta \simeq 0$ (quaternary carbons). Pulse delays ranging from five to ten times the longest relaxation time T_{1C} are loosely recommended for this purpose in the literature (48). This point will be examined more thoroughly in the next subsection.

Another technique to suppress the NOE effect, which has the simultaneous advantage of shortening the relaxation times T_{1C} and T_{1H}, is the addition of well-chosen paramagnetic species, the so-called *relaxation reagents*, as shown below in section III-4.

3 - Optimum conditions for non-destructive quantitative analysis (47)

If we assume a correct phase setting, a constant noise level of the spectrometer and a given total time of operation, then the final signal-to-noise ratio (eqns.8 and 9) will be proportional to

$$(S/N)_f \propto M_{zc}.(t_2+t_3)^{-1/2}.\sin\theta \quad (\underline{12})$$

This expression goes through a maximum for a given value θ_E of θ, the so-called Ernst angle (49) :

$$\theta_E = \arc\cos(\exp(-(t_2+t_3)/T_{1C})) \quad (\underline{13})$$

This condition effectively ensures a maximum S/N ratio to the examined resonance, but the sensitivity thus obtained is not uniform throughout the whole spectrum since several resonances are usually observed in the same spectrum, each with its own T_{1C} value. In quantitative analysis, condition (10) should be satisfied as explained above. Under this condition : $M_{zc} \simeq M_{oc}$, and we are then faced with the problem of finding the optimized set of instrumental parameters which ensure a maximum value to the function :

$$y = (t_2+t_3)^{-1/2}.\sin\theta \quad (\underline{14})$$

subject to condition (10), for all possible sets of the nuclear parameters : T_{1C}, T_{1H} , n_H and η.

The acquisition time t_2 is determined by the expected digital resolution $r = 1/t_2$, and, in practice, by the computer memory area which is reserved for data acquisition. Typical values are t_2=0.6 to 1.3 s for spectra at 80-300 MHz. We are then left with the problem of adjusting θ and t_3 so as to satisfy conditions (10) and (14). This requires in turn to establish the relationship between z_f and the instrumental parameters.

This problem is solved by computing the time evolution of ^{13}C and ^{1}H longitudinal magnetizations after each RF pulse. This is done by adapting Solomon's equations relative to a two spin systems (I-S) (50) to the case of one spin I (^{13}C) coupled to n_H equivalent spin S (^{1}H) (47). The results of this study can be stated as follows.

(a) For a given set of isochronous ^{13}C nuclei, the longitudinal magnetization M_{zc} at the beginning of each pulse assumes a steady-state value after a certain number of scans n_S, depending upon the flip angle : n_S = 0 to 10 when θ is varied from 90° to 30°. Let us assume that this steady-state is reached in the following.

(b) In a next step, the minimum waiting-time t_3 which is necessary to ensure simultaneously conditions (10) and (14) is computed as a function of θ. This function goes through a minimum which represents the optimum waiting-time t_{3m}. *An important point is that this optimum is reached in all cases for a value θ_m of the flip angle quite close or equal to 90°*. This is at strong variance with the values generally recommended in qualitative analysis ($\theta \simeq$ 10-20°) and even in quantitative analysis, typically θ = 30°.

Let us assume $\theta = 90°$ in the rest of this discussion.

(c) In the general case, in which the spectrum consists of several lines, the last problem is to evaluate the largest value t_{3M}, of the set of parameters t_{3m} obtained for each individual signal. This computation can be performed after a prior inspection of the range of values possibly assumed by the nuclear parameters T_{1C}, T_{1H}, η and n_H. This point is a difficult one, owing to the paucity of relaxation data in the literature. Rough limits only can be estimated for these parameters. The suggested limits for pairs of parameters T_{1C} and η are represented in the graph of Fig.1 which mainly applies to the case of saturated and aromatic

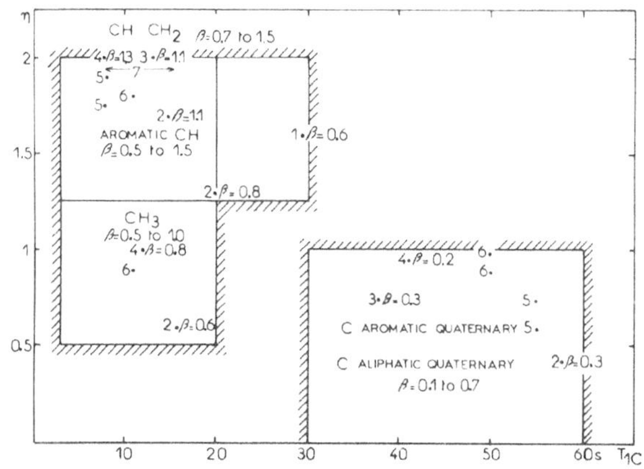

Figure 1. A schematic graph of the possible nuclear parameters (η, T_{1C}, n_H, $\beta = T_{1H}/T_{1C}$) for systems of pratical importance : (1) benzene (2) toluene (3) *O*-xylene (4) mesitylene (5) hemimellitene (6)isodurene (7) *n*-alkanes. From reference 47 by permission.

hydrocarbons. Rectangular areas are assigned to methyl carbons ($n_H = 3$), $0.5 \leqslant \eta \leqslant 1.98$ and $3 \leqslant T_{1C} \leqslant 20$ s ; aromatic unsubstituted carbons ($n_H = 1$), $1.5 \leqslant \eta \leqslant 1.98$ and $3 \leqslant T_{1C} \leqslant 30$ s ; and quaternary aliphatic and aromatic carbons (the exact value of n_H is not known), $0 \leqslant \eta \leqslant 1$ and $30 \leqslant T_{1C} \leqslant 60$ s. A horizontal segment is assigned to methylene and methine carbons, $\eta = 1.98$ and $0.5 \leqslant T_{1C} \leqslant 20$ s. Simultaneously T_{1H} may vary in each case ; the ratio $\beta = T_{1H}/T_{1C}$ is assumed to vary from 0.5 to 1.0 ; 0.1 to 0.7 and 0.5 to 1.5, respectively. We shall simplify the treatment of all these cases by gathering methyl, methylene, methine and aromatic unsubstituted carbons into one domain (I) for which $0.5 \leqslant \eta \leqslant 2$; $0.5 \leqslant T_{1C} \leqslant 30$ s ; $\beta = 0.5$ to 1.5 and $n_H = 3$ or 2 or 1. This domain may be further dissected into two subdomains, an aliphatic/aromatic domain I_A, where $0.5 \leqslant T_{1C} \leqslant 20$ s, and $n_H = 1$ to 3, and a purely aromatic domain (I_B), where $20 \leqslant T_{1C} \leqslant 30$ s, and $n_H = 1$. A second domain (II) will be considered for quaternary aromatic or aliphatic carbons $0 \leqslant \eta \leqslant 1$, $30 \leqslant T_{1C} \leqslant 60$ s, $\beta = 0.1$ to 0.7, $n_H = 1$ to 3 (the exact value of n_H will be shown to be of little importance).

The dependency of t_{3m} upon the nuclear parameters T_{1C}, β, η and n_H is expressed as plots of the ratio $\alpha = t_{3m}/T_{1C}$ as a function of T_{1C}, computed for series of values of the parameters β and η. Final sets of results are displayed in the form of graphs of Fig.2A and 2B. For carbons of domain I, a series of curves $C^{\beta}_{n_H}$

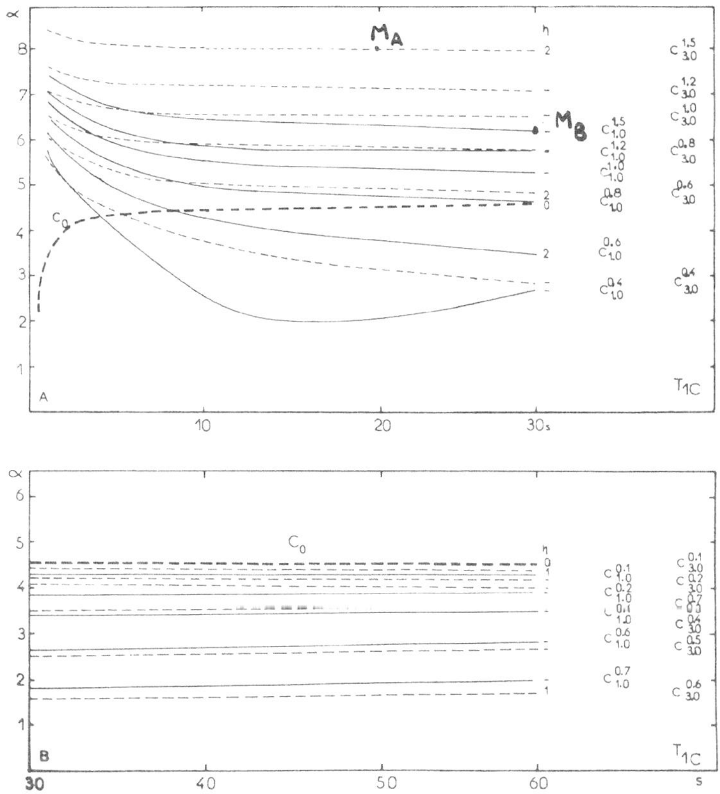

Figure 2. Plots of the convenient pulse delays t_{3m} (expressed as the ratios $\alpha = t_{3m}/T_{1C}$) as a function of T_{1C}. Curves $C^{\beta}_{n_H}$ correspond to different sets of parameters : $n_H = 1$ (solid line) and $n_H = 3$ (dashed line) ; and β (as shown in the figure). Two graphs are given (A) and (B) (above and below, respectively) according as $T_{1C} < 30$ s or $T_{1C} > 30$ s ($t_2 = 1.3$ s, $\theta = 90°$). From reference 47, by permission.

represents α as a function of T_{1C} for η = 1.98 and for various values of β (β = 0.4, 0.6, 0.8, 1.0, 1.2 and 1.5) and n_H (n_H = 1 and 3) (Fig.2A). The upper boundary of these graphs will yield the desired t_{3M} value. It can be shown that this upper boundary is always determined by the curves pertaining to the value 1.98 of η, which alone are represented for this reason in Fig.2A. The appropriate waiting time t_{3M} is therefore determined from the highest possible value α_M of α. This value is itself determined from the upper curves of Fig.2A, taking their right-hand extremities, i.e. points M_A and M_B of curves $C_3^{1.5}$ and $C_1^{1.5}$ in the subdomains I_A and I_B, respectively, for which α_M = 8 ; t_{3M} = 160 s, and α_M = 6.2 ; t_{3M} = 200 s.

For carbons of domain II, a similar series of curves C_{nH}^{β} is shown in Fig.2B for η = 1, β = 0.1 to 0.7, and n_H = 1 or 3. In fact all these curves lie below the unique curve C_o corresponding to η = 0, which is an alternative limiting curve to be used to determine t_{3M}. This curve is a horizontal straight line α = 4.55 corresponding to a classical value of the waiting time t_3=4.55 T_{1C} when NOE and cross-relaxation effects are absent. The maximum waiting time t_{3M} computed for the highest value of T_{1C} (60 s) is therefore t_{3M} = 270 s, a value to be used for a quantitative analysis of quaternary carbons.

In conclusion, the following conditions are recommended for performing a quantitative non-destructive carbon analysis of a complex sample :

(i) discard the first ten scans, if $\theta \neq 90°$

(ii) choose a flip angle θ = 90°

(iii) choose delay-time $t_3 \simeq 6.5\ T_{1C}$, when $T_{1C} \leqslant 30$ s, and $t_3 = 4.6\ T_{1C}$, when $T_{1C} \geqslant 30$ s, or alternatively,

(iiii) choose t_3 = 200 s for non quaternary aromatic or aliphatic carbons atoms and t_3 = 270 s for quaternary carbon atoms with unkonwn relaxation times.

The latter values are intermediate between 5 T_{1C} - which is a classical value for obtaining an equilibrium magnetization after a 90° pulse when no attention is paid to the relaxation of coupled protons - and 10 T_{1C}, which is a value recommended for measuring the NOE factor (48).

4 - Relaxation reagents (18)

These pulse delays will result in a considerable decrease of the number of scans in a given length of time T, and consequently, of the final S/N ratio, thus seriously limiting the accuracy of carbon titrations. This explain the necessity of using relaxation reagents which, without shifting resonance frequencies, strongly decrease both carbon and proton relaxation times T_{1C} and T_{1H} and cancel the NOE factor η. The nature and the concentration of the reagent Q are the factors which govern its effectiveness. The reagents most used at the present time are the acetylacetonates of chromium (III) and iron (III) : $Cr(acac)_3$ and $Fe(acac)_3$. More or less heavily substituted acetylacetonates have been tried so as to increase their solubility in organic solvents, but this advantage was more than counterbalanced by an unfavourable reduction in their effectiveness due to an increased distance of approach of reagent molecules to sample molecules. Relaxation reagents form labile complexes with sample molecules and thus induce paramagnetic relaxation rates approximately proportional to the concentration $[Q]$ of the reagent (51). The observed relaxation rates in the absence of reagent $1/T_1^\circ$ are thus increased by a large paramagnetic contribution $1/T_1^e$ after addition of the reagent

$$1/T_1^{obs} = 1/T_1^\circ + 1/T_{1e} \simeq 1/T_{1e}$$

Pulse delays t_3 thus decrease as more reagent is added to the sample and finally reduce to zero as soon as $t_3 = 4.55\ T_{1C}$. This limiting situation is reached for a concentration $[Q_M] \simeq 0.1$ M.

Calibration curves yielding t_3 as a function $f[Q]$ of $[Q]$ are therefore necessary in quantitative analysis. We used toluene as a model compound for hydrocarbons because of a wide range of relaxation times and NOE factors. For a given concentration $[Q]$, the waiting time t_3 was progressively decreased until the intensity ratios of the carbon signals were coincident with theoretical values (1:2:2:1:1 for carbons C_1, C_2, C_3, C_4 and CH_3) within experimental precision (standard deviation : 2.5 %). This limit yields the corresponding t_3 value to be plotted on curve (a) of Fig.3. Curves for the two relaxation reagents $Cr(acac)_3$ and $Fe(acac)_3$ were found to be coincident within experimental erros, thus showing approximately identical intrinsic effectiveness of both reagents. This calibration curve was checked using two other reference compounds : tetralin and 2-ethylnaphthalene. Typical waiting times are : 16, 6, 2 and 0 s for concentrations of 0.025, 0.050, 0.075 and 0.10 M, respectively. No pulse delay seems to be necessary for concentration $[Q] \simeq 0.1$ M, except however for samples containing highly condensed aromatic structures (asphalts) where a waiting time of *ca.* 10 seconds may be preferable.

Concentrations larger than the limiting value $[Q_M] = 0.1$ M are unnecessary since they do not result in a further decrease of the operation time, and, on the contrary, brings about an unwanted degradation of the resolution. Paramagnetic relaxation effectively induces linebroadenings r which were also measured as a function of $[Q]$ (curve (c) in Fig.3). Typical linewidths are : r = 2, 3.5, 5 and 6.5 Hz when $[Q]$ = 0.025, 0.050, 0.075 and 0.1 M, respectively.

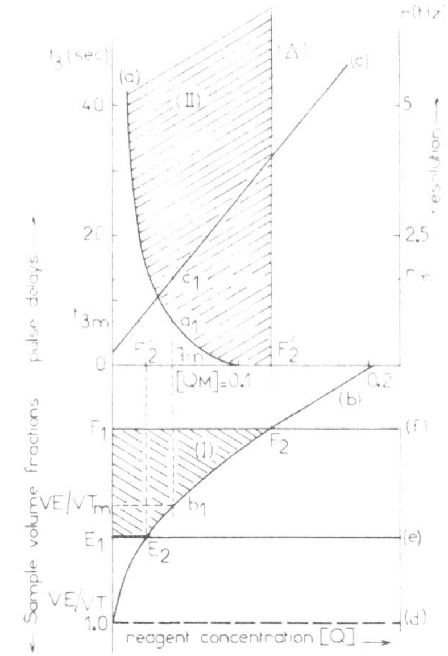

Figure 3. Schematic plots of the possible sample volume fractions VE/VT (shaded area I) and pulse delays t_3 (area II) as a function of the relaxation reagent concentration /Q/ ($mol.dm^{-3}$). Other notations are explained in the text. From reference 18, by permission.

Optimum conditions for such a sample may then be defined in either of two ways (i) *The resolution,* i.e. the mean linewidth r, should not be larger than a given value r_m, in which case curve (c) yields the corresponding concentration [Q], and, in turn, curve (a) the necessary waiting time t_3 (ii) *The time of operation* T should be reduced to a minimum (for a given final signal-to-noise ratio), in which case [Q] = 0.1 M and t_3 = 0 s.

These conclusions must be somewhat amended in the case of heavy oil fractions owing to the necessity for *diluting the sample* of volume VE with a convenient volume VS of an appropriate solvent so as to dissolve the required concentration [Q] of the relaxation reagent. A great variety of procedures have been described (11,12, 15,16) in the literature to solve this problem. It seems therefore useful to make some comparisons between the various solvent-reagent systems regarding the efficiency of the procedure. Deuterium chloroform $CDCl_3$ is the most commonly used solvent in quantitative analysis at the present time. We however selected 1,1,2,2-tetrachloroethane-d_4 which has the net advantages of a higher boiling point (E = 146° instead of 60°C) and better dissolving properties towards hydrocarbons. The sample volume fraction VE/VT, where $VT \simeq VS + VE$, may vary from 0 (pure solvent) to 1 (neat sample), as shown on the graphs of Fig.3 (abscissae axis and line (d)). The possible dilution ratios VE/VT are limited on this graph by the solubility curve $VE/VT = g[Q]$ of the reagent (curbe b) and by two instrumental constraints (a) $(VE/VT) < E_1$ (line e) so as to maintain a deuterium signal from the solvent sufficienty intense to allow a long-term field-frequency stabilization (lock) ; E_1 depends on the stability of the spectrometer which is used for these experiments, e.g. $(VE/VT) = 0.25$ in our experiments. (b) $(VE/VT) > F_1$ (line f) to avoid strong baseline distortion (cf.section III-6) due to an intense ^{13}C signal from the solvent ; $F_1 = 0.66$ with our spectrometer. The variation domain of (VE/VT) is thus defined by the shaded area I in Fig.3 enclosed between the ordinate axis (segment E_1F_1) and lines $b(E_2F_2)$, $e(E_1E_2)$ and $f(F_1F_2)$.

These graphs are now used to show the way for obtaining the desired optimum conditions.

(i) To achieve a given *resolution*, the desired minimum linewidth is again reported on curve (c) (point C_1), this determines the corresponding value q_m of $[Q]$, the waiting time t_3, from curve (a) (point a_1), and finally the necessary dilution ratio $(VE/VT)_m$ from the solubility curve (b) (point b_1).

(ii) If a *minimum time* of operation is desired ensuring a given final S/N ratio, the beneficial effect of adding relaxation reagents (until $t_3 = 0$ as in the above toluene sample) is now partially counterbalanced by the parallel necessity of increasing the dilution ratios (VE/VT). The total time of operation is now proportional to the quantity

$$V = \left(\frac{VE}{VT}\right)^2 \frac{1}{(t_2+t_3)} = \frac{g[Q]^2}{t_2+f[Q]} \qquad (15)$$

The optimization variable V effectively goes through a maximum for a reagent concentration $[Q_{max}]$ which was found in fact to be nearly coincident with the above optimum value $[Q_M] = 0.1$ M. Best results are obtained with the $Fe(acac)_3/C_2D_2Cl_4$ couple (due to a higher solubility of the reagent) which allows to reduce the time of operation by a factor of 3 to 5 as compared with the usual $Cr(acac)_3/CDCl_3$ pair.

5 - Experimental procedures (18,19,52)

Proton spectra were taken on a Cameca 250 superconducting spectrometer operating at 250 MHz in the continuous wave mode. Carbon-13 Fourier transform spectroscopy was performed with the same apparatus at 62.86 MHz. Experimental conditions were : flip angle, 80° ; pulse width, 10 µs ; acquisition time, 1.3 s ; pulse delay : 0 to 270 s ; sweep width, 12500 Hz (≃200 ppm) ; digital resolution, 0.76 Hz ; memory capacity used for data acquisition, 32K words of 20 bits ;

delay time, 40 µs ; frequency filter, 12400 Hz ; no. of FID's, 100-10000 ; sample temperature from 25 to 70°C ; sample tube diameter, 8 mm ; deuterium internal lock using the solvent signal.

Oil samples except asphalts were prepared to obtain a ^{13}C resolution of 2 Hz and a time of operation of 13.5 h, corresponding to an overnight experiment. These requirements demand in turn : concentration of reagent [Q] = 0.025 M ; pulse delay $t_3 \geq 16$ s ; dilution ratio $\frac{VE}{VT} \sim 0.6$. Typical solutions are prepared by mixing 1.75, 0.25 and 1.00 cm^3 of oil, an internal standard (D : dioxane or HMDS : hexamethyldisiloxane) and solvent ($C_2D_2Cl_4$), respectively. This solution is divided into two aliquots of 1.5 cm^3, one for ^1H spectroscopy, and the other for ^{13}C spectroscopy after the addition of 13 mg $Fe(acac)_3$. The sample for ^1H spectroscopy may be conveniently diluted with 2 to 10 volumes of $C_2D_2Cl_4$ to prevent self-association of aromatic components which otherwise could produce chemical shift deviations of up to 0.6 ppm (from the values of Table I) (52). This procedure, however, cannot be applied to heavy oil ends, e.g. asphalts, on account of a decreased solubility of the sample, by a factor of about 2, event in the better solvent $C_2D_2Cl_4$. Higher dilutions of the sample ($VE/VT \leq 0.3$) are therefore necessary which require in turn higher concentrations of relaxation reagent (up to the saturation limit) to keep the time of operation within reasonable limits. A compromise between these conflicting requirements was found by using a 0.24 volume fraction of sample, i.e. 2.25 cm^3 $C_2D_2Cl_4$, 0.50 cm^3 standard, 0.70 cm^3 or *ca.* 0.8 g of asphalt (a dark solid at 25°C) and a 0.075 M concentration of $Fe(acac)_3$ (38 mg in 1.5 cm^3). These conditions correspond to a resolution of 5 Hz and a time of operation of 18 h, which can be reduced to about 3-5 hours on more recent spectrometers.

Using more sensitive spectrometers also allows to introduce pulse delays longer than mentioned above, thus improving the accuracy of measurements, especially for highly condensed aromatic carbons.

Coal liquids generally show better dissolving properties towards relaxation reagents, which then may be added to the neat sample (typically, in this event : [Q] = 0.05 M, and $t_3 \sim 30$ s) (52).

6 - Digital corrections of the signal

Raw data obtained from Fourier Transformation of the FID signal should be submitted to some additional mathematical treatments before being integrated. Three main corrections are necessary to eliminate the effect of phase misadjustments, unequal amplification of signals and baseline distortions. Finally, there may be computation errors arising from the digital integration procedure itself.

(a) Digital phase corrections

Unlike spectra run in the CW mode, phase errors in FT-NMR are frequency dependent. This is due to flip angle variations across the spectrum (Section III-1), to delays between the end of each RF pulse and the start of data acquisition, and to the phase shifts introduced by the frequency filter (46). Phase corrections are performed mathematically by application of eqn.8 used point by point, generally assuming a linear variation of the phase angle $\emptyset$ with the frequency ν (53) :

$$\emptyset = a + b\nu \qquad (16)$$

In current practice, the unknown coefficients a and b are obtained from the knowledge of the phase angle $\emptyset_A$ and $\emptyset_B$ for two reference signals A and B. Angles $\emptyset_A$ and $\emptyset_B$ are determined consecutively by trial and error until an undistorted signal is computed for each resonance A and B. Thse angles are introduced as parameters into a routine allowing automatic computation of the phase-corrected spectrum over the entire frequency range through a linear interpolation of $\emptyset$ between $\emptyset_A$ and $\emptyset_B$.

However it should be pointed out that linear phase corrections are not able to remove phase distortions for resonances which are far from the reference signals A and B. Higher degree polynomials are necessary to compensate the non-linear phase shifts introduced by Butterworth or Bessel frequency filters. This requires in turn more reference lines, A, B, C... to compute the polynomial coefficients. We found a good compromise by using a quadratic form (54) :

$$\emptyset = a + b \quad + c\nu^2 \qquad (17)$$

and three reference compounds : hexamethyldisiloxane (HMDS), dioxane and CS_2.

(b) Digital intensity corrections

As mentioned above, the intensity of signals is attenuated all along the spectrum as a monotonic function of the frequency offset, because of filtering effects and, possibly, of a non-uniform amplification level. However the attenuation is reproducible and standardization is possible using calibration curves (10). Again the effect is linear only as a first approximation, and correction curves $f(\nu)$ should be represented by more complicated expressions (54), e.g. :

$$f(\nu) = a + (b+c\nu)\exp(d\nu) \qquad (18)$$

where parameters a, b, c and d depend upon the spectrometer. Eqn.18 allows an easy automatic intensity correction of the above phase-corrected spectrum, point by point.

(c) Digital baseline-flattening

The evaluation of integrated line intensities is often complicated by the presence of slightly distorted baseline (10,55-57). These distortions generally have U (10) -or S- shape (56) forms and are usually assigned to "pulse feed through", i.e. to ringing phenomenons in the receiver circuitry induced by RF pulses, and also to invalid data sampled early in the free induction decay. These artifacts should be digitally removed to improve the accuracy of data analysis. Quadratic forms have been recommended for describing U-shape distortions of the baseline (57). For S-shape distortions generally encountered in ^{13}C NMR, it is necessary to consider polynomial expansions P(ν) up to the third power (54) :

$$P(\nu) = a + b\nu + c\nu^2 + d\nu^3 \quad (19)$$

Polynomial coefficients are determined from the knowledge of at least four baseline points. Four signal-free regions are thus necessary for this purpose, to be chosen for each individual spectrum. This allows a fully automatic computation of P(ν) and of the baseline-corrected spectrum by subtraction. The efficiency of these corrections can be appreciated by recording the spectrum of a test sample constituted of a mixture of reference compounds. Spectra of actual oil samples are then phase-, intensity-, and linebase-corrected using three reference lines (CS_2, D and HMDS) and four baseline points (Fig.4).

(d) Digital integration

There are two main sources of errors possibly introduced by digital integration in the frequency spectrum.

(i) Limited integration domain. Signal areas may be underestimated since integration is in fact performed between finite boundaries ±(L(Hz) on either side of the signal. The corresponding error $(\Delta A/A)_D$ (in %) amounts to $\frac{200}{\pi}$ arctan (2L/W) for a Lorentzian curve. Integration domains L as large as thirty times the linewidth W are necessary to reduce these errors to less than 1 %. This means that linewidths of 2 to 5 Hz require an integration domain of ±100 Hz (*ca.* 1.5 ppm at 62.86 MHz). Spectra of oils are often composed of broad resonances which, however, should be considered as the continuous envelope of overlapping lines. This means that the above boundary values are still valid, i.e. the integration domain should be extended by *ca.* 100 Hz on each side of the examined envelope.

(ii) Truncation of the signals. In FT-NMR, the frequency spectrum is defined at discrete frequencies with a readout interval of $r = 1/t_2$ (which may be further decreased by zero-filling of the FID data, if computer memory area is still available). Severe

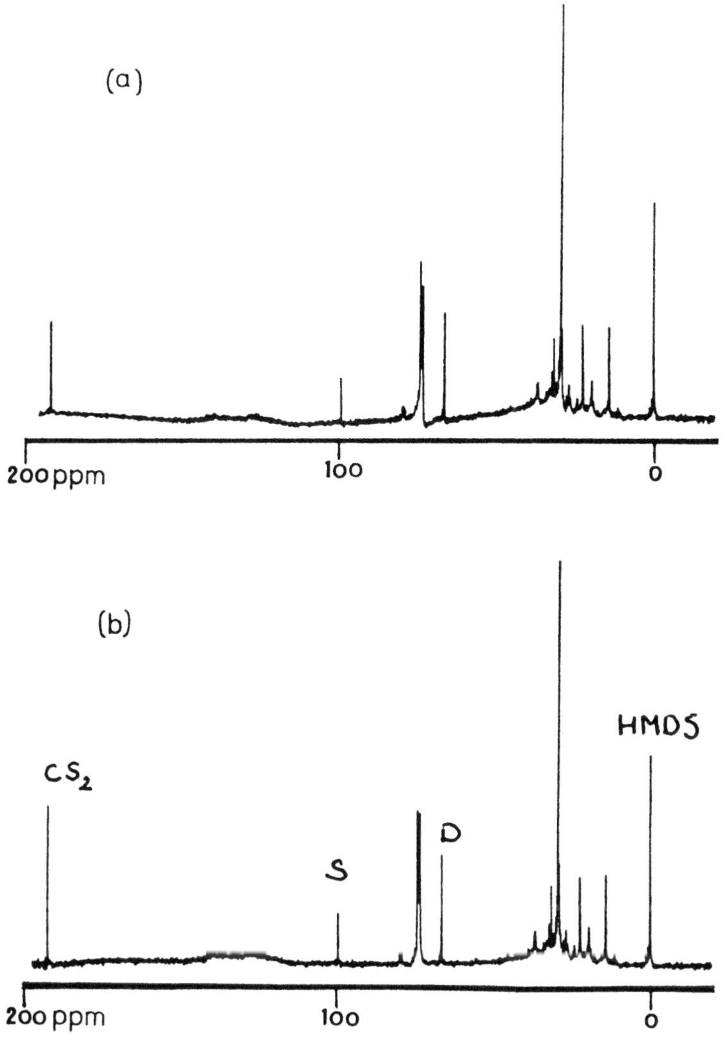

Figure 4. ^{13}C FT-NMR of an oil sample (Safaniya vacuum residue) at 62.86 MHz (a) raw spectrum (b) quadratic phase- and baseline-corrected spectrum.

line truncations may occur if digital resolution is large compared to the linewidth of signal. The maximum peak height error and the maximum frequency error which result from these truncation effects have been estimated (58).They are at a minimum when the maximum of the signal happens to be exactly on one of the discrete frequencies of the spectrum, and at a maximum when the summit of the peak lies halfway between two consecutive discrete frequencies. Truncation effects are more difficult to predict as far as signal areas are concerned. They can be estimated by performing the digital integration of a Lorentzian absorption curve and placing the maximum of the curve in either of the two above mentioned positions (59). Numerical errors of integration ($\Delta A/A$) thus obtained can be considered as the sum of the above error $(\Delta A/A)_D$ and of the truncation error $(\Delta A/A)_T$) (whose contribution to ($\Delta A/A$) is obtained by subtraction of $(\Delta A/A)_D$. Truncation errors are negligible for broad lines (W = 5 Hz), and they become critically large as soon as $r \geqslant W$. A resolution r = 0.76 Hz (our spectrometer) is still sufficient for a linewidth of 2 Hz, a limiting value commonly encountered in ^{13}C NMR. Doubling r to 1.5 Hz results in an untolerable increase of truncation error from 0.05 to 3 % . Trapezes method was used for integration ; no improvement was obtained by using parabolic interpolation procedure (Simpson method).

A control was performed using a synthetic mixture of hydrocarbons and integrating over the aliphatic and the aromatic protonated or unprotonated regions. Dioxane was added to test the validity of using this coumpound as an internal standard. The accuracy of 4 to 5 % formerly obtained by using calibration curves, linear phase correction, analogic integration and no baseline correction, was reduced to less than 1 % in these recent experiments (59). Digital corrections are therefore an important factor for obtaining a high accuracy in quantitative FT-NMR.

IV - PROTON FT-NMR SPECTROSCOPY

As many NMR spectrometers are equipped exclusively for FT operations at the present time, the use of FT spectroscopy is now extending to proton analyses of oil samples. Owing to the much higher receptivity of hydrogen atoms, one-scan operations are feasible for most samples. If it is not the case, a limited number of scans is generally sufficient for oil analyses. Sharp optimization of the pulse delay t_3 and addition of relaxation reagents are consequently useless. Generous waiting times $t_3 \sim 100$ s are recommended since the complete recovery of magnetization after each pulse still requires a time of 4.6 T_{1H}, i.e. of 80 s for a typical hydrocarbon, benzene. Flip angles of 90° still represent the best value for quantitative analysis (47,60). Digital resolutions of at least 0.1-0.2 HZ are necessary to sample peaks with linewidths of 0.2 to 0.5 Hz. If sharper lines are present in the spectrum, exponential multiplication of the FID can be used to artificially degrade linewidths up to the above values. This unpleasant manipulation may be necessary due to the computer memory area, whose capacity is primarily devised to allow ^{13}C operations and not ^{1}H operations. Finally ^{1}H signals should be digitally corrected along the same lines as those described for ^{13}C NMR. The linewidths of 0.2 to 0.5 Hz mentioned above require an integration domain of 5 to 15 Hz only, on either side of a sharp line or of a broad resonance.

V - CORRELATIONS BETWEEN CARBON AND HYDROGEN ANALYSES

Integrated ^{13}C or ^{1}H line intensities (ICI or IHJ) are proportional to the number (CI) of carbons CI of type I or (HJ) of hydrogens HJ of type J, respectively. The classification and nomenclature of carbon and hydrogen atoms are given in Tables I and II, e.g. CI may denote aromatic or saturated carbon atoms, CA and CS, respectively. The exploitation of ^{13}C and ^{1}H spectra results in the production of carbon or hydrogen number ratios, expressed as the ratio of carbon (CI) or of hydrogens (HJ) to the total number of carbons (CT) or of hydrogen (HT), respectively :

$$\% \ CI = 100(CI)/(CT) = 100 \ ICI/ICT$$
$$\% \ HJ = 100(HJ)/(HT) = 100 \ IHJ/IHT \qquad (\underline{20})$$

Correlations between carbon and hydrogen contents require the addition of a common internal standard R, dioxane (D), HMDS or TMS, in which the carbon to hydrogen ratio (CR)/(HR) is known a priori (CR)/(HR) = 1:2 or to 1:3, respectively. This allows in turn to compute any atomic carbon to hydrogen ratio

$$(CI)/(HJ) = (CR)/(HR).(IHR/IHJ).(ICI/ICR) \qquad (\underline{21})$$

The total carbon to hydrogen ratio, simply denoted as C/H, may thus be estimated as

$$C/H = (CT)/(HT) = 1/2.(IHR/IHT).(ICT/ICR)$$

(using D as a reference). This ratio can be converted from the atomic to the weight scale

$$(C/H)(weight) = 12(C/H) \qquad (\underline{22})$$

thus permitting comparisons between C/H ratios determined either by NMR or by elemental analysis (19). Eqn.21 allows also the determination of C/H ratios restricted to specific parts of the spectrum : (CA)/(HA)(aromatics), (CS)/(HS)(saturates)... The determination of these ratios does not require the knowledge of the amounts of either the sample or the reference, which would however be necessary to obtain the absolute carbon and hydrogen percentages, denoted as CI % and HJ %, as shown by Wooton and coworkers(12), i.e. the total elemental contents of carbon and hydrogen C % and H %.

Another possibility is the direct computation of the relative carbon percentage % CI from the total C/H ratio and from the relative % HI of hydrogens HI which are structurally related to carbons CI. An important example is the aromatic protons HA, which must be in equal number to the unsubstituted aromatic carbon atoms CAH. The equality (HA) = (CAH) results in the equation :

$$\% \text{ CAH} = \% \text{ HA}/(\text{C}/\text{H}) \qquad (\underline{23})$$

This formula is most useful in the present case where HA lines stand quite apart in the ^{1}H spectrum, thus allowing a reliable integration, while CAH lines may overlap with those of other types of aromatic carbons in ^{13}C spectrum. Similar formulae may be used for a specific methyl or methylene group, introducing a factor of 3 or 2 respectively in eqn.23.

Tables of characteristic atomic ratios can then be set up on the basis of the chemical shift ranges displayed in Tables I and II. Assignment of ^{13}C resonances is complicated by the existence of overlapping shift ranges. Inspection of Table II led us to the following conclusions (19). Besides the well-known possibility of distinguishing saturated (CS) and aromatic (CA) carbons, integration over the shift range 137-160 ppm yields the percentage of aromatic alkyl-substituted carbons (% CAL or, alternatively FA = % CAL/% CA). The rest of the spectrum (118-137 ppm) yields the overall sum : (CAH)+(CAA)+(CAAA)+(CAN)+($CACH_3$). The contribution (CAH) to this sum is obtained separately from the aromatic protons (see above) and the contribution ($CACH_3$) from the number of methyl carbons attached to an aromatic ring (which are themselves mostly located between 14.6-16.9 and 18.5-21-0 ppm (28,29). The sum Σ = (CAA)+(CAAA)+CAN) is thus obtained by difference.It is an important factor because it yields the percentage of non-peripheral aromatic factor and a compactness factor FC = Σ/CA. A poorly resolved problem is the separation of saturated carbons (CS) into their naphthenic (CN) and aliphatic (CL) parts. A rough estimate in the case of heavy fractions only is obtained from the difference between the integrated intensity of saturated carbons (CS) and of the infinite straight-chain pattern (falsely assumed to represent the totality of aliphatic carbons).

Such sets of average structural factors are displayed in Table III for two typical samples : a vacuum distillate VD (E = 370-535°C) from an Arabian Light crude oil (Elf Refining Company) (20), and a SRC recycle oil (DOE 417 R, Gulf Company) (52).

Finally, as mentioned in the introduction, these atomic fractions (% CI or % HJ) may be converted into *average molecular numbers* ($\overline{CI}$ or $\overline{HJ}$), if the average molecular weight $\overline{MW}$ and the total elemental contents of carbon and hydrogen C % and H % are known :

$$\overline{CI} = (\% \text{ CI}/100)(\text{C }\%/100)(\overline{MW}/12)$$
$$\overline{HJ} = (\% \text{ HJ}/100)(\text{H }\%/100)\overline{MW} \qquad (\underline{24})$$

TABLE III. Carbon and hydrogen percentages and average structural parameters for two typical samples VD (26) and SRC (52) (see the text and Tables I and II for nomenclature)

	VD	SRC		VD	SRC
% HA	5.0	39.0	$(C/H)_{tot.}$	0.58	0.93
% HS	95.0	61.0	$(C/H)_{alip.}$	0.46	0.49
% HS_{α}	10.0	32.0	$(C/H)_{aro.}$	2.83	1.62
% HS_{β}	65.0	21.0	FA=CA/C	0.24	0.68
% HS_{γ}	20.0	8.0	FAH=CAH/CA	0.34	0.62
			FAL=CAL/CA	0.20	0.10
% CA	24.4	68.0	$FACH_3=CACH_3/CA$	0.10	0.08
% CAL	4.8	7.1	$FC=\Sigma/CA$	0.36	0.19
% CAH	8.2	42.0			
% $CACH_3$	2.6	5.3			
Σ	8.8	13.6			
% CS	75.6	32.0			
% $CH_{3\gamma}$	10.9	1.5			

The raw average molecular formulae thus obtained can give rise to various average molecular models on the basis of additional structural assumptions (1,4,8,61-65).

VI - CONCLUSION

As far as the quantitativity in oil and coal liquids NMR analysis is concerned, we may note that significant progress have been achieved since the earlier investigations in this field. Using appropriate instrumental parameters, calibration and standardization procedures, efficient relaxation reagents and solvents, and modern digital treatments of the signals allows a *precision* of up to 1 %. The *accuracy* of the method, which is hoped to be of the same order of magnitude, can be checked using known mixtures of reference compounds representative of hydrocarbons found in petroleum or coal products. Such a control is indeed impossible with petroleum or coal products themselves. Three important cross-checkings are good indications of the reliability of the method: (i) A satisfactory agreement between the C/H ratios obtained either from elemental analysis or from ^{1}H and ^{13}C spectra (ii) A consistency between the carbon and hydrogen percentages of complementary oil fractions, e.g. a parent vacuum residue and two daughter fractions : the *n*-pentane solubles and the asphalt (20) (iii) Satisfactory comparisons between analyses perfomed by different laboratories using a common bank of samples.

ACKNOWLEDGMENTS

Much of this work is part of the theses of Dr. Gillet, Le Roux and Daubenfeld. The author gratefully acknowledges the participation of his coworkers, Dr. Boubel, Nicole and Rubini, and the active collaboration of Dr. J-C. Escalier and P. Valentin (Centre de Recherches Elf-Solaize) and of Dr. P. Chiche and Prégermain (CERCHAR) in the good achievement of these investigations. This work was supported by the Centre National de la Recherche Scientifique (CNRS - GRECO Charbon), the Elf Society, and the Ministère de l'Industrie et de la Recherche.

REFERENCES

1. Williams, R.B. ASTM *Spec. Techn. Publ.* 1958, 224, 168.
2. Brown, J.K., and Ladner, W.R. *Fuel* 1960, 39, 87.
3. Friedel, R.A., and Retcofsky, H.L. *Chem. Ind.* 1966, 455.
4. Knight, S.A. *Chem.Ind.* 1967, 1920.
5. Oelert, H.H. *Z.Anal.Chem.* 1967, 231, 105.
6. Retcofsky, H.L., and Friedel, R.A. *Fuel* 1968, 47, 487 and 1976, 55, 363.
7. Hirsch, E., and Altgelt, K.H. *Anal.Chem.* 1970, 42, 1330.
8. Clutter, D.R., Petrakis, L., Stenger, R.L., and Jensen, R.K. *Anal.Chem.* 1972, 44, 1395.
9. Katayama, Y., Hosoi, T., and Takeya, G. *Nippon Tagaku Kaishi* 1975, 1, 127 and 692.
10. Thiault, B., and Messerman, M. *Org.Magn.Resonance* 1975, 7, 575 and 1976, 8, 28.
11. Shoolery, J.N., and Budde, W.L. *Anal.Chem.* 1976, 48, 1458.
12. Dorn, H.C., and Wooton, D.L. *Anal.Chem.* 1976, 48, 2147.
13. Deutsch, K., Jancke, H., and Zeigan, D. *J.Prakt.Chem.* 1976, 318, 177 and 1977, 319, 1.
14. Kerkeni, M., Rubini, P.R., and Bouguerra, M.L. *Bull.Soc.Chim. France* 1977, 531.
15. Retcofsky, H.L., Schweighardt, F.K., and Hough, M. *Anal.Chem.* 1977, 49, 585.
16. Wooton, D.L., and Coleman, W.H. *Fuel* 1978, 57, 17.
17. Hajek, M., Sklenar, V., Sebor, G., Lang, I., and Weisser, O. *Anal.Chem.* 1978, 50, 773.
18. Gillet, S., Delpuech, J-J., Valentin, P., and Escalier, J-C., *Anal.Chem.* 1980, 52, 813.
19. Gillet, S., Rubini, P., Delpuech, J-J., Escalier, J-C., and Valentin, P. *Fuel* 1981, 60, 221.
20. Gillet, S., Rubini, P., Delpuech, J-J., Escalier, J-C., and Valentin, P. *Fuel* 1981, 60, 226.
21. Mashimo, K., Nakayama, I., Kiya, K., and Wainai, T. *Fuel* 1982, 61, 193.
22. Seshadri, K.S., Bacha, J.D., and Albaugh, E.W. *Fuel* 1982, 61, 1095.

23. Takegami, Y., Watanabe, Y., Suzuki, T., Mitsudo, T., and Itoh, M. *Fuel* 1980, 59, 253.
24. Dereppe, J-M., Moreaux, C., and Castex, H. *Fuel* 1978, 57, 435.
25. Alger, T.D., Grant, D.M., and Paul, G.E. *J.Amer.Chem.Soc.* 1966, 88, 5397.
26. Jones, A.J., Alger, T.D., Grant, D.M., and Lichtman, W.M. *J. Amer.Chem.Soc.* 1970, 92, 2386.
27. Stothers, J.B. "Carbon-13 NMR Spectroscopy", Academic Press, New York, 1972, and references therein.
28. Woolfenden, W.R., and Grant, D.M. *J.Amer.Chem.Soc.* 1966, 88, 1496.
29. Dalling, D.K., Ladner, K.H., Grant, D.M., and Woolfenden, W.R. *J.Amer.Chem.Soc.* 1977, 99, 7142.
30. Fischer, P., Stadelhofer, J.W., and Zandler, M. *Fuel* 1978, 57, 345.
31. Ernst, L. *Tetrahedron Letters* 1974, 3079.
32. Gillet, S. *Thesis*, University of Nancy, 1979.
33. Wilson, N.K., and Stothers, J.B. *J.Magn.Reson.* 1974, 15, 31.
34. Caspar, M.L., Stothers, J.B., and Wilson, N.K. *Can.J.Chem.* 1975, 53, 1858.
35. Buchanan, G.W., and Ozubko, R.S. *Can.J.Chem.* 1975, 53, 1829.
36. Seshadri, K.S., Ruberto, R.G., Jewell, D.M., and Malone, P.H. *Fuel* 1978, 57, 111.
37. Ozubko, R.S., Buchanan, G.W., and Smith, I.C.P. *Can.J.Chem.* 1974, 52, 2493.
38. Stothers, J.B., Tan, C.T., and Wilson, N.K. *Org.Magn.Reson.* 1977, 9, 408.
39. Maekawa, Y., Yoshida, T., and Yoshida, Y. *Fuel* 1979, 58, 864.
40. Retcofsky, H.L., and Friedel, R.A. "Spectrometry of Fuels" (Friedel, R.A. Ed.), Plenum, New York, 1970, p.90.
41. Bendall, M.R., Doddrell, D.M., and Pegg, D.T. *J.Amer.Chem.Soc.* 1981, 103, 4603.
42. Bendall, M.R., Pegg, D.T., Doddrell, D.M., and Thomas, D.M. *J.Magn.Reson.* 1982, 46, 43.
43. Bendall, M.R., Doddrell, D.M., Pegg, D.T., and Hull, W.E. "DEPT", Bruker Special publication, Karlsruhe, 1982.
44. Gerhards, R. *Proceedings of the International Conference on Coal Science* p.55, Dusseldorf, 1981.
45. Gerhards, R., this volume, p. 377.
46. Martin, M.L., Delpuech, J-J., and Martin, G.J. "Practical NMR Handbook", Heyden-Wiley, London, 1980, Chapt. 3, 4 and 9.
47. Gillet, S., and Delpuech, J-J. *J.Magn.Reson.* 1980, 38, 433.
48. Canet, D. *J.Magn.Reson.* 1976, 23, 361.
49. Ernst, R.R. "Advances in Magnetic Resonance" (J.S. Waugh, Ed.), Vol.2, p.1, Academic Press, New York, 1966.
50. Abragam, A. "The Principles of Nuclear Magnetism", pp.295, 333, Oxford Univ. Press (Clarendon), London, 1961.
51. Shoolery, J.N. "Progress in Nuclear Magnetic Resonance Spectroscopy" (Emsley,J.W., Feeney,J., and Sutcliffe, L.H. Eds), Vol.11, Pt.2, 79, PergamonPress, 1977.

52. Le Roux, M. *Thesis*, University of Nancy, 1983.
53. Gerhards, R. *Erdoel Kohle Erdgas Petrochem.* 1980, 33, 93.
54. Daubenfeld, J-M., Boubel, J-C., and Delpuech, J-J., to be published.
55. Schaefer, J., and Stejskal, E.O. *J.Magn.Reson.* 1974, 15, 173.
56. Pearson, G.A. *J.Magn.Res.* 1977, 27, 265.
57. Dietrich, W., and Gerhards R. *J.Magn.Reson.* 1981, 44, 229.
58. Comisarow, M.B., and Melka, J.D. *Anal.Chem.* 1979, 51, 2198.
59. Daubenfeld, J-M., Boubel, J-C., Helsens, S., and Delpuech, J-J., to be published.
60. Cookson, D.J., and Smith, B.E. *Anal.Chem.* 1982, 54, 2591.
61. Yen, T.F., Erdman, J.G., and Hanson, W.E. *J.Chem.Eng.Data* 1961, 6, 443.
62. Hirsch, E. *Anal.Chem.* 1970, 42, 1326.
63. Altgelt, K.H. *A.C.S.,Div.Petrol.Chem.,Prepr.* 1968, 13, 37.
64. Haley, G.A. *Anal.Chem.* 1972, 44, 580.
65. Speight, J.G. *Fuel* 1970, 49, 76.

^{13}C-NMR MULTI-PULSE SEQUENCES FOR THE ANALYSIS OF COAL AND PETROLEUM PRODUCTS

Reinhard Gerhards

Bergbau-Forschung GmbH, Franz-Fischer-Weg 61,
D-4300 Essen 13

Abstract:
The technological properties of a product made from coal or petroleum depend to a great extent on the distribution of specific chemical groups in the mixture. In the case of hydrocarbons most of the structural variations are directly related to the number of hydrogen atoms bonded to the carbon atoms. To discriminate between the four possible types of CH_x-groups (x : 0,1,2,3) four modern NMR multi-pulse techniques can be used. Two of them have been developed from conventional spin-echo sequences (Doddrell's method and PCSE), the other ones are polarization transfer methods (INEPT and DEPT). These methods are explained in detail in terms of a vector model. Some applications to coal and petroleum derived products are presented. The effect of varying coupling constants of the different carbon atoms within the same sample is discussed. Subspectra editing is used to produce spectra that contain signals of either CH_x-group exclusively.

1. Introduction

Since coal and petroleum are used as fuels and as raw materials for the chemical industry many methods have been developed to characterize the various products. Several of these methods have been established empirically by deducing simple relations between structural properties of model compounds and some simply accessible global parameters. One of these 'classical' methods is the ndm-procedure (1), occasionally still used today: The refractive index n, the density d, and the mean molecular weight m are correlated to the aromaticity of the sample. All the 'classical' methods have been designed for a specific class of products. Consequently they often fail if greater product variations occur. Mainly two reasons are responsible for the variations of

L. Petrakis and J. P. Fraissard (eds.), Magnetic Resonance. Introduction, Advanced Topics and Applications to Fossil Energy, 377–407.

the compositions of today's products. The first is that fossil fuels are produced at more and more different places. Accordingly they have been formed under very different geological, chemical and biological conditions thus their composition varies considerably. This is especially valid for oil shales and tar sands that have been explored increasingly during the last decade. The second reason is that the number of technologies to process coal and petroleum has grown rapidly. While coking and distillation have been used nearly exclusively before, nowadays numerous modern processes are available and are able to produce more valuable products from the raw materials. Examples are the various cracking processes, as well as the many coal liquefaction and gasification technologies.

Corresponding to this the need for improved analytical characterization methods became obvious. Especially designing and optimising conditions of a technical process depends very much on many and reliable analytical data.

Coal and petroleum derived products usually consist of at least several hundred compounds. A complete product analysis inclusive the identification of each individual compound and its quantitative determination can be done for low boiling products only. The 200°C+ fractions generally are too complex to be separated completely. On first sight this seems desadvantageous with respect to the aim of characterizing those products. But regarding the fact that these products are normally used as mixtures, it is plausible that they can be better described with statistical-type data than with the sum of the properties of its components. While the number of values is drastically reduced by the restriction to statistical data, these data however should be detailed as possible.

The technological properties of products made from coal or petroleum are strongly correlated to quantitative structural data. Some of these are, e. g., the aromaticity, the amount of atoms in an olefinic or aliphatic environment , the degree of condensation in ring systems, the type and the amount of branching in aliphatic chains, the mean length of the non-branched parts of these chains, or the distribution of heteroatoms in the molecules.

In the search for techniques that allow the determination of one or more of these data various types of spectroscopic and chromatographic methods have been investigated. Chromatography gives good results, provided a good separation of the compounds on a chromatographic columns can be achieved. This is generally possible for mixtures containing only small molecules. Higher boiling samples usually are more complex, so separation becomes poorer and less detailed information is only available. The major disadvantage of chromatographic methods however results from the fact that the molecules are generally affected as entireties in the separation procedure. This restricts the separation to only one single specific attribute of the molecules. However larger molecules usually have many different structural features at the same time, so it is hardly possible to track a specific one by chromatography.

Spectroscopic methods are better suited for supplying structural information. These allow the detection of substructures in molecules within a mixture. UV spectroscopy detects the system of π-electrons in the molecules selectively, while IR-spectroscopy shows the vibrations of groups of atoms in a molecule. With both methods a lot of different substructures in the molecules can be determined no matter how the rest of the molecules looks.

Unfortunately the intensities of the signals to be observed in these spectra depend extremely on the specific substance (absorption coefficient) that causes the signal. So quantitative results can only be obtained if the qualitative composition of the sample is known before. In the investigation of complex mixtures this problem normally prevents the determination of the concentration of different substructures even if these can be identified in the spectra. The measured signal in this case usually is the result of a superposition of a great number of lines, each produced by a different substance with a different absorption coefficient. Only if very similar products are investigated routinely a calibration is practicable (Oelert(2)). There is hardly a chance however to obtain correct quantitative results for a variety of coal and petroleum products always using the same calibration.

NMR-spectroscopy is another method that allows the discrimination of structural groups. But in this case quantitative results can be obtained even from spectra of complex mixtures. Principally in NMR the signal intensities are directly proportional to the concentration of the species that cause the specific signals. There is no different absorption coefficient for any signal within the same NMR spectrum. Only some practical problems have to be solved caused by the apparatus (3) or by some specific experimental conditions (see for example Delpuech et al. (4): If it is possible to measure a spectrum with a sufficiently high signal to noise ratio (S/N) without accumulating several spectra and if only the observing frequency is transmitted into the sample as in the case of ^{1}H-NMR spectroscopy, correct quantitative results can directly be obtained. If this is not possible or not recommended as in the case of ^{13}C-NMR spectroscopy some precautions have to be taken to avoid intensity destortions by relaxation effects and by interactions between different nuclei within a molecule, for example produced by the nuclear Overhauser enhancement (NOE).

These problems are discussed in more detail in the paper by Delpuech,published in the proceedings of this ASI meeting. He shows that by means of some easy to realize experimental techniques it is generally possible to produce spectra with correct intensity relations between the signals observed.

The importance of ^{13}C-NMR spectroscopy for the analysis of coal and petroleum products has steadily grown since its first use for this kind of samples by Knight (5) in 1967. Compared to ^{1}H-NMR spectroscopy, in ^{13}C-NMR spectroscopy the spectral width in which signals can be observed is expanded by a factor of 20. Moreover the signals in ^{13}C-NMR are not splitted into multiplets provided the C-H interactions are totally decoupled by broadband decoupling, a technique that can not be applied in ^{1}H-NMR spectroscopy. Both facts show in a much better resolution of ^{13}C-NMR spectra compared to ^{1}H-NMR spectra. Because of the more detailed structure of the ^{13}C-NMR spectra more informations can be yielded. Even considering that ^{13}C-NMR signals of natural occuring materials are by three orders of magnitude smaller than the corresponding ^{1}H-NMR signals (benzene for example) does not prevent the use of ^{13}C-NMR spectroscopy. This applies especially since the introduction of the Pulse-Fourier-Transform technique and the use of superconducting magnets with higher fields.

2. Discrimination of CH_3, CH_2, CH and C_{quart}-groups by conventional ^{13}C-NMR methods

^{13}C-NMR spectra of typical products from coal and petroleum normally contain a large number of signals. The positions of these signals depend mainly on the distribution of the electrons around the carbon atoms within a molecule. Different electron densities are caused in turn by different chemical bonds between the atoms. This includes as well the type of the bonds (single, double, aromatic, etc.) as the nature of the atom bonded. Usually the electron density close to a nucleus is affected by at least the next four neighbour atoms, but with decreasing influence. The greater the difference in the electron distribution near the nuclei of interest the more the corresponding signals separate from another in the observed spectrum. In ^{13}C-NMR spectra of hydrocarbons two well resolved groups of aromatic and aliphatic carbon signals can be distinguished. Both spectral regions generally display pronounced fine structure. However the assignment to specif structural elements is not always obvious. This is demonstrated in fig. 1, in which the ^{13}C-NMR signals of some pentane, hexane and heptane isomers are shown. The signals are displayed in the four groups representing CH_3, CH_2, CH, and C_{quart} signals, respectively. The classifying of the individual signals was done subsequent to the actual measurements, since with conventional NMR techniques all groups of signals can only be detected alltogether in one single spectrum.

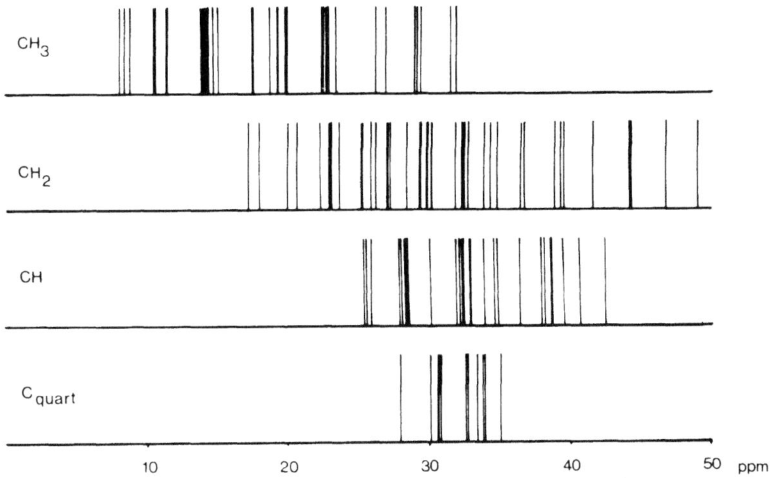

Fig. 1: ^{13}C-NMR signal distribution of pentane, hexane and heptane isomers

Apparently a straightforeward assignement of signals to distinct CH_x groups is possible for a small number of signals only. In the larger part of the spectral region shown the signals may be produced by either CH_x group.

If there was a possibility to differentiate between the signals of carbon atoms connected to different numbers of hydrogen atoms a major part of the analytical information wanted could be obtained. Nearly all structural features mentioned earlier, like branching, condensation or substitution that are influencing the technological behaviour of a hydrocarbon mixture are clearly related to the number of hydrogen bonded to carbon atoms.

Recognizing this relationship also some NMR methods have been discussed to discriminate between the different CH_x-groups. They all make use of the interaction between the ^{13}C nuclei and the ^{1}H nuclei in the same molecule. This interaction can be observed directly in ^{13}C-NMR spectra that are measured when only the ^{13}C-NMR frequencies are transmitted into the sample and no other NMR frequency is applied. In this case where the C-H interactions are not decoupled, a CH_3-signal is splitted into a quartet of peaks with the intensity relations1 : 3 : 3 : 1, a CH_2-signal into a triplet (1 : 2 : 1), and a CH-signal into a doublet (1 : 1), whereas a C_{quart}-signal remains unsplitted. The distance between the lines in the multiplets is a criterion for the strength of the C-H-interaction. It is measured in Hertz and is called coupling constant J.

So the most simple ^{13}C-NMR experiment could yet be used to differentiate between the individual CH_x-groups. However it has no importance in practice. Compared to the commonly used ^{13}C-NMR experiment, where all C-H-interactions are decoupled by transmitting all ^{1}H-NMR frequencies into the sample while measuring the ^{13}C-NMR spectrum (broadband decoupling), the quality of non-decoupled spectra is almost poor. In broadband decoupled ^{13}C-NMR spectra every CH_x-signal shows as a singlet. The signal to noise ratio in such spectra is much better than in non-decoupled spectra since the individual peaks of a multiplet collapse into a single line at the centre of the multiplet summing up all intensities of the splitted signal in this line. Additionally the line width in broadband decoupled spectra is much narrower than in the decoupled spectra; broadening is caused by long range C-H-interactions within the same molecule. Thus the signal amplitude again is increased because the signal area remains constant. Finally another effect in broadband decoupling, the NOE (Nuclear Overhauser Enhancement) can enhance the ^{13}C-NMR-signals by a factor up to 3.

All these facts together produce a 10 to 20 fold better signal to noise ratio in broadband decoupled spectra. Consequently this method is preferred in routine ^{13}C-NMR spectroscopy.

On the other hand the possibility to determine the multiplicity of ^{13}C-NMR-signals could be an incentive to use non-decoupled ^{13}C-NMR spectra. However in practice it turns out that these spectra can be evaluated only if they are displaying a small number of signals. The line broadening and the splitting into multiplets generally cause such a severe overlapping of signals that even in simple cases the individual multiplets can be assigned with very much experience only. An example is shown in fig. 2.

Comparing the non-decoupled with the broadband decoupled ^{13}C-NMR spectrum obviously signal overlapping complicates the interpretation of the spectrum extremely.This problem is even much more severe when products from coal and petroleum are investigated as it is demonstrated in fig. 3.No essential informations no longer can be derived from the non-decoupled spectrum.

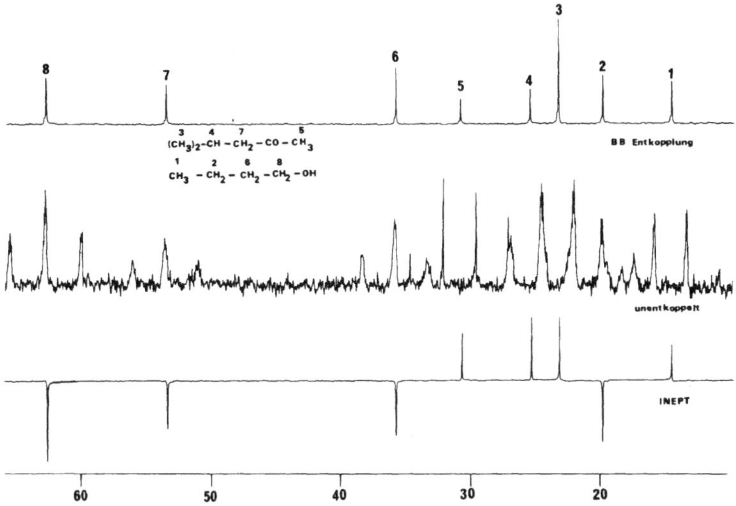

Fig. 2: Non-decoupled and broadband decoupled spectra of a mixture of isobutylmethyl ketone and n-hexanol

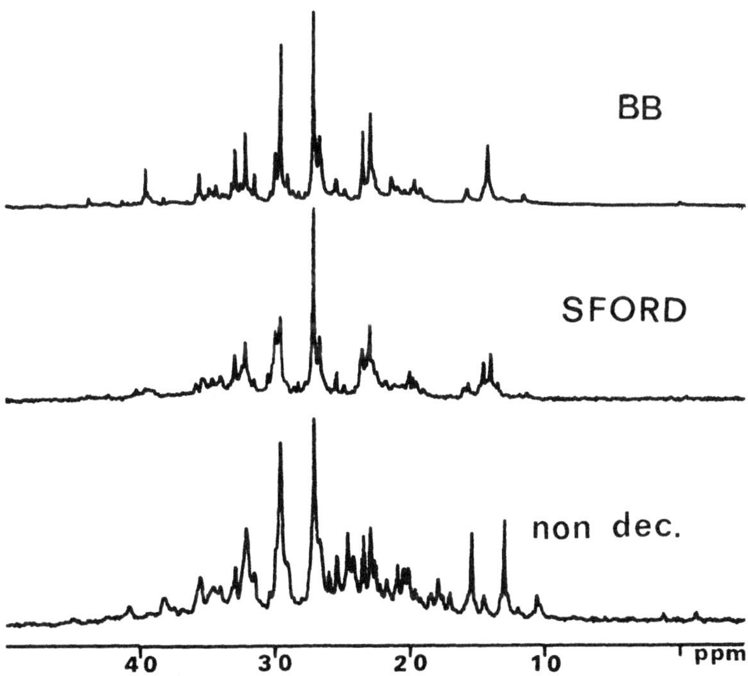

Fig. 3: Off-resonance decoupled, non-decoupled and broadband decoupled ^{13}C-NMR spectra of a coal derived oil

This still applies to the third spectrum shown in fig. 3 that has been obtained by using the SFORD technique (single frequency off resonance decoupling). In this case the ^{13}C-NMR spectrum is acquired while at the same time one single ^{1}H-NMR frequency is fed into the sample. This frequency is some 100 Hz displaced compared to the actual ^{1}H-resonances in the sample to avoid a total decoupling of a specific C-H-interaction. As a result of this kind of excitation the C-H-interactions are partially disturbed, long range interactions then can almost be neglected and the interaction of directly bonded carbon and hydrogen atoms is reduced. In the registered spectrum the signals become narrower than without decoupling and the peaks in the multiplets are shifted to the centre of the multiplet indicating the reduction of the interaction by a smaller coupling constant J.

Although the spectral resolution in SFORD spectra is better than in non-decoupled spectra it is not sufficient to allow a clear identification of all types of CH_x-signals in the spectra of complex mixtures. Consequently neither the non-decoupling nor the off-resonance-decoupling technique have been of great importance for the analysis of those products.

3. Discrimination between CH_3, CH_2, CH, and C_{quart}-groups by multi-pulse sequences

In the last five years several new techniques have been proposed to discriminate between the signals of the four possible CH_x-groups. In all cases multi-pulse sequences have been used for the excitation of the nuclei. Two of these methods are modified spin-echo techniques and two other are polarization transfer methods. All of them will be described in detail.

To make the explanations more clear some simplifications are made:

1. All relaxation effects during the multi-pulse sequences are ignored. In practice this is allowed because the duration of the sequences are in the order of some milliseconds whereas typical relaxation times of products discussed here are in the order of seconds.

2. The applied pulses are considered to be δ-pulses. Their actual lenghts are 10 to 30 us.

3. All long range C-H-interactions within a molecule are ignored.

4. The frequencies involved are described as rotating vectors. Their behaviour during the multi-pulse sequences will be explained in a rotating frame. Its rotation frequency is assumed to be equal to the actual frequency of the examined signal in the decoupled spectrum. When no decoupling is applied the centre frequency of the corresponding multiplet determines the

rotation frequency of the rotating frame. Thus both frequencies can be described as stationary vectors in such a rotating frame.

3.1 The partially coupled spin-echo technique (PCSE)

The most simple method to discriminate between carbon atoms attached to a different number of hydrogen atoms was proposed by Brown et al. (6) and by Le Cocq and Lallemand (7). First applications in the analysis of complex mixtures have been described by Snape (8, 9). A slightly modified experiment was used by Cookson et al. (10, 11).

The designed multi-pulse sequence is shown in fig. 4.

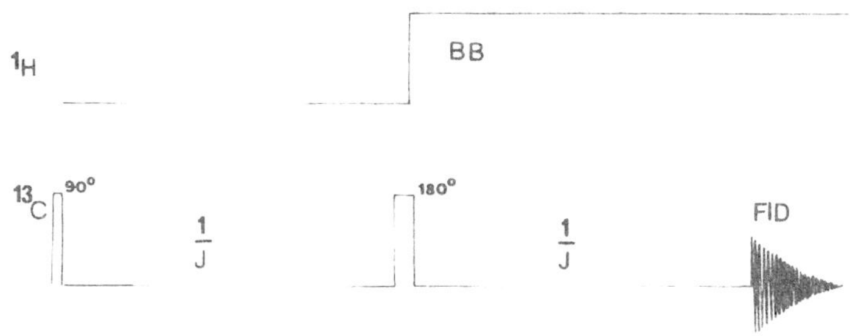

Fig. 4: The PCSE pulse sequence

In this case all events during the pulse sequence happening to the individual CH_x-signals are explained in every detail. Most of the considerations used here can easily be adapted to the other methods. Therefore it seems reasonable to explain them for the simpliest case.

The ^{13}C-NMR frequencies involved are shown in fig. 5 as vectors at different moments during the applied pulse sequence. Every CH_x-group is treated individually.

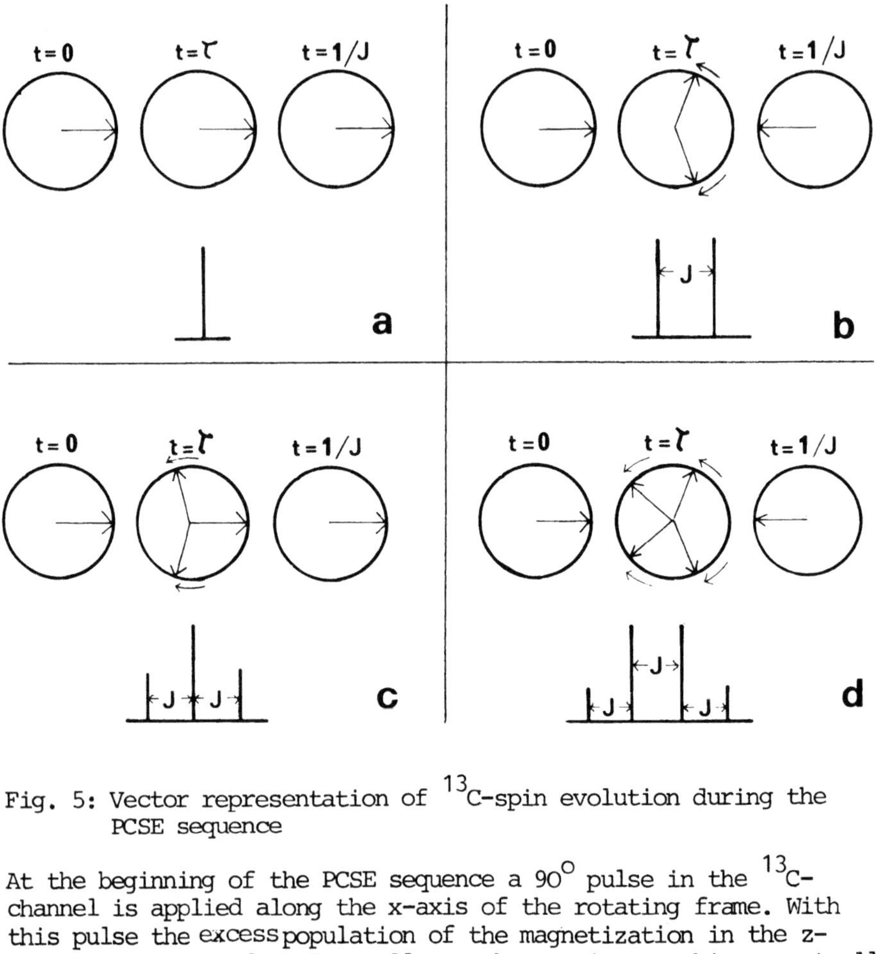

Fig. 5: Vector representation of ^{13}C-spin evolution during the PCSE sequence

At the beginning of the PCSE sequence a 90° pulse in the ^{13}C-channel is applied along the x-axis of the rotating frame. With this pulse the excess population of the magnetization in the z-direction is transferred totally to the y-axis. At this moment all frequencies of all signals in the spectrum are in phase. The evolution of the frequency vectors however then is different for every CH-group.

Starting with the signal of a quarternary carbon atom it is known that no interaction with any hydrogen atom is possible. So the corresponding signal is not affected by the fact that no ^{1}H-decoupling is applied during the evolution period. In the rotating frame the vector representing the C_{quart}-signal is stationary since it was assumed to rotate with the same frequency (fig. 5 a).

In any other case there are interactions between carbon and hydrogen atoms. For a CH-group in the non-decoupled spectrum a doublet is observed, the frequency of one line +J/2 Hz the other -J/2 Hz shifted from the centre of the multiplet. These two signals can be explained by two vectors in fig. 5 b. One vector rotates faster than the frame itself, the other slower. In the evolution period therefore one frequency vector rotates clockwise and the other anticlockwise in this frame. Since both frequencies are shifted by the same amount from the centre of the multiplet the positions of the corresponding vectors are symmetric to the y-axis at any moment. As it is shown in fig. 4 the duration of the evolution period is exactly 1/J seconds. After this time interval both vectors have completed exactly one half of a circle and superimpose as at the beginning of the evolution period, but now they are pointing to the opposite direction. At this moment the broadband decoupler is switched on (fig. 4) and the C-H-coupling is no longer possible. The dephasing of the two frequency vectors for the CH-group stops and they remain superimposed for the rest of the sequence. The C_{quart}-frequency vector in its rotating frame at this moment is exactly opposite to the superimposed vectors representing the CH-group. If a spectrum could be taken immidiately following the evolution period, non-splitted signals would be observed for C_{quart}-groups as well as for CH-groups but their amplitudes would be opposite to each other.

Because there is some inhomogeneity of the static magnetic field a second part in the pulse sequence is needed. By the 180° pulse in the ^{13}C-channel (fig. 4) a spin-echo is induced refocussing all effects resulting from inhomogeneity. The refocussing is perfect when a time intervall as long as the evolution period has passed. So this refocussing period again lasts 1/J seconds.

What is happening to the frequency vectors of the CH-group in this time interval? If the C-H coupling was still allowed, both vectors would continue to rotate independently and join again, exactly at the same position where the C_{quart}-frequency vector can be found. C_{quart} and CH-signal then no longer could be distinguished. Indeed however the broadband decoupler is switched on at the end of the evolution period. Then the CH-group too is represented by one single frequency that by convention is the frequency of the rotating frame. The superimposed vectors in this frame thus can no longer dephase and remain opposite to the C_{quart} vector all over the entire refocussing period. So the differentiation between both kinds of signals already produced during the evolution period is preserved for the rest of the sequence.

Similar considerations also apply to CH_2 and CH_3-groups: A CH_2-signal splits into a triplet if the C-H-interactions are not decoupled. The centre line has the same frequency as the rotating frame, the outer lines are J Hertz apart. In the rotating frame the latter start to move clockwise and anticlockwise as discussed before. Since they are shifted J Hertz from the centre of the multiplet they can finish both a full circle during the evolution period 1/J. There they superimpose the frequency vector of the centre line that remained stationary during the whole period. Thus the resulting vector is in phase with the frequency vector of C_{quart}-signals, in the acquired spectrum CH_2-signals therefore will have the same sign as the C_{quart}-signals.

CH_3-signals will show inverted amplitudes like CH-signals. In the CH_3-quartet the inner lines are J/2 Hertz apart from the centre of the multiplet and thus they can be treated like the CH-doublet. The outer lines are separated by 3•J/2 Hertz from the centre. The corresponding vectors can each rotate exactly 1.5 circles in the rotating frame during the evolution period and then they meet the vectors of the inner lines at the same position. All four vectors representing each one line in the quartet then superimpose. The resulting vector again is inverted compared to the starting orientation.

Following the same path of explanation it can be shown that by reducing the length of the evolution period as well as of the refocussing period by a factor of 2 to a value of 1/(2J), C_{quart}-groups can be observed in the acquired spectrum exclusively. For every other CH-group the corresponding frequency vectors then will exactly cancel in the rotating frame and thus prevent the appearance of these signals in the spectrum.

3.2 Doddrell's sequence

Very similar to the previously described technique is a pulse sequence proposed by Doddrell et al. (12). Again it is a spin-echo method. In this case the broadband decoupler is not used during the sequence itself but only during data acquisition (fig. 6). Broadband decoupling used in the PCSE sequence is replaced by a 180° pulse in the ^1H-frequency channel that is applied at the same moment as the 180° ^{13}C-pulse is transmitted (sequence Doddrell-1). The 180° ^1H-pulse inverts the populations of the ^1H-energy levels.

A ^{13}C-spin coupled to a ^1H-spin in the higher energy level before, after the 180° ^1H-pulse is coupled to the same spin that now is in the lower energy level, and vice versa. In the vector representation of the frequencies in the rotating frame this causes an inversion of the rotation direction of the vectors. A clockwise rotating vector after the 180° ^1H-pulse rotates anticlockwise and an anticlockwise rotating vector turns clockwise.

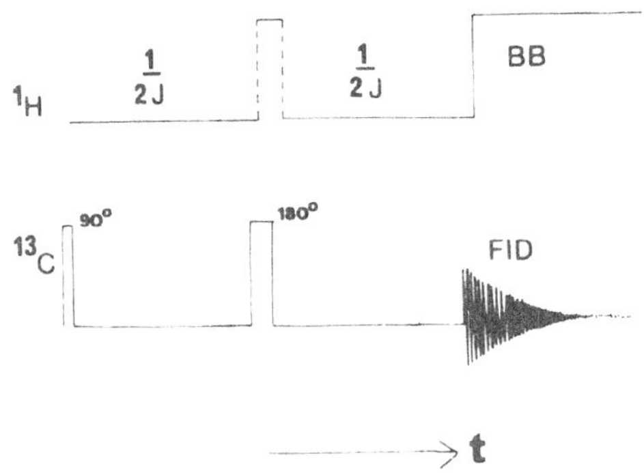

Fig. 6: Doddrell's pulse sequence

The length of the period from the beginning of the sequence to the 180° pulses is 1/(2J), half as long as the evalutation period in the PCSE technique. Consequently the frequency vectors can rotate only half of the angle in the rotating frame they did in the PCSE sequence. In Doddrell's sequence however the frequency vectors are allowed to continue their dephasing even after the 180° pulses because in this case the broadband decoupling is not applied and C-H-coupling is still possible.

The 180°-pulse in the ^{13}C-frequency channel is necessary for two reasons in Doddrell's sequence. On one hand it has to produce a spin-echo allowing the observation of narrow lines after the refocussing of the spins in the not totally homogeneous magnetic field. On the other hand the 180° ^{13}C-pulse prevents the rejoining of all vectors at the same position. By the 180° ^{13}C-pulse the individual vectors are transferred to their mirror positions with respect to the direction of the applied pulse but without changing the sense of rotation of these vectors. A vector rotating clockwise before the pulse still rotates clockwise afterwards. In the actual case however, any off-axis vector is moved to the position of its mirror vector. Thus a vector rotating anticlockwise has been replaced by a vector moving in the other direction. So the inversion of the rotation direction by the 180° ^{1}H-pulse is again inverted and the frequency vectors can continue to dephase. It has to be pointed out that only the combination of both 180°-pulses allows as well to separate the CH-signals into groups as the observation of narrow lines.

In the recorded spectrum CH_3 and CH-signals are inverted compared to CH_2 and C_{quart}-signals. Wether one or the other group of signals shows positive signals depends only on the direction of the two ^{13}C-pulses in the sequence. If in one case CH_3 and CH-signals are negative, they are positive if the ^{13}C-NMR frequencies are transmitted along an axis perpendicular to the first experiment.

If the 180° ^1H-pulse is not applied (sequence Doddrell-2) a usual spin-echo spectrum can be observed with all signals having positive intensities. Since all other experimental conditions are the same in both sequences it is very promising to calculate linear combinations of spectra measured with both methods. The difference spectrum then contains only CH_3 and CH-signals whereas in the sum spectrum CH_2 and C_{quart}-signals are the only ones to be observed. Especially when the two spectra are accumulated alternatively there are no problems caused by different line widths that otherwise can produce non satisfying difference spetra.
Reducing the delays in the pulse sequence by a factor of 2, again quartenary carbons can be detected exclusively. This corresponds exactly to the PCSE sequence.

In any case with Doddrell's method the same informations can be obtained as with the PCSE sequence. The most important difference between both methods is the doubled length of the PCSE sequence compared to the other one.

3.3 The INEPT sequence

The INEPT technique (insensitive nuclei enhancement by polarization transfer) has been proposed by Morris and Freeman (13) in 1979. The original INEPT pulse sequence then was prolonged by an additional refocussing pulse sequence described by Morris (14) and Burum and Ernst (15). This was designed to remove signal splittings from the observed spectrum due to C-H-couplings. Today it is common parctice to use the name INEPT for this longer pulse sequence as entirety.

The INEPT sequence (fig. 7) starts with a so called polarization period. At the beginning the ^1H-spins are excited by a 90° pulse. In the rotating frame then the vectors representing the frequencies of the CH-doublet start to dephase. It is important to recognize that for the ^1H-spins coupled to ^{13}C-spins always a doublet results, no matter if the hydrogen atom is in a CH_3, CH_2 or CH-group. In every case the ^1H-nucleus is coupled to one carbon only.

After the time interval of 1/(4J) seconds 180° pulses are applied in the ^1H frequency channel as well as in the ^{13}C-frequency channel. As it has been demonstrated in the previous section such a combination of 180° ^1H and 180° ^{13}C-pulses simply allows to continue the dephasing of the two frequencies but at the same time

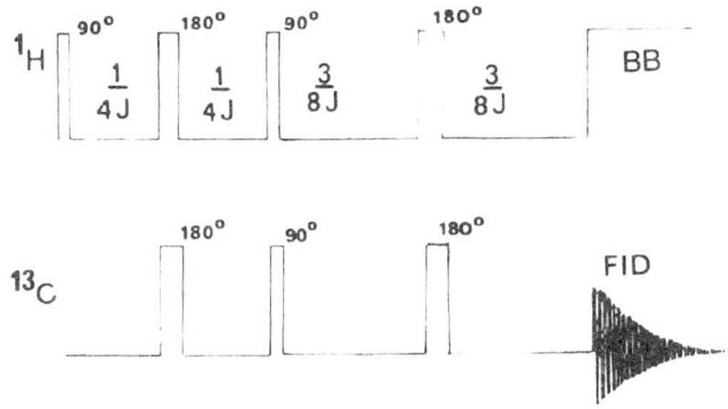

Fig. 7: The INEPT sequence

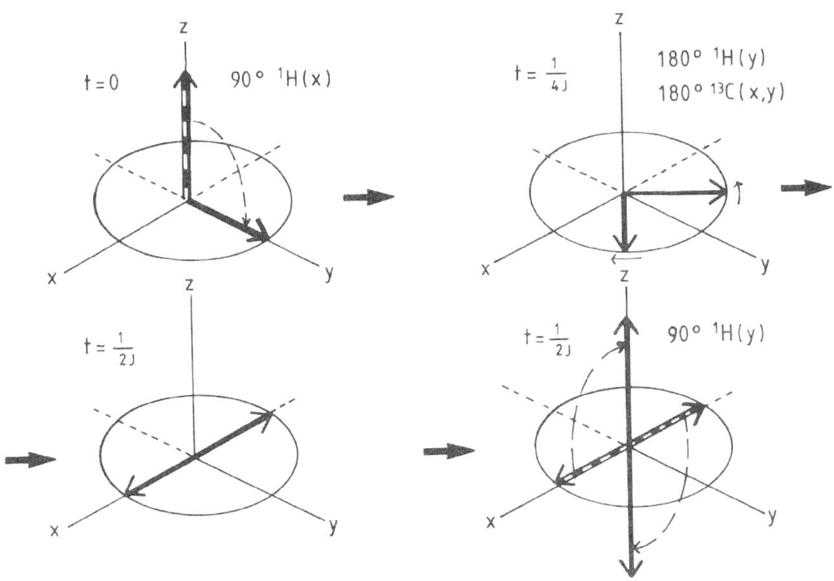

Fig. 8: Vector representation of the ^{1}H-frequencies of a hydrogen atom coupled to one ^{13}C-atom

cancels the field inhomogeneity (fig. 8).
In the first 1/(4J) period the two frequency vectors can each accumulate a 45° rotation clockwise and anticlockwise relative to their starting direction. After the next 1/(4J) seconds following the 180° pulses, they each are perpendicular to the starting direction and thus are exactly opposite to each other. Supposing the primary 90° pulse has been applied along the x-axis now the frequency vectors are aligned with the -x and +x direction. A 90° pulse in the ^1H frequency channel along the y-axis then will rotate these vectors into the +z and -z direction (fig. 8). This corresponds to a population inversion across one of the two ^1H-transitions (-z axis) but equilibrium population across the other (+z-axis). Because ^1H and ^{13}C spins share common energy levels the ^{13}C spins now must have population differences appropriate to the ^1H spins. Appropriate means that a negatively polarized ^1H spin will also produce a negatively polarized ^{13}C spin, in other words the ^1H-NMR signal as well as the corresponding ^{13}C-NMR-signal will be negative. Taking this into account the multiplets to be observed for the different CH_x-groups then will show as -1 : 1 doublets (CH), or -1 : 0 : 1 triplets (CH_2), or -1 : -1 : 1 : 1 quartets (CH_3). The intensities of the signals are enlarged by a factor of γ_H/γ_C since by the same amount of energy a γ_H/γ_C fold number of ^{13}C spins can be excited compared to the number of ^1H spins.

These effects can be observed if after applying a 90° pulse to the ^{13}C spins a spectrum is recorded immediatly, as it was proposed in the first description of the INEPT technique. For the purpose discussed here - the discrimination of different CH-signals without signal splitting - the following part in the INEPT sequence is as much important as the first one. In this so called refocussing period the splitting of the different multiplets can be cancelled when appropriate time intervals are choosen. The 180° pulses in the middle of this period again are only necessary to prevent signal broadening due to inhomogeneity of the magnetic field. They could be omitted of the field was completely homogeneous.

If the total length of the refocussing period is 1/(4J) (INEPT-1) all CH_x-signals can be observed non-splitted and all signals having positive amplitudes. Changing the length to 1/(2J) (INEPT-2) only CH-signals appear in the spectrum. If at last the duration of the refocussing period is 3/(4J) seconds (INEPT-3) CH and CH_3-signals will have positive amplitudes, whereas CH_2-signals have negative ones. All three types of signals are singlets.

INEPT-2 and INEPT-3 are those sequences that again allow to discriminate between the different CH_x-signals. Comparing these techniques with both spin-echo methods discussed earlier, two important differences have to be considered:

1. The signal to noise ratios in INEPT spectra are much better if the same total instrument time is used. This is caused by the mechanism of polarization transfer from ^{1}H to ^{13}C and by the fact that INEPT spectra accumulation needs shorter delays between the acquisition of subsequent spectra, since the effective relaxation time to be considered is that of the fast relaxing ^{1}H nuclei and not that of ^{13}C spins.

2. INEPT spectra show no quarternary carbon signals. Due to the mechanism of exciting the carbon atoms, only those carbon atoms can be polarized that are strongly coupled to hydrogen atoms. The polarization transfer mechanism is very sensitive to a proper setting of the interval times that depend directly on the coupling constants J. Only those carbon atoms thus can be excited whose actual J does not differ too much from the value used to calculate these periods. Principally it would also be possible to use long range couplings to polarize the quarternary carbon atoms, but the interval times needed then would be very long and the transverse relaxation could no longer be neglected. Nevertheless from the point of view to discriminate between the individual signals, even the non-observability of quarternary carbon atoms in INEPT spectra can be used as information. Comparing such a spectrum with a conventional ^{13}C-NMR spectrum of the same sample, in the latter the signals from quarternary carbons can easily be assigned.

3.4 The DEPT sequence

Another polarization transfer sequence was recently proposed by Doddrell et al (16). Again (see fig. 9) the carbon atoms are excited indirectly by transferring the polarization from the hydrogen nuclei to the carbon nuclei. After applying a 90° pulse to the hydrogen atoms in the sample the free evolution of the nuclear spins is allowed for a period of 1/(2J) seconds. In this time interval the two components of the ^{1}H doublet representing those ^{1}H atoms coupled to a bonded ^{13}C atom evolve to opposite directions of the frequency vectors in the rotating frame. This corresponds to a situation yet discussed for the INEPT sequence. In the INEPT sequence this polarization is transferred to the carbon atoms by applying a 90° pulse in the ^{1}H frequency channel and a 90° pulse in the ^{13}C-channel at the same time. In the DEPT sequence the 90° ^{1}H-pulse is replaced by a 180° ^{1}H-pulse. In the following period a multiple quantum precession takes place. This cannot be explained any longer in terms of the simple vector model used until now. An exact treatment of the changes for the different spin populations is only possible with the density matrix approach. In this context this treatment would go too much into details and therefore is relinquished. (A good description of the DEPT sequence using density matrices is published by Soerensen and Ernst (17)).

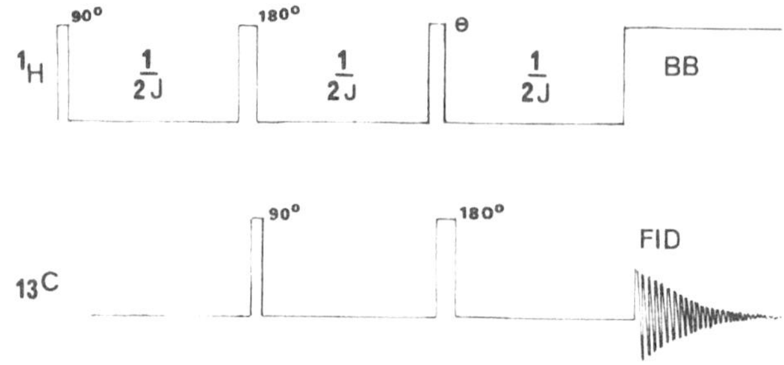

Fig. 9: The DEPT sequence

At the end of the second 1/(2J) period a 180° pulse is applied to the ^{13}C-atoms. This pulse is necessary to refocus the diverging spins by initializing a spin-echo whose second half is recorded after a third delay of 1/(2J) seconds. At the same moment the 180° ^{13}C-pulse is irridiated, in the ^{1}H-channel another pulse is applied. This is called Θ-pulse. Doddrell et al. could show that by applying this sequence in any case unsplitted signals for every CH_x-group can be obtained. The intensities of the individual CH_x-signals are modulated by the length of the Θ pulse. The signal amplitudes are proportional to sinΘ for CH-signals or proportional to sin 2Θ for CH_2-signals, or they follow the function sinΘ + 3·sinΘ in the case of CH_3-signals.

Knowing these relations it is easy to find that CH-signals are the only ones to be observed when Θ is 90°, in any other case the recorded spectrum will display all signals at once but each type of CH_x-signal modulated according to the actual value of Θ. By the calculation of appropriate linear combinations of spectra recorded for $\Theta = 45°$, $\Theta = 90°$ and $\Theta = 135°$ it is possible to produce spectra containing only one kind of CH_x-signals. The CH_2-subspectrum results when the spectra obtained with $\Theta = 45°$ and $\Theta = 135°$ are substracted from each other. The CH_3-subspectrum is calculated by adding these two spectra and subtracting the spectrum yielded with $\Theta = 90°$ scaled down by a factor of $\sqrt{2}/2$.

In practice the compensation of the signals to be surpressed is not complete if the subspectra are calculated as indicated before. This is caused by the inhomogeneity of the ^{1}H pulse that in common instruments usually is irridiated via the decoupling coil of the probeheads. By the use of some small correction factors the 'purity' of the CH_x-subspectra can be optimized. These factors have to be determined only once, because the homogeneity of the ^{1}H-pulse is specific for every probehead with its individual coil geometry. They are constant for all signals in the spectrum and for different samples.

As in INEPT spectra in DEPT spectra no signals of quarternary carbon atoms can be observed. The big advantage of DEPT is the possibility to obtain subspectra of every individual CH_x-group. Favourable is also the smaller number of pulses needed in the DEPT sequence, since the quality of pulse shapes and pulse lengths are most critical to the resulting spectra. If,as in practice pulses are not perfect,a method using less pulses than the other usually is preferred.

3.5 Summary

In table 1 all techniques to discriminate between CH_x-signals are summarized. In the first column the different sequences discussed are named individually with numbers to shorten the description in the following chapter. In the second column the most important parameter for the specific sequence is listed. The results of the experiments are given in the next column. For every individual CH_x-signal +a indicates a positive signal, -a a negative signal, and 0 is printed if the corresponding signal is surpressed. The spectra are named by capital letters in the last column of table 1. In some rows the characteristics of calculated spectra, resulting from linear combinations of recorded spectra are given. These are also named by capital letters. The mathematical procedure to calculate them from measured spectra are there given too.

4. Use of multi-pulse sequences to analyse coal and petroleum derived products

4.1 Experimental conditions

All pulse sequences described have been tested using the Bruker CXP-200 spectrometer with resonance frequencies of 200 MHz for ^{1}H nuclei and 50.3 MHz for ^{13}C nuclei, respectively. The pulse sequences have been programmed for the pulse generator Z 17 C using the DISCXP program of the ASPECT 2000 computer. To avoid errors due to non-ideal pulses the phases of convenient pulses in the different sequences have been shifted through a cycle. The data measured have been added or substracted in an appropriate cycle. The main goal of this procedure is to cancel quantitatively residual signals that are caused by not exactly matched pulse widths.

method	most important parameter	CH_3	CH_2	CH	C_{quart}	produced spectrum
Doddrell-1	$(180^{o}\ ^{1}H), t = 1/(2J)$	-a	+a	-a	+a	A
Doddrell-2	$(no\ ^{1}H\ ), t = 1/(2J)$	+a	+a	+a	+a	B
Doddrell-3	$(180^{o}\ ^{1}H), t = 1/(4J)$	0	0	0	+a	C
		0	+a	0	+a	D = (A + B)/2
		+a	0	+a	0	E = (B - A)/2
PCSE-1	$t = 1/(2J)$	0	0	0	+a	F
PCSE-2	$t = 1/J$	+a	-a	+a	-a	G
Echo	$t = 1/J$	+a	+a	+a	+a	H
		+a	0	+a	0	I = (G + H)/2
		0	+a	0	+a	J = (H - G)/2
INEPT-1	$t = 1/(4J), \tau = 1/(4J)$	+a	+a	+a	0	K
INEPT-2	$t = 1/(4J), \tau = 1/(2J)$	0	0	+a	0	L
INEPT-3	$t = 1/(4J), \tau = 3/(4J)$	+a	-a	+a	0	M
DEPT-1	$t = 1/(2J), \Theta = 45^{0}$	+a	+a	$+ a/\sqrt{2}$	0	N
DEPT-2	$t = 1/(2J), \Theta = 90^{0}$	0	0	+a	0	O
DEPT-3	$t = 1/(2J), \Theta = 135^{0}$	+a	-a	$+ a/\sqrt{2}$	0	P
		0	+a	0	0	Q = (N - P)/2
		+a	0	0	0	$R = (N+P-\sqrt{2} \cdot O)/2$

Tab.1 Comparison of multi-pulse sequences used to discriminate between CH_x groups

The investigations described have been performed with the broad-band probehead (20 - 85 MHz) using the decoupling coil to transmit the ^{1}H-pulses. The pulse widths determined were 18 μs for the 90° ^{13}C-pulse and 28 μs for the 90° ^{1}H-pulse. Both pulse widths are short enough to excite signals in the full bandwidths where signals can occur. The spectra have been accumulated using 16 k of computer memory. Fourier-transformation was performed on 32 k after zerofilling of the last 16 k. The spectral width observed was 12.5 kHz, so digital resolutions was 0.76 Hz per data point. During spectra accumulation a delay of ten seconds followed every data acquisition. This period has been proven to be long enough to restore the Boltzmann distribution between the energy levels involved.

The products investigated generally were dissolved in the same volume of $CDCl_3$. Dissolved oxygen was not removed and so could help to relax the excited nuclei.

The experiments discussed in detail later in this paper were planned to show the chances and the limits of the four pulse sequences with respect to their use in the analysis of complex mixtures. One paragraph deals with the effect of varying coupling constants of different carbon atoms within one sample. In another section the possibilities of evaluating quantitative data from spectra obtained with the individual techniques are investigated. Finally those methods that allow subspectra generation with only one type of CH_x-signals present, are examined in detail.

4.1 The effect of varying coupling constants

The results of all methods described depend directly on the knowledge of the coupling constants J of the carbon atoms within the sample under investigation. These constants are necessary to calculate the optimal delays in the pulse sequences. But even the knowledge of the J values does not help when there are signals with different coupling constants in the same spectrum. The delays in one experiment can principally fit to one coupling constant only. Every coupling constant deviating from the actually used value will cause the corresponding signal to show another intensity than predicted by theory. The frequency vectors in the rotating frame then will no longer exactly superimpose nor cancel. Vector addition then produces either shorter vectors corresponding to reduced signal intensities, or non-zero vectors corresponding to non-zero signal intensities in the case a special type of CH_x-signals should be surpressed in the recorded spectrum.

To give an impression of the effects to be expected in practice a pure substance has been investigated with both PCSE techniques: campher. The coupling constants J for the carbon atoms bonded to hydrogen in this molecule are: C_3:129.8 Hz, C_4:144.3 Hz, C_5:133.1Hz

C_6: 130.5 Hz, C_8: 125.0 Hz, C_9: 125,0 Hz, and C_{10}: 125,0 Hz. The numbering of the carbon atoms is given in fig. 10 additionally to the two PCSE spectra. The bottom spectrum was obtained with the delay 1/J between the pulses of the sequence PCSE-1 and the top spectrum using the delay 1/(2J) (PCSE-2). So the PCSE-2 spectrum shows CH and CH_3-signals with amplitudes opposite to CH_2 and C_{quart}-signals. In the PCSE-1 spectrum theoretically only signals from quarternary carbon atoms should be observed. The J value assumed for the calculation of the delays was 125.0 Hz.

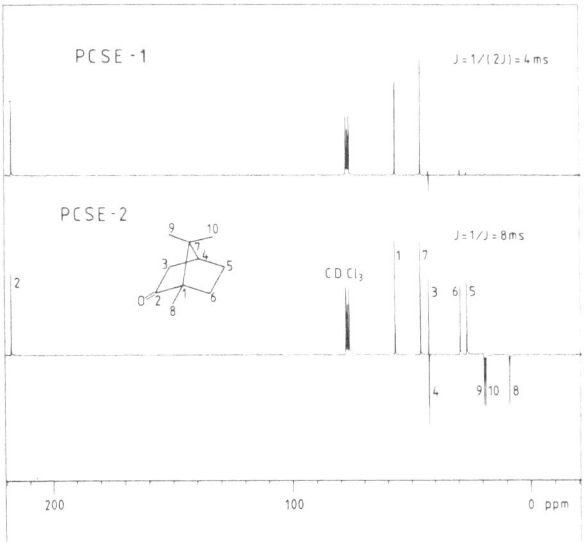

Fig. 10: The PCSE spectra of campher using J = 125 Hz

Obviously ths signals of quarternary carbon atoms in campher and in the solvent $CDCl_3$ produce the intense signals in the PCSE-1 spectrum. There are however some smaller peaks resulting from not completely surpressed signals that can be easily assigned to the different protonated carbon atoms in camphor. In this case in which a pure substance is investigated the residual signals do not make the interpretation of the spectrum more difficult. Using the symmetry of the molecule it can be deduced that the small signals can be observed in the PCSE-1 spectrum because the signal surpression is not complete. If the spectrum recorded was the result of an investigation of a mixture the interpretation would be not as clear. Then these peaks possibly also could be produced by quarternary atoms of a diluted substance in the mixture. A clear decision then is only possible by comparing the PCSE-1 spectrum with another ^{13}C-NMR spectrum that contains the signals of hydrogen bonded carbon atoms too. For example the PCSE-2 spectrum in fig. 10 can be used for this comparison. Then it is evident that the intensities of signals 3 - 6 are strongly reduced and thus can be assigned to non-quarternary carbon atoms.

The intensities of the residual signals in the PCSE-1 spectrum are closely related to the deviation of the actual J-value from the assumed 125.0 Hz: The greater the deviation the greater the observable intensity of the residual signal. Corresponding to its high J-value of 144.3 Hz the intensity of the C-4-signal in campher is less reduced than any other CH or CH_2-signal produced by this molecule. Methyl-signals on the other side are completely surpressed since their coupling constants are exactly 125.0 Hz.

Fortunately the J-values of substances in coal and petroleum derived products generally are very close together. Data in literature (18) show for pure hydrocarbons that are the main components of these products, J-values between 155 and 170 Hz for carbon atoms in aromatic bonds and between 120 and 130 Hz for those in aliphatic bonds.

Corresponding to these data two independent experiments generally should be performed with products made from coal or petroleum. One with the assumption J = 160 Hz giving the best results for the aromatic carbon atoms, and another with J = 125 Hz for the aliphatic carbon atoms.

This is however necessary only if the intensities of the recorded signals are to be evaluated quantitatively or if a technique is used to surpress one or the other CH_x-signal by cancellation of the individual frequency vectors in the rotating frame. The surpression of quarternary carbon signals by both polarization transfer methods is sufficient anyway if a J-value greater than 100 Hz is used since then there is no possibility of polarization transfer to the C_{quart}-atoms. The signals of quarternary carbon atoms sometimes observed in polarization transfer spectra generally are the result of a direct excitation of these nuclei by imperfect ^{13}C-pulses in the pulse sequence. By applying an appropriate phase cycling of the pulses however these signals can be surpressed completely.
If the experiment performed is only used to determine the nature of the individual signals, usually one single experiment with assuming an average coupling constant J = 140 Hz needs to be recorded. The actual coupling constants of aromatic and of aliphatic carbon atoms then are still close enough to differentiate between the CH_x-groups. The signal intensities in this case however are no more reliable. Moreover then some phase anomalies in the obtained spectra are observed, however they are generally interpretable and do not prohibit the assignment of the signals to the different CH_x-groups.

4.2 Quantitative results

As pointed out in the last section quantitative results can be evaluated only if the J-value used for the calculation of delays is very close to the actual coupling constants. In both experiments performed independently for the aliphatic and the aromatic carbon atoms in coal and oil products the variation width of the actual J-values however is ± 10 Hz with respect to the assumed one.

Again the vector model can help to explain what these deviations mean to the intensities of the individual signals in the recorded spectrum. As example the consequences to a CH_2-signal in a PCSE-2 experiment are calculated, assuming an actual coupling constant of 135 Hz and a value of 125 Hz determining the delays in the pulse sequence: Instead of exactly one cycle the two frequency vectors representing the outer lines of the CH_2-triplet in the time interval 1/J = 1/125 seconds will pass 135/125 cycles and then will be 28.8^{o} away from the starting position in the rotating frame. At that position the frequency vector of the centre line has remained all over the evaluation period. The total signal intensity then can be calculated by adding up all frequency vectors. Due to the symmetric distribution of the vectors with respect to the centre vector all vector components perpendicular to its direction exactly cancel. Assuming unit length for the frequency vectors representing the outer lines of the triplet, the length of their components pointing in the same direction as the frequency vector of the centre line turns out to be the cosine of 28.8^{o} which is 0.88. The vector sum then is 2 (for the double intensity of the centre line) plus 2·0.88 (for both outer lines) resulting to 3.76. Normalizing to a standard intensity 1 that would be observed if the actual and the assumed J-values are the same, in this case the intensity is only 0.94. With the same assumptions the intensity of a CH-group would be reduced to 0.97 and that of a CH_3-group to 0.91. The same values are valid also for signals in spectra obtained with Doddrell's sequences.

The intensity reduction introduced by a missetting of J that cannot be avoided if mixtures are studied, seems to be severe. Investigating coal and petroleum derived products however the observed reductions in signal intensity are tolerable. To show this, difference spectra have been calculated resulting from the subtraction of spectra obtained with one of the methods discussed in this paper from 'quantitative' ^{13}C-NMR spectra of the same sample. These 'quantitative' ^{13}C-NMR spectra have been recorded using the inverse gated decoupling technique and a delay of 10 seconds between the acquisition of the individual FIDs during spectra accumulation. The delay was proven to be long enough for this type of samples in an independent investigation. For reasons of abbreviation this method will be called C13QUANT from now on.

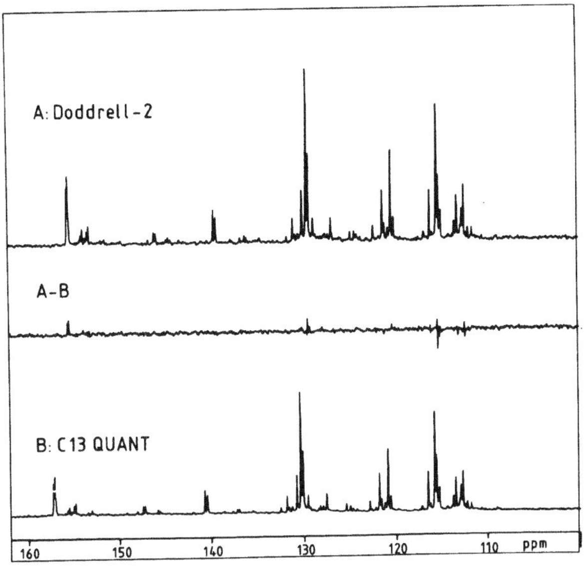

Fig. 11: Difference spectrum obtained by subtracting spectra obtained from a C13QUANT and a Doddrell-2 experiment

To produce significant results the spectra could not be subtracted directly but only after a normalization to compensate the different signal to noise ratios in the different spectra.

Comparing the qualities of the resulting difference spectra both spin-echo methods are superior to the polarization transfer methods with the respect to the display of correct quantitative signal intensities. This corresponds to theoretical calculations published by Soerensen and Ernst (19). If the spectra produced by multi-pulse sequences show correct signal intensities these have to cancel the intensities in C13QUANT spectra exactly after subtraction (or addition when negative signals are concerned) of the two spectra compared. In ideal cases the difference spectrum should show noise only.

One example of the calculated difference spectra is reproduced in fig. 11. In this case a Doddrell-2 spectrum is compared with the C13QUANT spectrum. Only the aromatic spectral region is shown. A J-value of 160 Hz was used to calculate the delays in the Doddrell-2 sequence.

The difference spectrum in the middle of fig. 11 shows a nearly complete compensation of the signals. The still observable residuals are only partly caused by the deviation of the actual J-values from the assumed 160 Hz. Another reason for these residuals are relaxation effects that can cause different line widths in both spectra. These different line widths however have not been taken into account when the difference spectrum was calculated.

The residual signals in the difference spectrum seem to be tolerable, at least if it is considered that the results of two completely different experiments had to be identical to obtain a complete compensation of all signals.

The Doddrell-2 spectrum was choosen as example for showing the quality of the method for C_{quart} and CH-signals in one single difference spectrum. The same results can be obtained using the Doddrell-1 spectrum or a linear combination of both Doddrell spectra.

The PCSE-spectra showed as good quantitative results as Doddrell's spectra. Remembering the fact however that the PCSE sequence lasts twice as long as Doddrell's sequences there might be a negative effect caused by short transverse relaxation times. In all cases investigated up to now, however this effect did not be obvious, probably because the size of the molecules in these samples is still small enough to allow slow transvers relaxation. Doddrell's sequence is preferrable for another reason: In the PCSE sequence the broadband decoupler is switched on some milliseconds before data acquisition starts. In this period the NOE could come into effect and change signal intensities. This effect is not very severe but again Doddrell's sequence has not to deal with this problem.

In practice however no big difference between both spin-echo methods was observed and it seems to be convenient for users of older spectrometers that do not allow to pulse the ^{1}H frequencies independently of the ^{13}C frequencies, to use the PCSE sequence as well to obtain quantitative data.

4.3 Subspectra editing

The four different techniques discussed in this paper each allow to discriminate between individual CH_x-signals in ^{13}C-NMR spectra. In most cases the differentiation is made by analysing the orientation of signal amplitudes within the same spectrum; positive signals represent one group of CH_x-signals, negative ones another.

This form of differentiation usually is clear enough to get the necessary information. In special cases however, things would be more simple if spectra could be produced containing signals of one type of CH_x-groups only. There are possibilities to obtain one of these 'pure' spectra directly by using special delays in the pulse sequences presented: Shortening the delay between the pulses in both spin-echo experiments to a value half as great as it is used to discriminate between all CH_x within the same spectrum, only signals of quarternary carbon atoms will be observed. One example is given in the middle of fig. 12: The PCSE-1 sequence has been applied to a phenolic fraction of a coal oil using the delay 1/2J= 1/(2·160) seconds. Below a Doddrell-1 spectrum of this sample displays CH and C_{quart}-signals alltogether. Generally the C_{quart}-signals can be assigned in both spectra as easy. Some small signals of quarternary carbon atoms however that absorb in a spectral region with very intense CH-signals can be better identified in the PCSE-1 spectrum than in the Doddrell-1 spectrum. Another 'pure' spectrum can be produced by either the INEPT-2 or the DEPT-2 experiment, both yield a CH-subspectrum. In fig. 12 the INEPT-2 spectrum is displayed as example.

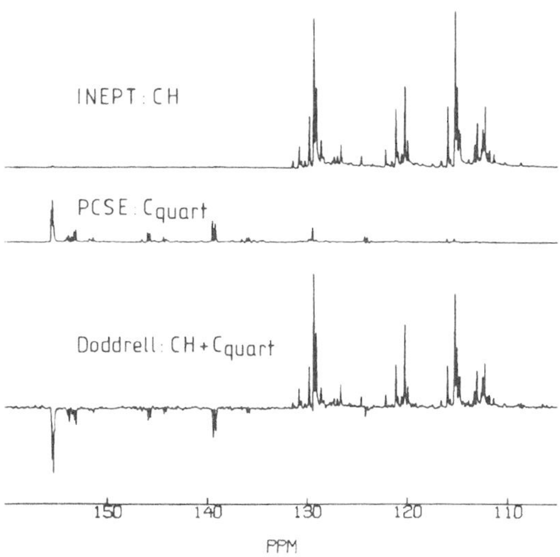

Fig. 12: CH and C_{quart}-subspectra of a phenolic coal oil fraction obtained directly using the INEPT-2 sequence and the PCSE-1 sequence respectively. The Dodrell-1 spectrum is shown for comparison.

So here are methods that in the aromatic region of a spectrum of coal and petroleum products can produce individual spectra for CH and C_{quart}-signals. Because CH and C_{quart}-groups are the only ones to be observed in those spectra, a clear separation can be achieved in this case. Usually both spectra, PCSE-1 and INEPT-2 spectra can be used only to assign the peaks within a conventional ^{13}C-NMR spectrum, but it is not possible to compare the signal intensities directly. Even the estimation of quantitative relations between C_{quart} and CH-signals is not possible when only these two spectra are known.

This problem can be managed if a so called spectra editing is possible with spectra showing the same signal intensities. Spectra editing is the result of a mathematical operation applied to several spectra with the aim to produce a calculated spectrum containing a reduced number of CH_x-signals only. Looking into table 1 for nearly every technique some of these mathematical operations are obvious.

If for example the spectra resulting from a Doddrell-1 experiment and from a Doddrell-2 experiment are added, all CH_3 and CH-signals should exactly cancel and only CH_2 and C_{quart}-signals should remain observable. Subtracting both types of Doddrell spectra the opposite result is achieved, only CH_3 and CH-signals are present in the resulting spectra.

Thus if only the aromatic spectral region is considered, again a clear separation of CH and C_{quart}-signals is possible. This is demonstrated in fig. 13. The two top spectra show edited C_{quart} and CH-spectra (in this case with negative amplitudes)

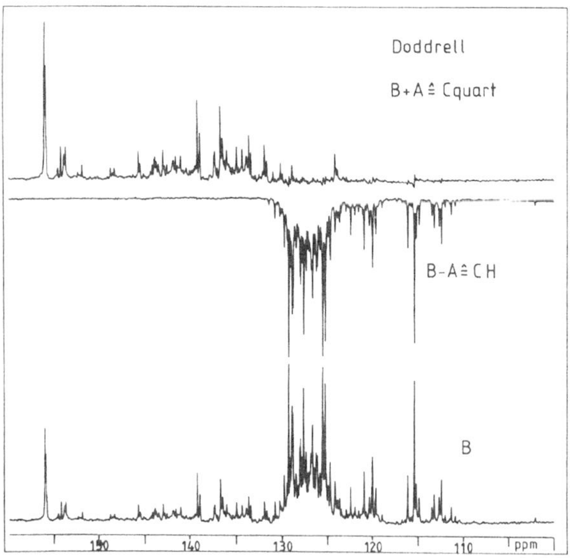

Fig. 13: Edited C_{quart} and CH-subspectra of a coal product using Doddrell's pulse sequences

To show the advantages of spectra editing at the bottom of fig.13 the Doddrell-2 spectra is displayed where as in conventional broadband decoupled ^{13}C-NMR spectra no differentiation between both types of signals is possible. Both edited spectra can be evaluated quantitatively independent of each other. The results can be compared directly since they are obtained from spectra produced with the same technique.

Similar results can be produced by combining PCSE-2 spectra with conventional spin-echo spectra. These can be obtained with the same pulse sequence as it is used in the PCSE-2 technique but broadband decoupling all over the sequence. Again signals produced by carbon atoms bonded to an even number of hydrogen atoms can be separated from the other carbon signals. Each group of signals then forms an extra spectrum.

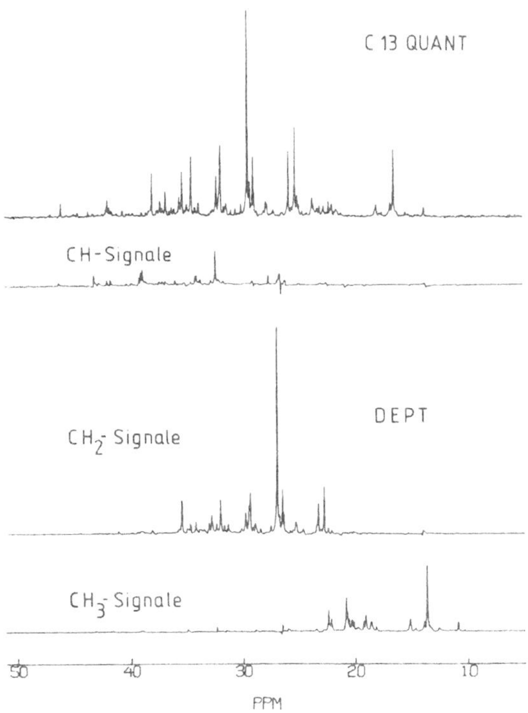

Fig. 14: Subspectra editing using the DEPT technique

The editing of individual CH_x subspectra using the spin-echo methods is clear for the aromatic region of a spectrum only because products made from coal and petroleum there show CH and C_{quart}-signals exclusively. In the aliphatic spectral region however all four types of CH_x-signals are present. CH_2 and C_{quart}-signals theoretically can be differentiated by combining the $CH_2 + C_{quart}$ subspectrum, calculated as indicated before, with the C_{quart} subspectrum obtained directly by the use of the appropriate pulse delays in the PCSE-1 or Doddrell-3 sequence. There is however no chance to discriminate between CH_3 and CH-signals using either of the spin-echo sequences.

The only method that can be used to separate these types of signals is the DEPT technique. Combining the three DEPT spectra resulting from $\Theta = 45^o$, $\Theta = 90^o$ and $\Theta = 135^o$ as described before, pure CH_3, CH_2 and CH subspectra can be calculated. One example is shown in fig. 14 for the aliphatic region of the ^{13}C-spectrum of a low boiling coal oil fraction. The delays in the pulse sequences have been calculated using a J-value of 125 Hz.

The separation of the different groups is acceptable and allows to identify every type of CH_x-group. The surpression of theoretically not allowed signals in the individual subspectra is sufficient. Quantitative results can be estimated in every DEPT subspectrum but it is preferable to use the C13QUANT spectrum which is proven to show the correct intensity relations. DEPT spectra however then are needed for a clear signal assignment.

5. Conclusion

Multi-pulse sequences developed in the last four years can be used succesfully to analyse ^{13}C-NMR spectra of complex mixtures. The possibility of subspectra editing to produce spectra containing signals of one single type of CH_x-group exclusively is very helpfull for the identification of chemical structural groups in coal and petroleum derived products. Quantitative results can be estimated from every of these spectra but the precision of the data is sufficient only if spin-echo spectra are evaluated. The best way to get correct quantitative results however, is to evaluate a conventional C13QUANT spectrum and to assign the portions of every signal to a specific CH_x-group by comparing the C13QUANT spectrum with one of the edited subspectra. Following this procedure both polarization transfer methods are faster than the spin-echo methods.

Subspectra editing resembles to a preseparation in chromatography. After applying these techniques, signals that before have been hidden under other signals now can be identified clearly. Using the previous known information again, in this case the chemical shift value, it is possible to come to new conclusions.

The described multi-pulse sequences thus allow a more detailed description of the hydrocarbon-type structure in molecules and by this is complementary to those NMR techniques using chemical derivatization that help to identify structural groups containing heteroatoms. Both developments together will give more data for a detailed analysis of structural groups in coal and petroleum derived products.

Acknowledgement
Part of the investigations discussed in this paper have been published previously (19,20). The author acknowledges the permission of both publishers to reuse some figures yet produced in those papers

Literature

1. K. v. Nes, H. A. v. Westen, "Aspects of the constitution of mineral oils", Elsevier, New York (1951)
2. H. H. Oelert, Z. Anal. Chem., 231, 81 (1967)
3. R. Gerhards, DGMK-Bericht No. 239, Hamburg (1979)
4. a) S. Gillet, J.-J. Delpuech, J. Magn. Reson., 38, 443 (1980)
   b) S. Gillet, J.-J. Delpuech, P. Valentin, J.-C. Escalier, Anal. Chem., 52, 813 (1980)
   c) S. Gillet, P. Rubini, J.-J. Delpuech, J.-C. Escalier, P. Valentin, Fuel, 60, 221 (1981)
5. S. A. Knight, Chem. Ind., 1920 (1967)
6. D. W. Brown, T. T. Nakashima, D. L. Rabenstein, J. Magn. Reson., 45, 302 (1981)
7. C. Le Cocq, J. Y. Lallemand, J. C. S. Chem. Comm., 150 (1981)
8. C. G. Snape, Fuel 61, 775 (1982)
9. C. G. Snape, Fuel 61, 1164 (1982)
10. D. J. Cookson, B. E. Smith, N. J. White, J. C. S. Chem. Comm., 12 (1981)
11. D. J. Cookson, B. E. Smith, Fuel, 62, 34 (1983)
12. M. R. Bendall, D. M. Doddrell, D. T. Pegg, J. Am. Chem. Soc., 103, 4603 (1981)
13. G. A. Morris, R. Freeman, J. Am. Chem. Soc., 101, 760 (1979)
14. G. A. Morris, J. Am. Chem. Soc., 102, 428 (1980)
15. O. P. Burum, R. R. Ernst, J. Magn. Reson., 39, 163 (1980)
16. D. M. Doddrell, D. T. Pegg, M. R. Bendall, J. Magn. Reson., 48, 323 (1982)
17. O. W. Sørensen, R. R. Ernst, J. Magn. Reson., 51, 477 (1983)
18. W. Bremser, E. Franke, L. Ernst, R. Gerhards, H. Hardt, "Carbon-13 NMR-Spectral Data", Verlag Chemie, Weinheim (1981)
19. R. Gerhards, Erdöl u. Kohle, Erdgas, Petrochemie, 36, issue July 1983
20. R. Gerhards, Z. Anal. Chem., 316, issue September 1983

COAL AROMATICITY

Bernard C. Gerstein

Ames Laboratory, US Department of Energy and
Department of Chemistry, Iowa State University
Ames, Iowa 50011

The extent to which coal may be considered an "aromatic" material is considered in terms of some old, and more recent criteria of aromaticity. These criteria include: a) chemistry, including solvent extraction, extraction, b) interatomic distances as inferred from density and bulk properties, and from x-ray diffraction, c) inferences from optical properties, especially infrared spectroscopy, d) proton and carbon NMR of solvent extracts and of the solid coals. The concept of coal as a polymer is developed within the formalism of VanKrevelen, and of Brown and Ladner. These ideas are used to develop models for some average chemical structures within coals, using inferences from some recent developments in high resolution solid state NMR of carbon in coals.

I. INTRODUCTION

The elemental composition and macroscopic features of coals have been extensively studied for decades. However, if a chemist were presented with an unknown organic substance and asked to produce it from, e.g., nylon, he would also want to know "What functional groups does it contain, and in what concentrations?". The present lecture deals with carbon and hydrogen aromaticity, and, by inference, carbon and hydrogen aliphaticity in coals.

Even the most peripheral glance at the concept of aromaticity itself, however, immediately reveals a further problem. "Aromaticity" is not a sharply defined concept. We therefore proceed with a brief review of the traditional concepts of aromaticity (Clar, 1964; Badger, 1969; Bergman and Pullman, 1971; Garratt, 1971). We then investigate the extent to which these concepts apply to coals.

L. Petrakis and J. P. Fraissard (eds.), Magnetic Resonance. Introduction, Advanced Topics and Applications to Fossil Energy, 409–439.

II. AROMATICITY

The accepted beginning of the concept of aromaticity were with Kekulé's description of the structure of benzene (Kekulé, A. (1869)). The power of this original idea was enormous. Today, more than 115 years later, the extent to which nuclear magnetic resonance chemical shifts of hydrogens in polycyclic aromatic rings correlate with connectivity (Randić, M., and Gerstein, B. (1980)) may be understood in terms of Kekulé structures.

Originally the concept of aromaticity development as a way of characterizing certain organic molecules that readily underwent substitutions but were disinclined to addition reactions and were thermally stable (Robinson, R. (1958)).

An earlier definition of aromaticity then was that a compound was aromatic if it had a chemistry like benzene. That is, reactivity (which encompasses both thermodynamics and kinetics) was used as a criterion for aromaticity. A more modern measure of aromaticity is that which defines it in terms of the standard energy of formation (Peters, D. (1960)). The quantum mechanical justification for this definition appeared in the form of calculations of the ground state "resonance energy" of a conjugated system (Craig, D.P. (1959)).

These criteria are not, however, without ambiguity, as evidenced by the example of the fulvenes, e.g., the cyclopentadienyl anion, $C_5H_5^-$. This compound is the prototype of non-benzenoid aromatic hydrocarbons, having a standard enthalpy of formation in line with "aromatic" character but it is also much more reactive than benzene. The difficulty had led to a proliferation of descriptions, e.g., quasi-aromaticity, anti-aromaticity and homo-aromaticity.

A rule widely used to predict aromaticity in non-benzenoid hydrocarbons is the Hückel 4n+2 rule (1931), for which the theoretical basis was developed by Coulsen and Longuet-Higgins (1948/1949), and by Dewar (1952). Inherent in all of the above criteria used to infer aromaticity is the concept of electron delocalization.

A feature inherent in the delocalization of electrons in aromatic rings is that of ring currents (Streitweiser, Jr., A. (1961)). This concept was of enormous importance for purposes of the present talk because it led directly to the use of isotropic chemical shifts in the nuclear magnetic resonance (NMR) spectra of aromatic compounds to diagnose aromaticity. Three papers of seminal importance in this are these of Waugh and Fessendon (1957), Pople and Unch (1966), and Haddon (1972). Another parameter related to the ring current concept is that of the isotropic value, and anisotropy, of the magnetic susceptibility (Dauben et al. (1969)). Bond lengths between carbons on the periphery of aromatically conjugated systems, (Julg and Francois (1967)), as well as a description of delocalization in terms of bond order (Trindle

(1969)) are two further parameters used to index aromaticity. Finally, the electronic absorption spectra and electron spin resonance spectra (Vincow (1971)) of molecules has been used as a tool to probe aromaticity.

When we ask about the application of all of the above criteria to the question of aromaticity in coals, it is immediately clear that for the most part we are on the horns of a dilemma. An implicit assumption, e.g., in the use of a criterion line "reasonance energy", to judge aromaticity is that one has the pure compound on which to perform the appropriate calorimetry, and/or to make calculations at some level of approximation. In fact, of all of the above criteria mentioned, only those capable of experimentally probing matter, the structure of which is assumed to be completely unknown and returning a more-or-less yes/no result are applicable.

The classic compilation of information from which aromaticity in coal has been inferred prior to 1960 is that in "Coal" (VanKrevelen (1961)).

At this point it is not inappropriate to remark however that one of the most obvious indicators of the possibility of aromaticity in coal is the atomic (H/C) ratio, available simply from major constituent analyses of the dry, mineral free portion. This ratio varies from 1 for lignites to 0.3 for anthracites. Clearly, unless there are extensive contributions from polycondensed diamond-like structures in coals, as suggested by Charkabartty and Kretschmer (1974) on the basis of degradative oxidation by OCl^-, but challenged by many (e.g., Aczel et al. (1975)), it would seem reasonable to conclude that coals do contain aromatic ring systems, and that the extent of such systems increases with increasing carbon content.

III. AROMATICITY AND CHEMISTRY

Since chemical reactivity was among the first of many criteria used to infer aromaticity, it is appropriate to start our considerations of aromaticity in coals with a discussion of reactivity. The simplest reaction, that of solution formation by chemical components with similar chemical functionality, might appear to be an appropriate place to begin. However, it has been said of coals that they are almost insoluable in almost everything. Where appreciable "solubility" has been achieved under conditions of temperatures and pressures high compared to normal laboratory conditions, the question of chemical degradation always arises. A monograph by Wise (1971) has reviewed solvent treatment of coals, and the solubility of various solvents in coals as a function of the donar-acceptor number concept of Guttman (1977) has been discussed by Marzac et al. (1979). Hombach (1980) has discussed the distinction between "extractability" and "solubility" in solvent treatment of coals. He has used the Higgins treatment of irregular mixtures (Higgins (1943)) to explain why mixtures of solvents are more effective in solubilizing coals than single solvents. It is pointed out that maximization of solubility occurs when the enthalpic contributions to the solubility are equal for solvent and solute. These contributions are equal to the cohesive energy density, which is calcuable from the heat of vaporization. A problem with inferring a particular chemical functionality from solubilities is the wide varieties of chemical functionality within coals. The fact that molecules as diverse as pyridine, benzene, methanol, and water are all imbibed to a certain extent by dried, powdered coals is indicative of such functional group diversity.

A fact not generally appreciated is that the halftimes of solvent uptake by even finely powdered, (50 μm) dried coal samples can be of order of days, at room temperature. A preliminary study of the kinetics of solvent uptake for a few selected solvents, and self diffusion constants of solvents as a function of fraction absorbed (in m moles g^{-1}) has been performed in the author's research group. (Figure III-1 and Tables III-1 and III-2 exhibit these data on coals ground to a size between 25 μm and 50 μm, and dried at 110°C for eight hours at a pressure

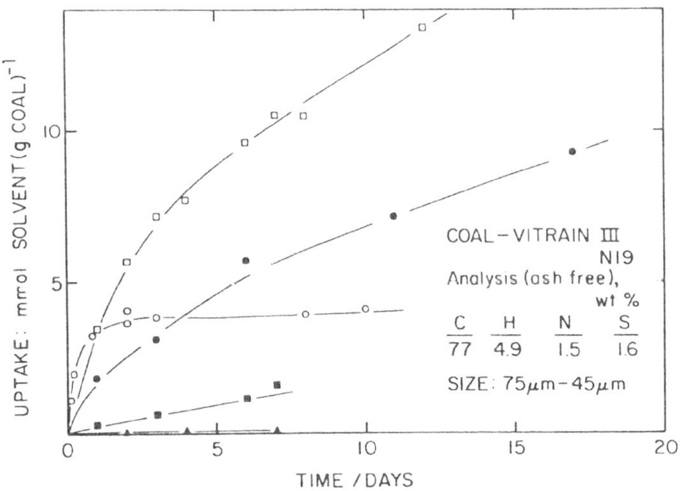

Fig. III-1. Uptake of solvents in an Iowa Coal, m Mol g^{-1} ((□) water, (o) benzene, (•) pyridine, (■) xylene, (▲) decalin)

of 0.1 Torr. The processes represented by these data could be as diverse as physical absorption (e.g., N_2 on charcoal) to solution of a solvent in a polymer (e.g., imbibing of toluene by polystyrene). For the data exhibited, solvent uptake was by the coal in a closed system exposed to the saturated vapor pressure of the solvent at room temperature. Self diffusion constants were determined using a multiple pulse NMR multiple pulse gradient technique developed in the author's laboratory (Silva-

Crawford et al. (1980)). These data indicate that for equivalent coverage pyridine is much less mobile in coals than the other molecules studied. The comparison of the value for Py with benzene indicates the relative lack of mobility to be associated with the nitrogen functionality. The other trends in the data do not appear to be relatable to aromaticity in coals in any direct manner.

Table III-1. Self Diffusion Coefficient, $D/cm^2\ sec^{-1}$, and Traverse Relaxation Time, T_2^*, of Protons in Pyridine and Methanol Imbibed by Star #2 Vitrain

Uptake Time/Days	Uptake/m mol g^{-1}	$T_2^*/\mu sec$	$10^8 \times D/cm^2 sec^{-1}$
	Solvent = Methanol		
2	4.70	406.2	---[a]
4.2	7.05	538.3	1.43
5.2	7.87	597.3	2.5
7.2	8.14	644.3	90
17	9.26	810.0	98
	Solvent = Pyridine		
1	0.9	64.8	---[a]
3	2.5	93.5	---[a]
4.2	3.44	136.5	---[a]
5.2	4.67	140.4	---[a]
7	5.65	144.6	---[a]
10.2	8.24	270.3	0.39
17	9.14	380.05	86

[a]Too small to be observed.

Table III-2. Self Diffusion Constants, $D/cm^2 sec^{-1}$, of Selected Solvents at Selected Uptake Times in an Iowa Vitrain (Vitrain III)

Uptake Time/Days	Solvent	$10^7\ D/cm^2 sec^{-1}$
55	H_2O	49
17	pyridine	31
2	benzene	31

With respect to "bond breaking" chemical reactions on coals which probe chemical functionality (VanKrevelen (1961)) and involve reactions as simple as silylation of hydroxyl groups and as complicated as degradative oxidation and high temperature pyrolysis there is an awesome wealth of data. Fortunately, there have been recent reviews on this subject (Gorbaty and Ouchi (1981), Bevkowitz (1979), Whitehurst et al. (1980), and Hayatsu et al. (1981)).

A reaction which appears to yield quantitative information about coal aromaticity is flash pyrolysis. Solomon (1981) has inferred f_a in 43 coals from infrared absorption spectra in the Fourier transform mode (vide infra) and found a linear correlation with the concentration of "fixed carbon", F.C., following the suggestion of such a correlation by VanKrevelen and Schuyer (1957). Fixed carbon is obtained from a pyrolytic decomposition, and defined as FC = 100 - weight percent of volitile matter. Solomon's model for flash pyrolytic decomposition of coals is one in which coal is described as a polymeric mixture of n-mers of varying molecular weight, and rapid thermal treatment vaporizes the lower molecular weight fragments while maintaining much of the original coal structure is the distillate. This model will be discussed more fully in Sections VII and IX. Degradative oxidation has also shown great promise as a diagnostic chemical tool. It appears that there exist oxidation procedures utilizing acid $Cr_2O_7^{2-}$ which are capable of selectively attacking aliphatic structures while preserving aromatic ring systems (Hayatsu et al. (1981)). It also appears that it may be possible to selectively oxidize aromatic rings with peroxide in perfluoroacetic acid while preserving aliphatic structures (Deno et al. (1978), (1981)) and references in Hayatsu et al. (1981)). It is perhaps not inappropriate at this point to paraphrase some of the conclusions from Hayatsu et al. (1981) on inferences from oxidative degradation. "Organic macromolecular material in coal is essentially a complex mixture of benzenes, phenols, naphthalenes, cycloalkanes, tetralins, furans, thiophenes, pyridines, etc." "These compounds are tied together by methylene bridges, alicyclics, ethers, etc." "Although much work has been done, there still are not enough data available to permit the comprehensive understanding necessary to provide detailed information concerning the structure origin, and transformation of coal molecules". We summarize by stating that classical organic chemistry appears to unequivocally demonstrate the existence of polycondensed aromatic structures in coals, and that for all coals except anthracites, the number of polycondensed rings does not appear to be much larger than three. At the present writing, however, there does not yet seem to be sufficient data from such studies alone upon which to quantitate the amount, or condensation index of polynuclear aromatics in coals.

IV. AROMATICITY AND BULK PROPERTIES; DENSITY AND RELATED PARAMETERS

As indicated in the introduction, the carbon-carbon bond distance is one parameter used to infer the presence of aromaticity. This distance may be empirically related to the density of condensed polymeric matter. Density in turn is a parameter involved in the molar sound velocity, and in the molar refraction of such bulk matter. In the present section we summarize results obtained from densities in particular. On the basis of connectivity arguments for polymers consisting of C, H, N, S, and O (see Section VI), VanKrevelen has provided a relation between the hydrogen to carbon ratio corrected for the presence of heteroatoms $(H/C)_{corr}$, the carbon aromaticity,

$f^c_{ar}=C_{ar}/C$, and the "ring condensation index", $2(R-1)/C$, where R is the total number of rings per carbon:

$$2 - \left(\frac{H}{C}\right)_{corr} = f^c_{ar} + 2\,\frac{(R-1)}{C}\ . \qquad [IV-1]$$

Using: 1) major constituent analyses, 2) inferences about functionality of N, S, and O from standard qualitative organic analytical probes on powdered coal samples, and 3) an additivity analyses of functional group contributions to molar volumes, VanKrevelen was able to use densities of macerals compared with graphs on model compounds ("graphical densities") to infer f^c_{ar} and the ring condensation index for individual macerals. f^c_{ar} of fusinites is found to be constant at a value near 1 for carbon contents between 80 and 95%. The values of f^c_{ar} and ring condensation index increase from 0.70 to 0.30, respectively, for a vitrinite containing 70.5% C to 1 and 0.77 for vitrinites containing 94.2 to 96% C. Micronites are more highly aromatic, and exinites less aromatic than vitrinites, but both approach the f^c_{ar} of vitrinites for %C greater than 85.

V. AROMATICITY AS INFERRED FROM X-RAY DIFFRACTION

X-radiation produces coherent diffraction patterns when scattered from matter having electron density with periodicity of order of the wave length of the X-rays. The well-known Bragg equation relating X-ray wave length, λ, to interplane spacing, d, and scattering angle, θ, is $\lambda = 2d \sin \theta$. The coherence length of the electron density periodicity necessary to produce sharp diffraction patterns in powdered samples is of order of 100 Å. Samples with smaller coherence lengths will produce increasingly more diffuse diffraction maxima. Coals are found to produce rather diffuse powder X-ray diffraction patterns, the extrema in the patterns becoming sharper and more similar to those exhibited by graphite as the carbon content of the coal increases (Franklin (1950), Hirsch (1960), Diamond (1958), Ergun and Tiensuu (1959)). From information at a wide range of scattering angles, and from analyses of radial distribution functions which reflect average - nearest and next nearest neighbor distances even in materials as disordered as liquids, a model has emerged for coals with carbon content from 65-95% which features: a) carbon-carbon distances similar to those of graphite, with C-C bands ~ 1.40 Å in length, b) interlayer distances between lamallae similar to those of graphite ~ 3.54-3.43 for coals with carbon content between 84-94%), c) polynuclear aromatic rings ranging from two to four condensed benzenoid structures in coals ranging in carbon from 65-90%. It seems reasonable to conclude from the results of X-ray diffraction on coals that lamellar structures similar to graphite exist, that the size and ordering of these structures increase with carbon content, and that if by "aromatic" is meant "with interatomic and interlayer distances similar to graphite", then coal becomes increasingly aromatic with increasing carbon content.

VI. OPTICAL PROPERTIES AND INFRARED SPECTROSCOPY

The use of reflection measurements to determine refractive indicies of macerals has been discussed by VanKrevelen (1961). The molar refraction, obtained from the refractive index is one of the parameters which may be modeled in terms of additive contributions from atomic and functional fragments of organic compounds, with data inferred from results on model compounds. Again, using an additive relation between molar refractivities and aromaticity, a knowledge of major constituents, and inferences about the major forms of functional groups associated with heteroatoms, VanKrevelen (1961) was able to infer f_{ar}^{c} for vitrinites. Those values increased smoothly from 0.7 to 1 for vitrinites varying in carbon content from 70.5 to 93.4%, respectively. About %C = 93.4, the values of f_{ar}^{c} were found to be 1.

In contrast to refractive index, which is a bulk property and relates to aromatic functionality via empirical modeling, infrared spectroscopy, probes the stretching and bending motions of specific molecular fragments. Wexler (1970) has reviewed the application of IR spectroscopy to identification of aromatic fragments in hydrocarbons. The frequencies with which motions sensitive to detection by IR occur can be directly correlated with chemical functionality, i.e., it is possible to distinguish a methylene C-H (2912 cm^{-1} and 2860 cm^{-1}) stretch from an out of plane aromatic C-H (850-870 cm^{-1}). A serious problem is that oscillator strengths are in general unknown, so the technique is at best semiquantitative in nature without recourse to very careful calibration procedures. Early spectral work (VanVucht et al. (1955)) suffered from low intensity, and poor resolution due to overlap of bands, but indicated the presence of polycyclic aromatic rings in coals, inferred from absorption bands at 850, 810 and 750 cm^{-1}. In addition, the presence of aromatic and aliphatic C-H bands was indicated, and of carbonyl and carboxyl in lignites. Notably absent were indications of olefinic or acetylenic linkages. More recent work by Retcofsky (1977) reports a comparison of the integrated intensities of the out of place C-H bonding vibrations with proton nuclear magnetic resonance (NMR) of pyridine and CS_2 soluble extracts of coals. This comparison calibrates the IR intensities on a "per aromatic hydrogen" basis (the transition probabilities of the NMR lines are the same for all spins; vide infra). The calibrated integrated IR intensities of the extracts were compared with the integrated intensities of the vitrains from which the extracts were obtained. It was found that for both extracts and coals, the integrated intensities in the aromatic region increased with increasing carbon content. This observation, coupled with the fact that the elemental hydrogen contents within a lithotype decrease with increasing carbon content suggests that the higher carbon materials are more aromatic and contain larger polynuclear condensed aromatic ring systems.

A discontinuous step forward in the use of infrared spectroscopy to infer chemical functionality in coals was that provided by Fourier

Transform infrared techniques (Griffiths (1975)). Rapid signal averaging with computer performed spectral substraction and least squares fitting has led to resolution and sensitivity enhancement previously unavailable. Painter and Snyder (1980) have described how the use of FTIR combined with acylation to remove hydroxyl groups and spectral substraction of the acylated from the native sample can be used to provide qualitative, and perhaps quantitative analyses of alkyl and aryl hydroxyl groups in coals. Solomon (1979) (1981) has suggested a procedure for quantitating the aliphatic hydrogen in coals using FTIR, and then quantitating aliphatic carbon using the stoichiometry H_{al}/C_{al} = 1.8. Aromatic carbon is then calculated by difference. The aromatic hydrogen values determined by Solomon on a number of coals with carbon content varying from 65.2-86.2% were in agreement with values obtained by VanKrevelen (1957) and by Mazumdar et al. (1957). The model of coals proposed by Solomon (1981) from infrared studies of coals and their pyrosylates is that discussed in Section II, i.e., that of a mixture of polymerized monomers of varying molecular weights. This model is an appropriate introduction to the next section.

VII. COAL AS A POLYMER

"The organic materials from which coal has formed are all polymers" (VanKrevelen (1961)). Thus starts a discussion which may be one of the most important points related to analyses of chemical functionality in coals. With the exception of anthracites, coals of any rank exhibit some degree of solubility in various solvents. To the extent that material extracted from coals is representative of the original coal structure, the problem of analyzing for chemical functionality becomes enormously simplified, simply because standard analytical techniques for probing chemical functionality are so amenable to detection of species in solution.

The early infrared studies of Cannon (1953), Roy (1957) and Brown et al. (1958) indicated that some coals may be represented as a polymeric solution of varying sized n-mers, with the lower molecular weight components extractable by appropriate solvents. The evidence for this view was that the infrared spectra of the extract, the residue, and the native coal were quite similar. This picture has been reinforced by studies of chemical degradation of coals by e.g. hydrogenolysis (Glen (1955)). From studies of the proton NMR of coals depolymerized using phenol-BF_3 under relatively gentle conditions, Heredy et al. (1957) concluded that analyses of soluble fragments by NMR yielded results "which could be extrapolated to the native coal with considerable confidence". The implication of this work was that under the depolymerization conditions described, the process of depolymerization involved breaking of methylene bridges. Weinberg and Yen (1980) have also investigated solubility parameters in coals and coal liquefaction products, and have come to the conclusion that molecules found in liquefaction products already exist in the virgin coal before liquefaction. Subsequent work by Zilm et al. (1979) coal derived liquids obtained by short residence contact with $ZnCl_2$, utilizing carbon 13 NMR on parent and liquid, and work by the authors on high resolution solid state NMR of ^{13}C in coals, extacts, and insoluble fractions under gentle pyridine extraction lends strength to the picture that coals may be represented as a mixture of n-mers of varying degrees of polymerization. The lower molecular weight portions appear to be extractable without minor disruption of the structure, and may be a fair representation of a major portion of the solid material before extraction. The minor differences between spectra of coal, extract, and residue, of course, are indicative of changes related to the depolymerization process.

If coal may be viewed as a polymer, a statistical model which uses connectivities of C, H, O, S and N with known functionalities may be developed and has been referred to earlier (VanKrevelen (1961)). As a basis for discussion of data to be presented in later sections, we summerize VanKrevelen's development:

1. For an aliphatic chain: $H = 2C + 2$ [VII-1]

2. If the chain closes to form a ring, two hydrogens are lost per closure;

$$H = 2C + 2 - 2R \qquad [VII-2]$$

where R is the number of rings formed by the chain (e.g., 2 for naphthalene).

3. For each aromatic carbon, one hydrogen is lost relative to the aliphatic structure;

$$H = 2C + 2 - 2R - C_{ar} \qquad [VII-3]$$

4. If the hydrocarbon units described by equation [3] "bridge together" to form polymerized structures, with the "degree of polymerization", P, (which is the number of monomeric units in a particular polymeric fraction), then two hydrogens are lost for each bridge per monomer (or per "average structural unit"). Denoting by the subscript u the H, C, C_{ar} and R in an average structural unit (i.e., an average monomer);

$$H_u = 2C_u + 2 - 2R_u - C_{ar,u} - 2b \,. \qquad [VII-4]$$

If the average "structural unit" were tetralin (tetrahydronaphthalene, $C_{10}H_{12}$) the degree of polymerization were P = 2, and there was one bridge, b = 1, a possible structure per structural unit, for the dimer could be (I) for which $C_u = 10$, $R_u = 2$, $C_{ar,u} = 6$, $b = 2/2 = 1$ and

$$H_u = 2C_u + 2 - 2R_u - C_{ar,u} - 2b$$
$$= 2 \times 10 + 2 - 2 \times 2 - 6 - 2 = 10,$$

the number of hydrogens in the monomer fragment (II).

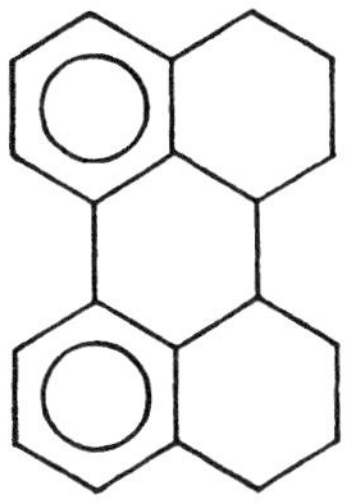

$P = 2$

$b = \frac{2}{2} = 1$

I

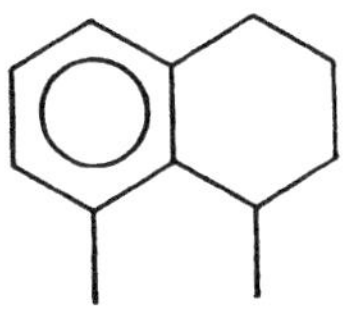

II

Note that the formation of b bridges can lead to extra rings per monomer unit, r_u in number. In the above example, $r_u = 1/2$. The relation between b, called the "polymerization intensity", the degree of polymerizaton, P, and the extra rings per average structural unit, r, is

$$b = \frac{P-1}{P} + r \qquad [VII-5]$$

For our above example, Eq. [5] yields

$$1 = \frac{2-1}{2} + \frac{1}{2}$$

The conditions for "netook formation" are that: a) the degree of polymerization be infinite:

$$\frac{P-1}{P} = 1 \qquad [VII-6]$$

and that both b and r be greater than unity.

Equations [4] and [5] may be rearranged to yield

$$\frac{2(R_u + r_u - 1/P)}{C_u} = 2 - \frac{C_{ar,u}}{C_u} - \frac{H_u}{C_u} \qquad [VII-7]$$

For P>>1, the values of H/C and H_u/C_u become the same (i.e., the composition of the monomer fragment represents the composition of polymer). With $f^C_{ar} = C_{ar,u}/C_u$, Eq. [7] may be written as

$$2\frac{R'_u}{C_u} = 2\left(\frac{R}{C}\right)_u = 2 - f^c_{ar} - \frac{H}{C} \qquad [VII-8]$$

Here $R'_u = R_u + r_u$, the total number of rings associated with a monomer fragment, f^c_{ar} is the fraction of aromatic carbon and the ratio $(R/C)_u$ is the total number of rings associated with a monomer unit. In our example using tetrahydronaphthalene as a monomer, one possible polymer might be (III)

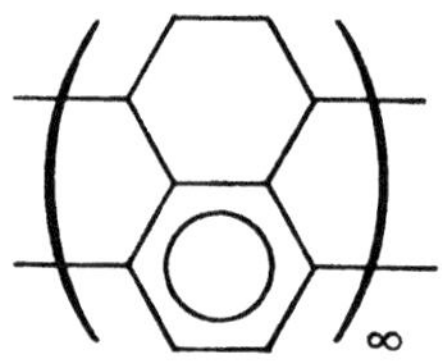

III

for which $R'_u = R_u + r = 2 + b\text{-}1 = 1\text{+}b$. Here, b = 2, and we find, with $C_u = 10$, $C_{ar,u} = 6$, and $H_u = 8$, that Eq. [8] becomes

$$2(3/10) = 2 - 6/10 - 8/10$$

Therefore, if the structure consists only of carbon and hydrogen and the carbon aromaticity is known, the number of total rings per carbon may be derived from the total hydrogen to carbon ratio via Eq. [7]. If there are heteroatoms present, the value of (H/C) must be corrected to account for their functionality; for example if an average structural unit were a tetralin carboxylic acid, (IV), with b = 2 as before,

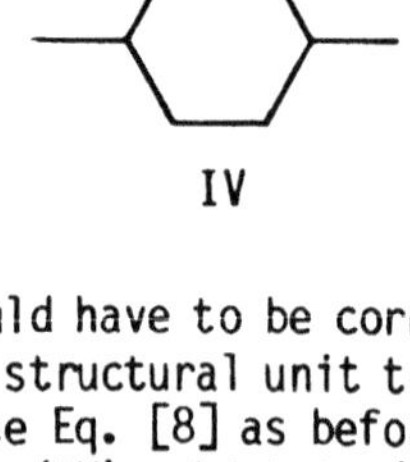

IV

then the value of (H/C) would have to be corrected to reduce the formulation of the average structural unit to that not containing the extra carbon in order to use Eq. [8] as before, i.e., the analytically determined value of H/C for (IV), which is (8/11) would have to be corrected back to (8/10). We therefore write Eq. [8] as

$$2(R/C)_u = 2 - f^c_{ar} - (H/C)_{corr} \qquad [VII-9]$$

to account for added heteroatom functionality, VanKrevelen (1961) points out that a minimum value of f^c_{ar} may be inferred from $(H/C)_{corr}$ determined by major constituent and functional group analyses, and the maximum value of $(R/C)_u$, which can be approximated

$$(f^c_{ar})_{min} = 1.23 - (H/C)_{corr} \cdot \qquad [VII-10]$$

for coals ranging from high volatile bituminous to anthracite, the value of $(f_{ar})_{min}$ thus calculated by VanKrevelen ranged from 0.43 to 0.93. As we have seen earlier, values inferred from bulk and spectroscopic measurements are generally higher than the minimum values just quoted.

It is instructive to examine the value of $2(R/C)_u$ for our example of the tetralin monomer, compared to that of the naphthalene monomer using b = 2, P = ∞. We find $2(R/C)_u = 2 - f_a - H/C = 2 - 0.6 - 0.8 = 0.6$ for the tetralin monomer, and a value of 2 - 1 - 0.4 = 0.6 for the naphthalene monomer. It is transparent that the value of $2(R/C)_u$ cannot be used to distinguish between hydroatomatic and aromatic rings. However if it is known that $C_{ar,u} = 6$, then for the tetralin monomer, with b = 2, P = ∞, $2(R_u/C_{ar,u}) \doteq 2(R/C)_u \times C_u/C_{ar,u} = 1$, and for the naphthalene monomer with $b \doteq 2$, P = ∞, $2(R/C_{ar,u}) = 0.6$. A knowledge of the value of $C_{ar,u}$ may therefore be used, in conjunction with f_a and elemental analyses to infer something about the number of rings per aromatic carbon in the representative with monomer.

On the other hand, if there exists in formation about the fractions of aromatic and aliphatic carbon and hydrogen and about the number of aliphatic hydrogens α to aromatic rings, it is possible to relate the carbon aromaticity, f^c_{ar}, to the C/H ratio, to the degree of substitution of aromatic rings, σ_a, and average sizes of polynuclear aromatic rings (Brown and Ladner (1960)). This information is only obtained, however, with some assumptions about heteroatom functionality, about the aliphatic hydrogen to carbon ratio, and about connectivity of aromatic rings. The Brown-Ladner relations are

$$f^c_{ar} = [(C/H) - (H^*_{\alpha}/x) - (H^*_{o}/y)]/(C/H) \qquad [VII-11]$$

$$\sigma_a = [(H^*_{\alpha}/x) + (O/H)]/[(H^*_{\alpha}/x) + (O/H) + H^*_{ar}] \qquad [VII-12]$$

$$\left(\frac{H}{C}\right)_{ar} = [(H^*_{\alpha}/x) + H^*_{ar} + (O/H)]/[(C/H) - (H^*_{\alpha}/x)-(H^*_{o}/y)] \qquad [VII-13]$$

Here, $H^* = H_\alpha/H$, and H_α is the number of aliphatic hydrogens α to an aromatic ring, a value obtainable from high resolution NMR of ^1H in liquid samples. $H^*_o = H_o/H$, and H_o is the remainder of aliphatic hydrocarbons. X is the ratio of aliphatic hydrogen to aliphatic carbon for α hydrogens (usually assumed to be 2), and y is the ratio of aliphatic hydrogen to aliphatic carbon for the remainder of aliphatic hydrogens. It is assumed that no direct aryl-aryl linkages exist (such as biphenyl), and that all oxygens are substituted on aromatic rings. $(H/C)_{aru}$ is the aromatic hydrogen to carbon ratio of the unsubstituted aromatic rings in the structure which gives a direct measure of the size of the polynuclear aromatic (PNA) rings; *e.g.*, this value is 1 for benzene, 0.8 for naphthalene, 0.6 for benzo-a-pyrene, etc. At this point, it is important to mention that a knowledge of the number of aliphatic hydrogens α to the aromatic rings is necessary to determine $(H/C)_{aru}$. Another way of saying this is that the connectivity of the aromatic rings must be known to derive $(H/C)_{aru}$. This concept has been discussed by the author (Gerstein, Ryan and Murphy (1979)). If the number of α aliphatic hydrogens is not known, but the ratio of $(H/C)_{aru}$ on the substituted aromatic ring is, a connectivity between average PNA units must be assumed to estimate the aromatic condensation index. For example, a monosubstituted benzene unit could have a value of $(H/C)_{aru}$ = 5/6 = 0.83, and a di-substituted naphthalene would have a value of $(H/C)_{aru}$ = 6/10 = 0.60. If "network formation" is to be assumed, we have seen (Eq. VII-6) that the minimum number of substituents per unit must be greater than two, which equivalent to a value greater than unity for b and for r. The reader will immediately recognize that a value of $(H/C)_{aru}$ = 0.65 could correspond to an average PNA size of unity (*i.e.*, to benzene rings in a network).

VIII. AROMATICITY AND ELECTRON SPIN RESONANCE

The electron spin resonance (ESR) experiment is dependent upon the presence of unpaired electrons in the sample under observation. In the presence of an oscillating macrowave field, resonance absorption occurs when the condition

$$\Delta E_{elec} = g\ \beta_e\ B$$

is met, with ΔE_{elec} being the separation of Zeeman levels of unpaired electrons in a static magnetic field, B. g is the electron g factor, and β_e is the Bohr magneton. Unpaired electrons may be associated with transition metal ions (*e.g.*, Fe^{+3}) and with stable free radicals in heterogeneous materials such as coals. The presence of stable free radicals in hydrocarbons is frequently associated with structures which are "resonance stabilized". Another way of making this statement is that free radicals in hydrocarbons are associated with polynuclear aromatic rings. With a fixed frequency of *ca.* 9 x 10^9 Hz, the field at which ESR can occur for unpaired electrons in transition metal ions may vary between 0.1 and 2T. In other words, the g factor can be very far from 2.0023, which is the value for a "free electron". In addition the contribution of orbital angular momentum to the total electronic angular momentum of unpaired electrons in transition metal ions frequently leads to broadening of spectral lines associated with longitudinal relaxation; the ESR of many transition metal ions is only observable at temperatures sufficiently low to severely attenuate this relaxation, and these temperatures can be as low as 2K. Unpaired electrons in organic free radicals, on the other hand, are noted for being characterized by a g factor very close to 2, and with an associated spin lattice relaxation time as to be observable at room temperatue.

It has been known almost since the discovery of ESR that coals gave rise to strong ESR signals at room temperature with g near 2 (Jugram et al. (1954), Ubersfeld (1954), Austen et al. (1959) and references in Austen et al. (1967)); also see the reviews by Tschamlev and Ruster (1963), in "Coal" (VanKrevelen (1961), and by Retcofsky et al. (1978)). The number of unpaired electron spins per gram in vitrains varies between 10^{18} and 10^{20} g^{-1}. This number increases with carbon content between 55 and 90%C, and then sharply decreases above 95%C. The number of unpaired spins in fusains is constant at about 10^{19} g^{-1}. The spin concentration for vitrains and fusains is found to increase with temperature, but for resinites, increased temperature leads to decreased spin concentration (Petrakis and Grandy (1981)). Linewidths of ESR spectra of free radicals in the liquid state can be very small relative to 10^{-4} T. These lines frequently show a rich nuclear hyperfine structure from which, e.g. spin densities in aromatic rings, and the presence of heteroatoms may be inferred. Linewidths of ESR signals in coals, however, are frequently between 2 x 10^{-4} T (for fusinites) and 8.6 x 10^{-4} T for vitrinites and exinites. It has been inferred from the g values, (Petrakis and Gandy (1978), Retcofsky et al. (1978)) relaxation parameters, and increase in electron spin concentration with heating that the dominant source of unpaired electron spin in coals is organic free radicals. The dominant source of broadening is said to be nuclear hyperfine interactions (Retcofsky et al. (1975)). The face that the spin concentration increases with "coalification" (i.e., generally with carbon content) has been taken to provide evidence of increased formation of polynuclear aromatic rings with increased "coalification". Electron g values, decrease from 2.0033 for brown coal and peat, to 2.0027 for medium volitile bituminous coals through anthracite. This variation is taken as evidence that in the younger coals there is spin orbit coupling to heteroatoms such as oxygen (Retcofsky et al. (1968)). As coalification proceeds, decarboxylation is known to occur, and the lowered g values are taken as evidence for decreased contact between electron and heteroatom in polynuclear aromatic rings.

More recent evidence that the identity of free radicals in coals is associated with the hydrocarbon portion of the coal has been provided by the technique of electron-nuclear double resonance (ENDOR). Schlick et al. (1978) have determined, from ENDOR signals on a coal, and a solvent refined derivative coal, that the closest protons to the observed unpaired electron are at a distance of 2.6±0.2 Å.

In summary, ESR and ENDOR have provided qualitative inferential evidence for polynuclear aromatic structures in coals via relaxation times, trends in g values in comparison with model polynuclear aromatic structures containing heteroatoms, spin concentrations as a function of temperatures and nuclear-electron contact distances.

IX. NUCLEAR MAGNETIC RESONANCE (NMR)

A. Aromaticity

NMR is the resonant absorption of radio frequency (5-500 MHz) radiation by nuclei with nonzero spin in the presence of a static magnetic field. Since there are isotopes of hydrogen, carbon, oxygen, and nitrogen possessing nonzero nuclear spin, coal is in principle a rich source of information via NMR. Because isotopic abundances are well known, and transition probabilities, at least for spin 1/2 nuclei, may be arranged to be essentially unity, the intensity of the NMR signal may be used, even in solid samples, as a quantitative analysis for the nuclear species under measurement (Gerstein and Pembleton (1977)). In the liquid state with sufficiently homogeneous static magnetic fields, it is possible to achieve NMR spectral linewidths of .01 parts per million (PPM). Molecules in many liquids are tumbling sufficiently rapidly that the symmetry of the electronic charge cloud about nuclei in such molecules is effectively spherical. The nuclei in such molecules experience two major interactions. The first is the static magnetic field supplied by the laboratory magnet which is produced by circulating currents in the magnet coils. The second may be viewed as the effective magnetic field produced by the interaction between the circulating electronic charge cloud about a nucleus in question and the static field. The latter is called the "shielding" field, and leads to shifts in NMR frequencies associated with local bonding environments which are not larger than 15 ppm for protons, and 200 ppm for e.g., ^{13}C in molecules in the liquid state. These "chemical shifts" for ^{1}H and ^{13}C have been tabulated for an enormous number of molecules, and form a broad data base upon which the practicing chemist working with solution species draws upon almost daily. In solids, on the other hand, NMR spectra can be as broad as 600 ppm with few features to aid the chemist. For spin 1/2 nuclei, the principal source of this broadening is nuclear dipolar interactions, which are averaged to zero by isotropic motion in a liquid. In well crystallized materials, where there are relatively few unique internuclear vectors within unit cell, the width of this broad resonance absorption can be used to validate a proposed crystal structure via the second moment of the absorption line (Van Vleck (1948)). If the width of the absorption line is dominated by dipolar interactions (which depend upon the inverse cube of internuclear distances) than the second moment is a reflection of the inverse sixth power of internuclear distances. Attempts have been made to infer average proton-proton distances, and thus aromativity from average carbon-carbon distances thus inferred (Newman et al. (1955), Smidt et al. (1959), Oth and Tschamler (1963)). Initial efforts in this direction utilizing ^{1}H NMR spectra of solid coals did not appear sharply diagnostic because of questions regarding sample purity, and, in the authors' opinions, because of lack of knowledge of the contribution of intermolecular interactions in materials as heterogeneous and randomly ordered as coals. Thus, prior to the use of transient pulse techniques in NMR to attenuate proton-proton nuclear dipolar broadening in coals

(Gerstein et al. (1977), Gerstein (1979)), the major source of information about coals from NMR of ^1H was obtained from inferences of liquified products. An important example of this type effort is the work by Brown et al. (1960), and Brown and Ladner (1960). In the latter work the high resolution NMR of chloroform and methylene chloride soluble products of distillates of coals from low temperature (~ 380°C) vacuum pyrolysis was reported. The coals studied ranged in carbon content from 77.7 to 90.2. The distillates represented roughly 10% of the solid coal, with H/C ratios generally higher than in the coal. The proton NMR spectra of such distillates exhibited a broad, clearly defined peak about 2.5 ppm in width with little observable structure in the region where "aromatic", olefinic, acetylinic, and hydroxyl protons absorb. There is also a well defined "aliphatic" region with three discernable overlapping peaks. The broad peak in the "aromatic" region was associated with phenolic OH and aromatic CH. The low field peak in the aliphatic region (at +2.8 ppm relative to H_2O) was associated with hydrogen on carbon atoms α to aromatic rings (and to olefinic linkages and carbonyl groups if present). The upfield aromatic peak at 3.8 ppm relative to water was associated with "other" saturated hydrogens. These data indicated that "with increasing rank of the parent coal, the fraction of aromatic hydrogen increased regularly". Further, more detailed information about the chemical structure of the distillate was inferred utilizing information about (H/C) in the non-aromatic fraction of the coal as indicated in Section VII. A carbon aromaticity somewhat lower than that inferred for coals of similar carbon content was inferred for the distillation products, and ranged from 0.58 for the product from a 77.7% C coal to 0.78 for the product from a 90.2%C coal. The degree of aromatic ring substitution, σ_a, varied from 0.56 (77.7%C coal) to 0.31 (90.2%C coal). The value of $(H/C)_{ave}$, the hydrogen to carbon ratio on the unsubstituted unit model polynuclear aromatic ring, which decreases with increasing ring condensation (unity for benzene, 0.8 for naphthalene, etc.), varied from 0.86 (77.7%C coal) to 0.70 (90.2%C coal).

Retcofsky and Friedel (1970) have provided a summary of aromaticity values of coals as a function of carbon content, inferred from high resolution ^1H NMR on coal extracts with the use of the Brown-Ladner relation for f_{ar}^C (Eq. VII-10). It is important to recall, at this point, that the value of f_{ar}^C obtained from the Brown-Ladner relation is not a direct measurement of f_{ar}^C. It involves a determination of proton aromaticity, an assumption about the ratio of aliphat hydrogen to aliphatic carbon $(H/C)_{ali}$, and assumption about how aromatic ring structures are connected. A direct measure of f_{ar}^C in coal derived liquids is available via ^{13}C NMR, since the aliphatic and aromatic regions of the ^{13}C spectra are well separated, and under high resolution conditions in the liquid state, it is possible to identify and account for the contribution of the carboxyl carbon to the aromatic region. Retcofsky and Friedel (1970) have reviewed ^{13}C chemical shifts in selected compounds which might model structure in coals. Retcofsky et al. (1977) have also provided us with one of the first comparisons of

f_{ar}^{C} determined via ^{13}C NMR and via proton NNR and the Brown-Ladner relation on coal derived liquids. It was found that the Brown-Ladner relation yielded values of f_{ar}^{C} which were in excellent statistical agreement with that obtained directly via ^{13}C NMR. The implication of these results is that the values of 2 for x and y in Eq. VII-10 is reasonable for many (but not necessarily all) coal derived liquids and perhaps also for the native coals. Illustrative examples of the use of both ^{1}H and ^{13}C NMR to characterize aromaticity and hypothetical average structures of petroleum derived liquids have also been provided by O'Donnell et al. (1980) and Dickensen (1980), and in coal derived asphaltenes by Dereppe et al. (1978) and by Tokono et al. (1978).

A problem continually present in the use of information from coal extracts to guess aromaticity parameters in coals is that of the similarity of the extract and the native coal. It would be highly desirable to have the value of f_{ar}^{C} from the solid coal itself. Following the development of enhancing the NMR spectra of ^{13}C in solids by proton cross polarization and strong proton decoupling, with rare spin observation (Pines et al. (1972)), it indeed became possible to observe relatively narrow lines for ^{13}C in solid coals (VanderHart and Retcofsky (1976), Retcofsky and VanderHart (1978)). The dominant broadening interaction on these ^{13}C spectra of whole coals was the anisotropy of the molecular electronic environment about a given carbon nucleus, i.e., the "shielding anisotropy" (Mehring (1976)). This anisotropy for a given carbon can be as large as 300 ppm, i.e., as large as the total range of isotropic chemical shifts observed for carbon in the liquid state. Nevertheless, the spectrum exhibited for ^{13}C in a solid coal under strong proton decoupling exhibits a two peak structure which could be associated with "aliphatic" and "aromatic" (and unsaturated) regions of coals and from which f_{ar}^{C} could be estimated. Subsequently, the addition of magic angle spinning to remove shielding anisotropies (Andrew (1971)) was added to the experiment Bartuska et al. (1977)). Similar experiments were described by Maciel et al. (1979), Zilm et al. (1979) and Gerstein et al. (1979). In all of these "high resolution" NMR spectra of ^{13}C in coals, there is a region downfield about 120 ppm from TMS, and about 50-70 ppm in width. This region is associated with aromatic, carboxyl, carbonyl, olefinic and acetylenic carbon. Cleanly separated from the "aromatic" region is a peak about 30 ppm downfield of TMS, and roughly 50 ppm wide associated with aliphatic carbon (Fig. IX-1). These spectra, while providing a clean separation between aliphatic and "aromatic" regions, do not necessarily provide a quantitative measure of f_{ar}^{C}. While the conditions for the combined use of cross polarization, strong proton decoupling, and magic angle spinning (CP-SD-MAS) to obtain quantitative information in cleanly defined model compounds are well understood, there is some disagreement about whether those conditions are satisfied in the application of such combined procedures to coals. These conditions have been discussed by VanderHart and Retcofsky (1976) and Gerstein (1979). It is the authors' view, from spin counting of ^{13}C under single pulse excitation of ^{13}C with strong decoupling of ^{1}H, that all ^{13}C spins may not be seen in coals under "high resolution" conditions, i.e., conditions of CP-SD-MAS. At least two possibilities exist if not all carbons are detected. One is that those carbons which are detected are representative of those which are not, such that the values of f_{ar}^{c} inferred from ^{13}C NMR are valid. This might be the case, for example, if those carbons not detected are in or near polynuclear aromatic ring structures with unpaired electrons, i.e., free radicals. In this case, the longitudinal relaxation time of many carbon spins could be sufficiently short such that lifetime broadening would prevent detection under high resolution conditions. Another possibility is that the longitudinal relaxation times of undetected carbons (if such exist) is too long for detection under the usual experimental conditions of signal averaging to improve the ^{13}C signal to noise ratio. Such might be the case if there are extensive regions of diamond-like adamantane structures in coals. It is the authors' belief that NMR spectra of ^{13}C in coals are quantitatively representative of carbon in coals, and it is with this belief, and caveat, with which we shall proceed.

The high resolution on solid state NMR spectra of ^{13}C has been determined in thirteen coals (courtesy of P. Given, Penn. State; H. Retcofsky, P.E.T.C.; and D. Biggs, Iowa State University) in the author's research group. Major constituent analyses and f_{ar}^{c} of these coals is given in Table IX-1.

Table IX-1. Elemental Analysis of Aromatic Fractions of the 13 Coals Studied in this Research

Rank	PSOC[a] #	%C[b]	%H	%O	f_{ar}^{c} ±0.03
Anthracite	871	95.2	2.1	1.5	0.97
Low vol. bit.	782	90.2	4.3	3.0	0.86
Medium vol. bit. (mvb)	801	85.0	5.2	3.0	0.79
High vol. A bit. (hvAb)	821	83.3	5.7	6.9	0.70
High vol. A bit (hvAb)	808	82.8	5.6	6.4	0.73
High vol. B bit. (hvBb)	854	81.0	5.9	11.1	0.63
High vol. A bit. (hvAb)	739	80.9	5.6	8.9	0.70
High vol. B bit. (hvBb)	773	79.1	5.5	11.0	0.75
High vol. C bit. (hvCb)	548	77.9	5.2	15.0	0.64
High vol. C bit. (hvCb)	866	75.2	5.3	17.7	0.64
Subbituminous C	785	71.9	5.3	20.4	0.54
Lignite A	791	70.4	5.5	21.9	0.55
Lignite	833	69.4	5.1	23.8	0.68

[a]Pennsylvania State Office of Coals.

[b]All samples dry, ash free (daf). The remainder of the percentage includes N and S.

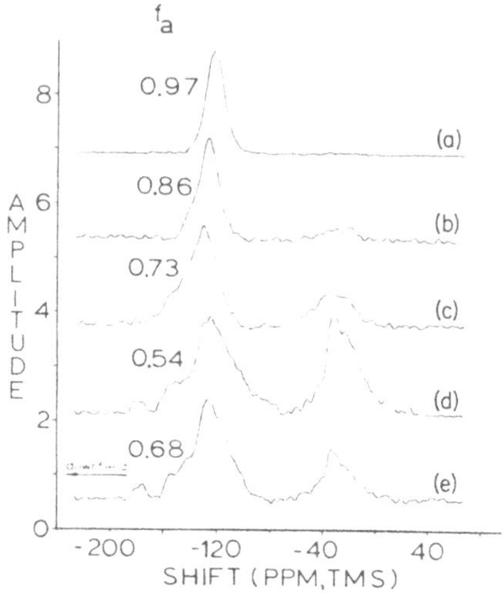

Fig. IX-1. High Resolution Solid State NMR of ^{13}C in Whole Coals as a Function of Rank

In addition to the study of carbon aromaticity reported above, the authors have studied the high resolution ^{13}C and ^{1}H NMR of eighteen coals ranging in carbon content from 75%-95%. Carbon spectra of representative samples are shown in Figure IX-1. Carbon aromaticities are indicated in Table IX-2.

Table IX-2. Carbon Aromaticity Values of (f^c_{ar}) in Coals Determined by Solid State ^{13}C NMR

Coal	% C	f^c_{ar} (NMR)
Anthracite	93.0	0.95
Pocahontas #4	90.3	0.86
PSOC 268	86.5	0.76
Powellton	85.1	0.77
PSOC 124	83.8	0.54
PSOC 351	83.5	0.77
PSOC 501	83.4	0.78
PSOC 103	82.9	0.75
PSOC 640	82.7	0.79
PSOC 330	82.0	0.74
PSOC 170	82.0	0.78
Bumik 40660	81.3	0.77
Upper Mich	81.0	0.70
PSOC 212	79.4	0.77
PSOC 155	77.9	0.38
Star	77.0	0.71

Unlike proton NMR spectra obtained under high resolution conditions in coal derived liquids (Brown and Ladner (1960)), the high resolution NMR spectra of protons in solid coals do not exhibit well resolved aromatic and aliphatic regions. Instead, all spectra characteristically show a structure which can be fitted with a superposition of two Lorentzian peaks. Characteristic spectra of protons in solid coals under high resolution conditions at a static field of 1.3T are shown for two coals in Figure IX-2.

Table IX-3 reports the values of f_{ar}^{h} determined from these data. It is observed that f_{ar}^{h} shows considerable scatter about a value of approximately 0.3 up to about 85 percent carbon, at which point it rises rapidly.

Using the values in Table IX-4 and results for the carbon aromaticity in the same series of coals measured by ^{13}C NMR, it is possible to calculate the mole ratios of aromatic hydrogen to aromatic carbon (H_{ar}/C_{ar}) in the coals. These values are reported in Table IX-4.

Table IX-3. Hydrogen Aromaticity Values (f_{ar}^{h}) in Coals Determined by Solid State ^{1}H NMR

COAL	% C	f_{ar}^{h} (NMR)
Anthracite	93.0	0.68
Pocahontas #4	90.3	0.53
PSOC 268	86.5	0.33
Powellton	85.1	0.43
PSOC 124	83.8	0.18
PSOC 351	83.5	0.34
PSOC 501	83.4	0.33
PSOC 103	82.9	0.39
PSOC 640	82.7	0.32
PSOC 330	82.0	0.33
PSOC 170	82.0	0.36
Bumik 40660	81.3	0.30
Upper Mich	81.0	0.31
PSOC 212	79.4	0.31
PSOC 155	77.9	0.16
Star	77.0	0.33
PSOC 308	76.6	0.34
Lovilia	75.0	0.29

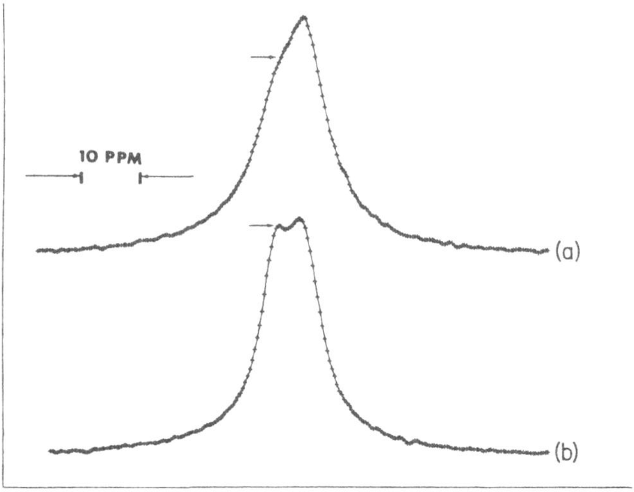

Fig. IX-2. High Resolution Proton NMR Spectra of Two Coals Taken Under Conditions of Strong Proton Decoupling and Magic Angle Spinning

From the major constituent analyses, and ^{1}H and ^{13}C NMR, and material balance, it is possible to calculate the H/C ratio in the aliphatic portion of the coals, givenin Table IX-4. Finally, on the basis of the following data, which we now have at hand: i) the weight percent of carbon and hydrogen in the coals, ii) f_{ar}, f_{ar}, f_{oh}, and iii) h_{ar}/C_{ar}, it is possible to place some limits on the average size of the polynuclear aromatic clusters in the coals (values of f_{oh}^{h} were supplied by P. R. Solomon from FTIR results on these coals). In particular, it is at least possible to determine the minimum allowable ring size consistent with the above information in view of the lack of information about α proton concentrations. These numbers are reported in the last column of Table IX-4.

Table IV-4. Structural parameters of coals

COAL	% C	H_{ar}/C_{ar}	H_{al}/C_{al}	Minimum Avr. Aromatic Cluster Size
Anthracite	93.0	0.23	2.06	30
Pocohontas #4	90.3	0.35	1.91	6
PSOC 268	86.5	0.36	1.75	3
Powellton	85.1	0.36	1.38	3
PSOC 124	83.8	0.31	1.69	1
PSOC 351	83.5	0.36	2.42	3
PSOC 501	83.4	0.32	2.31	3

PSOC 103	82.9	0.38	1.59	3
PSOC 640	82.7	0.31	2.34	3
PSOC 330	82.0	0.33	1.74	3
PSOC 170	82.0	0.34	2.14	3
Bumik 40660	81.3	0.35	2.11	3
Upper Mich	81.0	0.34	1.45	2
PSOC 212	79.4	0.31	1.91	3
PSOC 155	77.9	0.42	1.32	1
Star	77.0	0.34	1.89	2
PSOC 308	76.6	0.36	1.78	2
Lovilia	75.0	0.33	1.48	2

In order to effect this latter calculation, the following procedure was employed. Peri condensed atomic ring structures connected by aliphatic chains with average stoichiometry CH_x ($x = H_{al}/C_{al}$) were assumed as skeletal models for coal structure. Starting with benzene, these were increased in size until the minimum ring size consistent with the known data was determined.

From the number of aromatic carbon atoms, C_{ar}, in any model ring, the number of aliphatic carbon atoms, C_{al}, required in the model for consistency with a given coal was determined from the relation $C_{al} = \{\frac{1}{f^c_{ar}} - 1\}C_{ar}$, where f_{ar} is the NMR determined fraction of aromatic carbon in the coal. Next, the total molecular weight required of the model average molecular structure could be calculated according to M.W. = [(C_{ar} + C_{al}) x 12/wt. %C] x 100, with wt. %C being the weight percent of carbon in the coal. From the molecular weight for the average molecular structure, it is possible to determine the total number of hydrogen atoms required in the model from the relation H_{total} = M.W. x wt. %H/100. This total hydrogen could be further divided according to chemical types (H_{ar}, H_{al} and H_{oh}) on the basis of the average of the NMR and FTIR hydrogen results. The values generated in this manner are consistent with what is known about the coals. For example, for Lovilia vitrain, if the average ring size is presumed to be unity, the average molecular structure for the aromatic portion will be represented by benzene, which has six aromatic carbon atoms. The number of aliphatic carbons in the model, C_{al}, is then calculated to be C_{al} = (1/0.69 - 1)6 = 2.7, and the molecular weight is (8.7 x 12/75) x 100 = 139.2. The total number of hydrogen atoms is given by 139.2 x 4.77/100 = 6.6, and of those H_{ar} = 0.30 x 6.6 = 2.0, H_{al} = 0.60 x 6.6 = 4.0, and H_{oh} = 0.10 x 6.6 = 0.6.

At this point, the model average molecular structure may be fully constructed. The minimum possible ring size (minimum average polynuclear aromatic cluster size in the coal) is obtained if all aliphatic carbon is alpha to aromatic rings; i.e., only peripheral CH_3 groups and alkyl or naphthenic bridges with two carbon atoms were considered. To check for self-consistency, the H_{ar}/C_{ar} ratio of the model is then compared with that of the coal, as inferred by NMR. For example, for one ring on average in Lovilia vitrain, the average repeating unit may be represented by

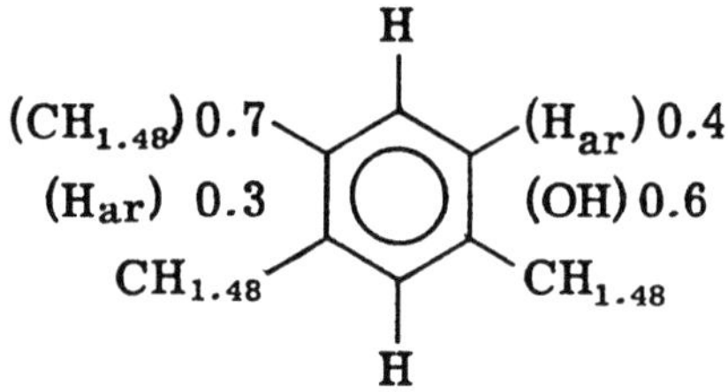

$C^{ar}_{6.0}C^{al}_{2.7}H^{ar}_{2.0}H^{al}_{4.0}(OH)_{0.6}$, or as the unit and $(H_{ar}/C_{ar})_{model} = 0.45$. This value is greater than the experimentally determined 0.33 for Lovilia, and, therefore, the average cluster size in that coal must be greater than 1.

The process described above is continued, employing progressively larger aromatic backbones, until $(H_{ar}/C_{ar})_{model} < (H_{ar}/C_{ar})_{coal}$. Values for the model which are less than that of the coal can be reconciled by considering some aliphatic bridges to be greater in length than two carbon atoms. The minimum possible cluster size becomes the smallest polyaromatic ring structure for which the above inequality is satisfied.

Table IX-4 lists the results of this type of analysis for the 18 coals examined in this work. While it is not possible to conclude that the condensed aromatic clusters indicated are necessarily the actual average ring sizes in the coals, it is interesting to observe that most of the coals have smallest possible ring sizes of at least two, and many are three or more. Further, the aromatic clusters appear to be smallest for coals of less than about 85 percent carbon. Above 85%C the size of the clusters increases significantly, with coal becoming nearly graphitic by 93% carbon.

Also interesting is the fact that the ring sizes shown in Table IX-4 are in generally good agreement with those determined by other methods despite the assumption of only relatively short aliphatic attachments to the aromatic backbones, and our ignoring of heteroatom functionalities except for OH. As has been previously mentioned Deno et al. (1978) have been able to study selectivity the aliphatic portion of coals by oxidation with aqueous hydrogen peroxide in trifluoroacetic acid. Those workers have determined that much of the aliphatic material in coals is present in relatively short structures, including particularly $ArCH_2CH_2Ar$.

Further, oxygen existing in coals is for the greater part considered to occur mostly in hydroxyl groups, ether groups, and heteroatom functionalities, while nitrogen and sulfur are relatively minor components of coal. This explains how the H_{ar}/C_{ar} ratio and the average cluster sizes in coals may be nearly equal to that of the hydrocarbon with identical average molecular structure, except with the heteroatoms eliminated (i.e., our average structural model).

In the previous discussions, we have not specified a particular maceral in determining f_{ar} or size of PNA rings. Very recently, Maciel et al. (1982) have measured ^{13}C NMR in a number of macerals using standard high resolution solid state NMR. These results are given in Table IX-5, and confirm VanKrevelen's earlier inferences that fusinite is almost completely aromatic, resinite is highly aliphatic, and the other macerals are greater than 2/3 aromatic.

Table IX-5. ^{13}C CP/MAS Results on Coal Macerals

Maceral	f_a'	$\Delta\nu_a^a$, ppm*
Fusinite		
**Illinois #6	0.91	24.6
(Hernshaw)	0.92	24.6
Vitrinite		
(FTC-50)	0.78	26.2
(Pitt. #8 seam)	0.81	28.9
(Hernshaw)	0.78	33.1
Micronite (Hernshaw)	0.78	23.1
Attrites Spore rich (Hernshaw)	0.73	22.3
Exinite (Hernshaw)	0.67	22.8
Resinite FTC 30	0.15	20.0

*Width at half height of the aromatic band, ± 5 ppm.

**This sample was provided by Dr. Jack Crelling, Southern Illinois University, Carbondale, Illinois.

B. Beyond "Just Aromaticity" Chemical Functionality

Thus far in this section on the use of NMR to infer aromaticity, the results on solid coals, with appropriate caveats about olefin content and quantitative character of the measurements, have been used to estimate f_{ar}^c and average PNA ring size with carbon content. It is possible to do a bit better than "just" f_{ar}^c from carbon NMR on solids utilizing a suggestion by Alla and Lippmaa (1976) which was subsequently followed by work of Opella et al. (1979), and by the present authors

(Murphy, Cassady and Gerstein (1982) and Murphy, Gerstein, Weinberg and Yen (1982)). The basic behind the experiments is that the transverse relaxation time of a ^{13}C directly bonded to a hydrogen, e.g., in aliphatic methylene or methyne groups or an aromatic ring hydrogen on a perimeter aromatic ring carbon, is about 12 μsec. The transverse relaxation time of carbons not directly bonded to hydrogens (e.g., quanternary aliphatic carbons, carboxyl and carbonyl carbons, and carbons in the interior of aromatic rings) is of order of 200 μsec. Carbons in methyl groups are a special case because of motion, and have T_2^*'s of roughly 100 μsec under conditions of magic angle spinning. The differences in these T_2^*'s in the absence of strong proton decoupling can be used to allow the carbons with shorter T_2^* to relax before starting proton decoupling and observation of carbons with larger T_2^*. The result is that, with appropriate corrections for the decay of signals from the zero of time, the signals from the two different types of carbons may be separated. Figure IX-3 shows the result of such a dipolar dephasing experiment on an anthracite coal. The interpretation of the difference spectrum (bottom) between the spectrum of the parent coal (top spectrum) and that obtained using a dipolar dephasing time of 60 μsec (center spectrum) is simple in this case. Since the anthracite is predominantly aromatic (f_{ar}^{c} = 0.95), the parent spectrum represents perimeter plus interior aromatic ring carbons. The depolar dephased spectrum (center) represents interior ring carbons in PNA rings (there is no evidence of carboxyl, carbonyl or phenotic functionality in this case). The difference spectrum represents the perimeter carbons. If a pericondensed ring system is assumed, and all carbons are detected, these data lead to a value of 33 for the number of condensed benzenoid rings. This compares

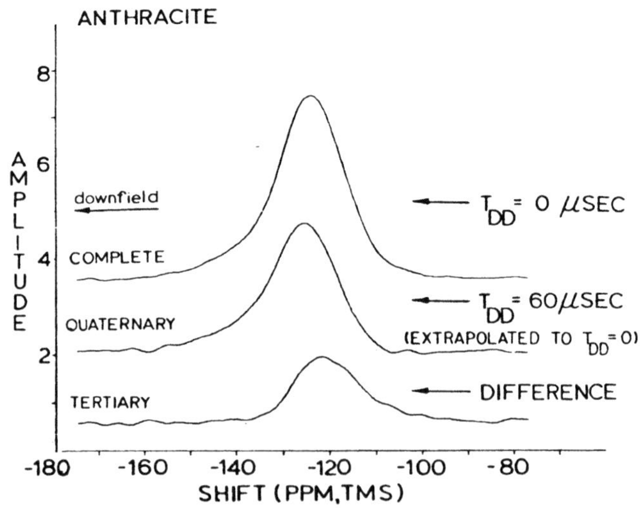

Fig. IX-3. Chemical Shifts Relative to TMS, for an Anthracite (top), the Interior Quaternary Aromatic Carbons (middle), and the Perimeter Carbons (bottom)

well with the value of 30 obtained earlier from ^{13}C and ^{1}H NMR on a solid anthracite, and is in agreement with estimates from X-ray diffraction on anthracites (Hirsch (1954)).

X. SUMMARY

With respect to the question of "aromaticity" in coals, we have seen that the primary criteria used to infer aromaticity have been the similarity of C-C bond distances, infrared spectra, NMR and ESR spectra, and selective chemical behaviour of coals and model "aromatic" compounds. Each of these views of coals vs. model compounds can be independently criticized. C-C bond distances in a disordered material such as coals may not at all yield a density similar to "regular" polymers such as polyethylene or polystyrene. Infrared spectra are notoriously nonquantitative. ESR spectra represent a very small portion of the sample, and may be associated with noncarbon centers. NMR spectra of extracts may not be representative of native coals. ^{13}C NMR spectra of solid coals may not detect all carbons. Coals are such a complicated mix of chemical functionality that standard organic reactions probing functionality may yield anomalous results because of competing reactions. In spite of these criticisms, we have seen that from each criterion used to judge, or estimate aromaticity in coals, a remarkably consistent picture arises. Densities of macerals are those estimated for aromatic polymers. Infrared spectra of solid coals clearly show major absorptions associated with aromatic C-H ring bending modes. ESR spectra support the presence of aromatic ring systems capable of stabilizing free radicals. Similarities of NMR and IR spectra of coals and extracts indicate native coals to contain appreciable aromatic ring systems. Carbon and proton NMR spectra of solid coals clearly indicate major contributions from aromatic ring systems. In anthracite in particular, there is good agreement between ring condensation index inferred from X-ray diffraction and dipolar dephasing difference experiments with ^{13}C NMR. Coals are consistently estimated to range from 60% aromatic for 70% carbon coals, to 90% aromatic for 95% carbon coals. Selective oxidative degradations, both of aromatic structures and of aliphatic structures are consistent with a picture of coal as a structure containing a few condensed rings attached by short aliphatic chains.

In 1961, VanKrevelen concluded that "the average structural unit in coals from the lignitic stage to the low volitile bituminour stage contain about 20 C atoms and about 4-5 rings. The aromaticity in this range increases from 0.7 to 0.9. In the anthracite stage the size of the clusters, and hence the number of rings per clusters, increases rapidly; the aromaticity goes up at the same time from 0.9 to 1".

In the work subsequent to 1961, there has been presented no strong evidence to oppose these conclusions, and much evidence to reinforce them. We now have a somewhat better idea of the number of <u>aromatic</u>

rings condensed to form an average structural unit, and feel somewhat firmer about the values of f^c_{ar} from solid state NMR. We also have developed some tools with which to move delicately, and perhaps more quantitatively probe chemical functionality in coals than existed in 1961. The basic picture of coal as a highly aromatic material stands firm.

ACKNOWLEDGEMENT

The author's work is supported by the Office of Basic Energy Sciences, Chemical Sciences Division, of the U.S. Department of Energy; Contract Number W-7405 ENG-82.

REFERENCES

Aczel, T., Gorbaty, M. L., Maa, P. S., and Schlosberg, R. H. (1975) Fuel, 54, 295.

Alla, M., and Lippmaa, E. (1976) Chem. Phys. Lett. 37, 260-264.

Andrew, E. R. (1971) "Progress in NMR Spectroscopy; the Narrowing of NMR Spectra of Solids by High-Speed Specimen Rotation, and the Resolution of Chemical Shift and Spin Multiplet Structures of Solids", Vol. 8, Pergamon, Oxford.

Austen, D.E.G., Ingram, D.J.E., Given, P. H., Binder, C. R., and Hill, L. W. (1967) in "Coal Science", Elsevier, (Amsterdam), pp. 344-362.

Badger, G. M. (1969) "Aromatic Character and Aromaticity", Cambridge University Press.

Bartuska, V. J., Maciel, G. E., Schaefer, J. and Stejskal, E. O. (1977) Fuel, 56, 354-358.

Bergman, E. D. and Pullman, B. (1971) in Vol. III "Aromaticity, Pseudoaromaticity, and Antiaromaticity" (Proceedings of an International Symposium held in Jerusalem, 31 March - 3 April 1970). The Israel Academy of Sciences and Humanities.

Berkowitz, N. (1979) "An Introduction to Coal Technology", Academic Press, N.Y.

Brown, J. K., Ladner, W. R. and Sheppard, N. (1960) Fuel, 39, 79-86.

Brown, J. K. and Ladner, W. R. (1960) Fuel 39, 87-96.

Brown, J. K., Dryden, I.G.C., Dunevein, D. H., Joy, W. K. and Parkhurst, K. S. (1958) J. Inst. Fuel 31, 259-273.

Cannon, C. G. (1953) Nature 171, 308.

Charkabartty, S. K. and Kretschmer, H. O. (1974) Fuel 53, 132-135.

Clar, E. (1964) "Polycyclic Hydrocarbons" Vol. I, Academic Press.

Coulsen, C. A. and Louguet-Higgins (1948/49) Proc. Roy. Soc. A195, 188-197.

Craig, D. P. (1959) in "Non Benzenoid Aromatic Hydrocarbons", 1-42 Ed. D. Ginsberg, Interscience.

Crawford, M. Mercedes-Silva, Gerstein, B. C., Kuo, An-L. and Wade, C. G. (1980) J. Am. Chem. Soc. 102, 3728-3732.

Dauben, H. J., Jr., Wilson, J. D. and Laity, J. C. (1969) J. Am. Chem. Soc. 90, 811-813.
Deno, N., Greigger, B. A. and Stroud, S. G. (1978) Fuel 57, 455-459.
Deno, N. C., Curry, K. W., Jones, Daniel A., Keegan, K. R., Rakitsky, W. G., Richter, C. A. and Minard, R. D. (1981) Fuel 60, 210-212.
Dereppe, J. M., Moreaux, J. M. and Castex, H. (1978) Fuel 57, 435-441.
Dewar, M.J.S. (1952) J. Am. Chem. Soc. 74, 3341-3363.
Diamond, R. (1958) Acta Cryst. 11, 129-138.
Dickensen, E. M. (1980) Fuel 59, 290-294.
Ergun, S. and Tiensuu, V. H. (1959) Acta Cryst. 12, 1050-1051.
Franklin, R. E. (1950) Acta Cryst. 3, 107-121.
Garratt, P. J. (1971) "Aromaticity", McGraw-Hill.
Gerstein, B. C., Ryan, L. and Murphy, P. DuBois (1979) in Am. Chem. Div. of Fuel Preprints 24, 90-95.
Gerstein, B. C. (1979) "Fingerprinting Solid Coals Using Pulse and Multiple Pulse NMR", Chapter 51, Vol. III, in "Analytical Methods for Coal and Coal Products", Ed. Clarence Karr, Jr., Academic Press.
Gerstein, B. C., Chow, C., Pembleton, R. G. and Wilson, R. C. (1977) J. Phys. Chem. 81, 565-570.
Gerstein, B. C. and Pembleton, R. G. (1977) Anal. Chem. 49, 75-77.
Glen, R. A. (1955) Fuel 34, 201-212.
Gorbaty, M. and Ouchi, K., Eds. (1981) "Coal Structure", Advances in Chemistry Series 192, American Chemical Society, Washington, D.C.
Griffiths, P. R. (1975) "Chemical Infrared Fourier Transform Spectroscopy", John Wiley and Sons, New York.
Guttman, Viktor (1977) Chemtech, April 1977, 255-263.
Hadden, R. C. (1972) Tetrahedron, 3613-3633.
Hayatsu, R., Scott, R. G. and Winans, R. E. (1981) "Oxidation of Coal", in "Oxidation in Organic Chemistry", W. S. Trahanovsky, Ed., Academic Press.
Heredy, L. A., Kostyo, A. E. and Neuworth, M. B. (1957), in "Coal Science", Elsevier, Amsterdam; pp. 493-502.
Hirsch, P. B. (1954) Proc. Roy. Soc. (London) 226A, 143-169.
Hirsch, P. B. and Cartz, L. (1960) Phil. Trans. 252A, 557-602.
Hombach, Hans Peter (1980) Fuel 59, 465-470.
Huckel, E. (1931) Z. Physik 70, 204-286.
Huggins, M. L. (1943) Ann. N.Y. Acad. Sci. 44, 431-443.
Huston, J. L. and Studier, M. H. (1981) Fuel 60, 1149-1152.
Ingram, D.J.E., Tapely, G. G., Jackson, R., Bend, R. L. and Murhaghan, A. R. (1954) Nature 174, 797-798.
Julg, A. and Francois, P. (1967) Theor. Chim. Acta (Berlin) 7, 249-259.
Kekulé, A. (1869) Ber. dtsch. Chem. Ges. 2, 362-365.
Larsen, J. W. (1975) NSF Workshop on the Fundamental Organic Chemistry of Coal, U. Tenn., Knoxville, TN.
Maciel, G. E., Sullivan, M. W., Petrakis, Leon and Grandy, D. W. (1982) Fuel (in press).
Maciel, G. E., Bartuska, V. J. and Miknis F. P. (1979) Fuel 58, 391-394.
Maekawa, Y., Yoshida, Tadashi and Yoshida, Yuji (1979) Fuel 58, 864-872.
Marzec, A., Juzwa, M., Betky, K. and Sobkowiak, M. (1979) Fuel Processing Technology 2, 35-44.

Mazdumar, B. K., Ganguly, S., Sauyal, P. K. and Lahivi, A. (1957) in "Coal Science", Elsevier, Amsterdam; pp. 475-492.
Mehring, M. (1976) "NMR Basic Principles and Progress; High Resolution NMR Spectroscopy in Solids", Vol. 11, Springer, New York.
Murphy, P. DuBois, Cassady, T. J. and Gerstein, B. C. (1982) Fuel 61, 1233.
Murphy, P. DuBois, Cassady, T. J., Trahanovsky, W. S. and Given, P. H. (1982) "NMR Spectra of 13-C in Solid Coals; First and Second Moments as a Function of Rank", Fuel (submitted).
Newman, P. C., Pratt, L., Richards, R. E. (1955) Nature 175, 645.
O'Donnel, D. J., Sigle, S. O., Berlin, K. D., Sturm, G. P. and Vogh, J. W. (1980) Fuel 59, 166-174.
Opella, S. J. and Frey, M. H. (1979) J. Am. Chem. Soc. 101, 5854-5856.
Oth, J.F.M. and Tschamler, H. (1963) Fuel 42, 467-468.
Painter, P. C. and Coleman, M. M. (1979) Fuel 58, 301-308.
Peters, D. (1960) J. Chem. Soc., 1274-1279.
Petrakis, L. and Grandy, D. W. (1981) Fuel 60, 115-119.
Petrakis, L. and Grandy, D. W. (1978) Anal. Chem. 50, 303-308.
Pople, J. A. and Unch, K. G. (1966) J. Am. Chem. Soc. 88, 4811-4815.
Radding, S. B. and Peters, H. M. (1978), "Proceedings of the 1978 Coal Chemistry Workshop", March 8-10, 1978, Stanford Research Institute, Menlo Park, CA.
Randić, M. and Gerstein, B. (1980) J. Mag. Res. 43, 207-212.
Retcofsky, H. L., Thompson, Gordon P., Raymond, Raphael and Friedel, Robert A. (1975) Fuel 54, 126-128.
Retcofsky, H. L., Stark, J. M. and Friedel, R. A. (1968) Anal. Chem. 40, 1699-1704.
Retcofsky, H. L., Schweighardt, F. K. and Haugh, M. (1977) Anal. Chem. 49, 585-588.
Retcofsky, H. L. and Friedel, R. A. (1970) in "Spectroscopy of Fuels", Plenum Press; Chapter 6, pp. 70-89.
Retcofsky, H. L. (1977) Applied Spectroscopy 31, 116-121.
Retcofsky, H. L. and Friedel, R. A. (1970) in "Spectrometry in Fuels", Ed., Friedel, R. A., Planum Press, NY; Chapter 7.
Robinson, R. (1958) Tetrahedron 3, 323-324.
Roy, M. M. (1957) Fuel 36, 249-252.
Schlick, S., Narayana, P. A. and Kevan, L. (1978) J. Am. Chem. Soc. 100, 3322-3326.
Smidt, J., Van Raayen, W. and VanKrevelen, D. W. (1959) Fuel 38, 539-540.
Solomon, P. R. (1981) Fuel 60, 3-6.
Solomon, P. R. (1979) A.C.S. Divison of Fuel Chemistry Preprints 24, #2, 184-195.
Streitweiser, A., Jr. (1961) "Molecular Orbital Theory for Organic Chemists", Wiley, New York, and references therein.
Trindle, C. (1969) J. Am. Chem. Soc. 91, 219-220.
Tingley, G. L. and Morrey, J. R. (circa 1973) "Coal Structure and Reactivity", a Battelle Energy Program Report, Battelle Pacific Northwest Laboratories, Richland, WA.

Tschamler, H. and DeRuiter, E. (1963) in "Chemistry of Coal Utilization", Suppl. Vol., H. H. Lowry, Ed., Wiley, NY.
Ubersfeld, J. (1954) J. Phys. Rad. 15, 126-127.
VanderHart, D. L. and Retcofsky, H. L. (1976) Fuel 55, 202-204.
VanKrevelen, D. W. (1961) "Coal", Elsevier, Amsterdam.
VanKrevelen, D. W. and Schuyer, J. (1957) in "Coal Science", Elsevier, Amsterdam.
VanKrevelen, D. W. and Schuyer, J. (1957) "Coal Science" Elsevier, Amsterdam.
Van Vleck, J. H. (1948) Phys. Rev. 74, 1168-1183.
Van Vucht, H. A., Rietveld, B. J., VanKrevelen, D. W. (1955) Fuel 34, 50-59.
Vincow, G. (1971) in "Aromaticity, Pseudoaromaticity and Antiaromaticity", Vol. III, Proceedings of an International Symposium held in Jerusalem, 31 March-3 April (1970), The Israel Academy of Sciences and Humanities.
Waugh, J. S. and Fessendon, R. W. (1957) J. Am. Chem. Soc. 79, 846-849.
Weinberg, V. and Yen, TenFu (1980) Fuel 59, 287-289.
Wexler, A. S. (1970) in "Spectrometry in Fuels", Ed., Friedel, R. A., Plenum Press, NY; Chapter 17.
Whitehurst, D. D., Mitchell, T. O. and Farcasiu, M. (1980) "Coal Liquefaction", Academic Press, NY.
Wise, W. S. (1971) "Solvent Treatment of Coal", Mills and Boon Monograph CE/2, London.
Yokomo, T., Miyazawa, K. and Sanada, Y. (1978) Fuel 57, 555-558.
Zilm, K. W., Pugmire, R. J., Grant, D. M., Wood, R. E. and Wiser, W. H. (1979) Fuel 58, 11-16.

E.P.R. IN FOSSIL CARBONACEOUS MATERIALS

J. CONARD

C.R.S.O.C.I. - C.N.R.S., 45045 - Orléans Cedex, France

1. STABLE FREE RADICALS PRODUCED IN ORGANIC MATTER.

Electron paramagnetic resonance (EPR or ESR) is a technique quite convenient for the detection and study of free radicals i.e. polyatomic structures having unpaired electrons.

In 1954, both Ingram [16] and Uebersfeld [12] used EPR to demonstrate the presence of stable free radicals in coals. During the following years this same technique has also shown the presence of free radicals in all types of carbons, either natural or artificial, and in the products of pyrolysis or irradiation of organic matter. They were found in kerogens by Marchand et al [18], and it can be said as a general fact that they appear in all degradation processes of organic or biological substances.

The free radicals formed in a solid may be very long-lived : their mobility is low, so they have a vanishing probability to get together and to recombine. Moreover theoretical chemistry considerations show that aromatic free radicals have a remarkably high stability.

The production of free radicals during carbonization has been extensively studied since 1954. For a given carbonization time, the free radical concentration increases with temperature of pyrolysis up to a maximum and then decreases. Under laboratory conditions this maximum depends upon the nature of the organic matter, the treatment time, and the nature of the surrounding atmosphere and pressure but remains essentially between 500 and 700° C, and reaches values of 10^{19} to 10^{20} free radicals per gram [14].

But the same concentration of free radicals may be obtained using a short time pyrolysis with a higher temperature, or a long time treatment with a lower one. Considering the very long times

L. Petrakis and J. P. Fraissard (eds.), Magnetic Resonance. Introduction, Advanced Topics and Applications to Fossil Energy, 441–459.

typical for a geological process, it is natural to expect a high free radical content in sedimentaty organic matter, even if treated at relatively low temperature. Such is the case of natural coals, and kerogens, which, during their evolution, have not been submitted to temperatures above 200° C.

Fruitfull tentative studies have been done by Singer [31] by heating solution in very stable solvant as metaquinquaphenyl with the purpose to keep the formed radical in solution. Then hyperfine structure resulting from interaction with protons or other nuclei is revealed and helps a chemical analysis. But the recombination is necessarily very different from that would occur in the solid state. And then the identified free radicals cannot be undoubtely obtained in a solid material.

2. PURITY-HOMOGENEITY

Purifying natural carbon can be the most difficult part of the investigation by EPR. Paramagnetic minerals are very often mixed with them and superimpose their own signals (sometimes with a hyperfine structure) to the one under study. Fig. 1 shows how complicated the EPR lines of non-purified kerogens may be. Even in less unfavorable cases, the impurity induced perturbation is such that it may be very difficult to make an estimation of the intensity or linewidth (Fig. 2).

We shall not go here into the detail of purification methods (which are described in [25]). They make use of solvent-extractions, acid treatment ($HC\ell$, HF) and washing followed by centrifugation. They also quite often vary somewhat from one laboratory to the other, which may make comparisons difficult.

Total purification is not always possible : pyrite especially can be so closely imbedded that its separation is practically impossible. The resulting EPR signals have an intensity or a width which can indeed be measured (Fig. 2 shows two examples), but it is not possible to assign a clear meaning to the results of such measurements : they must be used with extreme caution.

However purification is generally quite effective, as illustrated by Figs. 1a (before purification) and 1b (after purification). But it must be emphasized that the treatment may change the free radicals of the kerogen. A systematic investigation of the influence of the purification treatment in kerogens was undertaken by Durand et al. [7]. Their findings show that it is most desirable to keep the treatment at the minimum level, or to make use of various methods in order to compare their results. In any case it may be very dangerous to draw conclusions from the study of a single sample.

It should also be kept in mind that a natural sample, even one of very small dimensions (EPR utilizes only a few mg) is very seldom homogeneous : it is the product of numerous and diverse biological elements which may have had also, different pregeological histories.

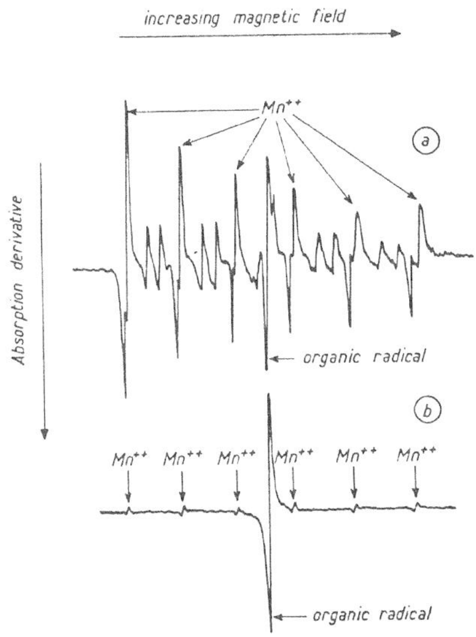

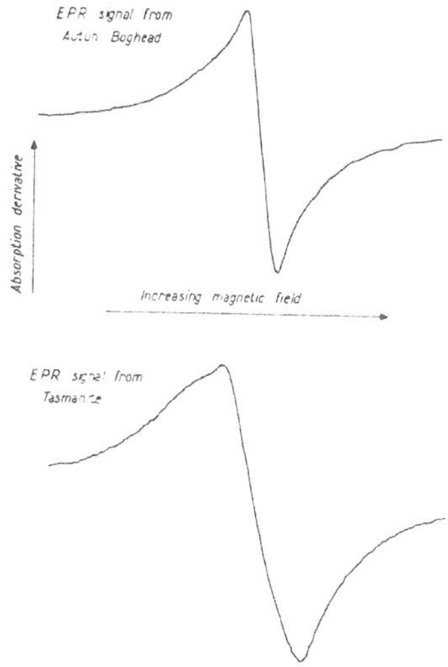

Fig. 1. E.P.R. signal of kerogen
a) before purification
b) after purification

Fig. 2. E.P.R. complex signals of natural samples. [25 p. 252]

3. INTENSITY OF THE EPR SIGNAL

Higher temperature, or longer duration of the degradation process, at least at the beginning, will result in larger free radical concentrations and increasing aromatic character of these radicals : it can then be expected that the history of a kerogen "catagenesis" is reflected in the intensity of its EPR signal.

That it is indeed so will be illustrated by the following typical examples where the variable used to define a maturation scale can be the coal rank, the depth or the treatment temperature.

The same maximum of spin concentrations is observed in the four series of samples. A monotonous but not necessarily linear relation exists between the variables able to scale the maturation process : essentially (paleo) temperature and time when they are known and if not, more accessible parameters as depth, rank i.e. carbon content, aromatization degree. Many others hidden variables certainly play an important role as parent materials, pressure, equilibrium with given off gases, catalysts and so on.

In the absence of a common, always accessible, parameter, we used intensively and fruitfully the comparative method and all possible correlations.

A very complete investigation was performed on a series of kerogens from the upper cretaceous of the Douala basin (Logbaba) : samples were studied down to a depth of 4 000 m.

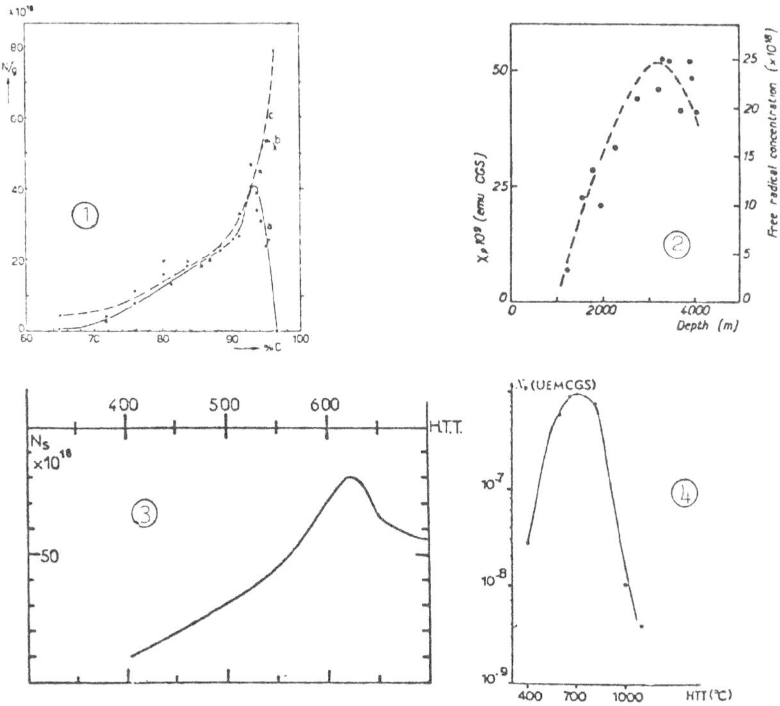

Fig. 3. Spin concentration vs maturation
1°-Coals: N_S vs rank, *in* a) air, b) vacuo, c) outgassed [26 p. 395]
2°-Kerogen: N_S vs depth, Douala Basin Cameroun [25 p. 255]
3°-Heat-treated coal: N_S vs treatment temperature, Faulquemont [27]
4°-Anthracene semi-coke: N_S vs treatment temperature GFEC [21].

The deeper a sediment is buried at a given location, the higher the temperature and the longer the duration of the heat-treatment it has undergone, so that depth of burial is an indication of the level of degradation ("catagenesis") of a kerogen. Fig. 3 shows the evolution of the paramagnetic susceptibility χ_p as a function of depth in the Logbaba series : after a steady increase, the paramagnetism reaches a maximum and then starts decreasing.

The maximum value corresponds to a concentration of 2.5×10^{19} free radical g of organic carbon. At such high concentrations the mean distance between radicals in the order of 30-40 Å, and small enough to allow direct recombinations : when the recombination rate grows larger than the creation rate of free radicals, their concentration starts decreasing. However the maximum concentration may be different for different types of organic or biological materials, since both the creation and the recombination rates depend on the chemical structure of the substance under study : it is well known, for instance, that oxygen-bridges between aromatic molecules inhibit the recombination of radicals and the further growth of the aromatic structures in artificial carbons [14].

The results illustrated by Fig. 4 show that the state of catagenesis of a kerogen can be characterized by the value of χ_p. It is important to compare this method with other techniques of characterization. The paramagnetic susceptibility of the Logbaba kerogens is plotted as a function of the H/C ratio (Fig. 4a), O/C ratio (Fig. 4b) and vitrinite reflectance (Fig. 4c) : arrows indicate increasing degradation of the material (increasing depth of burial). It is clear that there is no one-to-one correspondance between χ_p and the other criteria : the same paramagnetism may characterize two very different stages of catagenesis. For higly degraded kerogens χ_p is much more sensitive than the other parameters since it goes on decreasing while the H/C, O/C, and reflectance values remain nearly constant. At low catagenesis on the contrary all four criteria are equivalent since their variations are similar.

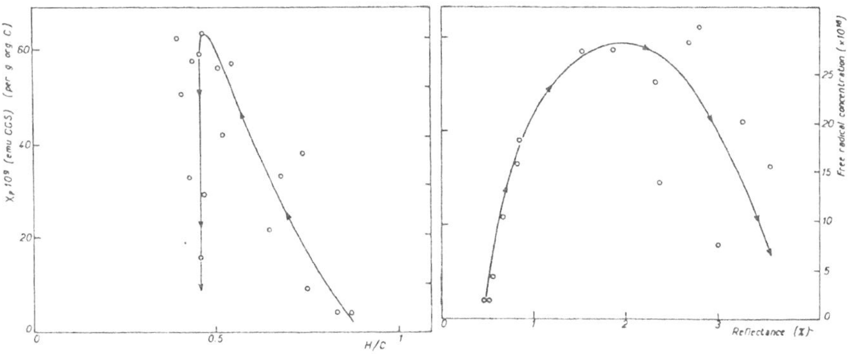

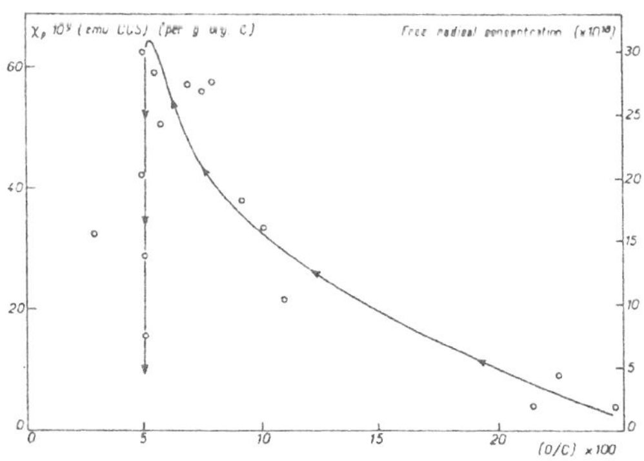

Fig. 4.
χ_p vs H/C, O/C & vitrinite reflectance
R. Douala kerogens [25, p. 256].

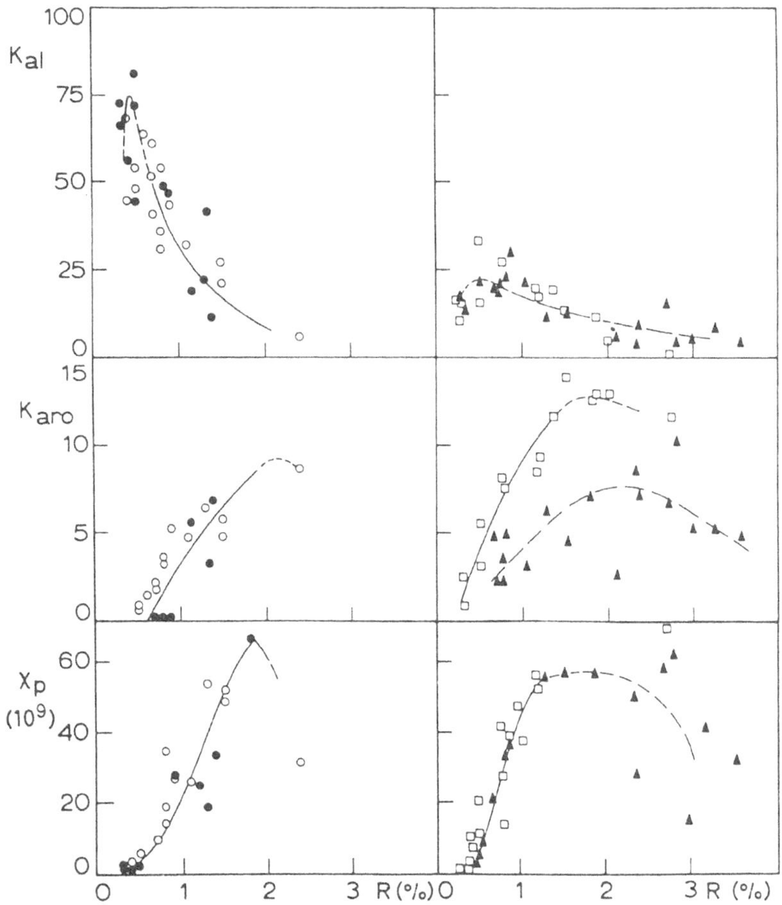

Fig. 5. I.R. absorption : Evolution of $K_{a\ell}$, K_{aro} and χ_p (emu GCS g^{-1}) as a function of reflectance R.
Kerogen series : ● I; o II; ▲ III. Coals : □ [25 p. 201]

IR absorption is another way of characterizing the evolution of a kerogen [25]. It will be seen that a very close correlation can be observed between χ_p and the intensity of the vibration bands of the aromatic CH groups (Fig. 5). There is also a good agreement of the variation of the 1 600 cm^{-1} aromatic band with the χ_p variation. These results were expected since the formation and stability of the free radicals are closely related to the aromatization of the kerogen structure.

4. ORIGIN AND NATURE OF FREE RADICALS IN CARBONACEOUS MATERIALS.

Practically all carbon materials, even those very recent or very little altered, give EPR signals. The paramagnetic susceptibility χ_p (per gram of organic carbon) varies approximately from 10^{-10} to $2\ 10^{-7}$ emu CGS at room temperature, which corresponds to free radical concentrations of $5\ 10^{16}$ to 10^{20} per g. [18,7,19,17].

As for other physical properties, the characteristic features of their EPR signal depends on the nature of the parent biological material and also on the "catagenetic process" this material has undergone. We will examine the influence of catagenesis.

The free radicals of artificial carbons have been the subject of a large number of investigations, which have shed some light on the origin and nature of these radicals. It may be thought that those in kerogens are similar.

During the pyrolysis of an organic material, as well as during the natural degradation of a biological substance, volatile products of small molecular mass are generated (CO_2, H_2O, light hydrocarbons). These are torn off the skeleton of the parent substance ; such a "peeling" leaves "scars" : one unpaired electron for each broken chemical bond. New electron pairings and bond rearrangements of course do occur, and hydrogen atoms replace some of the torn functional groups of the aliphatic chains or aromatic rings of the parent material. But a number of free radicals remain : they cannot migrate through the solid substance, they cannot recombine, and they may be stable for a quasi-infinite time.

A traditional presentation of this degradation process is the "Van Krevelen" diagram, where the O/C and H/C ratios (determined from an elementary chemical analysis of the solid residual substance) are plotted as abscissae and ordinates respectively. The chemical evolution of any organic material is illustrated in this diagram by a "trajectory" or "catagenesis path" along which the O/C and H/C values are simultaneously decreasing.

Studies of the carbonization process have shown that the production of volatile compounds, which starts below 200° C and can proceed up to 1000° C, is always accompanied by a progressive aromatization of the structure of the residual solid, which can be followed, for instance, through the increasing intensity of the 1 600 cm^{-1} IR band.

Moreover, quantum chemistry calculations show that unpaired electrons generated in one ring of an aromatic polynuclear structure are delocalized throughout the whole structure, with a resulting increase of stability of the radical.

X-rays and electron microscopy were used to estimate the size of aromatic clusters and to compare it with an average distance between paramagnetic centers. Simon [21] took advantage of a very well defined anthracene parent to obtain interference function with up to nine peaks for 800° C heat-treated anthracene semi coke. With the help of the Warren method, he proposed three

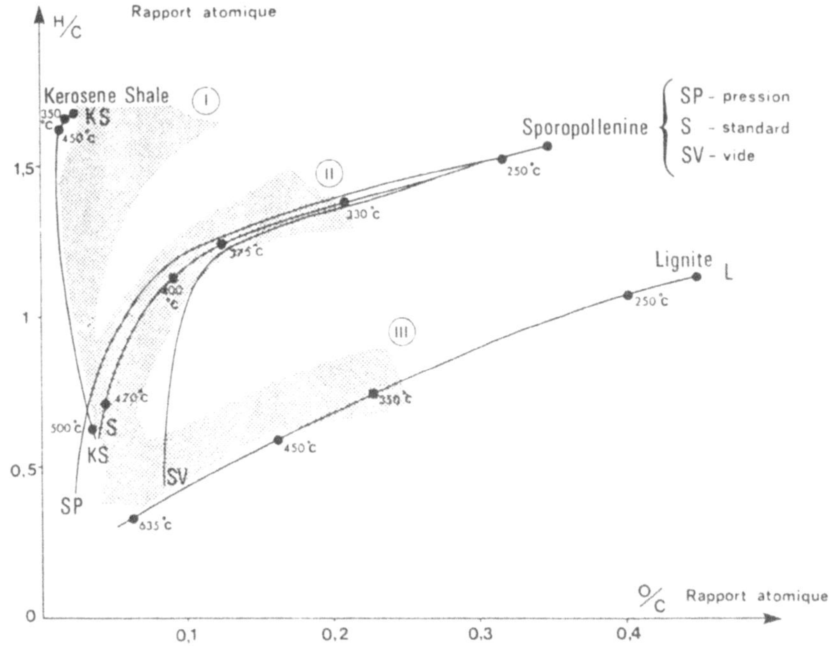

Fig. 6. Van Krevelen diagram (tinted area) and experimental paths for heated natural models of kerogens ; semi artificial models of kerogens [18].

approximate steps of growing corresponding to about 400, 600 and 800° C. Other EPR parameters agree with this analysis and especially 35 GHz EPR shows two kinds of free radicals (see g anisotropy). After these temperature range and cluster size, a conductivity process appears which drives the EPR intensity out of the Curie law (Fig. 7).

The maximum of spin concentration gives an average of 1 spin/ 10^3 A^3 in agreement with the size of the cluster II (HTT ∿ 700° C).

On the other hand in "artificial" kerogens, Viley [28] compared an electron microscopy estimation of elementary units of 4 to 10 cycles stacked by 2 or 3, with a loose bridging along a 50 to 2000 Å extension (Fig. 8), to spin concentrations of 10^{20}/g at the maximum. So the elementary unit is the good size for the free radicals ; but as the observed H/C ratio (0.05 to 0.6) would gives an aromatic unit of 20 to 30 carbon atoms, he deduced that not all the elementary units were actually free radicals.

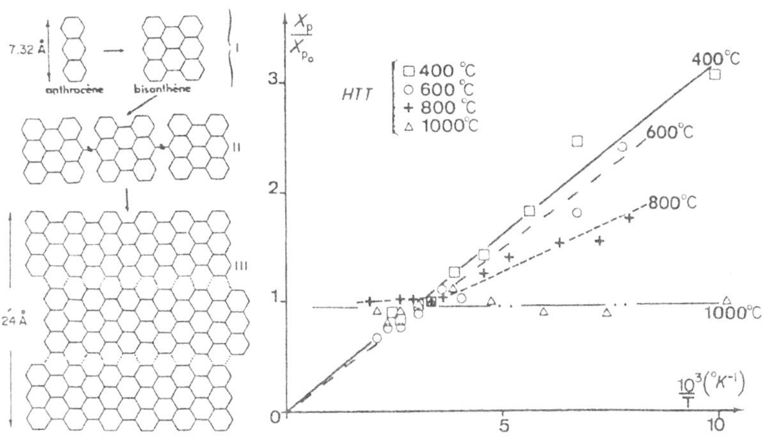

Fig. 7. Proposed cluster sizes for anthracene semi-cokes and susceptibility thermal variation (heat treatment as parameter) [21].

Another independant way to estimate the size of aromatic system is to study thermal variations of EPR intensity. If the odd electrons are quite localized they give a Curie law the intensity being proportional to H/T whereas free electrons give a temperature independant Fermi magnetism. The progressive extension of the aromatic system has been carefully checked in anthracene semi-coke between treatment temperatures of 400 to 1000° C (Fig. 7). In natural carbonaceous sediments the conducting state is not reached except, may, be in some anthracite coals.

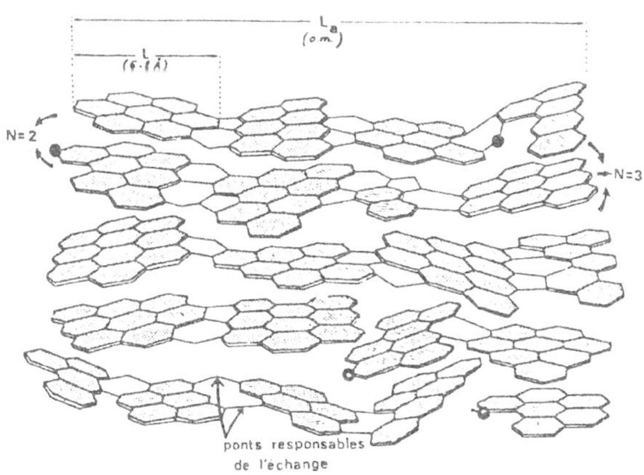

[25 p. 201]

Fig. 8. Proposed cluster assembling for artificial kerogens [28].

5. WIDTH AND LINESHAPE OF EPR

5.1. Width studies were, at the beginning, problematic because of the very frequent presence of paramagnetic impurities in samples. The more usual one is iron in oxydes and sulfides. Electron magnetic dipole is much larger than the nuclear ones and it imposes its local field at distances of the order of 10 Å. Then iron concentration as low as 10 ppm is enough to cancel any hope of using linewidth and shape as information sources.

Nevertheless, in a few case, expecially with synthetic kerogens, it as been possible to obtain a complete description of the physical phenomena which rule over these two easily measurable parameters. If one is lucky or very patient when purifying the sample, he can hope to use at least partially these results.

During the first steps of evolution a great number of protons are present and their magnetic dipoles produce a large local field in the range of ten gauss. As the electron is still quite localized, this dipolar interaction appears as static and the ESR line has approximately a gaussian shape. Resolved hyperfine structures are never perceptible in the solid state.

This step has been very seldom observed because the stabilization of free radicals is related to the creation of aromatic systems in which the odd electron describes already a large orbit keeping along a lot of tens of protons. Then a first process of averaging occurs, as the residence time of the electron near each proton becomes less than the reciprocal of the interaction value, measured in frequencies. The lineshape becomes lorentzian and the role of protons local field in the line width decreases continuously.

The next step is when the density of small aromatic systems grows so much that the distance between each other is less than 10 Å. Then, each one feels the local field produced by other electronic dipoles and the lineshape becomes gaussian as these intersystem dipolar interactions are time independant.

Then a new narrowing mechanism occurs in relation with the beginning of the overlap between the wave functions containing the odd electrons of neighbouring aromatic systems. This coupling is derived from quantum mechanics and is very similar to the exclusion Fermi principle. It says that when two odd electrons meet in the same region of the space (i.e. : they have almost the same space coordinates), they minimize their mutual energy by opposing their spins directions. This spin flip shortens the lifetime of the magnetic states and is a source of narrowing. It is called "exchange narrowing" because, after the interaction, it is not possible to distinguish the two electrons relatively to their origin.

They have "exchanged" one of their coordinates, including spin directions.

The last step of "low temperature" evolution occurs when the overlap of the two neighbouring aromatic systems is sufficient to give a chemical bonding. Then the two free electrons disappear from the line or if three systems are coupled together it remains only one odd electron with a quite extended wave fonction. Then we enter in the field of conductive carbon and graphite which is out of the field of this lecture. Paramagnetic molecules of oxygen can widen very much EPR lines in high porosity carbons and so these studies need to be done with samples under vacuum (cf.5.2).

We summarize with the listing of evolution steps :

1 - local field of protons (gaussian)

2 -averaging by "delocalization" inside the small aromatic system (lorentzian)

3 - dipolar fields of other aromatic systems (gaussian)

4 - exchange narrowing between aromatic system (lorentzian)

5 - chemical bonding : ESR signal intensity disappearence.

5.2. Oxygen effect

In numerous carbonaceous materials, the evaporation of small molecules produces a very important micropososity of the solid. Various methods from porosimetry [14] to NMR [33]were used to estimate size, distribution, extension and shape of this porosity. Specific area of hundred of square meter/g, size of a few Å and flat pores have been determined. Then, in air, oxygen molecules penetrates very deeply in the solid and produce two effects.

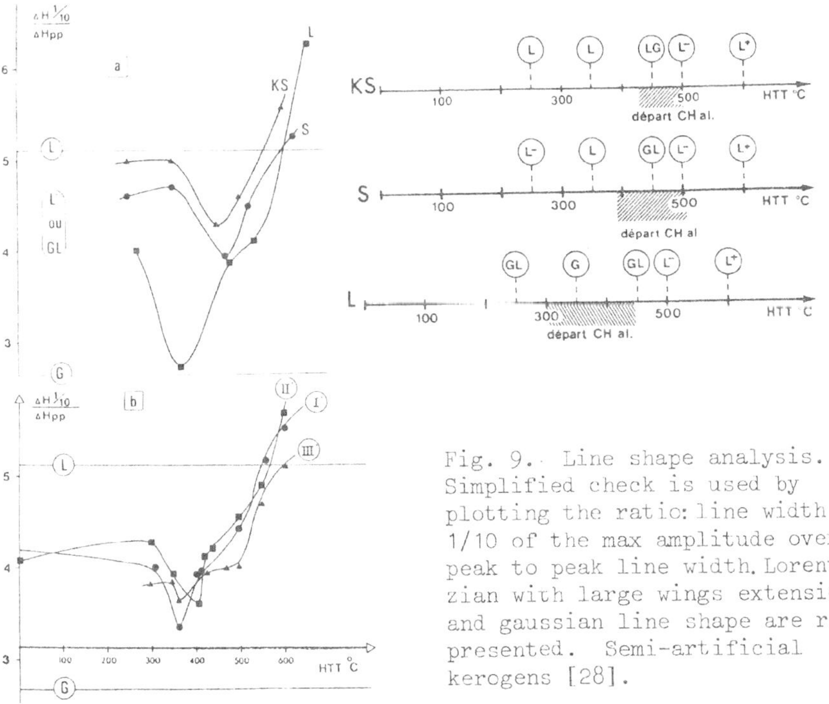

Fig. 9. Line shape analysis. Simplified check is used by plotting the ratio: line width 1/10 of the max amplitude over peak to peak line width. Lorentzian with large wings extension and gaussian line shape are represented. Semi-artificial kerogens [28].

1° - Chemical reaction with accessible free radicals. This effect is not reversible and produces firmly bonded CO or COOH groups. Then the corresponding free radical disappears out of the EPR line. This property is used to produce "activated carbons" with molecular sieve properties.

2° - Physical interactions between the dipolar magnetic moments of the oxygen molecules, which are paramagnetic, and the radicalar ones. This external local field widens very much the EPR line but is reversible. By putting the samples under vacuum a great quantity of physisorbed gases are evolved and the true line width can be seen. The oxygen effect upon anthracene char has been precisely measured and shows a maximum effect for treatment temperature around 600° C [Fig.9b] ; but in natural kerogens, without any heat treatment, no oxygen effect was found probably in relation with the high pressure during the diagenesis.

6. g VALUES

In general, g values very near the free electron one $g = 2.0023$ are found for the aromatic free radicals of natural coals and kerogens.

The two foreseeable sources of g variation are :

1 - the orbital contribution of aromatic ring currents. The first effect is also responsible for the susceptibility growing and anisotropy but has been clearly seen only for treatment temperature above 1000° C. That means probably that observed aromatic systems are too small to give enough effect [21].

2 - the contribution of spin orbit coupling in high Z atoms.

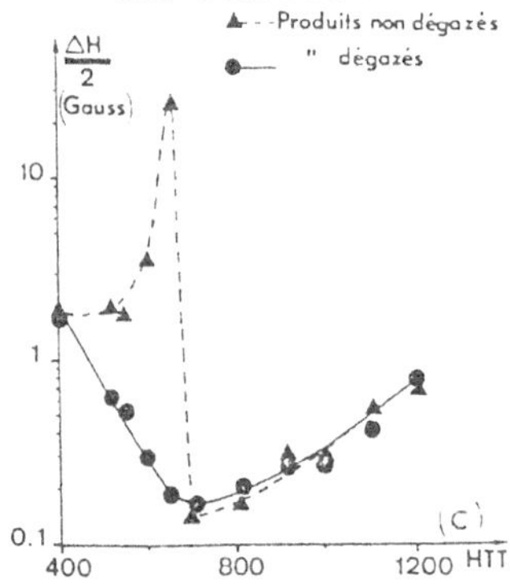

Fig. 9bis. Anthracene coke. Halfwidth, without g-anisotropy vs treatment temperature [21].

Representative values of ζ (zeta), the atomic spin-orbit coupling parameter.

Atom	ζ in cm^{-1}
H	∿ 0.5
C	28
N	70
O in OH	152
F	272
P	231
S in SH	382
Cl	587
Br	2460
I	4060

D.S. McClure, *J. Chem. Phys.*1949). 17, 905

The role of oxygen or sulfur atoms in g value has been clearly evidenced. If one plots g value in place of the O/C one in a Van Krevelen diagram, he obtains a very similar one [Fig.10].

The series one and two derived from algae, vegetal, waxes and toarcian of Paris basin are jointed together with a large H/C value, whereas the third series derived from higher plants with cellulose, go through the high g value region. The separation of the families in this diagram is so good that it could be used efficiently as identification method more safely than oxygen content analysis, always difficult to do.

The question arises to know where the heavy atom has to be, to give a shift of the g value. From liquid state studies we can compare anthracene anion (g = 2.00266) with 9-10 anthrasemiquinone anion (g = 2.00413) as an example-whereas coronene anion gives only (g = 2.00291) [29]. The highest values found in kerogen is g ∿ 2.0038 and in organic matter of phosphate g = 2.0050 so it is probably not possible to define very precisely the bondings

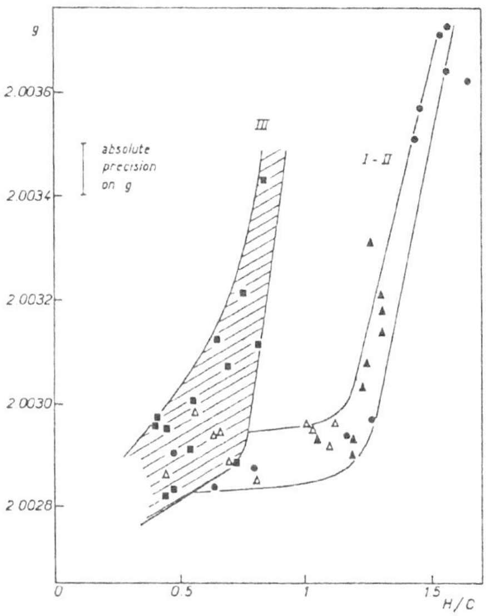

Fig. 10. Variation of the g factor as a function of atomic ratio H/C :
Series I : ●
Series II(Paris Toarcian): ▲
Series II(German Toarcian): Δ
Series III · ■ [25 p. 261]

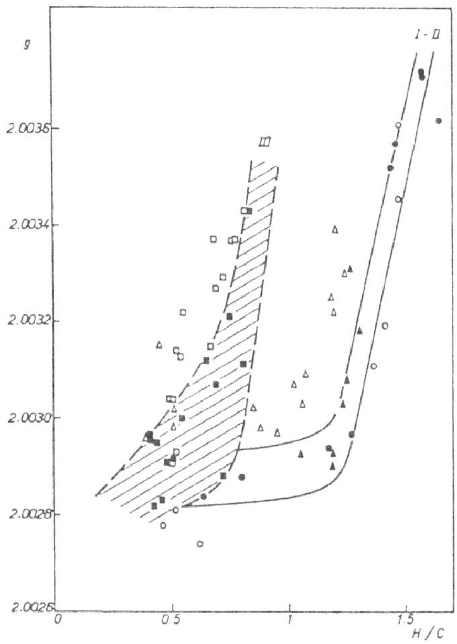

Kerogens of series I (natural: ●; artificial:o)
Series II (natural: ▲ and artificial: Δ)
Series III (natural: ■ and artificial: □).
[25 p. 264]

between aromatic systems and oxygen atoms, but what is sure is that the odd electron lives a certain amount of time on the oxygen atom, either directly, either through exchange mechanism.

Another remark can be made from these diagrams : there is

not a perfect superposition of the curves corresponding to natural kerogen and thermally treated one. In that case, the artificial series is obtained just by coaling the less evoluted one of the natural series. There is a shift of the "artificial series towards lower H/C content which could be related with the difference of equilibrium with evolved light gases as methane in the two processes.

In yet studied kerogen and coals only average g value is measured but in anthracene semicokes it was possible to obtain the two components of an axial g tensor.

In the low temperature domain, equivalent to natural carbonaceous sediments, $g_{\perp c}$ up to 2.004 are found whereas the $g_{/\!/ c}$ value is almost constant near 2.003. As these semicokes do not contain high Z atom, because they are prepared under argon gas, it is not very evident to find a reason for such a high g value.

Looking at structural results of ref. [21], indicates closed stacks of small aromatic systems with a relatively small plane to plane distance. At this state of organization, sp^3 orbitals probably contribute to "plane to plane" bonding and then we could think that an orbital interplane contribution is responsible for the high $g_{\perp c}$ value. In thermally simulated kerogens derived from Green River Shales (I') Toarcian (II') and Logbaba (III') sediments, the average g value decreases in the same temperature range and more markedly for the series (I) in agreement with such an interpretation but the simultaneous coupling with evolving S or O atoms cannot be excluded in these samples.

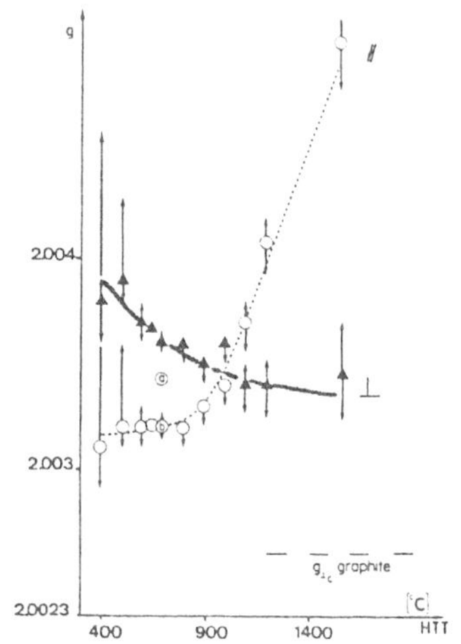

Fig. 11. g vs HTT in anthracene semi-cokes [21].

7. T_1 MEASUREMENT

The spin-lattice relaxation (T_1) can be easily measured through the saturation technique, because it is of the order of microseconds in aromatic free radicals. Indeed, this technique gives the product T_1 T_2 and some estimation of the spin-spin relaxation time T_2 has to be done. The reciprocal width can be used if no heterogeneous widening occurs. In general a too short T_2* is obtained because of present ^1H and paramagnetic impurities and a lower limiting T_1 value is then obtained by this way.

Two mechanisms at least help the relaxation, which are :

1 - phonon modulation of spin-orbit coupling

2 - exchange fluctuations.

The second one only is well documented in coals and anthracene series.

T_1 and T_2, vs N_S or HTT are given for Faulquemont, saccharose and anthracene semicokes in the Fig. 12 and 13. In the three cases when N_S reaches the value 10^{20}/g, $T_1 \to T_2$ which is characteristic of conduction electrons or high exchange frequencies.

When the number of radicals is growing, the exchange frequency which depends on the orbitals overlaps, is an exponential function of the distance between the aromatic systems and is responsible for a quick shortening of T_1, until $T_1 \sim T_2$ in the conducting region. The exchange frequency has been estimated in Faulquemont anthracite from the narrowing effect by the Anderson formula :

$$\omega_e = \frac{< \Delta\omega^2{}_{DIP} >}{\frac{1}{2} \quad \Delta\omega^{1\,2}} \text{ where we take } < \Delta\omega^2_{D18} > = 41.10^{-26} \; N_S \text{ cf.Fig. 14}$$

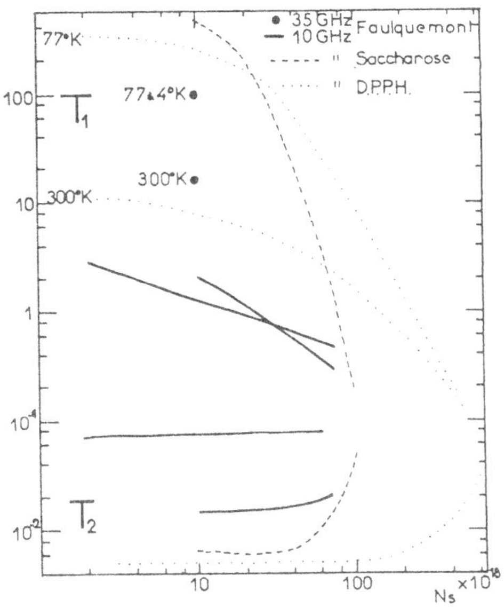

Fig. 12. T_1, T_2 vs N_S for heat treated Faulquemont's anthracite.

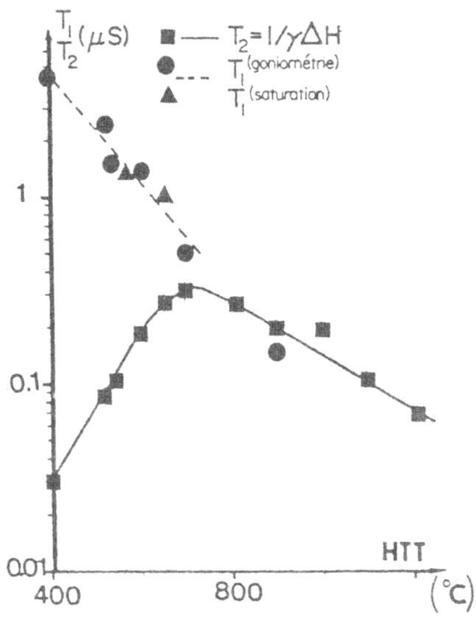

Fig. 13. T_1, T_2 vs N_S for anthracene semicoke [21].

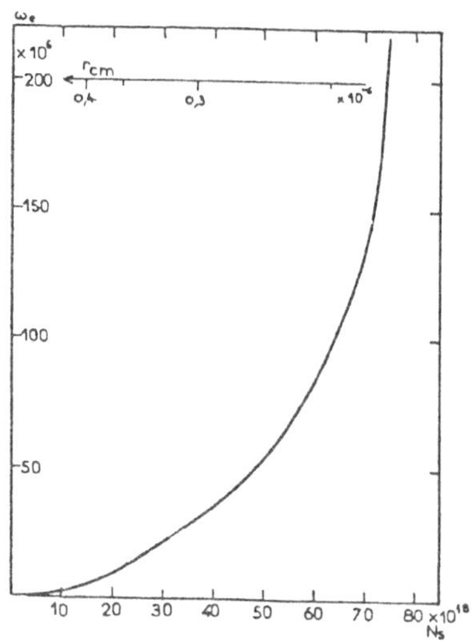

Fig. 14. Exchange frequency ω_e for heat treated Faulquemont's coals.

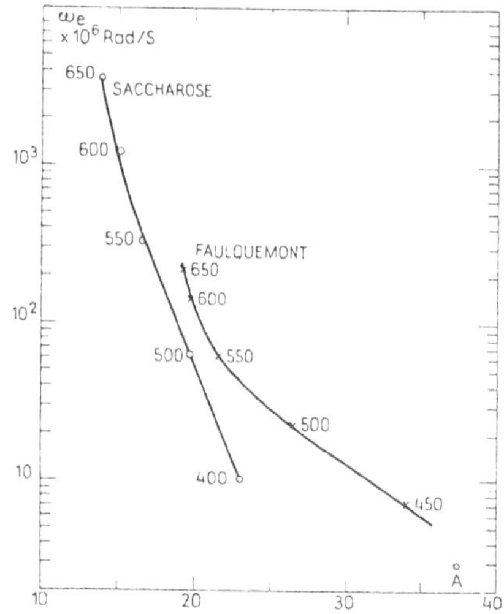

Fig. 15. ω_e vs distance. All are deduced from line width [27].

	S 400	S 500	S 550	S 600
F_{ext}	5.4	9.4	12.5	22
F_{int}	9	35	120	720

Table : exchange frequency f_e in MHz as deduced from T_1 direct measurement. (S = saccharose)

But in saccharose semicokes it was possible to have a direct T_1 measurement using a new goniometric method [32] without any use of the width. Then a series of fluctuation frequencies of much higher values was discovered from the T_1. We are now in the case where oxygen atoms contributes to the width through g anisotropy and also may be to the first relaxation mechanism. It was concluded that the found high frequency was related to the electron motion inside each aromatic radical -i.e. H local field averaging- and the lowest one to a true exchange between neighbouring radicals. To check the exchange frequency as a narrowing mechanism, we can use the idea that the orbital overlap depends exponentially on the distance between aromatic systems ($\omega_e \sim \exp^{-r}$). We must find a straight line in a ($\log \omega_e$) vs.r plot. That is almost true, except for the higher ω_e values. In fact the true distance parameter is the length between the bordering orbitals, something like $(r - \phi)$ were ϕ is the radicalar system diameter. For the Faulquemont coal, we could deduce from the curve Fig. 1 :

- ϕ before heating to be $\leqslant$ 20 Å,

- the distance (r - ϕ) to be in the range 15 → 1 Å during the phase of new centers creation 450 → 550° HTT,
- the curvature at high H.T.T. could result from the extension of aromatic systems size when the conductivity begins.

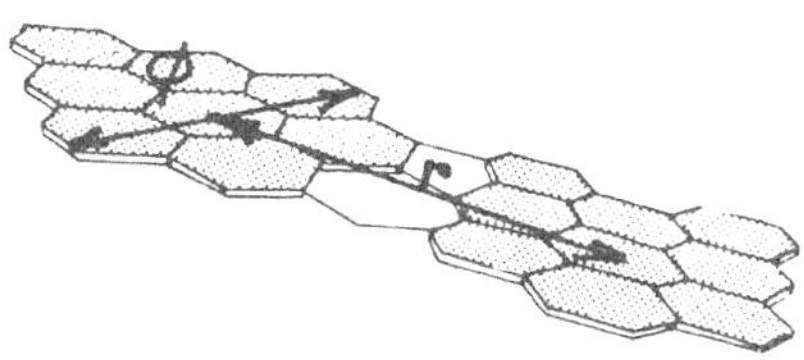

These results, obtained 15 years ago, agree with more re cent estimations using electron microscopy [28, 33] and a cautious comparison between the two techniques results let us assume that not all the small aromatic systems are indeed free radicals.

The role of remaining iron impurities in relaxation has to be cleared. From [27], it seems that it perturbs the exchange coupling and reduces its efficiency as a relaxation mechanism.

CONCLUSION

From the sensitive EPR methods, five parameters can be obtained which are related to the e-path in coals and kerogens. Two steps for this e-walk have been clearly evidenced ; what is occuring inside the aromatic system and the interactions from each other.

Even presented in very physical terms like exchange frequency, spin orbit coupling, they give informations upon bonding and chemistry at various length scale i.e. presence of oxygen, orbital overlap and so on.

In this field where very few method are very powerful, such E.S.R. informations are not at all negligible. But the main problem is to use them cautiously with a good understanding of physical and chemical concepts.

The most evident trap is to deduce a maturation estimation from a limited number of spins measurement only, as the related curve is multivalued. A continuous evolution path as deduced from depth depending series or coking at the laboratory has to be made. Careful sample preparation and competent examination of spectra must give to EPR method, an important role to play in the near future.

REFERENCES

[1] to [24] are refered with some detail in [25].

[1] - Albrecht, P., (1969), *Thèse*, Strasbourg.

[2] - Albrecht, P. and Ourisson, G., (1969), *Geochim. et Cosmochim. Acta*, 33, 138.

[3] - Alpern, B., Durand, G., Espitalié, J. and Tissot, B., (1972), *in : Advances in Organic Geochemistry* 1971, H.V. Gaertner and H. Werner ed., Pergamon Press, 1.

[4] - Bersohn, M. and Baird, J.C., (1966), "An Introduction to EPR", Benjamin, ed.

[5] - Carmona, F. and Delhaes, P., (1974), *in : Proceedings of Internat. Conference Industrial Carbons and Graphites*, London, 1974, preprint n° 120.

[6] - Dunoyer de Segonzac, (1969), *Thèse*, Strasbourg.

[7] - Durand, B., Marchand, A., Amiell, J. and Combaz, A.,(1977), *in : Advances in Organic Geochemistry* 1975, R. Campos and J. Goni ed., Enadimsa, Madrid, 753.

[8] - Durand B., Espitalié , J., Nicaise, G. and Combaz, A., (1972), *Rev. Inst. Franç. du Pétrole*, XXVII, 865.

[9] - Durand, B., Nicaise, G., Roucaché, J., Vandenbroucke, M. and Hagemann, H., (1977), *in : Advances in Organic Geoche-listry* 1975, R. Campos and J. Goni ed., Enadimsa, Madrid,601.

[10]- Erb., E., *Thèse de Doctorat d'Université*, Paris.

[11]- Espitallié, J. Durand, B., Roussel, J.C. and Souron, C., (1973), *Rev. Inst. Franç. du Pétrole*, XXVIII, 37.

[12]- Etienne, A., Cambrisson and Uebersfeld, J., (1954), *Nature*, 174, 614.

[13]- Conard, J., "Etude de la matière organique des phosphates par R.P.Æ.", *Géologie comparée des gisements de phosphate et de pétrole. BRGM n° 24 (1980).*

[14]- Groupe Français d'Etudes des Carbones (G.F.E.C.), (1965), *in : "Les Carbones"*, I. 469, Masson ed., Paris.

[15]- Hwang, P.T. and Pusey, W.C., (1973), U.S. *Patent n° 3740*, June 19, 1973.

[16]- Ingram, D.J.E. and Bennett, J.E., (1954), *Phil. Mag.* 45, 545.

[17]- Libert, P., (1974), *Thèse de Doctorat d'Etat*, Bordeaux.

[18]- Marchand, A., Libert, P. and Combaz, A., (1968), *C.R. Acad. Sci. Paris*, D.266, 2316.

[19]- Marchand, A., Libert, P. and Combaz, A., (1969), *Rev. Inst. Franç. du Pétrole*, XXIV, 3.

[20]- Salle, M., (1977), *Thèse de 3ème cycle*, Université d'Orléans.

[21]- Simon, C., Estrade, H., Tchoubar, D. and Conard, J., (1977), *Carbon*, 15.

[22]- Steelink, C., (1964), *Geochim. et Cosmochim. Acta*, 28, 10, 1615.

[23]- Tissot, B., Califet-Debyser, Y. Deroo, G. and Oudin J.L., (1971), *AAPG Bull.*, 55, 2177.

[24]- Uebersfeld, J. and Erb, E., (1959), *Proceedings 3rd Carbon Conference 1957, Pergamon*, 107.

[25] - Durand, B., (1980), *Kerogen*, Technip Paris.
[26] - Van Krevelen, D.W., (1961), *Coal*, Elsevier
[27] - Conard, J., Gradsztajn, (1967), *J. Phys.*, 28, 551-54
[28] - Villey, M., Estrade-Szwarckopf, H., Conard, J., *Rev. Inst. Franç. du Pétrole* (1981), 36, 3-16.
[29] - Oberlin, A., Villey, M.; Combaz, A. (1978), *Carbon*, 16, 73.
[30] - Poole, C., (1967), *Electron Spin Resonance*, Interscience.
[31] - Singer, L.S., Lewis, I.C., *Carbon*, 2, 115 (1964) & 5, 373, (1967).
[32] - Conard, J., Place, H., *C R A S Paris*, 265c, 1201.
[33] - Gradsztajn, S., Conard, J., Benoit, H. (1970), *J.P.C.S.*, 31, 1121 et 32, 1507.

APPLICATIONS OF DYNAMIC NUCLEAR POLARIZATION IN COAL RESEARCH

R.A.Wind, M.J.Duijvestijn, C.V.D.Lugt, J.Smidt and J.Vriend

Department of Applied Physics, Delft University of Technology
P.O.Box 5046, 2600 GA Delft, The Netherlands

ABSTRACT

The Nuclear Magnetic Resonance signals of coal can be enhanced to an appreciable amount by means of Dynamic Nuclear Polarization (DNP).

A brief outline is given of the three possible mechanisms, namely the Overhauser, Solid State and Thermal Mixing effect. The ^1H and ^{13}C DNP enhancement P_H respectively P_C have been investigated of 32 coal samples with a volatile matter percentage between 7 and 60%. The results are: (i) At high microwave powers (8W) P_H is mainly governed by the Solid State effect, whereas P_C is determined by a combination of the Solid State and Thermal Mixing effect; (ii) Both P_H and P_C decrease with decreasing coal rank, and become close to unity when VM% > 45%. The use of DNP in ^{13}C spectroscopy in coal has been investigated and examples of applications are given.

1. INTRODUCTION

For more than three decades coal has been subject of many Magnetic Resonance investigations, and many parameters have been measured to characterize the coal, to obtain information about its molecular structure and to investigate coal conversion processes (see e.g. reviews 1-5 and references cited therein).
During the last eight years much attention has been given to ^{13}C spectroscopy, because in principle this technique provides a way to measure in a non-destructive way a.o. the aromaticity of coal, one of the key parameters which characterizes the coal structure.

L. Petrakis and J. P. Fraissard (eds.), Magnetic Resonance. Introduction, Advanced Topics and Applications to Fossil Energy, 461–484.

Usually the ^{13}C spectra are measured using the ^{1}H-^{13}C cross-polarization (CP) technique (6) in order to keep the measuring time of a ^{13}C spectrum within acceptable limits. However, in many cases this measuring time can still be long (several hours is no exception), and it would be very worthwhile if this time could be reduced.

It is well known that coal contains free electrons (in fact, the first Magnetic Resonance experiments on coal were ESR measurements (7,8)), and it is also well known that in solids containing free electrons the NMR signal can be enhanced by irradiating at or near the electron Larmor frequency (9 ch.9 ; 10 ch.7 ; 11; 12 ch.6 and references cited therein). This is the so-called Dynamic Nuclear Polarization (DNP) effect. Recently DNP has been combined with ^{13}C spectroscopy in coal (13,14), and for some coals the measuring time of a ^{13}C spectrum could be reduced to minutes . This opens a.o. the possibility of a fast coal characterization via its aromaticity, whereas also more detailed information about the molecular structure can be obtained than was possible before. The purpose of this paper is threefold:

(i) A summary will be given of the different mechanisms which can cause a DNP effect in solids. It will be investigated which of these mechanisms determine the ^{1}H and ^{13}C DNP enhancement in coal.

(ii) It will be investigated for which coal rank DNP gives a non negligible signal enhancement, and which parameters determine this enhancement.

(iii) The use of DNP in ^{13}C spectroscopy in coal will be discussed. It will be shown that the shape of the ^{13}C spectra depends on the way DNP has been applied.

2. THE DNP EFFECT

In principle three DNP effects can be distiguished in a spin system containing nuclei and free electrons, namely the Overhauser effect, the solid state effect and the thermal mixing effect. Which of these effects dominates, depends on the nature of the interaction between the nuclei and the electrons. In general the hamiltonian of the spin system is given by

$$\hat{H} = -\omega_S \hat{S}_z - \omega_I \hat{I}_z + \hat{H}_{IS} + \hat{H}_{II} + \hat{H}_{SS} \qquad (1)$$

The first two terms represent the Zeeman interaction of the electrons and nuclei, respectively. $\hat{S}_z$ and $\hat{I}_z$ are the z-components of the electron and nuclear spin operators, $\omega_S = \gamma_S H_o$ and $\omega_I = \gamma_I H_o$, where γ_S and γ_I denote the gyromagnetic ratios of the electrons and nuclei respectively and H_o is the external magnetic field strength. $\hat{H}_{IS}$, $\hat{H}_{II}$ and $\hat{H}_{SS}$ represent the nucleus-electron interactions, nucleus-nucleus interactions and electron-electron interactions, respectively. Fig. 1 shows the energy-level scheme

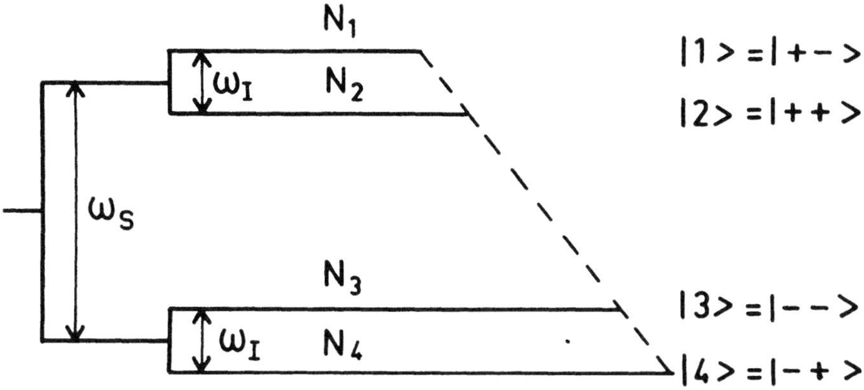

Fig.1. The Zeeman energy-level scheme of nucleus-electron spin pairs. N_1 to N_4 denote the population distribution in case of thermal equilibrium.

due to the first two terms in eq (1) for the case of nucleus-electron spin pairs. We confined ourselves to nuclear spin systems with spin 1/2 and a positive value of γ_I (γ_S is negative). Also the corresponding wave functions $|+\ -\rangle$ (S spin up, I spin down), etc... are given. We define the nuclear polarization α_I as

$$\alpha_I = \frac{(N_2 + N_4) - (N_1 + N_3)}{N_I} = \frac{N_I^+ + N_I^-}{N_I} \qquad (2)$$

N_1 to N_4 are the populations of the levels 1 to 4, and N_I^+ and N_I^- denote the number of nuclear spins in the + respectively in the minus state. $N_I = N_I^+ + N_I^-$ is the total number of nuclear spins. In thermal equilibrium the populations of the Zeeman levels are governed by the Boltzmann distribution and $\alpha_I \equiv \alpha_I^o$ is proportional to $N_I \omega_I T^{-1}$, where T is the temperature of the spin system, which in this case equals the lattice temperature. DNP is obtained if microwave irradiation is applied with a frequency ω equal or close to ω_S. Then electron transitions occur, possibly accompanied by nuclear transitions, resulting in a change of the populations of the Zeeman levels and α_I may become different from α_I^o. We define the DNP enhancement factor P_I as

$$P_I = \alpha_I / \alpha_I^o \qquad (3)$$

P_I is equal to the DNP enhancement of the NMR signal, as this signal is proportional to α_I. For the three DNP effects mentioned above we shall briefly consider which values of P_I can be expected and the conditions under which P_I becomes maximal. We assume that each nucleus is interacting with one electron only, and we take into account the possibility that the number of electrons N_S

may be different from the number of nuclei N_I. This means that each electron has to "service" $C_I = N_I/N_S$ nuclei. Finally, for simplicity we first suppose that the interactions between the nuclei and the electrons are the same, so that the transitions resulting from these interactions are equal for the different nuclei. Later on some remarks will be made about the change in results in case of a distribution of interactions.

2.1. The Overhauser effect

The Overhauser effect occurs when the interaction term $\hat{H}_{IS}$ in eq (1), which may represent both scalar and dipolar interactions between nuclei and electrons, is time-dependent, and if the correlation time τ_c describing this time-dependence is comparable to or smaller than ω_S^{-1} (15). Overhauser effects have been observed in samples with mobile electrons, like in liquids, conductors or semi-conductors (9 ch.9). About 25 years ago Overhauser enhancements have already been reported in graphite and in a high rank coal, probably a meta-anthracite (16). (We note that an Overhauser effect also occurs if the electron relaxation times are comparable to ω_S^{-1} (9 ch.9) , but in our case this possibility can be neglected.) Due to $\hat{H}_{IS}(t)$ in general relaxation transitions are induced between all Zeeman levels of the combined nucleus-electron spin system, see fig.2.

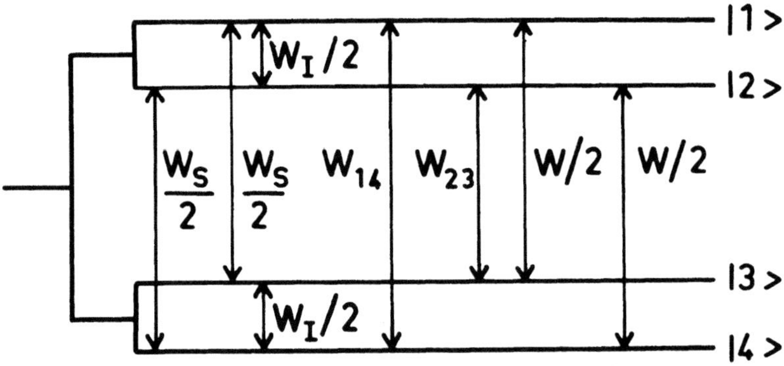

Fig.2 Possible relaxation transitions due to $\hat{H}_{IS}(t)$. W is the induced transition probability due to irradiation.

DNP is obtained by irradiation with a frequency ω near the electron larmor frequency. This gives rise to an induced transition probability W/2 between the electron Zeeman levels, with $W = \pi\gamma_S^2 H_1^2 g(\omega-\omega_S)$. H_1 is the amplitude of the microwave field and $g(\omega - \omega_S)$ the normalized absorption line-shape function of the electrons. Depending on W and the various relaxation transitions the populations of the Zeeman levels are re-arranged, resulting in a change in the nuclear polarization. A straight-forward calculation gives as result that in the stationary situation the enhancement factor $(P_I)_{OV}$ is given by

$$(P_I)_{OV} = 1 - \frac{W(W_{14} - W_{23})}{4C_I W_{14} W_{23} + W_I C_I (W_{14} + W_{23}) + (W + W_S)(W_{14} + W_{23} + W_I)} \frac{\gamma_S}{\gamma_I} \quad (4)$$

We found that for those coals that show a DNP enhancement, the approximation can be made that $W_S >> C_I W_I$, whereas it is also plausible that $W_S >> C_I W_{14}$, $C_I W_{23}$. Moreover, we separate the nuclear Zeeman relaxation rate into two terms: $W_I = W_I^S + W_I^X$, where W_I^S denotes the contribution due to nucleus-electron interactions only and W_I^X the contribution due to other interactions, e.g. the time-dependent I-I interactions. It follows in that case eq (4) can be rewritten as

$$(P_I)_{OV} \simeq 1 - \frac{W(W_{14} - W_{23})}{(W + W_S)(W_{14} + W_{23} + W_I^S)} \cdot \left[1 - \frac{W_I^X}{W_{14} + W_{23} + W_I} \right] \cdot \frac{\gamma_S}{\gamma_I} \quad (5)$$

From equations (4) and (5) it follows that $|P_I|_{OV}$ becomes maximal when $\omega = \omega_S$. To illustrate the maximum values of $|P_I|_{OV}$ we shall consider separately the cases that the nucleus-electron interactions are purely scalar and dipolar, respectively.
(i) scalar interaction. It can be shown that in this case $W_{23} = W_I^S = 0$. Hence for large values of W $(P_I)_{OV}$ becomes given by

$$(P_I)_{scalar} \simeq 1 - \left[1 - \frac{W_I^X}{(W_I^X + W_{14})} \right] \frac{\gamma_S}{\gamma_I} \quad (6)$$

Hence, if $W_I^X << W_{14}$ an enhancement of $1 - (\gamma_S/\gamma_I)$ would be obtained, e.g. a factor + 660 for protons and a factor + 2640 for ^{13}C. In fig. 3 the Overhauser effect for scalar interactions is illustrated.

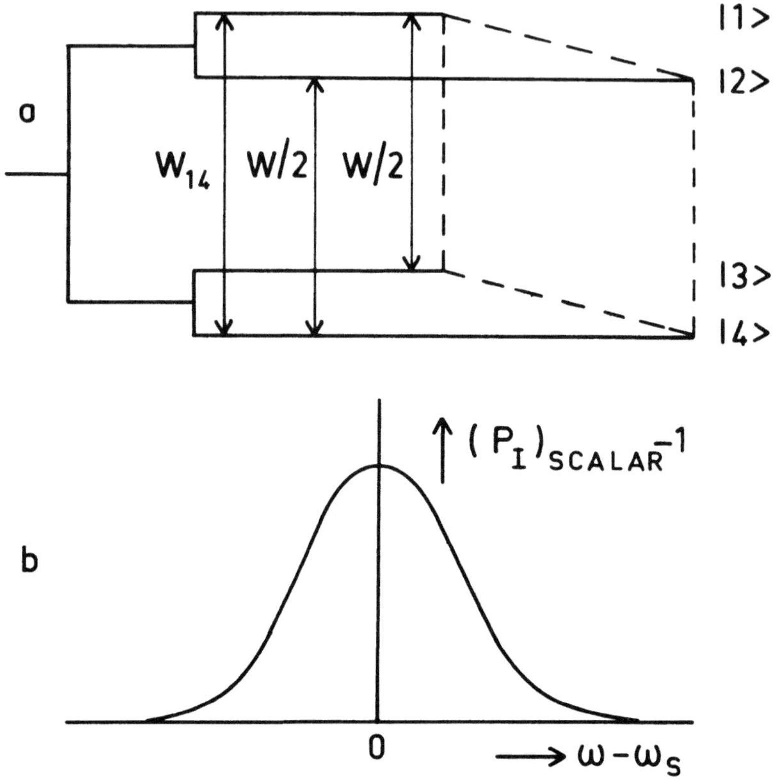

Figure 3. The Overhauser effect due to scalar nucleus-electron interactions. a: the population distribution of the Zeeman levels after strong irradiation; b: $(P_I)_{scalar}$ as a function of $(\omega - \omega_S)$

(ii) dipolar interactions. Here W_{14} , W_{23} , $W_I^S \neq 0$. It can be shown that under extreme narrowing conditions, where $\omega_S \tau_C << 1$, W_{14} : W_I^S : W_{23} = 1 : 3 : 6. It follows from eq (5) that for $W >> W_S$, $(P_I)_{OV}$ is given by

$$(P_I)_{dip} \simeq 1 + 1/2 \left[1 - \frac{W_I^x}{W_{14}+W_{23}+W_I} \right] \frac{\gamma_S}{\gamma_I} \tag{7}$$

Hence if $W_I^x << W_{23}$ an enhancement of $1+1/2(\gamma_S/\gamma_I)$ would be found, e.g. -330 for protons and -1320 for ^{13}C.
In practice both types of interactions will be present, which means that $(P_I)_{OV}$ can be anywhere between $1-(\gamma_S/\gamma_I)$ and $1+1/2(\gamma_S/\gamma_I)$, depending to the relative strengths of the scalar and dipolar interactions and the value of W_I^x.

2.2. The Solid State effect

The Solid State effect occurs when $\hat{H}_{IS}$ contains time-independent nucleus-electron dipolar interactions. This situation is met in solids containing fixed paramagnetic centra. It can be shown (9 ch. 9, 10 ch. 7) that due to these interactions the wave-functions, corresponding to the Zeeman levels, become mixed. When $|\gamma_S| >> |\gamma_I|$ then in first order the wavefunctions of fig. 1 are given by
$|1\rangle = |+-\rangle + q|++\rangle$; $|2\rangle = |++\rangle - q^*|+-\rangle$;
$|3\rangle = |--\rangle - q|-+\rangle$ and $|4\rangle = |-+\rangle + q^*|--\rangle$, with

$$q = 3/4 \frac{\gamma_S \hbar}{r^3} \cdot \frac{1}{H_o} \cdot \sin\theta \cos\theta \quad \exp(-i\phi) \tag{8}$$

r, θ and φ are the polar coordinates of the electron-nucleus vector if $\vec{H}_o$ is along the z-axis. It follows that if irradiation is applied at a frequency $\omega = \omega_S \pm \omega_I$, the normally forbidden induced transitions between the levels 1 and 4 respectively 2 and 3 become nonzero, with a transition probability W' given by

$$W'(\omega-\omega_S \pm \omega_I) = 2\pi |q|^2 \gamma_S^2 H_1^2 g(\omega-\omega_S \pm \omega_I) = 2|q|^2 W(\omega-\omega_S \pm \omega_I) \tag{9}$$

As for coal the average value of $|q|^2$ is typically 10^{-7} to 10^{-8}, it follows that for a given H_1 field $W' << W$.
If we neglect the relaxation transitions between the levels with a frequency difference $\omega_S \pm \omega_I$, it can be calculated that the stationary DNP enhancement factor $(P_I)_{SS}$ due to the solid state effect is given by

$$(P_I)_{SS} = 1 \pm \frac{W_S[W'(\omega-\omega_S \mp \omega_I)]}{W_S W_I + (W_S + C_I W_I)[W'(\omega-\omega_S \mp \omega_I)]} \left[\frac{\gamma_S}{\gamma_I} - 1 \right] \tag{10}$$

If we assume again that $W_S >> C_I W_I$ and if $|\gamma_S| >> |\gamma_I|$, eq (10) becomes

$$(P_I)_{SS} \simeq 1 \pm \frac{[W'(\omega-\omega_S\mp\omega_I)]}{W_I + [W'(\omega-\omega_S\mp\omega_I)]} \left|\frac{\gamma_S}{\gamma_I}\right| \quad (11)$$

It follows that $|(P_I)_{SS}|$ becomes maximal when $\omega=\omega_S\pm\omega_I$ and that for $W' >> W_I$ $(P_I)_{SS}$ becomes given by $-\gamma_S/\gamma_I$ when $\omega=\omega_S-\omega_I$ and $+\gamma_S/\gamma_I$ when $\omega=\omega_S+\omega_I$. Fig. 4a shows the solid state effect for nucleus-electron pairs for the case that $\omega=\omega_S-\omega_I$ and fig. 4b shows $(P_I)_{SS}$ as a function of ω.

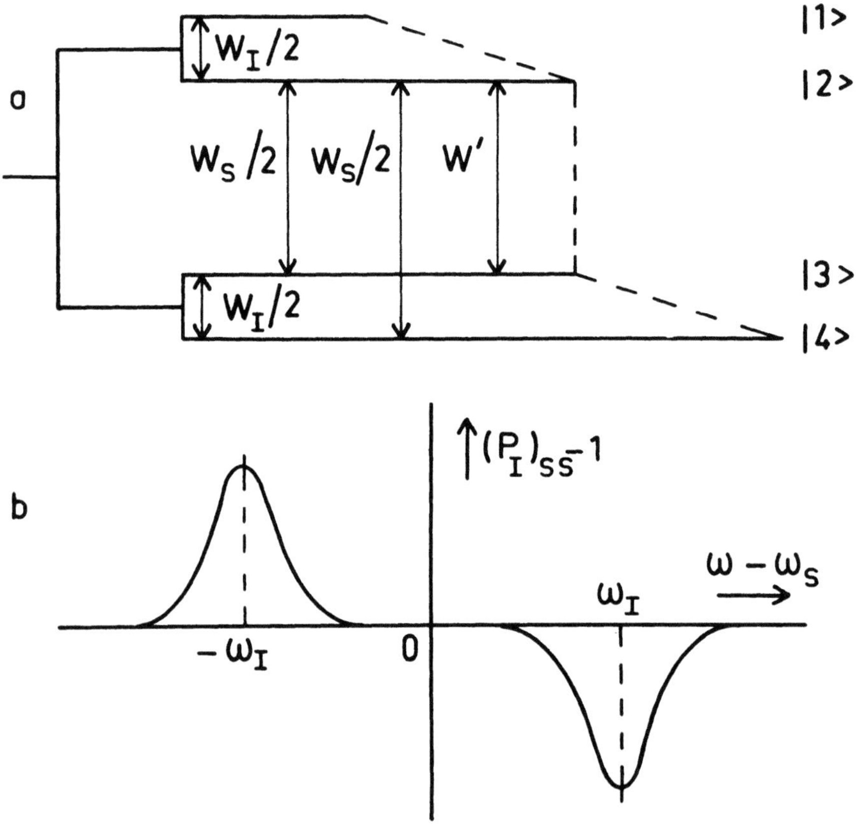

Figure 4. The solid state effect. a: the population distribution of the Zeeman levels after strong irradiation at $\omega=\omega_S-\omega_I$; b: $(P_I)_{SS}$ as a function of $(\omega-\omega_S)$

2.3. The Thermal Mixing effect

The thermal mixing effect occurs in solids containing such large concentrations of fixed paramagnetic centres that the homogeneous broadening of the ESR line of the free electrons due to the static electron-electron dipolar interaction can no longer be neglected (11,12 ch.6). As an example : as a rule of thumb (9 p.126) the average local field $\sqrt{H_d^2}$ due to this interaction is of the order of $N_S.10^{-19}$ Oe, where N_S is the number of free electrons per cm^3. In coal N_S is maximal a few times 10^{19} cm^{-3}, so that $\sqrt{H_d^2}$ is maximal a few Oe. This is comparable to the linewidth $\Delta H_{1/2}$ of the ESR line, which varies from ~ 7 to 13 Oe in the coal samples we investigated, and which is closely related to the total local field $\sqrt{H_L^2}$, including both the homogeneous and inhomogeneous broadenings. Then the energy E_L corresponding to the total local field can be characterized by a single temperature T_{LOC} via : $E_L \sim H_L^2 . T_{LOC}^{-1}$ (10:p.160;17). In thermal equilibrium T_{LOC} equals, of course, the lattice temperature T_L. According to the Provotorov theory (18) it is possible to change T_{LOC} by a few orders of magnitude by irradiating with a frequency ω close to, but not exactly at the electron Larmor frequency ω_S. Then energy quanta $\hbar\omega_S$ are absorbed by the electron Zeeman system, whereas the quanta $h(\omega-\omega_S)$ are either absorbed ($\omega > \omega_S$) or emitted ($\omega < \omega_S$) by the local electron system E_L, resulting in a negative (if $\omega > \omega_S$) or positive (if $\omega < \omega_S$) enhancement of T_{LOC}^{-1} (10 ch.4).
If, besides the electron-electron dipolar coupling also a static electron-nuclear dipolar coupling exists, an energy exchange can take place between theelectron local system E_L and the nuclear Zeeman energy. If the energy exchange rate is fast enough, the temperature T_{ZI} describing the nuclear Zeeman energy will become equal to T_{LOC}, so that the nuclear polarization α_I, which is proportional to T_{ZI}^{-1}, can be changed by irradiating near the electron larmor frequency. This has been called the thermal mixing effect. The thermal mixing effect is illustrated in fig.5. Irradiation with an induced transition probability W/2 causes a connection between the electron Zeeman system E_{ZS} and the electron local system E_L, resulting in a change of T_{LOC}; the electron-nuclear interactions cause a connection between E_L and the nuclear Zeeman system E_{ZI}, characterized by the rate $W_{I,L}$. W_S, W_L and W_I represent the spin-lattice relaxation rates of E_{ZS}, E_L and E_{ZI}, respectively.

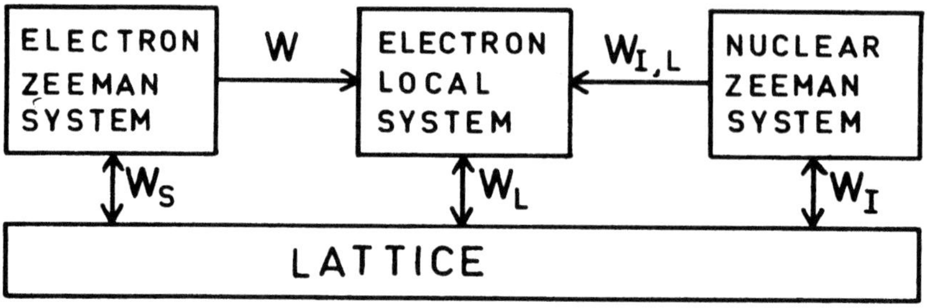

Figure 5. The thermal mixing effect.

It can be calculated that the thermal mixing effect causes a DNP enhancement $(P_I)_{TM}$ of the nuclear polarization, given by

$$(P_I)_{TM} = 1 + \frac{W_{I,L}}{(W_{I,L}+W_I)} \cdot \frac{(\omega-\omega_S)\omega_S \, W W_S}{W[B+(\omega-\omega_S)^2 W_S]+BW_S} \tag{12}$$

$$\text{with } B = \gamma_S^2 H_L^2 W_L + C_I \frac{W_I \cdot W_{I,L}}{W_I+W_{I,L}} \cdot \omega_I^2 \equiv B'+B'' \tag{13}$$

It can be shown that in all practical cases $B' >> B''$. It follows from eq (12) that for $W>>W_S$ $(P_I)_{TM}$ can be approximated by

$$(P_I)_{TM} \simeq 1 + \frac{W_{I,L}}{(W_{I,L}+W_I)} \cdot \frac{(\omega-\omega_S)\omega_S W_S}{B'+(\omega-\omega_S)^2 W_S} \tag{14}$$

$|P_I|_{TM}$ becomes maximal when $|\omega-\omega_S| \equiv |\Delta\omega_m| = \sqrt{\gamma_S^2 H_L^2}\sqrt{W_L/W_S}$ and is given by

$$\left| |P_I|_{TM} \right|_{max} \simeq 1+1/2\frac{W_{I,L}}{(W_{I,L}+W_I)} \cdot \frac{|\omega_S|}{|\Delta\omega_m|} = 1+1/2\frac{W_{I,L}}{(W_{I,L}+W_I)} \cdot \frac{|\omega_S|}{\sqrt{\gamma_S^2 H_L^2}}\sqrt{\frac{W_S}{W_L}} \tag{15}$$

Fig. 6 shows an example of $(P_I)_{TM}$ as a function of ω.

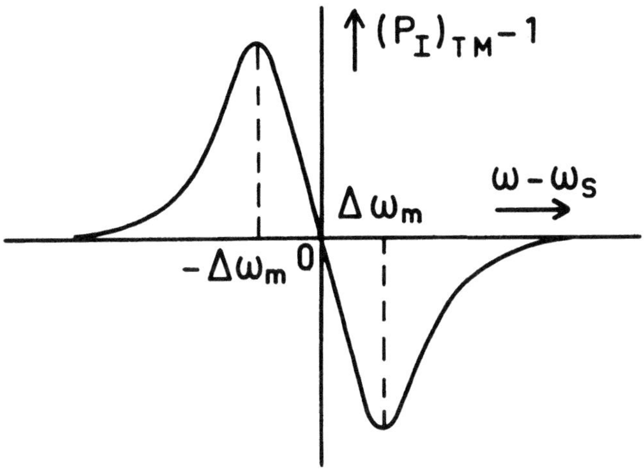

Figure 6. $(P_I)_{TM}$ as a function of $(\omega - \omega_S)$.

The maximum value of $|P_I|_{TM}$ can be larger or smaller than $|\gamma_S|/|\gamma_I| = |\omega_S|/|\omega_I|$, depending on the values of $W_{I,L}$, W_I and $|\Delta\omega_m|$.

2.4. Final remarks.

The question is, of course, which enhancement factors can be expected in practice. According to the above these factors depend on many, for the greater part unknown, parameters, which makes it impossible to predict these factors quantitatively. However, there is one exception : according to eq (11) the solid state effect gives rise to an enhancement factor $|\gamma_S|/|\gamma_I|$ provided that the quasi-fordidden transition probability W' can be made large compared with W_I. Hence it is not unrealistic to expect that for smaller values of the microwave field H_1 the Overhauser and/or thermal mixing effect, which occur due to the allowed transition probability W which is $1/2\,|q|^{-2}$ larger than W' , prevail, and that the solid state effect becomes important for larger values of H_1. As we shall see, this has indeed been observed in practice.

Finally, the following factors play also a role in the ultimate enhancement :

(i) Spin diffusion. In practice there exists a distribution of enhancements factors, depending on the magnitude and orientation of the inter electron-nuleus vector. Hence for dilute nuclear spin species such as ^{13}C, among which spin diffusion can be neglected, the final enhancement is found by integrating the individual enhancements over the sample. In this case it is also important whether the Zeeman relaxation rates W_I are also determined by other interactions than $\hat{H}_{IS}$, because this decreases the enhancements of the nuclei.

In case of abondant spin species like H the inter-nuclear dipolar interactions give rise to spin diffusion, which results in an average nuclear polarization enhancement over the sample, provided that the spin diffusion is fast enough (9.ch.9). Hence the enhancements are given by the eq (4), (10) and (12) for the different DNP effects, where the different parameters W_{14}, W_{23} etc. are now averaged over the sample volume.

(ii) The sample size. In our case the diameter of the sample tube and filling height are comparable to the wavelength of the microwave field.

Hence a distribution of enhancements over the sample exists even if spin diffusion is present.

(iii) the nuclear larmor frequency ω_I and the linewidth $\Delta H1/2$ of the ESR signal. If $\omega_I \lesssim 1/2\,\Delta H1/2$ the solid-state enhancement is suppressed due to the so-called unresolved solid state effect (10ch.7), where as for the thermal mixing effect $W_{I,L}$ decreases if $\omega_I/1/2\,\Delta H1/2$ increases (11), resulting in a decrease of the enhancement factor if $W_{I,L} < W_I$ see eq (12).

3. EXPERIMENTAL

We investigated the DNP-enhancements of a series of 32 coal samples of different rank. The samples were provided by the Exxon coal bank in Baytown, USA and by the Dutch coal bank SBN. Table 1 gives the weight percentages of volatile matter (VM), C, O and H of the different coals, as far as they are available. The Exxon coal samples have been preserved under argon, the DE series under nitrogen gas and the EB series under air. For the measurements the samples were powdered to a grain size $<$ 0,21 mm and put into pyrex tubes of 4 mm i.d. to a height of ca 1 cm. Then the tubes were evacuated at room temperature for at least two hours and sealed off. It was found that evacuation did increase the DNP enhancements considerably for the EB and DE series : after evacuation the proton enhancements factors of the EB series were 3 to 4 times larger than before. This is due to a decrease of about the same amount of the proton Zeeman relaxation rate, which without evacuation is probably governed by paramagnetic oxygen. This is partly caused by the fact that the EB series has been preserved under air : for the DE series the changes in the enhancements factors and relaxation rates were about a factor 2 (In this respect it is worth noting that for 6 DE it has been found that keeping the sample under air for a couple of months resulted in an decrease of the enhancement factor and an increase of the relaxation rate of about 30 % , even when the sample was evacuated for a long time. Obviously some paramagnetic oxygen has been trapped permanently in the coal). We did not investigate the influence of evacuation on the Exxon coal samples. The grain size did not affect the DNP enhancement or relaxation rates, provided that the size was not too large (for grain sizes of a few mm the DNP enhancements decreased with 30 to 50 % , may be because the oxygen could not be removed properly or because of penetration depth problems for the microwave field).

All measurements have been performed at room temperature. The used external magnetic field was 1.4 T , corresponding to a proton frequency of 60 MHz, a ^{13}C frequency of 15 MHz and an ESR frequency of 40 GHz. The microwave power P_e, incident on the cavity , could be varied and was maximal 8 W, in our set-up corresponding to a H_1 field of about 0,3 Oe (averaged over the sample). It was found that many samples were heated when irradiated with maximum microwave power , resulting in a decrease of the enhancement factors. The reason for this phenomenon is unknown. The decrease could be reduced by using ample air cooling, but still an enhancement reduction up to 30 % was found in some cases. The experimental set-up has been described elsewhere (19) . As the ESR frequency was fixed at ≃ 39.46 GHz, the H_o field had to be varied in order to measure the DNP enhancement as a function of the ESR off-set frequency. Still, in the next section the enhancements will be given as a function of the ESR frequency, because this renders the comparison between experimental and theoretical results more easily.

The following experiments have been performed:

a. For two samples of the DE series and three samples of the EB series the ^1H and ^{13}C DNP enhancements P_H and P_C, respectively, have been measured as a function of H_o for different values of P_e. It was found that qualitatively the results were very similar, and also hold for the other samples. In the next section the results of one coal will be shown. P_H was measured by comparing the magnitudes of the decay after a 90° r.f. pulse applied at the proton Larmor frequency, with and without microwave irradiation; P_C was determined by comparing the (proton decoupled) spectrum obtained after Fourier transformation of the ^{13}C free induction decay, with and without ESR irradiation.

b. For all coal samples the proton DNP enhancement P_H is measured for the value of H_o where the enhancement is maximal (this value was found to be the same for all samples). P_e was kept constant at 8 W. The values of P_H were measured a few seconds after the turning on of the microwave power, thus avoiding the heating problems mentioned before. For the few samples where also the ^{13}C enhancement has been measured, it has been found that this quantity varied in a very similar way as the ^{1}H enhancement factor.

c. The number of free electrons, N_S has been measured. This is done because a.o. N_S determines the proton DNP enhancement due to the solid state effect if $W' < W_I$ (it can be calculated that the mixing factor $|q|^2$, and according to eq (9) also W', is proportional to N_S). N_S has been determined using a home-built 9 GHz ESR spectrometer. The ESR signals of all coal samples consisted of a line with a width at half height varying from ca 7 to 13 Oe. For some samples also a narrow peak was detected almost on top of the other one. The origin of this narrow component is not well understood (3). As relatively few electrons contribute to the narrow line, and no correlation has been found between the DNP results and this line, this has not been further investigated.

d. The proton Zeeman spin-lattice relaxation rate W_H has been determined for all samples. This is done because this rate determines the DNP enhancement factor too. The Zeeman relaxation has been measured using a standard $180°-\tau-90°$ pulse sequence. For many samples it is found that the relaxation cannot be described by the exponential function $M_o-M_z(t)=(M_o-M_z(o))\exp(-\frac{t}{T_{1H}})$, ($M_z(t)$ denotes the magnetization of the protons parallel to H_o and M_o its thermal equilibrium value). This appears especially when the relaxation is fast and is probably due to the fact that in those cases even within the proton system the spin diffusion is not fast enough to maintain a single relaxation rate (20). Therefore W_H has been defined as $W_H=(t1/2/\ln 2)^{-1}$, where t1/2 is the time in which $M_o-M_z(t1/2)$ is equal to $1/2(M_o-M_z(o))$.

4. RESULTS

4.1 P_H and P_C as a function of H_o

Fig. 7 shows the proton DNP enhancement P_H as a function of the ESR offset frequency of the low volatile bituminous coal 6DE, for P_e=8W and P_e=1.1W. We observe that all three DNP effects are present in coal: a positive Overhauser effect, a solid state effect and a thermal mixing effect. In fig.7 these three contributions are shown separately. We shall briefly discuss the results: (i) the Overhauser effect gives rise to a positive enhancement of about 6.5 for P_e=8W and $\omega=\omega_S$. This means firstly, that the electrons cannot be regarded as completely fixed but must possess some mobility on a time scale $\approx \omega_S^{-1} \approx 10^{-11}$ to 10^{-12} sec, and secondly, that the time-dependent electron-proton scalar interactions are dominating the time-dependent electron-proton dipolar interactions. The absolute value of the enhancement cannot be explained because too many unknown quantities play a role, cf section 2.1.
We also observe that about the same enhancement is found for

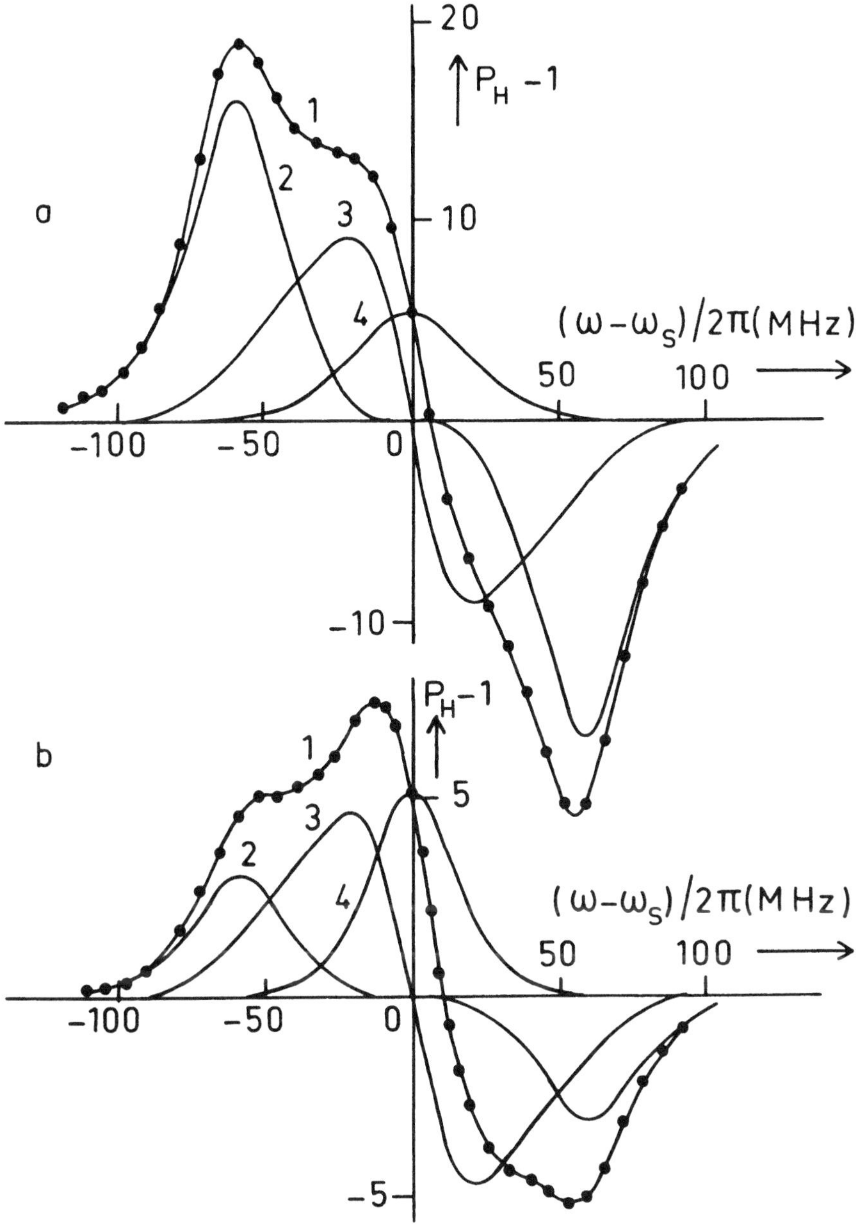

Figure 7.(P_H-1) of coal 6DE as a function of (ω-ω_S). a:P_e=8W; b: P_e=1.1 W. curve 1: experimental; curve 2: the solid state effect; curve 3: the thermal mixing effect; curve 4: the Overhauser effect.

P_e=1.1 W, so that the enhancement is rather close to the value which would have been obtained for infinite microwave power.

(ii) As predicted, the solid state effect causes an optimum enhancement for an ESR offset of ± 60 MHz. The enhancement is almost proportional to P_e (this is the more true if one realizes that at 8 W the enhancement is suppressed by about 20 % due to heating problems, as the microwave power has been left on continuously during this experiment). This means that $W' << W_I \equiv W_H$, see eq (11). Using the known values of $P_H \simeq +16$ for P_e = 8W and $(\omega-\omega_S)/2\pi=-60$ MHz and $W_H=2.5$ s^{-1} (table 1), it follows from eq (11) that $W'(o)=5.7$ 10^{-2} s^{-1}. It is known that P_e=8W corresponds to a microwave field with an average amplitude H_1=0.3 0e. If we assume $g(\omega)$ to be lorentzian it can be shown that at 40 GHz $g(o) \simeq 2.7.10^{-9}$ sec. It follows from eq. (9) that $W(o) \stackrel{\sim}{\sim} 2.4.10^5 s^{-1}$ and that $|q|^2 \stackrel{\sim}{\sim} 1.2.10^{-7}$, a realistic value. We conclude that eq (11) describes the proton solid state effect satisfactorily.

(iii) The thermal mixing effect causes an optimum enhancement at an ESR off-set frequency of about 21 MHz. The enhancement is less than proportional to P_e which means that for P_e=8W already some degree of saturation has been reached. It was not possible to investigated whether eq (12) predicts the correct value of $(P_I)_{TM}$ because this value depends on many unknown parameters. We therefore confine ourselves to noting that for low microwave powers the enhancement is mainly due to the thermal mixing effect and the Overhauser effect, whereas for high microwave powers the solid state effect dominates. Hence the choice of ESR offset to get a maximal enhancement depends on the microwave power. Figure 8 shows the ^{13}C enhancement P_C of coal 6DE as a function of $\omega-\omega_S$ for P_e=8W. We observe that a maximal enhancement of $\stackrel{\sim}{\sim}$ 75 is obtained for an ESR offset of ± 17 MHz. As the solid state effect is expected to give maxima at ± 15 MHz and the thermal mixing effect at ± 21 MHz (see fig.7) the observed enhancement is due to both. As an experimental separation of the two effects is not possible and as for a dilute spin system like ^{13}C the enhancements are given by more complicated functions than the equations given in section 2, we did not further analyze the ^{13}C enhancement. It also follows from fig.8 that the enhancement curve is almost completely anti-symmetrical, which means that the Overhauser enhancement can be neglected for the ^{13}C nuclei. We finally note that for small P_e a very similar curve is observed as given in fig.8, only the magnitude decreases from 75 to about 60 for P_e=1.1W, which means that already at 8W a considerable degree of saturation has been reached. This is probably due to the fact that the ^{13}C Zeeman relaxation rate W_C is much smaller than W_H ($W_C \simeq 5.10^{-2}s^{-1}$).

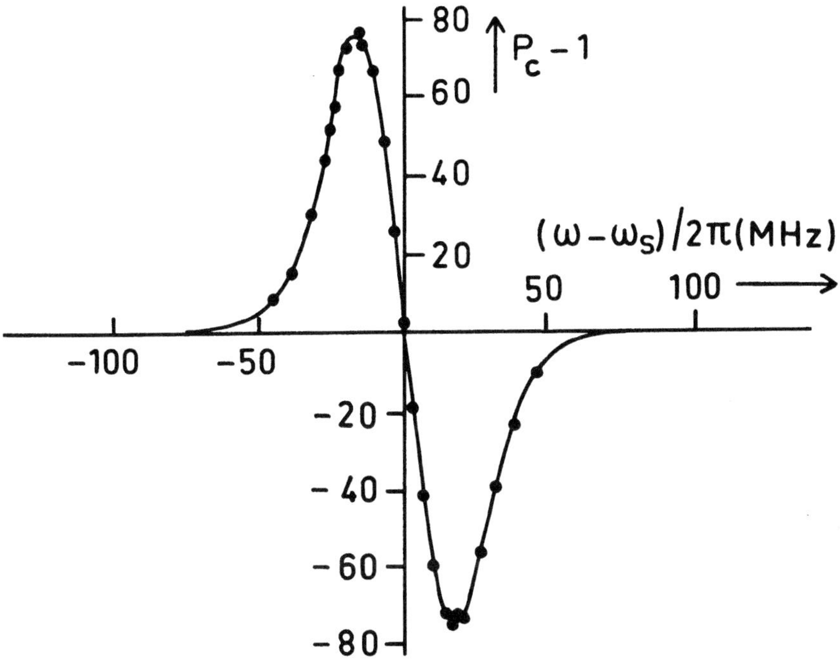

Figure 8. (P_C-1) of coal 6DE as a function of $(\omega - \omega_S)$. P_e = 8W.

4.2. The proton enhancement as a function of coal rank.

In table 1 the proton enhancement factors P_H are given for the different coal samples. They were measured with P_e=8W and an ESR off-set of -60 MHz, so that the enhancement is mainly due to the solid state effect (for all coals this effect appeared to dominate for P_e=8W). We observe that the enhancement reaches a maximum at about 20 % VM. In order to explain this phenomenon we consider eq (11) again. It follows that for $W' << W_I \equiv W_H$ and $\omega = \omega_S - \omega_H$ $|P_H - 1|$ is proportional to $W'/W_H = 2\pi |q|^2 \gamma_S^2 H_1^2 g(o)/W_H$ (eq (9)). It can be shown that $|q|^2 \sim N_S$. It follows that $|P_H - 1|$ is proportional to $N_S g(o)/W_H \sim N_S (\Delta H_{1/2})^{-1} (W_H)^{-1}$, where $\Delta H_{1/2}$ denotes the width of the ESR absorption line. In table 1 also N_S (per cm^3), W_H and $\Delta H_{1/2}$ are given ($\Delta H_{1/2}$ is obtained from the ESR first derivative line at 9 GHz assuming a lorentzian line shape function). It follows that N_S decreases with decreasing coal rank, in accordance with the literature (3,21) (an exception is 13 DE, where too low a value for N_S is found). W_H becomes large for lower coal rank. This means that in any case fot low coal rank the proton relaxation is not determined by the electrons governing the solid

coal	location	% V M 1)	% C 1)	% O 1)	% H 1)	$N_S x10^{-19}$ (cm^{-3})	$\Delta H_{1/2}$(Oe)	$W_H(s^{-1})$	P_H	A_1	A_2	f_a
20	PA	19	90.8	2.7	4.8	2.6	10.0	3.2	19.0	19.0	19.0	0.89
52	PA	23.9	89.6	2.8	4.9	2.3	10.6	2.4	20.5	21.0	21.0	0.89
24	CO	35.6	87.3	5.1	5.5	1.9	10.3	2.9	8.4	15.1	10.9	0.82^5
45	WV	39.4	85.5	6.0	5.6	1.4	10.3	2.4	5.8	13.5	8.9	0.83
01	JAP	40.8	85.8	6.1	6.1	1.4	11.8	2.9	6.3	10.0	6.5	0.78
07	KS	43.7	83.5	6.4	6.0	0.9	11.5	3.6	3.5	5.8	3.7	0.79
48	PA	37.4	84.7	7.5	5.6	1.3	11.2	3.4	6.9	8.5	6.0	0.83
60	KY	43.1	82.1	8.9	5.7	1.3	10.3	5.3	2.8	6.3	4.1	0.80
08	MO	41.1	81.9	9.3	5.4	1.2	10.9	5.9	2.1	5.1	3.5	0.84
22	CO	38.6	82.4	9.9	5.6	1.7	11.8	12.5	2.5	3.5	2.6	0.80
54	IL	45.2	79.4	10.5	5.6	0.7	11.8	12.5	1.5	2.0	1.6	0.80
27	UT	41.6	80.6	11.9	5.6	1.9	12.3	20.0	2.1	2.7	2.0	0.79
41	MM	45.6	80.0	12.0	5.8	1.6	11.2	17.2	1.5	2.8	2.0	0.72
56	CO	40.2	79.1	13.3	5.2	1.4	12.3	43.5	1.2	1.6	1.4	0.79
31	WY	38.5	79.0	14.7	4.7	1.6	10.9	23.3	1.4	2.4	1.9	0.85
33	ARI	45.7	77.7	15.1	5.6	1.2	11.8	45.5	1.1	1.5	1.3	0.74
59	CO	42.1	78.0	15.7	5.1	1.3	12.1	43.5	1.1	1.5	1.3	0.79
34	WY	39.4	76.2	17.5	4.8	0.9	11.8	57.1	1.1	1.2	1.1	0.87
02	ARK	58.6	74.0	18.4	5.9	0.2	11.2	66.9	1.0	1.1	1.0	0.61
21	WY	50.2	73.4	19.7	5.1	0.5	12.6	141	1.0	1.1	1.0	0.69
36	ND	45.2	71.7	22.3	4.9	0.5	10.9	140	1.1	1.1	1.0^5	0.77
65	GER	50.6	69.1	25.5	4.6	0.2	8.9	111	1.0	1.0	1.0	0.81
EB060	BEL	15	83.8	4.0	4.2	3.5	9.2	4.5	21.0	19.7	19.7	0.93^5
EB059	BEL	21	82.0	5.2	4.6	2.8	9.2	2.9	21.0	23.2	23.2	0.91
EB058	BEL	23	82.7	5.4	4.2	2.0	9.5	3.0	19.0	16.5	16.5	0.90
EB057	BEL	27	79.8	5.9	4.8	1.9	9.5	3.0	16.0	15.7	14.6	0.88
EB056	BEL	31	80.0	6.7	5.0	1.9	9.8	3.0	12.0	15.3	12.5	0.87
EB054	BEL	37	77.6	10.0	5.0	1.7	10.0	4.5	7.0	9.3	6.6	0.85
1 DE	GER	7	92.7			4.3	6.9	11.1	9.0	13.4	13.4	1.00
13DE	GER	11	82.5			1.8	8.9	4.2	21.0	11.6	11.6	0.96^5
6DE	GER	20	80.0			2.3	9.2	2.5	23.0	23.1	23.1	0.91
7DE	GER	32	67.7			1.9	9.8	2.6	12.0	17.5	13.9	0.86^5

Table 1. Some relevant parameters of the investigated coal samples; The EB and DE series are provided by the Dutch coal bank SBN, the rest by the Exxon coal bank, Baytown USA. 1): dmmf for the Exxon coals, daf for the SBN coals. For the meaning of the symbols see text. The accuracy in the values of N_S, $\Delta H_{1/2}$, W_H and P_H are 10 %, the accuracy in f_a is 0.01

state effect. As W_H strongly increases if the oxygen percentage exceeds 10 %, W_H then possibly is determined by paramagnetic oxygen, trapped in the coal (this is supported by the fact that W_H also increases when the coal has been exposed to air, cf section 3). $\Delta H1/2$ decreases for increasing coal rank, in accordance with previous observations (3). In order to see whether P_H can be estimated, the quantity $A_1 = 1 + \text{const.}N_S(\Delta H1/2)^{-1}(W_H)^{-1}$ is given in table 1, where the constant $= 2.21.10^{-17}$ has been given such a value that A_1 equals P_H for coal 20. We observe that, though A_1 follows the over-all behavior of P_H rather well, especially for the Exxon coals with values of the volatile matter between 35-45% A_1 is larger than P_H. The reason is not known. It could be caused by a change in the mobility in the coal, resulting in a decrease of $|q|^2$ and hence in P_H. Whether this is the real explanation remains to be investigated. A good agreement with the experimental results is obtained if the following empirical formula is used

$$A_2 = 1 + 2.21.10^{-17}.\frac{N_S}{W_H \Delta H1/2}\cdot\frac{25}{VM}, \quad VM > 25\% \qquad (16)^A$$

$$A_2 = 1 + 2.21.10^{-17}.\frac{N_S}{W_H \Delta H1/2}, \quad VM < 25\% \qquad (16)^B$$

VM denotes the volatile matter percentage. It follows clearly from table 1 that A_2 follows P_H much closer than A_1.

5. ^{13}C SPECTROSCOPY IN COAL USING DNP

One of the major application of DNP is the fast measurement of ^{13}C spectra. In principle this can be done in two ways: firstly, by enhancing the ^{13}C signal directly and measuring the (proton decoupled) free induction decay after 90° rf pulse (the DNP-FID experiment); and secondly, by enhancing the ^{1}H signal and transferring this enhanced polarization towards the ^{13}C spin system by means of cross-polarization (the DNP-CP experiment). Both methods have some disadvantages: if the ^{13}C spins are enhanced directly, the enhancement needs not to be uniform, due to the lack of spin diffusion among these spins, which may result in spectrum distortions. This problem is much less serious in the DNP-CP experiment because the spin diffusion among the abundant protons will probably result in a single average enhancement of the proton signal, and therefore also of the ^{13}C signal after cross-polarization. However, a draw-back of this method is that only

^{13}C atoms in the vicinity of protons will be observed after cross-polarization (22), which also can lead to spectrum distortions. In order to illustrate this in fig.9, the ^{13}C FID, DNP-FID, CP and DNP-CP spectra are given of the high volatile bituminous coal 01.

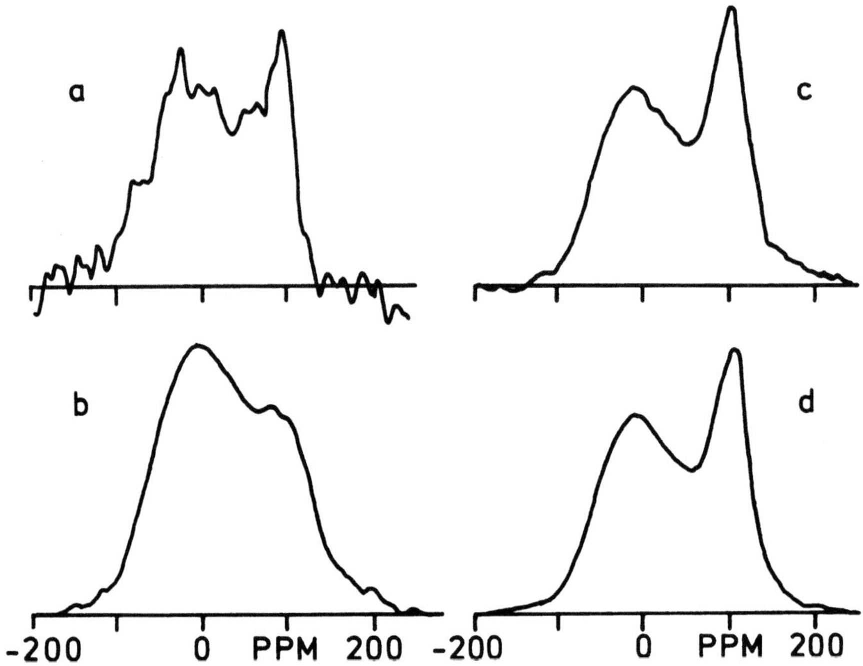

Figure 9. The ^{13}C spectrum of coal 01, measured with different techniques. a: the FID experiment. Number of scans = 6100, recycle delay = 60 sec, measuring time = 100 hrs; b: the DNP-FID exp. Number of scans = 100, recycle delay = 10 sec, measuring time = 17 min, ESR off-set = -21 MHz, P_e=8W; c: the CP exp. Number of scans = 30,000, recycle delay = 0.8 sec, measuring time = 6.7 hrs, matching time = 0.9 ms; d: the DNP-CP exp. Number of scans = 3,000, recycle delay = 0.8 sec, measuring time = 40 min, matching time = 0.9 ms, ESR off-set = -60 MHz, P_e = 8W.
During the acquisition of the ^{13}C signal a proton decoupling field of 10 Oe was applied.

First of all we observe the large improvement in measuring time due to DNP: both the ^{13}C DNP-FID and DNP-CP spectra with an excellent signal-to-noise ratio were obtained in a couple of minutes. Secondly we see that the ^{13}C CP and DNP-CP spectra are equal in shape, indicating that indeed the proton DNP enhancement is uniform. However the (DNP)-CP spectra differ considerably from the DNP-FID spectrum: the peak height due to the aliphatic carbon atoms in the DNP-FID spectrum is much less than that in the (DNP) -CP spectrum. This can be explained in two ways: (i) the DNP enhancement of the aliphatic carbon atoms is less than that of the aromatic carbon atoms. This may occur if the aliphatic carbon atoms are far away from the radicals, or if the relaxation rate of these atoms is larger than that of the aromatic carbon atoms. (ii) A part of the aromatic carbon atoms is not observed in the (DNP)-CP spectrum because of a very weak ^1H-^{13}C coupling. In order to investigate which of these statements is true we also measured the ^{13}C FID spectrum of coal 01 without DNP enhancement. In principle this should give the most reliable results. Fig.9 shows the result. The high frequency peak due to aliphatic carbon atoms seems larger than in the DNP-FID spectrum and smaller than in the DNP-CP spectrum, so that both effects may play a role. However, the signal-to-noise ratio of the FID spectrum is too small to draw definite conclusions.

For completeness, also the apparent ^{13}C aromaticities are given in table 1 of the different samples, following from the (DNP)-CP experiments. Here the aromaticity of the anthracite 1DE has been defined as unity, so that the values display the change in aromaticity compared to 1DE. We observe that the aromaticity decreases with decreasing coal rank, in accordance with the literature (21,22). However, it follows from the above that it might be dangerous to take these numbers as absolute, may be the values are too small.

6. CONCLUSION

Especially for higher rank coals considerable ^1H and ^{13}C DNP enhancements have been found. The enhancement factors depend on several parameters: (i) coal parameters like the number of free electrons, $\Delta H_{1/2}$, the nuclear Zeeman relaxation rate, and the volatile matter percentage; (ii) experimental parameters like the magnitude of the microwave field, the sample volume, the degassing and the powdering of the sample. For the used sample size and experimental set-up the maximum ^1H enhancements of the different coals varied between 1 and 23, and the ^{13}C enhancements between 1 and 75. For our coal samples with a VM% > 7% the proton maximum enhancement is due to the solid state effect, whereas the ^{13}C enhancement is probably due to a combination of the solid state and thermal mixing effect. (We note that for coals with a VM% of a few percent or less mainly a positive Overhauser effect is observed, see also ref (16)). We emphasize that especially the proton enhancement, which was found to be proportional to the microwave power, can be further enlarged if the average microwave field over the sample could be increased (this was confirmed by the fact that for coal

13 DE an increase in the enhancement factor of a factor 2 was obtained by decreasing the inner diameter of the sample from 4 to 2 mm Moreover, it was recently found that the use of quartz instead of pyrex sample tubes did increase the enhancement factor by about 15% due to the lesser damping of the quartz). Investigations are in progress to enhance the microwave field.

For lower coal ranks, typically when %VM > 45%, %C > 80% or %O > 10% the enhancements become small (though it should be kept in mind that even an enhancement of 1.4 already gives a reduction 2 in measuring time). The main reason is the increase of the nuclear relaxation rates with decreasing coal rank, and as this increase is probably due to the increasing amount of oxygen, the %O may be the key parameter for the use of DNP.

DNP can be used for several purposes:

(i) DNP reduces the measuring time of parameters which can be used to characterize the coal, like the apparent ^{13}C aromaticity from a DNP-CP experiment (for high coal ranks it takes only a few minutes to obtain a ^{13}C DNP-CP spectrum), but also parameters such as the proton relaxation rate which is also closely related to the coal rank. Moreover, the enhancement factors themselves can also be used as a means of coal characterization. In fact, in favourable cases an on-line check of the coal quality seems possible (this is the more true because we found that a degassing of the sample during a few minutes gives almost the same result as a few hours).

(ii) DNP can also be combined with magic angle spinning of the sample if more resolution in the ^{13}C spectra is wanted (19). This might e.g. help in detection of molecular groups of low concentration, like carbonyl or carboxyl groups.

(iii) DNP can be used to reduce the measuring time of experiments performed to obtain information about the molecular structure of the coal, like a ^{13}C DNP-CP experiment with a variable matching time (this provides e.g. information about the number of direct CH bonds in the aromatic and aliphatic parts in the coal), or a ^{13}C DNP-FID experiment which provides information about the regions where the radicals are located; but also recently developed Magnetic Resonance techniques like 2 D-NMR or MQT-NMR (24), which have been applied successfully to liquids to resolve complicated structures, now also become feasible for coal.

(iv) So far DNP has been investigated for ^{1}H and ^{13}C only. However, coal contains many other rare spin species like ^{33}S, ^{35}Cl, ^{17}O, ^{14}N and ^{15}N which seem to play an important role in some coal conversion processes. Normally the number of such nuclei is too small to be measured by means of Magnetic Resonance, but may be DNP can assist here also. This will be investigated in future.

ACKNOWLEDGEMENT

The investigations were supported by a grant from the Project Office for Energy Research of the Netherlands Energy Research Foundation ECN, within the framework of the Dutch National Coal Research Program. One of us (MD) was supported by the Netherlands Foundation for Chemical Research (SON) with financial aid from the Netherlands Organization for the Advancement of Pure Research (ZWO).

REFERENCES

1.Bartle,K.D. and Jones,D.W.,1978, Anal.Meth.for Coal and Coal Products (ed.C.Karr), Acad.Press,NY,Vol.II,pp.104-162.
2.Retcofsky,H.L. and Link,T.A., ibid pp. 162-210.
3.Sharkey,A.G. and McCartney,J.T., 1981, Chemistry of coal Utilization (ed.M.A.Elliott), John Wiley & Sons, NY 1981, second supplementary volume, pp. 220-252.
4.Retcofsky,H.L., 1982, Coal Science Vol.I (Ed.M.L.Gorbaty,J.W. Larsen and I.Wender), Acad.Press, London, pp. 43-82.
5.Miknis,F.P., 1982, Magn.Res.Rev.7, pp. 87-121.
6.Pines,A., Gibby,M.G. and Waugh,J.S., 1973,J.Chem.Phys.59, p.569.
7.Uebersfeld,J., Etienne,A. and Combrisson,J., 1954, Nature (London) 174, p.614.
8.Ingram,D.J.E., Tapley,J.G., Jackson,R., Bond,R.L. and Murnahgan, A.R., ibid, p.797.
9.Abragam,A., 1961, The Principles of Nuclear Magnetism, Oxford University Press, London.
10.Goldman,M., 1970, Spin Temperature and NMR in Solids, Oxford University Press, London.
11.Abragam,A and Goldman,M., 1978, Rep.Progr.Phys. 41, p.395.
12.Abragam,A. and Goldman,M., 1982, Nuclear Magnetism: Order and Disorder, Clarendon Press, Oxford.
13.Wind,R.A., Trommel,J. and Smidt,J., 1979, Fuel 58, p.900.
14.Wind,R.A., Trommel,J. and Smidt,J., 1982, Fuel 61, p.398.
15.Trommel,J., 1978, Thesis, Technical University, Delft.
16.Abragam,A., Landesman,A. and Winter,J.M., 1958, CR Acad.Sci. 247, p.1852.
17.Wind,R.A., van Baren,B.A., Emid,S., Kroonenburg,J.A. and Smidt, J., 1973, Physica 65, p.522.
18.Provotorov,B.N., 1962, Sov.Phys. JETP 14, p.1126.
19.Wind,R.A., Anthonio,F.E., Duijvestijn,M.J., Smidt,J., Trommel, J. and De Vette,G.M.C., 1983, J.Magn.Res. 52, p.424.
20.Tse,D. and Hartmann,S.R., 1968, Phys.Rev.Lett. 21, p.511.
21.Smidt,J. and Van Krevelen,D.W., 1959, Fuel 38, p.355.
22.Vander Hart, D.L. and Retcofsky,H.L., 1976, Fuel 55, p.202.
23.Miknis,F.P., Sullivan,M., Bartuska,V.J. and Maciel,G.E., 1981, Org. Geochem. 3, p.19.

24. Bax, A., 1982, Two dimensional NMR spectroscopy in liquids, Reidel, Hingham.

INFORMATIONS ON THE CHEMICAL STRUCTURES OF LIQUID FUELS AND COALS BY COMBUSTION PROCESS STUDIES.

BEN-AÏM Roger I.

Professor - Université Pierre et Marie Curie
4 Place Jussieu 75230-PARIS CEDEX-05 (FRANCE)

ABSTRACT

We develop three cases in which studies on combustion of a fossil fuel give information on its composition, structure or texture.

- In an engine, the octane number is related to the ratio of primary to secondary carbon atoms of the gasoline.

- During coal devolatilization, the early volatiles are formed from the functional groups located at the periphery of lamellae. The activation temperature indicates the nature of the adjoining sites.

- Oxidation of lamellae begins by adsorption of oxygen molecules on active centres. Desorption gives rise to carbon monoxide or carbon dioxide according to the nature of the centres.

Structure and reactivity are strongly related. Usually many experimental techniques are used to investigate the structure of fossil fuels and to predict the reactions that these chemical compounds undergo. As the main use of fossil fuels is combustion, the studies on their compositions and structures finally aim at improving their oxidation processes. In this paper, we develop the reverse point of view : we emphasize some cases in which the combustion in a given device or the products formed or the kinetic law of reaction give important information on the composition, the structure of the texture of a fuel.

L. Petrakis and J. P. Fraissard (eds.), Magnetic Resonance. Introduction, Advanced Topics and Applications to Fossil Energy, 485–500.

I. *LIQUID FUELS*

I.1. Octane Number of gasoline

A very important fraction of liquid hydrocarbons (ranging from C_5 to C_9) is used as a fuel in internal combustion engines. In the case of the commanded ignition motor, the quality of gasoline is represented by the Octane Number. Specialists distinguish between the M.O.N. (Motor Octane Number) and the R.O.N. (Research Octane Number). This property, determined by use of a typical motor (C.F.R.) in standard running conditions, represents the ability of preventing the occurrence of knock.

In 1920 it was noted that some additives such as lead alkyls increase R.O.N.; tetramethyl lead and tetraethyl lead were use as anti-knock agents. The practical problem was solved, although completely unexplained.

The explanation appeared progressively as follows. In 1930, LAFFITTE (1)(2)(3) and Coll. discovered the anomalous oxidation of hydrocarbons called "Cool Flames". In 1948, MILLER (4) took high speed photographs in a cylinder and showed that engine knock results when an anomalous flame appears at one point before the flame front ignited by the plug, reaches this point. In 1948, TOWNEND and Coll. (5) report complete studies on cool flames. A discussion of this paper by MILLER establishes the relationship between knock and cool flame.

But what causes cool flames? This question is strongly related to mechanism of oxidation and has been the object of a large number of papers. For solving the problem it is not necessary to enter into this much debated question. Simple remarks are sufficient:

n C_7H_{16}	:	R.O.N. = 0
2.2.4 trimethyl pentane	:	R.O.N. = 100

In a gasoline, secondary carbon atoms decrease R.O.N., while primary or aromatic ones increase R.O.N.

The refining industry has now two means of increasing the R.O.N. of a gasoline: adding lead alkyl or reforming the fuel so as to transform normal hydrocarbons into aromatic or branched molecules. However, lead compounds cause pollution while reforming requires new plants and important investments. Table I gives regulations in different countries (after to C & EN April 1983, p.13).

Country	Lead content $(g.l^{-1})$
Belgium	0.4
Denmark	0.15 (regular), 0.4 premium moving to 0.15 premium in 1984
France	0.4
West Germany	0.15
Greece	0.4 moving to 0.15 in 1984
Japan	Unleaded regular more than 90% of total sales
U.S.	Unleaded gasoline 54% of total sales; moving to 0.29
U.K	0.4 moving 0.15 in 1984 and no lead in 1990

Table I

It is evident that the trend is to lower lead percent in gasoline and consequently a better understanding of engine knock is needed in order to lower the cost of reforming the fuel. MEYERS and Coll. (6) made a very interesting attempt to relate R.O.N. or M.O.N. to the composition of gasoline and to the structure of the hydrocarbons. They proposed the empirical law:

R.O.N. = A + B (isoparaffin index) + C (aromatics) + D (lead) + E (Sulfur)

Coefficients A, D, E show influence of composition of gasoline.
Coefficients B, D show influence of structure of hydrocarbons.
Isoparaffin index is a measure of the amount of branching, that is, the ratio of CH_3/CH_2 for a given paraffin or a mixture of paraffins. It is determined experimentally by N.M.R. The above authors quoted the Chemical shift for different structures (Table II).

Proton type	Chem. shift
Methyl (paraffins)	0.6 to 1.0 ppm
Methylene (paraffins)	1.0 to 1.5 ppm
Ring aromatic	6.6 to 8.0 ppm

Table II

Other studies have attempted to establish a direct relation between engine-knock properties of a combustible and its cool flame domain (7) or its low temperature slow oxidation (8). As these phenomena are simpler and better defined than engine-knock it might be possible to deduce them from composition or structure of the gasoline. Semi-experimental studies exist of the stability of peroxydic compounds (that play an important part in low temperature oxidation and cool flames) and of the comparative reactivity of carbon groups, according to their place in the molecule.

I.2. Consequences

From this example we can see how strongly the structure of a fuel and its behaviour in a combustion device are related, even if the combustion process includes so many parameters that an explicit relation is difficult to point out.

The study of MYERS and Coll. cannot give quantitative results in all cases due to complexity of various gasolines (content of olefins, of Pb, of S) and due to the numerous physico-chemical problems in an engine (boiling point of gasoline,heterogeneous oxidation or inhibition). However, it provides a useful guide for obtaining a better value of R.O.N. by reforming the fuel.

The evidence of a strong interaction between the structures of the hydrocarbons and their engine-knock properties gave an opportunity to Industry to solve an important practical question. The reforming units in refineries are now able to produce a gasoline which conforms with regulations as well as with specifications of car makers.

II. COAL DEVOLATILIZATION

Combustion of coal is a very complex experimental operation including homogeneous and heterogeneous chemical reactions and transport phenomena. It is very interesting to connect some of the combustion processes to the chemical composition or the chemical structure of the coal sample.

We emphasize two stages among all those occuring during combustion of a coal particle: devolatilization and heterogeneous reaction with oxygen.

II.1. Structure of coal

Let us recall that coal is now presented as a compound mainly formed by polycondensed aromatic rings (9). Carbon atoms included in these rings are 88 to 93 percent of the total number. This is refered as the aromaticity number. These aromatic structures (called lamellae or foils) are tied together by aliphatic brigdes. Tetrahedral bonds of these aliphatic parts give coal its three dimensional structure. At the periphery of lamellae exist the functionnal groups usually encountered in coal (alcohols, ether-oxides, carboxylic acids, amines, thiols, thiophene, ...). Figure 1 shows a typical structure of a lamella and its symbolic representation.

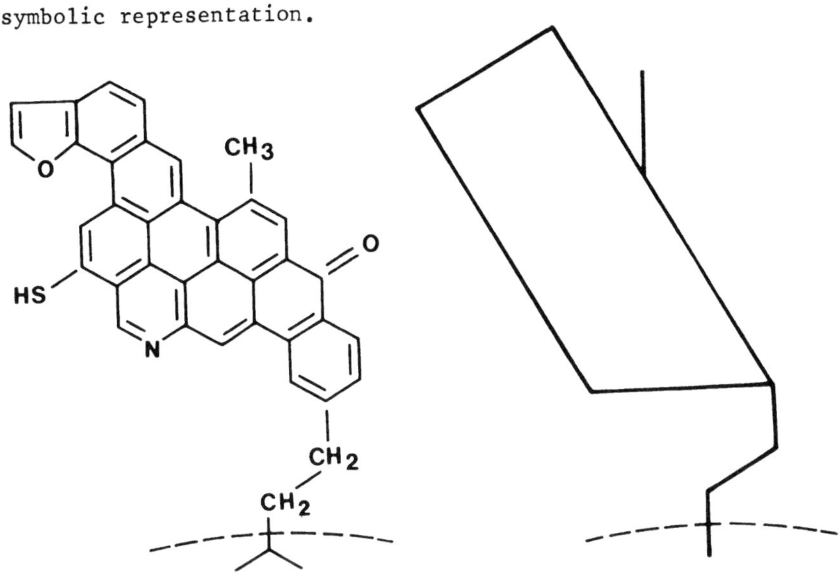

Figure 1

Arrangement of the rings in a foil with the lateral aliphatic groups and the functional groups (left). Representation of the same foil in the form of a quadrilateral with the aliphatic chains (right).

The number of rings in each foil increases from lignites to sub-bituminous coals and to anthracites. Figure 2 shows the mean number of cycles for each lamella as a function of the carbon content of the coal sample.

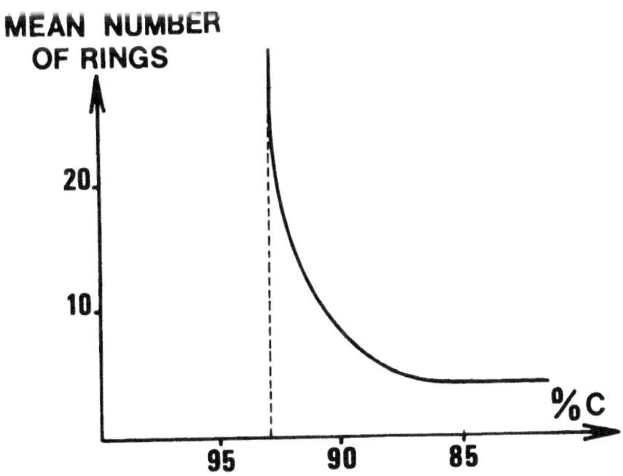

Figure 2. The mean number of rings is between 5 and 25 according to the rank of the coal sample.

As the mean number of rings increases, the arrangement of lamellae becomes better ordered. Figure 3 shows disposition of lamellae in a sub-bituminous coal and in an anthracite.

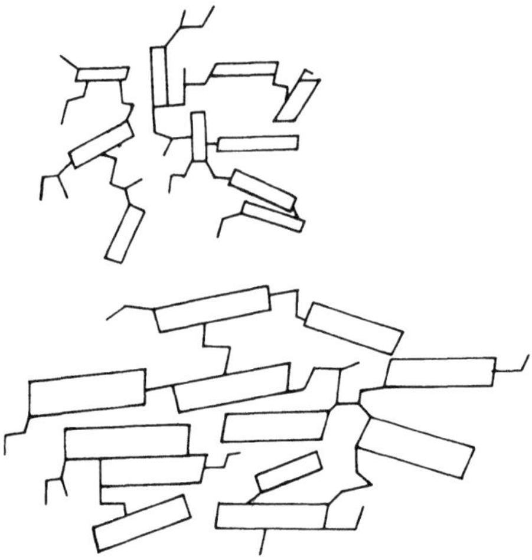

Figure 3. In a sub-bituminous sample, foils are random distributed (up). In an anthracite sample, the foils are larger and are preferentially oriented.

II.2 Products of devolatilization

In a burner, a coal particle is heated very rapidly ($dT/dt \sim 10^4$ $K.s^{-1}$). Gases are emitted and the process is called devolatilization. We distinguish three periods:

- the first one correspond to physical desorption of gases formed during the process of carbonization (i.e. formation of coal from wood). For example we observe ethane.
- the second one corresponds to the pyrolysis of the functional chains of the edge of the lamellae. They are called the early volatiles.
- the third period corresponds to the formation, from the aliphatic chain of liquid compounds and tars which undergo pyrolysis and produce some gases called late volatiles.

It is obvious that the second period can give information on the structure of coal. So, for the standard test of devolatilization (7 minutes at 950°C) the weight loss is in accord with the percent of atoms involved in rings or in the aliphatic chains or functional groups. This result has been clearly shown by SOLOMON and COLKET (10).

The aliphatic chains give rise to hydrocarbons (CH_4, C_2H_6, C_2H_2, C_2H_4, ...).

The ether-oxide function (-C-O-C-) produces CO.

The carboxylic function forms CO_2

$$\emptyset - COOH \longrightarrow \emptyset H + CO_2$$

The alcohol function forms H_2O according to

$$\emptyset - OH + \emptyset' H \longrightarrow \emptyset - \emptyset + H_2O$$

or : $$\emptyset - OH + \emptyset' OH \longrightarrow \emptyset - O - \emptyset + H_2O$$

The thiol function (-C-SH) produces the sulfur compounds (SH_2 , CS_2).

Nitrogen is mainly involved in cycles (pyridine, pyrrole,...) and does not appear in the early volatiles (or at most in very small amount).

II.3 Kinetic law of devolatilization

Formation of the early volatiles includes solid phase reaction, pyrolysis of liquids, desorption and transport phenomena through the coal sample.

Many studies have been devoted to the kinetic law of devolatilization but the results are not yet satisfactory. So we shall not try to attempt an interpretation but we limit to description of some partial results.

II.3.1 Activation temperature

The usual form of the constant rate k for an elementary reaction is given by the Arrhenius law

$$k = A \exp(- E/RT)$$

R = gas constant
E = activation energy per mole.

We shall use the more convenient form:

$$k = A \exp(- \theta /T)$$

= activation temperature (in Kelvin).

Table III gives relationship between E and θ :

E	θ
20 $kcal.mol^{-1}$ = 83,6 $kJ.mol^{-1}$	10.000 K
40 " = 167,2 "	20.000
100 " = 418 "	50.000

Table III

DE SOETE (11), using the data of ref. 12 and 13 has given the activation temperatures for the pyrolysis of some bituminous coals in table IV.

% of volatiles in coal.	Products							
	CH_4	C_2H_6	C_3H_6	C_3H_8	$n\text{-}C_4H_{10}$	H_2O	CO	CO_2
10				31200		16800 à 17100	13600	13300
19		20600 à 21600		29900		16100 à 16800	13300	13000
29				29300		15800 à 16700	12800	11600
39,5	26900 à 34700	30000 à 31200	30200	28400	29800	15600 à 16600	11100	11000

Table IV

It is noticeable that the activation temperatures corresponding to hydrocarbons (27.000 $< \theta <$ 35.000) is higher than those corresponding to formation of oxygenated compounds (11000 $< \theta <$ 17000). This agrees with the fact that the strength of C-C bond is higher than the C-O bond.

The scattering of the values of θ comes from the different sites where the lateral chains are bound. However, the experimental results are not so precise that the nature of the sites could be now determined from the corresponding value of θ .

II.3.2 Rate constant of devolatilization

It depends on the chemical reaction, on the transport phenomena, and on the rate of heating. According to (14) and (15), the minimum diameter d_m for which the rate of devolatilization is mainly imposed by chemical kinetics is related to the rate of heating and given in table V.

Rate of heating ($K.s^{-1}$)	d_m (μm)
10^3	1800 - 7000
10^4	400 - 1500
10^5	110 - 400
10^6	20 - 90

Table V

So, it appears that for pulverized coal (d $\sim$ 80 µm) in an industrial burner (heating $\sim 10^4.K.s^{-1}$) or in Laboratory conditions, the chemical regime is rate determining. For blocks of coal transport phenomena are rate determining.

II.4 Consequences

Devolatilization involves the side chains and the functional groups fixed at the edge of lamellae but brings little changes to the foils. Composition of early volatiles compared to the composition of coal gives informations on the distribution of atoms. For example table VI has been established by (16)for a bituminous coal.

	coal	coke	corrected values
% volatiles	39.9	2.4	
H/C	0.73	0.095	0.13
O/C	0.202	0.0027	0.0038
N/C	0.019	0.0135	0.019

Table VI

The volatiles still contained in coke (2,4%) indicate pyrolysis of tar and formation of late volatiles. Corrected values have been calculated by us and correspond to the initial ratio of atoms in the part of coal that constitutes coke after devolatilization. It appears that:

- 6/7 of hydrogen atoms are located at the periphery of lamellae
- Oxygen atoms are mainly present in lateral chains
- Nitrogen atoms are only present in ring structures, that is, in pyridinic or pyrrolic groups.

From this example, it clearly appears that analysis of early volatiles can help to establish the chemical structure of a coal sample.

III. OXIDATION OF COKE AND CHAR

Devolatilization of coal leaves a solid residuum called "coke" or "char", the composition, structure and texture of which depend on the coal sample, size of the particle, heat gradient, time of heating; thus, oxidation of coke by a gaseous mixture varying with time (O_2, CO, CO_2, H_2O,...) is a heterogeneous reaction that offers the maximum complexity.

III.1 Composition of coke and char

From the results of paragraph II, it is evident that coke corresponds to lower values of H/C and O/C and to higher values of N/C than in the corresponding coal. As a first approximation, it can be stated that char is mainly constituted of carbon atoms forming foils; coke corresponds to partly fused char.

As there remain very few lateral groups, we can assume that each lamella is a cristallite of graphite. The cristallites are formed by different numbers of aromatic rings and overlap.

The mean number of rings and the mean angular orientation and distances of the graphitic planes (or eventually the distribution laws, corresponding to these data) may represent the texture of . However another parameter plays a prominent part in the oxidation reaction: the arrangement of the carbon atoms at the edge of each lamella. The distances between the atoms shown on figure 4 are indicated below :

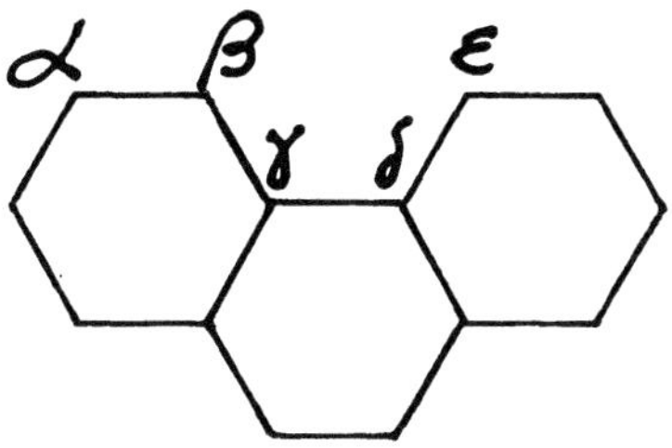

Figure 4 : distance (in Å) of
α - β : 1.42
β - δ : 2.46
β - ε : 2.84
Distance of oxygen atoms in a molecule : 1.21 Å

III.2 Oxidation reaction

We shall try to describe the oxidation reaction in a simple manner, assuming that coke is formed by regular graphitic planes reacting with molecular oxygen. It is evident that only the carbon atoms of the edge will react, whereas the surface is inert. BONNETAIN (17) showed that only 5% of the total surface of a graphite sample (measured by B.E.T. method) is covered by surface oxides. Some authors found much larger quantities of oxygen retained by graphite. This may be attributed to insertion of O_2 between two graphitic planes.

III.3 Surface oxides

In 1863, SMITH (18) discovered that oxygen is chemisorbed on charcoal and could be recovered by heating only as carbon dioxide. Since then, many studies have been devoted to this subject; a review was given by BOEHM (19) and does not confirm the results of SMITH.

Molecular oxygen is adsorbed physically and chemically by charcoal. The irreversible uptake starts at - 40°C; there is only reversible physical adsorption at lower temperatures.

Chemisorption may occur on one active centre (without dissociation of oxygen) or on two adjoining sites (leading to dissociation of oxygen). The higher is the temperature, the more important is the second type of chemisorption. HENNIG (20) found by observations of single crystal of graphite, that after oxidation with dry oxygen the "armchair" conformation of the periphery resulted, (fig. 5). In the presence of small amounts of water, the "zigzag" conformation resulted (fig. 5).

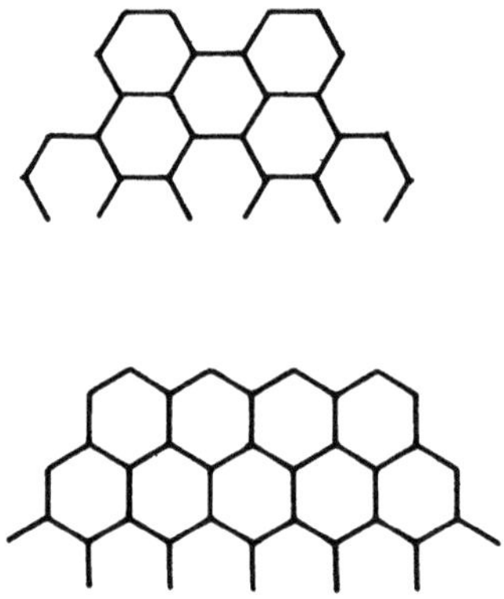

Figure 5 : The rings of the periphery of graphite may adopt the disposition of armchair (up) or zigzag (down).

Oxidation leads to different carbonyl groups of the surface (fig. 6).

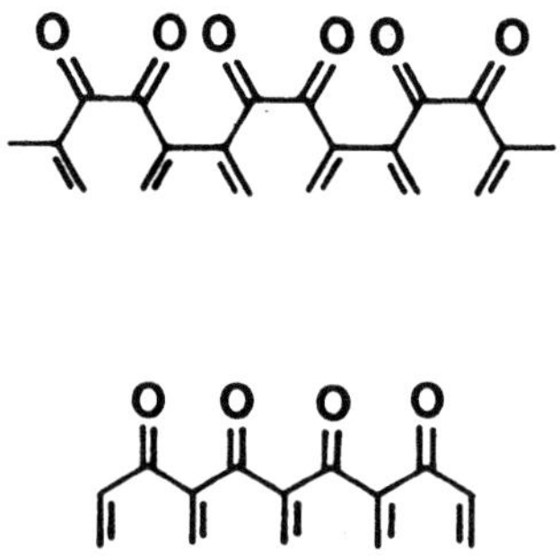

Figure 6 : Formation of oxide at the periphery of foils. Oxides corresponding to the armchair disposition (up) or to zigzag disposition (down).

III.4 Products

Two reactions ways of desorption of surface oxides can be imagined:

$$(-CO) \longrightarrow CO \qquad (1)$$
$$2(-CO) \longrightarrow CO_2 + (-C) \qquad (2)$$

The first case corresponds to the formation of carbon monoxide and the second to the formation of carbon dioxide.

The experimental activation temperature of desorption has been estimated to be 20000 - 40000 K. After systematic studies RHEAD and WHEELER (20) concluded that CO and CO_2 are primary products of combustion.

DE SOETE (11) thinks that in the case of flames (that is at

high temperature), reaction 1 is predominent; CO undergoes subsequent oxidation in gaseous phase.

An attempt here is made to relate the products formed to nature of the active centre.

Figure 7 shows that the armchair configuration can lead to CO by reaction 1.

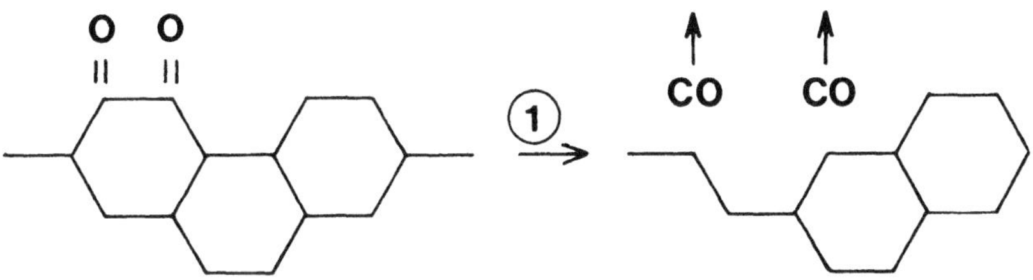

Figure 7
Formation of two CO necessitates to open only one ring.

Figure 8 shows an other armchair configuration with a more complex situation. Desorption with formation of two CO molecules would need breaking of two aromatic rings and is improbable. So desorption with formation of CO_2 is certainly easier.

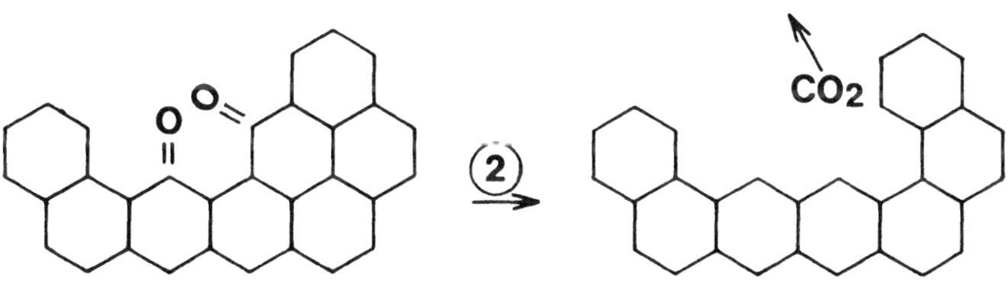

Figure 8
This special armchair configuration can lead to CO_2

It is more difficult to decide what products result from desorption of oxide in the zigzag configuration.

It is very important to remark that in the general case of armchair disposition, ablation of the first range of oxide leads to the same arrangement.

All these considerations are made with the assumption that periphery of coke is perfectly well defined. Evidently this is not the case and it is the reason why kinetic data, such as order of reaction, activation temperature, ratio CO/CO_2 , suffer from a large uncertainty.

LAINE, VASTOLA and WALKER (22) have determined the ratio CO/O_2 in the gases produced by oxidation of Graphon, at different temperatures. This ratio increases considerably with temperature as shown in table 7.

t° (°C)	CO/CO_2
575	3.2
625	4.8
675	6.3

Table 7

III.5 Consequences

The ratio CO/CO_2 obtained by combustion of coal in an industrial device cannot directly be related to the structure of the periphery of lamellae because four main competitive chemical reactions occur:
- adsorption and reaction of water
- gas phase oxidation of CO to form CO_2
- heterogeneous reaction of CO_2 to form CO
- catalytical reactions with metal atoms which are always present in coal.

It is the reason why, up to now, combustion of coal is mainly performed according to empirical methods; chemical considerations are of little importance.

CONCLUSION

Energy evolved is the most important datum looked for a fossil fuel; it depends mainly on its composition. The second important point is to obtain the most convenient mode of use with:
- better yield,
- less pollution,
- simpler practical device,
- higher performance,
- cheaper fuel,...

For this purpose, kinetic data are necessary, they depend mostly on the structure of molecules (or macromolecules) of the combustible as shown in two cases.

References

(1) Laffitte, P., Prettre, M., Dumanois, 1930, Comptes Rendus Ac. Sc., 191, p. 329, p.414.

(2) Prettre, M., 1931, Ann. Comb. Liq., 6, p. 269, p. 533.

(3) Prettre, M., 1932, Ann. Comb. Liq., 7, p. 699.

(4) Miller, C.D., 1948, Phot. Soc. Am., 14, p. 668.

(5) Spence, K., Townend, D.T.A., 1948, Third Symposium on Combustion, Baltimore, p. 404.

(6) Myers, M.E., Stollsteimer Jr, J., Wims, M., 1975, Anal. Chem. 47, p. 2301.

(7) Antonik, S., Delfosse, L., Baillet, C., 1973, Bull. Soc. Chim. de France, p. 1887.

(8) Karbassian, A., Ben-Aïm, R.I., Chamboux, J., Viossat, V., 1981, Rev. Inst. Fr. du Pétrole, 36, p. 801.

(9) Given, P.H., 1960, Fuel, 39, P. 147, "The distribution of hydrogen in coals and its relation to coal structure".

(10) Solomon, P.R., Colket, M.B., 1979, 7th Symposium (International) on Combustion. Pittsburgh, p. 131.

(11) De Soete, G.G., 1982, Rev. Inst. Fr. du Pétrole, 34, p.403; p. 503.

(12) Jüntgen, H., Van Heek, K.H., 1979, Fuel Processing Technology, 2, p. 261.

(13) Klein, J., 1971, Thesis. Aix-La-Chapelle.

(14) Schwandter, D., 1971. Thesis. Aix-La-Chapelle.

(15) Koch, V., 1968, Thesis.Aix-La-Chapelle.

(16) Smoot, L.D., Pratt, D.T., 1979, "Pulverised-coal combustion and gazification". Plenum Press. New-York.

(17) Bonnetain, L., 1961, J. Chim. Phys., 58, p. 34.

(18) Smith, A., 1863, A 12, p. 424.

(19) Boehm, H.P., 1966, Advances in Catalysis Academic Press, p. 179.

(20) Henning, G.R., Proc. of 5th Symp. Conf. on Carbon, 1963, Pergamon. Oxford. Vol. I, p. 143.

Henning, G.R., 1962, Z. Electrochem., 66, p. 629.

(21) Rhead, T.F.B., Wheeler, R.V., 1912, J. Chem. Soc., 101, p. 831, p. 846; 1913, J. Chem. Soc., 103, p. 1210.

(22) Laine, N.R., Vastola, F.J., Walker, P.L., 1963, Proc. of 5th Symp. Conf. on Carbon. Pergamon, Oxford, p. 211.

TEACHING THE NEW NMR: A COMPUTER-AIDED INTRODUCTION TO THE DENSITY MATRIX TREATMENT OF DOUBLE-QUANTUM SPECTROMETRY

G. D. Mateescu and A. Valeriu

Department of Chemistry
Case Western Reserve University
Cleveland, Ohio 44106

Abstract. This paper constitutes a guide for the use of density matrix calculations necessary for the understanding of double-quantum coherence detection in NMR. In keeping with its didactic nature, the article follows in more detail than usual the evolution of the density matrix elements throughout single- and two-dimensional sequences applied to coupled carbon nuclei. The relation of the density matrix treatment with the vectorial representation is discussed.

1. INTRODUCTION

Frequently, multipulse NMR sequences are described in terms of the evolution of magnetization vectors. Such description is insufficient for sequences involving double-quantum coherence. Only the density matrix treatment can properly represent the evolution of the spin system in such cases.

These notes are mainly intended as a theoretical background for a computer-generated movie dedicated to the teaching of modern NMR experiments. However, they have been written in such a way as to constitute an independent study material. It is assumed that the student is familiar with basic FT NMR, and elementary matrix algebra. A summary of density matrix formalism is given in the Appendix. A "math reminder" for those who may need it can be found in reference 1.

The paper begins with an intuitive description of the density matrix and continues with the density matrix calculation of INADEQUATE (Incredible Natural Abundance Double QUAntum Transfer

L. Petrakis and J. P. Fraissard (eds.), Magnetic Resonance. Introduction, Advanced Topics and Applications to Fossil Energy, 501–524.

Experiment) and its two-dimensional extension, the carbon-carbon connectivity experiment.

2. THE DENSITY MATRIX

Consider an AX system of two coupled carbon-13 nuclei. The four energy levels of such a system and the allowed transitions are shown in Figure 1. We can visualize the relations between the

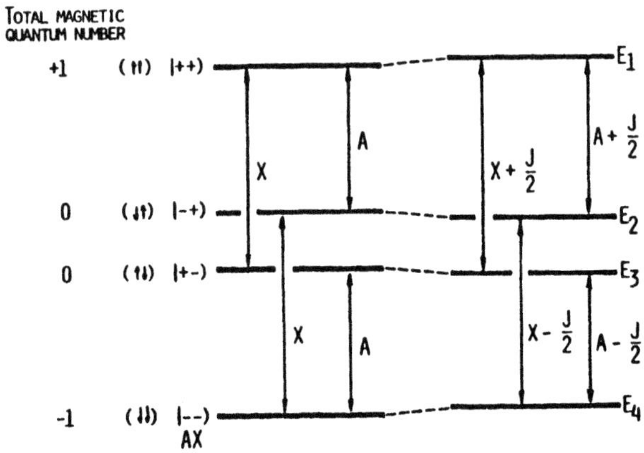

Figure.1 Energy levels of an uncoupled and coupled AX system. Transition frequencies and the coupling constant are in Hz.

four states by building a matrix as shown in Figure 2. The states, represented by "kets" |++), |-+), |+-), and |--), are in order of decreasing energies. The diagonal elements represent

	|+ +)	|- +)	|+ -)	|- -)
|+ +)	P_1	$1Q_A$	$1Q_X$	$2Q_{AX}$
|- +)		P_2	0Q	$1Q_X$
|+ -)			P_3	$1Q_A$
|- -)				P_4

↑ ↑
|A X)

Figure 2

the populations of the four levels (see Appendix for quantum mechanical basis). The off-diagonal elements, called "coherences" connect pairs of different states. They indicate possible transitions between corresponding levels. Single quantum (1Q) coherences connect states between which transitions are allowed ($\Delta m = \pm 1$). Each 1Q coherence is labeled according to the nucleus which is flipped. For example, in going from $|++\rangle$ to $|-+\rangle$ only the spin of A is inverted. Therefore the d_{12} element of matrix D [cf. relation (A15) of Appendix] is a $1Q_A$ coherence. All 1Q coherences correspond to transverse magnetizations which can be represented in vector form [see (A22)]. The element d_{14} is a double quantum coherence, $2Q_{AX}$, since both A and X spins are inverted from $|++\rangle$ to $|--\rangle$. The "zero-quantum" coherence d_{23} corresponds to a "flip-flop" transition (E_2-E_3). The name of the coherence is not a happy choice since the energy of the transition is not necessarily zero.

The matrix elements below the diagonal do not represent other coherences. They are mirror images (complex conjugates) of the elements above the diagonal:

$$d_{kj} = d^*_{jk}. \qquad (1)$$

3. DENSITY MATRIX DESCRIPTION OF INADEQUATE

The main goal of INADEQUATE[2] was to eliminate the strong signal of noncoupled C-13 nuclei in order to easily observe the 200 times weaker satellites due to C-C coupling. This is realized by exploiting the different phase responses of the coupled and noncoupled spin signals to a cycle of minimum two scans in which the observe pulse phase is varied by 90°. Then, the receiver is phase matched with the desired signal. It shall be seen that the different phase behavior of coupled nuclei is connected with their double-quantum coherence. The beauty of Inadequate resides in its basic simplicity (see Figure 3): only a two-step cycle is theoretically needed to eliminate the unwanted signal. That the real life sequences may reach 128 or more steps is exclusively due to hardware (pulse) imperfections whose effects must be corrected by additional phase cycling.

The essence of Inadequate can therefore be understood by following only the basic sequence shown in Figure 3.

<u>Preparation (equilibrium populations)</u>

At thermal equilibrium the four energy levels shown in Figure 1 are populated according to the Boltzmann distribution law:

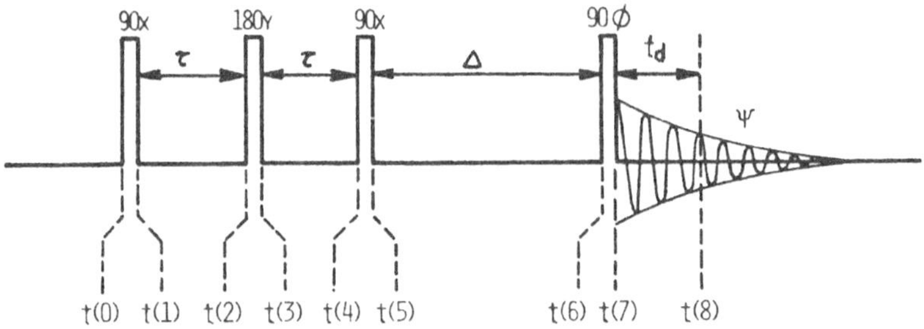

Figure 3. The INADEQUATE sequence (proton decoupling is applied throughout the experiment).

$$\frac{P_i}{P_j} = \frac{\exp(-E_i/kT)}{\exp(-E_j/kT)} = \exp\left|\frac{E_j - E_i}{kT}\right|. \tag{2}$$

Taking the least populated level 1 as reference we have:

$$P_2/P_1 = \exp[(E_1 - E_2)/kT] = \exp[h(A + J/2)/kT]. \tag{3}$$

Since at today's spectrometer fields transition frequencies are at least five orders of magnitude larger than coupling constants (J), and four orders of magnitude larger than chemical shifts (δ), we can neglect both J and δ when calculating populations. Of course they will not be neglected when calculating peak frequencies. Thus, $hA/kT \approx hX/kT = p$. A first order series expansion of the exponential in (3) is justified since $p \ll 1$:

$$P_2/P_1 = \exp(hA/kT) = 1 + (hA/kT) = 1 + p, \tag{4}$$

$$P_3/P_1 = \exp(hX/kT) = 1 + (hX/kT) = 1 + p, \tag{5}$$

$$P_4/P_1 = 1 + h(A+X)kT = 1 + 2p. \tag{6}$$

We now normalize the sum of populations:

$$\begin{aligned} P_1 &= P_1 \\ P_2 &= (1 + p)P_1 \\ P_3 &= (1 + p)P_1 \\ P_4 &= (1 + 2p)P_1 \\ \hline 1 &= (4 + 4p)P_1 = P_1 S \end{aligned} \tag{7}$$

Hence,

$$P_1 = 1/S$$

$$P_2 = P_3 = (1 + p)/S$$

$$P_4 = (1 + 2p)/S,$$

and the density matrix at equilibrium is:

$$D(o) = \begin{vmatrix} P_1 & 0 & 0 & 0 \\ 0 & P_2 & 0 & 0 \\ 0 & 0 & P_3 & 0 \\ 0 & 0 & 0 & P_4 \end{vmatrix} = \frac{1}{S} \begin{vmatrix} 1 & 0 & 0 & 0 \\ 0 & 1+p & 0 & 0 \\ 0 & 0 & 1+p & 0 \\ 0 & 0 & 0 & 1+2p \end{vmatrix} =$$

$$= \frac{1}{S} \begin{vmatrix} 1 & 0 & 0 & 0 \\ 0 & 1 & 0 & 0 \\ 0 & 0 & 1 & 0 \\ 0 & 0 & 0 & 1 \end{vmatrix} + \frac{p}{S} \begin{vmatrix} 0 & 0 & 0 & 0 \\ 0 & 1 & 0 & 0 \\ 0 & 0 & 1 & 0 \\ 0 & 0 & 0 & 2 \end{vmatrix} . \qquad (8)$$

It is seen that the first term of the sum above is very large, compared to the second term, because of the small value of p. However, containing the unit matrix, the first term is not affected by any evolution operator and, as such, it does not play a part in the evaluation of NMR observables. Though much smaller, it is the second term which is important because it contains the population differences. From now on we will work with this term only, ignoring the constant factor p/S and taking the license to continue to call it D(o):

$$D(o) = \begin{vmatrix} 0 & 0 & 0 & 0 \\ 0 & 1 & 0 & 0 \\ 0 & 0 & 1 & 0 \\ 0 & 0 & 0 & 2 \end{vmatrix} . \qquad (9)$$

In order to compare the results of the density matrix treatment with those of vectorial representation, we will calculate, for every step of the sequence, the magnetization components, using the Appendix relations (A19) - (A24). One can check that at thermal equilibrium [D = D(o)]

$$M_{zA} = (M_o/2)(d_{11} - d_{22} + d_{33} - d_{44}) =$$
$$= (M_o/2)(\ 0 \ - \ 1 \ + \ 1 \ - \ 2\ ) = -M_o.$$

Since M_{zX} is also $-M_o$, the total magnetization is $-2M_o$ (see Figure 4a).

The transverse magnetization

$$M_{TA} = M_{TX} = 0.$$

First pulse: 90°x

At time t(o) a nonselective pulse 90xAX is applied. Since all pulses in this sequence are nonselective, the notation AX will be omitted. The density matrix D(1) after the pulse is calculated according to:

$$D(1) = R^{-1}D(O)R \tag{10}$$

which is explained in the Appendix. The rotation operator, R and its reciprocal R^{-1} for the nonselective 90x pulse are:

$$R_{90x} = \frac{1}{2}\begin{vmatrix} 1 & i & i & -1 \\ i & 1 & -1 & i \\ i & -1 & 1 & i \\ -1 & i & i & 1 \end{vmatrix} \; ; \; R^{-1}_{90x} = \frac{1}{2}\begin{vmatrix} 1 & -i & -i & -1 \\ -i & 1 & -1 & -i \\ -i & -1 & 1 & -i \\ -1 & -i & -i & 1 \end{vmatrix}.$$

First we postmultiply D(o) by R:

$$D(o)R = \frac{1}{2}\begin{vmatrix} 0 & 0 & 0 & 0 \\ i & 1 & -1 & i \\ i & -1 & 1 & i \\ -2 & 2i & 2i & 2 \end{vmatrix} \; ;$$

then we premultiply the result by R^{-1}:

$$D(1) = R^{-1}[D(o)R] = \frac{1}{4}\begin{vmatrix} 4 & -2i & -2i & 0 \\ 2i & 4 & 0 & -2i \\ 2i & 0 & 4 & -2i \\ 0 & 2i & 2i & 4 \end{vmatrix} = \frac{1}{2}\begin{vmatrix} 2 & -i & -i & 0 \\ i & 2 & 0 & -i \\ i & 0 & 2 & -i \\ 0 & i & i & 2 \end{vmatrix} \tag{11}$$

We note that the 90° pulse equalizes the populations and creates single quantum coherences. The longitudinal magnetization is null while

$$M_{TA} = M_o(d^*_{12} + d^*_{34}) = M_o(\frac{i}{2} + \frac{i}{2}) = iM_o,$$

$$M_{TX} = M_o(d^*_{13} + d^*_{24}) = Mo(\frac{i}{2} + \frac{i}{2}) = iM_o,$$

$$M_{TAX} = 2iM_o.$$

We note that the transverse magnetization is imaginary. Since $M_T = M_x + iM_y$, it results that $M_x = 0$ and $M_y = 2M_o$ (Figure 4b).

So far, the vector representation would have been much simpler to use. Let us see, though, what happens further.

Evolution from t(1) to t(2)

The standard formula describing the (laboratory frame) time evolution of the density matrix elements in the absence of a pulse is:

$$d_{mn}(t) = d_{mn}(o)\exp(-i\omega_{mn}t). \tag{12}$$

d_{mn} is the matrix element and $\omega_{mn} = (E_m - E_n)/\hbar$ is the angular frequency of transition $m \to n$. Note that $d^n_{mn}(o)$ is the starting point of the evolution immediately after a given pulse. In our case the elements $d_{mn}(o)$ are those of D(1).

If the evolution is described in a frame rotating at transmitter frequency ω_o equation (12) becomes:

$$d_{mn}(t) = d_{mn}(o)\exp(-i\omega_{mn}t)\exp[i(m_m - m_n)\omega_o t], \tag{13}$$

where m_m and m_n are the total magnetic quantum numbers of states m and n (see Appendix). Let us apply (13) to our particular case ($m_1 = 1$; $m_2 = m_3 = 0$; $m_4 = -1$). We observe that during the evolution the diagonal elements d_{mm} are invariant since both exponentials are equal to 1. All single quantum coherences above diagonal have $m_m - m_n = 1$. Hence,

$$\begin{aligned} d_{mn}(t) &= d_{mn}(o)\exp(-i\omega_{mn}t)\exp(i\omega_o t) = \\ &= d_{mn}(o)\exp[-i(\omega_{mn} - \omega_o)t] = \\ &= d_{mn}(o)\exp(-i\Omega_{mn}t), \end{aligned} \tag{14}$$

where Ω_{mn} is the difference between the precession frequency and the rotating frame frequency. The rotating frame treatment is useful not only for better visualization of vector evolution but, also, because the detection is actually made at the resulting low

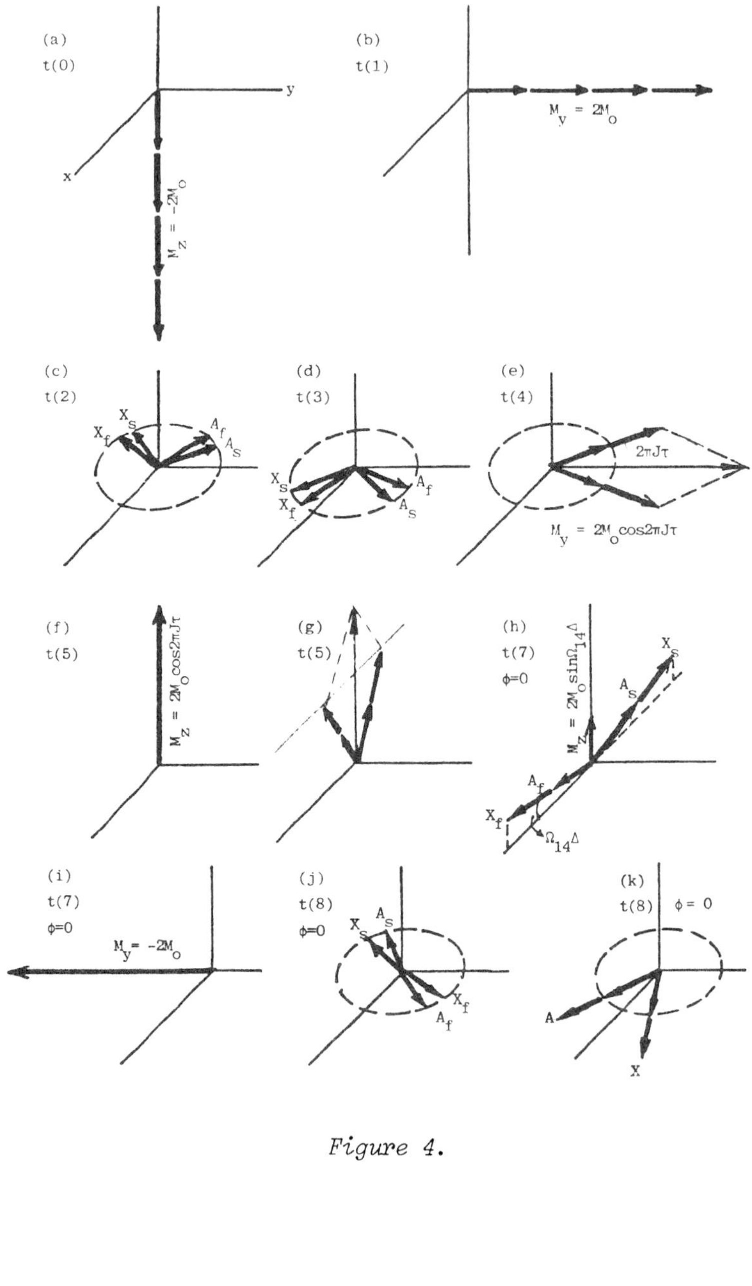
(a)
t(0)
y
x
$M_z = -2M_o$
(b)
t(1)
$M_y = 2M_o$
(c)
t(2)
X_f
X_s
A_f
A_s
(d)
t(3)
X_s
X_f
A_f
A_s
(e)
t(4)
2πJτ
$M_y = 2M_o \cos 2\pi J\tau$
(f)
t(5)
$M_z = 2M_o \cos 2\pi J\tau$
(g)
t(5)
(h)
t(7)
φ=0
$M_z = 2M_o \sin\Omega_{14}\Delta$
X_s
A_s
A_f
X_f
$\Omega_{14}\Delta$
(i)
t(7)
φ=0
$M_y = -2M_o$
(j)
t(8)
φ=0
X_s
A_s
X_f
A_f
(k)
t(8)
φ = 0
A
X

Figure 4.

(audio) frequencies.

For the double-quantum coherence

$$d_{14}(t) = d_{14}(o)\exp(-i\omega_{14}t)\exp[i(1 + 1)\omega_o t] =$$

$$= d_{14}(o)\exp(-i\Omega_{14}t), \qquad (15)$$

where $\Omega_{14} = \omega_{14} - 2\omega_o$. We note that both single- and double-quantum coherences evolve at low frequencies in the rotating frame.

The zero-quantum coherence is not affected by the rotating frame $(m_2 - m_3 = 0)$:

$$d_{23}(t) = d_{23}(o)\exp(-i\omega_{23}t). \qquad (16)$$

The zero-quantum coherence evolves at low frequency in both the laboratory and rotating frames.

We now want to calculate D(2) at time t(2). Inspecting D(1) we see that only single quantum coherence elements evolve. For instance

$$d_{12}(2) = d_{12}(1)\exp(-i\Omega_{12}\tau) = (-i/2)\exp(-i\Omega_{12}\tau) \qquad (17)$$

To save space we let $(-i/2)\exp(-i\Omega_{mn}\tau) = B_{mn}$ and at t(2) we have:

$$D(2) = \begin{vmatrix} 1 & B_{12} & B_{13} & 0 \\ B_{12}^* & 1 & 0 & B_{24} \\ B_{13}^* & 0 & 1 & B_{34} \\ 0 & B_{24}^* & B_{34}^* & 1 \end{vmatrix} . \qquad (18)$$

The z-magnetization is still zero (relaxation effects are neglected). The transverse magnetization components are:

$$M_{TA} = M_o(B_{12}^* + B_{34}^*) = \frac{iM_o}{2}[\exp(i\Omega_{12}\tau) + \exp(i\Omega_{34}\tau)], \qquad (19)$$

$$M_{TX} = M_o(B_{13}^* + B_{24}^*) = \frac{iM_o}{2}[\exp(i\Omega_{13}\tau) + \exp(i\Omega_{24}\tau)]. \qquad (20)$$

We see that there are four vectors rotating with four different angular velocities in the equatorial plane. We can identify (see Figure 1):

$$\Omega_{12} = \omega_{12} - \omega_o = \omega_A + 2\pi\frac{J}{2} - \omega_o = \Omega_A + \pi J,$$

$$\Omega_{34} = \Omega_A - \pi J.$$

$$\Omega_{13} = \Omega_X + \pi J,$$

$$\Omega_{24} = \Omega_X - \pi J.$$

In order, these vectors are labeled A_f, A_s, X_f, and X_s, where f and s mean fast and slow (Figure 4c).

Second pulse: 180°y

The rotation operator for this pulse,

$$R_{180AX} = \begin{vmatrix} 0 & 0 & 0 & 1 \\ 0 & 0 & -1 & 0 \\ 0 & -1 & 0 & 0 \\ 1 & 0 & 0 & 0 \end{vmatrix} = R^{-1}_{180AX}. \quad (21)$$

At time t(3) the density matrix is:

$$D(3) = R^{-1}D(2)R = \begin{vmatrix} 1 & -B^*_{34} & -B^*_{24} & 0 \\ -B_{34} & 1 & 0 & -B^*_{13} \\ -B_{24} & 0 & 1 & -B^*_{12} \\ 0 & -B_{13} & -B_{12} & 1 \end{vmatrix}. \quad (22)$$

Two important changes have been induced by the 180° pulse (Figure 4d). First, all single quantum coherences were conjugated and changed sign. This means that all M_x-components changed sign while the M_y-components remained unchanged:

$$M_T = M_x + iM_y,$$

$$-M^*_T = -M_x + iM_y.$$

This shows, indeed, that all four vectors rotated 180° around the y-axis. Second, coherences corresponding to fast precessing nuclei were transferred in "slots" corresponding to slow evolution. This means the vectors also changed labels (Figure 4d).

Evolution from t(3) to t(4)

As it happened during the first delay τ, only single quantum coherences evolve here. According to (14) and (20)

$$d_{12}(4) = d_{12}(3)\exp(-i\Omega_{12}\tau) = -B^*_{34}\exp(-i\Omega_{12}\tau) =$$

$$= (-i/2)\exp(+i\Omega_{34}\tau)\exp(-i\Omega_{12}\tau) =$$

$$= (-i/2)\exp[-i(\Omega_{12} - \Omega_{34})\,] = (-i/2)\exp(-i2\pi J\tau)\,. \tag{23}$$

$$d_{13}(4) = (-i/2)\exp(-i2\pi J\tau) = d_{12}(4) = U \tag{24}$$

$$d_{24}(4) = d_{34}(4) = (-i/2)\exp(+i2\pi J\tau) = V\,. \tag{25}$$

Hence,

$$D(4) = \begin{vmatrix} 1 & U & U & 0 \\ U^* & 1 & 0 & V \\ U^* & 0 & 1 & V \\ 0 & V^* & V^* & 1 \end{vmatrix}\,. \tag{26}$$

Let us calculate the corresponding magnetization vectors:

$$M_{zA} = M_{zX} = 0\,,$$

$$M_{TA} = M_{TX} = M_o(U^* + V^*) = iM_o\cos 2\pi J\tau\,. \tag{27}$$

We see (Figure 4e) that, while the chemical shifts refocussed, the coupling continues to be expressed, due to the label change.

Third pulse: 90x

Using the same operators as in (10) applied to D(4) we obtain:

$$D(5) = \begin{vmatrix} 1-c & 0 & 0 & s \\ 0 & 1 & 0 & 0 \\ 0 & 0 & 1 & 0 \\ s^* & 0 & 0 & 1+c \end{vmatrix}\,, \tag{28}$$

where $c = -\cos 2\pi J\tau$ and $s = -i\sin 2\pi j\tau$. D(5) tells us that the single quantum coherences vanished, a double quantum coherence was created, and the magnetization was transferred entirely along the z-axis (Figure 4f). For the "vector" enthusiast we can still accept the intuitive picture of Figure 4g. But our interest is in the double-quantum coherence. We want to maximize it, making $\tau = (2k+1)/4J$. Then, $c = 0$ and $\sin 2\pi J\tau = \pm 1 = (-1)^k$, where k = integer. In this particular case the z-magnetization due to coupled carbons is also null. All information resides in s. Of course the noncoupled carbons preserve their large z-magnetization ($J = o$, hence $\cos 2\pi J\tau = 1$).

Evolution from t(5) to t(6)

Only the double-quantum coherence element s evolves, according to (15):

$$D(6) = \begin{vmatrix} 1-c & 0 & 0 & w \\ 0 & 1 & 0 & 0 \\ 0 & 0 & 1 & 0 \\ w^* & 0 & 0 & 1+c \end{vmatrix}, \qquad (29)$$

where

$$w = s \exp(-i\Omega_{14}\Delta) = -i \sin 2\pi J\tau \exp(-i\Omega_{14}\Delta).$$

D(6) tells us that the magnetization components experience no change during the evolution Δ. This is difficult to match with the vector representation. If the vectors in Figure 4g would evolve preserving their labels at t(5), a transverse magnetization would appear. This is not consistent with the rigorous density matrix treatment, which suggests that there is a "scrambling" (total loss of coherence) in the equatorial plane. The fourth pulse (vide infra) will give D(7), so rich in information.

Fourth pulse 90°ϕ

This pulse is phase cycled, i.e., it is applied successively along the x, y, -x, and -y axes. For clarity, we will discuss separately the coupled and the noncopuled (isolatėd) carbon situations.

In the first case, with $\tau = (2k + 1)/4J$,

$$D(6) = \begin{vmatrix} 1 & 0 & 0 & w \\ 0 & 1 & 0 & 0 \\ 0 & 0 & 1 & 0 \\ w^* & 0 & 0 & 1 \end{vmatrix}. \qquad (30)$$

We observe that the populations are equalized and there still are no single quantum coherences. All information is contained in the w elements. The vectorial approach cannot properly represent this situation.

Applying the rotation operator given in (A33) we obtain

$$D(7) = \begin{vmatrix} 1+a^2F & -a^*G & -a^*G & F \\ aG & 1-a^2F & -a^2F & a^*G \\ aG & -a^2F & 1-a^2F & a^*G \\ F & -aG & -aG & 1+a^2F \end{vmatrix}, \tag{31}$$

where

$$\begin{aligned} a &= i\,\exp(-i\phi)\,, \\ F &= (w + w^*)/4 = -\tfrac{1}{2}(-1)^k \sin\Omega_{14}\Delta\,, \\ G &= (w - w^*)/4 = -\tfrac{i}{2}(-1)^k \cos\Omega_{14}\Delta\,. \end{aligned} \tag{32}$$

D(7) shows that the newly created single quantum coherences contain the double-quantum coherence information, Ω_{14}. The transverse magnetization is zero. A longitudinal magnetization proportional to $\sin\Omega_{14}$ appears. All this cannot be deduced from the vector representation. Yet, the density matrix treatment allows us to draw Figure 4h.

To save time and space we will treat the isolated (uncoupled) carbons as an AX system in which A and X belong to two different molecules. We can then use D(6) letting J = 0(c = -1; s = 0):

$$D(6)' = \begin{vmatrix} 2 & 0 & 0 & 0 \\ 0 & 1 & 0 & 0 \\ 0 & 0 & 1 & 0 \\ 0 & 0 & 0 & 0 \end{vmatrix}. \tag{33}$$

In this case D(7) will be

$$D(7)' = \begin{vmatrix} 1 & a/2 & a/2 & 0 \\ a^*/2 & 1 & 0 & a/2 \\ a^*/2 & 0 & 1 & a/2 \\ 0 & a^*/2 & a^*/2 & 1 \end{vmatrix}. \tag{34}$$

D(7)' shows only single quantum coherences and equalized populations. The transverse magnetization is $2M_o$ (the orientation depends on the value of ϕ). We can now draw Figure 4i.

Detection

The magnetization at t(8), due to coupled (cpl) and noncoupled (ncp) nuclei can be calculated starting from the single quantum coherences in (31) and (34), respectively.

$$M_T(\text{cpl}) = \mathcal{M}(\text{cpl})[\exp(i\Omega_{12}t_d) + \exp(i\Omega_{13}t_d) - \exp(i\Omega_{24}t_d) - \exp(i\Omega_{34}t_d)], \quad (35)$$

where

$$\mathcal{M}(\text{cpl}) = -M_o aG^* = (M_o/2)\exp(-i\phi)(-1)^k \cos\Omega_{14}\Delta.$$

These are the four peaks of the coupled AX system. In vector representation, they are shown in Figure 4j.

$$M_T(\text{ncp}) = \mathcal{M}(\text{ncp})[2\exp(i\Omega_A t_d) + 2\exp(i\Omega_X t_d)], \quad (36)$$

where $\mathcal{M}(\text{ncp}) = M_o a^*/2 = (-iM_o/2)\exp(i\phi)$.

These are the two peaks of the uncoupled nuclei (cf. Figure 4k). Each of them is 200 times more intense than each of the four peaks in (35).

The culminating point of INADEQUATE is the selective detection of M_T(cpl). We note that M_T(cpl) and M_T(ncp) depend in opposite ways on ϕ [cf. (35) and (36)]. They can be discriminated by cycling ϕ and properly choosing the lock-in detector phase ψ (the detected signal $S = M_T\exp[-i\psi]$). The table below shows that two cycles are sufficient to eliminate M_T(ncp).

Cycle	ϕ	ψ	ncp $\exp(i\Phi)\exp(-i\psi)$	cpl $\exp(-i\phi)\exp(-i\psi)$
1	0	0	1	1
2	90	-90	-1	1
Accumulated signal:			0	2

Even small imperfections of the rf pulse will allow leakage of the strong undesired signal into the resultant spectrum. This makes it necessary to apply one of several invented cycling patterns consisting of up to 256 steps, which attempt to cancel the effects of too long, too short, or out of phase pulses.

4. CARBON-CARBON CONNECTIVITY

A significant extension of INADEQUATE is its adaptation for two-dimensional experiments.[3] The second time-domain (in addition to t_d) is created by making Δ a variable. All we have to do now is to discuss (35) and (36) in terms of the two time variables.

As seen in Figure 5, all four peaks of $M_T(cpl)$ will be aligned along the same frequency, Ω_{14}, in domain two ($\Omega_{14} = \Omega_A + \Omega_X$). The great advantage of the 2D display consists of the fact that every pair of coupled carbons will exhibit its quartet along its own Ω_{14} frequency. This allows one to "trace out the carbon skeleton of an organic molecule."[3] The student is invited to determine the carbon connectivities of the molecule whose 2D spectrum is shown in Figure 6.

It should be noted that $M_T(ncp)$ does not depend on Δ [cf. (36)]. This means that, when imperfectly eliminated, the peaks of isolated carbons will be "axial" (dotted circles on the zero frequency line of domain two).

We also note that $M_T(cpl)$ is phase modulated with respect to t_d and amplitude modulated with respect to Δ. This means that mirror-image peaks will appear at frequencies $-\Omega_{14}$. This reduces the intensity of the displayed signals and imposes restrictions on the choice of transmitter frequency, increasing the size of the data matrix. A modified sequence has been proposed[4] to obtain phase modulation with respect to Δ, which is the analog of a quadrature detection in domain two.

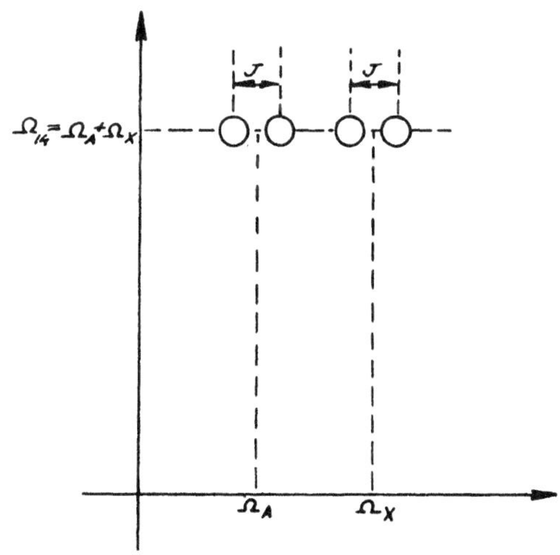

Figure 5

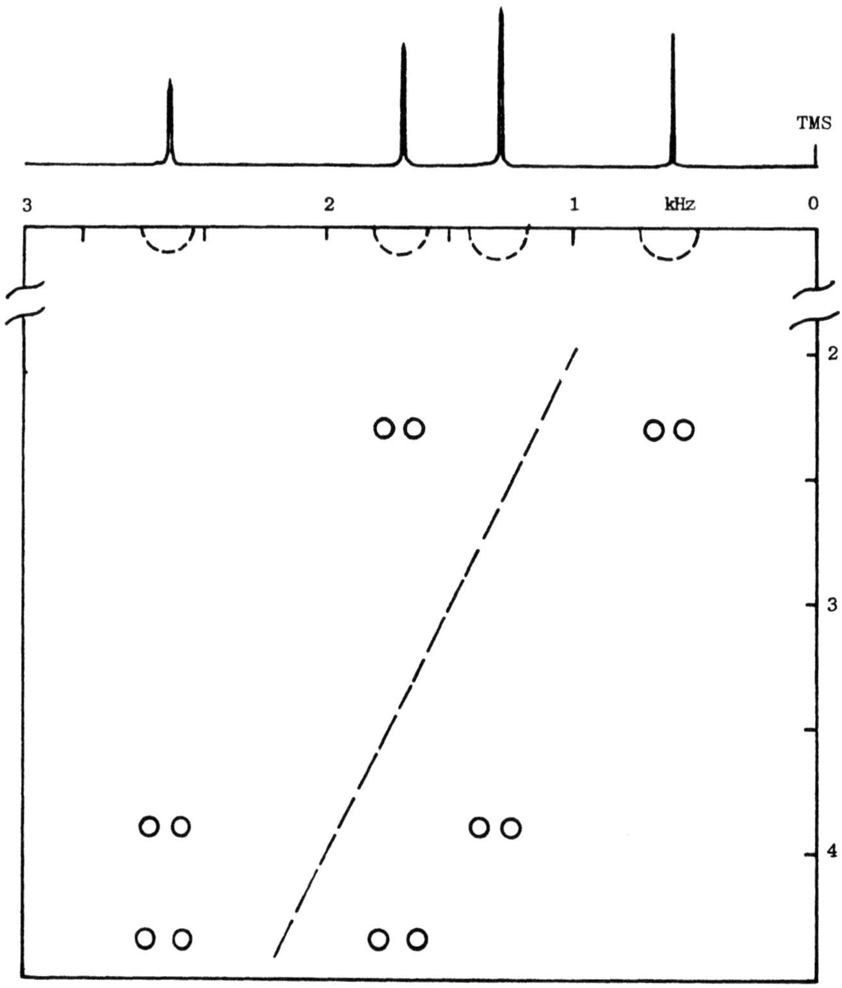

Figure 6. The 50 MHz carbon-carbon connectivity spectrum of a molecule with MW = 137

APPENDIX: DENSITY MATRIX FORMALISM

The density matrix is a tool used to describe the state of a spin ensemble as well as its evolution in time. It allows the passage from the probabilistic treatment of a system of a few spins to the statistical treatment of a large ensemble of such systems.

Since we are interested in the magnetization we want to describe this observable in terms of the wave function ψ of the system:

$$\langle\mu_x\rangle = (\psi|\mu_x|\psi) = \gamma\hbar(\psi|I_x|\psi) \quad . \tag{A1}$$

$\langle\mu_x\rangle$ is the expectation value of the x-component of the magnetic moment and I_x is the x-component of the angular momentum operator of the system. In order to calculate the macroscopic magnetization, the ensemble average is to be taken:

$$M_x = N_o\overline{\langle\mu_x\rangle} = N_o\gamma\hbar\overline{(\psi|I_x|\psi)} \;, \tag{A2}$$

where N_o is the number of spins per unit volume.

Since I_x is an invariant operator, the time dependence is contained in ψ which, in turn, may be expressed as a linear combination of the eigenstates $|n)$ of the system:

$$\psi = \sum_{n=1}^{n} c_n|n) \quad . \tag{A3}$$

The Schrödinger equation

$$-\frac{\hbar}{i}\frac{\partial\psi}{\partial t} = H\psi \tag{A4}$$

becomes:

$$-\frac{\hbar}{i}\Sigma\frac{dc_n}{dt}|n) = \Sigma\, c_n E_n|n) \;, \tag{A5}$$

where E_n is the eigenvalue corresponding to the state $|n)$ in the time independent equation:

$$H|n) = E|n) \quad . \tag{A6}$$

Rearranging (A5) gives:

$$\Sigma(c_n E_n + \frac{\hbar}{i}\frac{dc_n}{dt})|n) = 0. \tag{A7}$$

Due to the orthogonality of the eigenfunctions, (A7) is satisfied if and only if each term of the sum is null:

$$c_n E_n + \frac{\hbar}{i}\frac{dc_n}{dt} = 0 \quad . \tag{A8}$$

Hence

$$\frac{1}{c_n}\frac{dc_n}{dt} = -\frac{iE_n}{\hbar}$$

or

$$\frac{d}{dt}(\ell n c_n) = -\frac{iE_n}{\hbar} \quad . \tag{A9}$$

Integrating (A9) yields:

$$\ell n c_n = -\frac{iE_n}{\hbar}t + C \quad . \tag{A10}$$

$$c_n = \exp(-iE_n t/\hbar)\exp(C) \quad . \tag{A11}$$

The integration constant C may be related to the value of c_n at time t = 0

$$c_n(o) = \exp(C)$$

and (A11) becomes

$$c_n = c_n(o)\exp(-iE_n t/\hbar) \quad . \tag{A12}$$

Knowing the evolution of all c_n's will allow us to predict the time variation of ψ in (A2), hence, the time variation of the magnetization M_x.

We now calculate $(\psi|I|\psi)$ where I may be I_x, I_y or I_z:

$$(\psi|I|\psi) = |c_1^* \, c_2^* \ldots c_N^*| \times \begin{vmatrix} I_{11} & \cdots & I_{1N} \\ \cdot & & \\ \cdot & & \\ \cdot & & \\ I_{N1} & \cdots & I_{NN} \end{vmatrix} \times \begin{vmatrix} c_1 \\ c_2 \\ \cdot \\ \cdot \\ c_N \end{vmatrix} =$$

$$= |c_1^* c_2^* \ldots c_N^*| \times \begin{vmatrix} \sum^m I_{1m}c_m \\ \sum^m I_{2m}c_m \\ \vdots \\ \sum^m I_{Nm}c_m \end{vmatrix} = \sum\sum^{nm} I_{nm}c_m c_n^* . \tag{A13}$$

Since I_{nm} are constant throughout the ensemble, only $c_m c_n^*$ have to be averaged in (A2):

$$M = N_o \gamma \hbar \sum^{nm}\sum I_{nm} \overline{c_m c_n^*} \quad \text{(A14)}$$

Here M may be taken as M_x, M_y, or M_z using the corresponding matrix elements I_{nm} of the operators I_x, I_y, or I_z.

We notice that the only variable elements in (A14) are the averaged products $\overline{c_m c_n^*}$. There are N^2 such products which, arranged in a square table, form the density matrix:

$$D = \begin{vmatrix} d_{11} & d_{12} & \cdots & d_{1N} \\ d_{21} & & & \\ \cdot & & & \\ \cdot & & & \\ \cdot & & & \\ \cdot & & & \\ d_{N1} & \cdots & \cdots & d_{NN} \end{vmatrix}, \quad \text{(A15)}$$

with $d_{mn} = \overline{c_m c_n^*}$.

It follows that (A14) can be written

$$M = N_o \gamma \hbar \sum^{nm}\sum I_{nm} d_{mn} = N_o \gamma \hbar \sum^{nm}\sum I_{nm} d^*_{nm}, \quad \text{(A16)}$$

since D is Hermitian.

Therefore, the magnetization can be calculated by multiplying every element of I with the complex conjugate of the corresponding element of D and summing all products.

A more "elegant" but less practical avenue is calculating the trace (sum of diagonal elements) of the product I x D, e.g.:

$$M_x = \mathrm{Tr}(I_x D). \quad \text{(A17)}$$

Note that in (A17) the constant factor $N_o \gamma \hbar$ has been omitted. This is general practice since one is interested only in relative intensities.

However, at this point we want to make a quantitative evaluation of the magnetization components for our particular C_A-C_X case. We therefore have to take $N_o \gamma \hbar$ and p/S[cf(8) & (9)] into account:

$$M = (N_o \gamma \hbar p/S) \sum^{nm}\sum I_{nm} d^*_{nm}. \quad \text{(A18)}$$

Since $\frac{P}{S} = \frac{\hbar|\gamma|H_o}{4kT}$, the parenthesis in (A18) is $\frac{\gamma}{|\gamma|} M_o$ where

$$M_o = \chi_o H_o = N_o \gamma^2 \hbar^2 \frac{I(I+1)}{3kT} H_o,$$

is the equilibrium magnetization due to either nuclei A or X.

We now need the longitudinal (I_z) and transverse ($I_T = I_x + iI_y$) components of the angular momentum operator:[1]

$$I_{zA} = \frac{1}{2}\begin{vmatrix} 1 & 0 & 0 & 0 \\ 0 & -1 & 0 & 0 \\ 0 & 0 & 1 & 0 \\ 0 & 0 & 0 & -1 \end{vmatrix} ; \quad I_{zX} = \frac{1}{2}\begin{vmatrix} 1 & 0 & 0 & 0 \\ 0 & 1 & 0 & 0 \\ 0 & 0 & -1 & 0 \\ 0 & 0 & 0 & -1 \end{vmatrix} ;$$

$$I_{TA} = \begin{vmatrix} 0 & 1 & 0 & 0 \\ 0 & 0 & 0 & 0 \\ 0 & 0 & 0 & 1 \\ 0 & 0 & 0 & 0 \end{vmatrix} ; \quad I_{TX} = \begin{vmatrix} 0 & 0 & 1 & 0 \\ 0 & 0 & 0 & 1 \\ 0 & 0 & 0 & 0 \\ 0 & 0 & 0 & 0 \end{vmatrix} .$$

Applying (A18) we find, for $\gamma > 0$:

$$M_{zA} = (M_o/2)(d_{11} - d_{22} + d_{33} - d_{44}) \quad \text{(A19)}$$

$$M_{zX} = (M_o/2)(d_{11} + d_{22} - d_{33} - d_{44}) \quad \text{(A20)}$$

$$M_{zAX} = M_o(d_{11} - d_{44}) \quad \text{(A21)}$$

$$M_{TA} = M_o(d^*_{12} + d^*_{34}) \quad \text{(A22)}$$

$$M_{TX} = M_o(d^*_{13} + d^*_{24}) \quad \text{(A23)}$$

$$M_{TAX} = M_o(d^*_{12} + d^*_{34} + d^*_{13} + d^*_{24}). \quad \text{(A24)}$$

It is instructive to note that the transverse magnetization is given exclusively by single quantum coherences, while the longitudinal magnetization depends on populations only. Relations (A19) - (A24) take into account our convention to orient H_o in the direction of $-z$ axis for nuclei with $\gamma > 0$.

Now, having established the relation between magnetization and density matrix, we discuss the time evolution of the matrix from the thermal equilibrium throughout the pulse sequence.

At equilibrium the nondiagonal elements are null because of the random phase distribution of the complex coefficients:

$$d_{mn} = \overline{c_m c_n^*} = \overline{|c_m||c_n|\exp[i(\alpha_m - \alpha_n)]} \quad .$$

(A complex number, $\exp[i(\alpha_m - \alpha_n)]$, described as a vector in the complex plane, may be oriented in any direction defined by the phase difference $\alpha_m - \alpha_n$; hence, the average of a multitude of such vectors will be null). The diagonal elements represent populations since

$$d_{mn} = \overline{c_m c_m^* = |c_m|^2}$$

(In quantum mechanics $|c_m|^2$ is the probability of finding the system in the state m).

According to (A12) the evolution of the matrix elements is given by

$$d_{mn}(t) = d_{mn}(o)\exp[-i(E_m - E_n)t/\hbar] =$$

$$= d_{mn}(o)\exp(-i\omega_{mn}t). \qquad \text{(A25)}$$

The populations are invariable because $E_m - E_m = 0$ (relaxation processes are neglected throughout this paper).

A more general form of the evolutaion equation is:

$$\frac{dD}{dt} = \frac{i}{\hbar}(DH - HD) \quad ,$$

the solution of which is:

$$D(t) = \exp(-iHt/\hbar)\ D(o)\ \exp(iHt/\hbar). \qquad \text{(A26)}$$

The advantage of (A26) is that it allows us to see the effect of a pulse, i.e., the effect of a field H_1, perpandicular to H_o. In the rotating frame the Hamiltonian

$$H = |\gamma|\hbar H_o I_z$$

may be replaced by $|\gamma|\hbar H_1 I_x$. (We establish the following convention: the applied field H_o is oriented along the positive z-axis for $\gamma < 0$, and along the negative Z-axis for $\gamma > 0$). Then,

$$i\frac{H}{\hbar}t = i|\gamma|H_1 I_x t = iI_x\alpha.$$

Hence, for a pulse $\alpha = |\gamma|H_1 t$,

$$D(f) = \exp(-iI_x\alpha)\ D(i)\ \exp(iI_x\alpha) =$$
$$= R^{-1}D(o)R$$

where R is the rotation operator; D(i) and D(f) denote the density matrix before and after the pulse.

Let us calculate $R_{x\alpha}$ for a spin ½. From angular momentum tables

$$I_x = \frac{1}{2}\begin{vmatrix} 0 & 1 \\ 1 & 0 \end{vmatrix}$$

$$R_{\alpha x} = e^{iI_x\alpha} = 1 + i\alpha I_x + \frac{(i\alpha)^2}{2!} I_x^{\ 2} + \frac{(i\alpha)^3}{3!} I_x^{\ 3} + \ . \ . \ . \qquad \text{(A27)}$$

The powers of I_x may easily be calculated if one notices that

$$I_x^{\ 2} = \frac{1}{4}\begin{vmatrix} 1 & 0 \\ 0 & 1 \end{vmatrix},$$

$$I_x^{\ 3} = \frac{1}{4} I_x = \frac{1}{8}\begin{vmatrix} 0 & 1 \\ 1 & 0 \end{vmatrix},$$

$$I_x^{\ 4} = \frac{1}{16}\begin{vmatrix} 1 & 0 \\ 0 & 1 \end{vmatrix}.$$

In general, for n = even we have:

$$I_x^{\ n} = \frac{1}{2n}\begin{vmatrix} 1 & 0 \\ 0 & 1 \end{vmatrix}, \qquad \text{(A28)}$$

and for n = odd,

$$I_x^{\ n} = \frac{1}{2^n}\begin{vmatrix} 0 & 1 \\ 1 & 0 \end{vmatrix}. \qquad \text{(A29)}$$

Introducing (A28) and (A29) in (A27) and separating the even and odd terms gives:

$$R_{\alpha x} = [1 - (\frac{\alpha}{2})^2 \frac{1}{2!} + (\frac{\alpha}{2})^4 \frac{1}{4!} - \ldots] \times \begin{vmatrix} 1 & 0 \\ 0 & 1 \end{vmatrix} +$$

$$+ i[\frac{\alpha}{2} - (\frac{\alpha}{2})^3 \frac{1}{3!} + \frac{\alpha}{2}^5 \frac{1}{5!} - \ldots] \times \begin{vmatrix} 0 & 1 \\ 1 & 0 \end{vmatrix} . \qquad \text{(A30)}$$

The two parentheses are the cosine and sine series. Therefore

$$R_{\alpha x} = \cos \frac{\alpha}{2} \times \begin{vmatrix} 1 & 0 \\ 0 & 1 \end{vmatrix} + i\sin \frac{\alpha}{2} \begin{vmatrix} 0 & 1 \\ 1 & 0 \end{vmatrix} =$$

$$= \begin{vmatrix} \cos \frac{\alpha}{2} & 0 \\ 0 & \cos \frac{\alpha}{2} \end{vmatrix} + \begin{vmatrix} 0 & i\sin \frac{\alpha}{2} \\ i\sin \frac{\alpha}{2} & 0 \end{vmatrix} = \begin{vmatrix} \cos \frac{\alpha}{2} & i\sin \frac{\alpha}{2} \\ i\sin \frac{\alpha}{2} & \cos \frac{\alpha}{2} \end{vmatrix} \qquad \text{(A31)}$$

For example

$$R_{90x} = \frac{1}{\sqrt{2}} \begin{vmatrix} 1 & i \\ i & 1 \end{vmatrix} \qquad \text{(A32)}$$

For the sequences discussed here, we need the rotation operators of nonselective 90° and 180° pulses applied successively along the x, y, -x, and -y axes (phase cycling).

The general expressions for such operstors are:

$$R_{90\phi AX} = \frac{1}{2} \begin{vmatrix} 1 & a & a & a^2 \\ -a^* & 1 & -1 & a \\ -a^* & -1 & 1 & a \\ a^{*2} & -a^* & -a^* & 1 \end{vmatrix} ; \quad R_{180\phi AX} = \begin{vmatrix} 0 & 0 & 0 & a^2 \\ 0 & 0 & -1 & 0 \\ 0 & -1 & 0 & 0 \\ a^{*2} & 0 & 0 & 0 \end{vmatrix} \qquad \text{(A33)}$$

where ϕ is the angle between the x-axis and the H_1 direction and $a = i\exp(-i\phi)$.

The x, y, -x, and -y operators are obtained by making $\phi = 0, 90°$, 180°, and 270°.

Since the effects of rf pulses are treated in the rotating frame, we have to describe the evolution between pulses also in the same system. The net result of rotating the x and y axes around the z-axis is a "reverse" rotation with angular velocity ω_o. This can be expressed as follows:

$$D_R = \exp(i\omega_o t I_z) D_L \exp(-i\omega_o t I_z), \qquad (A34)$$

where D_R and D_L are the density matrices for the rotating and laboratory frame, respectively. I_z is a diagonal matrix with $(I_z)_{jj} = m_j$ (the magnetic quantum number of state j). The exponential is also diagonal; its j-th element is: $\exp(\pm i\omega_o m_j t)$.

According to (A34), in the rotating frame [cf. (A25)]

$$d_{mn}(t) = d_{mn}(o)\exp(-i\omega_{mn}t)\exp[(m_m - m_n)\omega_o t]. \qquad (A35)$$

REFERENCES AND RECOMMENDED READING

1. Mateescu, G. D. and Valeriu, A. "Teaching the New NMR: A Computer-Aided Introduction to the Density Matrix Formalism of Multipulse Sequences", NATO Adv. Studies Institute on the Spectroscopy of Biological Molecules, C. Sandorfy and T. Theophanides, Eds., Reidel: Dordrecht, 1983.
2. Bax, A.; Freeman, R.; Kempsell, S. P., J. Am. Chem. Soc., 1980, 102, 4849.
3. Bax, A.; Freeman, R.; Frenkiel, T.A.; Ibid., 1981, 103, 2102, and references cited therein.
4. Bax, A. "Two-Dimensional NMR in Liquids"; Reidel: Dordrecht, 1982, p 120. Bax, A; Frenkiel, T. A.; Freeman, R.; Levitt, M. H., J. Magn. Reson. 1981, 43, 478.
5. Jeener, J., Ampère International Summer School, Basko Polje, Yugoslavia, 1971.
6. Wokaun, A.; Ernst, R. R., Chem. Phys. Lett., 1977, 52, 407.
7. Bodenhausen, G.; Vold, R. L.; Vold, R. R., J. Magn. Reson. 1980, 37, 93.
8. Waugh, J. S.; Huber, L. M.; Haeberlen, U., Phys. Rev. Lett., 1968, 20, 180.
9. Vega, S.; Pines, A.; J. Chem. Phys., 1977, 66, 12, 5624.
10. Schäublin, A.; Höhener, A.; Ernst, R. R., J. Magn. Reson., 1974, 13, 196.
11. Slichter, C. P. "Principles of Magnetic Resonance", 2nd Edition, Springer-Verlag: Berlin, 1980.

INSTRUMENTATION CONSIDERATIONS IN SOLID STATE NMR ANALYSIS

Mark J. Sullivan

JEOL (USA), Inc.

ABSTRACT: Two major improvements in NMR instrument capability for applications to solid samples have recently been made commercially available through JEOL and CHEMAGNETICS. It is now possible to obtain variable temperature magic-angle spinning operation on a broad band solid-state probe system. In the present article we present the results of the application of multinuclear NMR to the study of two important classes of fossil fuel catalysts, zeolites and aluminophosphates. The significance of facile changing of nuclei with respect to sample throughput is discussed. Data showing the role of variable temperature magic-angle spinning C-13 NMR in the quantitative analysis of solid fossil fuels is shown. These results reveal changes that occur in the spectra as a function of temperature that have an important bearing on the validity of the measurement of aromaticity in the vacuum residue of a heavy oil.

INTRODUCTION

In 1979 JEOL addressed the emerging demands of the NMR market to provide a solid state spectrometer by introducing the FX-60QS equipped with a CHEMAGNETICS magic-angle spinning probe system. These instruments have been used primarily in the area of solid fossil fuels (i.e., coal and oil shale) characterization. In fact, the quantity apparent aromaticity that is obtained on these types of samples through solid state NMR has become one of the most significant parameters in determining their value as fuels and fuel sources for conversion to other products.

The advent of superconducting magnets that has revolutionized the field of high resolution NMR in liquids has had perhaps an

L. Petrakis and J. P. Fraissard (eds.), Magnetic Resonance. Introduction, Advanced Topics and Applications to Fossil Energy, 525–533.

even greater effect on the NMR of solids. It is not only possible to look at smaller quantities of material due to the higher sensitivity that is gained at high field, but SCM's also make it feasible to observe a wide variety of quadrupolar nuclei in the solid state. JEOL responded to these advances in technology by introducing an SCM based MAS probe built by CHEMAGNETICS for the FX-200 and FX-270 spectrometers in 1981. Again, as the the probe technology continued to improve, CHEMAGNETICS designed and built a variable-temperature magic-angle spinning probe system for wide-bore superconducting magnets. Two versions of the VT/MAS probe, one for high-band nuclei (i.e., Si-29 to P-31) and a second for low band nuclei (i.e., N-15 to Si-29) are currently available.

Future plans include several improvements in solid-state probe technology. Work is underway at CHEMAGNETICS to complete the development of a variable angle sample spinning (VASS) probe for observation of quadrupolar nuclei. With this probe that was first revealed at the Experimental NMR Conference in 1983, it is possible to program the variations in the spinning angle by the use of a software controlled stepping motor. In addition, CHEMAGNETICS is continuing to move to higher field solid state probes for use on the JEOL GX-400 and 500 MHz spectrometers. As always, the commitment to improving the basic operation of the spinning system is continuing. This development work is intended not only to upgrade performance, but to improve reliability as well.

APPLICATIONS OF MULTINUCLEAR NMR TO CATALYSIS

Zeolites

While the case structures that give rise to the catalytic properties of natural and synthetic zeolites are well known due to x-ray techniques, the microstructure of these compounds could not easily be determined from these results. In a recent review, Fyfe et. al. (1) have summarized the information available from the application of Si-29 and Al-27 solid-state NMR to the study of zeolites. An example of the Si-29 and Al-27 NMR spectra of Na-Y zeolite taken on a JEOL 200 MHz spectrometer is shown in Figure 1. The multiple resonances that are seen in the Si-29 spectrum correspond to distinct structural groups of the type Si(OAl)n(OSi)4-n (n=0,1,..,4). In combination with the elemental analyses, it is possible to determine the ordering in the Si/Al framework. In the case of Al-27, tetrhedrally and octahedrally coordinated sites can be distinguished. This is a very significant result since this information can be used to determine the extent of incorporation of aluminum into the lattice framework in relation to that which is present in the cavities.

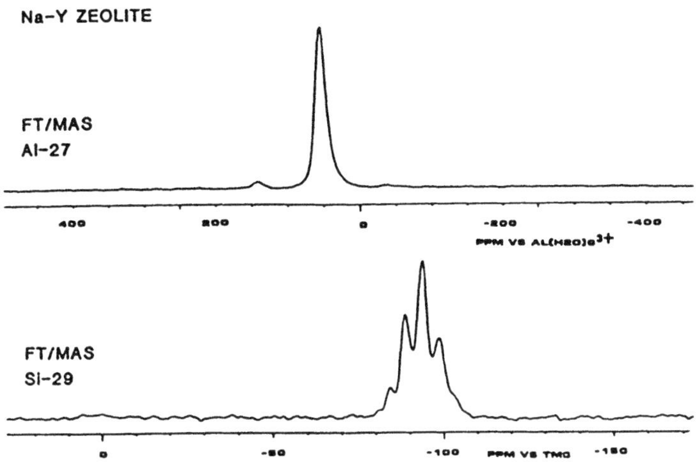

Figure 1: FT/MAS spectra of Na-Y zeolite; top Al-27 bottom Si-29.

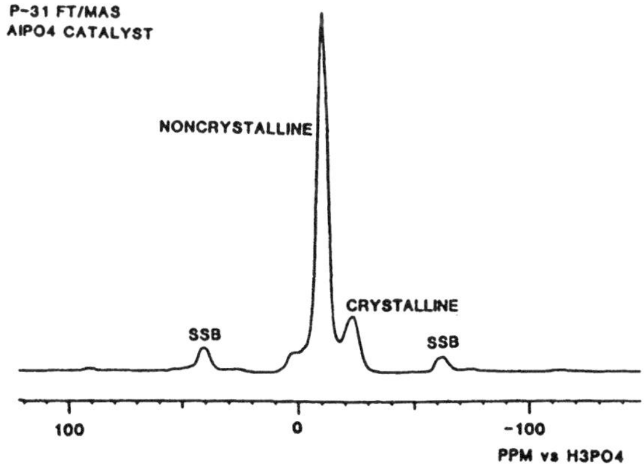

Figure 2: P-31 FT/MAS spectrum of an aluminophosphate catalyst. SSB = spinning sideband.

Aluminophosphates

A new class of catalytically active microporous solids called aluminophosphates has recently been discovered. (2) These materials display a number of the same attributes as zeolites including the property to withstand the high temperatures encountered during their regeneration. Like zeolites, the aluminophosphates can be readily studied by solid-state NMR techniques. As will be shown below, the use of P-31, Al-27 and C-13 NMR serve as sensitive probes for the design and synthesis of AlPO4 catalysts.

A solid-state P-31 FT (i.e., Bloch decay) magic-angle spinning spectrum of an aluminophosphate is shown in Figure 2. The most salient feature of this spectrum is the appearance of two lines located at about -10 and -13 ppm, respectively, versus H3PO4. Based upon the relative P-31 spin-lattice relaxation times for these two resonances, the lower field peak is assigned to the non-crystalline portion of the sample and the higher field peak to the crystalline portion. The degree of crystallinity as determined by P-31 NMR was 17 percent which agrees favorably with the percent crystallinity of 20 found by x-ray diffraction techniques.

Crystallization of aluminophosphates is seeded by the presence of an organic template that is typically a tertiary amine. Prior to its thermal decomposition as the final step in preparing these catalysts, the template can be examined by C-13 NMR. A carbon-13 CP/MAS spectrum of the tertiary n-propyl amine template in an AlPO4 is displayed in Figure 3. For this sample that contains only a few percent Carbon by weight, we chose to use a Delrin rotor that in turn gives rise to the two peaks thus identified in the figure. In most cases, for C-13 observation, Kel-F rotors that have no CP/MAS C-13 signal are preferable. Due to the much better dimensional stability of Delrin, however, these rotors can be machined with thinner walls than for Kel-F rotors and thus can contain larger sample volumes. In cases where sensitivity is critical and spectral overlap does not cause an interference, Delrin offers an important alternative. Both types of rotor materials are available for CHEMAGNETICS probes.

Aluminum-27 spectra of an aluminophosphate are shown in Figure 4. Both spectra were recorded with a 50 kHz sweep width at a resonance frequency of 51.9 MHz. Spectrum A was obtained using a normal single pulse sequence with a 4μs 90 degree pulse width and a 50μs receiver deadtime. A phase alternated spin echo pulse sequence (3) was used to generate the upper spectrum B. Both spectra were the result of 64 scans with a pulse delay time of 0.5 s. Two benefits are realized by using the spin echo pulse sequence. It eliminates baseline roll by canceling out spurious signals due to breakthrough to the receiver during the collection of the first few points of the FID. In addition, the spin echo permits a shorter delay time to be inserted between the end of

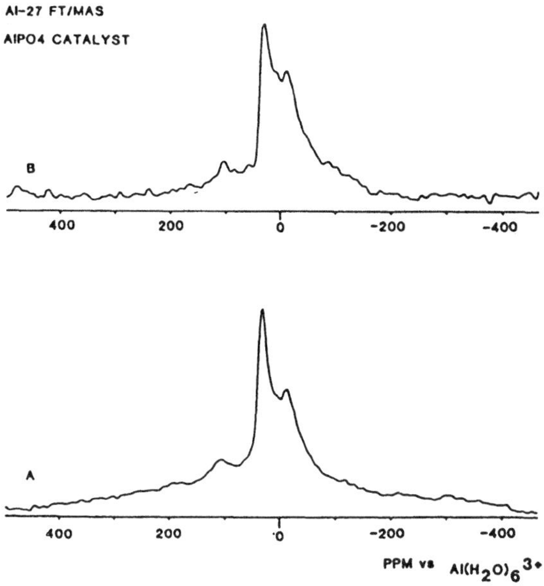

Figure 3: Al-27 FT/MAS spectra of an alumino-phosphate catalyst. A single pulse spectrum. B phase alternated spin echo.

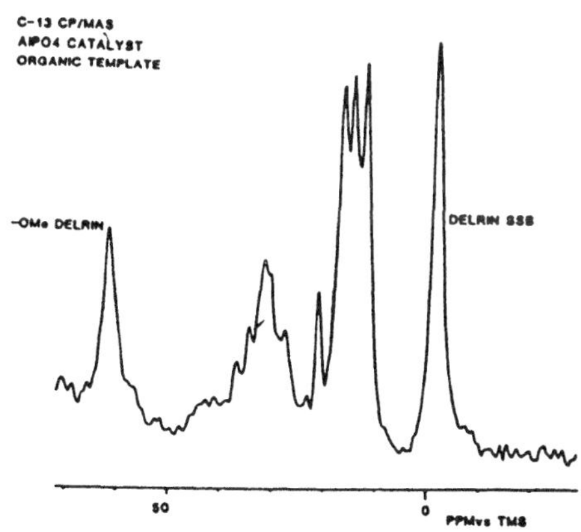

Figure 4: C-13 CP/MAS spectrum of the organic template in an aluminophosphate catalyst. The Delrin peaks are due to the rotor.

the observe pulse and the beginning of the collection of the NMR signal. Both of these points can be seen in a comparison of spectra A and B. The broad intensity that is distributed across the entire spectral range in A is absent in B. The relative intensities of the two peaks that bracket the center of each spectrum differ in A and B. It appears that the upfield resonance is broader than the downfield resonance. Therefore, the longer receiver delay time that was required in A decreases its intensity relative to the narrower downfield peak. This type of distortion leads to very different relative intensities in the two experiments. It should be noted that Delrin rotors were used in these experiments. Aluminum oxide rotors give rise to broad signals that cover a large portion of the Al-27 chemical shift range and cannot be used for aluminum observation without subsequent interference.

MEASUREMENT OF AROMATICITY IN THE VACUUM RESIDUE OF A HEAVY OIL

Aromaticities of chloroform soluble vacuum residues, commonly referred to as resids, are normally obtained by solution NMR. These numbers are quite reliable when run under quantitative conditions. In the present study, we have performed a variety of experiments to test the analytical reliability of solid-state experiments on the same materials.

CP/MAS Spectra

Several points must be considered at the outset of any quantitative cross-polarization magic-angle spinning study on a new sample. This type of analysis requires a fairly complete knowledge of the spin dynamics of the sample in question. The first criterion that must be met to insure quantitative results is that the cross relaxation time< contact time< proton T1rho. This determination can be made from a variable contact time experiment. In this case, it was found that the resid did not strictly adhere to this prerequisite since the proton T1rho was rather short, on the order of 2msec. The second criterion that must be satisfied in order to determine apparent aromaticity is that there be no overlap between the spinning sidebands of the aromatic peak and the aliphatic peak. With the spinning speeds used in this study of roughly 2000 to 2500 Hz at a carbon frequency of 50 MHz, it was necessary to invoke the TOSS spinning sideband elimination technique developed by Dixon (4). In order for the TOSS technique to retain quantitative information, however, either the T2's in the sample must be long in relation to the TOSS spin echo delay time, or the T2's in the sample must be uniform. The Hahn spin echo data given in Figure 5 show a marked change in the overall shape of the aliphatic band and almost a total loss of the aromatic resonance. This finding effectively eliminates the possibility for obtaining accurate quantitative information from the TOSS

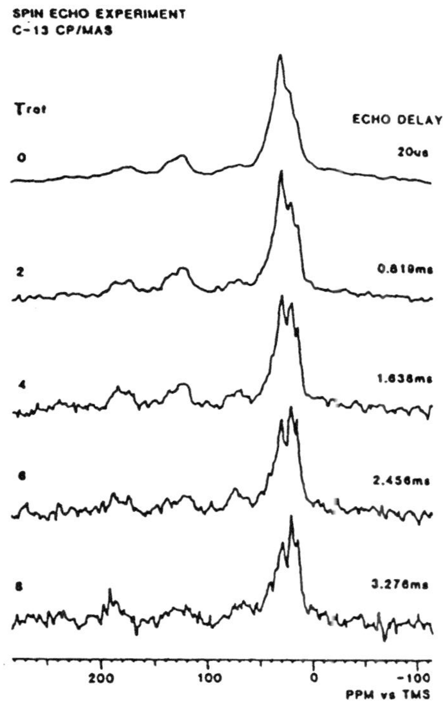

Figure 5: C-13 CP/MAS spectra of a heavy crude oil residue obtained with a spin echo pulse sequence.

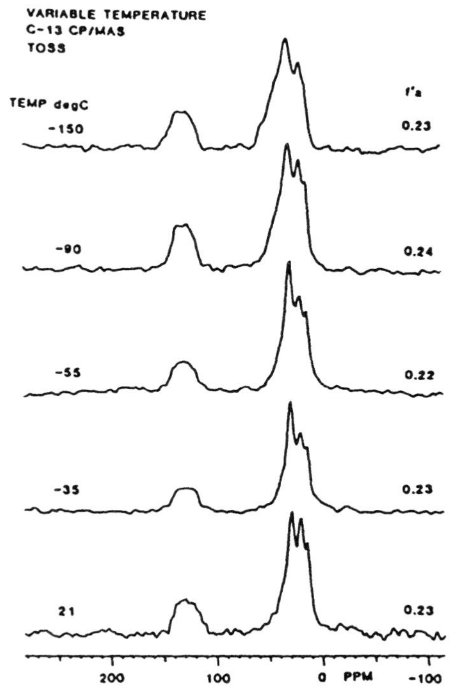

Figure 6: Variable temperature C-13 CP/MAS spectra of a heavy crude oil residue obtained with the TOSS sequence.

experiment. On the other hand, the qualitative information content is improved under these conditions since three distinct peaks are revealed in the aliphatic region.

Variable Temperature Magic-Angle Spinning C-13 NMR

A determination of the effects of temperature variation on the C-13 CP/MAS spectrum of the resid was run using the TOSS experiment. The reason for doing so was to obtain the improved resolution referred to above. Also, even though the aromaticities measured under these conditions are not quantitative in an absolute sense, changes in the relative aromaticity can be monitored. The results of these experiments that were carried out over a temperature range from ambient down to -150 deg C are shown in Figure 6. Clearly, a number of changes are occurring in these spectra with decreasing temperature. The relative intensities of the three distinguishable aliphatic peaks are altered and the overall width of the aliphatic envelope increases as the temperature is lowered. The shape of the aromatic peak appears to remain unchanged within the limits of the signal to noise obtained in this experiment. Close inspection shows that the increase in the width of the aliphatic peak envelope is due to the growth of a shoulder on the low field side of that peak. This additional peak is isolated in Figure 7 in which the difference spectrum between the -150 deg and -35 deg spectra is displayed. What is remarkable about these variable temperature experiments is that the apparent aromaticity does not change despite the changes in line shape that do occur.

VT/MAS System Description

The variable-temperature magic-angle spinning solid system operates over a temperature range from ambient down to -150 deg C. Cooling is obtained by passing the driving gas through a copper coil immersed in a bath of liquid nitrogen. The low temperature gas is then warmed to the desired temperature by an external heater. The sample chamber in the probe is insulated with a sliding plastic dewar arrangement that provides easy access for changing rotors. A platinum resistance sensor located at the gas inlet to the sample chamber is used to monitor the temperature. The magic-angle spinning system employs a double bearing air drive that is capable of variable speed spinning up to 5 kHz. The probe is equipped with separately adjustable drive and bearing gas supplies that provide stable spinning at all temperatures.

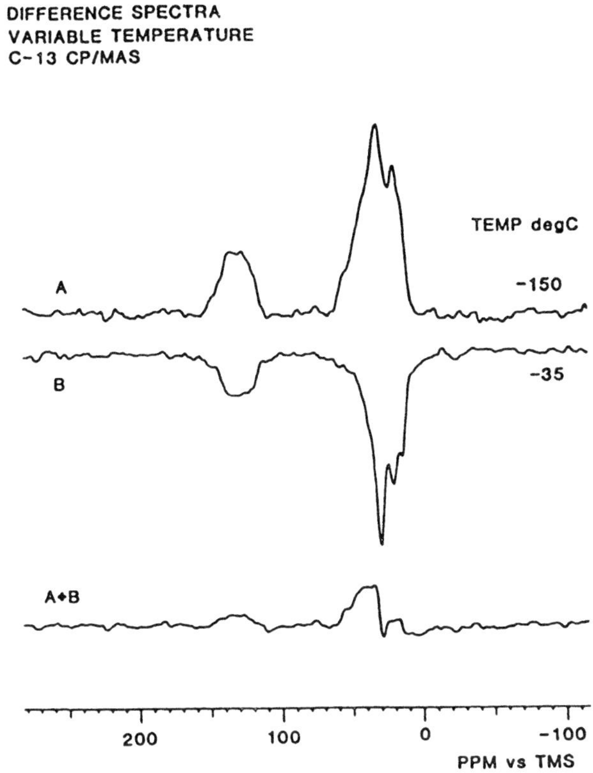

Figure 7: Difference spectrum of two C-13 CP/MAS spectra of a heavy crude oil residue obtained at -150 and -35 deg.C.

References

1. C.A. Fyfe, J.M. Thomas, J. Klinowski and G.C. Gobbi, Angew. Chem. Int. Ed. Engl.,22, 259-273(1983).
2. S.T. Wilson, B.M. Lok, C.A. Messina, T.R. Cannan and E.M. Flanigen, A.C.S. Symposium Series,#218,Chap. 5, 1983, pp 79-106.
3. M.Rance and R.A. Byrd, J. Magn. Reson.,52, 221-240(1983).
4. W.T. Dixon, J. Chem. Phys.,77, 1800-1809(1982).

QUANTITATIVE ASPECTS OF STRUCTURAL RESOLUTION METHODS IN CP.MAS ^{13}C NMR

J.M. DEREPPE

University of Louvain

Louvain-la-Neuve (Belgique)

Solid state ^{13}C NMR studies of materials such as coals, kerogens, asphaltenes are usually made in order to characterize and quantify chemical identities of the various constituents. Unfortunately these spectra do not exhibit a high level of resolution. They usually consist of two broad bands corresponding to aromatic/olefinic carbons and to aliphatic carbons. These aromatic and aliphatic bands sometimes show shoulders suggesting the presence of certain structure moeities. It has been shown(1) that the line width is primarily due to the heterogeneous nature of the sample, e.g. due to a multitude of similar but slightly different chemical shifts. Resolution enhancing methods, based on differential relaxation behavior, are therfore useful for distinguishing structural differences that would not appear in the usual CP.MAS spectra. The general problem of the analytical reliability of the results obtained by the CP.MAS method has been studied in various ways(2,3). The purpose of this work is to explore the quantitative aspects of the results obtained by some structural resolution methods. The approach used here is to compare these results with data obtained by the more classical methods of high resolution on liquids.

EXPERIMENTAL

* Sample. This work was carried out on an asphaltene isolated from a Venezuelian crude (80.98% C, 7.32% H, 1.07% O, 0.93% N, 8.37% S).

L. Petrakis and J. P. Fraissard (eds.), Magnetic Resonance. Introduction, Advanced Topics and Applications to Fossil Energy 535–544.

* Solution NMR measurements. The asphaltene (0.3gr) was dissolved in 3 ml of deuterochloroform. NMR measurements were carried out on a Bruker WM 250 in various conditions described in the pulse sequences section. ^{13}C chemical shift were measured from deuterochloroform but were recorded with respect to TMS.

* Solid state NMR measurements. The solid state ^{13}C spectra were obtained on a Bruker CXP 90 spectrometer operating at 22.64 MHz with a ^{19}F external lock. The spinners were made of perdeuterated polymethylmethacrylate (PMMA) with an internal volume of approximately 0.4 ml and were spun with air at 2.5 K Hz. In some cases difference spectroscopy was used to take into account the background signal corresponding to the rotor signal. The Hartman -Hahn match for CP experiments was for ^{13}C and ^{1}H respectively of 44 G and 11 G. The proton field was maintained at that level for high power proton decoupling. Spectra were usually recorded by using 2 K data points and a spectral width of 10 K Hz.

* Pulse sequences. The sequences are depicted in fig.1.

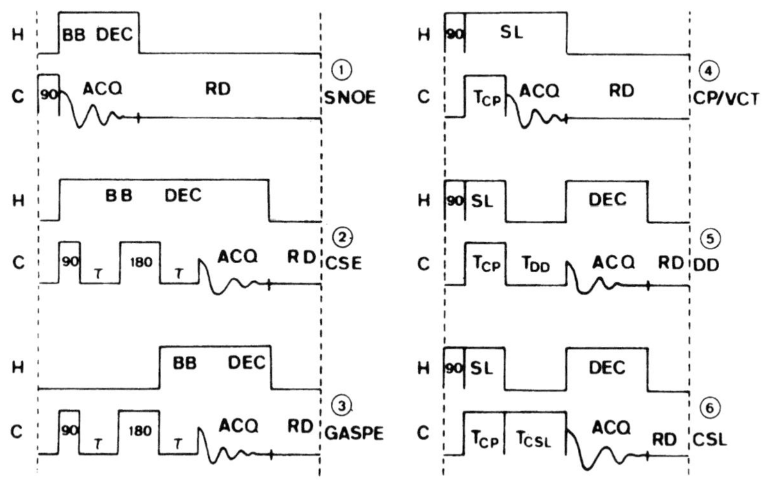

FIG 1

For high resolution work, they consist of (1) Suppressed NOE or inverse gated decoupling (SNOE) ; (2) Conventional spin echo (CSE) ; (3) Gated spin echo (GASPE). For solid state work the sequences are (4) single contact cross polarization (CP) or variable contact time experiment (VCT) ; (5) Dipolar Dephasing (DD) and (6) Carbon spin lock (CSL) experiment. These sequences are fully described in ref 4 to 7.

RESULTS AND DISCUSSION

Solution studies

Quantitative ^{13}C NMR of oil and low molecular weight compounds is not a straight-forward procedure because of long T_1 spin lattice relaxation time and nuclear Overhauser enhancement. Several studies have been devoted to the determination of the best experimental conditions(8). Work on higher molecular weight materials is not so demanding(9). It is actually known that asphaltenes are paramagnetic due to the presence of organic radicals and that they have inherent rapid relaxation. A relaxation delay of a few seconds is usually convenient for quantitative work provided the inverse gated decoupling technique (SNOE) is used. The ^{13}C spectra, recorded by using the SNOE sequence with relaxation delays varying between 5 Sec. and 15 Sec., showed no significant variations and were considered as representative of the carbon distribution. The aromaticity factor, f_A, was determined as the ratio of the integrated intensity of the aromatic carbon band (C_A : 160 to 110 ppm) to the integrated intensity of the aromatic and aliphatic carbon (C_S : 80 to 0 ppm) bands. The aromatic carbons (C_A) are usually divided into unsubstituted, $C_{A(CH)}$, and substituted or bridge carbons, $C_{A(C)}$, in order to estimate the maximum size of the aromatic sheet. This division, usually made on a chemical shift basis does not take into account the fact that bridge carbons in condensed aromatic ring systems also appear in the 125 - 135 ppm region(6). The number of unsubstituted and non bridge aromatic carbon can also be obtained either by combining the results of ^{13}C and ^{1}H spectroscopy and elemental analysis or by using one of the recently introduced pulse sequence used for peak assignment according to the number of H substituents.

The ratio of peripheral unsubstituted aromatic carbons, $C_{A(CH)}$, to the total number of aromatic carbons, C_A, in a polycondensed aromatic sheet if given by

$$C_{A(CH)}/C_A = H_A/f_A.(H/C)_{at}$$

where $(H/C)_{at}$ is the atomic hydrogen to carbon ratio and H_A is the fraction of aromatic protons given by the ^{1}H spectrum.

The measurement of the $C_{A(CH)}/C_A$ ratio by using the GASPE pulse sequence was also considered in order to confirm the previous evaluation and also to test the application of the spin echo method to heavy material such as asphaltenes. Actually, some potential problems related to short T_2 relaxation time for these compounds have been pointed out(10). Our approach was essentially experimental as we compared the relative intensities obtained from SNOE spectra and CSE spectra recordedwith τ delays ranging from 1 to 4 mS. Apart from a slight decrease in the S/N ratio, no noticeable difference was found between CSE and SNOE spectra for this sample. The selection between protonated and non proto-

nated aromatic carbon was then made by using the GASPE sequence with $\tau = 3.125$ mS which represents one halve of the inverse of the $^{13}C - ^{1}H$ coupling constant for aromatic carbons. The spectra show that some non protonated C can be observed in the protonated carbon band. The $C_{A(CH)}/C_A$ ratio is calculated by simply measuring the intensities in the absolute mode for an equal number of scans from the CSE and GASPE spectra.
Finally, the amount of C in methyl groups, $C_{S(CH_3)}$, is measured by the integrated intensity of the 0 to 22.5 ppm band and the amount of methyl groups involved in linear or branched mobile chains(11) is approximated by the integrated intensities of the bands centered on 14 and 19.5 ppm.
The information extracted from the high resolution work is as follow : $f_A = 0.5$; $C_{A(CH)}/C_A = 0.17$ (from ^{13}C and ^{1}H spectra) and 0.20 (from CSE and GASPE ^{13}C spectra) ; $C_{S(CH_3)}/C_S = 0.26$; $C_{S(CH_3)}$ in linear or branched chains/$C_S = 0.08$.
The ratio of protonated to non protonated aromatic carbons obtained from these data is $C_{A(CH)}/C_{A(C)} = 0.20$. The fraction of quaternary saturated carbons is rather small for compounds having an atomic H/C ratio for the saturated part of the molecule close to 2. The contribution of the saturated quaternary carbons will be neglected in the following analysis and the ratio of mobile CH_3 carbons to CH and CH_2 type carbons is calculated as $C_{S(CH_3)}/C_{S(CH+CH_2)} = 0.09$.

Solid state work

Most of the resolution enhancement techniques try to exploit differences in relaxation behavior of the ^{13}C nuclei in different environment to produce a resolution that is not observed in the usual experiment.
In a normal CP.MAS experiment(12), the proton spin polarization is transferred to the ^{13}C spin system during the cross polarization time T_{CP*} at a rate characterized by the apparent cross relaxation time T_{CH}. Fig.2 shows a typical dependance of the amplitude of the ^{13}C signal obtained in a CP experiment as a function of T_{CP}. The initial slope for short contact time is determined by the relaxation time T_{CH}. At long contact time the ^{13}C signal decreases and this is associated with the decreases of the ^{1}H magnetization spin locked along the ^{1}H RF field. This decay of the ^{1}H spin magnetization is characterized by a relaxation time $T_{1\rho H}$. The magnetization curve for the ^{13}C spin systems can be represented as a function of time by(13)

$$M^*(t) = M\{1 - \exp(-t/T^*_{CH})\}.\exp(-t/T_{1\rho H})$$

where $$T^*_{CH} = T_{CH}\{1 + (T_{CH}/T_{1\rho C}) - (T_{CH}/T_{1\rho H})\}^{-1}$$

T_{CH} is the cross relaxation time and $T_{1\rho C}$ is the ^{13}C relaxation time in the rotating frame. Mo is the magnetization of the ^{13}C spin system that would be observed at long cross polarization time and constant magnetization of the H spin system. This limit value only is representative of the number of carbon in resonance.

Hence, it should be possible to estimate T^*_{CH} and $T_{1\rho H}$ by systematically varying the cross polarization time in a variable contact experiment (VCT). The cross polarization time T_{CH} for the ith carbon mainly depends on the C_i-H dipolar interaction which in turn depends on the sixth power of the C_i-H distances. A distribution of T_{CH} values can be expected and a large difference should occur between carbon with no attached hydrogen atoms and those who have one or more attached hydrogen. The quaternary carbons should polarize slowly and protonated carbons should polarize much faster. Nevertheless the methyl carbons should polarize more slowly than other protonated carbons due to the usually fast methyl group rotation which reduces the C-H dipolar interaction. The VCT experiment may thus differentiate between carbon types.

A close inspection of the spectra obtained for various VCT shows that the line shapes for aromatic and aliphatic carbons gradually change as the polarization time is increased. At short T_{CP}, mostly protonated aromatic and non methyl aliphatic carbons respectively centered at 130 and 29 ppm are observed. The aromatic lineshape becomes more and more skewed towards the low field side and a similar line shape change is seen on the high field side of the aliphatic band as T_{CP} is increased indicating the observation of non protonated aromatic carbons and methyl group carbons which are slow to polarize. The magnetization curves, M^*_A and M^*_S measured by the integrated areas of the resonance bands, for aromatic and saturated carbons are given fig.2.

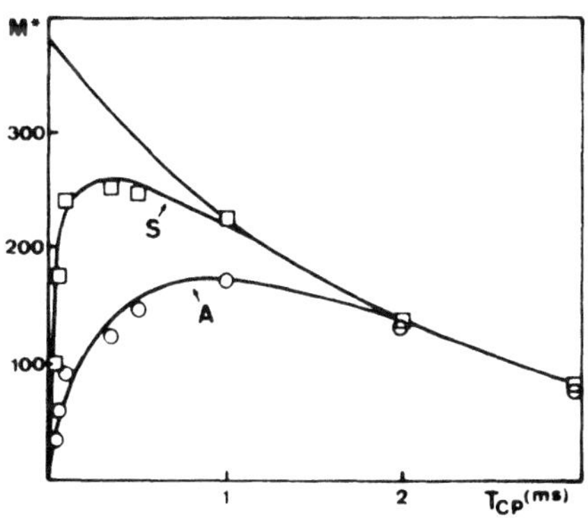

FIG 2

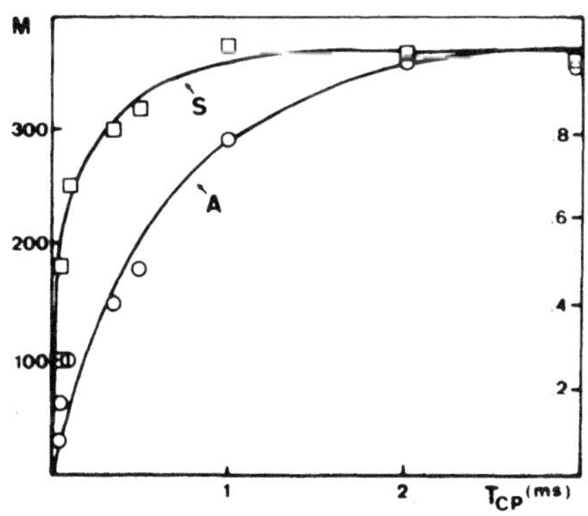

FIG 3

The maximum polarization obtained after ~0.25 mS for the aliphatic carbons and ~1 mS for the aromatic carbons then slowly decreases for both components at a rate governed by the $T_{1\rho H}$ relaxation time. Previous studies have shown that proton $T_{1\rho}$ are not exactly the same for aromatic and saturated carbon regions. For pure organic solids this kind of multiple $T_{1\rho}$ behavior is not observed. This small difference found for coals indicates that the aromatic and aliphatic protons are not in intimate contact. In this treatment, all the protons were considered as having a single relaxation behavior in the totating frame, i.e., all protons are characterized by the same $T_{1\rho H}$. The last three points of the magnetization curve for saturated carbons were used to estimate $T_{1\rho H}$. The 1.9 mS found here correlate well with the 1.7 mS observed in ^1H NMR. In that case the magnetization curves can be corrected for the adverse effects of the rather fast decrease of the proton magnetization in the rotating frame.

$$M_A(t) = M_A^*(t).\exp(t/T_{1\rho H}) \text{ and } M_S(t) = M_S^*(t).\exp(t/T_{1\rho H})$$

The corrected magnetization M_A and M_S are given fig.3. This plot clearly shows than carbons polarize at very different rates. This behavior is not unexpected : actually it has been shown by solution studies that most of the aromatic carbons are quaternary carbons which should polarize rather slowly. On the contrary, most of the saturated carbons are CH and CH_2 type carbons characterized by short polarization time. This type of experiment clearly shows that the usual 1 mS cross polarization time could lead, for some samples, to non quantitative measurements. The "true" intensity ration $C_A/C_S = 1$ given by the HR study is only obtained for cross polarization times longer than 2 mS (when the S/N ratio is not the best). A variable contact time experiment is thus very helpfull to ascertain the experimental conditions for quantitative work. Each magnetization curve can be divided into two components having different cross relaxation time T^*_{CH}.

$$M_A(t) = M_{A(C)}.\{1-\exp(-t/T^*_{CH(C)})\}+M_{A(CH)}\{1-\exp(-t/T^*_{CH(CH)}\}$$

where $M_{A(C)}$ is the limit magnetization associated with the quaternary i.e. substituted or bridge aromatic carbons and $M_{A(CH)}$ is associated with unsubstituted, i.e. protonated, aromatic carbons. $T^*_{CH(C)}$ and $T^*_{CH(CH)}$ are the associated cross relaxation times. A similar expression can be written for the saturated carbons.

$$M_S(t) = M_{S(CH+CH_2)}\{1-\exp(-t/T^*_{CH(CH+CH_2)})\} + M_{S(CH_3)}\{1-\exp(-t/T^*_{CH(CH_3)})\}$$

where $M_{S(CH+CH_2)}$ and $M_{S(CH_3)}$ are the magnetization associated with $CH + CH_2$ and CH_3 type carbons. In principle, the experimental data could be fitted to these formula to determine the limit magnetization and the relaxation time associated with each type of carbon. In this case, the amount of protonated aromatic and methyl carbons is quite low and the discrimination should be poor. The data were analyzed by using the carbon ratio obtained in HR work, in order to find the cross relaxation times T^*_{CH}. Values are aromatic (C) : 650 μS ; aromatic (CH) : 100 μS ; saturated $(CH + CH_2)$: 85 μS ; saturated (CH_3) : 500 μS.

The dipolar dephasing experiment (DD) is similar to a CP.MAS experiment except for the introduction of a delay T_{DD} before the acquisition under high power decoupling. During this period, the ^{13}C signal will decrease with a characteristic decay constant T_2^* which depends mainly on C-H heteronuclear dipolar interaction but also on molecular motions, MAS rotation speed and eventually spin diffusion(3). In a DD experiment, the carbons strongly coupled to protons, e.g. unsubstituted aromatic, methylene and methyne dephase more quickly than quaternary and methyl carbons and this type of experiment may also differentiate between carbon types. It has been shown that the amplitude decay for a given carbon type is usually approximated as a first order exponential

$$M(t) = M \exp(-t/T_2^*)$$

The DD spectra clearly show a fast decrease of the low field component of the aliphatic band and a decrease of the high field shoulder of the aromatic band. This behavior can be related to the fast dephasing of the protonated carbons compared to the unprotonated or rotating methyl group carbons. The M_A and M_S magnetization from the integrated areas of the resonance bands for aromatic and saturated carbons are given fig.5 and a logarithmic plot is given fig.5. These values were obtained with a standard cross polarization time of 1 mS. In that case all carbons are expected to be fully polarized except the quaternary aromatic carbons and the methyl group carbons. This fact will be taken into account later on.

Each curve could be represented as a sum of components characterized with different dephasing times T_2^*. The beginning of the M_S magnetization curve clearly represents to a large extent the evolution of the non methyl protonated carbons : these carbons should dephase rather quickly and they represent ~90% of the saturated carbons. The first few points could be used to estimate the mean dephasing time of these carbons, which is 35 μS. The mean dephasing time of the methyl carbons could be similary determined by using the last few points of the magnetization curve. Nevertheless at the longest dephasing time used here, 100 μS, the CH and CH_2 carbons still represents ~5% of the magnetization. As, inferred from HR results, the methyl carbons represent ~10% of

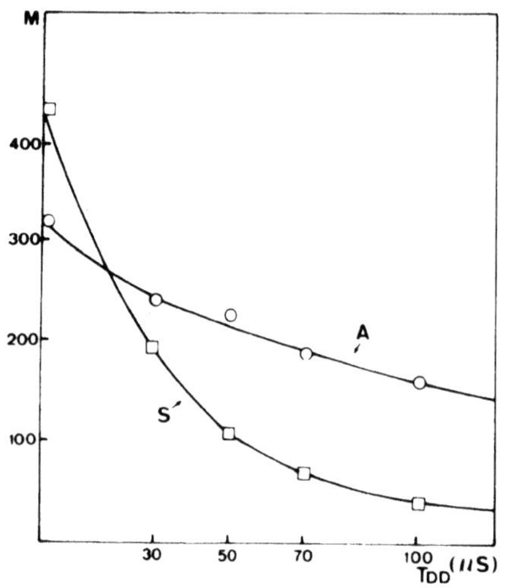

FIG 4

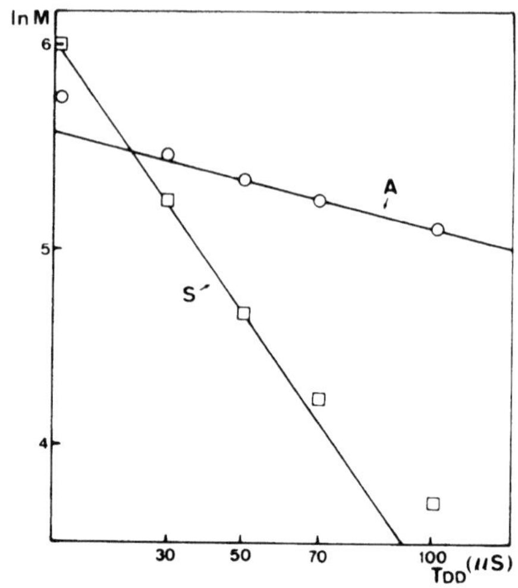

FIG 5

the magnetization and the dephasing time was not calculated here but taken from the literature(7). A meaningfull determination of T_2^* for CH_3 carbons should require measurements at longer dephasing time where the S/N ratio is very poor. The M_S magnetization curve was then represented by the sum of two exponentials, one for the CH and CH_2 carbons with $T_2^* = 35$ μS and one for the CH_3 carbons with $T_2^* = 80$ μS.

$$M_S(t) = M_{S(CH+CH_2)} \exp(-t/\ 35\ \mu S) + M_{S(CH_3)} \exp(-t/\ 80\ \mu S)$$

The best fit to the experimental values gives $M_{S(CH+CH_2)} = 389$ and $M_{S(CH_3)} = 51$ (in arbitrary units).
Similary the end of the magnetization curve mostly represents the non protonated aromatic carbons which should dephase slowly. The last points could be used to estimated the mean dephasing time of these carbons. The 200 μS value here obtained is the same as that obtained by Gerstein. Due to the lack of experimental data at short dephasing times, the T_2^* value for protonated aromatic carbons was also taken from the literature(7). The M_A magnetization curve was represented by the sum of two exponentials, one for the protonated carbons with $T_2^* = 17$ μS and one for the non protonated carbons with $T_2^* = 200$ μS.

$$M_A(t) = M_{A(CH)} . \exp(-t/\ 17\ \mu S) + M_{A(C)} . \exp(-t/\ 200\ \mu S)$$

The best fit to the experimental value gives $M_{A(CH)} = 55$ and $M_{A(C)} = 270$. The true ratios between the various types of carbons could be obtained provided the above values are corrected by factors taking into account the fact that the equilibrium magnetization is not obtained at 1 mS cross polarization time for non protonated aromatic and methyl carbons.

$$C_{S(CH\ )}/C_{S(CH+CH\ )} = 51(1.16)/389 = 0.15$$

$$C_{A(CH)}/C_{A(C)} = 55/270(1.28) = 0.16$$

The correcting factors 1.16 and 1.28 are obtained from the VCT experiment. The carbon ratios obtained here are in satisfactory agreement with those obtained in HR work considering the fact that both methods are subject to fairly large errors.

Schaefer and Stejskal introduced the CP pulse sequence for the measurement of the apparent ^{13}C rotating frame spin lattice relaxation time $T_{1\rho C}$. In this sequence, a variable relaxation period, T_{CSL}, is inserted between the cross polarization period and the acquisition period. During the carbon spin lock period, T_{CSL}, the ^{13}C magnetization decays at a rate measured by the apparent $T_{1\rho C}$. The magnetization curves, obtained with 1 mS cross polarization time , are given fig.6. The aromatic band appears to decay homogeneously with a 44 mS time constant. On the contrary, the aliphatic band appears to decay inhomogeneously with two time constant : 43% of the magnetization decays with a short 8 mS time constant and 57% of the magnetization decays with a 41 mS time constant quite similar to the value found for the aromatic carbons. These values are close to those reported for coal and kerogen(14). The apparent ^{13}C $T_{1\rho}$ can be related to spin - spin or spin - lattice or more probably to a complex interplay of spin - spin and spin - lattice processes. The relative importance of these two processes were analysed for polymers by Vanderhart and Garroway(15) and Schaefer, Stejskal and Buckdael(16). The discrimination between these two types of relaxation mechanism is not obvious for coals and related compounds. When the observed relaxation time is dominated by a spin - lattice process it can be used to probe the density of molecular motion in the tens of KHz region. This type of interpretation has been suggested(6) for coal specimens.

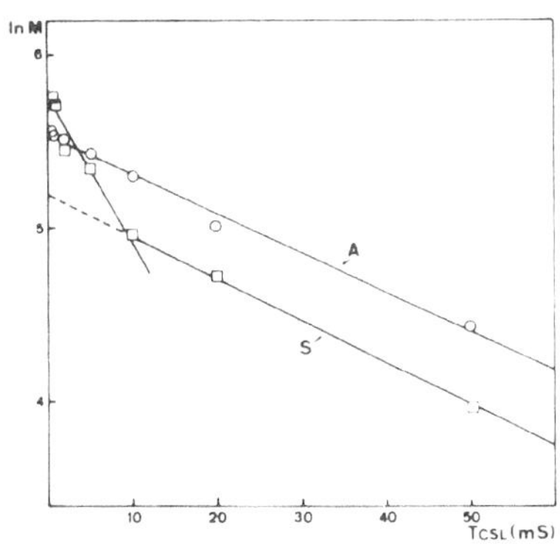

FIG 6

Carbon $T_{1\rho}$ data for coals and fossil fuels are quite scarce and not sufficiently well understood to be used as discriminating parameters in structural resolution studies.

CONCLUSIONS

Carbon 13 CP/MAS NMR appears to be a useful quantitative technique to study the organic structures in solid fossil fuels provided the experimental conditions are carefully selected. Special techniques such as dipolar dephasing and carbon spin lock appear to be useful methods to simplify complicated spectra and to obtain quantitative informations. Prospects of these methods are encouraging, nevertheless further work will be necessary to explore the full potential of these approaches

REFERENCES

(1) A.R. Palmer and G.E. Maciel, 1982. Anal.Chem., 54, p.2194
(2) G.E. Maciel, V.J. Bartuska, F.P. Miknis, 1979. Fuel, 58, p.391
(3) P. Du Bois Murphy, T.J. Cassady and B.C. Gerstein, 1982, Fuel, 61, p.1233
(4) D.J. Cookson and B.E. Smith, 1981. Org.Magn.Reson., 16, p.111
(5) A. Pines, M.G. Gibby and J.S. Waugh, 1973. J.Chem.Phys., 59, p.582
(6) M.T. Sullivan and G.E. Maciel, 1982. Anal.Chem., 54, p.1606
(7) P. Du Bois Murphy and B.C. Gerstein, 1982. Anal.Chem., 54, p.522
(8) S. Gillet and J.J. Delpuech, 1980. J.Magn.Res., 38, p.433
(9) M.A. Wilson, P.J. Collin, R.J. Pugmire and D.M. Grant, 1982. Fuel, 61, p.959
(10) D.J. Cookson and B.E. Smith, 1982. Fuel, 61, p.1007
(11) T. Suzuki, 1982. Fuel, 61, p.402
(12) J. Schaefer and E.O. Stejskal, 1976. J.Am.Chem.Soc., 98, p.1031
(13) A.N. Garroway, W.B. Moniz and H.A. Resing in "Carbon-13 NMR in Polymer Science", Ed W.M.Baska. Am.Chem.Soc.Symp.Ser.103
(14) A.R. Palmer and G.E. Maciel, 1982. Anal.Cehm., 54, p.2194
(15) D.L. Vander Hart and A.N. Garroway, 1979. J.Chem.Phys., 71, p.2773
(16) J. Schaefer, E.O. Stejskal and R. Buchdal, 1977. Macromolecules, 10, p.384

SOME EXAMPLES OF ^{13}C NMR APPLICATIONS TO OIL SHALES

F. P. Miknis and G. E. Maciel

University of Wyoming Research Corporation, Laramie, Wyoming and Colorado State University, Fort Collins, Colorado

ABSTRACT

Some examples are given of the use of solid state ^{13}C NMR techniques to study oil shales. These include the use of ^{13}C NMR for oil shale resource evaluation and for oil shale thermal decomposition. In addition, the use of ^{13}C NMR for evaluating petroleum source rock potential illustrated. NMR measurements of the fraction of aliphatic carbon in an oil shale are shown to correlate with the fraction of carbon converted to oil upon Fischer assay. Similar measurements on oil shales as a function of depth of burial can be used to assess the thermal maturation of the source rock. Interrupted decoupling experiments have been conducted on oil shales heated to various temperatures. The results of these measurement are discussed.

INTRODUCTION

Since the introduction of the solid state ^{13}C NMR techniques of cross-polarization (CP) with high-power decoupling and magic-angle spinning (MAS), there has been a steady increase in the applications of these techniques to the study of fossil fuels. Indeed, some of the earliest measurements illustrating the use of the techniques were applied to coals (1,2) and oil shales (3,4). The main impetus for this was that the CP/MAS techniques provide information about the carbon functionality in insoluble organic materials such as coals and oil shales. Moreover, the measurements are direct, non-destructive and can be applied to whole coal or oil shale samples ie. organic concentrates do not have to be prepared.

L. Petrakis and J. P. Fraissard (eds.), Magnetic Resonance. Introduction, Advanced Topics and Applications to Fossil Energy, 545–555.

In this paper we present some examples of the application of CP/MAS techniques to the analysis of oil shales. The applications range from oil shale resource evaluation to the study of oil shale thermal decomposition. Because oil shales are organic-bearing sedimentary rocks they can also be considered as source rocks for the generation of petroleum. An example of the use of NMR to assess an oil shale for source rock potential is included in this report.

APPLICATIONS

Oil Shale Resource Evaluation

Probably the most frequent analysis of an oil shale is the Fischer Assay which reports the oil yielding potential of an oil shale in terms of gallons of oil per ton of shale. The Fischer Assay is simply a specification test which measures the amount of oil produced from a 100 gram sample of oil shale heated at a prescribed rate to a final temperature (12°C/min to 500°C, isothermal at 500°C for 40 min). It is widely used to measure the commercial viability of a deposit, although it provides no information about the type of organic matter in the oil shale.

It is well known that the quality or type of organic matter in an oil shale largely determines what fraction of the original organic matter will be converted to oil upon heating (5). For example, oil shales of a predominantly algal origin (highly aliphatic) produce a greater fraction of liquids upon heating than do oil shales of predominantly terrestrial origin (highly aromatic) for an equivalent amount of organic carbon. The reason for this is that the aliphatic chemical structures contain a greater proportion of hydrogen, which is critical for producing liquids from organic materials during pyrolysis. Thus while the total quantity of organic matter in an oil shale deposit is an important factor for producing quantities of oil sufficient for commercial development, the quality of the organic matter is an important factor for determining the conversion characteristics of the oil shale deposit.

CP/MAS ^{13}C NMR techniques have proven useful for oil shale resource evaluation because these techniques provide a direct assessment of the quality of an oil shale and hence, its conversion characteristics upon heating. We have applied CP/MAS techniques to measure the organic carbon distributions in oil shales from various worldwide deposits (6) and to quantitate that portion of the carbon functionality which is responsible for producing liquids(7). Some of the results are shown in Figure 1 which shows a plot of the oil yields produced by Fischer Assay versus the amount of the aliphatic carbon in these samples.

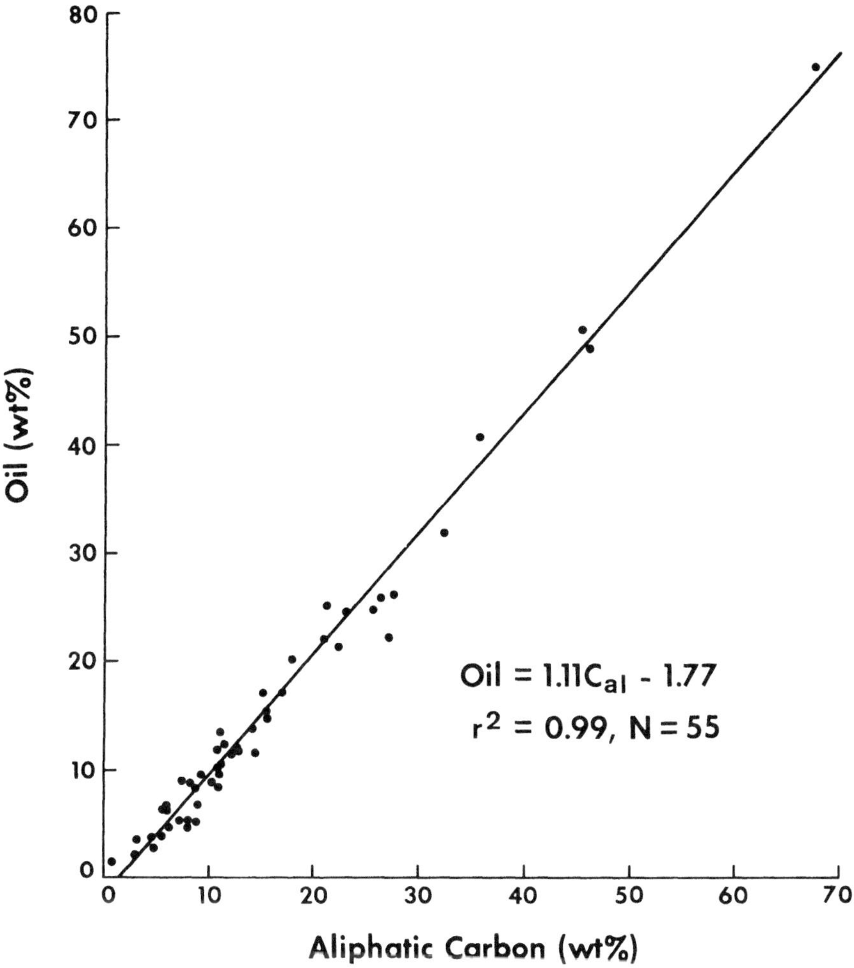

Fig. 1. Plot of Oil Yield vs. Aliphatic Carbon Content

The weight percent of aliphatic carbon is determined by multiplying the total organic carbon in the sample by the fraction of aliphatic carbon determined by NMR. Linear regression parameters are also shown for the fifty-five samples used in this study. The correlation clearly establishes the importance of aliphatic carbon structures for producing oils from oil shales regardless of the shales geographic origin, geologic age or depositional environment. The importance of aliphatic structures for producing liquids from oil shales is not new; however, the CP/MAS techniques now offer a method to quantitate the amounts of these structures in raw oil shales.

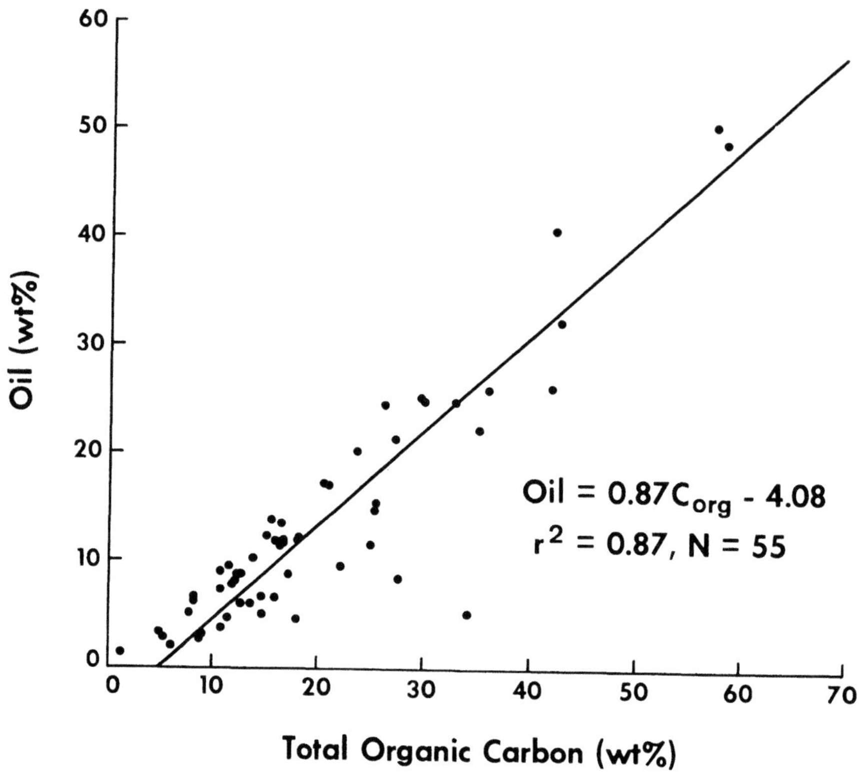

Fig. 2. Plot of Oil Yield vs. Total Organic Carbon

To further illustrate the importance of aliphatic materials for producing liquids from oil shales, Figure 2 shows a plot of Fischer Assay oil yields versus total organic carbon contents of the shales. This correlation is significantly poorer than that shown in Figure 1 and further illustrates that total quantity of organic matter is not a sufficient indicator to assess an oil shales potential to produce oil. Clearly, the type of organic matter in the deposit is an important factor. The scatter in Figure 2 is attributable largely to the fact that the total organic carbon contains contributions from the oil producing aliphatic carbons and from the non-oil producing aromatic carbons and in differing proportions among the samples.

Petroleum Source Rock Evaluation

As mentioned in the introduction an oil shale can be a petroleum source rock. However, for this to happen it must be subjected to a sufficient depth of burial for a sufficient period of geologic time to allow for the thermal maturation processes associated with depth of burial to effect the transformation of the organic matter to liquids and gases. By using NMR to measure changes in the organic carbon distribution of oil shales as a function of depth of burial it is possible to observe changes associated with thermal maturation.

We have applied CP/MAS techniques for source rock evaluation by studying oil shales from the Permian Phosphoria Formation in eastern Idaho and southwestern Montana (8). In brief, the

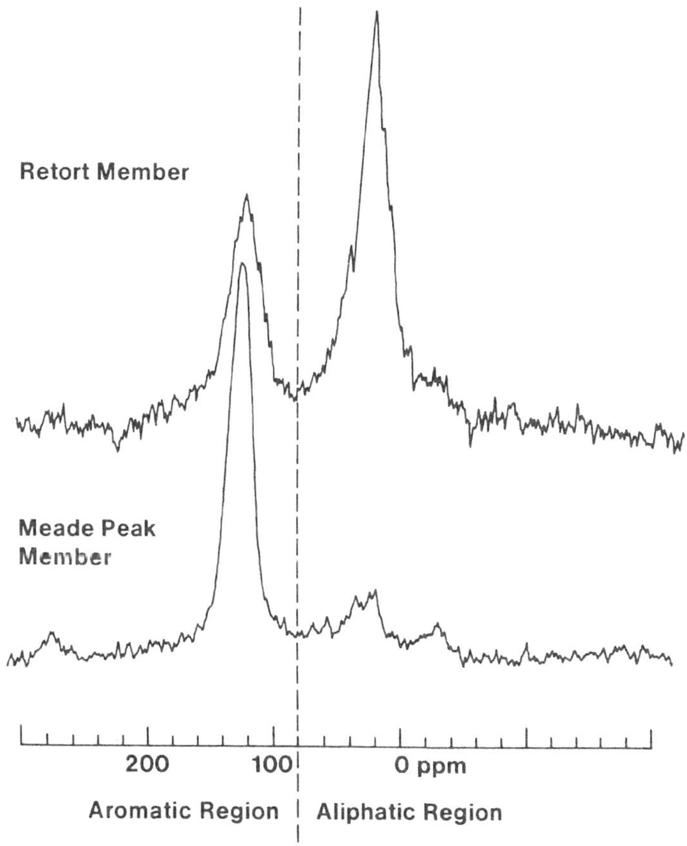

Fig. 3. CP/MAS ^{13}C NMR Spectra of Phosphoria Formation Oil Shales, illustrating changes in carbon distribution with depth of burial.

Phosphoria Formation consists of two organic-bearing shale members - the Retort and Meade Peak members. During Jurassic time, due to tectonic overthrusting, the Meade Peak member became more deeply buried (up to 9 kilometers in some places) than the Retort member. Geologic and geochemical studies (9) have concluded that the Phosphoria oil shales are the source rock for much of the oil found in upper Paleozoic rocks of the northern and central Rocky Mountain region. Our results which are shown in Figure 3 support these conclusions. The Retort member shale which was not subjected to a great depth of burial has a high aliphatic carbon fraction and represents a good petroleum source rock. The Meade Peak member which was subjected to a greater depth of burial was thermally altered as evidenced by its depleted aliphatic carbon fraction and probably produced oil.

A present disadvantage of using NMR to evaluate source rock potential is that often source rock measurements are made on samples containing less than one percent organic carbon. As a result, CP/MAS measurements on such samples become prohibitively time-consuming. However, the techniques of dynamic nuclear polarization presently being developed by Wind and coworkers (10,11) offer hope that in the future NMR measurements on samples having small amounts of organic carbon will be feasible.

Oil Shale Thermal Decomposition

Solid State ^{13}C NMR techniques are promising for the study of oil shale thermal decomposition and reaction kinetics. A major reason for this is that the techniques provide more insight into the chemistry of the decomposition reaction. Most oil shale thermal decomposition studies are based on the scheme:

$$\text{kerogen} \rightarrow \text{bitumen} \rightarrow \text{oil} + \text{gas} + \text{residual carbon}$$

or variations thereof. While such a scheme is useful for the "global" modelling of a retorting process for engineering purposes, the scheme does not provide my fundamental information about the chemistry of the reaction. With solid state ^{13}C NMR techniques it is possible to monitor the fate of the various carbon functionalities during heating.

A new model of kerogen decomposition has been proposed (12). The basis for the model is that aliphatic carbons are largely responsible for oil production, while aromatic carbons are largely responsible for the residual carbon. A subsequent NMR study (13) has shown that part of the aliphatic carbon "aromatizes" during heating, while a small, undefined portion of the aromatic carbon contributes to the oil production.

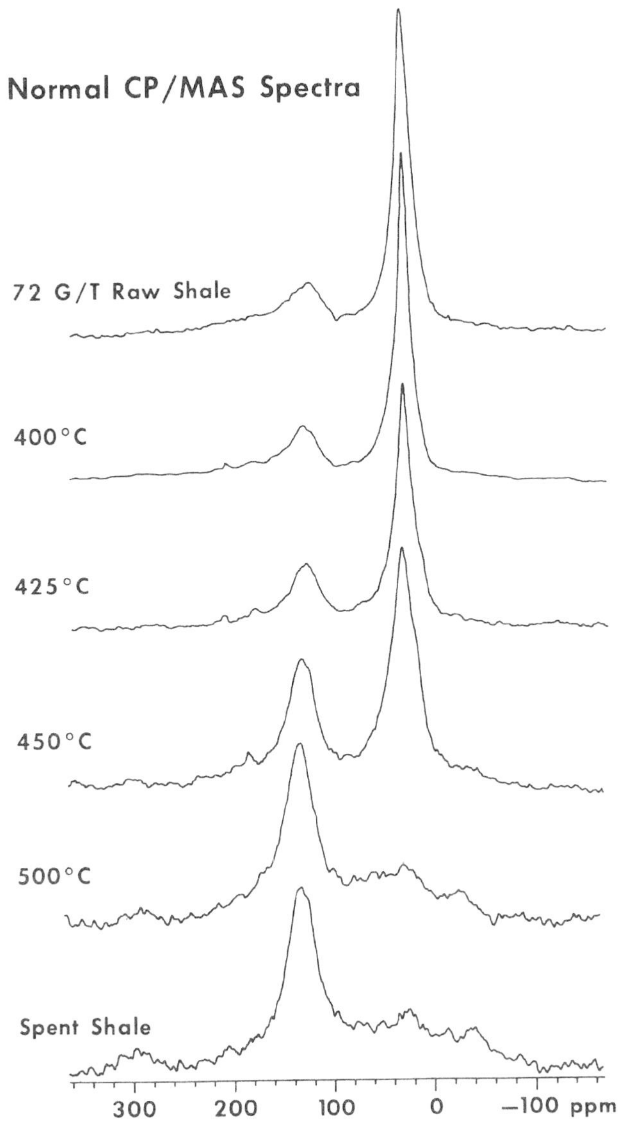

Fig. 4. CP/MAS ^{13}C NMR Spectra of Heated Colorado Oil Shale

Changes in the carbon distribution that occur upon heating can be easily measured by ^{13}C NMR. Examples are shown in Fig. 4. The six spectra represent six different samples of a 72 G/T Colorado oil shale heated in a Fischer assay retort to the final temperatures shown in Fig. 4. When the retort reached the desired temperature it was shut off and allowed to cool. No provisions were made for rapidly quenching the reaction. As a result, no kinetic information was obtained for these samples. Instead, the samples were generated for the purpose of obtaining a set of samples of varying carbon distribution to obtain a qualitative assessment on the use of interrupted decoupling techniques for oil shale kinetic experiments. The interrupted decoupling spectra are shown in Fig. 5 and are discussed later.

Some observations can be made about the CP/MAS spectra in Fig. 4. The aliphatic carbon fraction decreases with increasing temperature. This is expected since the aliphatic carbons are primarily responsible for the produced oils. At 450°C there is some "broadening" of the aliphatic band and the appearance of a shoulder at ≈ 19 ppm. This suggests that the aliphatic carbon remaining in the sample contains more short chain alkyl substitutes than the parent oil shale. Some aliphatic carbon remains even after Fischer assay. In a previous study, it was shown that the residual carbon after Fischer assay is almost exclusively aromatic carbon. The 72 G/T oil shale is a rich oil shale and showed a tendency to coke during the Fischer assay. The residual aliphatic carbon may be due to incomplete coking, although coking is usually considered an "aromatization" reaction.

The aromatic carbon band appears to narrow upon heating. This suggests that at the higher temperatures, the aromatic carbon remaining in the sample is less substituted, and more condensed. Similar observations have been made for coals as a function of rank (14, 15). The heated oil shales are analogous to oil shales of different "rank". In this case the "rank" is artificially produced by heating to high temperatures for short periods of time as compared to the slow heating over geologic time as in the case for coals.

The interrupted decoupling spectra are shown in Fig. 5. These spectra were recorded at an interrupt time of 75 μs and were not scaled to zero time as suggested by Murphy et al. (16, 17). Our purpose for these experiments was to determine whether interrupted decoupling techniques can provide additional information about the carbon distribution in heated oil shales.

Some qualitative observations can be made about the spectra in Fig. 5. The carbonyl region (>170 ppm) is enhanced over the corresponding region of the normal CP/MAS spectra. Some phenolic carbons are suggested by the shoulders at ≈ 155 ppm. The center

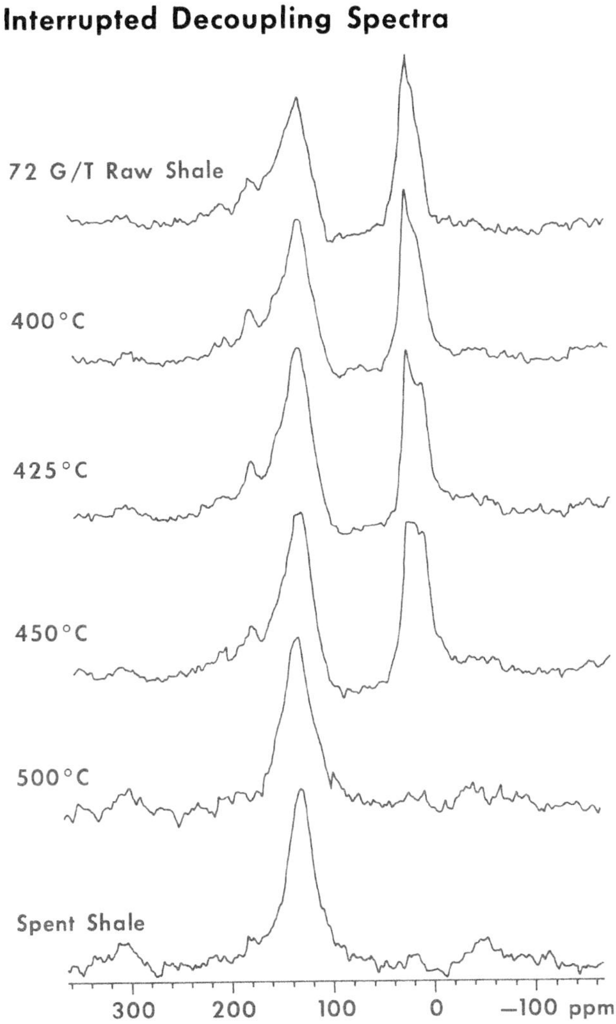

Fig. 5. Interrupted Decoupling Spectra of Heated Colorado Oil Shale. The interrupt time was 75μs.

of the aromatic head is shifted more upfield with increasing temperature and also becomes "narrower". These observations suggest a trend from substituted quaternary aromatic carbon at low temperatures to more polycondensed quaternary aromatic carbon at higher temperatures. The residual aliphatic carbon at 500°C has been removed in the interrupted spectra, suggesting that these carbons were secondary and tertiary aliphatic carbons.

While the data for the heated shales are preliminary and mostly qualitative, the chemical information obtainable by using CP/MAS and interrupted decoupling techniques make solid state ^{13}C NMR techniques promising for the study of oil shale reaction kinetics. More detailed studies are in progress.

REFERENCES

1. VanderHart, D.L. and Retcofsky, H.L. 1976, Fuel 55, 202.

2. Bartuska, V.J., Maciel, G.E., Schaefer, J. and Stejskal, E.V. 1977, Fuel 56, 354.

3. Resing, H.A., Garroway, A.N. and Hazlett, R.N. 1978, Fuel 57, 450.

4. Maciel, G.E., Bartuska, V.J. and Miknis, F.P. 1978, Fuel 57, 505.

5. Tissot, B.P. and Welte, D.H. "Petroleum Formation and Occurrence", Springer-Verlag, New York, 1978.

6. Miknis, F.P., Maciel, G.E. and Bartuska, V.J., 1979, Org. Geochem. 1, 169.

7. Maciel, G.E., Bartuska, V.J. and Miknis, F.P., 1978, Fuel 57, 505.

8. Miknis, F.P., Smith, J.W., Maughan, E.K. and Maciel, G.E. 1982, Am. Assoc. Petrol. Geol. Bull. 66, 1396.

9. Claypool, G.E., Love, A.H. and Maughan, E.K. 1978, Am. Assoc. Petrol. Geol. Bull. 62, 98.

10. Wind, R.A., Trommel, J. and Smidt 1982, Fuel 61, 398.

11. Wind, R.A., Anthonio, F.E., Duijvestijn, M.J., Smidt, J., Trommel, J. and De Vette, G.M.C. 1983, J. Magn. Reson. 52, 424.

12. Fausett, D.W. and Miknis, F.P., 1981, Proc. of 14th Oil Shale Symposium (ed J. Gary) Colo. Sch. of Mines, Golden, CO., p. 154.

13. Hershkowitz, F., Olmstead, W.N., Rhodes, R.P., and Rose, K.D., in "Geochemistry and Chemistry of Oil Shales" ACS Symposium Series 230 (eds F.P. Miknis & J.F. McKay)(in press).

14. Miknis, F.P., Sullivan, M., Bartuska, V.J. and Maciel, G.E., 1981, Org. Geochem. 3, 19.

15. Dereppe, J.M., Boudou, J.P., Moreaux, C. and Durand, B., 1983, Fuel 62, 575.

16. Murphy, P.D., Gerstein, B.C., Weinberg, V.L., and Yen, T.F., 1982, Anal. Chem. 54, 522.

17. Murphy, P.D., Cassady, T.J. and Gerstein, B.C., 1982, Fuel 61, 1233.

COUNTING CARBON SPINS IN CP-MAS SPECTRA OF FOSSIL FUELS

J.R. Kalman

CSIRO Division of Fossil Fuels
P O Box 136
North Ryde, NSW 2113 Australia

INTRODUCTION

Carbon aromaticity measurements by CP-MAS NMR form a logical basis for deductions about chemical changes during fossil fuel maturation or processing. An example is in oil shale retorting, where it has been found that the mass of aromatic carbon in the residue equals that in the starting shale (1). Similarly work by colleagues at CSIRO indicates that during hydrogenation of coal in tetralin solvent, there is no change in the total amount of aromatic carbon (2). Wherever aromaticity values are being used to follow changes in the mass of some component of the carbon in a material, it is important to be sure that all the carbons in the sample are contributing equally to the spectrum. Although the reliability of aromaticity values derived from CP-MAS spectra has been the subject of much discussion in the literature (3 - 7), there has been little recent mention that the total intensity of a spectrum should be related to the mass of carbon in the sample (3).

L. Petrakis and J. P. Fraissard (eds.), Magnetic Resonance. Introduction, Advanced Topics and Applications to Fossil Energy, 557–567.

EXPERIMENTAL

NMR spectra were obtained on a Bruker CXP-100 spectrometer, operating at 22.63 MHz for carbon, equipped with an iron magnet and a cross-polarisation/magic angle spinning (CP-MAS) probe designed for Andrews-type rotors. Rotors consisted of a Kel-F air-bearing cone joined by a screw thread to a boron nitride barrel, which served as sample holder. Routine CP-MAS spectra were obtained using a single-contact CP pulse program (fig.4 in the text). Contact times were optimised to maximum signal, recycle times were 1 second, and spin rates between 2.6 and 3.2kHz.

RESULTS AND DISCUSSION

Recently we have examined a number of samples, derived from both coals and oil shales, in which very noticeable differences could be observed in signal-to-noise ratio, despite the use of optimum contact times and the presence of similar amounts of carbon in each sample. This prompted a quantitative examination of the amount of signal obtained from samples of known weight and carbon content.

Spin Counting

In principle, with the spectrometer conditions fixed, a sample should generate a signal proportional to its carbon content (we have confirmed that signal is directly proportional to both sample weight and number of scans). Comparison of measured intensity of the spectrum with the known carbon content of a sample, and reference to a suitable standard should show whether all the carbons of the sample are contributing to the spectrum. Harris and Packer (9) have applied this method of "spin counting" to polymers; VanderHart and Retcofsky (3) applied it to coal in 1976, but felt their results were not conclusive because of possible differences in probe tuning.

The method was calibrated using samples of alumina containing known amounts of hexamethylbenzene; the estimated accuracy is 20%. Within this experimental error, Liddell and Walloon bituminous coals, and highly aliphatic kerogens from Rundle oil shale, give carbon CP-MAS spectra of the expected intensity. For some samples, however, this is not the case.

Effect of Heat Treatment on Carbon Spin Count.

Fig.1 shows CP-MAS carbon spectra of an oil shale kerogen from the Julia Creek deposit (Queensland), and the demineralised residue after retorting of the raw shale at 495°C. The spectra have been normalised to the same mass of carbon in each sample, and the same number of scans (0.2ms contact times were optimum for each sample). The residue gives a spectrum about 15% of the expected intensity. In fig.2 are spectra of a series of solid residues derived from a Yallourn brown coal (Victoria). Samples of coal were suspended in water and heated in an autoclave for 3h at the indicated temperatures, as part of an experiment on accelerated maturation. The intensities of the spectra are directly comparable as before. Note that with increasing treatment temperature, the spectra become less intense. Similar results have been observed with residues from the hydrogenation of Liddell coal (N.S.W. bituminous coal) in tetralin (fig.3). The experiments, part of a CSIRO study conducted by Dr Neil Foster, were performed by rapidly injecting a coal slurry into preheated tetralin, followed by rapid quenching after a given time. In the case of fig.3, the spectra were referenced to hexamethylbenzene. The integrated intensities of 3b and 3d were respectively 80% and 45% of the expected values.

Each of figs.1,2 and 3 depict a suite of samples derived from the same starting material, and in each, the same trend is evident: that a decrease in signal intensity occurs with increased severity of treatment. In fig.2, a monotonic decrease occurs over a range of samples; the Julia Creek example in fig.1 is quite striking.

In order to discuss possible reasons for the reduced signal in these samples, we first present a simple picture of the evolution of carbon signal during the cross-polarisation process.

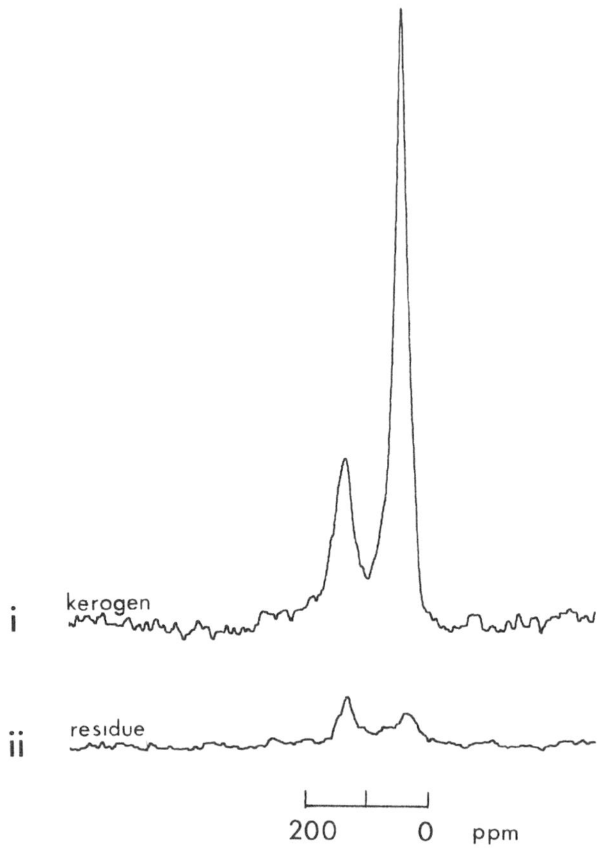

Fig.1. C-13 CP-MAS spectra of (i) the kerogen obtained on demineralising a sample of oil shale from Julia Creek (Queensland), and (ii) a demineralised sample at the residue obtained after retorting the shale at 495°C.

In adjusting the relative amplitude of the spectra the number of scans, and the weights and carbon analyses of the samples have been taken into account so as if the same amount of carbon had been present in each sample.

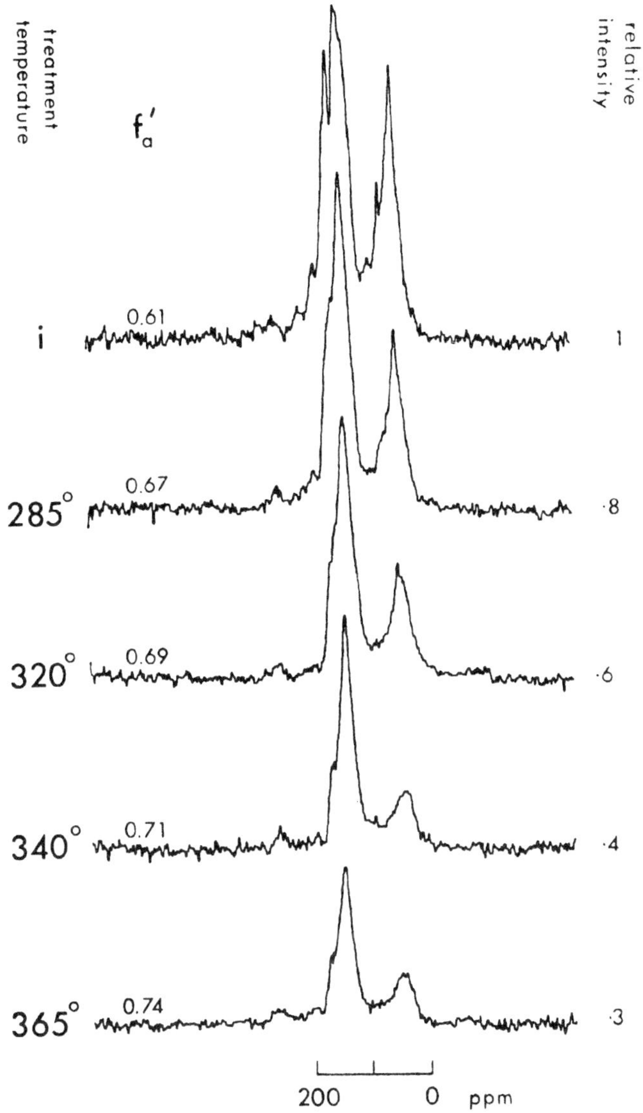

Fig.2 C-13 CP-MAS spectra of (i) the acid-exchanged form of Yallourn brown coal (H^+), and samples derived from it by hydrothermal treatment. The acid form coal was sealed in an autoclave with excess water in a N_2 atmosphere, and rapidly heated to the indicated temperature which was maintained for 3h.

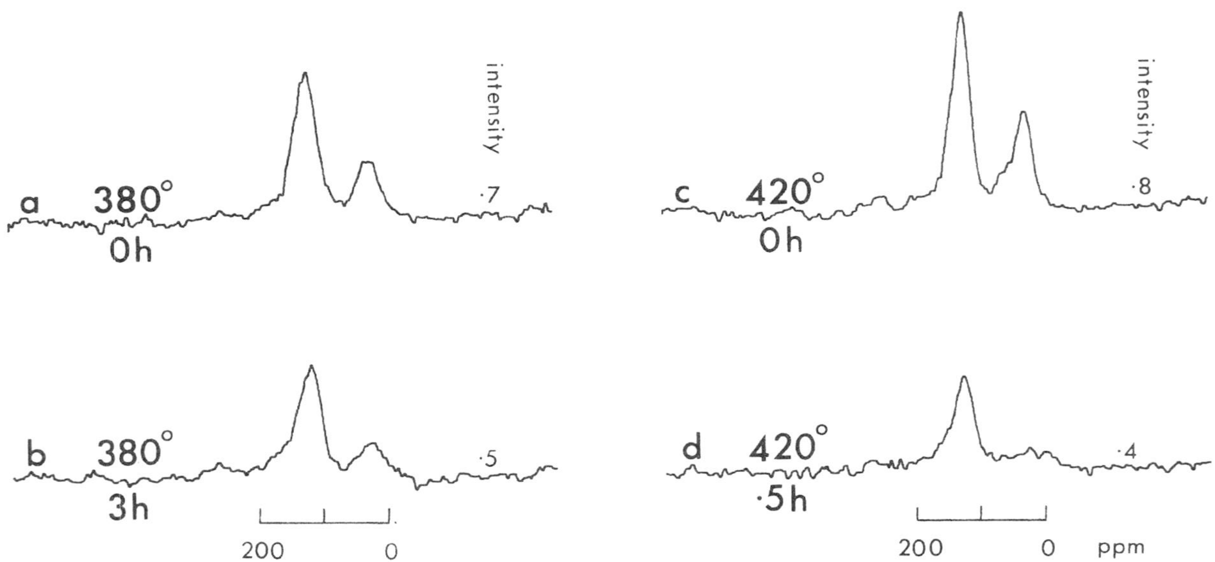

Fig.3 C-13 CP-MAS spectra of four samples of tetrahydrofuran insoluble Liddell coal residues. The samples were rapidly injected as a coaltetralin slurry into an autoclave containing preheated tetralin at 380 or 420°. Two control reactions (a and c) were then quenched immediately to below 350°, while (b) and (d) were quenched after the indicated reaction time.

The spectral amplitudes were adjusted as for Fig.1, and additionally referred to a spectrum of hexamethylbenzene. The intensities (relative to hexamethylbenzene = 1)are indicated to the figure.

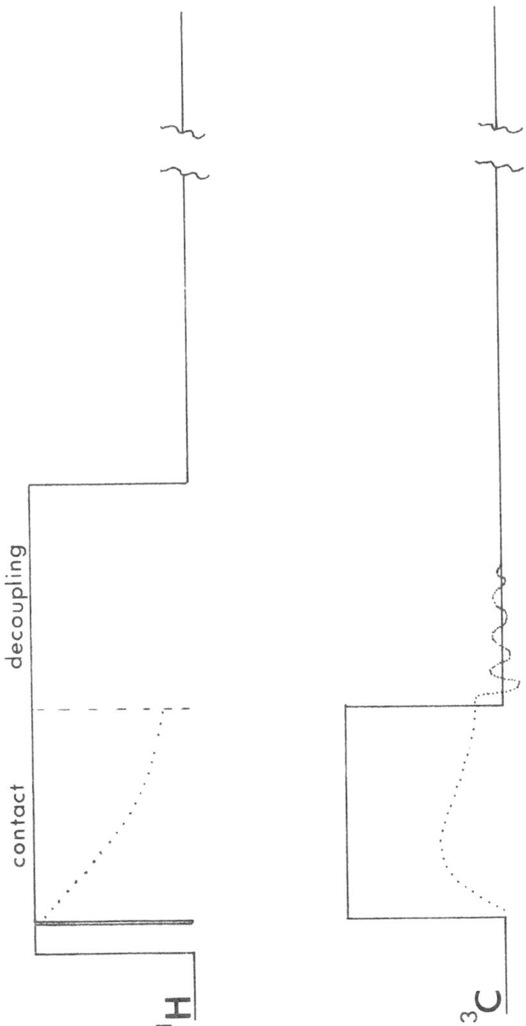

Fig.4 Cross-polarisation pulse sequence. The possible evolution of ^1H and ^{13}C magnetisation during Hartmann-Hahn contact is indicated by the dotted curves.

Cross Polarisation Kinetics

The familiar cross polarisation pulse diagram is reproduced in fig.4. To a first approximation, the level of carbon magnetisation following Hartmann-Hahn contact depends on (i) the initial value of the spin locked ^1H magnetisation, and (ii) the rate constants T_{CH} and $T^H_{1\rho}$ which characterise the build up of carbon magentisation and the decay of the proton spin locked magnetisation, respectively. When build up and decay are simple exponential processes, the evolution of carbon magnetisation during spin-locked contact can be expressed by equation 1.

$$M = M_o \exp(-t/T^H_{1\rho}) \, [1-\exp(-t/T_{CH})] \quad \underline{1}$$

The carbon signal is predicted to reach the maximum intensity at $t = T_{CH} \ln(1 + T^H_{1\rho}/T_{CH})$. The ratio $T^H_{1\rho}/T_{CH}$ thus governs both the optimum contact time (maximum signal) and the signal intensity at this maximum. It is instructive to examine the values predicted by this simple equation.

$T^H_{1\rho}/T_{CH}$	optimum contact	intensity
0.1	$0.1T_{CH}$	$0.03M_o$
1	$0.7T_{CH}$	$0.25M_o$
10	$2.4T_{CH}$	$0.72M_o$
100	$4.6T_{CH}$	$0.95M_o$
1000	$6.9T_{CH}$	$0.98M_o$

Typical values of $T^H_{1\rho}$ for coals and oil shales lie in the range 1 to 10ms (10); carbons with a proton attached have T_{CH} about 50μsec, but for non-protonated carbons this value can be considerably greater (5,10). If the average value for aromatic carbons is 100μsec, and $T^H_{1\rho}$ for aromatics is around 6msec, then $T^H_{1\rho}/T_{CH}$ is ∿60, and the intensity of the carbon signal at its maximum has dropped to $0.91M_o$. The table shows how sensitive this effect might be to the relative values of $T^H_{1\rho}$ and T_{CH}. It should be noted that T^H_1 and $T^H_{1\rho}$ of coals and oil shales are generally not simple exponential decays (11,12) when measured directly by ^{1}H nmr. If the different components of the relaxation correspond to physically different parts of the structure of the sample, then the predictions in the table above may apply to these regions individually.

Factors Leading to Diminished Carbon Signal

Now we can state some of the possible reasons for loss of carbon signal.

(a) Some carbons may have long T_{CH} due to their remoteness from protons, or possibly, in mobile samples, to motional averaging of dipolar interactions.

(b) $T_{1\rho}$ may be unusually short.

(c) The proton line may be very broad (the pulse width should ideally be less than 1/(4xfull proton line width) (13). This would result in a lower intensity of spin-locked ^{1}H, which in turn would influence the carbon signal.

Both (b) and (c) are possible consequences of paramagnetism (organic or inorganic) in the sample.

Only for the first of these factors are we able to present any direct evidence. All the spectra of heat-treated samples show some increase in aromaticity over the starting material. On removal of alkyl groups, carbons are on average further from protons, and an increase in average T_{CH} is expected. Condensation reactions during heating would have the same effect.

No dramatic changes in $T^{H}_{1\rho}$ upon heat treatment, are observed, as judged by the contact time dependences of the carbon signals. However this is not a reliable guide since only the protons affecting the observed carbon signal are being measured. Rapidly-relaxing protons cannot be observed through carbon; conclusions about whether a greater proportion of protons are relaxing rapidly in the heat treated materials must await direct ^{1}H measurements, which are now in progress. Paramagnetic centres created by heating may be trapped in the structure of the solid, creating a region of rapidly-relaxing protons linked by spin diffusion. If $T_{1\rho}^{H} < 0.1\ T_{CH}$, no carbon signal at all would result from such a region in the sample. If $T_{1\rho}^{H} \sim T_{CH}$, the predicted carbon signal would be 1/4 of that expected on the basis of the number of carbon atoms present.

Implication for Aromaticity Measurements

Low carbon spin count does not necessarily imply an incorrect analysis of carbon functionality. However in cases where aromatic and aliphatic signals have different, multi-exponential proton-rotating frame relaxations, an incorrect result is more likely, and here a C13-FID experiment can be valuable as a check (7). The value of routine spin counting is that it can alert us to a sample needing special attention.

In conclusion, we have shown that some samples of interest in Fossil Fuel research, such as certain heat treated coals and oil shales, give CP-MAS spectra which are less intense than expected. Carbon spin-counting can be a convenient way to establish this for a particular sample. In these cases, due caution must be exercised in performing quantitative analyses of carbon functionality.

ACKNOWLEDGEMENTS

The author thanks Dr J.D. Saxby and Dr N. Foster for supplying respectively the demineralised Julia Creek shale and spent shale, and the Liddell hydrogenation residues.

REFERENCES

(1) Miknis, F.P., Szeverenyi, N.M. and Maciel, G.E. 1 982, Fuel 61, p. 341.

(2) Wilson, M.A., Pugmire, R.J., Vassallo, A.M., Grant, D.M., Collin, P.J. and Zilm, K.W. 1982, Ind.Eng.Chem.Prod.Res.Dev. 21, p.497.

(3) VanderHart, D.L. and Retcofsky, H.L. 1976, Fuel 55, p. 202.

(4) Maciel, G.E., Bartuska, V.J. and Miknis, F.P. 1979, Fuel 58, p. 391.

(5) Wemmer, D.E., Pines, A. and Whitehurst, D.D. 1981, Phil.Trans.R.Soc.Lond.A 300, p. 15.

(6) Dudley, R.L. and Fyfe, C.A. 1982, Fuel 61, p. 651.

(7) Wind, R.A., Trommell, J. and Smidt, J. 1982, Fuel 61, p. 398.

(8) Hagaman, E.W. and Woody, M.C. 1981, Proc.Int.Conf. Coal Sci.Dusseldorf, p. 807.

(9) Aujla, R.S., Harris, R.K., Packer, K.J., Parameswaran, M., Say, B., Bunn, A. and Cudby, M.E.A. 1982, Polymer Bull. 8, p. 253.

(10) Sullivan, M.J. and Maciel, G.E. 1982, Anal.Chem. 54, p. 1615; Palmer, A.R. and Maciel, G.E., 1982, Anal.Chem. 54, p. 2194.

(11) Gerstein, B.C., Chow, C., Pembleton, R.G. and Wilson, R.C. 1977, J.Phys.Chem. 81, p. 565.

(12) Kalman, J. Unpublished results.

(13) Farrar, T.C. and Becker, E.D. 1971, "Pulse and Fourier Transform NMR; Introduction to Theory and Methods", Academic Press, NY., p. 67.

^{13}C NMR INVESTIGATION OF THE CHEMICAL STRUCTURES OF COKING AND NON-COKING COALS IN THE ORIGINAL AND REDUCTIVELY ALKYLATED SOLID STATES

Gaye Erbatur and Oktay Erbatur, Department of Chemistry, Çukurova University, P.K. 171 Adana, Turkey; and Abdullah Coban, Iskenderun Steel Company, Turkey; and Mark F. Davis and Gary E. Maciel, Department of Chemistry, Colorado State University, Fort Collins, Colorado 80523, U.S.A.

ABSTRACT

Chemical structures of four Turkish coals in their original and reductively alkylated forms were investigated in the solid state by ^{13}C NMR, using cross polarization and magic-angle spinning. Changes in ^{13}C NMR intensity accompany reductive alkylation, and these changes are useful in establishing structural distributions in the coals.

INTRODUCTION

A thorough understanding of the detailed relations between chemical structure and coking properties of coals is not yet available (1). The complex, heterogeneous structures of coals and the complex nature of chemical reactions taking place during carbonization make it extremely difficult to elucidate these relations. In this respect, the developing technique of ^{13}C NMR using cross polarization and magic-angle spinning (CP/MAS) (2-6) is promising for chemical characterization of coals in the solid state (5-13).

Ultimate analysis of coking coals give the following range of values on a daf (dry ash free) basis:(14) 84-91% C, 4-4.5% H, 4-8% O, 1-1.5% N and 0.6-0.9% organic S. The highest content of alicyclic structures has been observed in coking coals in which most of the aromatic carbons are unsubstituted (15). Ether linkages are believed to be the dominant functional groups in coking coals (16,17). The average number of hexagonal units in

L. Petrakis and J. P. Fraissard (eds.), Magnetic Resonance. Introduction, Advanced Topics and Applications to Fossil Energy, 569–574.

a condensed structural unit is assumed to be somewhere in between 3 and 10 in coking coals, and these condensed units are believed to be linked to each other either directly or *via* ether or aliphatic bridges (18).

Reductive alkylation of coals effectively cleaves ether bridges linking aromatic units; aromatic structures are also reduced and alkylated to a certain degree (19,20). It has also been observed that aliphatic bridges linking aromatic units are cleaved to a lesser degree (21-23), but this happens only when glymes are used as solvent. There is no effective fission of methylene and ethylene bridges when THF is solvent (24,25). Therefore, in this work reductive alkylation has been carried out in THF to permit the investigation of the effect of ether bridge cleavage and the effect of partial reduction and alkylation of aromatic structures on the coking properties of coals.

The present paper presents spectroscopic data on four Turkish bituminous and subbituminous coals. The derived structural information is related to coking characteristics. A later publication will deal with pyridine extracts and residues of these coals.

EXPERIMENTAL

The coals that were investigated in this work were labelled as C1, C2, C3, and C4. They were ground to pass 100 mesh (USA), dried at 80°C under reduced pressure and preserved under an inert atmosphere. C1 and C2 were from the Zonguldak district, C3 from the Amasra district and C4 from the Armatcuk district; all three districts are in the western Black Sea region of Turkey.

Reductive alkylations were carried out according to Sternberg's procedure (18,19) with ethyl iodide as the alkylating agent. After alkylation THF and excess ethyl iodide were separated from the reaction mixture in a rotary evaporator under reduced pressure, and the alkylated coal was precipitated by the addition of distilled water. The filtered alkylated coal was washed several times with 30% aqueous ethanol and water and then dried at 80 °C under reduced pressure. The dried alkylated coals were stored under an inert atmosphere.

^{13}C CP/MAS spectra were obtained at 15.0 MHz, using a JEOL FX-60QS spectrometer, with a MAS spinning speed of about 2.5 KHz. Hartmann-Hahn match conditions (3) involved radiofrequency field strengths of about 12 gauss and 48 gauss for ^{1}H and ^{13}C, respectively. A contact time of 1 ms and a repeat time of 1 s were employed.

RESULTS AND DISCUSSION

Figures 1 and 2 show the CP/MAS ^{13}C NMR spectra of the original and reductively alkylated coals. The fractions of aromatic and aliphatic carbons were obtained by the integration of these spectra.

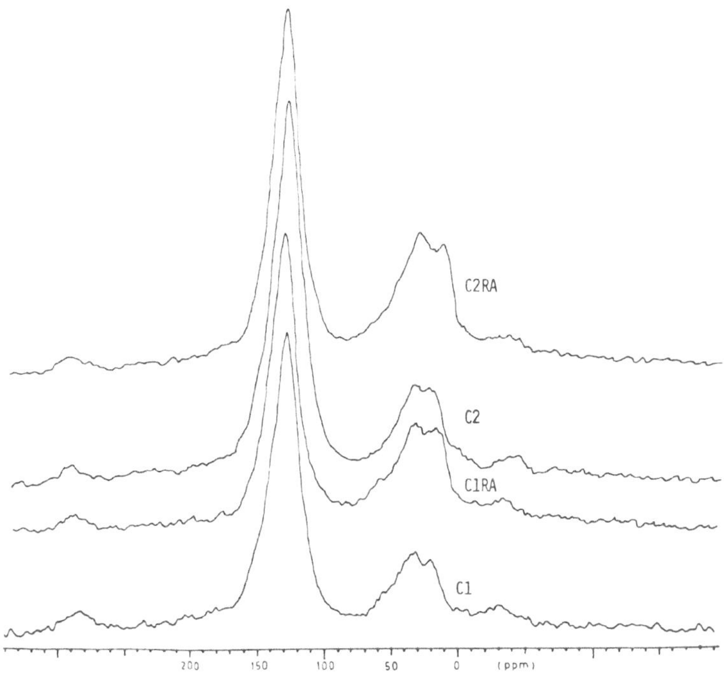

Figure 1. ^{13}C CP/MAS NMR spectra of coals C1 and C2 in their original and reductively alkylated (RA) solid states.

Various schemes have been used to represent the distribution of organic carbon in coal samples, as determined by ^{13}C CP/MAS NMR. In the present case, the intensity between 80 ppm and 0 ppm is taken to represent aliphatic (sp^3) carbons. The remaining intensity in the spectrum is for present purposes identified with aromatic (sp^2) carbons, from which the apparent aromatic fraction (f_a') is calculated. Of course, this approach does not explicitly differentiate carbonyl (including carboxyl) carbons, in the chemical shift range from about 210 to 165 ppm, from aromatic carbons. However, for the spectra shown in Figures 1 and 2, it is clear that the carbonyl contents are very small. Accordingly, f_a' values determined in this way do not have large errors (say, ±0.03) and can be considered to be sufficiently accurate for the purposes of this study.

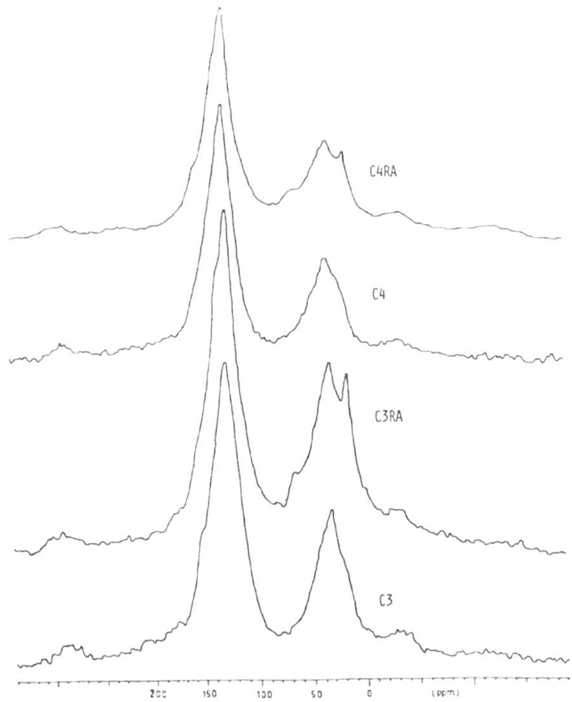

Figure 2. ^{13}C CP/MAS NMR spectra of Coals C3 and C4 in their original and reductively alkylated (RA) solid states.

The ^{13}C NMR spectra of the coking coals, namely C1 and C2, are similar to each other. Intensities due to unsubstituted aromatic carbons (128 ppm) are high, and there is substantial intensity in the region, 139 to 145 ppm, indicating the presence of alkyl substituents to a certain degree in both of the coals. Intensities in the region for phenolic carbons (155 to 163 ppm) are very weak for both of the coals, Aliphatic carbon resonances are broad, and show essentially two maxima. These observations are consistent with a distribution of short and straight alkyl substituents (22 to 23 ppm) and branched chains and/or alicyclic structures (33 to 45 ppm). When we combine these observations with the IR findings (not presented here) (26) we conclude that the aliphatic parts of C1 and C2 are composed of short and straight alkyl substituents and alicyclic structures.

The non-coking coals, C3 and C4, have only single maxima in the aliphatic region of the ^{13}C NMR spectra, in contrast to the double maxima of coking coals. Resonance intensity at about 32 to 35 ppm indicates that the aliphatic parts of C3 and C4 are made up largely of branched chains and/or alicyclic structures. A pronounced shoulder at about 155 ppm in the ^{13}C NMR spectrum of C3 shows that this coal has a significant phenolic carbon content in comparison to the others.

The f_a' values for C1, C2, C3 and C4 are 0.79, 0.78, 0.73 and 0.73, respectively. After reductive alkylation the f_a' values decreased to 0.68, 0.66, 0.65 and 0.66, respectively. These decreases are due both to the incorporation of ethyl groups and to the partial reduction of aromatic structures. These f_a' decreases also indicate that C1 and C2 are the most reactive coals toward reductive alkylation. After reductive alkylation peaks due to methyl groups (15 to 18 ppm) appear in the ^{13}C NMR spectrum of each coal. After reductive alkylation there are also significant increases in the ^{13}C NMR intensity of alkyl carbons bound to ether oxygen (55 to 70 ppm). A plausible explanation for this is that, during reduction with potassium, Ar-O-Ar ether linkages are cleaved and the ArO^- ions formed are subsequently alkylated *via* ethyl iodide to form Ar-O-C_2H_5. During reduction phenolic groups also form ArO^- ions, and subsequent alkylation also produces Ar-O-C_2H_5. From the increases in relative intensities in the methyl and ether-substituted alkyl regions of the ^{13}C NMR spectrum, it appears that formation of $-OCH_2CH_3$ structures upon reductive ethylation is more prominent for the noncoking coals (C3 and C4) than for the coking coals. The increase in 55-to-70 ppm intensity is most prominent in reductively alkylated C3, because of the high content of phenolic groups in C3.

ACKNOWLEDGMENTS

The authors gratefully acknowledge partial support of this research by the U.S. Department of Energy (Laramie Energy Technology Center).

REFERENCES

1. Wandless, A.M., 1971, J. Inst. Fuel, 44, 531.
2. Andrew, E.R., 1972, "Prog. Nucl. Magn. Reson. Spectroscopy", 8, 1.
3. Pines, A.; Gibby, M.G.; and Waugh, J.S., 1972, J. Chem. Phys., 59, 569.
4. Schaefer, J. and Stejskal, E.O., 1979, "Topics in Carbon-13 NMR Spectroscopy", (Ed. G.C. Levy), Vol. 3, Wiley, New

York, p. 283-324.
5. Bartuska, V.J.; Maciel, G.E.; Schaefer, J. and Stejskal, E.O., 1977, Fuel, 56, 354.
6. Maciel, G.E.; Bartuska, V.J. and Miknis, F.A., 1979 Fuel, 58, 391.
7. Maciel, G.E.; Sullivan, M.J.; Szeverenyi, N.M.; Miknis, F.P. 1981, "Chemistry and Physics of Coal Utilization-1980"; Cooper, V.R., Petrakis, L., Eds.; American Institute of Physics: New York, pp 66-81.
8. Maciel, G.E. and Sullivan, M.J., 1982, ACS Symp. Monograph 191, G.C. Levy, Ed., p. 319.
9. Zilm, K.W.; Pugmire, R.J.; Larter, S.R.; Allan J. and Grant, D.M., 1981, Fuel, 60, 717.
10. Hagaman, E.W. and Woody, M.S., 1982, Fuel, 61, 53.
11. Havens, J.R.; Koenig, J.L. and Painter, P.C., 1982, Fuel, 61, 393.
12. Sullivan, M.J. and Maciel, G.E., 1982, Anal. Chem., 54, 1606, 1615.
13. Maciel, G.E.; Sullivan, M.J.; Petrakis, L and Grandy, D.W., 1982, Fuel, 61, 411.
14. Jasienko, S., 1978, Fuel, 57, 131.
15. Van Krevelen, D.W., 1961, "Coal", Elsevier, Holland
16. Blom, L., 1960, Ph.D. Thesis, Univ. of Delft.
17. Wachowska, H. and Pawlak, W., 1977, Fuel, 56, 422.
18. Dryden, I.G.C., "Chemistry of Coal Utilization", Ed. H.H. Lowry, Suppl. Vol., Wiley, New York, 1961.
19. Sternberg, H.W.; Delle Donne, C.L.; Pantages, P.; Moroni, E.C. and Markby, R.E., 1971, Fuel, 50, 432.
20. Sternberg, H.W., and Delle Donne, C.L., 1974, Fuel, 53, 172.
21. Hombach, H.-P., and Niemann, K., Comp. 1977/78, Suppl. Vol. Erdol, Kohle, Erdgas Petrochem. verein. Brennstoff Chem., p. 295.
22. Schanne, L. and Haenel, M.W., 1979, Tetrahedron Lett., 4245.
23. Collins, C.J.; Hombach, H.-P.; Maxwell, B; Woody, M.C. and Benjamin, B.M., 1980, J. Am. Chem. Soc., 102, 851.
24. Erbatur, O. and Gaines, A.F., unpublished data.
25. Erbatur, O. Ph.D. Dissertation, 1981, Çukurova University.
26. Erbatur, G., Erbatur, O., Coban, A., Davis, M.A. and Maciel, G.E., Fuel, submitted.

SOLID STATE ^1H AND ^{13}C NMR STUDIES OF PITCHES

J. A. MacPhee and B. N. Nandi

CANMET/Energy Research Laboratories
Energy, Mines and Resources Canada
Ottawa, Canada K1A 0G1

and

J. A. Ripmeester and R. E. Hawkins

Division of Chemistry
National Research Council of Canada
Ottawa, Canada K1A 0R9

ABSTRACT

The aromaticity of a number of whole and quinoline solvent-fractionated pitches was determined by means of CP/MAS ^{13}C NMR spectroscopy. Several factors which had to be taken into account include spinning sideband intensity, and the variation of aliphatic and aromatic band intensities with cross-polarization time. ^1H NMR lineshapes and spin-lattice relaxation times are presented as well.

INTRODUCTION

Pitches from a variety of sources are currently used as binders in the production of metallurgical coke from poor coking coals (1) and as an essential ingredient in the formation of carbon anodes for the aluminum industry (2). The desirability of certain properties such as chemical composition, rheology, quinoline insoluble fraction, benzene soluble fraction etc. depends to a large extent on the use for which a given pitch is destined. Until recently relatively few studies of a more fundamental nature that attempt to relate chemical structure to

L. Petrakis and J. P. Fraissard (eds.), Magnetic Resonance. Introduction, Advanced Topics and Applications to Fossil Energy, 575–586.

technological behaviour have been available (3). A publication dealing with the formation of mesophase from asphaltenes from different bitumens (4) and one treating the formation of mesophase from various coal derived liquid fractions (5) demonstrate that carbon aromaticity (f_a) is one of the important factors in determining the carbonization behaviour (mesophase texture, bonding ability etc.) of coal and petroleum derivatives.

The role played by aromatcity and other factors in determining the technological properties of pitches will become clearer as more results become available. As a step towards a more complete understanding of the carbonization behaviour of coal and petroleum derivatives two pitches were selected for study: one of these, a coal tar pitch (Aluminum Company of Canada) and the other a pitch from a processed petroleum residue (Kureha Chemical Co., Japan). These two pitch samples were separated into quinoline soluble (QS) and insoluble fractions (QI), also used in this study. In the past the amount of quinoline insoluble material in the pitch was taken as an indication of pitch aromaticity. This is not always satisfactory. Since the determination of aromaticity by chemical means is a complicated and time consuming process, we have explored the possibility of using solid state ^{13}C NMR as a rapid method of determining pitch aromaticity.

Experimental Methods

^{13}C spectra were obtained by means of cross-polarization and magic angle spinning at a frequency of 45.28 MHz with a Bruker CXP-180 NMR spectrometer. A single cross-polarization contact (6) was used with flip-back of the proton magnetization (7). Radio frequency field amplitudes were 45 kHz. Typically, 400-1000 free induction decays (fids) were added at ≈2 or 4 sec repetition rates. The fids, consisting of 500 points digitized at a rate of 25 μsec/point, were zero filled to 4K before Fourier transformation. In order to obtain spectra for ^{13}C nuclei coupled weakly to protons, a delay of 40 μsec with the high power decoupler off was inserted before data acquisition (8) (dipolar dephasing). Kel-F rotors of the Andrew type were used to achieve magic angle spinning rates of ≈3.3 kHz.

Proton NMR spectra were obtained at 180 MHz by Fourier transformation of the fids. Proton spin-lattice relaxation times were measured at 60 MHz on a Bruker SXP pulsed NMR spectrometer by means of $90°-\tau-90°$ pulse sequences.

Elemental analyses (C, H, N) were performed using the Perkin Elmer 240 Analyser and are given in Table 1.

Table 1

	N%	C%	H%	H/C	Ash(%)	QI(%)
1. Coal Tar Pitch	0.94	93.95	4.06	0.52	0.22	16.1
2. Processed Petroleum Pitch	1.03	85.74	6.69	0.93	0.086	8.6

Quinoline Insolubles

	N%	C%	H%	H/C
1.	0.57	94.72	2.11	0.27
2.	1.24	85.62	4.68	0.65

Quinoline Solubles

	N%	C%	H%	H/C
1.	0.68	93.05	4.44	0.57
2.	1.02	86.74	6.84	0.94

RESULTS AND DISCUSSION

Proton NMR

Proton NMR spectra of the whole pitches are shown in Figure 1 where it is seen that the lines are quite broad, ≈40 kHz width at half height. There is thus relatively little motional freedom of the protons in these samples.

Proton spin-lattice relaxation times (T_1) for the whole pitches and quinoline soluble (QS) fractions are shown in Table 2. For the processed petroleum pitch the T_1 values increase slightly (10%-30%) whereas for the coal tar pitch T_1 decreases by a factor of almost 4 on going from the whole pitch to the QS fraction. Since the relaxation mechanisms for the pitches are not well understood, it is difficult to interpret these changes in T_1 values. However, from an empirical relationship between otpical texture of cokes and ^{1}H T_1's of the parent materials (9) one may expect considerable differences in coking properties for the whole coal tar pitch and its QS fraction.

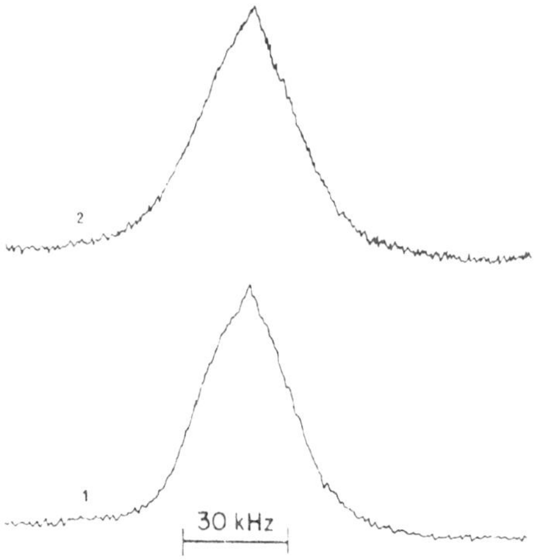

Figure 1 - ^1H NMR spectra of solid pitches

Table 2

^1H Spin-Lattice Relaxation Times

	T_1/Sec Whole Pitch	Quinoline Soluble Fraction
1.	0.960	0.254
2.	0.295	0.328

Quantitative Aspects of CP/MAS ^{13}C NMR

The intensity of lines in ^{13}C NMR spectra obtained by cross-polarization techniques depends on the cross-polarization time τ_{cp} which controls the growth of the signal, and the ^1H rotating frame relaxation time $T_{1\rho}$ which depends on the motional correlation times, the type of motion, and perhaps free radical content and distribution.

Another aspect which must be dealt with is the significant intensity of spinning sidebands in the ^{13}C NMR spectrum. At the ^{13}C NMR frequency of 45.28 MHz, the usual spinning rate of the Kel-F rotors of ≈3.3 kHz leads to spinning sidebands some 70 ppm to high and low field from spectral lines due to aromatic and carbonyl carbons.

The importance of the above factors can be illustrated by examining the results obtained for several pure compounds. Figure 2 shows the spectrum of the ammonium salt of p-toluic acid. Carbon types can be divided into four groups: methyl, quaternary aromatic, protonated aromatic and carboxyl.

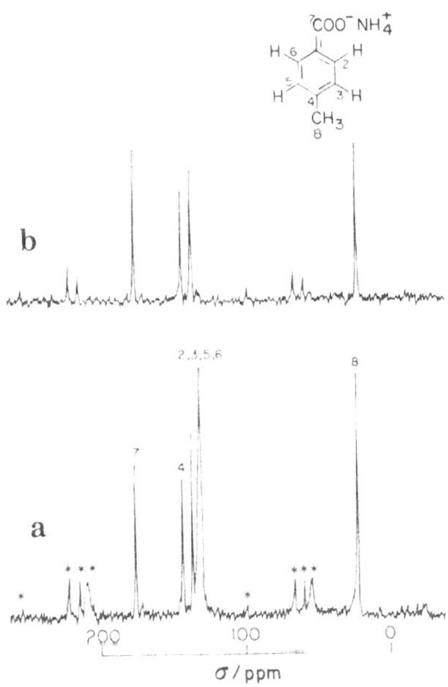

Figure 2 - CP/MAS ^{13}C NMR spectra of the ammonium salt of p-methyl benzoic (p-toluic) acid without a) and with b) dipolar dephasing. Spinning sidebands are starred.

Figure 3 shows the variation of integrated line intensity as a function of cross-polarization time. Equilibrium values of the CP intensities are not reached because of an efficient relaxation (T_{1p}) process controlled by ammonium ion reorientation. Cross polarization (growth) times (τ_{cp}) vary in the order $CH<CH_3<C_{quat.}<CO$. Correct intensity ratios for the 4 types of carbon can be obtained at τ_{cp} values of ≈3-4 msec, well away from the region where maximum signal intensities can be obtained. In systems without an efficient relaxation mechanism, such as the parent p-toluic acid, the correct intensities can be simply obtained from the equilibrium values reached at long τ_{cp}, Figure 4.

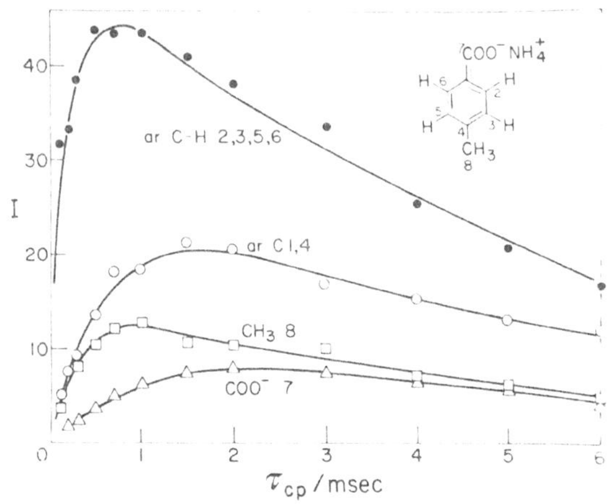

Figure 3 - Signal intensities as a function of cross-polarization time for different carbon types in p-methyl benzoic acid ammonium salt.

From Figure 2 it can be seen that the spinning sidebands of the aromatic bands have ≈10%-25% of the intensity of the corresponding centre bands, and in quantitative work account must be taken of the total integrated intensity of sidebands plus central band. In Figure 2 it can also be seen that the first order spinning sidebands on the high and low field sides of the centre band are essentially equal in intensity. For the pitches studied here, the high field sideband of the aromatic carbons overlaps the aliphatic carbon band.

To correct the integrated aromatic and aliphatic carbon intensities it was always assumed that the high and low field side bands were of equal intensity. In the case of the coal tar pitch, Figure 7, where both side bands are resolved, this can be seen to be a reasonable assumption.

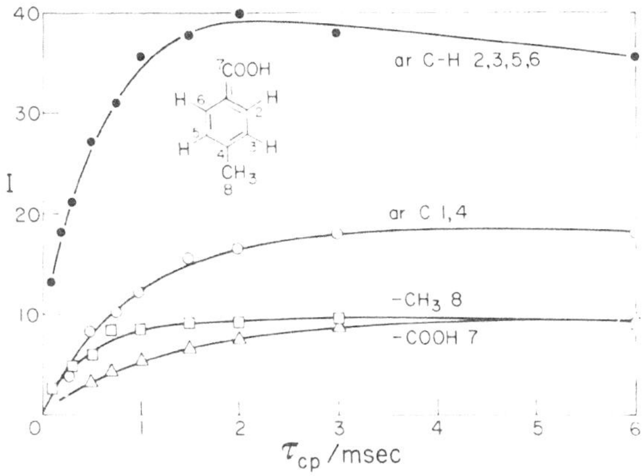

Figure 4 - Signal intensities as a function of cross-polarization time for different carbon types in p-methyl benzoic acid

^{13}C NMR SPECTRA

^{13}C NMR spectra of the whole pitches are shown in Figure 5. The pitches have some features in common, including well separated aliphatic and aromatic carbon bands with maxima at ≈30 and 128 ppm respectively. The coal tar pitch (1) has very little intensity in the aliphatic region, and quite a sharp aromatic band. Dipolar dephasing can be used to obtain spectra of carbons only weakly coupled to protons (quaternary carbons, carbons involved in motion averaging such as methyl groups and long chain methylenes). Then, by subtracting spectra obtained by means of the dipolar dephasing technique from the normal CP/MAS spectrum, a spectrum results for carbons strongly coupled to protons. Figure 6 shows spectra obtained in this manner for the processed petroleum pitch (2).

Bands at ≈13 and ≈20 ppm can be assigned to methyl groups, the band at ≈30 ppm to methylene carbons in aliphatic chains. The aromatic band is now resolved into two bands centered at ≈125 ppm and ≈135 ppm and correspond to protonated and quaternary aromatic carbons respectively.

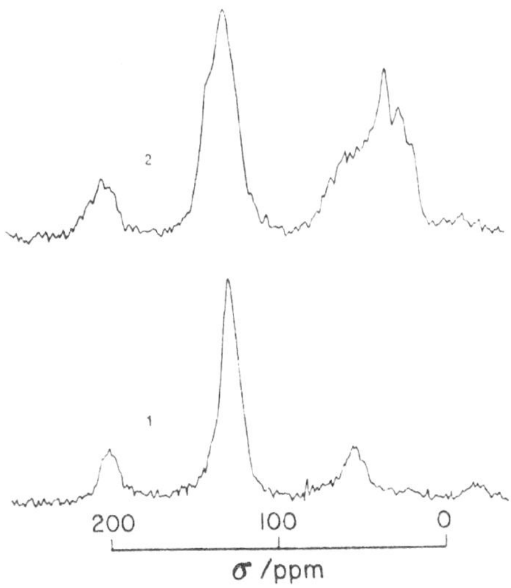

Figure 5 - ^{13}C CP/MAS NMR spectra of solid pitches

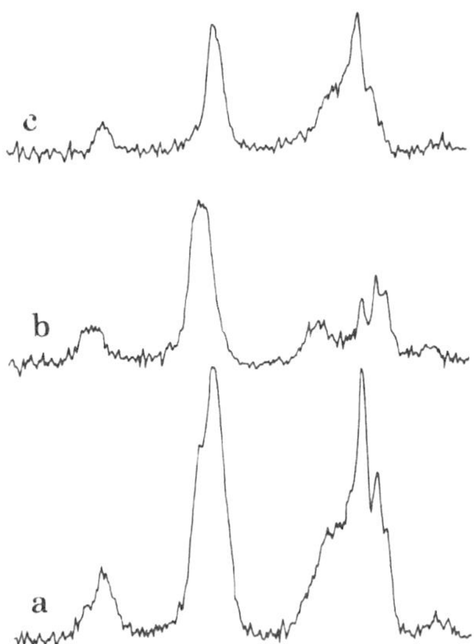

Figure 6 - a) ^{13}C CP/MAS NMR spectrum of processed petroleum pitch 2, b) with dipolar dephasing, c) difference spectrum (a-b)

Figure 7 shows spectra obtained for the coal tar pitch (1). In this instance the protonated and quaternary aromatic bands occur at 125 ppm and 130 ppm. The difference in location of the quaternary aromatic bands for the two pitches, 130 ppm and 135 ppm, indicates that 1 has more carbons in condensed aromatic structures, whereas 2 has more peripheral carbons bonded to aliphatic groups.

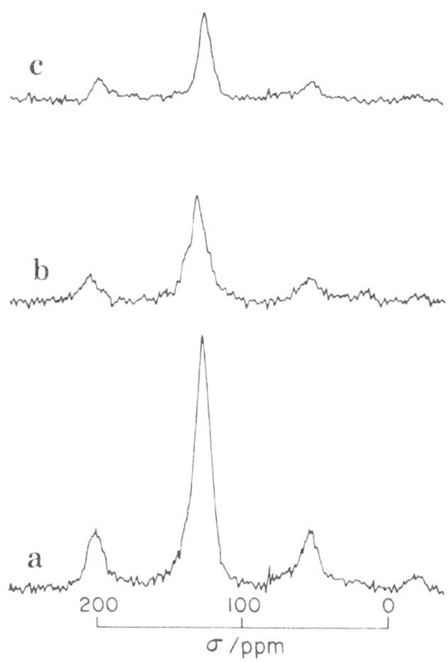

Figure 7 - ^{13}C CP/MAS NMR spectrum of coal tar pitch 1, b) with dipolar dephasing, c) difference spectrum (a-b).

Figure 8 shows the variation of integrated band intensities as a function of cross-polarization time for pitches 1 and 2. The intensity ratio of the aliphatic and aromatic bands becomes constant at τ_{cp} values of ≈1.5-3 msec, Figure 9, and aromaticities for the pitches derived from τ_{cp} values in this range are given in Table 3.

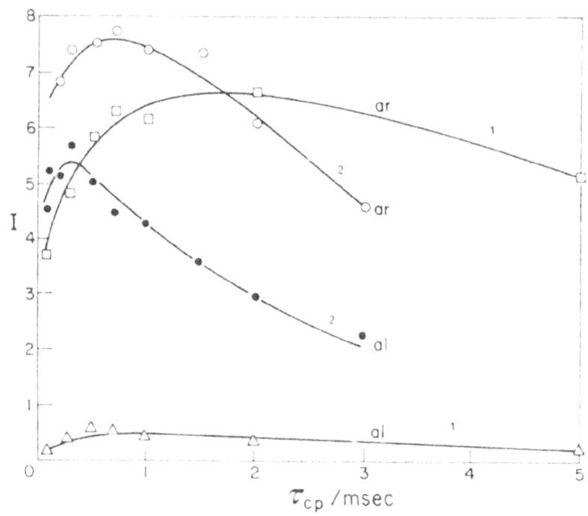

Figure 8 - Signal intensity as a function of cross-polarization time for Pitches 1 and 2.

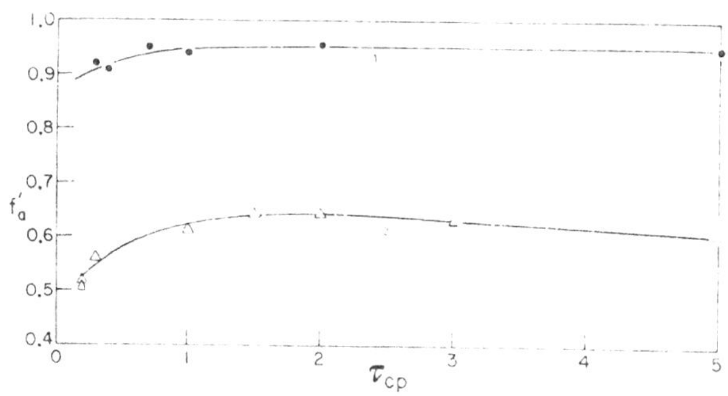

Figure 9 - Apparent aromaticity as a function of H/C ratio for pitches and pitch fractions

Table 3

Apparent aromaticities[a]

	Whole Pitch	QS	QI
1.	0.94 (0.96)	0.82	0.94
2.	0.61 (0.65)	0.62	0.68

[a]values obtained at τ_{cp} value of 1 msec; values in brackets are from Figure 9.

For the fractionated pitches, aromaticities were estimated from spectra obtained at a single τ_{cp} value, 1 msec, Table 3. Compared to aromaticities derived at slightly higher τ_{cp} values, the 1 msec aromaticities are slightly underestimated. In general, quinoline insoluble pitch fractions are slightly more aromatic than whole pitches, and quinoline soluble fractions are slightly less aromatic, although in some instances the difference is quite small and within experimental error.

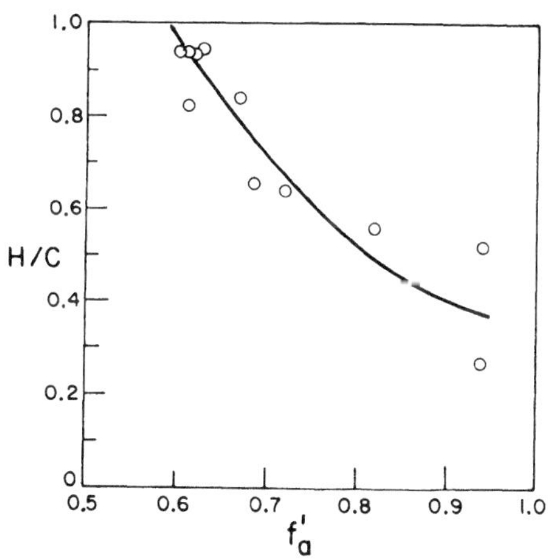

Figure 10 - Apparent aromaticity (f'_a) as a function of H/C ratio for pitches and pitch fractions

One parameter which can be expected to correlate roughly with pitch aromaticity is the H/C ratio, obtainable from Table 1. Figure 10 shows a plot of this ratio as a function of apparent aromaticity, and indeed a reasonable correlation exists. Some scatter must be expected, especially at high aromaticity, as the H/C ratio depends greatly on the degree of condensation of the aromatic ring structures.

CONCLUSION

The rapid and reproducible determination of pitch aromaticities is possible by means of ^{13}C CP/MAS NMR spectroscopy. This provides a convenient parameter which, along with more conventional techniques, should provide information relating to carbonization behaviour.

REFERENCES

1. Nandi, B. N., Belinko, K., Ciavaglia, L. A., Pruden, B. P. and Denis, J. M. US Patent 4,234,387; Nov 18, 1980.
2. Jungten, H., Klein, J., Knoblauch, K., Schroter, H. J and Schultze, J. "Chemistry of Coal Utilization"; Ed Martin A. Elliot, Wiley-Interscience, Ch 30; 1981.
3. Miyazawa, K., Yokono, T., Chiba, T. and Sanada, Y. "Coal Structure"; Eds. Gorbaty, M. L. and Ouchi, K. ACS 192; 1981; Ch 5; p 59.
4. Nandi, B. N., Ciavaglia, L. A. and MacPhee, J. A. Extended Abstracts; 16th Biennial Conference on Carbon; American Carbon Society; July 18-22, 1983; San Diego, CA.
5. Weinberg, V. A. and Yen, T. F.; Carbon; 21; 1983; pp 39-45.
6. Pines, A., Gibby, M. C. and Waugh, J. S. J. Chem. Phys.; 59; 509; 1973.
7. Tegenfeldt, J. and Haeberlen, U. J. Magn. Reson.; 36; 453; 1979.
8. Opella, S. J. and Frey, M. H. J. Am. Chem. Soc.; 101; 1265; 1979.
9. Miyazawa, K., Yokono, T., Sanada, Y. and Marsh, H. FUEL; 58; 691; 1979.

A MAGNETIC RESONANCE STUDY OF NATIVE AND MODIFIED ILLINOIS COAL

Edward P. Duliba and R. B. Clarkson

School of Chemical Sciences, University of Illinois
505 S. Mathews Ave., Urbana, IL 61801

ABSTRACT

Several experimental techniques, including electron paramagnetic resonance, electron spin echo spectroscopy, solid state ^{13}C nuclear magnetic resonance, Mossbauer spectroscopy, and elemental analysis, were applied to a series of coal chars of an Illinois #6 coal in order to develop a method of following correlations between sulfur and iron chemistry in low-temperature charring processes.

INTRODUCTION

In an effort to identify a suitable system of chemical and instrumental techniques for the study of low-temperature coal chars and as a mechanism to follow possible desulfurization processes, we are conducting a series of studies on a high-sulfur content Illinois coal. Multi-frequency electron paramagnetic resonance (EPR), solid state ^{13}C nuclear magnetic resonance (NMR), Mossbauer spectroscopy, and elemental analysis of charring products were employed to examine coal chars and attempt to follow organic sulfur chemistry during low-temperature charring. The possibility of iron acting as a catalytic site for hydrodesulfurization is discussed.

EXPERIMENTAL

All coal examined in this study is an Illinois #6 coal, registry number C-21722, from the mine Old Ben #26.

L. Petrakis and J. P. Fraissard (eds.), Magnetic Resonance. Introduction, Advanced Topics and Applications to Fossil Energy, 587–596.

Table 1. Elemental Analysis of Coal Charred in Vacuo

Char Temperature (°C)	%C	%H	%N	%S	S:C Ratio	% Other Elements (by difference)
Untreated	69.39	4.54	1.24	3.43	4.9	21.30
200	70.96	4.43	1.75	3.30	4.7	19.56
250	69.49	3.94	1.47	2.78	4.0	22.32
300	70.14	3.85	1.37	2.96	4.2	21.68
350	67.03	2.79	1.79	3.12	4.7	25.27
400	70.84	2.45	1.65	2.93	4.1	22.13
450	49.43	1.00	2.36	0.92	1.9	46.29

Char Analysis

Coal samples were heated in vacuo and analysis of the resultant chars performed by the University of Illinois Microanalytical Laboratory. The coal (ground to less than -100 mesh) was evacuated to at least 10^{-5} torr with a diffusion pump, heated to 50°C below the desired temperature for 24 hrs, then heated to the final temperature for 24 hrs while the volatiles are collected in a liquid nitrogen-cooled trap. The chars are cooled in vacuo to prevent oxidation. The analyses are given in Table 1. Analysis of the coal distillate products for sulfur content and chemical functionality is currently underway using techniques developed by other workers (19-21).

Electron Paramagnetic Resonance

A variety of experiments employing EPR over a wide range of frequencies and microwave power levels have yeilded significant information concerning the nature of the organic radicals in coal. Untreated coal displays a single line in its EPR spectrum about 0.7 mT wide at g = 2.00277. Based on observations by other workers (7,10) of anomalies in EPR spectra due to charge conduction in pure coal samples, coal in this work was thoroughly mixed with approximately 50% v:v KCl or NaCl to eliminate these effects. An additional set of six weak peaks was found at g = 2.00192 with A = 266.4 MHz. A search of higher and lower fields failed to find additional peaks. EPR spectra were recorded on a Varian E-9 spectrometer. The magnetic fields were monitored with a Bruker B-NM20 tracking NMR oscillator.

A study of the effect of particle size on the unpaired spin concentration for unevacuated untreated coal was conducted. The results are shown in Table 2. Spin concentrations were calculated by comparison of peak areas to a solution of diphenylpicrylhydrazyl (DPPH) of known concentration.

Table 2. Spin Concentration of Untreated Coal at 9.137 GHz

Coal Particle Size	Spin Concentration (10^{19} spins g^{-1})	Freshly Ground (10^{19} spins g^{-1})
Greater than -20 mesh	1.8	
-20 to -30 mesh	1.9	
-30 to -50 mesh	1.8	2.1
-50 to -60 mesh	2.1	
-60 to -80 mesh	2.1	
-80 to -100 mesh	2.3	2.5
Less than -100 mesh	2.3	

The EPR linewidth is quite dependent on the presence of oxygen. The EPR spectrum of evacuated coal shows two overlapping lines (see Figure 1). After several days, the EPR spectrum of the same sample, if left undisturbed, will display the same signal as the original unevacuated sample.

A parallel-mode X-band cavity was utilized to observe evacuated untreated coal. No additional peaks, other than the residual allowed peak, were observed.

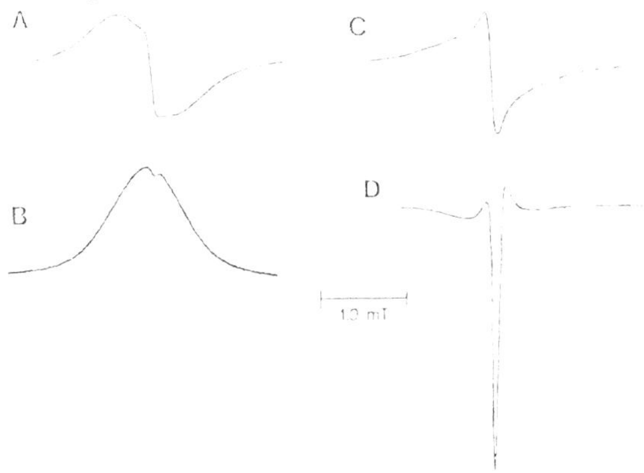

Figure 1. EPR spectrum of evacuated untreated coal at ν = 9.13821 GHz, scan range = 3.0 mT, modulation amplitude = 0.05 mT, scan time = 8 min, time constant = 0.003 sec, temperature = 77K, gain = 400, scan centered at 0.3240 T.

A. Standard EPR, microwave power = 0.5 mW.
B. Second harmonic, microwave power - 0.5 mW.
C. Standard EPR, microwave power = 50 mW.
D. Second harmonic, 90 out-of-phase with respect to B, microwave power = 50 mW.

Table 3. EPR Power Saturation Studies of Untreated Coal

Band (GHz)	g	Saturation Power (mW)	T_1 (10^{-3} sec)	T_2 (10^{-9} sec)	Linewidth (mT)
L(1.2) (non-evacuated)	2.00279	2.0	1.5	10.9	0.60
S(3.6) (non-evacuated)	2.00280	17.6	1.6	10.1	0.65
X(9.1) (non-evacuated)	2.00277	20.2	1.6	10.1	0.65
X(9.4) (evacuated, second harmonic, broad line)	2.00277	0.7			1.23
X(9.1) (evacuated, second harmonic, narrow line)	2.00262	4.8			0.048
Q(36.5) (non-evacuated)	2.00276	20.0	1.6	9.9	0.73

Electron spin echo (ESE) spectroscopy, performed on evacuated untreated coal, showed only proton modulation at 12.5 kHz. An ESE spectrum is shown in Figure 2.

Power saturation studies of untreated coal were made at four frequencies with results given in Table 3. Calculations of relaxation times were made using $T_1 = 1.97 \cdot 10^{-7}\ \Delta H_{pp}/g(2H_1)^2$ and $T_2 = 2/\gamma\Delta H_{1/2}$.

Irradiation of untreated coal at room temperature by high-energy X-rays failed to cause a change in the EPR spectrum or spin concentration.

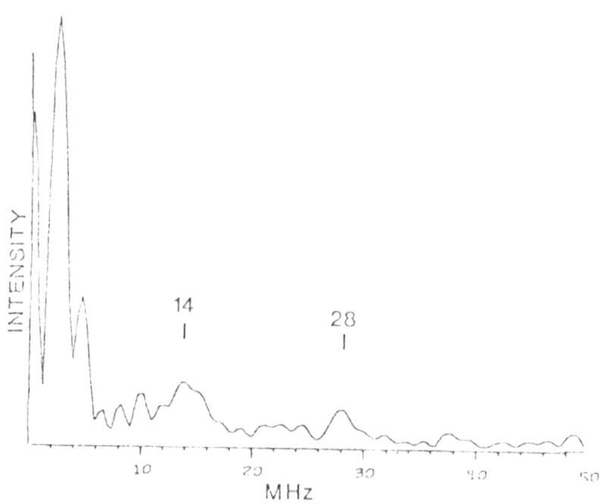

Figure 2. Electron spin echo spectrum of coal:

Table 4. Solid State ^{13}C NMR Results on Coal Chars

Char Temperature	Aliphatic %	Aromatic %	Aromatic:aliphatic Ratio
Untreated	38.02	61.98	1.63
200°C	34.20	65.80	1.92

Estimated uncertainties are ±0.5%.
Additional chars are currently in progress.

Second harmonic detection methods clearly exhibit two lines in the EPR spectrum compared to the single unresolved peak found by standard 100 kHz modulation methods. One peak is fairly broad, 1.23 mT, and is easily saturated. Setting the detector ninety degrees out-of-phase shows a narrow line, 0.048 mT, which does not saturate until much higher power is applied. Figure 1 shows the effect of different power levels on the two lines using the second harmonic technique.

Nuclear Magnetic Resonance

A series of coal chars were examined by solid state ^{13}C nuclear magnetic resonance with 2.5 kHz magic angle spinning used in conjunction with strong cross polarization and multiple pulse techniques, usually denoted as proton enhanced nuclear induction spectroscopy (11-12,16). Time averaging of a large number of pulse sequences is used to enhance the signal-to-noise ratio. Samples completed are given in Table 4. An example spectrum is shown in Figure 3.

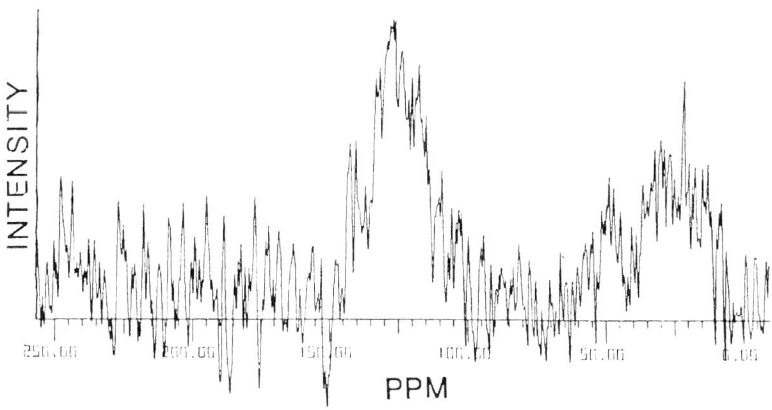

Figure 3. Solid state ^{13}C NMR spectrum of 200° C char. Spectrum based on 1600 scans, 2 sec between pulses, cross polarization time 2.0 msec, temperature = 20°C.

Table 5. Isomer Shift and Quadrupole Splitting Data for Mossbauer Spectra of Several Coal Chars

Temperature of Char Treatment (°C)	Temperature of Mossbauer Spectrum (°K)	Isomer Shift (mm/sec) (±0.02)	Quadrupole Splitting (mm/sec) (±0.02)
Untreated	300	0.30	0.63
	4.2	0.38	0.62
250	300	0.30	0.63
	4.2	0.39	0.62
300	300	0.30	0.63
	4.2	0.40	0.63
350	300	0.31	0.64

Each spectrum consists of two nonoverlapping peaks due to the aliphatic and aromatic portions of the coal structure. The aliphatic region shows a shift of 31±2 ppm and the aromatic region a shift of 127±2 ppm. Some additional unresolved structure is apparent in the aliphatic region, at least two overlapping peaks.

Mossbauer Spectroscopy

Mossbauer spectroscopy is being used to examine a series of chars. The initial data are given in Table 5. Mossbauer spectroscopy has not been used extensively in the past for coal analysis (18). Further analysis of additional Mossbauer spectra is currently underway.

DISCUSSION

Char Analysis

The char analyses demonstrate relatively little change in the chemical composition below 400°C. Above 400°C, the coal structure begins to depolymerize, occurring along with a substantial loss of organic material. There is a steady decrease in the hydrogen percentage, indicating a loss of aliphatic side chains. This observation is supported by the NMR data which also shows a preferential degradation of the aliphatic organic material.

The sulfur content is relatively constant, as demonstrated by the sulfur:carbon ratio, up to 400°C. Therefore, there is no desulfurization effect below 400°C through charring alone. Above 400°C, the sulfur:carbon ratio decreases, but with a concomitant volatilization of a large part of the organic matrix. Our investigation is currently pursuing higher charring temperatures in conjunction with various chemical treatments.

Electron Paramagnetic Resonance

The unpaired spin density of coal is dependent on the size of the sample particle size. Fresh grinding breaks additional chemical bonds and can increase organic radical concentration up to 20%. The line broadening of the EPR signal is also dependent on the presence of oxygen. Since newly evacuated coal shows a narrower linewidth for one component, but over a period of weeks returns to the original unevacuated value, the change is possibly due to slow oxygen release from inner pores and absorption on the coal surface. This corresponds to observations made by previous workers (2-10).

The six weak peaks are undoubtedly due to the presence of a small quantity of manganese(II) in the coal. The absence of any iron peaks indicates that all iron(II) in the coal is high spin (S = 3/2), which ordinarily relaxes too rapidly at 77 K to be observed.

The absence of any additional signals in the parallel mode EPR spectrum shows that any non-organic radicals present are in too low a concentration to be detected. The electron spin echo spectrum shows modulation solely by protons at 14 MHz and its overtone frequencies. This further validates the use of cross polarization in the NMR experiments as a means for reducing relaxation rates.

The key result of the EPR experiments is a novel method via second harmonic detection for the separation of the peaks due to the vitrain and fusain portions of the coal. The vitrain portion reveals a very narrow EPR line which does not saturate below 20 mW. The fusain portion is much more easily saturated and displays a broader signal.

The power saturation studies demonstrate very little inhomogeneous broadening. The method used to calculate T_1 gives only an upper limit. However, values in the millisecond range are quite typical for organic free radicals.

Nuclear Magnetic Resonance

The solid state ^{13}C NMR data, when completed, will provide a mechanism for observing the relative proportions of aliphatic and aromatic carbon. The preliminary data indicate a preferential loss of aliphatic side chains, as shown by the increasing aromatic:aliphatic carbon ratio as temperature is increased. One caveat which must be mentioned are that only carbon atoms bonded to hydrogen are detected since the associated hydrogen atoms are used as a thermal spin reservoir to increase the ^{13}C relaxation rate. Our initial data is in general agreement with results found by other workers (11-17).

Mossbauer Spectroscopy

The initial Mossbauer results are indicative of primarily pyrites and a smaller percentage of illite. Below a char temperature of 350°C, there is essentially no change in the isomer shift and quadrupole splitting which indicates little change in the iron environment. This observation is consistent with the elemental analysis data. A more detailed interpretation will be possible when the complete range of chars has been finished.

Further investigation of the role of iron as a catalytic site for hydrodesulfurization would appear to be in order (21). The use of Mossbauer spectroscopy to follow the evolution of the iron chemistry under the low-temperature charring should provide interesting additional insight. High-temperature EPR will also allow one to follow the changes in g value as a function of heteroatom content as determined by the other techniques discussed here. A high frequency (100 GHz), high magnetic field EPR apparatus is currently under development at the University of Illinois in an attempt to resolve possible hyperfine structure in the EPR spectrum

ACKNOWLEDGEMENTS

This work was supported by a grant from the Center for Research on Sulfur in Coal/Illinois Coal Board. We would like to thank R. Haidle of Northwestern University for the NMR spectra. We would also like to thank Dr. P. DeBrunner of the University of ILlinois for the Mossbauer spectra and assistance with their interpretation. We would also like to thank Dr. M. Bowman at Argonne National Laboratory for the electron spin echo spectra.

REFERENCES

1. C. W. Kruse and N. F. Shimp, Coal Processing Technology, Vol. III, Chemical Engineering Process Technical Manual, AIChE 1981, pp. 124-134.

2. H. L. Retcofsky, J. M. Stark, and R. A. Friedel, Anal. Chem. 40, 1699 (1968).

3. H.-D. Engelmann and H. vonHarpe, Z. Anal. Chem. 267, 37 (1973).

4. L. Petrakis and D. W. Grandy, Anal. Chem. 50, 303 (1978).

5. L. Petrakis and D. W. Grandy, Geochemica et Cosmochimica Acta 44, 763 (1980).

6. L. Petrakis and D. W. Grandy, Phil. Trans., R. Soc. Lond. A300, 83 (1981).

7. H. L. Retcofsky, "Magnetic Resonance Studies of Coal," Coal Science, Vol. I, M. L. Borbaty, J. W. Larsen, and I. Wender, eds., Academic Press: New York, 1982, pp. 43-82.

8. H. L. Retcofsky, G. P. Thompson, R. Raymond, and R. A. Friedel, Fuel 54, 126 (1975).

9. C. L. Kwan and T. F. Yen, Anal. Chem. 51, 1225 (1979).

10. L. M. Vasil'eva, V. F. Anufrienko, and A. A. Shklyaev, Kinetika i Kataliz 12, 1310 (1971).

11. D. L. VanderHart and H. L. Retcofsky, Fuel 55, 202 (1976).

12. H. L. Retcofsky and D. L. VanderHart, Fuel 57, 42 (1978).

13. V. J. Bartuska, G. E. Maciel, J. Schaefer, and E. O. Stejskal, Fuel 56, 355 (1977).

14. H. A. Resing, A. N. Garroway, and R. N. Hazlett, Fuel 57, 450 (1978)

15. G. E. Maciel, V. J. Bartuska, and F. P. Miknis, Fuel 57, 505 (1978); ibid., 58, 155 (1979); ibid., 58, 391 (1979).

16. R. M. Davidson, "Molecular Structure of Coal," Coal Science, Vol. I, M. L. Gorbaty, J. W. Larsen, and I. Wender, eds., Academic Press: New York, 1982, pp. 83-156.

17. T. Yokono and Y. Sanada, Fuel 57, 334 (1978).

18. F. E. Huggins and G. P. Huffman, "Mossbauer Analysis of Iron-Bearing Phases in Coal, Coke, and Ash," Analytical Methods for Coal and Coal Products, Vol. III, C. Karr, ed., Academic Press: New York, 1978, Chap. 50.

19. A. Attar, "Sulfur Groups in Coal and Their Determinations," Analytical Methods for Coal and Coal Products, Vol. III, C. Karr, ed., Academic Press: New York, 1978, Chap. 56.

20. J. H. Karchmer, ed., The Analytical Chemistry of Sulfur and Its Compounds, Vol. I, Wiley Interscience: New York 1970.

21. R. Cypres and C. Soudan-Moinet, Fuel 59, 48 (1980).

TEMPERATURE-DEPENDENT PROTON T_1 AND $T_{1\rho}$ STUDY OF OIL SHALES

J. W. Harrell, Jr. and Taro Kohno

Department of Physics and Astronomy, University of Alabama, University, Al. 35486

Proton T_1 and $T_{1\rho}$ measurements have been made over a wide temperature range on oil shales from Eastern and Western United States. The T_1 values for the different shales differ considerably but they all exhibit a similar temperature dependence. A single minimum occurs above room temperature and almost no low-temperature variation occurs. All the Eastern shales have almost identical $T_{1\rho}$ values, with almost no temperature dependence observed over the entire temperature range. The $T_{1\rho}$ values for the Wyoming shale shows a broad low-temperature minimum and a well-defined high-temperature minimum. Preliminary measurements suggest that T_1 is determined in part by relaxation due to paramagnetic centers and that $T_{1\rho}$ is affected by slow spin diffusion. ESR measurements of spin concentration have been made on shales and compared with T_1, with the hydrogen NMR signal, and with extractable oil. The ESR signal can be separated into component signals with different g-factors.

INTRODUCTION

NMR has proven to be a useful technique for characterizing oil shales. Some of the most important recent developments have been made using CP/MAS C-13 NMR (1,2). However, conventional broadline pulse NMR has an advantage in that temperature dependent relaxation times can be easily measured. Relaxation times are dependent in part upon the mobility of the kerogen components and may provide useful information relating to the thermal decomposition of the kerogen and the extraction of oil from the shale.

L. Petrakis and J. P. Fraissard (eds.), Magnetic Resonance. Introduction, Advanced Topics and Applications to Fossil Energy, 597–606.

In this study both hydrogen T_1 and $T_{1\rho}$ measurements have been measured as a function of temperature in several Eastern and Western U.S. shales. ESR measurements have also been made to try to determine the contributions to the relaxation due to paramagnetic centers. While no correlation could be found between the ESR measurements and the T_1 measurements, the ESR results were found to be of interest apart from the T_1 results.

EXPERIMENTAL

Oil shale samples and Fischer assays were supplied by the Mineral Resources Institute of the University of Alabama. Untreated samples (with associated Fischer assays) were from the following locations: Madison Co., Alabama (12.3 GPT), Limestone Co., Alabama (16 GPT), Dekalb Co., Alabama (~1 GPT), Franklin Co., Tennessee (9.5 GPT), Powel Co., Kentucky (16.9 GPT), and Wyoming (60 GPT). Madison Co., AL and Wyoming samples which were prepared with different kerogen concentrations by the beneficiation method (3) had Fischer assays in the ranges 0.3 GPT to 27 GPT and 5 GPT to 91 GPT, respectively. Madison Co. spent shales were also used. The Alabama and Tennessee shales were from the Chattanooga formation, the Kentucky shale was from the New Albany formation, and the Wyoming shale was from the Mahogany Zone formation. The Dekalb Co. shales were deformed shale from a geologically folded region.

The pulse NMR system consisted of a 36 MHz Spin-Lock CPS-2 spectrometer with a homebuilt programmer. Free-induction-decay signals were recorded on a Nicolet Explorer III digital oscilloscope which was interfaced to a microcomputer for signal averaging of weak signals. Temperature control was achieved by a gas flow system with an electronic regulator. T_1 measurements were made using a saturation-τ-90° pulse sequence, and $T_{1\rho}$ measurements were made using a 90° pulse followed by a 90° phase-shifted spin-locking pulse of variable length.

ESR measurements were made on homebuilt X-band and K-band systems. Spin concentration measurements were made by digitally integrating the first harmonic signal. Signals were compared with the signal from a fixed amount of Cr^{3+} which was added to the shale and which had a sufficiently different g-factor so that almost no signal overlap with the shale occured. Saturation measurements of the two separate oil shale signals were made at X-band using second harmonic detection because of the better resolution as compared to the first harmonic signals. G-factor measurements were made at K-band using second harmonic detection.

RESULTS

A. Relaxation Time Measurements

Both T_1 and $T_{1\rho}$ measurements were nonexponential. For some samples the T_1 relaxation was characterized by a large signal fraction which had a well-defined T_1 value and by a small signal fraction which had a distribution of short T_1's. For some samples no well-defined T_1 component could be extracted. Since individual T_1 components could not be reliably determined for all samples, effective T_1 values were determined by the relationship $T_{1eff} = t_{1/2}/\ln 2$, where $t_{1/2}$ is the time for half recovery of the magnetization after a saturation sequence. $T_{1\rho\, eff}$ values were similarly determined from the length of a spin-locking pulse for which the signal was 1/2 that for a 0.1 ms pulse.

The temperature dependence of T_1 and $T_{1\rho}$ for five different raw shales is shown in figures 1-5. The general features of the data are described as follows. The T_1 results for all samples are qualitatively similar. Essentially no temperature dependence was found below room temperature except for a small dependence for the Tennessee shale. All samples show high-temperature minima in the range from 20° to 85°C. The increase in T_1 above the minimum is more pronounced for the samples which have short T_1 minima. The

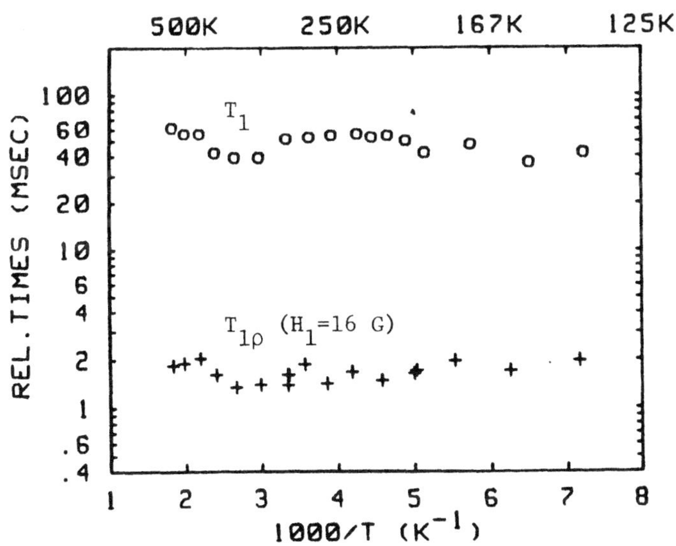

Figure 1. T_1 and $T_{1\rho}$ for Madison Co., AL shale.

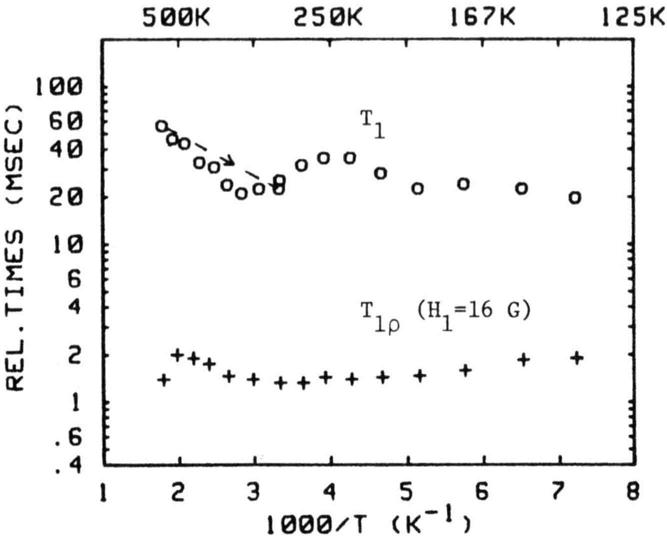

Figure 2. T_1 and $T_{1\rho}$ for Tennessee shale.

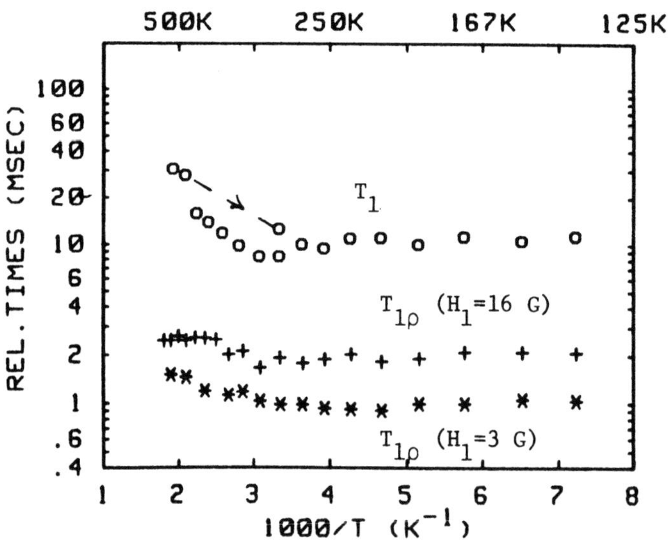

Figure 3. T_1 and $T_{1\rho}$ for Limestone Co., AL shale.

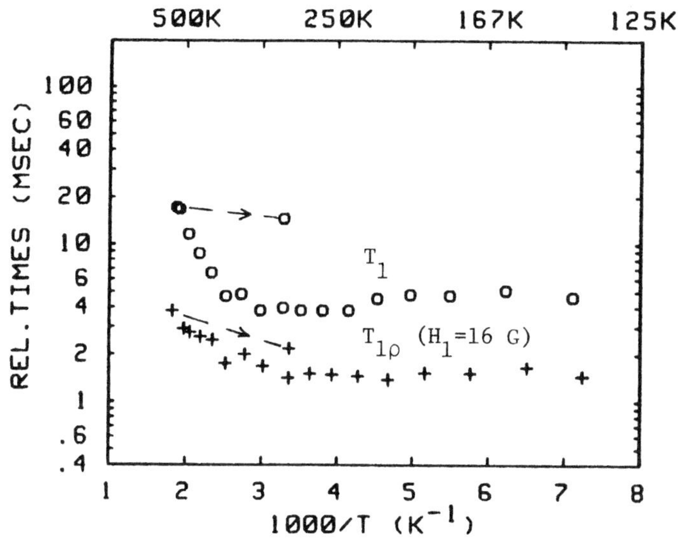

Figure 4. T_1 and $T_{1\rho}$ for Kentucky shale.

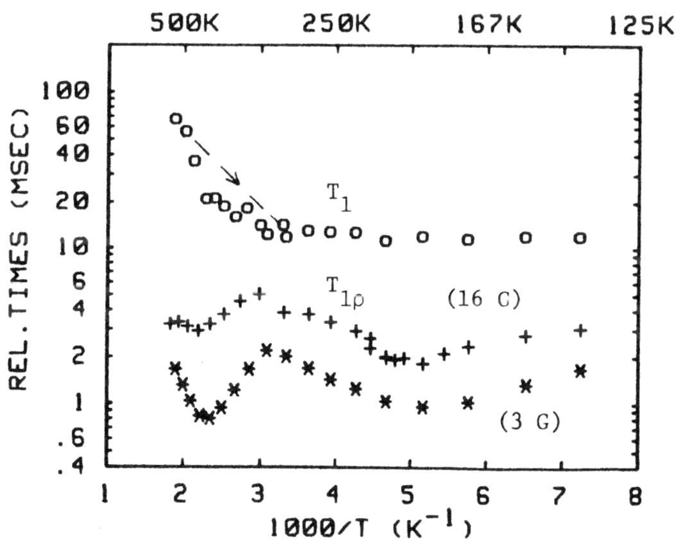

Figure 5. T_1 and $T_{1\rho}$ for Wyoming shale.

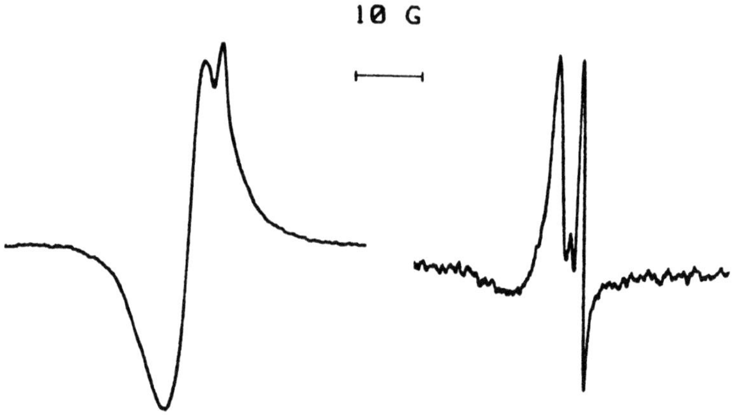

Figure 6. X-band 1st and 2nd harmonic Kentucky shale ESR spectra.

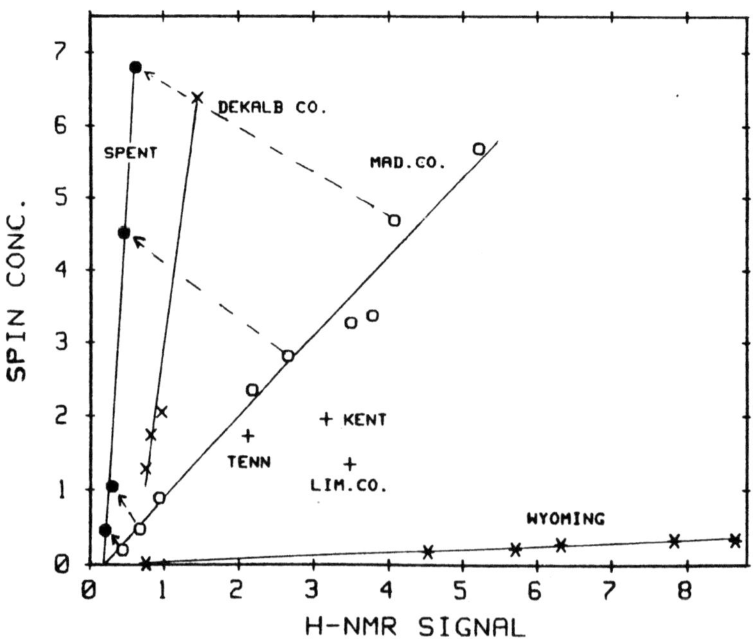

Figure 7. Spin concentration versus intensity of proton NMR signal.

range of T_1 minima observed among the different shales differs by a factor of about 10, from a low of 4 ms for the Kentucky shale to a high of 40 ms for the Madison Co., Al shale. Partial sample decomposition may have affected the T_1 values near the highest temperature attained (280°C) for the Kentucky and Limestone Co., Al shales, which had low T_1 minima; however, the minima could be repeated for samples which were not excessively heated.

The $T_{1\rho}$ data for the four Devonian shales is nearly identical, especially below room temperature where essentially no temperature dependence is obtained. Slight increases in $T_{1\rho}$ occur above room temperature with the most pronounced change occurring for the Kentucky shale. For the Limestone Co., Al shale the measurements for H_1 = 3 G were uniformly smaller than the measurements for H_1 = 16 G by a factor of about 2 over the entire temperature range.

The most interesting $T_{1\rho}$ results were obtained for the Wyoming shale. For H_1 = 16 G a broad low-temperature minimum was obtained at about -75° and a high-temperature minimum was obtained at about 182°C. Below room temperature the 3 G measurements were smaller by a factor of about 2, as was the case for the Limestone Co. shale. However, the high-temperature measurements showed a much stronger dependence upon H_1, with the high-temperature minima differing by a factor of 3.7, and with a greater H_1 dependence on the low-temperature side of the minimum. The high-temperature minimum could be reproduced after heating the sample to 270°C.

B. ESR

All samples exhibited ESR signals characteristic of organic free radicals. The Eastern shale signals were of particular interest is that they consisted primarily of two separate signals which could be partially resolved at X-band and almost fully resolved at K-band. An X-band Kentucky shale signal is shown in figure 6. The two component signals partially overlap when first harmonic detection is used, but much better resolution is obtained when using second harmonic detection. In fact, a weak third component signal could also be seen when using second harmonic detection and small modulation. The second harmonic K-band signals are almost completely separated due to g-factor differences. Preliminary g-factor assignments are 2.0030 and 2.0006 for the two strongest components. The sharper signal to the right, which has the smaller g-factor, is easily saturated compared to the broader signal. The Wyoming ESR signal, which was much weaker than the Eastern shale signals, had only one discernible component.

Relative spin concentrations were determined for the shales of figures 1-5 and for a number of other shales as well. These additional shales included Madison Co., Al and Wyoming shales with a range of concentrations of kerogen which were obtained by the beneficiation technique from the raw shales of figures 1 and 5, deformed shales from Dekalb Co., Al, and spent Madison Co., Al shales. Relative spin concentrations were determined per unit mass of sample, per unit of extractable oil, and per unit of hydrogen NMR signal. No correlation between these spin concentrations and room temperature T_1 values could be seen. Other trends in the measurements, however, were of interest.

Figure 7 shows spin concentrations per unit mass plotted against the intensity of the hydrogen NMR signal per unit mass. The Madison Co., Al and Wyoming shales which have varying concentrations of kerogen each showed a very linear relationship with a positive intercept on the NMR signal axis. Similar relationships were observed for the spent Madison Co., Al shales, which have gone through the Fischer assay treatment for extracting oil but which still have small amounts of hydrogen, and for the Dekalb Co., Al shales. When comparing raw, untreated shales, a nearly inverse relationship exists between the spin concentration and the amount of hydrogen. The kerogen-poor Dekalb Co., Al shales have the greatest spin concentrations and the kerogen-rich Wyoming shale has the lowest spin concentration.

DISCUSSION

The high-temperature T_1 data suggests some type of motion common to all the shales, the nature of which cannot be determined. The low T_1's observed below room temperature, the very weak temperature dependence, and the wide range of T_1's observed among the different samples suggests that relaxation due to paramagnetic impurities may be important. The lack of an observed correlation between the T_1 values and the spin concentrations does not necessarily rule out this possibility. The oil shales are complex, heterogeneous substances and there may be significant differences in how the spin concentrations are distributed within the different shales.

The nearly identical $T_{1\rho}$ values observed for the Eastern shales, the almost nonexistent temperature dependence, the weak dependence upon H_1, and the size of the $T_{1\rho}$ values suggest that the intrinsic $T_{1\rho}$ relaxation mechanisms may be obscurred by slow spin diffusion. Spin diffusion may explain why the motion giving rise to the high-temperature T_1 minima does not produce similar low-temperature $T_{1\rho}$ minima in the Eastern shales. While slow spin-diffusion may also be important in the Wyoming shale, especially below room temperature, it does not completely mask the motional relaxation mechanisms. The broad low-temperature minimum is possibly due to the same motion which produces the high-temperature T_1 minimum. The high-temperature minimum is due to a motion which cannot possibly be seen in the T_1 data. It is interesting that this motion shows up in the Wyoming shale and not the Eastern shales, although the Eastern are known to show evidence of thermal decomposition at lower temperatures (4).

The oil shales are too complex to accurately determine the relaxation mechanisms involved. Some insight may be obtained by making comparisons with simpler model systems such as the n-alkanes (5) and polymers. There are similarities in some of the results presented here and in those found for poly(vinylidine fluoride) (PVF_2) in which slow spin diffusion was found to be important in explaining the $T_{1\rho}$ results (6).

The nature of the ESR spectrum observed for the Eastern shales is different in several respects from the ESR spectrum observed for bituminous coals, for example, from Alabama (7) and from Pennsylvania (8). The coal spectra also consist of a combination of sharp and broad signals; however, the component signals have very similar g-factors and the broad signal saturates more easily than the narrow signal, unlike what is observed for the shales. Also, the coal signals are about 10 times more intense than the shale signals.

The linear relationships observed in figure 8 for a particular type of shale suggest that the ESR signal is due entirely to the kerogen, and the intercept suggests that some of the hydrogen in the shale is in some form other than kerogen.

The high spin concentrations measured in the spent shales and in the Dekalb Co., Al shales are consistent with our findings that heat-treating normal shale samples increases their spin concentrations.

ACKNOWLEDGEMENTS

This project is supported by a grant from the School of Mines and Energy Development at the University of Alabama. The authors would like to thank Dr. H. S. Hanna for supplying the oil shale samples and Fischer assays, and Drs. I. Miyagawa, C. Alexander, and Hanna for helpful discussions.

REFERENCES

1. Maciel, G. E., Bartuska, V. J., and Miknis, F.P. 1978, "Correlation between Oil Yields of Oil Shales and C-13 Nuclear Magnetic Resonance Spectra," Fuel 57, pp. 505-506.
2. Maciel, G.E., Bartuska, V.J., and Miknis, F.P. 1979, "Improvement in Correlation between Oil yields of Oil Shales and C-13 NMR Spectra," Fuel 58, pp. 155-156.
3. Lamont, W.E. and Hanna, H.S. 1982, "Upgrading of Eastern Oil Shale by Beneficiation," Proceedings 1982 Eastern Oil Shale

Symposium, Oct. 11-13, Lexington, Kentucky, Pub. IMMR, to appear.
4. Hanna, H.S., personal communication.
5. Douglass, D. C. and Jones, G.P. 1966, "Relaxation of n-Alkanes in the Rotating Frame," J. Chem. Phys. 45, pp. 956-963.
6. McBrierty, V.J., Douglass, D.C., and Weber, T.A. 1976, "Nuclear Magnetic Relaxation and Molecular Motion in Poly(vinyldine Fluoride)," J. Polym. Sci., Polym. Phys. Ed. 14,pp. 1271-1286.
7. Miyagawa, I., Shibata, K., and Alexander, C. 1981, "ESR-ENDOR Study of Coal," 15th Biennial Conference on Carbon, Philadelphia (abstract), pp. 464-465.
8. Das, U., Doetschman, D.C., Jones, P.M., and Lee, M. 1983, "Electron-Spin Echoes and Saturation in Coal," Fuel 62, pp. 285-288.

PROTON SPIN-LATTICE RELAXATION IN COALS OF VARYING RANK

Mark J. Sullivan,[‡] Nikolaus M. Szeverenyi and Gary E. Maciel[*]
Department of Chemistry,
Colorado State University
Fort Collins, Colorado 80523

and

Leon Petrakis[†] and D.W. Grandy
Gulf Research and Development Company
Pittsburgh, Pennsylvania 15230

ABSTRACT

Proton NMR T_1 measurements were made on a series of 10 coals of varying rank at 60 MHz, using an indirect ^1H-^{13}C cross polarization method. Aromatic and aliphatic proton T_1's were distinguished by this technique. Electron spin concentrations and g values of the naturally occurring free radicals in these samples were determined by epr. These results show relatively smooth curves for the proton T_1's as a function of rank, with an apparent maximum at a value of 85% carbon content (maf). A correlation between increasing unpaired electron spin concentration and increasing rank was found, in agreement with previously published results. These data suggest that the proton T_1 relaxation process in coal is dominated by molecular motions, rather than solely by paramagnetic dipolar fluctuations. The combination of proton T_1 and apparent aromaticity (f_a') data provides an empirical classification of coals with a wide dynamic range.

‡ Present address: JEOL (USA), Cranford, NJ 07016.
* To whom correspondence regarding the NMR measurements should be addressed.
† To whom correspondence regarding the EPR measurements should be addressed.

L. Petrakis and J. P. Fraissard (eds.), Magnetic Resonance. Introduction, Advanced Topics and Applications to Fossil Energy, 607–616.

INTRODUCTION

Proton spin-lattice relaxation measurements by pulsed nmr techniques have been shown to be useful for characterizing coals and coal products in the solid-state (1-4). In the present work, a proton spin-lattice relaxation (T_1) study was carried out on a series of coals, using a ^1H-^{13}C cross polarization/magic-angle spinning (CP/MAS) method(5,6). This technique involves the indirect observation of ^1H relaxation via ^{13}C, and offers an important advantage in resolution over the direct pulsed ^1H nmr technique. Since the aromatic and aliphatic resonances are resolved as two peaks in ^{13}C CP/MAS nmr spectra of coals (7), the relaxation times of the protons associated with these two carbon types can be determined separately.

Cross polarization is commonly used in solids to enhance the nmr sensitivity of rare, low-frequency nuclei (e.g., ^{13}C, ^{29}Si) in the presence of abundant, high-frequency nuclei (e.g., ^1H, ^{19}F) (8). In this experiment, the rare spin species becomes polarized at the expense of the more abundant nuclei through a dipolar spin-spin interaction that occurs under the Hartmann-Hahn condition (9). By the nature of this exchange, when the two spin reservoirs are allowed sufficient time to equilibrate, they attain a common spin temperature. This aspect of the cross polarization process is exploited in the indirect proton T_1 relaxation experiment. Relaxation is observed indirectly by preparing the proton magnetizaton in a non-equilibrium state with an inverting 180° pulse, followed by a variable relaxation delay (τ), as in the standard 180°- τ -90°-observe experiment (10). At the end of this relaxation period, the resultant magnetization is tipped into the transverse plane with a 90° pulse and spin locked. By turning on the carbon transmitter under the appropriate matching conditions, ^{13}C magnetization is generated through cross polarization. The extent of recovery of the ^1H magnetization during τ is reflected in the sign and amplitude of the ^{13}C signal that is observed. The proton T_1 value is obtained by measuring the ^{13}C magnetization as a function of τ, and is determined from the same expression that is used in the standard inversion-recovery experiment (10). A comparison between the timing diagrams for the normal cross polarization experiment and the indirect proton T_1 sequence is shown in Figure 1.

EXPERIMENTAL

The coal samples were obtained from the Illinois State Geological Survey, and are part of a suite of 172 whole coals that have been studied by Gluskoter and coworkers (11). The initial sets of ^{13}C CP/MAS nmr spectra and epr spectra were run on these samples, as received, in sample holders that did not prohibit contact with air. A second set of nmr and epr measurements were made on samples of these coals that were dried for 6 to 12 hours at 80°C under a pressure of 1-2 mtorr in an Abderhalden drying pistol containing P_2O_5. For the nmr experiments, the dried samples were packed in the sample rotors under a dry nitrogen atmosphere in a glove bag. The sealed rotor was then quickly transferred into the nmr probe, which was purged throughout the sample run with dry nitrogen gas. The epr samples were sealed in glass tubes under vacuum after drying. The coals were powdered, but were not sorted according to particle size for any of these experiments.

The ^{13}C CP/MAS spectra were obtained at 15.0 MHz on a home-built spectrometer based on a 12-inch Varian 1.4-T electromagnet and a Nicolet 1180 data system. The probe was of a double-tuned, single coil design,[12] utilizing a bullet-type Kel-F rotor with a volume of 0.6 cm^3. Spectra were obtained at spinning speeds of 2,500 Hz. The proton and carbon-13 rf field strengths used for the Hartmann-Hahn match were 11 G and 44 G, respectively.

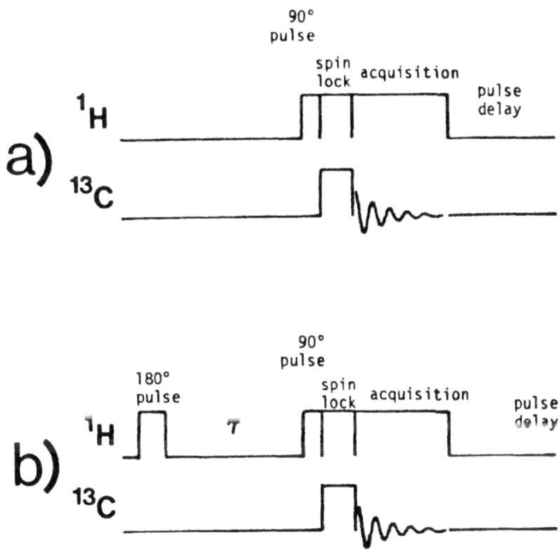

Figure 1. Timing diagrams for a) the simple cross polarization pulse sequence; b) the sequence for the measurement of T_{1H} *via* ^1H-^{13}C cross polarization.

The pulse delays between successive acquisitions in the proton T_1 measurements were chosen to be more than five proton T_1 values, in order to approach complete relaxation between pulse cycles. Delay times (τ values) were ordered at random, with several infinity values interspersed. Between 1000 and 2000 scans were signal averaged for each τ value in order to obtain adequate signal-to-noise ratios for T_1 calculations. Contact times of 1 ms were used. All signal intensities were measured from peak heights obtained with 30-Hz line broadening applied. Proton T_1 values were derived from the initial slopes of plots of the equation, $\ln(I_\infty - I_\tau) = \ln 2I_\infty - \tau/T_1$ (13).

The epr measurements were made in the X band on a Varian model E-100 epr spectrometer. This instrument was equipped with a dual cavity for accurate calibration in spin counting experiments, using as a secondary standard a bituminous coal that had been calibrated previously against a freshly prepared sample of DPPH;[14] accuracy of the spin concentrations is about ± 20 %. Measurement of g values employed a single-cavity probe, and is accurate to about ± 0.0001.

RESULTS AND DISCUSSION

The proton spin-lattice relaxation data obtained *via* ^{13}C for both the unaltered and evacuated coal samples are given in Table 1. This table also includes the percent carbon and hydrogen for each coal by elemental analyses, and the apparent aromaticity (f_a'), obtained from each ^{13}C nmr spectrum. As indicated above, separate ^1H T_1 values for the protons associated with the aliphatic and aromatic carbons can be distinguished by this technique and are recorded separately in Table I. Two major trends are drawn from the data presented in this table. First,

Table I. NMR Results on the Ten Coals Studied.

Sample No.[a]	ASTM Rank	Percent[b] Carbon	T_{1H}(ms) as received al	as received ar	evacuated al	evacuated ar	Percent[b] Hydrogen	f_a'[c]
C18433	lig.	70.22	14	8.5	40	28	4.42	71
C18450	subc	73.66	17	11	42	34	5.08	65
C18451	subb	74.62	16	10	49	24	5.10	69
C18462	suba	72.69	11	8	27	17	5.42	69
C17279	hvcb	77.23	35	28	80	70	5.56	69
C17016	hvcb	71.91	22	22	79	65	5.24	75
C14735	hvcb	80.98	70	66	131	120	5.86	75
C15331	hvcb	79.61	73	69	142	135	5.88	74
C18824	mvb	81.53	100	100	143	128	5.81	83
C18820	lvb	90.27	100	103	110	104	4.63	85

[a]Numbers designated by the Illinois State Geological Survey.
[b] maf.
[c]Determined by integration of the ^{13}C CP/MAS spectra (18). Values good to about ± 3 % (absolute) and were found to be essentially the same for degassed samples and samples used as received.

the proton T_1's are dependent upon sample treatment. Evacuation to remove oxygen and moisture yields an increase in T_{1H}. This is not unexpected, e.g., paramagnetic molecular oxygen trapped in the pores and surface regions of the powdered coal samples might contribute to proton spin-lattice relaxation (1,2). A second and perhaps more important feature that is seen in this table is the small, but significant difference between the T_{1H} values for the aromatic and aliphatic protons. This finding is indicative of the heterogeneous nature of coal.

In a typical organic solid, the proton T_1's measured for all chemically distinct carbon sites are identical, due to rapid proton spin diffusion. The spin diffusion mechanism is a spin-spin flip-flop, a T_2-type process in which nuclei that are coupled <u>via</u> the dipolar interaction exchange spin states; this mechanism does not contribute to a change in spin temperature. For a collection of strongly-coupled spins this is a very efficient process, and such a set exhibits a uniform spin-lattice relaxation time. The rate of spin diffusion has a strong inverse dependence on the proton-proton internuclear distance. Therefore, as the separation between protons increases, the proton spin diffusion rate drops dramatically. Also, as the rate and intensity of motions that average the dipolar interaction increase, the spin diffusion rate decreases.

On the basis of the proton T_1 results reported above, it can be concluded that coal consists of at least two separate domains that are, respectively, aromatic-rich and aliphatic-rich in character. Within the individual domains spin diffusion is rapid, as evidenced by the homogeneous decays of the aromatic and aliphatic peak envelopes as τ increases (spectra not given here). The extent of spin communication and the nature of the isolation between domains (i.e., the criteria for domains) cannot be determined directly from the data presented here. The similar magnitudes of the aromatic and aliphatic T_1 values in coal may indicate partial proton spin diffusion between domains; or this similarity may be accidental. Experiments, such as the Goldman-Shen sequence, have been used successfully to measure spin diffusion rates as a means for determining the dimensions of domains in solids.[15] By the nature of these experiments, however, the differences in the intrinsic ^1H T_1 values for the two domains must be large, and the spin diffusion rate must be fast on the time scale of the T_{1H} for the more slowly relaxing domain.

One can think of proton spin-lattice relaxation in a solid as occurring by means of the following processes: a) direct relaxation of individual protons by means of a time-dependent modulation of some local field (a nuclear or electron spin) and b) spin diffusion, in which the primary relaxation processes of a) due to interactions with relaxation sinks are spread out to all of the dipolar-coupled protons in the sample. Relaxation sinks could be paramagnetic centers or motionally active C-H pairs (e.g., in rotating methyl groups). *A priori* a plausible ^1H relaxation mechanism for coal would seem to be the functioning of free radicals as relaxation sources, interacting either directly with all or most of the protons in the sample (a) or indirectly *via* spin diffusion (b).

Table II gives the results of the esr studies of the ten coals of the present study. Both g values and unpaired electron concentrations are presented for evacuated coals, as well as the samples as received. These parameters are of interest for comparison with the nmr relaxation data.

Table II. EPR Data on the Ten Coals Studied.

Sample No.	Unpaired Spins Per Gram x 10^{-18} as received	evacuated	g Factors as received	evacuated
C18433	4.44	3.29	2.00352	2.00345
C18450	4.54	3.43	2.00362	2.00341
C18451	4.32	3.55	2.00344	2.00345
C18462	3.92	3.59	2.00353	2.00350
C17279	5.59	4.85	2.00307	2.00311
C17016	5.70	4.69	2.00293	2.00411
C14735	8.36	7.94	2.00290	2.00304
C15331	7.64	7.84	2.00293	2.00305
C18824	8.21	16.09	2.00288	2.00285
C18820	23.63	21.00	2.00274	2.00279

In a previous investigation the effect of the addition of a paramagnetic dopant on the proton T_1 in polystyrene was reported (16). For concentrations of unpaired electron spins of 8.5×10^{18} to 8.5×10^{19} per gram, a range typical of coals, T_{1H} was found to decrease linearly in proportion to the paramagnetic content. Similar changes in proton relaxation times have been reported for samples doped with other kinds of paramagnetics (17) and for γ-ray irradiated polymers (3). In contrast, the data in Tables I and II show that for the series of coal studied here, the proton spin-lattice relaxation times are larger for increasing unpaired electron spin concentrations. These data are plotted in Figures 2 and 3 for the evacuated and unaltered samples, respectively.

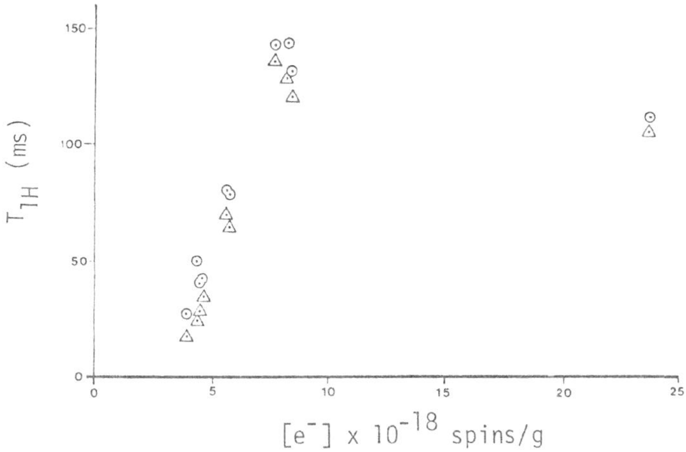

Figure 2. Plot of proton spin-lattice relaxation time _versus_ unpaired electron concentration in the evacuated coal samples listed in Tables I and II. ⊙ , aliphatic; △, aromatic.

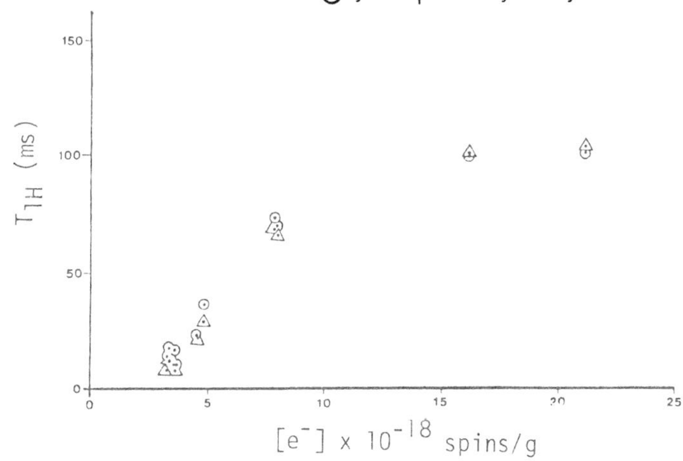

Figure 3. Plot of proton spin-lattice relaxation time _versus_ unpaired electron concentration in the coal samples, as received, listed in Tables I and II. ⊙, aliphatic; △, aromatic

The increase in free radical content that was found with increasing rank (Table 2) in these samples is, of course, accompanied by major changes in their chemical composition. In an earlier study on this suite of coals, a correlation was reported between coal rank and the apparent aromaticity (f_a') determined from solid-state ^{13}C nmr (18). Those results agree with the classical model for the coalification processes, in which the "typical coal molecule" becomes increasingly aromatic with geologic age. If we were to assume that the ^1H spin-lattice relaxation process were rate limited by proton spin diffusion, then one might conclude from the present data that the average proton-proton internuclear distance increases with rank, although one should note that for the present suite of coals the weight percent of hydrogen is very nearly constant. Furthermore, Gerstein _et al._ have pointed out that the proton second moment is dependent, not upon the average internuclear distance, but rather the distance between nearest-neighbor protons (1), and that the nearest proton-proton distance in the aliphatic structure, polyethylene (1.763 Å), is considerably less than in the aromatic compound, naphthalene (2.407 Å). This would appear to support the view that the lower efficiency of ^1H spin lattice relaxtion at higher rank (with higher f_a values) is associated with less efficient spin diffusion (assuming relaxing centers provide the major T_1 mechanism). However, inspection of Table 1 shows an inconsistency with this view, as the T_{1H} values associated with aliphatic sites are typically larger than those associated with aromatic sites.

Comparison of Tables 1 and 2, as well as Figures 2 and 3, shows that paramagnetic centers do not provide the dominant source of ^1H spin-lattice relaxation in this suite of coals, as the largest T_{1H} values are associated with the larger free radical concentrations. The free radical concentrations and the range in concentration in the coals studied here may be too small to have a significant effect on the T_{1H} values. If one uses typical values of 8.0 x 10^{18} spins/g, 80% carbon, and f_a value of about 0.70, and if one assumes that the average "aromatic cluster" in coals contains about 17 carbon atoms (say, four fused rings), then one concludes that there are 100 to 200 aromatic clusters per unpaired electron. If the ordering of clusters is random, rather than stacked as in a graphite structure, then the fraction of nuclei within an effective radius of the unpaired spin may be small. (Of course, electron spin delocalization must be taken into account in any serious assessment of this point.) ESR lines are known generally to narrow with increasing coal rank (14). If the observed ESR lines are envelopes of unresolved electron-proton hyperfine interactions, then the decreasing linewidth as the rank increases could be due to a lessening of proton-electron interactions. A more systematic investigation of a set of macerals treated over a wide range of temperatures should reveal more readily the expected electronic effects on relaxation times. The reason for this is that the free radical concentrations varies by four orders of magnitude among macerals and, in addition, high-temperature treatment can increase the radical concentraton by two orders of magnitude.

The generally larger T_{1H} values for evacuated samples, in comparison to samples as received, imply some role for O_2 or other adsorbed species (e.g., H_2O). Adsorbed molecular oxygen is known to broaden, as expected, ESR lines in the spectra of fusinates by a factor of four (14). Furthermore, the reported observation of large ^{13}C and ^{1}H signal enhancements in coals by dynamic nuclear polarization shows that nuclear spin/electron-spin dipolar interactions in coals are important (19). Clearly, many details remain to be answered (20) before a good understanding of ^{1}H relaxation in coals is complete.

Even without a complete, fundamental understanding of the factors determining T_{1H} in these coal samples, it is possible to draw the following useful (largely empirical) conclusions: 1) adsorbed species (e.g., O_2 or H_2O) contribute in some way to ^{1}H spin relaxation in coals; 2) the T_{1H} values measured for evacuated coals are roughly one order of magnitude smaller than what one typically observes for solid samples of small diamagnetic organics; 3) the protons on aliphatic centers are not in efficient thermal communication with the protons on aromatic centers, implying a degree of domain heterogeneity; 4) T_{1H} increases with rank, in spite of increasing free radical concentrations, implying that a mechanism in which a combination of primary relaxation by paramagnetic centers and ^{1}H spin diffusion does not dominate ^{1}H spin-lattice relaxation in these systems; 5) T_{1H}, especially in combinaton with f_a' values, provides a useful parameter for the characterization of a coal, related to its rank.

Acknowledgements

The authors gratefully acknowledge partial support of this work from the U.S. Department of Energy (Laramie Energy Technology Center) and the U.S. Geological Survey, Reston, VA.

REFERENCES

1. Gerstein, B.C., Chow, C., Pembleton, R.G. and Wilson, R.C., 1977, J. Phys. Chem., 81, 565.
2. Yokono, T. and Sanada, Y., 1978, Fuel, 57, 334.
3. Miyazawa, K., Yokono, T., Sanada, Y. and Marsh, H., 1979, Fuel, 58, 691.
4. Chow, C.K., 1981, Fuel, 60, 749.
5. Szeverenyi, N.M., Sullivan, M.J. and Maciel, G.E., presented at the 21st Experimental NMR Conference, Talahassee, FL, Apr. 1980.
6. Sullivan, M.J. and Maciel, G.E., 1982, Anal. Chem., 54, 1615.
7. Bartuska, V.J., Maciel, G.E., Schaefer, J. and Stejskal, E.O., 1977, Fuel, 56, 354.

8. Pines, A., Gibby, M.G. and Waugh, J.S., 1973, J. Chem. Phys., 59, 569.
9. Hartmann, S.R. and Hahn, E.L., 1962, Phys. Rev., 128, 2042.
10. Farrar, T.C. and Becker, E.D., "Pulse and Fourier Transform NMR", Academic Press, New York, 1971.
11. Gluskoter, H.J., Ruch, R.R., Miller, W.G., Cahill, R.A., Dreher, G.B., and Kuhn, J.K. 1977, Trace elements in coal: occurrence and distribution; Ill. State Geol. Surv. Circ., 499.
12. Bartuska, V.J. and Maciel, G.E., 1981, J. Magn. Reson., 42, 312.
13. Levy, G.C. and Peat, I.R., 1975, J. Magn. Reson., 18, 500.
14. Petrakis, L. and Grandy, D.W., "Free Radicals in Coals in Synthetic Fuels", U.S. Coal Science and Technology, Elsevier Publication, in press.
15. Cheung, T.T.P. and Gerstein, B.C., 1981, J. Appl. Physics, 52, 5517.
16. Sullivan, M.J. and Maciel, G.E., 1982, Anal. Chem., 54, 5517.
17. Brown, C.E., 1982, J. Am. Chem. Soc., 104, 5608.
18. Miknis, F.P., Sullivan, M., Bartuska, V.J. and Maciel, G.E., 1981, Fuel, 3, 19.
19. Wind, R.A., Trommel, J. and Smidt, J., 1982, Fuel, 61, 398.
20. Sullivan, M.J., FACCS Meeting, Philadelphia, 1982.

REACTION TIME ^1H NMR STUDIES OF THERMAL TRANSFORMATION PROCESSES OF FOSSIL FUELS

L.J. Lynch, D.S. Webster, N.A. Bacon and W.A. Barton

CSIRO Physical Technology Unit,
Ryde. NSW. 2112. Australia.

ABSTRACT

Information of great value in understanding the mechanisms involved in the conversion of fossil fuel materials to commercially useful products can be obtained by dynamic measurement techniques. Proton nuclear magnetic resonance (^1H NMR) can, in principle, detect many of the changes expected during thermal processing of fossil fuels, but its application as a dynamic measurement technique over the temperature range of importance presents many difficulties. Results of the application of dynamic ^1H NMR measurement techniques to the study of the pyrolysis of acenaphthylene, oil shale, brown coal, bituminous whole coals and macerals, and pyridine treated coal are presented and discussed.

INTRODUCTION

Because the proton nuclear magnetic resonance (^1H NMR) characteristics of a substance are determined by both structural and dynamical aspects of the molecular lattice and it is possible to measure NMR parameters which depend differently on these molecular properties, dynamic ^1H NMR measurements have considerable potential for monitoring the changes caused by thermal processing of fossil fuels. Specifically, alterations to the concentration and distribution of magnetic species (i.e. protons and unpaired electrons) and changes in the molecular mobility should be readily observed. Thus in the course of thermal processing of fossil fuels (brown coals, bituminous coals and oil shales) we expect to record such events as - (i) fusion transitions which are probably the basis of thermoplasticity in bituminous

L. Petrakis and J. P. Fraissard (eds.), Magnetic Resonance. Introduction, Advanced Topics and Applications to Fossil Energy, 617–628.

coals, and the subsequent resolidification processes; (ii) glass to rubber type transitions in the amorphous parts of the structures and the effects of progressive thermal activation of molecular mobility above the transitions; (iii) stable free-radical formation resulting from homolytic scission of covalent bonds and their quenching due to recondensation reactions, and (iv) changes in the hydrogen content and distribution due to decomposition of parts of the heterogeneous structure and evolution of low molecular weight products.

The time resolution of this dynamic NMR technique is limited by the combination of measurement sensitivity and spin-lattice relaxation rate, R_1, both of which will alter in the course of the thermal processing. This is the result of the temperature dependence of the macroscopic magnetization and response of the NMR detection circuit on the one hand, and the variations in the spin-lattice relaxation expected as a consequence of the chemical and physical changes in the specimen, on the other. Typically the room-temperature signal-to-noise ratio (S/N) of an open 500mg coal specimen ($\sim 2 \times 10^{22}$ protons) of $\sim$30:1 will be reduced to $\sim$2:1 at 800K whereas the R_1-limited repetition rate can change by a factor of 5 during heating to this temperature.

APPARATUS

The experiments are performed with a high power pulsed NMR spectrometer operating at 60MHz and using phase sensitive detection. The regulated electrically-heated high temperature NMR probe used and the methods of its construction have been described (1). The specimen probe has a tapped-series-tuned resonant circuit duplexed via a modified version of the technique introduced by Lowe and Tarr (2). Ring down time of the transmitter is limited by switched diode quenching circuits in the output resonant circuit of the transmitter (3). The specimen tuned circuit is damped to a Q of less than 20. The effective "dead-time" of the receiver is 4μs and 90° pulse lengths of between 1 and 2μs are easily attained using a modified Bruker SXP 4-100 transmitter. The oven is free of thermal gradients over the radio frequency (r.f.) coil and the specimen temperature is obtained by making a calibrated correction to the temperature sensed by the control thermocouple mounted below the specimen. Temperature control is effected by channelling the thermocouple output into a feedback amplifier operating in Class A mode with the heating element as its load. Coupling between the radially wound r.f. coil and heater circuit is minimized by the axial winding of the heater wire. Heating rates up to 20K min^{-1} have been used (4) and shock-heated specimens reach 95% of the test temperature in less than three minutes.

EXPERIMENTAL

Most of our ^1H NMR thermal analyses are based on observations aimed at quantitatively monitoring the ^1H NMR transverse relaxation signal I(t), but we also have been successful in dynamically measuring the spin-lattice relaxation of coal specimens during pyrolysis (5). In the former case the $90\text{-}\tau\text{-}90_{90}$ pulse sequence is used to stimulate the nominal solid-echo representation of the transverse relaxation (6). This is considered necessary in an attempt to overcome the "dead-time" problem, which is significant for the rapidly decaying signals encountered. The spin-lattice relaxation is monitored with the progressive saturation sequence.

The ^1H NMR measurements are made at regular intervals during a heating program - i.e. either while heating at a uniform rate or under isothermal conditions after prior heating to the test temperature. The amount of signal averaging possible is further limited by the necessity to maintain the tuning of the probe and adjust the 90° pulse conditions between measurements as the temperature and nature of the specimen change.

Specimens of 100-500mg are suitable for measurement - the actual size depending on their "richness" in hydrogen. These are usually finely ground just prior to measurement, oven dried at 378K under nitrogen and then either placed in open glass tubes or sealed in ∿20mm long Pyrex ampoules both 8mm O.D.. During measurement the specimen is positioned within the inner 2/3 of the r.f. coil so that there is uniform irradiation of the material. Thus for sealed ampoules all products of the pyrolysis are observed. Open tubes are flushed with dry nitrogen to ensure a non-oxidizing atmosphere and to quickly remove volatile products from the measurement zone. This presentation will be limited to the discussion of "solid-echo" transverse relaxation data for materials pyrolysed at 4K min^{-1} under nitrogen.

ANALYSIS

For a rigid-lattice pair-wise dipolar coupled population of proton spins the solid-echo signal is a full representation of the transverse relaxation and the peak of the echo is a measure of the total hydrogen content of the specimen (6). The proton populations of the fossil fuel specimens studied by us will deviate from this model in complex ways during pyrolysis. Deviations are expected because of the varying degrees of molecular mobility that occur, interaction of the protons with unpaired electrons and because of multiple proton dipolar interactions. How the nature of the recorded signals is affected by these factors has been considered to some extent by others (7,8), but we are unaware of any comprehensive theory. For rigid structures the solid-echo signal peak intensity will be attenuated by dipolar coupling between protons and unpaired electrons. For motionally "narrowed" signal components for which $T_2 >> \tau$, very little dephasing will have occurred before the 90°-phase-shifted second pulse, which will therefore have correspondingly little effect on the signal. It is the intermediate situation that presents the conceptual difficulties. In our experiments we use a value of $\tau = 6\mu s$ which is the minimum value for which the peak of the echo can be detected because of the "dead-time" of the receiver system. From our experience with these types of measurements, including comparisons with the results of measurements made using the single 90° pulse, we are of the opinion that the signal stimulated by the $90-\tau-90_{90}$ sequence, with $\tau = 6\mu s$, is a close representation of the true transverse relaxation signal for most of the conditions encountered in these experiments, except for the predictable influence of unlike spin coupling referred to above.

The "solid-echo" signals which dynamically monitor the transient states of the pyrolysing specimens can be analysed in a number of ways to provide parameters useful for describing the pyrolysis properties (9). We here consider two such parameters - (i) the signal intensity which is related to the residual hydrogen content of the specimen and (ii) a parameter characterizing the ^1H NMR transverse relaxation and which is computed as a truncated second-moment, M_2^*, of the power spectrum, $a(\nu)$, obtained by Fourier transformation of the recorded time domain signal, $I(t)$.

By treating the "solid-echo" signals as full representations of the transverse relaxation the echo peak height is nominally proportional to the proton spin population. An empirical correction to these measured intensities is necessary to allow for the temperature sensitivity of the ^1H NMR signals. This correction is based on a calibration of the probe using a brown coal char specimen sealed in a glass ampoule. This material previously had been pyrolysed at a temperature of greater than 950K and besides having a constant hydrogen content, appears relatively constant in its ^1H NMR properties over the temperature range of the calibration.

The power spectrum second-moment, M_2^*, is chosen as a parameter to characterize the ^1H NMR transverse relaxation in preference to the usual frequency spectrum second-moment, M_2, because of the large scatter in the M_2 values obtained for a thermal scanning experiment. This lack of precision in M_2 results from its critical dependence on the phase adjustment necessary subsequent to Fourier transformation - figure 1. The power spectrum $a(\nu)$, however, is independent of small phase shifts. Ejchardt and Wroblewski (10) have demonstrated the validity of using this power spectrum, rather than the frequency absorption spectrum, $u(\nu)$, for line shape analysis of high resolution spectra.

We have therefore computed the second-moment parameter of a(ν) as

$$M_2^* = \int_{\nu_o}^{\nu_x} a(\nu)\Delta\nu^2 d\nu \Big/ \int_{\nu_o}^{\nu_x} a(\nu) d\nu \ ,$$

where $a(\nu) = u'^2(\nu) + v'^2(\nu) = u^2(\nu) + v^2(\nu)$

and u' and v' are the quadrature components of the complex frequency spectrum, g(ν), which results from Fourier transformation of the "solid-echo" signal I(t). u and v are the pure

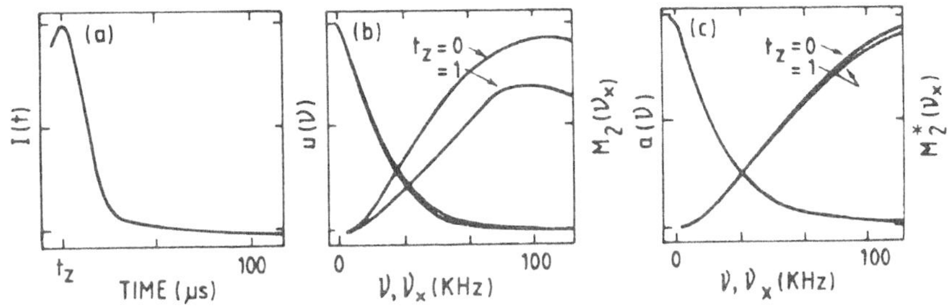

FIG. 1 The truncated second-moments of (b) the frequency absorption spectrum, M_2, and (c) the power spectrum, M_2^*, plotted against truncation frequency for two assumed zeros, shifted by 1μs, of the time domain data, I(t), as shown in (a).

absorptive and dispersive modes of g(ν). ν_o is the central resonance frequency, $\Delta\nu = (\nu - \nu_o)$ and the limit of integration ν_x is usually chosen in practice to be a frequency where $a(\nu_x) = a(\nu_o)/10$. Truncation by this method has been arrived at by trial as a compromise between increased random errors that accrue for truncation at higher values of ν_x and the increase in systematic errors for truncation at lower values. The computation of a(ν) also involves a subjective baseline adjustment. The sensitivities to "phase error" of the frequency and power spectra second moments are contrasted in figure 1. Here both second-moments computed for an extremely rapidly decaying signal of a coal char are plotted as a function of truncation frequency, ν_x, for two cases in which the zero of the time domain data differs by 1μs.

RESULTS AND DISCUSSION

Acenaphthylene: The carbonization properties of this polynuclear aromatic substance have been extensively studied (11) including its propensity to form thermotropic liquid crystalline structures and thence anisotropic cokes. The temperature dependence of M_2^* (here computed using a truncation for which $a(\nu_x) = a(\nu_o)/20$) during pyrolysis at 4K min^{-1} to 777K and subsequently at 777K for

125 minutes is compared to a thermogravimetric analysis at 10K $hour^{-1}$ (11) in figure 2. The many features in the ^{1}H NMR thermal analysis of acenaphthylene can be interpreted in the context of its known pyrolysis behaviour (11). A first order melting transition occurs at ∿368K, followed by polymerization in the melt above 480K to form polyacenaphthylene which fuses at ∿550K. Thermal depolymerization and successive dehydrogenative condensation reactions occur in the softened state between 610K and 777K with decacyclene the major product. The slow increase in M_2^* under isothermal conditions at 777K parallels the growth of thermotropic liquid crystalline domains and eventual solidification.

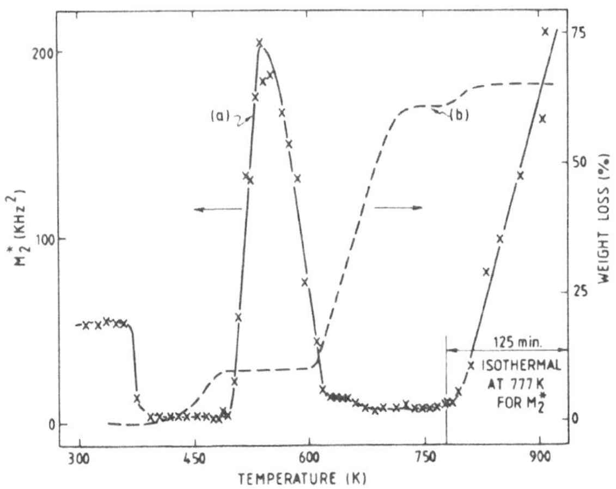

FIG. 2 Pyrolysis of acenaphthylene. (a) M_2^*, during heating at 4K min^{-1} to, and at the isothermal temperature of 777K. (b) TGA data of Fitzer et al. (11) at 10K $hour^{-1}$.

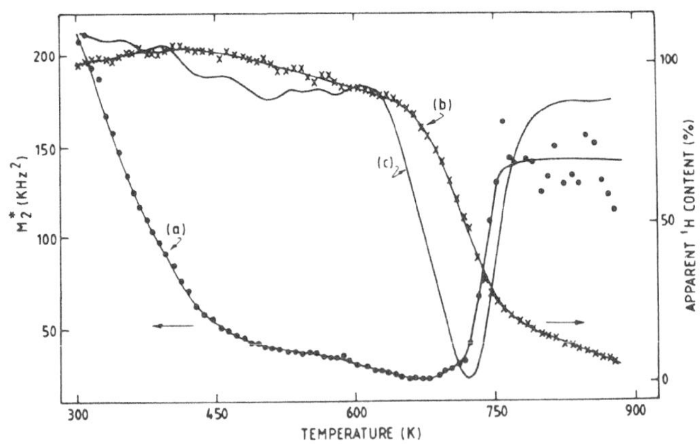

FIG. 3 Pyrolysis of Rundle oil shale at 4K min^{-1}. (a) M_2^*; (b) apparent hydrogen content, (c) differential hydrogen content.

Oil Shale: ^{1}H NMR thermal analysis is a convenient method for studying the pyrolysis of oil shale kerogen which is embedded in an inorganic matrix. The results of such an analysis of a highly aliphatic Rundle (Queensland, Australia) oil shale are shown in figure 3. Both the second-moment parameter, M_2^*, and the apparent hydrogen content from the "solid-echo" peak intensity, are plotted. The differential hydrogen content ($\Delta^1H/\Delta T$) obtained numerically from a spline fit to the hydrogen content data is also shown. We see from the apparent hydrogen content that there is little loss of volatile material below about 650K. We note also the anomalous increase in hydrogen content on heating between 300 and 450K. This phenomenon is characteristic of most shales we have studied and will be discussed later on. The room temperature value of M_2^* is typical of an amorphous polymer and similar to room temperature values we have measured for dried coals (see below). The rapid fall in M_2^* on heating above room temperature is characteristic of a diffuse glass to rubber type transition in an amorphous polymer of low cross-link density and the small value of M_2^* achieved shows that most of this highly aliphatic kerogen substance is involved. Above about 575K there is evidence of a second "softening" process which probably involves a minor, possibly aromatic, fraction of the kerogen. M_2^* reaches a very small minimum value at ∿680K consistent with almost all of the residual organic material being softened. Rapid pyrolytic decomposition, monitored here by the fall in apparent hydrogen content, commences at 650K before this minimum in M_2^* is reached and it is only after the decomposition is well advanced that M_2^* begins to increase, signifying the formation of a rigid-lattice residual coke. The constancy and value of M_2^* above 750K is consistent with a rigid aromatic organic residue.

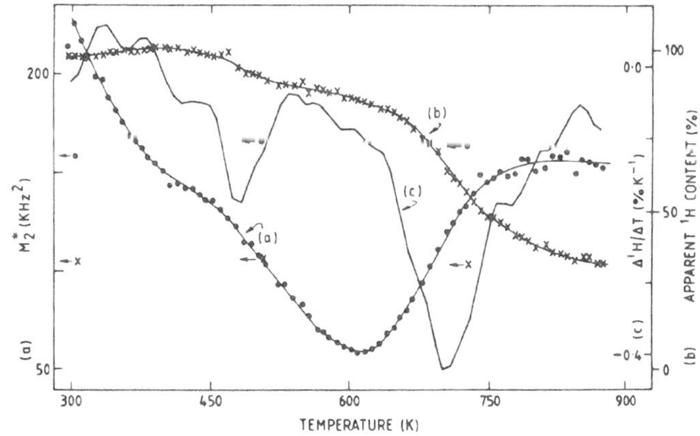

FIG. 4 Pyrolysis of Morwell brown coal at 4K min^{-1}. (a) M_2^*; (b) apparent hydrogen content, and (c) differential hydrogen content. Values measured during cooling are marked with arrows.

Brown Coal: Results of the ^1H NMR thermal analysis of a predried Morwell brown coal are shown in figure 4. This type of coal has a carbon aromaticity of 50-60%. Between room temperature and 615K there is a progressive increase in the average molecular mobility indicated by the fall in M_2^* while the apparent hydrogen loss is less than 15% of the original. Above ~615K the rate of hydrogen loss increases and the decreasing trend of M_2^* is reversed. The minimum value reached by M_2^* shows that a significant proportion of the structure remains rigid throughout the pyrolysis. The value of M_2^* attained above ~750K and its constancy during recooling are consistent with the residue above 750K being a highly aromatic rigid-lattice substance. These effects are interpreted as the results of thermal activation of molecular mobility in, and subsequent decomposition of, the aliphatic and functional group components of the structure, while the aromatic part of the coal remains largely unaffected and therefore constitutes most of the residue.

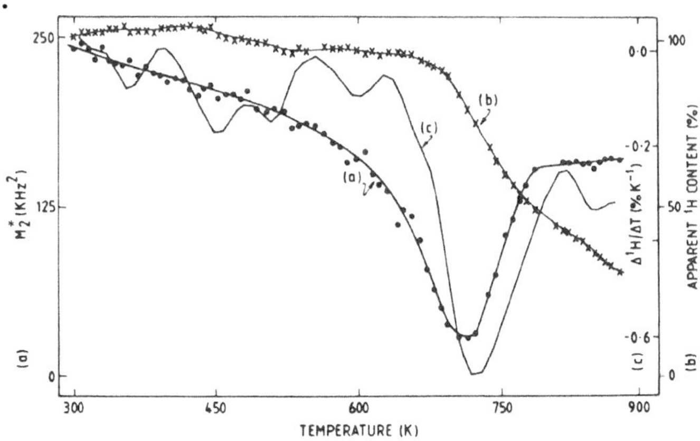

FIG. 5 Pyrolysis of high-volatile bituminous coal at 4K min^{-1}. (a) M_2^*; (b) apparent hydrogen content, and (c) differential hydrogen content.

Bituminous Coals: The ^1H NMR thermal analyses of a typical high-volatile vitrinite-rich bituminous coal and of a vitrinite concentrate of a different high-volatile bituminous coal are shown in figures 5 and 6 respectively. The general behaviour of the two specimens is similar and is characteristic of the reactive thermoplastic vitrinite macerals of these coals. The thermoplastic zone for these coals is clearly delineated by the M_2^* parameter, the minimum value of which occurs at temperatures very close to those corresponding to the maximum Giessler fluidities. The relatively small decrease in M_2^* between room temperature and 600K as compared with brown coal is no doubt a consequence of the much higher aromaticity of these bituminous coals. M_2^* begins to fall rapidly above 600K and, in contrast to the behaviour for brown coal

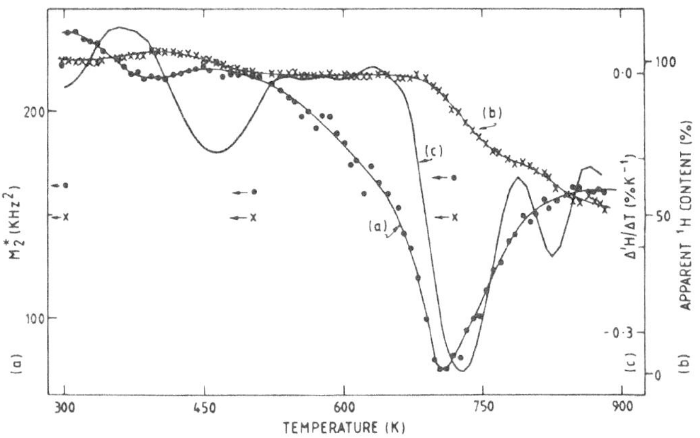

FIG. 6 Pyrolysis of vitrinite concentrate of Liddell high-volatile bituminous coal at 4K min^{-1}. (a) M_2^*; (b) apparent hydrogen content, and (c) differential hydrogen content. Values measured during cooling are marked with arrows.

(figure 4), it continues to decrease for a considerable period after rapid loss of hydrogen commences. Beyond its minimum, M_2^* rapidly increases to reach constant values that are similar for both specimens and typical for highly aromatic rigid-lattice structures. It is significant that the loss of hydrogen continues beyond the thermoplastic regions and does not appear to be closely correlated with the thermoplasticity. These ^{1}H NMR observations are consistent with the proposition that thermoplasticity, as defined for bituminous coals, is a manifestation of the molecular mobility that results from the thermal destabilization of aromatic structural zones of the vitrinite maceral and it would appear that this process is largely independent of the decomposition of the mainly non-aromatic parts of the structure to produce low molecular weight volatile materials.

The effects of the imbibing of pyridine, which appears to penetrate and swell the aromatic zones of bituminous coals, are under investigation in our laboratory (12). The ^{1}H NMR thermal analysis of a Liddell vitrinite containing 12.3% (on dry coal weight basis) of pyridine apparently intercalated in the specimen is shown in figure 7. This specimen was prepared by first swelling the coal with pyridine and then exhaustively evacuating it at 294K. These results can be directly compared to those for the normal vitrinite in figure 6. Several very interesting effects are apparent. Firstly, the low values of M_2^* reached near 400K show that a considerable fraction of the vitrinite structure has acquired a degree of molecular mobility, presumably because it is plasticized by the intercalated pyridine. Secondly, the apparent loss of hydrogen between ~400 and ~600K, which far exceeds

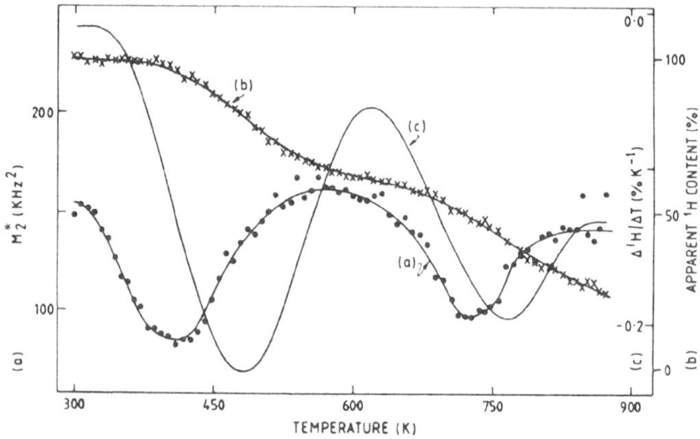

FIG. 7 Pyrolysis of Liddell vitrinite containing intercalated pyridine. (a) M_2^*; (b) apparent hydrogen content, and (c) differential hydrogen content.

that equivalent to the intercalated pyridine, indicates that the presence of the pyridine facilitates low temperature volatilization of a considerable fraction of the coal. Thirdly, there is a significant reduction in the average molecular mobility, indicated by the much higher M_2^* values, in the thermoplastic zone of the vitrinite. This reduction of the molecular mobility probably renders the specimen "non-plastic" according to such criteria as Giessler fluidity. It is well known that solvent extraction destroys the thermoplasticity of coals (13). It could therefore be the loss of material from the coal by volatilization between 400 and 600K that accounts for the subsequent reduction in mobility in the thermoplastic zone. An alternative explanation could be the disruption of order in the condensed aromatic zones as a result of solvent swelling.

The ^1H NMR thermal analysis data of a semi-fusinite concentrate of Bayswater bituminous coal are shown in figure 8. This information is definite evidence supporting the proposition that the inertinites of some Australian Permian coals are reactive in the sense that they have coking properties (14). Like other bituminous coals there is only a gradual fall in M_2^* and little loss of volatile material at lower temperatures. At 667K a rapid decrease in apparent hydrogen content commences and M_2^* begins to rise to reach a maximum value near 800K which is much greater than the original room temperature value for the coal. This suggests that the high values of M_2^* attained are not only the result of loss of the mobile non-aromatic parts of the coal but also the result of a build-up of stable free-radicals and therefore of proton-unpaired electron interactions. This is confirmed by studies of the residues of specimens heat treated to these

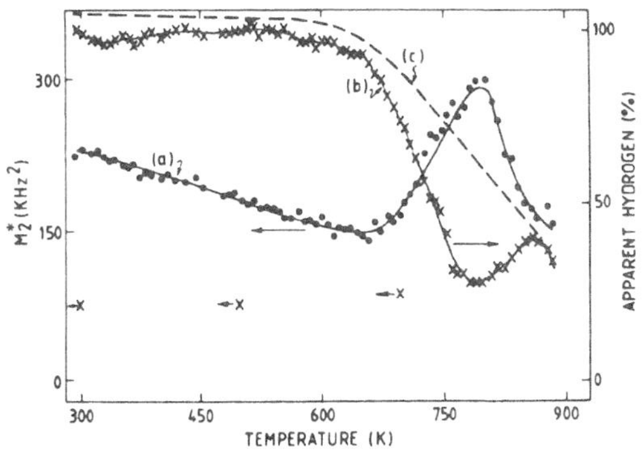

FIG. 8 Pyrolysis of Bayswater semi-fusinite concentrate. (a) M_2^*; (b) apparent hydrogen content, and (c) the suggested real variation of the hydrogen content during the pyrolysis.

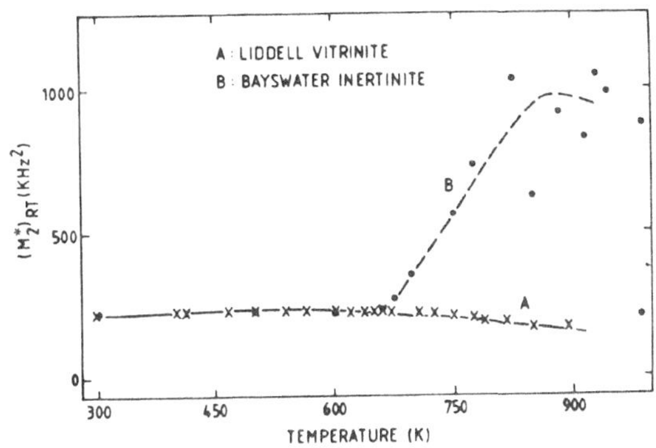

FIG. 9 M_2^* measured at room temperature, $(M_2^*)_{RT}$, as a function of maximum temperature reached during heating at 4K min^{-1} for (A) Liddell vitrinite, and (B) Bayswater inertinite.

temperatures (12). In fact there is a significant increase in M_2^* on recooling the residues, indicating some temperature dependence of the proton-unpaired electron coupling. This is not observed for the vitrinite. Room temperature values of the second-moment parameter, $(M_2^*)_{RT}$, plotted in figure 9 against maximum pyrolysis temperature for both the Bayswater inertinite and the Liddell vitrinite coals, illustrate this distinction.

The rapid drop in M_2^* between 800K and 870K which is accompanied by an anomalous increase in the apparent hydrogen content, is the result of a sudden softening or fusion of the aromatic zones of the residual material. We therefore attribute to the Bayswater inertinite reactive-softening properties, albeit at significantly higher temperatures, that are usually associated with the reactive vitrinite macerals of bituminous coals. The anomalous increase in apparent hydrogen content is the result of motional attenuation of the strong static coupling between protons and nearby free-radicals. In the rigid state that exists below 800K these strong interactions cause the coupled protons to be NMR "invisible". The decrease in apparent hydrogen content on recooling is likewise a consequence of the high free-radical concentration and the resolidification. A similar explanation is appropriate for the anomalous increases in apparent hydrogen content recorded at low temperature for most oil shales and also for brown coals impregnated with paramagnetic ferrous ions. The dotted line in figure 8 indicates a more likely trend for the true hydrogen content of the pyrolysing inertinite.

REFERENCES

1. Webster, D.S., Cross, L.F. and Lynch, L.J. 1979, Rev. Sci. Instrum. 50, 390.
2. Lowe, I.J. and Tarr, C.E. 1968, J. Phys. E. 1, 320.
3. El-Hannay, U. 1973, Rev. Sci. Instrum. 44, 1069.
4. Lynch, L.J. and Webster, D.S. 1981, Proc. Intern. Conf. on Coal Science, Dusseldorf, 7-9 Sept. 1981, D14, 675.
5. Webster, D.S. and Lynch, L.J. 1981, Fuel 60, 549.
6. Powles, J.G. and Strange, J.H. 1963, Proc. Phys. Soc. (London) 82, 6.
7. Cosgrove, T. and Barnett, K.G. 1981, J. Magn. Reson. 43, 15.
8. Mansfield, P. 1965, Phys. Rev. 137, A961.
9. Lynch, L.J. and Webster, D.S. 1983, Proc. Symp. on Geochemistry and Chemistry of Oil Shale, Seattle, March 1983, ACS Symposium Series, eds. F. Miknis and J. McKay. (To be published).
10. Ejchardt, A. and Wroblewski, K. 1980, J. Magn. Reson. 40, 469.
11. Fitzer, E., Mueller, K. and Schaefer, W. 1971, Chemistry and Physics of Carbon, Vol. 7, Ed. P.L. Walter, Marcel Dekker Inc., New York, 237.
12. Barton, W.A., Lynch, L.J. and Webster, D.S. (To be published).
13. Brown, H.R. and Waters, P.L. 1966, Fuel 45, 17.
14. Diessel, C.F.K. 1982, Final Project Report on NERDDP Project 78/2629, University of Newcastle, Australia.

ELECTRON SPIN ECHO AND ENDOR STUDIES OF COALS

H. Thomann, I.B. Goldberg, C. Chiu and L.R. Dalton

Department of Chemistry
University of Southern California
Los Angeles, California 90089-0482

Electron Nuclear Double Resonance (ENDOR) and Electron Spin Echo (ESE) techniques have been applied to the investigation of several coals. The results of these measurements are compared with analogous studies of pristine polyacetylene, composites of polyacetylene/polyethylene, and of a variety of hydrocarbon radicals. This comparison of ENDOR and ESE results provides insight into paramagnetic electron delocalization and into electron Heisenberg spin exchange interactions existing in coals.

1. ELECTRON SPIN ECHO STUDIES

1.1. Introduction

Electron spin echo (ESE) techniques afford the opportunity for measuring spectral parameters such as hyperfine and quadrupolar interactions and for measuring relaxation times with substantially improved resolution relative to conventional EPR techniques. The advantages of the ESE method are particularly apparent for the study of materials giving rise to inhomogeneously broadened EPR lines. Special pulse procedures avoid the destructive interference in the free induction decay (FID) of magnetization vectors associated with different resonance frequencies in the inhomogeneously broadened resonance line. When electron spin echo envelop modulation (ESEEM) patterns are analyzed, information about weak hyperfine and quadrupolar interactions can often be obtained (1). Moreover, the phase memory time measured by ESE techniques reflects the homogeneous spin packet width or "true spin-spin relaxation time" and it is this time that can be related by appropriate theory to relevant lattice dynamical processes (2). The inhomo-

L. Petrakis and J. P. Fraissard (eds.), Magnetic Resonance. Introduction, Advanced Topics and Applications to Fossil Energy, 629–638.

geneous linewidth or "effective spin-spin relaxation time" is often an unresolved combination of static (unresolvable hyperfine interactions) and dynamic line broadening contributions.

Electron spin echo techniques also afford the opportunity for the direct measurement of electron spin-lattice and phase memory relaxation times. For solid state materials, spin-lattice relaxation times are often substantially longer than phase memory relaxation times and are usually determined by different relaxation mechanisms. Measurement of both relaxation times often provides insight into different aspects of the lattice dynamics.

1.2. Electron Spin Lattice Relaxation Measurements

The temperature dependence of the electron spin-lattice relaxation rate (T_{1e}^{-1}) for many localized π-electron radicals can be described by the general expression (3)

$$T_{1e}^{-1} = AT^{n} + \sum_{i} B_{i} \text{ cosech } \left(\frac{\Delta i}{T}\right) + CT^{m}J_{m-1}(\theta/T) \tag{1}$$

where n = 1 or 2 and m = 7 or 9. This expression derives from Van Vleck, Raman and Orbach-Murphy relaxation mechanisms. Such mechanisms also predict either no dependence of relaxation rate upon applied dc magnetic field or that the relaxation rate increases with increasing magnetic field.

Recently, we have conducted an extensive investigation of electron spin-lattice relaxation for samples of pristine trans-polyacetylene and for composites of polyacetylene (PA)/low density polyethylene (PE). Representative results are shown in Fig. 1. In these materials, the paramagnetic defect has been shown by ENDOR to be an extensively delocalized π-electron radical. The temperature and magnetic field dependence of the measured electron spin-lattice relaxation rates is observed to be drastically different than that usually observed for localized π-electron radicals. For example, spin-lattice relaxation rates are observed to decrease with increasing magnetic field strength and the temperature dependence can be described by the expression

$$T_{1e}^{-1} = A' \int_{0}^{\frac{\varepsilon}{T}} \Delta \text{ cosech } \left(\frac{\Delta}{T}\right) d\Delta \tag{2}$$

We have shown that both temperature and magnetic field dependence can be derived from the TLS model of Phillips and of Anderson and coworkers which considers spin relaxation to occur via phonon assisted tunneling between energy minima (4). The precise magnetic field dependence suggests that this process may modulate electron dipolar and electron Heisenberg spin exchange interactions.

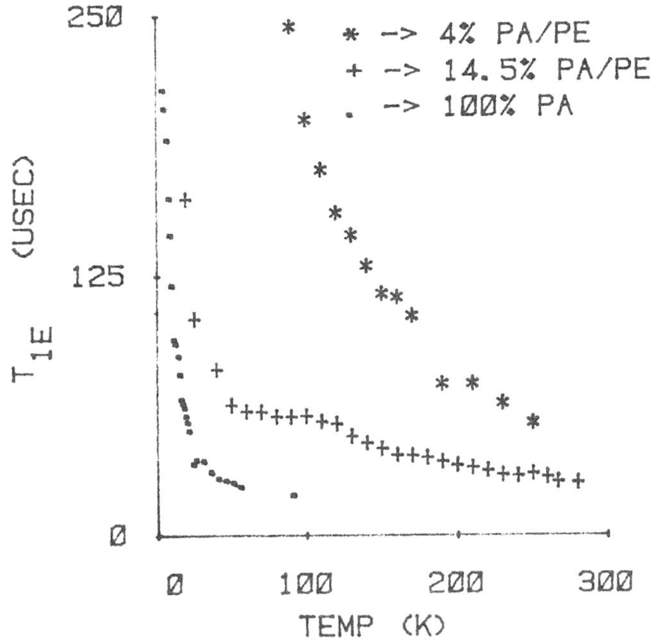

Figure 1. Electron spin-lattice relaxation times for trans-polyacetylene and composites of trans-polyacetylene/low density polyethylene are shown as a function of temperature. Compositions of samples are indicated in the figure.

This observation is also supported by the lack of dependence of measured spin-lattice relaxation times upon the isotopic (^1H, ^2H) composition of polyacetylene and by the observation of a dependence of electron spin-lattice relaxation times upon trans-polyacetylene concentration in composites of PA/PE.

1.3. Electron Phase Memory Relaxation Measurements

For localized π-electron radicals, measured phase memory relaxation times are observed to be essentially temperature independent. This is consistent with lattice motions effecting nuclear spin flips and thus modulation of electron nuclear hyperfine interactions.

For samples of trans-polyacetylene, experimentally determined phase memory times are observed to vary with temperature. Pristine samples of trans-polyacetylene yield phase memory times which become shortened as the temperature is increased from 3°K to 150°K then the times are observed to lengthen as the temperature is increased to 300°K (See Fig. 2). This minimum in the

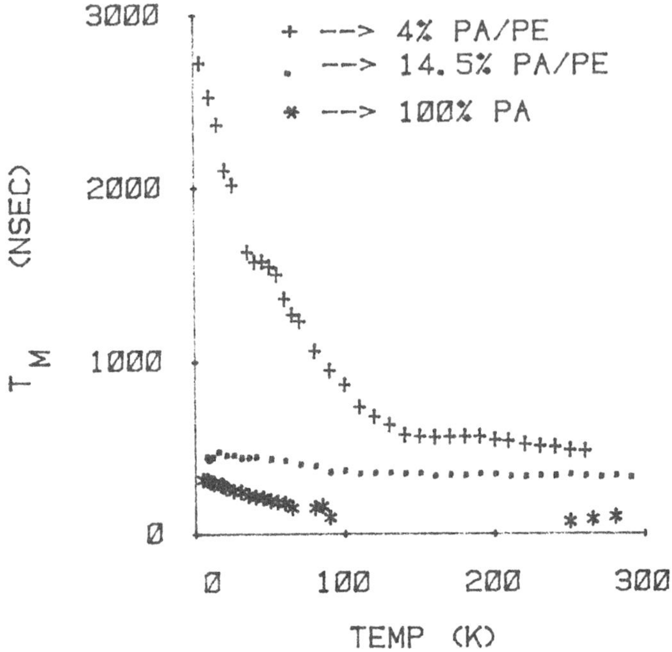

Figure 2. Electron phase memory relaxation times for trans-polyacetylene and composites of trans-polyacetylene/low density polyethylene are shown as a function of temperature.

graph of phase memory time versus temperature is consistent with spatial diffusion of the paramagnetic defect modulating hyperfine and electron Zeeman interactions. This observation is supported by the observed dependence of phase memory time upon the isotopic (^1H, ^2H) composition of the polyacetylenes studied and by the orientation dependence of phase memory times for samples of partially aligned polyacetylenes. The observed phase memory time behavior implies an activation energy of 400°K for the defect diffusion process.

The behavior for composites of PA/PE is somewhat different in that only the "slow motion" (motional frequencies less than the frequencies of modulated magnetic interactions) region is observed. (See Fig. 2) That is, phase memory times are observed to become shorter with increasing temperature with a minimum value reached at approximately 300°K. A temperature region for which phase memory times lengthen with increasing temperature is not observed. Phase memory times are observed to shorten with increasing trans-polyacetylene concentration suggesting that electron Heisenberg spin exchange may contribute to the measured phase memory time, i.e.,

$$(T_M)^{-1}_{expt} = (T_M)^{-1}_{other} + \omega_{ex} \quad (3)$$

where ω_{ex} is the Heisenberg spin exchange frequency.

1.4. Electron Spin-Echo Envelop Modulation (ESEEM)

Fourier transformation of ESEEM patterns often permits the measurement of weak hyperfine and quadrupolar interactions. As such analysis is the subject of other contributions at this conference, we will not consider such effects here.

2. ELECTRON NUCLEAR DOUBLE RESONANCE STUDIES

2.1. Introduction

Electron nuclear double resonance (ENDOR) provides a means of avoiding the reduction in spectral resolution resulting from the inhomogeneous broadening of EPR lines. For weak hyperfine interactions, only two ENDOR transitions are observed for each hyperfine interaction, with the ENDOR transition frequencies given by

$$\nu_{\pm} = \nu_N \pm \frac{A}{2} \quad (4)$$

Thus, ENDOR transition frequencies are linearly related to hyperfine interactions. Processes which modulate hyperfine interactions may in favorable cases be investigated by the effect that they have upon ENDOR transition frequencies and linewidths.

2.2. ENDOR Studies of Hydrocarbon Radicals

ENDOR studies of a variety of localized π-electron radical systems demonstrates that resolved ENDOR spectra are often obtained and that these can be interpreted by appropriate theory. A similar observation holds for delocalized π-electron radicals when hyperfine interactions are not modulated by defect diffusion or by Heisenberg spin exchange. Such an example is the ENDOR spectrum for cis-rich polyacetylene (5).

For pure trans-polyacetylene, the ENDOR spectrum consists of a single line at temperatures above 50°K with the linewidth decreasing as the temperature is increased. These observations are consistent with ESE phase memory time data and suggest that hyperfine interactions are modulated by defect diffusion. Typical results for composites of polyacetylene/polyethylene are shown in Fig. 3. These spectra suggest defect diffusion in the "slow motion" region and indicate that ENDOR spectra may be influenced by electron Heisenberg spin exchange as spectra are found to depend

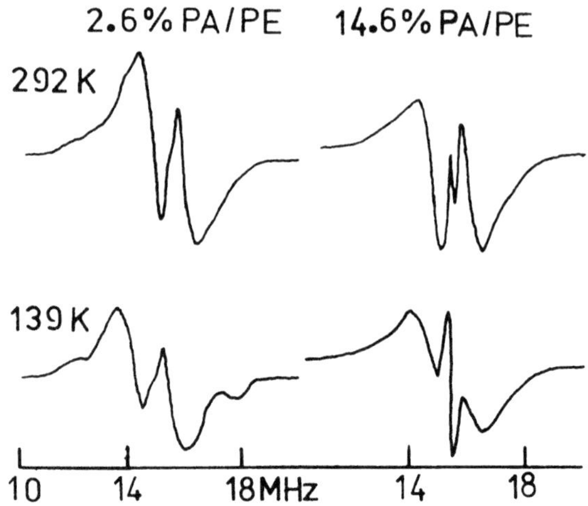

Figure 3. The ENDOR spectra of composites of trans-polyacetylene and low density polyethylene. Spectra are shown for two polyacetylene concentrates and for two temperatures.

upon trans-polyacetylene concentration in the PA/PE composites.

3. EXPERIMENTAL METHODS

The ENDOR and ESE spectrometers have been described elsewhere as has the preparation of polyacetylene samples (5). The subbituminous coal from the Rosebud mine of Montana was found to contain 1.07 × 10^{19} unpaired electron spins and by ^{13}C-cross polarization-magic angle spinning NMR was found to exhibit an aromaticity of 0.72 (6). Elemental analysis indicated that this coal contained 71.16% C, 4.45% H, 1.18% N and by difference 23.21% O and S (6). No spin concentration or chemical composition data is currently available on the other coals. These samples were degassed by pumping on a vacuum line and were sealed under vacuum in glass EPR tubes. Care was taken not to heat the samples above 300°K.

4. ELECTRON SPIN ECHO STUDIES

4.1. Electron Spin-Lattice Relaxation Measurements On Coals

The temperature dependence of measured spin-lattice relaxation times for two coal samples are shown in Fig. 4. The functional dependence of spin-lattice relaxation time upon temperature is essentially identical for the two coal samples; however, spin-lattice relaxation is observed to be consistently faster through-

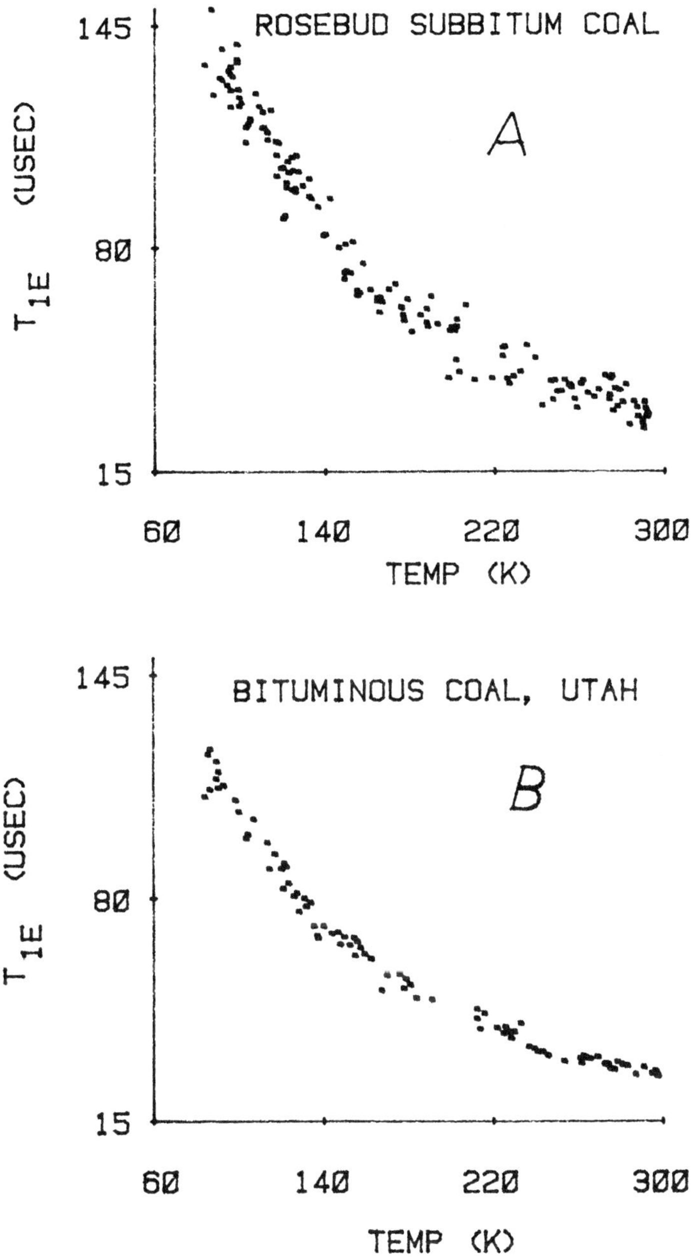

Figure 4. Spin-lattice relaxation times for two coals shown as a function of temperature. (A) Rosebud mine subbituminous coal. (B) Bituminous coal from Utah.

out the temperature range for the bituminous Utah coal. The temperature dependence is similar to that observed for polyacetylene/polyethylene composites for PA/PE concentrations between 4 and 15%. This may in turn suggest relaxation through a TLS mechanism; however, confirmation of this suggestion awaits the determination of the magnetic field dependence of electron spin-lattice relaxation rates.

4.2. Electron Phase Memory Time Measurements On Coals

Phase memory time measurements for two coal samples are shown in Fig. 5. Again the functional dependence upon temperature appears to be the same with consistently faster relaxation times observed for the bituminous coal from Utah compared to the Rosebud mine subbituminous coal.

The measured phase memory times are substantially longer than "effective spin-spin relaxation times" calculated from EPR linewidths. This observation is consistent with inhomogeneous broadening of the EPR resonance. The measured phase memory times are substantially shorter than normally observed for solids containing dilute concentrations of localized π-electron radicals. Indeed, the high spin concentrations in these samples suggest that the short phase memory times reflect strong electron Heisenberg spin exchange.

5. ELECTRON NUCLEAR DOUBLE RESONANCE (ENDOR) MEASUREMENTS ON COALS

Representative ENDOR spectra are shown in Fig. 6. In general, ENDOR spectra of coals are observed to consist of two components. The narrow component at the free proton Larmor frequency which may well be a distant ENDOR signal. The broader component likely arises from weak hyperfine interactions. Comparison of these results with the ENDOR spectra of PA/PE composites suggests that electron Heisenberg spin exchange influences the ENDOR spectra of coals.

6. CONCLUSIONS

Electron phase memory times, electron spin-lattice relaxation times, and ENDOR spectra of coals indicate the existence of electron Heisenberg spin exchange. It is possible that exchange frequencies may be quantitatively estimated from measured phase memory times; however, further experiments are necessary to verify this contention. Comparison of spin-lattice data for coals and for composites of PA/PE suggests at a TLS mechanism may dominate spin-lattice relaxation.

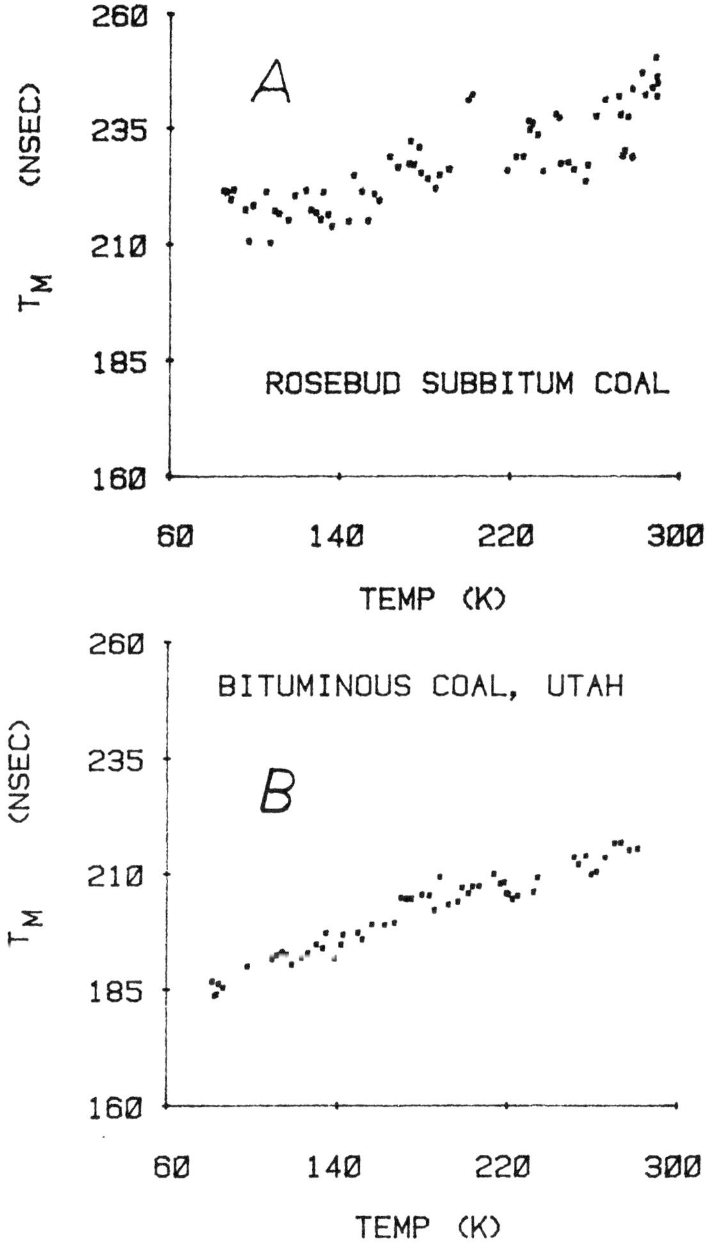

Figure 5. Phase memory relaxation times for two coals shown as a function of temperature. (A) Rosebud mine subbituminous coal. (B) Bituminous coal from Utah.

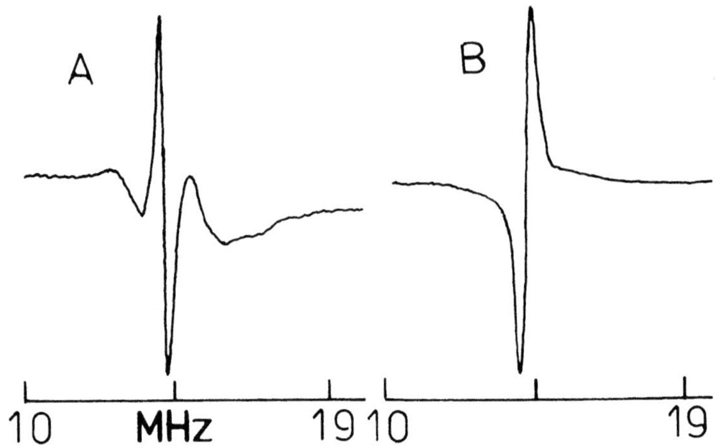

Figure 6. ENDOR spectra of two coals recorded at 250°K. Note that the spectra are each composed of two components.

Acknowledgement. We wish to acknowledge support from the Air Force Office of Scientific Research under Grant No. 82-NC-0184.

References

1. Narayana, P.A. and Kevan, L., 1983, Mag. Reson. Rev. 7, p. 239-274.

2. Thomann, H., Dalton, L.R. and Dalton, L.A., 1983, Biological Applications of Time Domain ESR, in: Biological Magnetic Resonance, ed. L.J. Berliner and J. Reuben (Plenum, New York).

3. Dalton, L.R., Kwiram, A.L. and Cowen, J.A., 1972, Chem. Phys. Lett. 17, p. 495-499.

4. Deville, A., Gaillard, B., Blanchard, C. and Livage, J., 1983, J. Physique 44, p. 77.

5. Thomann, H., Dalton, L.R., Tomkiewicz, Y., Shiren, N.S. and Clarke, T.C. 1983, Phys. Rev. Lett. 50, p. 533-536.

6. Goldberg, I.B., McKinney, T.M. and Ratto, J.J. 1983, Time Dependence of Radical Formation and Decay During Coal Hydrogenation, to be published in the Proceedings of the International Conference on Coal Science, Pittsburgh, PA., August, 1983.

THE ELECTRON SPIN ECHO CHARACTERISTICS OF SEVERAL BITUMINOUS COALS

David C. Doetschman and Devkumar Mustafi

Department of Chemistry, State University of New York, Binghamton, New York 13901 U.S.A.

Electron spin echo spectrometry was employed to isolate the vitrain maceral free radical spectrum in several bituminous coals by means of relaxation time differences between macerals. Two echo characteristics show that the free radicals are well-isolated species with S=1/2, (a) the microwave pulse length for maximum echo and (b) the single exponential time dependence of approximately T_1=200 μsec for the saturation recovery of the echo. The heterogeneous broadening of the echo spectra of about 10 G width is characterized with the approximately T_2=1 μsec echo decay times and the echo modulations of around 5 MHz. The modulations are consistent with the hyperfine interactions in a perinaphthyl or similar type of radical oriented in coal layers.

An electron spin echo study of the free radicals in a low sulfur Pittsburgh bituminous coal was reported recently by Das (1). The phenomenon was mentioned briefly in earlier accounts of other subjects by Schlick (2) and Narayana (3). It was found that the echo spectrum of the vitrain maceral component could be selected from composite EPR spectra. The coal EPR power and temperature dependence suggested that the echo selectivity is based on the facts that the vitrain and fusain macerals have different T_1 and T_2 times and that these times are highly temperature dependent. A modulation of the echo decay was observed and its frequency was equivalent to the hyperfine interaction of a proton about 0.36 nm from a point spin such as in the free radical structure proposed by Yen (4). In this paper some of the

L. Petrakis and J. P. Fraissard (eds.), Magnetic Resonance. Introduction, Advanced Topics and Applications to Fossil Energy, 639–644.

general characteristics of the vitrain maceral electron spin echoes will be discussed and examples will be given from studies of a range of coal samples.

The mechanism by which coal free radicals are formed is not fully understood but is expected to resemble the results of carbonization studies (5). Whether the radicals are isolated or are formed in proximate pairs is therefore of interest. In the past, magnetic susceptibility measurements were made on the materials to establish the spin multiplicity of the radicals. The Curie law behavior in most cases suggests that the radicals are isolated. However weakly coupled pairs could be missed in studies at insufficiently low temperatures. Two types of spin echo studies provide further evidence that the coal free radicals examined are well-isolated.

The following two echo experiments provide additional evidence that the coal radical spin multiplicity is a doublet, S=1/2. We have compared the microwave pulse duration at which the echo is maximum in coal samples with a standard sample of known multiplicity. Theory shows that the doublet pulse duration should be $\sqrt{2}$ times the triplet duration (6). The pulse duration for the maximum coal echo intensity in the low sulfur Pittsburgh bituminous coal is 1.48 ± 0.14 times longer than a known triplet S=1 molecule, fluorenylidene (7). Similar times have been observed in a high volatility "b" Illinois bituminous coal (PSOC 580) and in a standard high volatility "a" Pittsburgh bituminous coal (PSOC 704).

The second experiment that indicates a doublet coal radical spin multiplicity is a spin echo saturation recovery experiment. It is based on a demonstration that in general a triplet spin system relaxes to thermal equilibrium with a biexponential dependence on time (8) whereas a doublet relaxes exponentially provided there is no phonon bottleneck. Figure 1 shows a single exponential saturation recovery of the low sulfur Pittsburgh bituminous coal. The data indicate that T_1=225 ± 20 μsec. Similar results were found in the standard high volatility "a" Pittsburgh bituminous coal (PSOC 704), T_1=160 μsec, at 100 K and in the high volatility "b" Illinois coal (PSOC 580), T_1=194 μsec., at 90 K.

These experiments both indicate that the coal free radicals have doublet spin multiplicity. The free radicals are therefore isolated to the extent that exchange or dipolar coupling between them is less than the microwave magnetic field of about one Gauss. A dipole interaction of $\leq$ 1G requires a spin separation of $\geq$ 17Å.

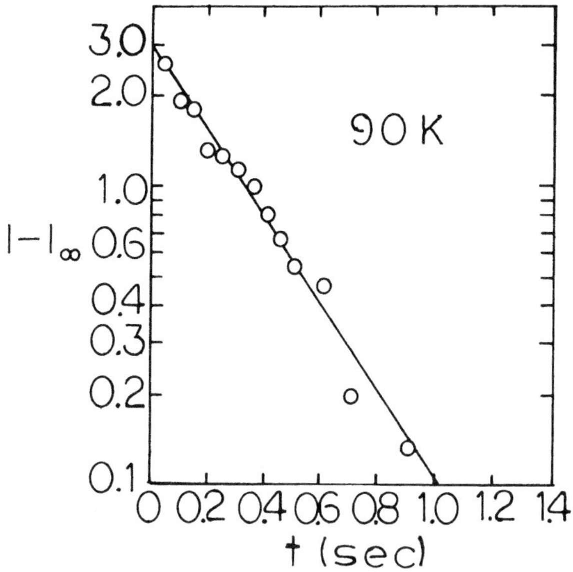

Figure 1. Low sulfur Pittsburgh bituminous coal electron spin echo saturation recovery at 90 K for the vitrain maceral component. The C.W. saturating pulse length is 1.8 msec, the echo pulse duration is 360 nsec and the echo pulse interval is 1.2 μsec. The magnetic field is 3280 G and the microwave frequency is 9192.5 MHz.

The intensity of the two-pulse Hahn echo decay of the low sulfur Pittsburgh bituminous coal was measured as a function of pulse interval. The echo decays with a time T_2=1.3 μsec at 90 K and drops to T_2=1.0 μsec at 120 K The echo decay is modulated at a frequency of 5.0 ± 0.5 MHz. The high volatility "b" Illinois bituminous coal (PSOC 580) decays with T_2=1.40 μsec at 85 K and with a similar but deeper and somewhat more complex modulation. The standard high volatility "a" Pittsburgh coal (PSOC 704) decays with T_2=0.96 μsec at 100 K without well-defined modulations.

These echo decay characteristics reveal how the approximately 10 G wide echo spectrum of the low sulfur Pittsburgh bituminous coal is heterogeneously broadened. Based on T_2, the individual components are about 0.3 G wide. The modulations indicate some approximately 1.7 G hyperfine structure components. These components are spanned (partially) by the rotating frame microwave magnetic field to generate the modulations. The standard Pittsburgh (PSOC 704) and Illinois (PSOC 580) coals have similar linewidths and decay characteristics and apparently have the same general type of heterogeneous line broadening.

Perinaphthyl and similar radicals are frequently formed in the process of carbonization (5,9). It has been suggested that a perinaphthyl type of structure is bonded into the partially aromatic coal layers (4). Perinaphthyl radicals have also been found in the volatiles from coal pyrolysis (10).

Figure 2 shows a model of perinaphthyl in a coal layer with spin densities based on the perinaphthyl solution EPR spectrum (5,9). The principal hyperfine constants shown for the meta and para protons are the products of the respective spin densities and the anisotropic hyperfine constants of the CH on the malonic acid radical.

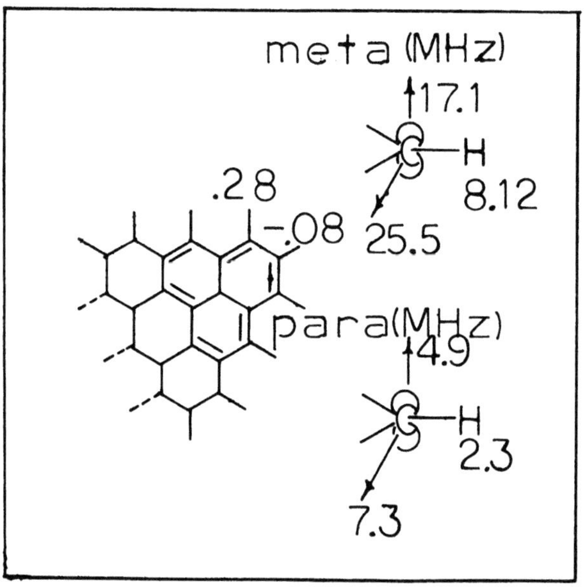

Figure 2. Perinaphthyl type of radical in a plane of the layered coal structure. Meta and para spin densities and principal hyperfine interaction frequencies on the CH are shown.

When the field is perpendicular to the plane of a given radical, the hyperfine interaction of all identical CH become equivalent, unlike any other field direction. Therefore para echo decay modulations around 4.9 MHz and meta modulations around 17.1 MHz will accumulate for fields near the aromatic plane normal. Modulations of various frequencies for other field directions will tend to cancel one another. Any tendancy for the aromatic planes of different radicals to be aligned with random in-plane orientation will further enhance the accumulation of modulation frequencies for the heme normal. This kind of alignment of partially aromatized layers is commonly believed to exist in coals. The predicted 4.9 MHz para modulation is in good agreement with 5.0 ± 0.5 MHz frequency observed in the low sulfur Pittsburgh bituminous coal. The 17.1 MHz frequency is presently outside the spectrometer bandwidth.

The more complex echo decay modulation in the Illinois coal (PSOC 580) could arise from related radicals. For example the solution spectrum of naphthanthryl (9) similarly predicts the accumulation of modulation frequencies, 2.2 and 3.5 MHz, as well as 4.8 MHz.

In summary, electron spin echoes in bituminous coals can isolate the vitrain maceral spectra. The echo decay characteristics give a general picture, of the composition of the heterogeneously broadened lines and show that the radicals are well-isolated. The echo modulation frequency observed is in agreement with the hyperfine of the para proton of perinaphthyl radical in microscopically aligned coal layers.

1. Das, U., Doetschman, D.C., Jones, P.M. and Lee, M., 1983, Fuel 62, pp. 285-288.
2. Schlick, S., Narayana, P.A. and Kevan, L., 1978, J. Am. Chem. Soc. 100, pp. 3322-3326.
3. Narayana, P.A., Massoth, R.J. and Kevan, L., 1982, Rev. Sci. Instr. 53, pp. 624-625.
4. Yen, T.F., Erdman, J.G. and Saraceno, A.J., 1962, Anal. Chem. 34, pp. 694-700.
5. Lewis, I.C. and Singer, L.S., "Electron Spin Resonance and the Mechanism of Carbonization", pp. 1-88, vol. 17 "Chemistry and Physics of Carbon", P.L. Walker, Jr. and P.A. Thrower, eds. Marcel Dekker, Inc., New York, 1981.
6. Sloop, D.J., Yu, H.-L., Lin, T.-S. and Weissman, S.I., 1981 J. Chem. Phys. 75, pp. 3746-3757.
7. Brandon, R.W., Closs, G.L., Davoust, C.E., Hutchison Jr., C.A., Kohler, B.E. and Silbey, R., 1965, J. Chem. Phys. 43, pp. 2006-2016.

8. Doetschman, D.C. and Botter, B.J., 1981, J. Magn. Res. 44, pp 348-354.
9. Singer, L.S., and Lewis, I.C., 1982, Applied Spectroscopy, 36, pp. 52-57.
10. Kim, S.S., Jarand, M.L. and Durai-Swamy, K., 1982, Fuel 51, pp. 1124-1126.

ESR OF CARBON RADICALS IN ISOLATED COAL MACERALS

B. G. Silbernagel, L. A. Gebhard
Exxon Research and Engineering Company
Linden, New Jersey 07036

G. R. Dyrkacz
Argonne National Laboratories
Argonne, Illinois 60439

ABSTRACT

ESR g-values, linewidths, and radical densities are compared with elemental analyses and other properties of a series of coal macerals isolated by the gravity centrifugation technique. Increases in g-values are attributed to the presence of phenolic oxygen pendant to the condensed ring structures of the coal. For vitrinites, all three ESR parameters vary with increasing coal rank: g-values fall, while linewidths and radical densities increase. Linewidths decrease and radical densities increase for more condensed inertinite forms having lower H/C values. Exchange between radicals is suggested to explain the observed narrowing phenomenon.

INTRODUCTION

The recent development of techniques for the separation of organic particles by density gradient centrifugation allows small (~3 μm) particles of coal to be separated with a density resolution of ~0.01 g/cm^3.[(1)] At this resolution it is possible to isolate samples which are almost entirely (>99%) comprised of a single coal maceral type. Such materials are of particular interest for the study of both coal maceral chemistry and the coalification process because these relatively homogeneous samples allow us to establish a less ambiguous relationship between the intrinsically microscopic ESR parameters and more global variables such as maceral type, elemental analysis, and aromatic carbon content. Previous experiments on mixed macerals did not allow this degree of definition, and information

L. Petrakis and J. P. Fraissard (eds.), Magnetic Resonance. Introduction, Advanced Topics and Applications to Fossil Energy, 645–653.

on minority maceral types was particularly ill defined. This paper will discuss an ESR survey of three isolated maceral types: vitrinite, inertinite, and exinite.[2,3] We find that each maceral type has its own "signature" as reflected in such ESR parameters as g-values, linewidths, and spin densities. We also find systematic changes in these properties for macerals of coals with different ranks.

The ESR observations described here were performed on ~10 mg samples at microwave frequencies of 10 GHz. A detailed description of the experimental conditions is contained in Reference 2. The present paper will deal with three major questions: (a) the variation of g-value with oxygen for vitrinites of different rank, (b) the relationship of radical density and ESR linewidth to the chemical properties of vitrinites, and (c) the interrelationship of linewidth, lineshape and radical density for inertinite samples, which show dramatic variations in these quantities.

G-VALUE VS. O/C FOR VITRINITES

The variation of g-value with atomic O/C ratio is shown in Figure 1. For high rank coals, especially low and medium volatile bituminous coals, the g-values observed approach g≃2.0028, as expected for carbon radical anions localized on aromatic molecules. The g values increase significantly to values near

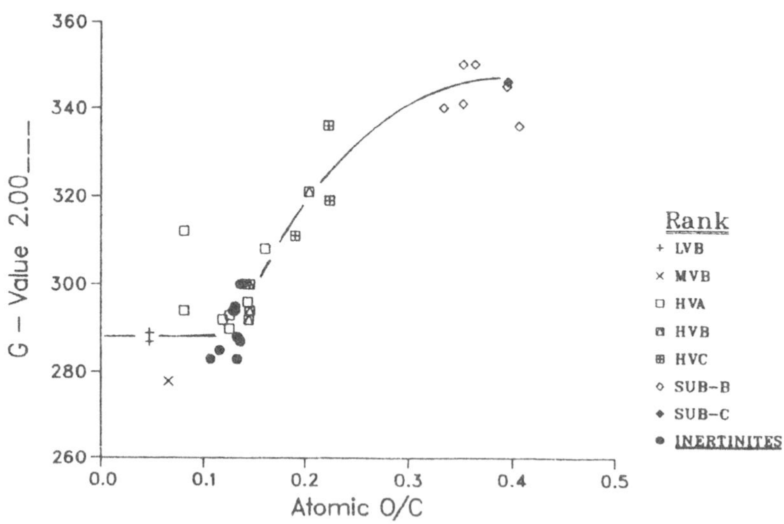

Figure 1. G Value vs. Atomic O/C.

= 2.0032 for high volatile coals. This increase generally tracks the coal rank and is nearly linear in O/C. The g-values are still higher for subbituminous coals (~2.0034) but inspection of Figure 1 shows that the increase is no longer linear in O/C ratio.

As observed by Petrakis and Grandy,[4] the presence of heteroatoms in the carbon radical host molecules often leads to higher g-shift values. While nitrogen and sulfur might also have an effect, the most obvious candidate for the observed g-value increase is oxygen, since it represents such a large fraction of the total number of heteroatoms. From Figure 1 it is clear that the heteroatom effect is decidedly nonlinear in the present case.

This nonlinearity can be explained by considering the different forms of oxygen which occur in the coal. A fraction of the oxygen can occur in heterocyclic forms or as etheric bridges, while other oxygens can be phenolic, with direct substitution of oxygen for hydrogen on the periphery of the aromatic molecule "cores" which occur in the coal structure. Oxygen can also occur as carboxylic functionalities pendant to these cores. In a series of controlled oxidations, Lewis and Singer demonstrated that condensed aromatic molecules were susceptible to oxygen attack and analysis of the hyperfine structure of the resultant products revealed oxygen substituted for hydrogen at the periphery of the molecule.[5] They also demonstrated that the magnitude of the g-shift associated with the oxidation was inversely related to the energy splitting of the lowest unoccupied orbital of the parent molecule. The latter observation implies that the detailed electronic structure of the core molecule - and not simply the ring size - dictates the magnitude of the g-shift.

We can rationalize the systematic g-value variations in terms of the coalification process. As systematic studies of oxygen functionalities in coal demonstrate,[6] a subbituminous coal can contain significant amounts of oxygen in the form of carboxyl, phenolic and etheric forms. By contrast, a high volatile bituminous coal has virtually no residual carboxyl species, since they were lost during the coalification process. The peripherial oxygen of the phenolic type is labile under the more stringent conditions encountered in transforming coal to higher ranks. We suggest that the heterocyclic and etheric oxygen forms do not contribute significantly to the shift in g-values, as indicated by the nearly constant g-values for $O/C \leq 0.15$. Phenolic forms seem to be particularly effective at causing g-shifts. Conversely, the carboxylic oxygen appears less effective, presumably because of the fact that the oxygen is less proximate to the molecular core.

ESR INTENSITY AND LINEWIDTH VARIATIONS FOR VITRINITES

Figure 2 shows the variation of the ESR linewidth, as represented by the splitting between maxima of the ESR absorption derivative (ΔH_{pp}), as a function of H/C ratio for vitrinites and inertinites. Vitrinite ΔH_{pp} values lie for the most part between 5G and 7G, with ΔH_{pp} showing a linear decrease as H/C increases. It is also apparent from an inspection of the symbols which denote coal rank, that linewidth increases for higher rank coals. The inertinites show a significantly different behavior. For samples closest in physical density to the vitrinites, $\Delta H_{pp} \sim$ 5-6G is observed. However, other inertinite samples show a sharp drop in ΔH_{pp} which roughly tracks falling H/C. We will return to this precipitous fall-off in H/C for inertinites in the next section.

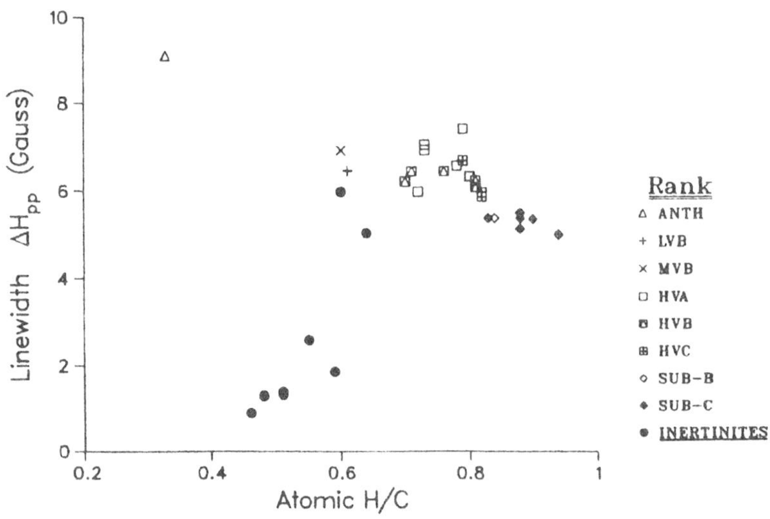

Figure 2: ESR Linewidth vs. Atomic H/C.

There are also significant variations in carbon radical density, as Figure 3 indicates. For vitrinites, the general trend is for the radical density to increase with increasing coal rank (up to anthracites), consistent with previous whole coal studies.[7] By contrast the inertinites show dramatic variations with H/C, ranging from values comparable to higher rank vitrinites to radical densities which exceed these by nearly a factor of five. This radical density variation is not monotonically related to H/C, suggesting that a variety of factors, including the chemical form of the resulting residue, determine the final radical density in these materials.

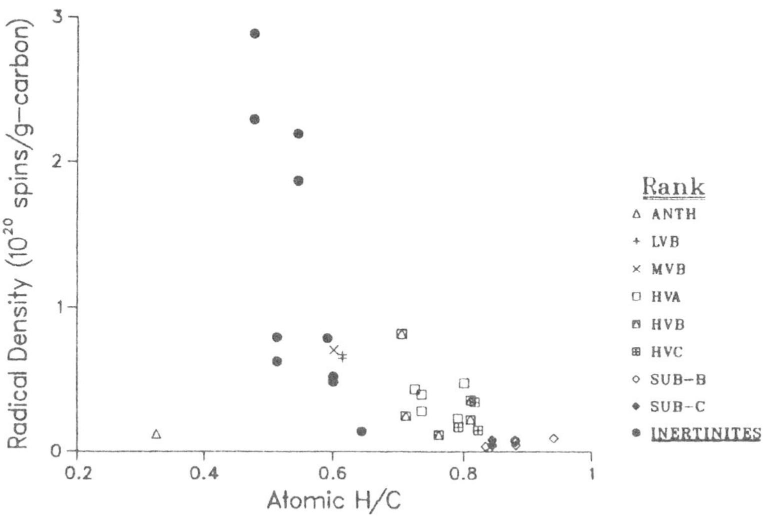

Figure 3: ESR Radical Density vs. Atomic H/C.

Figures 2 and 3 reveal some general trends for vitrinites of increasing coal rank. A greater extent of coalification leads to lower H/C values, higher carbon radical densities and broader ESR lines. These three observations are chemically self-consistent. The lower H/C of higher rank coals suggests that a greater fraction of the organic material is in a condensed, aromatic form, particularly desirable molecular hosts for carbon radicals. Furthermore, an increased radical density results in stronger magnetic interactions between the dipole moments of the carbon radicals as they become more densely packed in the material. At the low radical densities encountered in subbituminous coals, dipole interactions are relatively weak, and the observed linewidths (ΔH_{pp} ~5G) are associated with unresolved hyperfine couplings, i.e., interaction of the unpaired electron on the carbon radical with protons of the host and adjacent molecules. At higher densities, the interactions between the electrons themselves become more prominent. For a random distribution of radicals the linewidth and shape resulting from dipole interactions can be calculated explicitly.[8] The lineshape is approximately Lorentzian with a width of:

$$\delta = 2\pi^2\gamma\hbar n/3\sqrt{3} \qquad (1)$$

where γ is the gyromagnetic factor, $\hbar$ is Planck's constant and n is the spin density/cm^3. At $5x10^{19}$ spins/g-carbon, dipole contributions are on the order of 5 Gauss and should result in line broadening.

This behavior is illustrated in Figure 4, where the linewidth is plotted as a function of radical density. A linewidth increase is observed, but it is clear that there is no rigorous linear variation. Likewise in Figures 2 and 3, while linewidths and radical densities show clear trends in their variations with H/C, a considerable amount of "scatter" is inherent in the data. This reflects the fact that each of these quantities is determined by several variables, not all of which are completely accounted for here.

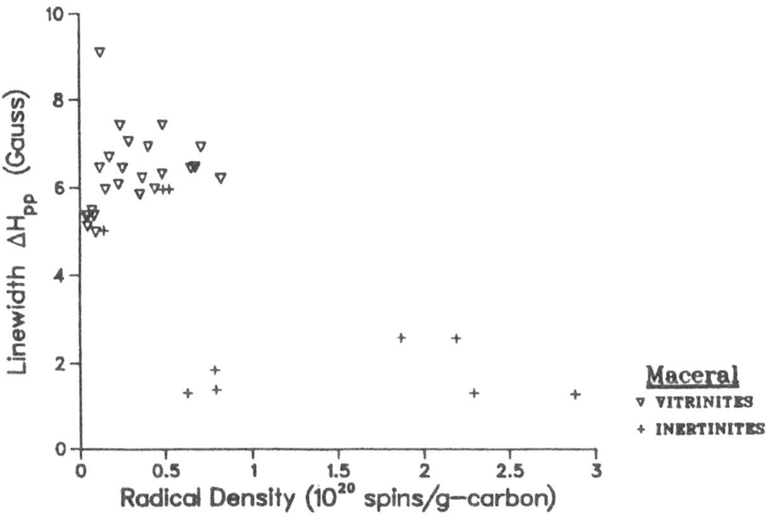

Figure 4. ESR Linewidth vs. Carbon Radical Density.

ESR PARAMETERS OF INERTINITES

The inertinite materials present a particularly difficult interpretational challenge because of their poorly defined history. Such macerals are often referred to as mineralized charcoal, and it is believed that they underwent significant chemical transformation prior to burial.(7) The present data suggest that the extent of such transformations depend on the details of this early metamorphosis. As Figure 1 indicates, the g-values for all inertinite samples lie close to 2.0028, suggesting that the radicals are associated primarily with condensed aromatics. Although significant quantities of oxygen exist in these inertinites, it must occur as heteroatoms or ethers since the oxygen does not influence the g-value. As Figures 2 and 3 suggest, there is a wide variability in inertinite behavior. Certain samples have linewidths and radical densities comparable to those of low and medium volatile bituminous coals. However, for samples that have undergone significant metamorphosis, narrow linewidths and high radical densities are observed.

The very narrow ESR linewidths observed for these latter samples cannot be explained using the arguments appropriate for vitrinites. The unresolved proton hyperfine interactions alone should produce a linewidth of ~5G, even if the H/C values, and hence the number of protons in the sample, are smaller than for vitrinites. Furthermore, there is no obvious relationship between inertinite linewidth and radical density of the type seen for vitrinites (Figure 4). The difference is even more pronounced when one examines the shape of the resonance line, as characterized by the ratio of the full width at half height of the ESR absorption ($\Delta H_{1/2}$) to ΔH_{pp}. While all vitrinites have $\Delta H/_{1/2}/\Delta H_{pp}$ values of ~1.4, as do the less transformed inertinites, there is a dramatic lineshape change for samples with lower H/C values (Figure 5). A comparison of ΔH_{pp} and

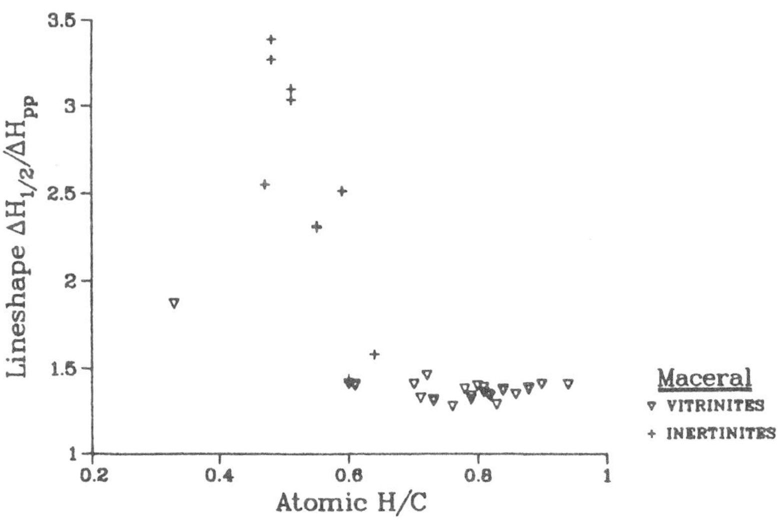

Figure 5. $\Delta H_{1/2}/\Delta H_{pp}$ vs. Atomic H/C.

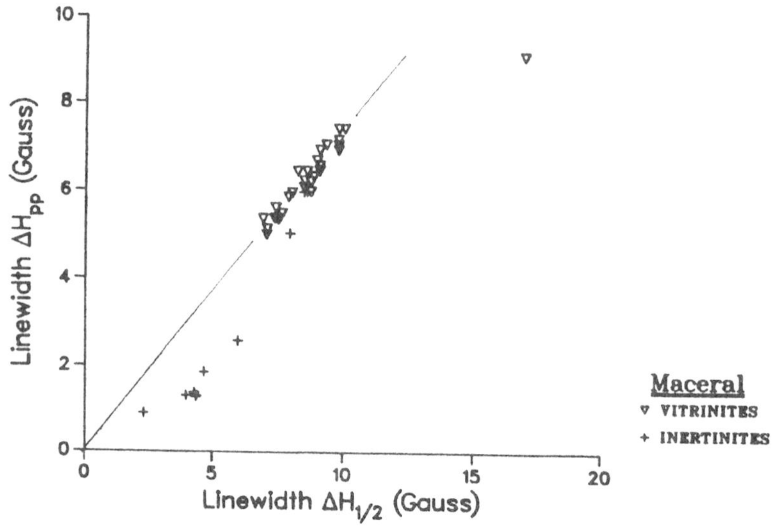

Figure 6. $\Delta H_{1/2}$ vs. ΔH_{pp}.

$\Delta H_{1/2}$ (Figure 6) shows the very uniform behavior for vitrinites and suggests that the high $\Delta H_{1/2}/\Delta H_{pp}$ ratio arises from a relatively rapid change of ΔH_{pp}.

We propose that this striking lineshape behavior arises from the onset of narrowing of the carbon radical spectra in these inertinites. Such narrowing could arise either from diffusional motion of the radical electrons from one molecular host to the next. Equivalently, exchange interactions between adjacent radicals could also cause such a narrowing phenomenon. Several lines of argument suggest that exchange narrowing may be operative in the present case. First, the linewidths of the inertinite spectra are independent of temperature from 300K-100K for all of the inertinites, which would not be expected if thermally activated diffusion were operative. Second, for substantial carrier diffusion, the ESR linewidths become asymmetric,(9) and the lines observed here remain symmetric. Finally, at spin densities on the order of 3×10^{20} spin/g carbon, carbon radicals should be in close proximity, with a mean interradical spacing of ~15Å. At this distance overlap of radical wave functions could lead to the narrowing observed. It is clear that the narrowing process, particularly if it arises from exchange overlap of radical wave functions, will not be equivalent for all radicals, since their positions relative to one another will exhibit a high degree of variability. Thus, the ESR absorption for inertinites is not expected to exhibit the Lorentzian lineshape expected for a set of equivalent exchange-narrowed spins. Evidence for this distribution should be contained in the lineshape. Since in these isolated macerals the inertinite ESR absorption is not mixed with spectra of other maceral types, the prospect for a more extensive analysis is good and is currently in progress.

CONCLUSION

The ultimate goal of magnetic resonance studies of coal should be a quantitative description of the ESR features and their relationship to the microscopic properties of the organic materials in which these radicals occur. It is important to explore the extent to which more detailed resonance analysis can be applied to these systems. These relatively homogeneous, well characterized samples provide an excellent opportunity for such research.

We wish to extend special thanks to M. L. Gorbaty for his constant encouragement and helpful comments.

REFERENCES

1. Dyrkacz, G. R., and Horowitz, E. P. 1982, Fuel 61, pp. 3-12.
2. Silbernagel, B. G., Gebhard, L. A., and Dyrkacz, G. R. 1983, Am. Chem. Soc. Fuel Division Preprints 28, pp. 118-124.
3. Silbernagel, B. G., Gebhard, L. A., and Dyrkacz, G. R. 198- ACS Symposium Series Volume on Chemistry and Characterization of Coal Macerals, to be published.
4. Petrakis, L., and Grandy, D. W., 1978, Anal. Chem 50, pp. 303-8
5. Lewis, I. C., and Singer, L. S. 1981, J. Phys. Chem 85, pp. 354-60.
6. Liotta, R. 1981, J. Am. Chem. Soc, 103, pp. 1735-42.
7. Retcofsky, H. L. 1982, in "Coal Science and Technology, vol. 1", Gorbaty, M. L., Larsen, J. W., and Wender, I., eds., Academic, New York, p. 43-82.
8. Abragam, A. 1961, "Principles of Nuclear Magnetism", Clarendon, Oxford, pp. 126-28.
9. Winter, J. 1971, "Magnetic Resonance in Metals", Clarendon, Oxford, ch. 10.

STUDY OF PARAMAGNETIC CENTERS IN NATURAL AND SOLVENT REFINED COAL BY MULTIFREQUENCY ESR, ENDOR AND ESE MODULATION SPECTROSCOPY

Shulamith Schlick* and Larry Kevan†

Departments of Chemistry, *University of Detroit, Detroit, Michigan 48221 and †University of Houston, Houston, Texas 77004

ABSTRACT

Electron spin resonance (ESR) measurements at 35 GHz have revealed two spectral components in natural bituminous coal. Comparison with results at 9 GHz together with lineshape analyses at both frequencies identifies one ESR component due to exchange narrowing. Detection of strong proton matrix ENDOR lines indicates interaction of radicals with nearby protons. ENDOR spectra are well simulated by assuming that the closest protons are situated at 0.26±0.02 nm. Differences in ENDOR signals of natural and solvent-refined coal are explained in terms of different electron and nuclear spin-lattice relaxation times. Electron spin echo measurements of natural and solvent refined coal show proton modulation with the modulation being much deeper in the solvent refined coal. For the solvent refined coal, 16 protons at 0.5 nm are indicated. These results clearly indicate more hydrogen within 0.6 nm of the paramagnetic centers in solvent refined vs. natural coal.

INTRODUCTION

Electron spin resonance (ESR) spectroscopy has been extensively used to characterize coals of different ranks and carbon content (1-3). The primary ESR spectral parameters are unpaired spin concentration, g-factors and linewidths. The variation of g-factors has been associated with carbon content and the presence of heteroatoms such as oxygen and possibly sulfur. In general, only a single line is observed and its width has been attributed to unresolved hyperfine interactions with

L. Petrakis and J. P. Fraissard (eds.), Magnetic Resonance. Introduction, Advanced Topics and Applications to Fossil Energy, 655–665.

surrounding protons or to exchange narrowing with other electron spins.

In some coal samples both broad and narrow ESR lines appear to be superimposed. The broad and narrow lines have sometimes been attributed to vitrain and fusain components. The g-values of these two components have been correlated with the carbon content of the coal, while the linewidths of vitrains have been correlated with hydrogen content (4).

The ESR work accomplished has not yet identified positively the nature or natures of the radical species in coals and how such species might be related to the structure of coal.

In this report we review results obtained by applying various electron magnetic resonance techniques to the study of a natural bituminous coal and the corresponding solvent refined coal.

Multifrequency microwave power saturation studies and linewidth analysis on the bituminous coal suggest that the narrow line detected at 35 GHz is due to an exchange-narrowed site. Analysis of the proton matrix ENDOR line reveals protons at a distance of 0.26 ± 0.02 nm from the unpaired electron, with an isotropic hyperfine constant of 1.3 ± 0.3 MHz. Electron Spin Echo (ESE) modulation patterns from natural and solvent refined bituminous coal have been observed and analyzed in order to obtain structural information on the proton environment around the unpaired electron and on the effect of the solvent refining process.

EXPERIMENTAL

The samples were Pittsburgh #8 bituminous coal (feed coal for solvent refining) and a corresponding solvent refined sample, (SRC), from a coal conversion pilot plant at Wilsonville, Alabama, both kindly supplied to us by Dr. L.T. Taylor of the Virginia Polytechnic Institute and State University. Analysis of the feed coal gave 86.0 wt. % C, 5.0 wt. % H and 2.2 wt. % N while for the solvent refined coal the analysis is 89.5 wt. % C, 6.8 wt. % H and 0.8 wt. % N. In solvent refining pulverized natural coal is mixed with a high boiling point aromatic solvent, hydrogen is added and dissolution occurs at $\sim 450^{o}$C. The results of several analytical techniques to partially characterize the solvent refined coal have been reported (5). Of interest to us here is that solvent refining lowers the radical concentration and increases the hydrogen concentration relative to the natural feed coal (5,6).

ESR spectra were taken at 9 GHz with a Varian V-4500 spectrometer and 6 kHz magnetic field modulation. At 35 GHz a Varian E-line spectrometer was used with 1 kHz magnetic field modulation. The microwave power was measured with a Hewlett Packard HP431C power meter. Below ambient temperature 35 GHz spectra were taken by immersing the cavity in a flow dewar cooled by cold helium gas. ENDOR spectra were recorded on a Varian V-4500 ESR spectrometer interfaced with a Varian E-700 ENDOR unit. The radio-frequency (rf) field is pulsed at 6 kHz and the magnetic field is modulated at low frequency (40 Hz). Electron spin echo spectra were recorded at 4.2 K using a home-built spectrometer (7). Samples had been stored in capped vials at room temperature and subsequently were evacuated to a pressure of 10^{-6} torr at room temperature and sealed in quartz tubes for the spectral measurements.

Additional experimental details have been published (8,9).

RESULTS AND DISCUSSION

1. Multifrequency ESR Studies

The ESR spectrum of powdered Pittsburgh #8 feed coal at 9 GHz (X-band) and ambient temperature consists of a single structureless line with a peak-to-peak derivative width (ΔH_{pp}) of 6.1 G at $\leq$1 mW microwave power which decreases to 5.5 G at 200 mW. The lineshape is essentially Lorentzian over the entire microwave power range.

At 35 GHz (Q-band) at ambient temperature and below ~ 0.1 mW (30 db) microwave power, as shown in Fig. 1, two superimposed ESR lines are observed, with H_{pp} values of ~ 2.5 and ~ 5 G. As the microwave power increases, the broad ESR line becomes less prominent and the narrow ESR line predominates above ~ 1 mW (20 db). However, even at the highest microwave power studies of ~ 100 mW (0 db) it can be seen that the observed line is broader than a single Lorentzian line. This implies a superposition of two lines as is clearly demonstrated by the lineshape at low microwave power.

Microwave power saturation curves were measured at 35 GHz by monitoring the peak-to-peak derivative height and are shown in Fig. 2 at different temperatures. The slight saturation is most probably indicative of saturation of the broad component.

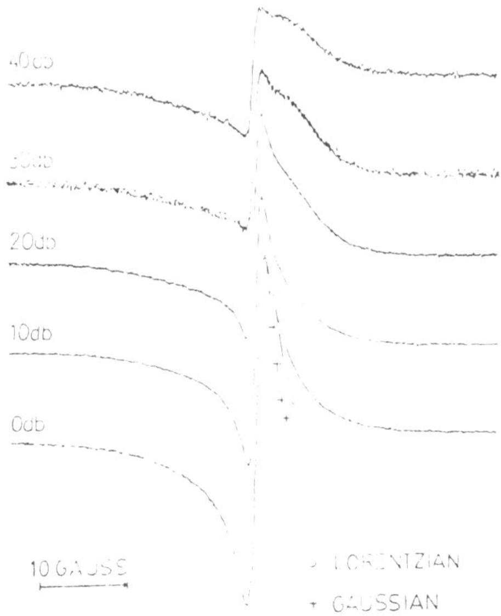

Fig. 1. Room temperature 35 GHz ESR spectra of Pittsburgh #8 feed coal at different power levels. 0 db corresponds to ~100 mW. Lorentzian and Gaussian lineshapes were calculated for lines having the experimental ΔH_{pp} and I_{max}.

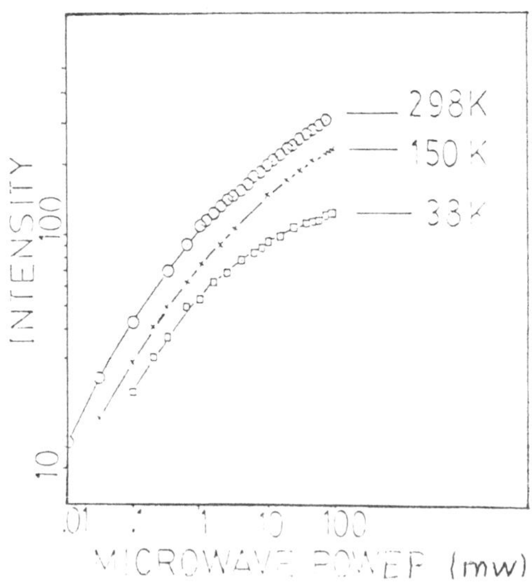

Fig. 2. Microwave power saturation curves at 35 GHz of Pittsburgh #8 feed coal at several temperatures.

Fig. 3 gives a saturation curve at 298 K for the feed coal at 9 GHz which also shows only slight saturation. However, SRC shows strong saturation at 298 K at 9 GHz and is easily distinguished from the feed coal by this criterion. The easier saturation of solvent refined coal is consistent with a decreased local radical concentration by the solvent refining process.

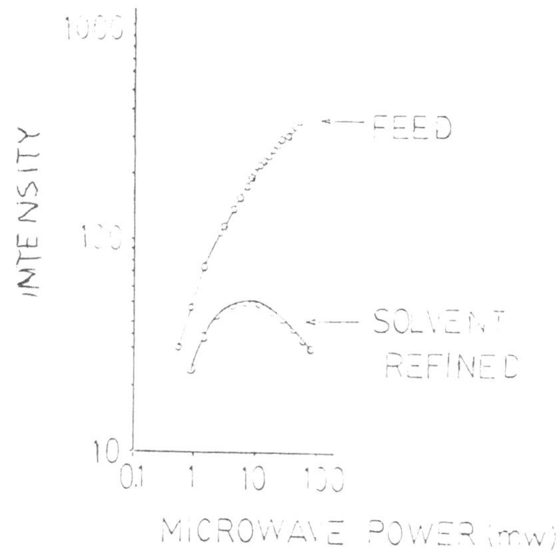

Fig. 3. Microwave power saturation curves at 9 GHz of Pittsburgh #8 feed and solvent refined coals at room temperature.

Results presented in Figs. 1-3 can be interpreted as follows. Saturation curves from the feed coal at 9 and 35 GHz are very similar. Spectra at 35 GHz, however, clearly show the presence of two components with linewidths differing by at least a factor of two. The narrow component at 35 GHz is interpreted as arising from an exchange narrowed species due to a high local radical concentration. This is supported by the Lorentzian lineshape, by the lack of much saturation and by the insensitivity of the limited saturation to temperature. The broad component at 35 GHz, whose lineshape is not well defined, is most probably due to unresolved hyperfine interaction with neighboring protons. In this case its lineshape should be Gaussian and its linewidth should be little changed by the microwave frequency of observation.

For systems where both dipolar and exchange interactions among electron spins are present, the observed linewidth ΔH depends on the relative magnitudes of the dipolar field H_d, the exchange field H_e and the observation field H_o, as seen in eq. (1).

$$\Delta H = \left(\frac{H_d^2}{H_e}\right)\left(1 + \frac{5}{3}\exp\left(-\frac{1}{2}\left(\frac{H_o}{H_e}\right)^2\right) + \frac{2}{3}\exp\left(-2\left(\frac{H_o}{H_e}\right)^2\right)\right) \quad (1)$$

This expression was deduced for isotropic g, $S = \frac{1}{2}$, a cubic lattice of unpaired spins and a Gaussian electron spin correlation function (10). Eq. (1) predicts that the linewidth ΔH is independent of the observation field H_o if $H_e \ll H_o$ or $H_e \gg H_o$. Results presented here as well as in previous studies (1,4) seem to indicate that the linewidth of the exchange narrowed signal is about the same at 9 GHz and 35 GHz. This result implies that $H_e \leq 2$ GHz or $H_e \gg 100$ GHz.

Practically all previous ESR studies on coals have been carried out at 9 GHz. Typically the lineshape has not been found to be purely Gaussian or purely Lorentzian although the trend is toward Lorentzian for higher rank coals. Based on our results at 35 GHz we suggest that there may generally be two superimposed components in most coals corresponding to two different environments with respect to radical concentration. If an exchange narrowed component can be identified at two sufficiently different microwave frequencies, then it should be possible to determine the exchange frequency by the method outlined here.

2. ENDOR Studies

Strong matrix ENDOR signals around 14 MHz were observed for SRC at room temperature, thus providing first spectroscopic evidence for the presence of protons interacting with the unpaired electron. No ENDOR signals are detected in natural coal at room temperature. ENDOR is seen, however, between 110-160 K, with less intensity than for SRC. The strongest signals occur near 150 K.

The matrix ENDOR line is interpreted by assuming predominantly dipolar interaction between the unpaired electron and surrounding protons (11) and contains information about the extent of delocalization of the unpaired electron. This information can be extracted by simulation of the matrix ENDOR line within the constraints of a model for the ENDOR response.

Interpretation of the ENDOR signal from SRC is based on the Hochman et al formulation of the ENDOR response (12) as modified by Narayana et al (13). This theoretical model explicitly introduces the dependence of the ENDOR response on the microwave (H_1) and radio frequency (H_2) magnetic fields, as well as on all relaxation paths operative in the interacting system of electron and nuclear spins.

Matrix ENDOR lines can be simulated from this model using experimentally measured or estimated values of H_1, H_2 and relaxation parameters, and compared with experimental spectra in order to determine r_{min}, the distance of the closest nuclei to the unpaired electron and a_{iso}, the average isotropic hyperfine constant of the closest nuclei. In addition, the ENDOR response vs. H_1 and H_2 can be calculated and compared with experiment. Simulation of the matrix ENDOR line at the microwave power which gives the maximum ENDOR response (4 mW) leads to the good agreement with experiment shown in Fig. 4, for r_{min} = 0.26 nm and a_{iso} = 1.3 MHz.

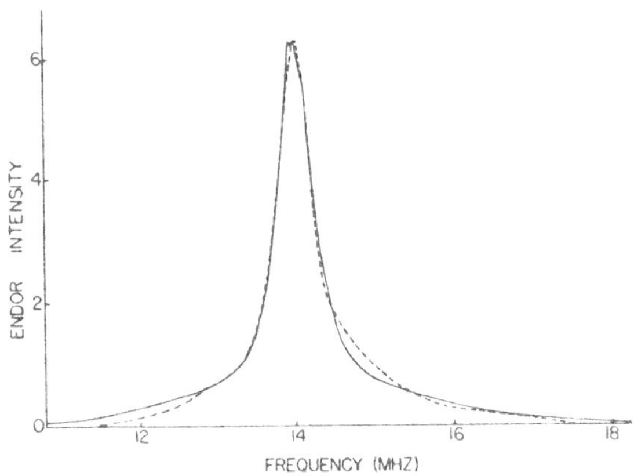

Fig. 4. Experimental matrix ENDOR line at 4 mW for SRC at room temperature (broken line) and simulated line (full line).

For aromatic ring protons in anthracene or tetracene a_{iso} is at least 3 MHz. The observed isotropic hf constant of 1.3 MHz rules out interaction of the unpaired electron with ring protons in SRC.

Different ENDOR intensities for natural and SRC are rationalized in terms of different electron spin-lattice relaxation times T_{1e}, the shorter T_{1e} being assigned to the natural coal. This result is in agreement with differences in saturation properties reported in the previous section.

Additional ENDOR studies on coals and coal-like materials have been published (14-16).

3. ESE Modulation Studies

The analysis of ESE modulation has been described in detail (17). Echoes are generated in pulsed ESR experiments. The 2-pulse sequence, 90-τ-180-τ-echo, has been used in coal studies. The 90° pulse flips the spins from the Z direction into the XY plane. For a time τ the spins precess in the XY plane and are subject to time dependent magnetic interactions. The 180° pulse flips the spins to the opposite direction in the XY plane so that the spins then precess together and refocus at a time 2τ after the original pulse to produce a burst of microwave energy called an echo. This echo intensity is then measured as a function of the time between the pulses to generate an echo decay envelope.

The echo decay envelope is a product of two functions, a relaxation function and a modulation function. The relaxation function decreases monotonically with time and carries information about all of the spin relaxation processes that cause the electron spins to dephase. The modulation function carries structural information with which we are primarily concerned. The general analysis procedure is to analyze the spin echo modulation pattern in terms of a number of equivalent nuclei n, located at a distance r from the paramagnetic species which have a small overlap of the unpaired electron wavefunction to produce a small isotropic hyperfine coupling, a_{iso}. Usually the physically reasonable assumption that n is integral is made. For reasonably good quality data it is then typically possible to uniquely determine the value of n to the nearest integer up to about $n = 10$. The parameter r can be determined to about ± 0.01 nm and a_{iso} to about $\pm 15\%$.

Strong electron spin echoes are obtained from both the natural and solvent refined coals at ambient temperature, 77 K and 4.2 K. Fig. 5 shows the experimental 2-pulse electron spin echo for SRC at 4.2 K. Clear proton modulation is observed with a repetition pattern of about 70 ns at 9 GHz. This modulation is identified as due to protons by its modulation period. Also shown in Fig. 5 is the calculated modulation pattern for the best fit parameters for a single shell model of 16 ± 2 protons at 0.5 ± 0.03 nm and zero isotropic hyperfine coupling. It is possible that more complex multishell models may fit as well. The regularity of the modulation pattern is also indicative of little or no isotropic hyperfine coupling (17). This suggests that the unpaired electron is not strongly delocalized over, for example, an aromatic system, in agreement with ENDOR results.

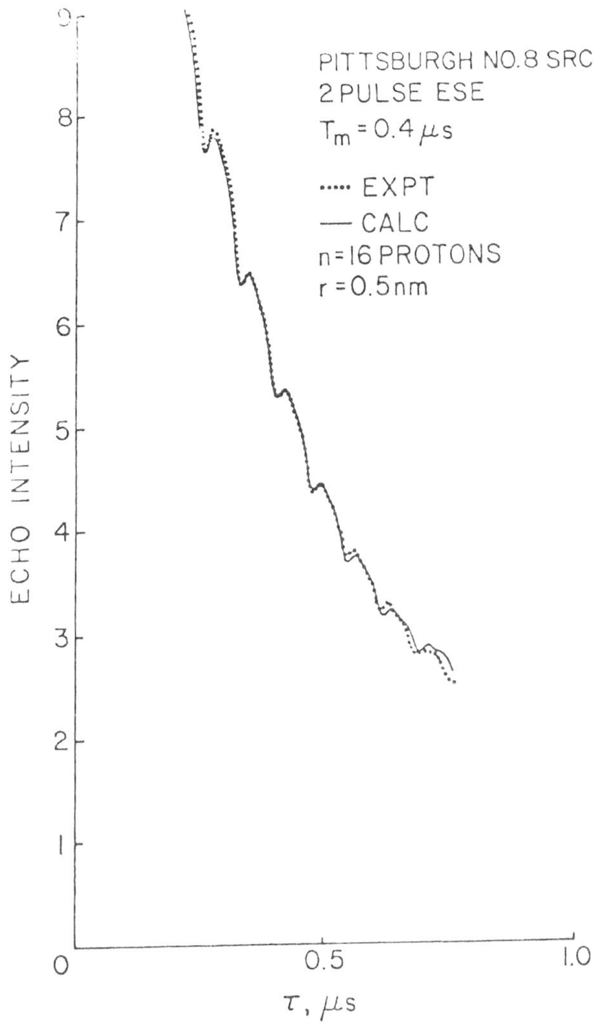

Fig. 5. Experimental and calculated two-pulse electron spin echo proton modulation in solvent refined coal at 4.2 K. The decay function is $g(\tau)=\exp(2.53 - 0.60\tau - 5.38\tau^2 + 4.66\tau^3)$.

The electron spin echo decay curve for the natural coal shows definite proton modulation but is very shallow. This is qualitatively similar to the weak proton modulation recently reported in a bituminous coal by Das et al. (18). Qualitatively this result indicates fewer interacting protons or a greater proton interaction distance or both compared to the solvent refined coal. The phase memory time of the echo decay (e^{-1} time) is 0.4 μs for both the natural and solvent refined coals.

These electron spin echo results clearly show the existence of hydrogen within 0.6 nm of the paramagnetic center in coal. In the solvent refined coal they provide the first structural model on both the number and distance of the surrounding hydrogen. The difference between natural and solvent refined coal indicates that there is probably a higher hydrogen concentration near the paramagnetic centers in the solvent refined coal. This is certainly consistent with the addition of hydrogen during the solvent refining process for coal. The large number of protons within 0.6 nm from the radical site might seem at odds with some models of coal macerals (4). However, our result is observed for the solvent refined coal, where the increased hydrogen content can be correlated with reduced carbon aromaticity. This correlation is consistent with the present ESE results on solvent refined and feed coal. Additional support for the existence of regions with high proton density in solvent refined coals is given by proton spin lattice relaxation studies (19). Variation of the proton T_1 with frequency indicates spin diffusion effects which are usually detected in spin systems with high proton density.

There is a contrast between the hydrogen seen by the ESE results on solvent refined coal with that seen by ENDOR. ENDOR detected some hydrogens at 0.26 nm with a significant isotropic hyperfine coupling. It is possible that these closest hydrogens do not produce ESE modulation because the isotropic coupling is too large. If the ESE microwave field is not large enough to excite both the allowed and forbidden ESR transitions, modulation is not generated (17). Under our experimental conditions we do not expect this, however, so the discrepancy is not completely explained at present.

REFERENCES

1. Retcofsky, H.L., Stark, J.M. and Friedel, R.A. 1968, Anal. Chem. 40, p. 1699.
2. Retcofsky, H.L., Thompson, G.P., Raymond, R. and Friedel, R.A. 1975, Fuel 54, p. 126.
3. Retcofsky, H.L., Thompson, G.P., Hough, M. and Friedel, R.A. 1978, In "Organic Chemistry of Coal", Amer. Chem. Soc. Symposium Series No. 71, Larsen, J.E., editor, p. 142.
4. Retcofsky, H.L. 1982, In "Coal Science", vol. 1, Gorbaty, M.L., Larsen, J.W. and Wender, I., editors, Academic Press, New York.
5. Wooton, D.L., Dorn, H.C., Taylor, L.T. and Coleman, W.M. 1976,

Fuel 55, p. 224.
6. Taylor, L.T., private communication.
7. Ichikawa, T., Kevan, L. and Narayana, P.A. 1980, J. Phys. Chem. 83, p. 3378.
8. Schlick, S., Narayana, P.A. and Kevan, L. 1978, J. Amer. Chem. Soc. 100, p. 322.
9. Schlick, S., Narayana, M. and Kevan, L. 1983, Fuel, 62, p. 0000.
10. Kubo, R. and Tomita, K. 1954, J. Phys. Soc. Japan, 9, p. 888.
11. Vedrine, J.C., Hyde, J.S. and Leniart, D.S. 1972, J. Phys. Chem. 76, p. 2087.
12. Hochman, V.L., Zevin, V.Ya. and Shanina, B.D. 1968, Fiz. Tverd. Tela 10, p. 337; 1968, Sov. Phys.-Solid State (Engl. Trans.) 10, p. 269.
13. Narayana, P.A., Bowman, M.K., Kevan, L. and Schwartz, R.N. 1966, J. Chem. Phys. 67, p. 1990.
14. Decaillot, M. and Uebersfeld, J. 1967, C.R. Acad. Sc. Paris 265, p. 155.
15. Miyagawa, I. and Alexander, C. Jr. 1979, Nature 278, p. 40.
16. Clarkson, R.B. 1979, In "Molecular Motion in Polymers by ESR", Boyer, R.F. and Keinath, S.E., editors, Harwood Press, New York, p. 263.
17. Kevan, L. 1979, In "Time Domain Electron Spin Resonance", Kevan, L. and Schwartz, R.N., editors, Wiley-Interscience, New York, chapter 8.
18. Das, U., Doetschman, D.C., Jones, P.M. and Lee, M. 1983, Fuel 62, p. 285.
19. Yokono, T., Miyazawa, K., Sanoda, Y. and Marsh, H. 1979, Fuel 58, p. 896.

EPR STUDY OF EXCITED TRIPLET STATES AND INTERACTIONS BETWEEN THE MOLECULAR PHASES IN COAL

Stanisław Duber* and Andrzej B. Więckowski

Institute of Molecular Physics, Polish Academy of Sciences, ul.Smoluchowskiego 17, 60-179 Poznań, Poland
* Department of Petroleum and Coal Chemistry, Polish Academy of Sciences, ul. 1 Maja 62, 44-100 Gliwice, Poland

ABSTRACT

EPR studies are performed for coal and its products of extraction with tetrahydrofurane /THF/, pyridine /Py/ and ethylenediamine /EtDA/. The spectra of the coal and the residues after extraction consisted of two lines, the one broad and the other narrow. Their intensities depend differently on the temperature of measurement. The system of paramagnetic centers related with the narrow line obeys the Curie law /S = 1/2 /, whereas that related with the broad line does not, and consists of free radicals /S = 1/2 / and of centers with singlet ground state /S = 0/ and lowest excited triplet state /S = 1/. Quenching of paramagnetic centers during extraction is found to be related with the destruction of intermolecular bonds of the electron donor-acceptor type between the two molecular phases in coal.

INTRODUCTION

The nature of the paramagnetic centers present in bituminous coal has been under investigation ever since coal was shown to give an EPR signal. The signal is generally thought to be due to organic free radicals.

Elofson and Schulz /1/, and Schwager and Yen /2/ put forward the as yet unconfirmed hypothesis that the paramagnetic centers of

L. Petrakis and J. P. Fraissard (eds.), Magnetic Resonance. Introduction, Advanced Topics and Applications to Fossil Energy, 667–676.

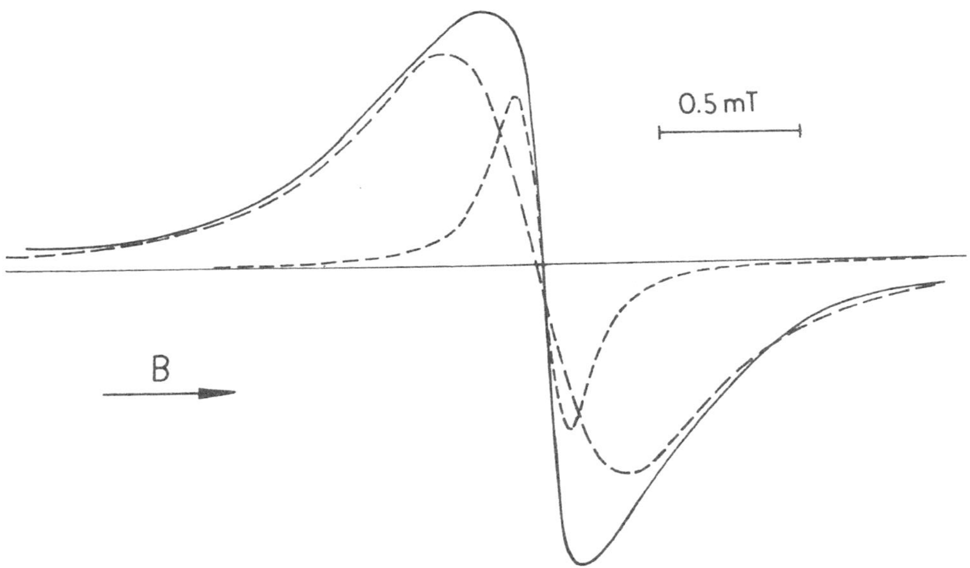

Figure 1. EPR spectrum of raw coal resolved into two components /continuous - experimental curve, dashed - narrow line and broad line/

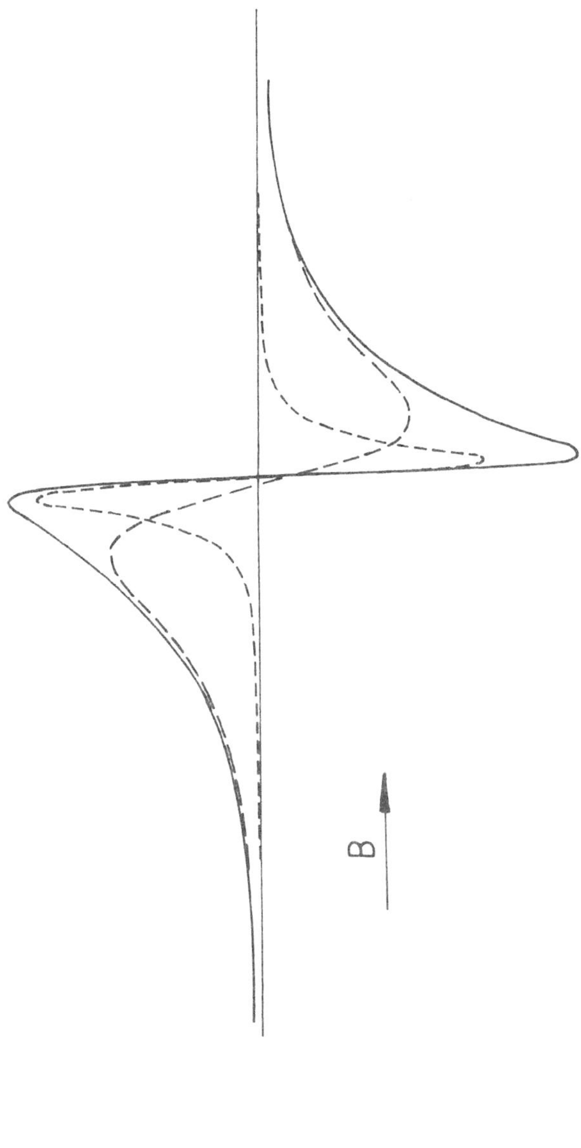

Figure 2. EPR spectrum of the residue on extraction with EtDA /notation as in Fig. 1/.

coal originate in electron donor-acceptor /EDA/ complexes. Many complexes are known to have a singlet ground state /S = 0/ and a triplet as their first excited state /S = 1/. If the singlet--triplet /S-T/ excitation energy J is comparable with the thermal energy kT, this should lead to a contribution of the thermally excited paramagnetic centers to the total EPR line intensity. Yen and Young /3/, when studying the dependence of the EPR line intensity on the temperature of measurement, have shown that excited triplet states exist in asphaltenes of petroleum origin. Retcofsky et al. /4, 5/, investigating coal under analogous conditions, found no excited triplet states.

We start from the following general assumptions concerning the structure of coal:

Coal consists of macromolecules, connected by cross bonds forming a three-dimensional network /6/. We will be referring to this part of the substance of coal as the macromolecular /MM/ phase.

Beside the macromolecules, the organic part of coal consists of a low-molecular phase. It undergoes extraction by strongly electron-donor solvents /7/. We refer to this as the molecular /M/ phase. The M phase is situated in pores and on the boundaries of the MM phase.

A. Marzec /8/, in her study of the interaction between solvents and coal, has shown the two phases to be connected by donor--acceptor bonds, leading to the formation of complexes.

The EPR spectrum of coal consists of two lines, the one broad and the other narrow. Our analysis of the relaxation times: spin--lattice T_1 and spin-spin T_2 /9/ led us to the conclusion that the paramagnetic centers related with the narrow line are attached to the latticed macromolecular phase MM, whereas the centers related with the broad line are distributed throughout the molecular phase M.

EXPERIMENTAL

The coal under study came from the mine Jankowice in Upper Silesia, Poland /7/. Extraction was carried out at room temperature. As solvents, use was made of tetrahydrofurane /THF/, pyridine /Py/ and ethylenediamine /EtDA/. The coal and the extraction residues were placed in quartz tubes and degassed during 5 h at about 1 mPa, and the tubes were vacuum sealed. The obtained X-band EPR spectrum was resolved by computer analysis into its two components /Fig. 1

and 2/, either of which was assumed lorentzian. The method had been used by us previously /9/ to analyze EPR spectra of natural coals. As spin concentration standard we applied a ruby crystal, situated outside the quartz dewar containing the sample.

RESULTS AND DISCUSSION

The EPR spectrum of the coal under investigation consists of two lines with different line widths and with slightly different g-factors /Fig. 1/. Similar spectra were found in the residues obtained after extraction with the organic solvents /Fig. 2/. The lower graph of Fig. 3 shows that the system of paramagnetic centers in dublet state related with the narrow line in coal behaves according to the Curie law /D-type centers/. The system of centers related with the broad line /upper graph of Fig. 3, and Fig. 4/ is a mixture of D-type centers and thermally activated cemters with singlet-triplet excitation /S-T type centers/. The intensity I of the broad line was found to vary with temperature according to the formula:

$$I.T = c + b\left(\exp(J/kT) + 3\right)^{-1} \qquad /1/$$

where J is the singlet-triplet excitation energy. On performing a Yen-Young /3/ fitting procedure the values J were determined /Table 1/. The continuous graphs of Fig. 5 show the variations of the product I.T calculated from Eq. /1/ for coal and for the residues after extraction. Clearly, our experimental points are in agreement with the calculated values.

During slow extraction with the organic solvents, some of the paramagnetic centers are quenched and the efficiency of quenching is the greater the higher is the extraction yield achieved with the solvent. The line widths of the components also change /Table 1, Figs 1 and 2/. Considering the non-destructive conditions of extraction /room temperature/ under which the solvents used do not destroy the intramolecular bonds but only affect the weak intermolecular electron donor-acceptor bonds, we are led to the conclusion that some of the paramagnetic centers in coal are stabilized by intermolecular interactions between the two molecular phases MM and M.

CONCLUSIONS

We have proved the existence of thermally excited triplet states /S-T type centers/ in coal by measuring the temperature dependence of the intensity of the two EPR line components. The broad line and the narrow line show a different behavior when the temperature is made to vary. The singlet-triplet excitation

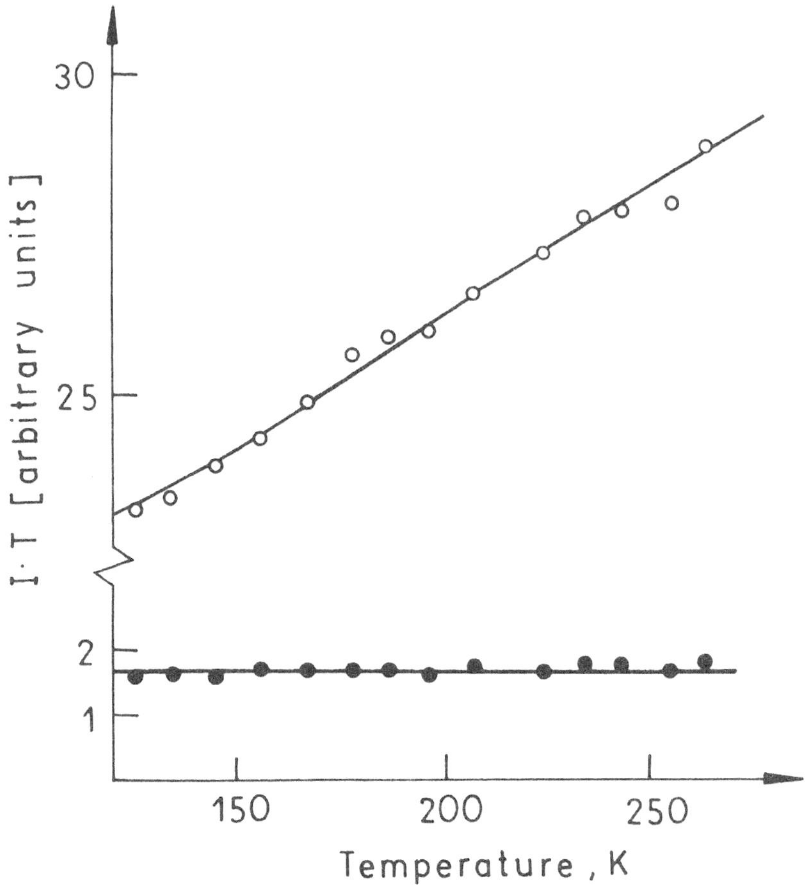

Figure 3. The product I.T /with I - the EPR line intensity and T - the temperature of measurement/ vs. temperature T for raw coal /upper - for the broad line; lower - for the narrow line/. The system of paramagnetic centers related with the narrow EPR line fulfills the Curie law. The continuous graph for the broad EPR line was calculated applying Eq. /1/ on insertion of the value J = 47 meV.

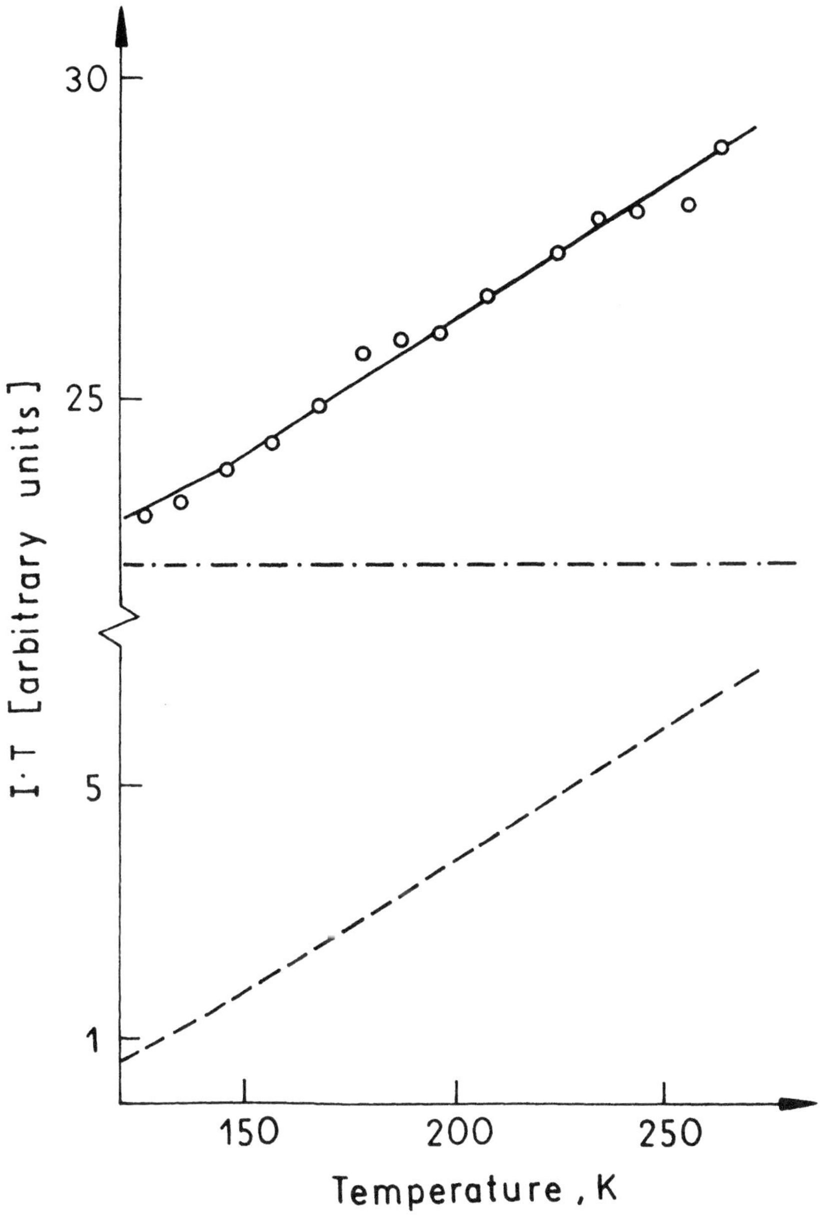

Figure 4. The product I.T of the broad line vs. temperature T for raw coal resolved into two components for D-type centers with S = 1/2 /upper line/ and for S-T type centers with S = 1 /lower line/.

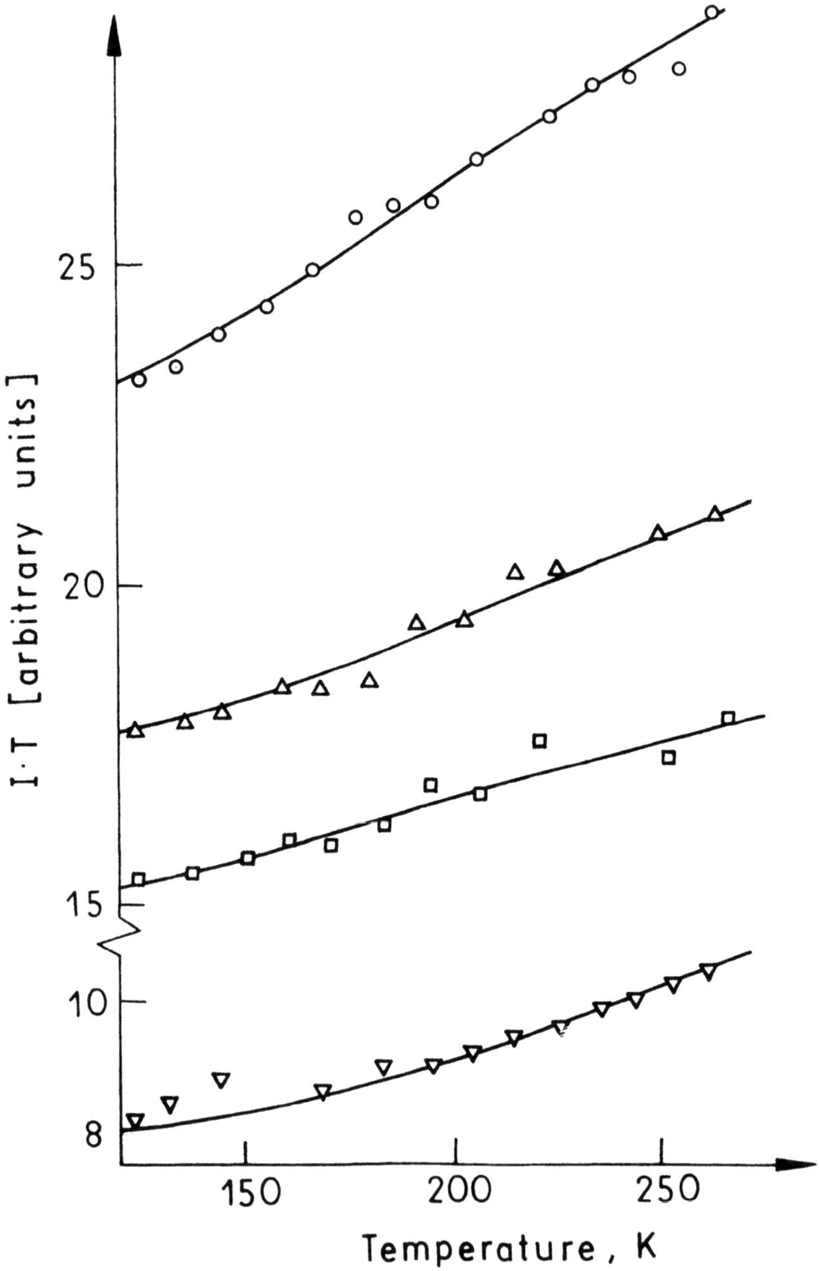

Figure 5. The product I.T of the broad line vs. temperature T for raw coal and for the residues on extraction with THF, Py and EtDA, respectively.

Table 1. Line width $2\Delta B_{1s}$ /mT/, g-factor, spin concentration N /10^{18} spin g^{-1}/, singlet-triplet excitation energy J /in meV, cm^{-1} and K, respectively/ and fraction of singlet-triplet states $f_{ST} = N_{ST}/(N_D + N_{ST})$.

Sample /between parantheses the extraction yield fraction is given/	Narrow line			Broad line				
	$2\Delta B_{1s}$	g	N	$2\Delta B_{1s}$	g	N	J	f_{ST}
Coal from Jankowice mine	0.17	2.0033	0.5	0.61	2.0035	8.7	47 380 550	0.44
Residue after extraction with THF /0.080/	0.18	2.0033	0.5	0.56	2.0034	6.9	53 430 620	0.41
Residue after extraction with Py /0.125/	0.16	2.0032	0.4	0.53	2.0034	6.7	46 370 530	0.33
Residue after extraction with EtDA /0.224/	0.12	2.0031	0.4	0.49	2.0033	4.2	64 510 740	0.62

energy J is comparable with the thermal energy kT of lattice vibrations at room temperature.

Our experimental results can be explained on assuming the model of molecular phases in coal proposed by Larsen /6/ and Marzec /8/. We have found that in the process of slow extraction part of the donor-acceptor type complexes, existing as a result of intermolecular interaction between the molecular phase M and the macromolecular phase MM, are destroyed. These complexes are responsible for the presence of the S-T type centers, some of which are quenched during extraction. The quenching of D-type centers during extraction is due to their recombination.

ACKNOWLEDGEMENT

The authors wish to thank Professor Jan Stankowski of Poznań and Professor Anna Marzec of Gliwice for their valuable discussions and interest throughout the present work.

REFERENCES

1. Elofson, R.M. and Schulz, K.F. 1967, Am.Chem.Soc. Div. Fuel Chemistry 11, p. 513.
2. Schwager, I. and Yen, T.F. 1976, Am.Chem.Soc. Div. Fuel Chemistry 21, p. 199.
3. Yen, T.F. and Young, D.K. 1973, Carbon 11, p. 33.
4. Retcofsky, H.L., Thompson, G.P., Hough, M. and Friedel, R.A., in "Organic Chemistry of Coal", Ed. J.W. Larsen, Am.Chem.Soc. Symposium Ser. No. 71, Washington 1978, p. 142.
5. Retcofsky, H.L., Hough, M.R., Maquire, M.M. and Clarkson, R.B., in "Coal Structure", Ed. M.L. Gorbaty and K. Ouchi, Am.Chem.Soc. Adv. in Chemistry Ser. No. 192, Washington 1981, p. 37.
6. Larsen, J.W. and Kovac, J., in "Organic Chemistry of Coal", Ed. J.W. Larsen, Am.Chem.Soc. Symposium Ser. No. 71, Washington 1978, p. 36.
7. Bodzek, D. and Marzec, A. 1981, Fuel 60, pp. 47-51.
8. Marzec, A., Juzwa, M., Betlej, K. and Sobkowiak, M. 1979, Fuel Processing Technology 2, p. 35.
9. Duber, S. and Więckowski, A.B. 1982, Fuel 61, pp. 431-436.

CHEMICAL STRUCTURE AND HYDROLIQUEFACTION REACTIVITY OF COAL

T.Yoshida, Y.Hasegawa, and Y.Maekawa

Government Industrial Development Laboratory, Hokkaido.
2-17 Tsukisamu-Hiagshi, Toyohira, Sapporo 061-01, Japan

Using Yallourn and five Japanese coals, chemical structures and their relation to the reactivity in hydroliquefaction and further to the structure of coal-derived oil were investigated. Carbon and oxygen distributions, and structural indices (fa and Haru/Car) for coal were obtained from CP/MAS ^{13}C NMR spectra. These informations led to propose average chemical structure for each coal and further suggested that the cleavage of ether linkages in the depolymerization of coal contributed to the formation of asphaltene or preasphaltene alone. Furthermore, the reactivity characteristics of coals were compared with their chemical structures and good correlation was found between oil yield and CH_2 content of coal. The structure of coal-derived oil from hydroliquefaction was also related closely with CH_2 content of coal.

The reactivity of coal in hydroliquefaction must be primarily influenced by the chemical structure. In the past, the reactivity of coal has been evaluated in terms of conversion obtained by the solvent extraction method, and compared with coal parameters (1,2) such as C%, atomic H/C ratio, volatile matter (V.M.), and volatile carbon (V.C.). However, these parameters do not directly reflect chemical structure of coal, and therefore do not necessarily correlate well with the reactivity. At present, coal is considered to contain relatively small polycyclic aromatic and hydroaromatic rings as structural units, which are combined each other by ether linkages or methylene bridges to form a macromolecular structure (3,4). However, these structural informations have been obtained from soluble components of coal and by no means contribute to the informations on bulk structure. In recent years, a new technique

L. Petrakis and J. P. Fraissard (eds.), Magnetic Resonance. Introduction, Advanced Topics and Applications to Fossil Energy, 677–688.

called cross-polarization (CP)/magic angle spinning (MAS) ^{13}C nuclear magnetic resonance (NMR)(5) has enabled direct determination of carbon distribution in coal. A number of reports have already appeared which applied this technique to the structural studies of coal(6-9).

In this work, we applied CP/MAS ^{13}C NMR technique to study structural characterisitics of various kinds of coals and these characteristics were correlated with the reactivities of coals in hydroliquefaction. The structural correlation between coal and oil was also discussed.

Experimental

Briquettes of Yallourn brown coal (from Australia) and five kinds of Japanese coals were used. The ultimate and proximate analyses of these coals are summarized in Table 1. Coals were pulverized to pass through a 100 mesh screen, and dried in vacuum at 107°C for 12 h before being used in autoclave experiments. The hydroliquefaction of coal was carried out at 400°C for 30 min under an initial hydrogen pressure of 9.81 MPa in a rotating 10 L autoclave. An well-shaken mixture of 200g of coal with 20g of Adkins catalyst was charged into the autoclave together with 20 stainless steel balls. In order to facilitate the analysis of reaction products, no vehicle oil was used. Non-gaseous products were separated into oil (soluble in hexane, HS), asphaltene (insoluble in hexane but soluble in benzene, BS), preasphaltene (insoluble in benzene but soluble pyridine, PS), and residue (insoluble in pyridine, PI) by the solvent extraction method. The oil was further separated into five fractions, namely saturates (Fr-P), monoaromatics (Fr-M), diaromatics (Fr-D), tri- and tetraaromatics (Fr-T), and poly- and polar aromatic compounds (Fr-PP) with a liquid chromatograph equipped with double-layered packed column of silica and alumina(10). CP/MAS ^{13}C NMR spectra of coal and solid products have been reported elsewhere(9,11), including conditions

Table 1 Ultimate and proximate analyses of coals

	Proximate analysis (wt%)			Ultimate analysis (wt%, dmmf coal basis)			Atomic ratio	
Coal	Volatile matter	Fixed carbon	LTA[a]	C	H	O	H/C	O/C
Yallourn	43.2	37.6	1.7	66.2	3.9	29.4	0.701	0.333
Soya-koishi	33.6	35.3	22.9	72.0	5.5	21.1	0.909	0.220
Taiheiyo	49.6	33.8	13.8	76.6	6.2	15.9	0.964	0.156
Horonai	47.9	46.1	7.0	81.1	6.5	10.9	0.954	0.101
Akabira	41.7	51.2	5.7	83.2	6.3	8.6	0.901	0.078
Shin-yubari	34.7	56.7	9.6	87.4	6.2	4.6	0.845	0.039

[a]LTA, low-temperature ash

of measurements, and assignments of various types of carbon. The group-type analysis of oil(12) was carried out based on the field-ionization (FI) mass spectra of fractions P, M, D, and T.

Results and discussion

Chemical structure of coal

CP/MAS ^{13}C NMR spectra of of coals are shown in Figure 1 and carbon distributions calculated from their spectra are summarized in Table 2. The contents of polar groups, especially carboxyl and carbonyl groups, decrease as the carbon content of coal increases. Aromatic carbon is least abundant in Soya-koishi coal and increases

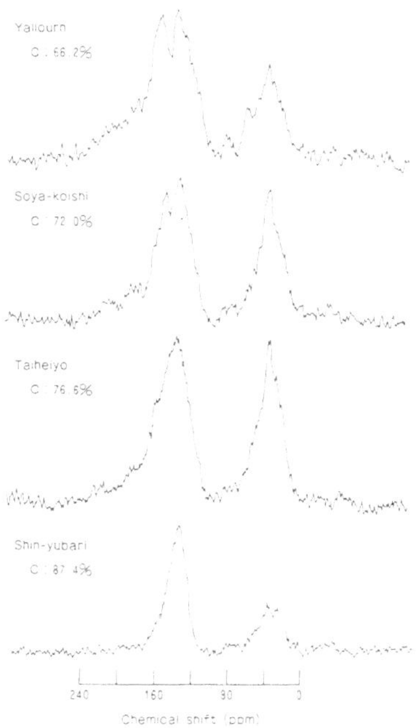

Fig.1 CP/MAS ^{13}C NMR spectra of various kinds of coal

Table 2 Distribution of carbon atoms in coals

Coal	Distribution (%)				
	Polar	Aroma	CH_2	α-CH_3	end-CH_3
Yallourn	22	57	12	5	4
Soya-koishi	18	51	19	7	5
Taiheiyo	13	54	21	8	4
Horonai	5	61	22	8	4
Akabira	6	66	15	7	5
Shin-yubari	6	71	12	6	5

with increasing carbon content in other coals. Yallourn coal contains relatively large amount of aromatic carbon despite the smallest carbon content among the coals studied. This unexpected observation may be interpreted as the result of dehydrogenation of naphthenic structural portions by heat treatment at the time of briquette preparation process, which leads to the increase in aromatic carbon. On the other hand, aliphatic carbons are most abundant in subbituminous coals. Taiheiyo and Horonai coals showed the highest content in CH_2 groups, whereas all coals studied except for Yallourn coal had almost the same level of CH_3 content. These observations agree with the high H/C ratios of subbituminous coals in Table 1.

We can calculate oxygen distribution from the oxygenated carbon distribution. However, two problems included in the spectral data prevent the calculation of true contents of various types of oxygen atoms: one is the overestimation of oxygenated carbon atoms, compared to non-oxygenated carbon ones, and another is no information on the amounts of hydroxy group and aromatic ether due to the insufficient resolution. It is therefore difficult to calculate the oxygen distribution of coal from the spectral data alone. In this work, the determination of various types of oxygen atoms in coal is based on two accurate data, namely oxygenated carbon distribution from CP/MAS spectrum and OH content obtained by the modified Brown's acetylation method(13). The method for the calculation of oxygen distribution is illustrated in Table 3. Multiplying the oxygen distribution by the atomic O/C ratio of coal, we can obtain true contents of various types of oxygen atoms. The results for the oxygen distribution of coal are summarized in

Table 3 Calculation of oxygen distribution from oxygenated carbon distribution for Soya-koishi coal

	-CHO =CO	-COOH	C_{Ar}-O			C-O-C
			Ar-OH	Ar-O-Ar	$-OCH_3$	
No. of oxygenated carbon atom per 100C	5.1	4.3		21.0		2.2
			(y	2x	3.5)	
No. of oxygen atom per 100C	5.1	8.6	y[a] (y=14.1)	x[a] (x=1.7)	3.5	1.1
Oxygen distribution (%)	15.0	25.2	41.3	5.0	10.3	3.2
True content of oxygen atom per 100C	3.3	5.5	9.1	1.1	2.3	0.7

[a] X is the number of oxygen atom in aromatic ether and y is the number of oxygen atom in OH group obtained from spectral data. The values of x and y are calculated from the following two simultaneous equations:

$$\begin{cases} \text{Carbon balance} & 2x + y + 3.5 = 21.0 \\ \text{Oxygen balance} & y/(x+y+5.1+8.6+3.5+1.1) = \dfrac{O_{OH}/C^{\,b}}{O_{total}/C^{\,c}} = 0.414 \end{cases}$$

$$\therefore \; x=1.7 \quad y=14.1$$

[b] O_{OH}/C is the OH content in coal obtained by acetylation method.

[c] O_{total}/C is the atomic ratio O/C of coal obtained by ultimate analysis.

Table 4 Oxygen distribution of coal[a]

Coal	O/C x 100 (%)					
	-CHO =CO	-COOH	Ar-OH	Ar-O-Ar	$-OCH_3$	C-O-C
Yallourn	7.9	7.7 (7.2)[b]	9.7	2.9	4.0	0.8
Soya-koishi	3.3	5.5 (5.6)	9.1	1.1	2.3	0.7
Taiheiyo	1.9	3.2 (2.0)	7.0	1.2	1.6	0.5
Horonai	0	1.7 (0.7)	5.6	0.2	2.3	0.2
Akabira	0	1.2 (0.4)	4.7	0.5_5	1.1	0.1
Shin-yubari	0	0 (0)	2.5	0.6_6	0.7	0

[a] Values obtained by assuming that alcohols are not present in coal.
[b] Values obtained by barium acetate method.

Table 4. It is particularly noteworthy that ether oxygens were present in much less concentration than expected in every coal and were distributed in aromatic ether rather than in aliphatic ether. The validity of the results obtained by above method is proved by comparison with COOH content determined by barium acetate method (14). We can find out that low-rank coals have a good agreement between our data and values obtained by barium acetate method in Table 4. The increase in the difference between two data with the increase in rank of coal would be probably caused by the tailing of NMR peaks.

In order to know the changes in average chemical structure of coal with increasing carbon content, two basic structural indices, aromaticity fa and atomic H/C ratio for the hypothetical unsubstituted aromatic nuclei Haru/Car, were calculated using the following equations:

$$fa = Car/\text{Total carbon} \qquad \text{---- (1)}$$

$$Haru/Car = \frac{H/C - 1/2(2-1) \times C_{CH_2}/C - 1/2 \times 2 \times C_{CH_2}/C - (3-1) \times C_{\alpha\text{-}CH_3}/C - 3 \times C_{\beta\text{-}CH_3}/C}{Car/C} + \frac{1/2(O_{OH}/C + O/C)}{Car/C} \qquad \text{---- (2)}$$

where aromatic carbon content Car was obtained from the integrated intensity of bands between 93 and 171 ppm. Three assumptions were made in the calculation of Haru/Car. First, alkyl groups contain only CH_3 group, and β-CH_3 group is assigned to that bound to naphthenic ring. Therefore, all CH_2 group belong to the naphthenic CH_2 and CH_2 bridge. Second, one half of all methylene groups are located in α-position to aromatic ring. Third, all phenolic OH and one half of the other oxygen atoms are bonded directly to aromatic ring. These structural indices are shown in Table 5. The fa value of sample coal generally increased with increase in the carbon content, and ranged between 0.51 and 0.71. Table 6 summarizes Haru/Car values of representative model structures obtained by using equation (2). This Table further indicates that Haru/Car values of model aromatic hydrocarbons are increased by the substitution of oxygen-containing polar groups. These materials allows us to present average chemical structures for the coals examined here. Yallourn, Soya-koishi, and Taiheiyo coals are dominated by hydroaromatic structures containing monoaromatic or diaromatic rings. In contrast, Horonai, Akabira, and Shin-yubari coals have hydroaromatic structure of chrysene and pyrene-types. The unit structure of coal is thus supposed to consist basically of relatively small hydroaromatic structures.

Hydroliquefaction characteristics of coal

At first, we evaluate the roles of ether oxygens in the depolymerization of coal. As shown in Table 4, the ether groups in coal were classified as either aliphatic or aromatic ethers. The aliphatic ether oxygen is present to the extent of less than 0.8% relative to carbon atom in Yallourn coal and does not exist in coals having more than 85% of carbon. Aromatic ether is most abundant again in Yallourn coal. Most of the aromatic ether oxygen in Yallourn coal, where the average number of aromatic ring is one, should be assigned to the diphenyl ether type. Ogata et al. (15) have recently reported that diphenyl ether was the most stable in the thermal treatment and that the decomposition of aromatic ethers was enhanced by the increase in the number of aromatic ring. Therefore, the ether linkages in high-rank coals must have been thermally cleaved with the increase in the number of aromatic ring during coalification to give hydroxy group. This may be the reason for the decrease in aromatic ether oxygen and the relative increase in hydroxy group with the increase in the carbon content of coal.

Table 5 Structural indices of coals

Coal	fa	Haru/Car
Yallourn	0.57	0.92
Soya-koishi	0.51	0.95
Taiheiyo	0.54	0.91
Horonai	0.61	0.69
Akabira	0.66	0.67
Shin-yubari	0.71	0.61

Table 6 Value of Haru/Car for model compounds

Structure	Haru/Car	Structure	Haru/Car
	1.0	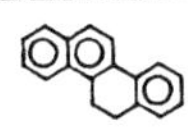	0.69
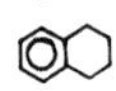	1.0	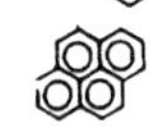	0.63
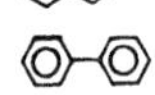	0.83		0.64
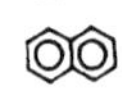	0.80	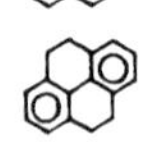	0.67
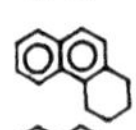	0.80	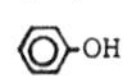	1.20
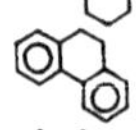	0.75	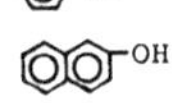	0.90
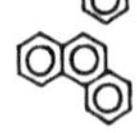	0.71	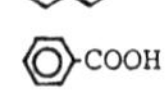	1.30
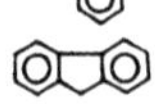	0.71	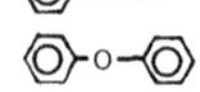	0.92
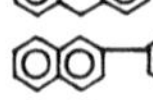	0.75	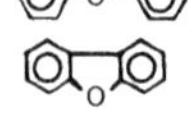	0.75
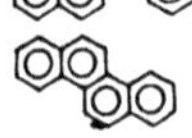	0.67		

We briefly discuss the contribution of ether oxygen to the depolymerization of coal. For example, Soya-koishi coal (fa=0.51, Haru/Car=0.95) has the average number of aromatic ring of one to two, corresponding to the number of carbon atoms of 12 to 20 in an average unit structure. Based on the results of ultimate analysis of the coal, the unit structure is formulated to be between $C_{12}H_{10.9}O_{2.6}$ (MW=197) and $C_{20}H_{18.2}O_{4.4}$ (MW=329). Hence there should be one ether oxygen per 2.8 to 4.6 structural units, namely per molecular weight of approximately 910. Thus, the cleavage of ether linkages in early stages of depolymerization of coal leads to the formation of asphaltene or preasphaltene in low-rank coals.

In the study of elucidating the correlation of hydroliquefaction reactivity with chemical structure of coal, an ideal reaction would be the one which adequately reflects any difference in the chemical structure of coal and still changes the structure as little as possible. We chose the hydrogenation under Adkins catalyst, which is known to hardly hydrogenate aromatic nuclei, but cleaves the bond like ether linkages connecting structural units(16). Table 7 presents the results obtained under the mild condition (400°C, 30 min). Small yields of hydrocarbon gases from all of the coal sample studied confirm our expectation that gasification due to the cleavage of C-C bond did not occur to significant extent. The yields of oil based on the extraction method are high for subbituminous coals such as Taiheiyo, Horonai, and Akabira coals, and low for bituminous and brown coals.

Quantitative parameterization of the reactivity of coal, as evaluated by the yield of oil, appears to be highly useful for the elucidation of the chemical structure of coal and the mechanism of hydroliquefaction. Although the possible relation between such coal parameters as C%, H/C, V.M., and V.C., and the reactivity of coal has already been suggested, it cannot be denied that these

Table 7 Experimental results for hydroliquefaction of coals[a]

	Yallourn	Soya-koishi	Taiheiyo	Horonai	Akabira	Shin-yubari
H_2 consump.	4.0	4.8	3.8	2.9	2.8	1.1
Gas	17.0	11.8	5.8	4.2	4.0	1.8
C_1-C_5	4.3	2.6	3.7	3.2	3.3	1.2
COx	12.7	9.2	2.1	1.0	0.7	0.6
Water[b]	9.1	9.2	6.3	4.0	3.3	1.6
Oil	28.8	21.9	54.7	52.7	43.0	11.1
Asp	16.7	14.1	15.6	15.5	20.7	35.7
Preasp	6.7	11.2	5.0	3.8	10.8	26.7
Residue + Loss	21.7	36.6	16.3	22.6	21.1	24.2

[a] wt%, dry mineral-matter free (dmmf) coal basis.

[b] Values obtained by distillation method.

previous parameters do not directly reflect the chemical constitution of coal. Even if some relation was found between one of these parameters and reactivity, it was impossible to know which part of coal structure is responsible for the observed reaction. In recent years, Maciel et al.(17) reported a good relation between the oil yield and the content of aliphatic carbon of oil shale based on solid state NMR studies. The depolymerization of coal is generally believed to proceed through the cleavage of ether linkages(18) and CH_2 bridges(19) connecting structural units. However, coal is essentially not rich in ether linkages and the cleavage contributes only to the formation of asphaltene or preasphaltene(20). Further degradation of coal to produce oil must involve the cleavage of CH_2 bridges. If this assumption is right, the content of CH_2 group rather than that of CH_3 group should correlate with the yield of oil.

Figure 2 shows the relation between the aliphatic carbon contents obtained from spectra and the oil yields obtained from extraction method for various coals. As shown in Figure 2a, plotting of oil yield against the total aliphatic carbon content (CH_2+CH_3) revealed significant correlation between these two quantities except for Soya-koishi and Shin-yubari coals. However, Figure 2b indicates that similar plotting against CH_2 content gave a better correlation. Namely, the oil yield correlated better with CH_2 content than with CH_2+CH_3 contentof coal. Figure 3 illustrates the dependencies of yields of PS, BS, and HS on the CH_2 content of coal. While the PS yields are independent of CH_2 content except for low-rank coals, the oil yields are dependent. In other words, the formation of intermediate like preasphaltene can be achieved by the cleavage of ether linkages and hence hardly depends on the CH_2 content of coal. On the other hand, the dependence of yield of extractable oil on the CH_2 content of coal could be interpreted as follows: 1) the promotion of depolymerization by the relative increase in the number of CH_2 bridge which is linked to naphthenic rings, 2) the increase in the solubility of oil components with the increase in the number of naphthenic ring, and 3) the increase in the amount of transferable hydrogen with the increase in the naphthenic CH_2 content. Thus, the yield of oil depends not only on the cleavage of CH_2 bridge, but on the amount of naphthenic CH_2 group in coal.

In order to obtain informations on the reaction mechanism from the change in carbon atom distribution of coal during hydroliquefaction, CP/MAS ^{13}C NMR spectra of coal and crude products after getting rid of water and gases are compared. Based on the yield, the carbon content, and the carbon atom distribution of the crude product, the changes in the CH_2, CH_3 and aromatic carbon contents during the hydroliquefaction of coals were calculated. Results are shown in Figure 4. Heat treatment

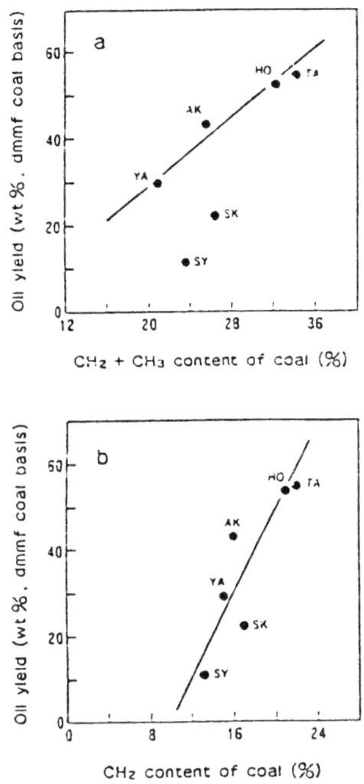

Fig.2 Relationship between oil yield and aliphatic carbon content of coal

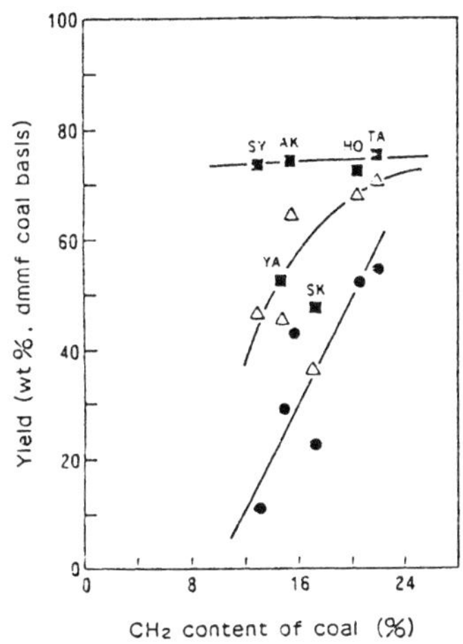

Fig.3 Relationship between yields of products and CH_2 content of coal

of coal, especially low-rank coals, readily brought about the marked decrease in CH_2 content and the increase in aromatic carbon content(21). These facts indicate that alicyclic structures of coals are dehydrogenated to convert to aromatic rings. Therefore, the hydroliquefaction at temperatures above 400°C would involve the dehydrogenation from alicyclic structure as well as the hydrogenation of aromatic rings. Indeed, Figure 4 indicates that the dehydrogenation from alicyclic structure rather than the hydrogenation of aromatic rings is predominent in low-rank coals, and the hydrogenation in high-rank coals. The considerable increase in CH_3 content in low-rank coals should be caused by the cleavage of CH_2 bridge between aromatic rings.

Structural correlation between coal and oil

Oil is the most extensively hydrogenated fraction in the hydroliquefaction of coal and contains hydroaromatic compounds as the predominant components. Hence its structural analysis must include the method of determining naphthenic structures

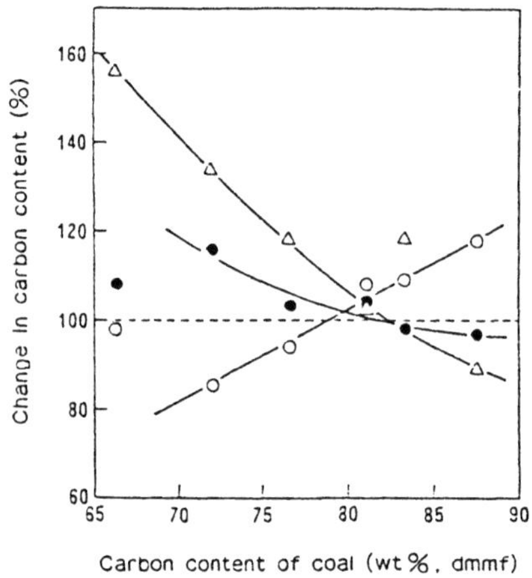

Fig.4 Changes in CH_2, CH_3 and aromatic carbon contents during hydroliquefaction of coal

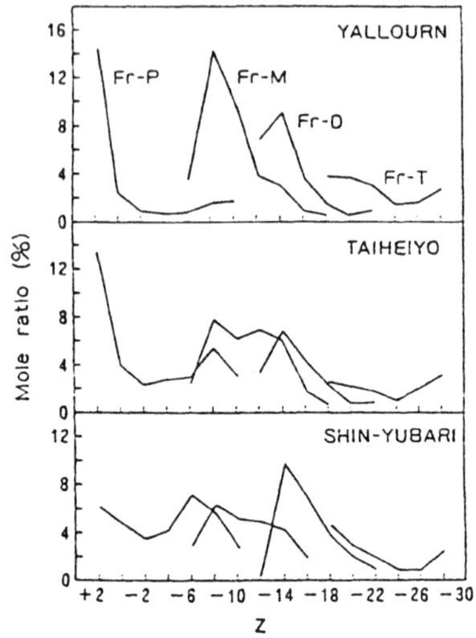

Fig.5 Z series distribution for ring-type fractions of oils

in addition to aromatic nuclei in hydroaromatic ring species. The informations obtainable from such analysis should be useful in assessing the distribution of structural types of oil and in elucidating the reaction mechanisms. Mass spectral analysis based on FI and FD techniques provides molecular weight distribution of oil and informations on the nucleus structures of oil components(12).

Figure 5 gives the distribution (z series distribution) of molar contents of various compounds in the ring-type fractions of oils from Yallourn, Taiheiyo, and Shin-yubari coals as obtained by the group-type analysis of FI mass spectra. The z series distribution of oils revealed that the most abundant component was always hydroaromatic hydrocarbons for all the aromatic ring fractions studied, and shows the following response of each fraction against the increase in coal rank. Paraffins (z=+2) decreased and various kinds of cycloparaffins increased in Fr-P, and major components acquired smaller z values, namely contained increased number of naphthenic rings in fractions M, D, and T. If we combine these observations with the fact that the major component (z=-8) in Fr-M decreases and the major component (z=-14) in Fr-D increases with the increase in coal rank, we are led to conclude that the average size of structural unit in coal increases with its rank. Encouraged by the discovery of the dependence of structural distribution in oil on coal rank, we proceeded to a quantitative study of the structural correlation

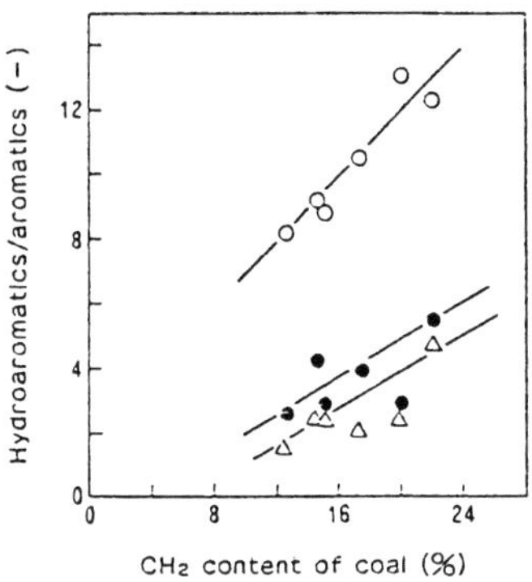

Fig.7 Plots of hydroaromatics/aromatics ratio against CH_2 content of coal

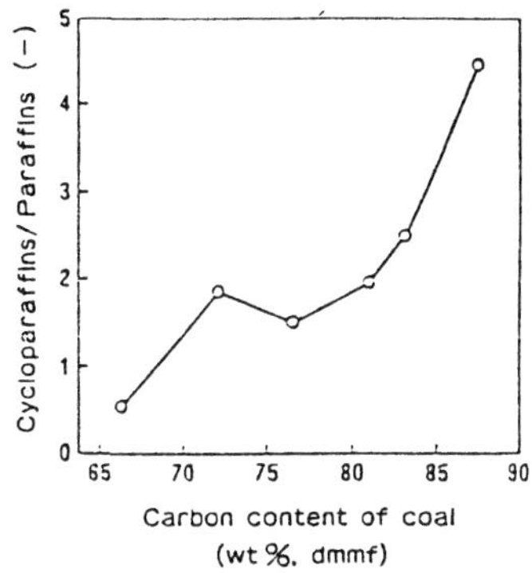

Fig.6 Change in cycloparaffins/paraffins ratio of Fr-P with carbon content of coal

between coal and oil.

In Figure 6, the ratios cycloparaffins/paraffins of Fr-P are plotted against the carbon content of coal. The positive relation obtained indicates possibilities that low-rank coals are rich in free paraffins, and that high-rank coals contain large amounts of cycloparaffins formed by hydrogenation of aromatic rings. In Figure 7, the ratios hydroaromatics/aromatics of fractions M, D, and T are plotted against the CH_2 contents of coal. On the whole, the ratio changed proportionately with the CH_2 content for all the ring-type fractions. These results suggest that the CH_2 groups of coal originate not from alkyl chains but alicyclic structure, and that coals rich in alicyclic structures produce oils also rich in naphthenic structures. It is now clear that correlations regarding chemical structures exist between coal and oil. In conclusion, we propose that not only the yields but also the structural characteristics of oil depend on the CH_2 content of coal.

References

1) Given,P.H., Cronauer,D.C., Spackman,W., Lovell,H.L., Davis,A. and Biswas,B. Fuel 1975, 54, 40
2) Whitehurst,D.D. 'Coal Liquefaction Fundamentals', ACS sympo. Series 139, Washington, D.C., 1980, Chap.4
3) Dryden,I.G.C. J.Inst.Fuel 1957, 30, 193
4) Wiser,W.H. 'Scientific Problems of Coal Utilization', Technical Information Center, U.S.Department of Energy, 1978, p.219

5) Pines,A., Gibby,M.G. and Waugh,J.S. J.Chem.Phys. 1973, 59, 569
6) VanderHart,D.L. and Retcofsky,H.L. Fuel 1976, 55, 202
7) Zilm,K.W., Pugmire,R.J., Grant,D.M., Wood,R.E. and Wiser,W.H. Fuel 1979, 58, 11
8) Maciel,G.E., Bartuska,V.J. and Miknis,F.P. Fuel 1979, 58, 391
9) Yoshida,T., Nakata,Y., Yoshida,R., Ueda,S., Kanda,N. and Maekawa,Y. Fuel 1982, 61, 824
10) Yokoyama,S., Tsuzuki,N., Kato,T. and Sanada,Y. J.Fuel Soc.Jpn. 1978, 57, 748
11) Yoshida,T., Maekawa,Y. and Fujito,T. Anal.Chem. 1983, 55, 388
12) Yoshida,T., Maekawa,Y., Higuchi,T., Kubota,E., Itagaki,Y. and Yokoyama,S. Bull.Chem.Soc.Jpn. 1981, 54, 1171
13) Hasegawa,Y., Tokuhashi,K. and Maekawa,Y. 'Proceedings of the 17th conference on coal science', Tokyo, 1980, p.15
14) Bhaumik,J.N., Mukherjee,P.N. and Lahiri,A. Fuel 1959, 38, 211
15) Ogata,H., Goto,K., Nomi,T. and Kamiya,Y. 'Proceedings of the 19th conference on coal science', Tokyo, 1982, p.153
16) Utoh,S., Oda,H. and Yokokawa,C. 'Proceedings of the 19th conference on coal science', Tokyo, 1982, p.222
17) Maciel,G.E., Bartuska,V.J. and Miknis,F.P. Fuel 1978, 57, 505
18) Takegami,Y., Kajiyama,S. and Yokokawa,C. Fuel 1963, 42, 291
19) Maekawa,Y., Shimokawa,K., Ishii,T. and Takeya,G. J.Fuel Soc. Jpn. 1967, 46, 927
20) Yoshida,T., Tokuhashi,K., Narita,H., Hasegawa,Y. and Maekawa,Y. Fuel 1984, in press
21) Yoshida,T., Yoshida,R., Maekawa,Y. and Miyazawa,M. 'Proceedings of the 20th conference on coal science', Tokyo, 1983, p.305

NATURALLY ABUNDANT ^{17}O NUCLEAR MAGNETIC RESONANCE AND THE DETERMINATION OF OXYGEN FUNCTIONALITIES IN SYNTHETIC FUELS

D. W. Grandy, L. Petrakis, D. C. Young, and
B. C. Gates*
Gulf Research & Development Company P.O. Drawer 2038
Pittsburgh, PA 15238
*University of Delaware, Department of Chemical
Engineering, Newark, DE 19711

We report the successful use of nuclear magnetic resonance (NMR) of naturally abundant ^{17}O to determine oxygen functionalities in synthetic fuels. Heteroatomic (N, S, O) species are a significant factor in the processability, stability, and value of natural and synthetic fossil fuels. Sulfur and nitrogen removal has received much attention, but oxygen is also quite significant, since it is the most abundant heteroatom in most coals and for its removal, hydrogen must be consumed. This can impact adversely the economics of coal liquefaction. Oxygen bonds (e.g., ethers) on the other hand may be the "weak link" in the depolymerization of coal. Also, oxygen functionalities affect the stability of fuels, and during desulfurization and denitrogenation, the same oxygen functionalities compete and may interfere with these reactions in a fuel upgrading process. Thus, there is need for the unequivocal identification and quantitative determination of oxygen functionalities, for their understanding could aid materially in the design of processes to produce and upgrade synthetic fuels in a cost-effective manner.(1,2)

The samples used in our study are the heavy distillates produced from Powhatan No. 5 coal using the SRC-II process (3) and three acidic and two basic fractions that have been prepared chromatographically from the heavy distillate.(4) The elemental analyses and yields of the various fractions are given in Table I. Additional data on these materials can be found in Reference 4.

^{17}O NMR has found only limited utility in the study of natural substances due to its low natural abundance (0.037%) and

L. Petrakis and J. P. Fraissard (eds.), Magnetic Resonance. Introduction, Advanced Topics and Applications to Fossil Energy, 689–698.

Table I

ELEMENTAL ANALYSES OF FRACTIONS
AND PERCENTAGE OF HEAVY DISTILLATE
(Wt% of Fractions)

Fraction	C	H	N	O	S	% of Total
Neutral Oils	91.3	7.2	0.1	0.7	0.6	72.8
Very Weak Bases	84.3	6.8	1.3	7.1	0.5	2.6
Weak Bases	76.7	6.5	3.8	8.4	2.2	1.8
Strong Bases	85.7	6.9	4.8	1.8	0.6	5.7
Very Weak Acids	82.6	8.6	0.1	8.9	0.1	1.0
Weak Acids	82.8	7.8	0.0	9.8	0.2	1.2
Strong Acids	76.0	5.6	0.0	17.8	0.6	0.2
Neutral Resins	79.5	6.8	0.8	6.8	0.4	1.1
Asphaltenes	85.4	6.4	3.2	4.5	0.5	8.4

the broad lines encountered due to its quadrupolar nucleus (I = 5/2). However, as advances are made in FT NMR instrument sensitivity, ^{17}O NMR may become very important as a tool in the study of oxygen functionalities in natural systems that until now have been thought to be too dilute in oxygen for successful study.(5) In spite of its difficulties, ^{17}O NMR has several important advantages. The chemical shift of ^{17}O has a wide range of about 800 ppm. The ^{17}O chemical shifts of many small (MW <100) compounds have been reported throughout this range.(6-10) The quadrupolar nature of the nuclei which broadens the observed lines also shortens the spin-lattice relaxation time (T_1) of the nuclei, such that very rapid pulsing is possible. More than a million transients can be recorded overnight. Determination of the relative amounts of the various oxygen functionalities can also be done, if ^{17}O spectra with sufficient signal-to-noise ratio are available and the various signals are resolved.

At present, the spectroscopic techniques used for the identification and quantitative analysis of oxygen-containing functional groups in coal liquids are IR, ^{29}Si NMR, and ^{13}C spectroscopy. IR spectroscopy is quite useful, however, it is often difficult to get good quantitative data from this method due to the different absorbtion coefficients of the various bonds and band overlap. Recently, ^{29}Si NMR has been applied to coal liquids, but this method is limited to protonated species and requires derivitizing these functional groups with silating agents.(11) ^{13}C NMR is very widely used to obtain structural information on organic molecules and mixtures. All of the samples, used in this study have been analyzed by ^{13}C NMR and the spectra of many of the fractions have been published.(12) Typically the ^{13}C NMR spectrum of the very weak acid shows a number of peaks in the 110 to 156 ppm (TMS) chemical shift

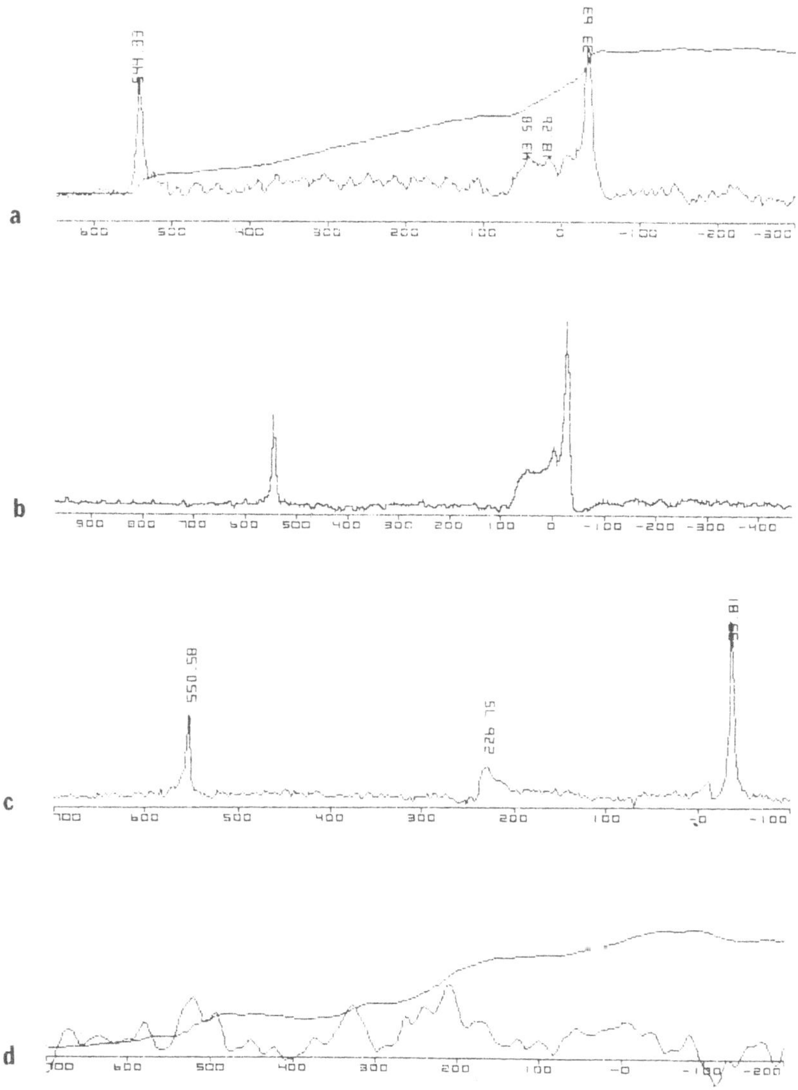

Figure 1

^{17}O NMR spectra of coal liquids, a. very weak acids (A1), b. weak acids (A2), c. strong acids (A3), d. total heavy distillate range. In this spectral region aromatic carbons and aromatic oxygen-containing groups appear, and the overlap does not allow the definitive differentiation of the relative contributions. Nevertheless these signals indicate that phenols, ethers, or perhaps esters may be present.(13)

The data presented here have been obtained with a Varian XL-200 spectrometer using a 14 μsec pulse (90°), a 13 msec acquisition time, and a 20 μsec delay. Using this sequence, about 30 transients/sec or 108,000/h can be obtained, yielding a good spectrum in a few hours. Deuterated methylene chloride was used as the solvent for acid and base fractions and was used in all samples as the field frequency lock solvent. From 3×10^5 to 2.7×10^6, transients were recorded before Fourier transformation. The ^{17}O spectra of the very weak acids (A1), weak acids (A2) at strong acids (A3) are shown in Figure 1. All three of these samples were spiked with 50 μl each of methanol and acetone to aid in the adjustment of the spectral phasing. Methanol appears as the prominent peak at about -35 ppm and acetone is the prominent peak at about 540 ppm. All three spectra also exhibit a peak near 0 ppm identifiable as water. The weak acid (A2) and the very weak acid (A1) spectra both display a large broad peak, resolvable in A1, in the -10 to 80 ppm region. This is characteristic of alcohols, phenols, and ethers.(6,7)

There appears to be some aldehyde and/or ketone functional groups in the very weak acid (top of Figure 1) as indicated by an upfield sholder on the acetone peak. Both the very weak and weak acids as well as their hydrotreated products have been exhaustively studied by GC/MS and GC methods.(14-16) The very weak acid is shown by these methods, to contain dimethyl benzofuranone, cyclohexylphenol isomers, ethylphenoxybenzene, methylcyclohexylphenol, trimethyldihydronaphthalenone, tetrahydronaphthol, and phenylphenol. The weak acid contains tetrahydronaphthol, phenylphenol, methyltetrahydronaphthol, methylphenylphenol, cyclohexylphenols, and dimethylphenoxybenzene. The ^{17}O spectrum of the strong acids (A3) shows only the prominent acetone and methanol peaks, a small peak around 0 ppm, probably due to water and a peak at about 230 ppm. This is the appropriate spectral region for furans and the carbonyl oxygen of carboxylic acids. The chemistry of the separation procedure and the IR data indicate that carboxylic acids are the most likely choice.(4)

The ^{17}O NMR spectrum of the total heavy distillate is shown in Figure 1d. Although 2.7 million scans were used, the spectrum is noisy. Integration measurements indicate about 19% of the oxygen is in the alcohol, phenol, or ether region of -50 to 100 ppm. Phenols are most likely. About 44% of the oxygen is

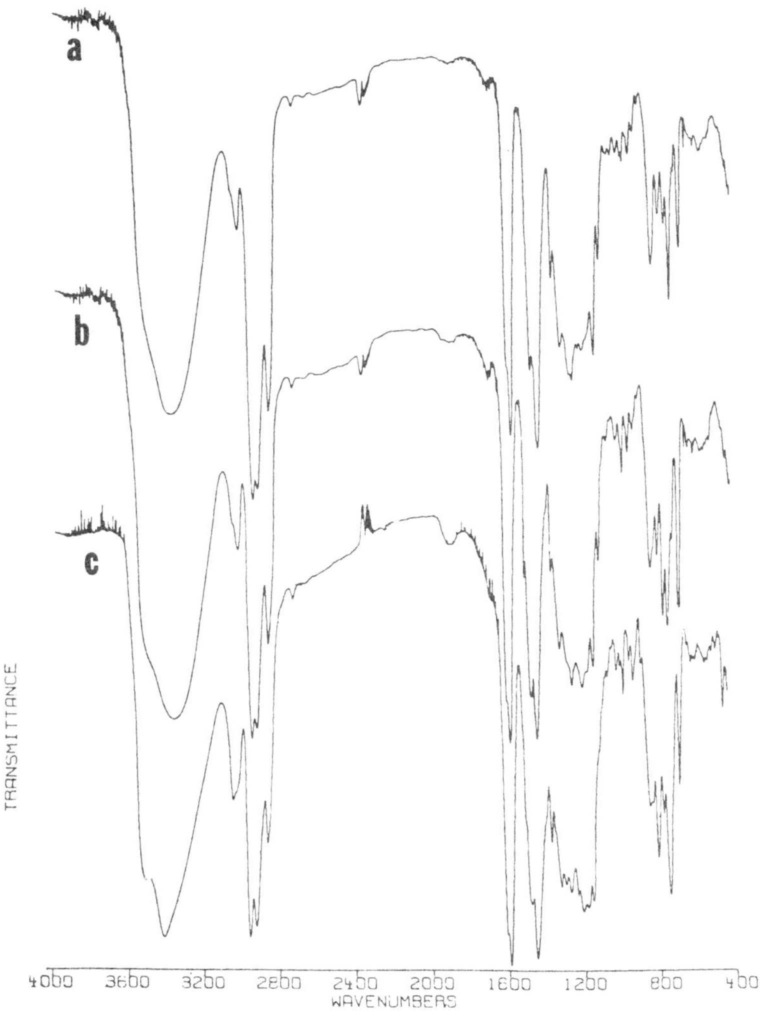

Figure 3

IR spectra of a. very weak acids, b. weak acids, and c. very weak bases in the carboxylic acid (C=O) and furan region from 150 to 280 ppm. The chemical shift range of 310 to 380 contains about 14% of the oxygen. Esters are the only reported compounds having ^{17}O NMR signals in this region; (6,7) however, many of the oxygen-containing aromatics that may be in coal liquids such as quinones and aromatic acids have unknown ^{17}O NMR shifts. The carbonyl region due to ketones and aldehydes (480 to 550 ppm) contains about 23% of the oxygen. (6,9)

A very favorable situation for quantitative analysis exists in the ^{17}O spectrum of the very weak acid (A1), shown at the top of Figure 1. From the spectrum integration, about 4% of the oxygen is in the ketone or aldehyde region of 500-550 ppm, 29% is in the phenol, alcohol, and ether region of -10 to 80 ppm and the remainder is in the 100 to 500 ppm region. If a sufficient catalogue of ^{17}O chemical shifts of coal-like compounds existed, one might be able to split this region into oxygen functionality sections. Overall, the integration values for the coal liquid signals and the methanol and acetone signals are consistent with the 50 µl of each pure compound and a 1-g sample of coal liquid acid containing about 9% oxygen by weight.

The ^{17}O spectra of the sample of very weak bases (B1) and that of the strong bases (B3) (Figure 2) have peaks at 0 ppm in B3 or slightly negative in B1 due to traces of water. A strong peak from 30 to 80 ppm in B1 is assigned to alcohols, ethers, phenols, and higher aromatic alcohols. Phenol is reported to have a chemical shift of 69 or 79 ppm. (6,7,8) The large peak in B3 at 250 ppm, also found in B1 as a smaller peak, may be due to furan-type oxygen. Furan is reported to have an ^{17}O chemical shift of 241 ppm. (6,7) The spectra of the strong bases (B3) lack any prominent ^{17}O peaks above 400 ppm. The strong resonance between 450 and 520 in B1 may be due to compounds such as acetophenones substituted by -OH or NH_2 groups. (9) The presence of phenolic amines are indicated in this fraction by other spectroscopic techniques. (12)

The IR spectrum of A1 and A2, the very weak and the weak acids, and B1, the very weak base, are essentially identical (Figure 3). They consist mainly of a strong absorption centered at 3400 cm^{-1} that is assigned to OH groups, a band between 2800 and 3050 cm^{-1} due to aromatic and aliphatic hydrogen, a very weak band at 1700 cm^{-1} in the acids that may be due to C=O, bands at 1600 and 1400 cm^{-1}, also due to C-H, a strong band centered at 1200 cm^{-1} due to OH and bands at 800 cm^{-1}, again due to aromatic C-H. Hydroxyl groups appear to be the only major nonhydrocarbon bands in the IR spectrum. In the ^{17}O NMR spectrum, hydroxyl groups are the major peaks in A1 and A2, although some carbonyl oxygen also appears in A1. The very weak base, B1, has some hydroxyl oxygen present, but also some other peaks,

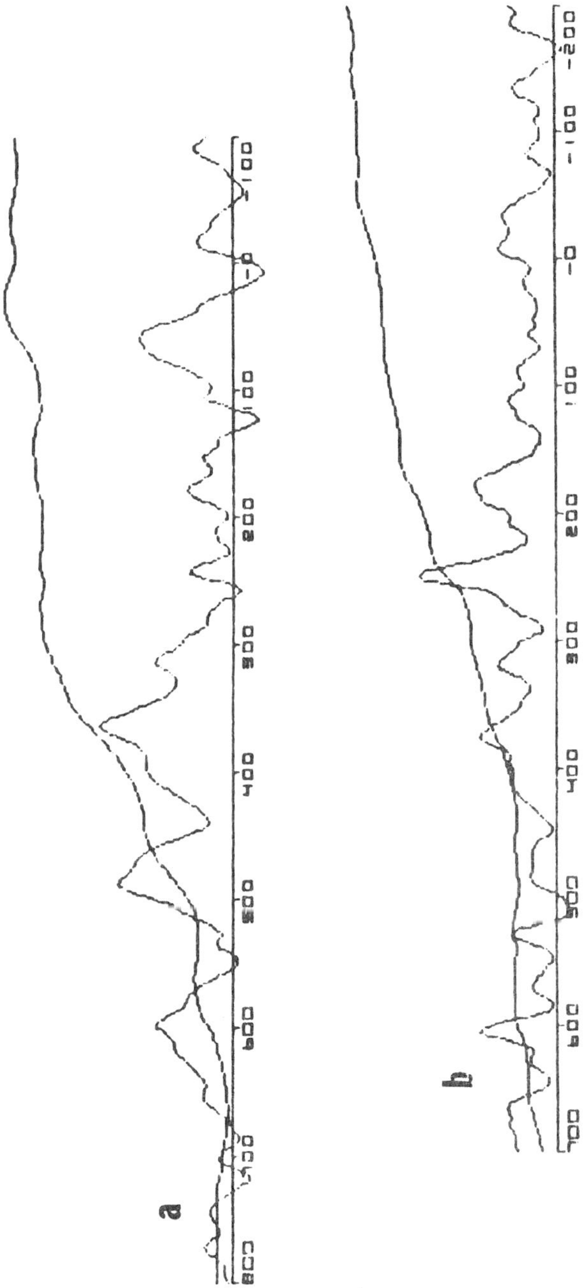

Figure 2

^{17}O NMR spectra of coal liquids, a. very weak bases,
b. strong bases

Table II

OXYGEN FUNCTIONALITY BY IR AND ^{17}O NMR

Fraction	IR Freq. (cm^{-1})	IR Function-ality	^{17}O NMR Chemical Shift (ppm)+	^{17}O NMR Functionality
Total Distillate			-50 --> 100	Alcohol, phenol, ether
			150 --> 280	Furans and carboxylic acids
			310-380	Esters
			480-550	Ketones and aldehydes
Very Weak Bases (B1)	3200-3500	OH	30 --> 80	Alcohols, phenols, ethers
			250-	Furans
			450-520	Phenones
Strong Bases (B3)	3200-3500	OH or NH	250	Furans
Very Weak Acids (A1)	3400	OH	-10 --> 80	Alcohols, phenols, ethers
	1200	OH		
Weak Acids (A2)	3400	OH	-10-80	Alcohols, phenols, ethers
	1200	OH		
Strong Acids (A3)	3400	OH	230	Furans or carboxylic acids
	3100	carboxylic acid		
	1700	carboxylic acids		
	1200	OH		
	900	carboxylic acids		

+ From water

such as those in the furan region, not observed in the IR spectrum of this fraction. The strong base, B3, has the same CH bands as the others (A1, A2, B1) but lacks the strong hydroxyl bands at 3400 and 1200 cm^{-1}. There are some weak absorbances from 3200-3500 cm^{-1} which may be due to amines or a low relative concentration of hydroxyl groups. The IR and ^{17}O NMR assignments are summarized in Table II. By ^{17}O NMR, as discussed above, there is evidence of much more oxygen functionality in B3. There is good evidence of acids and other hydroxyl functionality in the IR spectra of the strong acid, A3, in agreement with what was found in the ^{17}O spectrum.

On materials of low oxygen content (~1%) as many as 16 h of spectrometer time may be necessary to obtain usable spectra, an expense that may discourage ^{17}O NMR analysis of large numbers of routine samples. Since ^{17}O NMR is specific to one atom and the signal area is directly proportional to the concentration of the functional group, it offers an excellent method of quantitative analysis. Thus, ^{17}O NMR can be a very powerful tool for the investigation of oxygen functionalities in the complex mixtures present in synthetic fuels. The method will be even more useful as progress is made in the sensitivity of FT NMR spectrometers and the chemical shifts of whole classes of large, substituted oxygen-containing aromatic compounds such as quinones, aromatic acids, and phenyl ethers are catalogued. At this stage there is no doubt about the potential of the technique, including the possibility of doing magic angle spinning experiments in the solid state, routinely. However, much remains to be done in this field and a definative comparison of ^{17}O NMR with other techniques cannot be made until many more data have been obtained.

Acknowledgment

The authors acknowledge the support of the U.S. Department of Energy for the preparation of the samples and DOE Contract DE-AC22-F90ET14880.

REFERENCES

1. Gorbaty, M.L., Wright, F.J., Lyon, R.K., Long, R.B., Schlosberg, R.H., Baset, Z, Liotta, R., Silbernagel, B.G., and Neskora, D.R. 1979, Science, 206, p. 1029.
2. "American Physical Society Study Group Report," 1981, Rev. Mod. Phys., 53 (4), pp. 51-5168.
3. Schmid, B.K. and Jackson, D.M. 1981, Phil. Trans. R. Soc. Lond., A300, p. 129.
4. Petrakis, L., Young, D.C., Roberto, R.G. and Gates, B.C. 1983. I&EC Fundamentals, 22, p. 292.
5. Gerothanassis, I.P., Lauterwein, J. and Sheppard, N. 1982, J. Mag. Res., 48, p. 431.

6. Christ, H.A., Diel, P., Schneider, H.R. and Dahn, H. 1961, Helv. Chim. Acta., 44, p. 865.
7. Sugawara, T.,Kawada, Y. Katoh, M. and Iwamura, H. 1979, Bull. Chem. Soc. Japan, 52, p. 3341.
8. Crandall J.K. and Centeno, M.A. 1979, J. Org. Chem., 44, p. 1183.
9. St. Amour, T.E., Burgar, M.I., Valentine, B. and Fiat, B. 1981, J. Am. Chem. Soc., 103, p. 1128.
10. Kalabin, G.A., Kushnarev, D.F., Valeyev, R.B. Trofimov, B.A., and Fedotov, M.A. 1982, Org. Mag. Res., 18, p. 1.
11. Rose K.D. and Scouten, C.G. "^{29}Si NMR: A New Tool for Coal Liquids Characterization," in "Chemistry and Physics of Coal Utilization," Cooper B.R. and Petrakis, L. eds., Am. Inst. of Physics, NY, NY, 1981, pp. 82-100.
12. Petrakis, L., Young, D.C., Ruberto, R.G. and Gates, B.C. 1983, I&EC Fundamentals, 22, p. 298.
13. Jensen R.K. and Petrakis, L. 1972, J. Mag. Res., 6, p. 105.
14. Petrakis, L., Grandy, D.W. and Gates, B.C. in preparation.
15. Katti, S., Westerman, D.W.B., Gates, B.C., Petrakis, L., Grandy, D.W., Youngless, T., Proceedings, 1983 International Conference on Coal Science, Pittsburgh, PA, August 15-19, 1983.
16. Li, C.L., Xu, Z.R., Gates, B.C. and Petrakis, L. I&EC Process Fundamentals, submitted 1983.

FUNCTIONAL GROUP ANALYSIS OF COAL LIQUIDS

D. Allen[1], L. Petrakis[2], D. Grandy[2], G. Gavalas[1], and B. Gates[3]

[1]Department of Chemical Engineering, California Institute of Technology
[2]Gulf Research and Development Company
[3]Department of Chemical Engineering, University of Delaware

A new method of structural analysis is applied to a group of hydroliquefied coal samples. The method uses elemental analysis and NMR data to estimate the concentrations of functional groups in the samples. The structural characterization of these samples demonstrates that estimates of functional group concentrations can be used to provide detailed structural profiles of complex mixtures and to obtain limited information about reaction pathways.

INTRODUCTION

Characterizing the structures of fuels derived from coal is an important step in developing an understanding of the chemistry of coal reactions. The determination of exact chemical structures of coal liquids from elemental composition, Nuclear Magnetic Resonance (NMR) and possibly other analytical data is an underdetermined problem, however, because of the complexity and variety of the molecules in the liquids. Since molecular structures cannot be delineated fully, recourse must be made to approximate characterization methods. The data on which such characterizations are most commonly based are elemental analysis, proton NMR spectra, and less routinely carbon-13 NMR spectra.

Structural characterization most frequently takes the form of a representative molecular structure or a set of average parameters, such as carbon aromaticity, average alkyl chain length and the degree of aromatic substitution (1-5). In this work, we apply an alternative method of structural characterization, termed functional group analysis. The goal of

L. Petrakis and J. P. Fraissard (eds.), Magnetic Resonance. Introduction, Advanced Topics and Applications to Fossil Energy, 699–707.

functional group analysis is to calculate the concentrations of the functional groups present in the sample. The method is based on the premise that although the number of individual molecular species in a coal liquid is large, most of these species are composed of a limited number of functional groups. The relative concentrations of the functional groups determine the chemical behavior of the liquids. Thus, calculating the concentrations of the major functional groups is a useful structural characterization.

A set of functional groups appropriate for the characterization of hydrotreated coal liquids is given in Figure 1. This set is not a complete listing of all structures known to be present in coal liquids. Rather, it is a concise representation of the major structural features of hydrotreated coal liquids, i.e., hydroaromatic groups, aromatic clusters, alkyl chains, bridges and heteroatomic groups. The method of functional group analysis seeks to calculate the concentrations of these groups utilizing elemental composition, NMR and possibly other supplemental data. An extensive description of the methodology has been given (6). Briefly, the NMR and elemental analysis data are used to construct linear balance equations. In these balance equations, the concentration of each atomic species is expressed as a sum of the contribution from the functional groups, each with a suitable stoichiometric coefficient. For example, for the functional groups of Figure 1,

$$b_1=6y_1+10y_2+14y_3+16y_4 \tag{1}$$

where b_1 is the concentration of aromatic carbon, and y_i is the concentration of functional group i. The set of balance equations can be written succinctly in the matrix form,

$$\sum_{j=1}^{n} A_{ij}y_j = b_i \quad (i=1,\dots.m) \tag{2}$$

where y_j (j=1,...n) are the functional group concentrations, b_i (i=1,...m) are atomic species concentrations and A_{ij} are stoichiometric coefficients. The number of equations is generally smaller than the number of unknowns, $m < n$.

In addition to the constraints imposed by the balance equations, there are the constraints of non-negative concentrations.

$$y_j \geq 0 \tag{3}$$

Equations 2 and 3 define a space of feasible solutions. The structure of the coal liquid is characterized by selecting

Figure 1

Coal Liquid Functional Groups

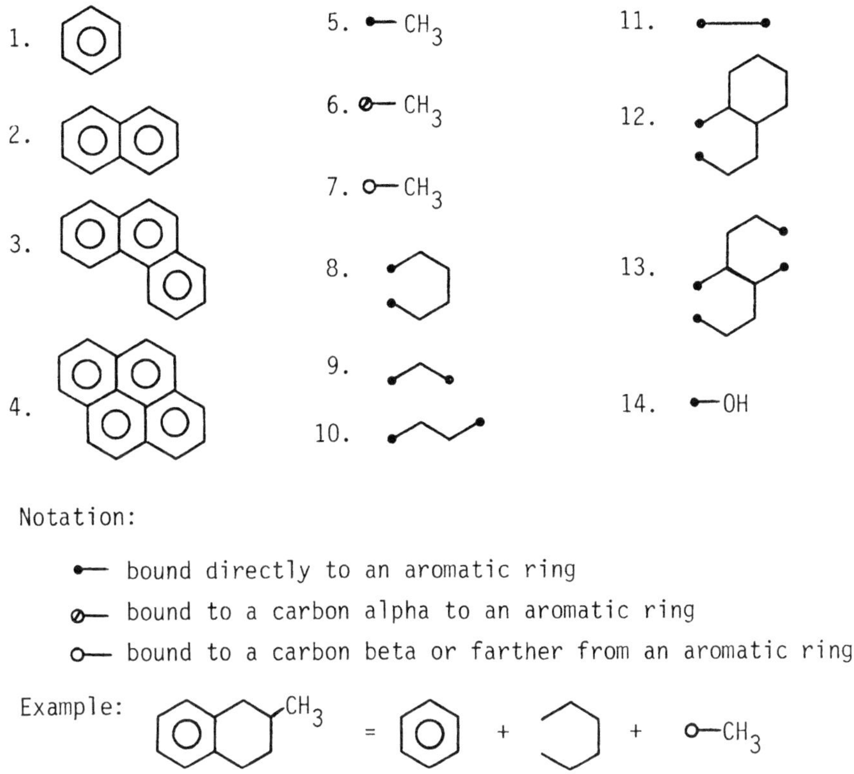

a solution from the feasible space. The set of concentrations $y_1,...y_n$ is selected such that a function $P(y_1,...y_n)$ is minimized. The form of the function depends on what, if any, data are available in addition to elemental analysis and NMR. For example, if high resolution mass spectra are available, equation 4 is minimized,

$$P = \sum_{i=1}^{n} (y_i - f_i)^2 \qquad (4)$$

where f_i is the concentration of functional group i predicted by the mass spectral data. If no data other than elemental analysis and NMR are available, then equation 5 is minimized,

$$P = \sum_{i=1}^{m} \sum_{j=1}^{n} \left(\frac{b_i}{A_{ij}N_i} - y_j \right)^2 \qquad (5)$$

where N_i is the number of functional groups containing species i. Equation 5 was constructed by assuming that the distribution of atomic species among the functional groups was random. Concentrations obtained by minimizing this equation subject to the constraints of equations 2 and 3 were found to be in reasonable agreement with functional group concentrations based on mass spectral data and with the results of other methods of structural analysis (6).

In this work, functional group concentrations will be estimated for a group of asphaltene and oil fractions obtained in a series of coal hydroliquefaction experiments. The functional group concentrations obtained for these samples will help to identify liquefaction pathways.

EXPERIMENTAL

A group of pentane soluble and toluene soluble/pentane insoluble extracts were obtained from the products of a series of autoclave experiments. The samples were prepared by shaking a reactor containing 5 g of a Powhatan #5 coal and 5 g of a liquefaction solvent in a heated fluidized sand bath. The reactor consisted of a 22 ml stainless steel bomb with a 1/4-in. stainless steel gas inlet tube. In each run, the reactor was pressurized to 11.0 MPa with hydrogen and was shaken in the sand bath for the specified time. Not included in the nominal reaction time was the reactor heating time, i.e., the time required for the reactor's internal temperature to reach within 1°C of the sand bath temperature. Heating time was found to be 4 minutes at both reaction temperatures.

After the necessary time had elapsed, the reactor was removed from the sand bath and cooled with an air jet. About 2 minutes were required to cool the sample to 300°C. After 7 to 10 minutes of cooling, the reactor was immersed in dry ice. The reaction products were removed from the reactor with a methylene chloride wash. The methylene chloride was then driven off at room temperature by a nitrogen purge. Pentane and toluene extracts were obtained by successive Soxhlet extractions. The pentane and toluene were driven off, along with the liquefaction solvent, by heating the extracts at 60°C under a nitrogen stream.

The reaction conditions were varied to determine the effect of solvent, temperature and residence time on product structure. Tetralin and Solvent Refined Coal Heavy Distillate (SRC-HD) were chosen as solvents to represent different hydrogen donating capabilities. Experiments were done at two temperatures and three residence times.

	Pentane Fraction				Toluene Fraction			
Sample	Yield*	%C	%H	%O	Yield**	%C	%H	%O
400S40	.117	89.5	7.4	2.2	.197	84.1	6.3	5.3
460S03	.106	88.2	7.2	2.4	.136	84.6	5.7	6.5
460S10	.154	89.0	7.0	2.2	.141	86.0	5.3	5.5
460S40	.103	90.3	6.6	2.0	.101	88.2	5.0	4.4
460T03	.322	82.2	7.4	5.4	.176	83.9	6.4	6.4
460T10	.423	85.0	7.2	4.5	.183	86.0	6.2	6.0
460T40	.491	87.8	7.0	3.6	.213	87.9	6.0	5.4

$$^{*}\text{Pentane Yield} = 1 - \frac{\text{Wt Pentane Insolubles}}{\text{Wt Ash-Free Coal}}$$

$$^{**}\text{Toluene Yield} = 1 - \frac{\text{Wt Toluene Insolubles}}{\text{Wt Ash-Free Coal}} - \text{Pentane Yield}$$

The sample number consists of the reaction temperature, solvent and reaction time, consecutively. The temperature is in degrees C and the time is in minutes.

Table 1: Extraction Yields and Elemental Analysis Data for Hydroliquefied Samples

Proton NMR, ^{13}C NMR and elemental analysis data were obtained for each of the samples. The results of the elemental analysis and the yields of the extractions are listed in Table 1. The NMR data are given in Table 2.

RESULTS AND DISCUSSION

The pentane and toluene extracts examined here contain liquefaction products of moderate molecular weight. Low molecular weight products are removed with the liquefaction solvent and very high molecular weight products tend to be insoluble in pentane and toluene. The purpose of the hydroliquefaction experiments was to help define the structures of moderate molecular weight products and obtain clues about reaction pathways.

We start by examining the data of Tables 1 and 2. The samples liquefied in tetralin and SRC-HD exhibit some straightforward trends. As the residence time increases, aromatic carbon increases, aliphatic carbon and hydrogen decrease. The samples are also deoxygenated. The changes in the content of aliphatic carbon and various types of aliphatic hydrogen are more pronounced in the samples generated in the solvent SRC-HD

Sample	Aromatic H[a]	Alpha H[b]	Beta H[c]	Gamma H[d]	Carbon Aromaticity
		Pentane Fractions			
400S40	45.6	25.9	21.2	7.2	.74
460S03	44.6	28.6	20.1	6.6	.72
460S10	48.4	29.5	17.3	4.8	.76
460S40	56.3	31.5	10.2	1.9	.81
400T40	23.1	38.1	29.5	9.3	.51
460T03	28.4	34.3	27.8	9.5	.54
460T10	29.4	33.9	28.3	8.5	.60
460T40	30.0	36.3	25.7	8.0	.69
		Toluene Fractions			
400S40	39.8	33.6	21.4	5.2	.79
460S03	48.7	32.4	14.9	4.1	.82
460S10	52.4	32.3	11.9	3.5	.84
460S40	66.7	24.2	7.5	1.7	.90
400T40	32.8	36.0	24.9	6.4	.67
460T03	33.9	38.0	23.7	4.4	.70
460T10	43.5	36.2	16.4	3.9	.75
460T40	53.6	29.7	13.3	3.4	.80

[a]Aromatic Hydrogen have chemical shifts between 9.0 and 5.0 ppm.

[b]Alpha Hydrogen have chemical shifts between 5.0 and 1.9 ppm.

[c]Beta Hydrogen have chemical shifts between 1.9 and 1.0 ppm.

[d]Gamma Hydrogen have chemical shifts between 1.0 and 0.5 ppm.

The sample number consists of the reaction temperature in ^{o}C, the solvent and the reaction time in minutes, consecutively.

Table 2: NMR Data

relative to the samples generated in tetralin. They are also more pronounced in the toluene fractions compared to the pentane fractions.

For each sample, the concentrations of the functional groups listed in Figure 1 were calculated. The functional group concentrations were obtained by minimizing equation 5 subject to the constraints of equations 2 and 3. The functional group concentrations listed in Tables 3 and 4 suggest the following interpretation of the analytical data. Partially hydrogenated aromatic structures are partially or completely dehydrogenated and the hydroaromatic portion of some of these structures is being cleaved. Multiple

Functional Group	Liquefied in Tetralin				Liquefied in SRC-HD			
	400^0 40 min	460^0 3 min	460^0 10 min	460^0 40 min	400^0 40 min	460^0 3 min	460^0 10 min	460^0 40 min
	.343	.320	.188	.188	.235	.270	.230	.247
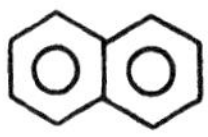	.096	.102	.112	.126	.142	.139	.141	.156
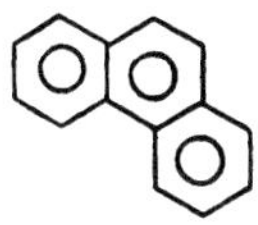	.040	.052	.079	.093	.099	.088	.100	.112
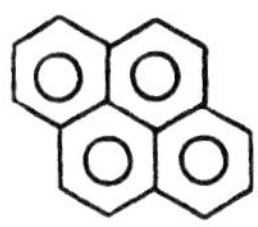	.002	.014	.067	.085	.083	.066	.086	.099
CH_3	.413	.283	.263	.427	.356	.302	.315	.323
 CH_3	.102	.062	.245	.406	.300	.140	.150	.074
 CH_3	.229	.234	.204	.187	.177	.158	.112	.042
	.199	.162	.148	.157	.144	.132	.114	.092
	.001	.000	.023	.204	.000	.000	.015	.075
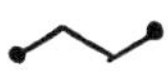	.074	.062	.137	.105	.083	.106	.124	.125
	.000	.000	.000	.159	.000	.000	.000	.000
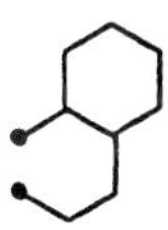	.158	.149	.074	.000	.020	.063	.028	.002
	.044	.069	.081	.035	.031	.040	.048	.033
 OH	.338	.338	.281	.225	.138	.150	.138	.125

Concentrations are in moles per 100 g of sample.

Table 3: Functional Group Concentrations for the Pentane Soluble Fractions

Functional Group	Liquefied in Tetralin				Liquefied in SRC-HD			
	400^0 40 min	460^0 3 min	460^0 10 min	460^0 40 min	400^0 40 min	460^0 3 min	460^0 10 min	460^0 40 min
	.176	.182	.200	.218	.151	.215	.223	.246
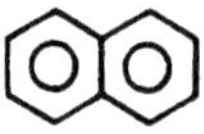	.114	.122	.134	.146	.116	.144	.150	.165
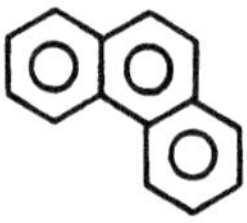	.083	.091	.100	.109	.145	.108	.111	.123
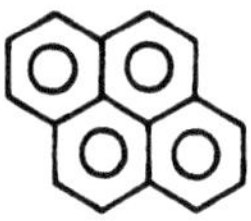	.075	.083	.091	.099	.090	.098	.102	.112
CH_3	.129	.276	.195	.128	.856	.410	.247	.212
 CH_3	.123	.235	.095	.051	.478	.199	.111	.072
 CH_3	.136	.094	.081	.068	.109	.078	.062	.028
	.101	.119	.083	.061	.006	.082	.069	.044
	.182	.237	.244	.201	.001	.128	.171	.116
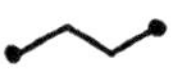	.133	.118	.123	.099	.001	.059	.084	.056
	.000	.063	.046	.105	.406	.384	.453	.568
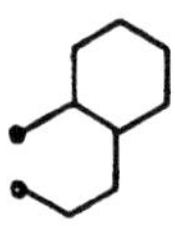	.077	.026	.032	.033	.000	.000	.000	.000
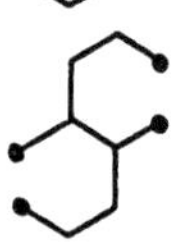	.084	.056	.057	.050	.000	.000	.021	.002
OH	.400	.400	.375	.338	.331	.406	.344	.275

Concentrations are in moles per 100 g of sample.

Table 4: Functional Group Concentrations for the Toluene Soluble Fractions ring hydroaromatic groups (functional groups 12 and 13) appear to be more labile than single ring hydroaromatics (group 8).

CONCLUSIONS

In general, functional group analysis is best suited for the structural analysis of complex liquids on which a large amount of data, from a variety of sources, is available. When applied, the method has several distinct advantages over other methods of structural characterization. It allows data from diverse sources such as NMR, IR and separation procedures to be incorporated into a single characterization; changes in structure are easily quantified by comparing concentrations; the types of structures present are easily visualized and the concentrations could provide a starting point for kinetic modeling.

REFERENCES

(1) Brown, J.K. and Ladner, W.R. 1960, Fuel, 39 pp. 87-96.
(2) Charlesworth, J.M. 1980, Fuel, 59 pp. 865-870.
(3) Clutter, D.R., Petrakis, L., Stenger, R.L. and Jensen, R.K. 1972, Anal. Chem., 44 pp. 1395-1405.
(4) Dickinson, E.M. 1980, Fuel, 59 pp. 290-294.
(5) Oka, M., Chang, H. and Gavalas, G.R. 1977, Fuel, 56 pp. 3-8.
(6) Allen, D.T., Petrakis, L., Gavalas, G.R. and Gates, B.C. 1983, Anal. Chem., in press.

ANALYSIS OF MIDDLE DISTILLATE FUELS BY FLOW LIQUID CHROMATOGRAPHY/PROTON NUCLEAR MAGNETIC RESONANCE

Robert N. Hazlett

Naval Research Laboratory, Washington, D. C., USA

and

Harry C. Dorn and Thomas E. Glass

Virginia Polytechnic Institute and State University, Blacksburg, Virginia, USA

Normal phase liquid chromatography (LC) is an appropriate technique for separating middle distillate fuels - jet and diesel - into major hydrocarbon classes. The use of a proton nuclear magnetic resonance (^{1}H NMR) instrument as a continuous flow LC detector, in conjunction with a halocarbon liquid phase, generates a proton spectrum for each separated fraction in the sample. Specific proton types are measured via appropriate integration windows for each hydrocarbon class. A detailed set of equations allows the integration data to be interpreted as an average molecular composition for each chromatographic fraction. The treatment is particularly useful with the mono- and di-aromatic fractions.

INTRODUCTION

The middle distillate fuels used by the U. S. Navy - diesel fuel marine (DFM) and JP-5 jet fuel - must meet many specification requirements. These requirements are related to properties which are important with respect to fuel use and handling. The most significant characteristics of these fuels relate to safety, stability, cold flow behavior and combustion phenomena. Safety aspects depend on physical properties for the most part but the other key characteristics are controlled by

L. Petrakis and J. P. Fraissard (eds.), Magnetic Resonance. Introduction, Advanced Topics and Applications to Fossil Energy, 709–720.

chemical composition. Trace nitrogen, sulfur, and oxygen compounds in fuel exert a major control on instability patterns and n-alkanes predominate the freezing and pour point phenomena. Aromatic hydrocarbons in middle distillate fuels influence the combustion phenomena, particularly in gas turbine fuels. Thus it is important to define the aromatics in Navy fuels and relate their concentration and character to combustion parameters.

Total aromatic hydrocarbons are determined by an empirical test in the military jet fuel specification (1). Fraction aromatic carbon by ^{13}C NMR has been suggested as an alternate approach to the definition of total aromatics (2). Mass spectrometry with atmospheric pressure column chromatography is sometimes tried to get a breakdown of aromatic hydrocarbons by classes (3). None of these techniques afford detailed, reliable information on the composition of middle distillate fuels.

A new analytical technique, combining high pressure liquid chromatography (HPLC) with high field proton NMR as a detector provides a powerful method for defining the aromatic hydrocarbons in fuels. This combined continuous flow separation/analysis scheme allows the definition of the average molecular composition of each chromatographic fraction and the quantitative determination of structural features of each class. The separation requires only 30-40 minutes.

EXPERIMENTAL

Separation of hydrocarbon classes was accomplished on a Whatman Magnum-9 PAC column (500 mm x 9 mm i.d.). The packing in this column is silica gel which has been derivatized to introduce amino and cyano functionalities to the surface. A non-protonated solvent, 97.5% 1,1,2 trichlorotrifluoro ethane and 2.5% deuterated chloroform, insured a proton NMR spectra free from solvent interferences. The chromatographic solvent also contained 0.05% (v/v) of hexamethyldisiloxane (HMDS) as a chemical shift and quantitation reference.

The column effluent flowed directly to a refractive index detector and on to a NMR flow cell located in a Jeol FX-200 nuclear magnetic resonance spectrometer. The NMR cell design and operating parameters have been described (4,5,6,7). Results with simple model hydrocarbon mixtures have been reported (7,8).

TREATMENT OF DATA

The proton NMR data were integrated for specific structural features in the various hydrocarbon classes. The monocyclic aromatic and dicyclic aromatic fractions were particularly amenable to the definition of average molecular parameters. The breakout of structural types for the aromatics is depicted in Figure 1. The integration windows corresponding to the specific proton types are listed in Table I along with other symbols used in this paper.

TABLE I - Definitions of Representative Symbols
(See Figure 1 for structural designations)

H^x_{AR}	^1H NMR integrated area for aromatic hydrogens in the monocyclic or dicyclic aromatic fraction (x = m or d), m window - 6.3-7.5 ppm; d window -6.7-8.5 ppm
$H^x_{\alpha CH_3}$	^1H NMR integrated area for aliphatic methyl hydrogen α to aromatic rings (e.g., methyl groups for toluene and/or β-methylnaphthalene) m window 1.95-2.40 ppm; d window 2.2-2.7 ppm.
$H^x_{\alpha CH_2}$	^1H NMR integrated area for aliphatic methylene hydrogen α to aromatic rings (e.g., methylene group in ethyl-benzene) m window 2.40-2.68 ppm; d window 2.7-2.9 ppm.
$H^m_{\alpha CH}$	^1H NMR integrated area for aliphatic methine hydrogen α to aromatic rings, m window 2.9-3.1 ppm.
$H^x_{>\alpha}$	^1H NMR integrated area for aliphatic hydrogen groups not adjacent to aromatic rings (i.e., greater than α), m window 0.50-1.95; d window 0.50-2.0.
$H^x_{\alpha tet}$	^1H NMR integrated area for aliphatic ring hydrogen α to aromatic rings, m window 2.68-1.9; d window 3.2-4.0.
ADS^x	the average degree of aromatic substitution for the monocyclic or dicyclic fractions.
MW^x	the number average molecular weight for the monocyclic or dicyclic fraction
f_a	the carbon aromaticity for the total sample (i.e., the total aromatic carbon ($C_{Ar(total)}$) divided by total carbon (C_{total})).
f_a^x	the partial carbon aromaticity contribution for the monocyclic aromatic or dicyclic aromatic fractions (i.e., $f_a = f_a^m + f_a^d$)
F^x_{total}	the fraction of total carbon in each chromatographic fraction (i.e, $F^a_{total} + F^m_{total} + F^d_{total} = 1$)

$*C^{x}_{un}$ number of unsubstituted aromatic carbons in hypothetical average molecule

$*C^{x}_{sub}$ number of substituted aromatic carbons

$*C^{x}_{\alpha CH_3}$ number of methyl carbons α to aromatic ring

$*C^{x}_{\alpha CH_2}$ number of methylene carbons α to aromatic ring

$*C^{x}_{\alpha tet}$ number of tetralin carbons α to aromatic ring

$*C^{x}_{CH_2 > \alpha}$ number of aliphatic methylene carbons not bonded to aromatic ring

$*C^{x}_{CH_3} > \alpha$ number of terminal methyl carbons not bonded to aromatic ring

$*C^{x}_{BH}$ number of bridgehead carbons for the dicyclic aromatic fraction

$*C^{x}_{T}$ total number of carbons in a given fraction

The proton data is converted to moles of carbon type by using the appropriate C/H ratios. The amounts of carbon types can be quantitated relative to the total proton signal of the sample or absolutely by comparison to the amount of HMDS reference. Equations have been derived (8,9) to convert the integration data into moles of substituted, unsubstituted and bridgehead aromatic carbons in a fraction; average degree of ring substitution; the carbon aromaticity for the total sample and for each fraction; the fraction of total carbon in each chromatographic fraction; and the number average molecular weight.

RESULTS AND DISCUSSION

The capability of the HPLC system is illustrated in Figure 2. The three major hydrocarbon classes for a JP-5 jet fuel (81-9) are well resolved in this trace for a refractive index detector. The aliphatic class peaks at 17 min, the monocyclic aromatic class at 22 min, and the dicyclic aromatic class at 34 min. The partially resolved shoulder on the mono-aromatic peak is due to tetralins.

As a consequence of the good separation on the HPLC column, the NMR can afford meaningful data on each separated fraction when it serves as the detector. The profiles for a DFM (81-6)

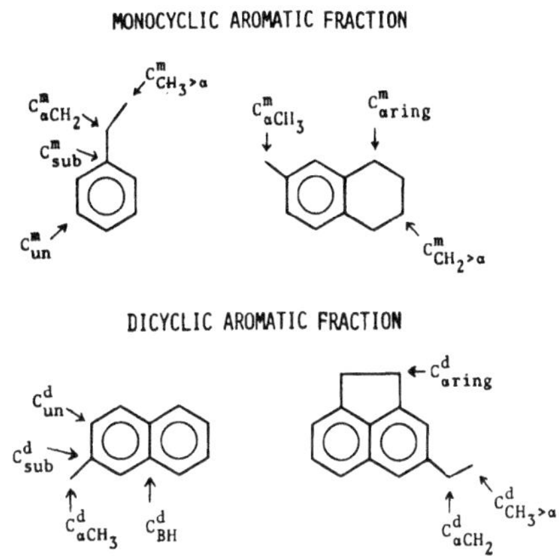

Figure 1. Representative carbon types for the monocyclic and dicyclic aromatic fractions

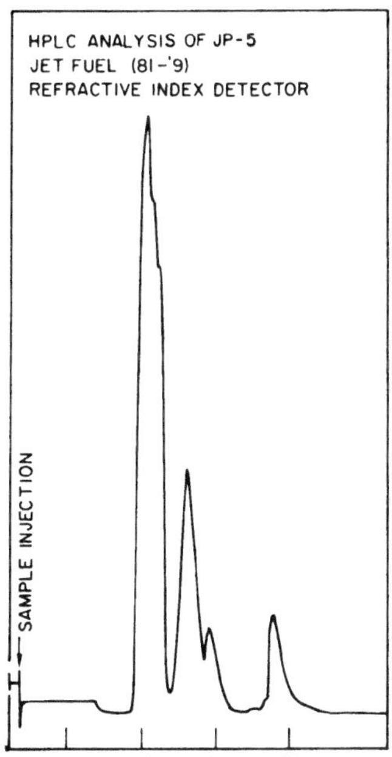

Figure 2. Typical chromatogram with refractive index detector. See text for separation conditions.

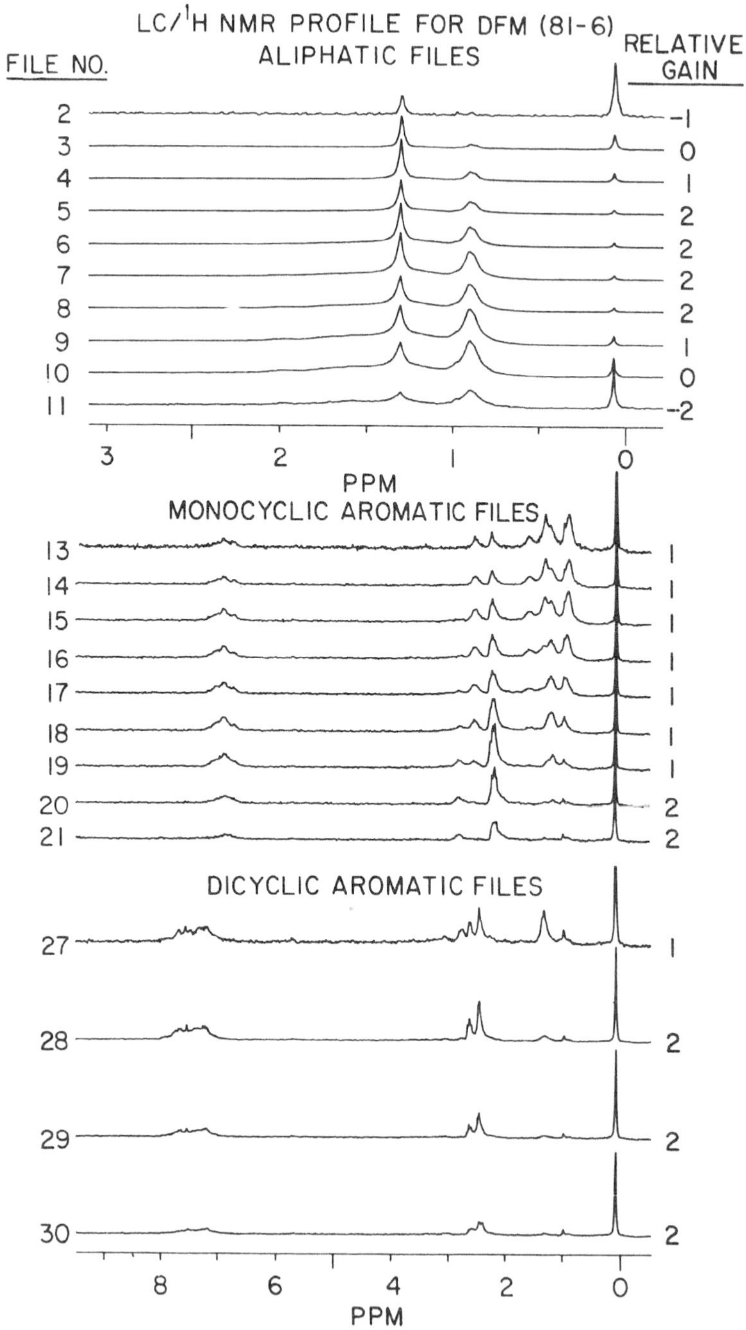

Figure 3. Proton NMR spectral files for a diesel fuel marine sample

are depicted in Figure 3. Note that the scale in the Figure for the aliphatic files, 2 - 11, covers a narrower range than the scale used for the monocyclic, files 13 - 21, and dicyclic aromatics, files 27 - 30.

The reliability of the LC/^{1}H NMR integration and quantitation procedures is illustrated by comparison with GC/MS data on toluene bottoms, a multicomponent chemical blending stock (8). The average molecular parameters for the monoaromatic fraction (99% of total) are listed in Table II. The two techniques give excellent agreement. It is clear that the LC/^{1}H NMR measurements are accurate and that the average composition equations are correct and fully applicable to the monocyclic aromatic hydrocarbons.

Table II. Comparison of LC-NMR and GC-MS for Monocyclic Fraction of Fuel 81-3

Property	NRL GC-MS	VPI LC-NMR
MW^m	116.3	116.5
F_a^m	0.677	0.682
$*C^m_{\alpha CH_3}$	2.12	2.07
$*C^m_{\alpha CH_2}$	0.291	0.278
$*C^m_{\alpha CH}$	0.009	0.0007
$*C^m_{\alpha tet}$	0.00	0.013
$*C^m_{CH_2>\alpha}$	0.031	0.108
$*C^m_{CH_3>\alpha}$	0.309	0.279
ADS	2.39	2.36

The flow LC/^{1}NMR procedure has been applied to over 60 samples - petroleum and shale derived jet and diesel fuels, aromatic and aliphatic blending stocks, and hydrocracked fractions. The measured and calculated average structural parameters are presented in Table III for two jet fuels and three DFM's.

Table III - Average Structural Parameters for Middle Distillate Fuels

Parameter	JP-5 (80-32) Petroleum	JP-5 (80-33) Shale	DFM (81-5) Petroleum Straight Run	DFM (81-6) Petroleum Cat-Cracked	DFM (D-11) Shale
(a) Fractional Aromaticity Data					
F^{a}_{total}	0.777	0.729	0.680	0.529	0.693
f_{a}^{m}	0.098	0.143	0.095	0.112	0.125
F^{m}_{total}	0.182	0.254	0.199	0.212	0.265
f_{a}^{d}	0.036	0.013	0.098	0.206	0.033
F^{d}_{total}	0.041	0.017	0.121	0.258	0.042
f_{a}	0.134	0.156	0.194	0.318	0.158
(b) Monocyclic Aromatic Fraction					
ADS^{m}	2.28	2.48	2.60	2.49	3.05
MW^{m}	150.4	142.6	169.3	153.3	171.0
$*C^{m}_{\alpha CH_3}$	1.48	1.39	1.49	1.46	1.13
$*C^{m}_{\alpha CH_2}$	0.665	0.665	0.727	0.684	0.902
$*C^{m}_{\alpha CH}$	0.019	0.065	0.056	0.068	0.470
$*C^{m}_{\alpha tet}$	0.118	0.365	0.329	0.282	0.549
$*C^{m}_{CH_2 > \alpha}$	2.21	1.43	3.26	2.16	2.28
$*C^{m}_{CH_3 > \alpha}$	0.684	0.728	0.784	0.752	1.37

(c) Dicyclic Aromatic Fraction

ADS[d]	1.30	1.55	1.62	1.71	2.08
MW[d]	145.8	151.4	160.0	163.6	166.7
$*C^{d}_{\alpha CH_3}$	1.22	1.28	1.50	1.57	1.62
$*C^{d}_{\alpha CH_2}$	0.08	0.272	0.123	0.146	0.469
$*C^{d}_{CH_2 > \alpha}$	0.00	0.12	0.524	0.668	0.195
$*C^{d}_{CH_3 > \alpha}$	0.08	0.272	0.123	0.146	0.469

Of the two jet fuels, the shale derived fuel (10) contains about 5% more total aromatics, 7% more monocyclic aromatics, 2% less dicyclic aromatics, and a 2% higher fraction aromaticity than the petroleum derived jet fuel. The shale monoaromatic class has slightly more average substitution but somewhat shorter side chains. The petroleum fuel contains only one-third the amount of tetralins compared to the shale. This is explained by the hydrocracking process used to produce the latter fuel. Naphthalenes are converted to tetralins in this high pressure hydrogenation. The dicycloaromatic fractions of the jet fuels consist primarily of methyl and dimethylnaphthalenes although the shale contains some 2 and 3 carbon sidechains.

The diesel fuels have more total aromatics than the jet fuels and can have a significantly higher fraction aromaticity. This is due, mainly, to much more dicyclic aromatics in the diesel fuels. Noteworthy is the more than 25% dicyclic aromatics in DFM 81-6, a product made by catalytic cracking. Significant instability problems have been encountered with this fuel. The diesel fuel fractions also exhibit higher average molecular weights, greater ring substitution, and longer side chains than the jet fuels.

Comparison of the two petroleum derived diesel fuels indicates that the straight run product is much lower than the catalytically cracked product in total aromatics, carbon fraction aromaticity, and dicyclic aromatics, but higher in average side chain length for the monocyclic aromatic fraction. Other structural features are quite similar for these two fuels.

The shale derived DFM (10) is low in dicyclic aromatics but high in mono-aromatics, average degree of substitution, and tetralins. The side chains of the mono-aromatic fraction of the shale DFM appear to have considerable branching.

We have explored several possible correlations of fuel structural information with the measured smoke point. The latter measurement is an empirical test in the jet fuel specification (1) to control combustion behavior in gas turbine engines. The carbon fraction aromaticity affords a reasonable relationship with smoke point and gives a correlation coefficient of 0.80 for a set of 18 fuels.

Another parameter of interest is the substituted aromatic carbon content. This has some theoretical significance since substituents would serve as points of intermolecular condensation to build the high molecular weight soot and smoke particles during combustion.

Based on some model hydrocarbon studies, an equation was developed to predict smoke points based on LC/^{1}H NMR derived molecular parameters.

$$(\text{smoke point})_{\text{predicted}} = 21.5 + 15X^a + X^m \left[\frac{(^*C^m_{sub})(-139.5)}{^*C^m_T} - 21.8\right]$$

$$+ X^d \left[\frac{(^*C^d_{sub} + {^*C^d_{BH}})(-139.5)}{^*C^d_T} - 31.2\right]$$

This equation includes factors for the hypothetical average molecule. These factors are the moles of substituted aromatic carbon for the monocyclic ($^*C^m$ sub) and dicyclic ($^*C^d$ sub) aromatic fractions, for the total carbon in the monocyclic ($^*C^m_T$) and dicyclic ($^*C^d_T$) fractions, and the moles of bridgehead carbon for the dicyclic aromatic fraction (i.e., $^*C^d_{BH} = 2.0$). The values for these parameters are obtained directly from the average molecule data obtained from the LC/^{1}H NMR results. In addition, the equation includes values for the mole fractions of the aliphatic (X^a), monocyclic (X^m), and dicyclic (X^d) aromatic fractions.

The constants 15, -139.5, -21.8 and -31.2 were all derived from the synthetic model study. The -139.5 value was determined from results with a series of substituted monocyclic aromatic hydrocarbons. Since data were not available for dicyclic aromatics, this same -139.5 value was used for them. The 21.5 constant was empirically derived.

Smoke point values calculated using this equation are plotted in Figure 4 versus the measured smoke point. The good relationship indicated is corroborated by the excellent correlation coefficient of 0.91. It must be noted that this relationship is limited to the smoke point range used in the graph.

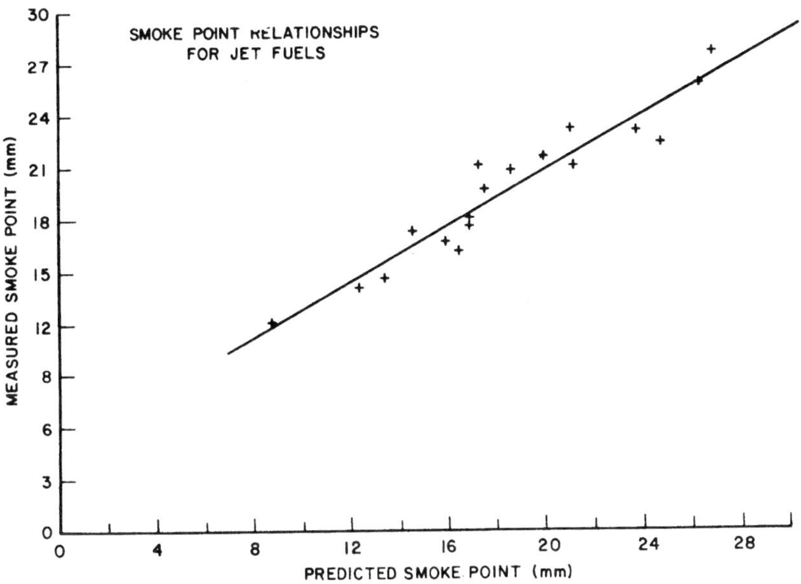

Figure 4. Smoke point comparisons. Predicted value calculated using equation in text.

Thus the LC/^{1}H NMR analysis of middle distillate fuels provides information which can be related to an important fuel property, namely combustion behavior. The use of the bridgehead carbon factor for dicyclics along with factors for degree of ring substitution is reasonable since this accounts for a condensation which is already accomplished.

CONCLUSIONS

The analysis of a middle distillate fuel by continuous flow LC/^{1}H NMR affords a wealth of structural information about the various hydrocarbon classes in a fuel. The treatment is particularly useful for the mono- and dicyclic aromatic fractions. The analysis method has been applied to over 60 samples.

Selected molecular parameters derived by this combined analytical procedure have been incorporated into an equation which correlates with measured smoke points.

REFERENCES

1. Turbine Fuel, Aviation, Grades JP-4 and JP-5, U. S. Military Specification, MIL-T-5624L, 18 May 1979.
2. Dorn, H. C., '^{1}H and ^{13}C Fourier Transform NMR Characterization of Jet Fuels Derived From Alternate Energy Sources,' VPI and SU, Final Progress Rpt., Feb. 1982.
3. American Society for Testing and Materials, Methods D2425-67, D2786-81, and D3239-81, Annual Book of Standards, Philadelphia, PA. 1982.
4. Dorn, H. C. and Wooton, D. L. 1976, Anal. Chem. 48, pp. 2146-2148
5. Haw, J. F., Glass, T. E. and Dorn, H. C. 1981, Anal. Chem. 53, pp. 2332-2336.
6. Haw, J. F., Glass, T. E. and Dorn, H. C. 1982, J. Magn. Reson. 49, pp. 22-31.
7. Haw, J. F., Glass, T. E., Hausler, D. W., Motell, E. and Dorn, H. C. 1980, Anal. Chem. 52, pp. 1135-1140.
8. Haw, J. F., Glass, T. E. and Dorn, H. C. 1983, Anal. Chem. 55, pp. 22-29.
9. Haw, J. F., Glass, T. E. and Dorn, H. C. 1981, Anal. Chem. 53, pp. 2327-2332.
10. 'Oil Shale, Tar Sands and Related Materials,' H. C. Stauffer, ed., (a) Wasilk, N. J., and Robinson, E. T., Ch. 15, pp. 223-236, (b) Solash, J., Hazlett, R. N., Burnett, J. C., Beal, E., Hall, J. M., Ch. 16, pp. 237-252, (c) Affens, W. A., Hall, J. M., Beal, E., Hazlett, R. N., Leonard, J. T., Nowack, C. J. and Speck, G., Ch. 17, pp. 253-266, A. C. S. Symposium Series No. 163, Am. Chem. Soc., Washington, D. C. 1981.

COMPLEMENTARY APPLICATION OF CHROMATOGRAPHY, MASS SPECTROMETRY AND N.M.R. SPECTROSCOPY TO THE STUDY OF U.K. COAL MACERAL EXTRACTS

John A. G. Drake* and Derry W. Jones**

* National Institute for Higher Education, Limerick, Republic of Ireland;

** School of Chemistry, University of Bradford, Bradford, W. Yorkshire, BD7 1DP, U.K.

ABSTRACT

In studies of solvent extracts (refluxed in pyridine and in carbon disulphide) of groups of macerals (Hirsch series) sorted from three bituminous U.K. coals, gas-liquid chromatography and Shpol'skii luminescence spectra indicate the presence of several individual polycyclic aromatic hydrocarbons of up to 10 rings. High-resolution ^{1}H NMR spectra are consistent with occupation by alkyl substituents of many of the peripheral aromatic-ring sites and with a dependence of aromatic content on maceral type. Molecular-ion peaks from field-desorption mass spectrometry of carbon-disulphide extracts correspond to the presence of at least some highly condensed aromatic systems of MM above 1000 and of long-chain alkanes (especially n=24 to 35).

INTRODUCTION

Like other fossil fuels, coal is a heterogeneous material formed over geological time from decomposition of mainly plant matter. The predominant non-mineral organic carbonaceous material occurs macroscopically in bands corresponding to lithotypes of somewhat different chemical compositions(1). Microscopically there are three main chemically and physically distinct groups of macerals: exinite (E) (most reactive, most volatile, and highest H/C ratio), vitrinite (V), and inertinite (I) (least reactive, least volatile, lowest H/C ratio). In general, for higher rank coals, chemical compositions of these maceral groups are much the same but for bituminous coals of rather lower rank the inertinites have the highest carbon contents.

L. Petrakis and J. P. Fraissard (eds.), Magnetic Resonance. Introduction, Advanced Topics and Applications to Fossil Energy, 721–728.

The wide range of magnetic resonance techniques is particularly sensitive to the presence of specific unpaired-electron species and chemical functional groups, though may be less effective at distinguishing between closely related individual members of a series. So-called average structures can be derived (2) though these should be regarded as representative of the structural types present rather than as well-defined molecular structures(3). With these limited aims, investigation of the complex structure of coals requires the combination of several spectrometric and spectroscopic techniques, each sensing different facets of molecular structure. In turn, these techniques are most effective when preceded and accompanied by differential solvent extraction, chromatography and other separation procedures(4). Size-exclusion chromatography (SEC) is emerging as a most effective means of determining MM distributions for coal derivatives(5), while gas-liquid chromatography (GLC) is particularly valuable for series such as the n-alkanes.

Electron-impact (EI) mass spectrometry (MS) of coals and coal derivatives yields peaks from fragment hydrocarbon ions as well as from molecular ions, with consequent excessive complexity of spectra. More gentle methods of ionization can yield predominantly parent ions. Field ionization (FI) (i.e. with very high electric-field gradients) has an advantage over chemical ionization (CI) in that it produces few quasi-molecular $(M-1)^+$, $(M+1)^+$, etc ions(6). Soft ionization MS methods of studying higher molecular-mass (MM) ranges in coal materials often involve experimental complexities though bombardment with fast noble-gas atoms (FAB) is promising as a way of obtaining MM profiles. Field-desorption (FD) MS, in which ions from the sample on a heated wire emitter are desorbed with the aid of a high electric field, has revealed the presence of cyclic alkanes in coal tar(7), and is now applied to coal macerals.

Whereas MS gives rise to MM distributions, Shpol'skii luminescence spectroscopy can reveal the presence of some individual polycyclic aromatic hydrocarbons (PAH). Simple ^{1}H NMR of solutions can enable comparisons to be made between the distribution of groups of aromatic and other types of hydrogen in different extracts.

EXPERIMENTAL

The samples of high-volatile, caking coals formed part of the well-characterized Hirsch series (but stored under nitrogen), manually sorted into vitrinite, exinite and inertinite groups of macerals, from three coalfields in England. Extraction was by refluxing finely ground maceral samples for 6h in (a) carbon disulphide and (b) pyridine. 100 MHz ^{1}H NMR spectra of carbon-disulphide and pyridine solutions were recorded on a Jeol MH-100 spectrometer. Shpol'skii luminescence spectra (370-480 nm) of carbon-disulphide extracts in cyclohexane/n-hexane solvents(8) were recorded at 77 K on a luminescence spectrometer(9) (Imperial College, London) with excitation by a xenon-arc or mercury-vapour lamp. Gas chromatography was on a column packed with 4% OV 101 supported on Celite AW, with nitrogen carrier gas over a programmed temperature range 423-593 K at 60 min^{-1} on a Pye Cambridge 104 chromatograph. FD mass spectra were recorded (from a sample deposited on a carbon filament) on a Varian CH5D mass spectrometer linked to a Varian Spectrosystem 100 (University College, Cardiff).

RESULTS AND DISCUSSION

For each maceral sample, much higher percentages were extracted by the more polar solvent pyridine (from 1% for two inertinites to 35% for Chislet vitrinite) than by carbon disulphide (0.3 for two inertinites to 6.3 for Chislet exinite). This is especially true for the three vitrinites, which contain more oxygen than the other macerals; presumably pyridine is better able to dissolve rather larger PAH and those with oxygen-containing groups. For the vitrinites, the proportion extracted by pyridine increased - 13, 16, 35% - as NCB coal rank code changed in the sequence 902, 702, 601, i.e. as coal rank increased. Retcofsky and Friedel (10) found pyridine extractions higher (∿21%) for US coals with m.a.f. 83-87% C than for coals of slightly lower or higher rank.

Differences between hydrogen-type indications from ^1H NMR spectra of corresponding extracts from the two solvents were greater than reported for US vitrains(10) and Japanese coals(11). For all the pyridine extracts, estimates of aromatic content derived from ^1H NMR spectra are much greater (e.g. 42% compared with 11% for Chislet inertinite) than for carbon-disulphide extracts. The carbon-disulphide extracts contain more H_β and H_γ (i.e. predominantly methylene and methyl groups β and γ to aromatic rings) but generally less aromatic hydrogen and hydrogens α to aromatic rings than the corresponding pyridine extracts; the solvent difference is smaller for the exinites than for other macerals. In carbon-disulphide extracts, percentages of H_α are in the sequence E > V > I, e.g. 16, 8 and 5% for Markham Main (Fig. 1), while aromatic hydrogen contents are greater for E than I.

Shpol'skii spectra of all the samples examined(8) contained lines assigned to the 7-ring PAH coronene (A). For the Chislet (Kent) coal, Shpol'skii spectra (Fig. 2) of all three macerals also contained lines attributed to ovalene (B), with ten rings, and perylene (C); overlap for these two at 445 nm can be removed by changing the matrix from n-hexane to n-heptane. The Markham

Main (E and V) and Cannock Wood E extracts all included benz[g,h,i]perylene (D), pyrene (E), and benz[a]pyrene (F) (for which the method seems particularly sensitive), while other individual PAH identified in these three extracts included chrysene (H), benz[a]anthracene (I), 9,10-dimethylanthracene (J) and dibenz[a,h]pyrene

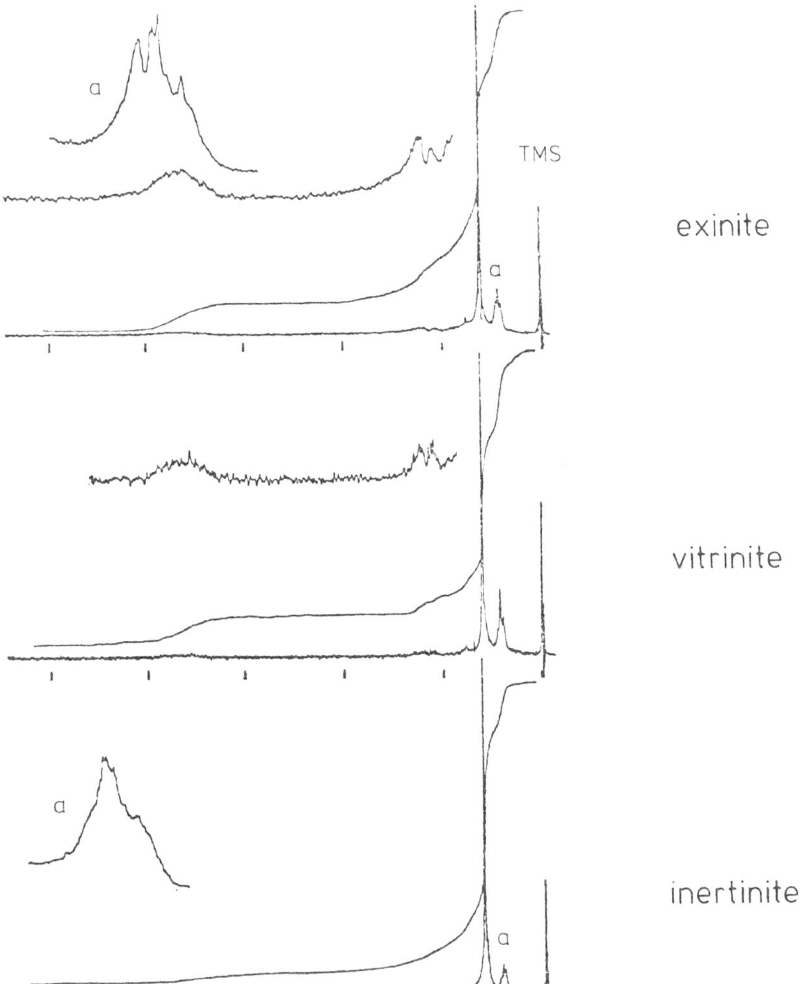

Figure 1. 100 MHz ^{1}H NMR spectra of carbon-disulphide extracts of macerals from Markham Main high-volatile coals; D.m.m.f. % C : E, 83; V, 82; I, 92.

(L). For all the PAH detected in the luminescene spectra, lines of appropriate MM were also observed in the corresponding FDMS. While positive identification of such large PAH in coal extracts that have been subjected only to rather mild treatment is surprising in view of the paucity of plausible naturally occurring precursors with so many rings, it should be emphasised that the quantities detected are small and in any case they relate only to the small carbon-disulphide-soluble fraction of the macerals.

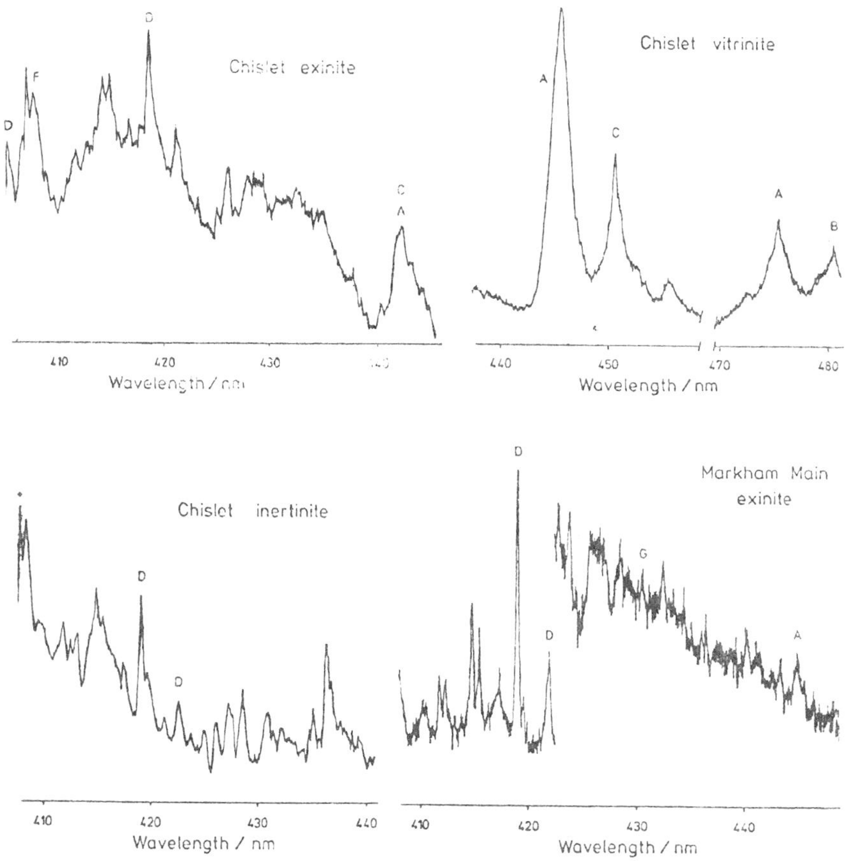

Figure 2. Shpol'skii luminescence spectra of carbon-disulphide extracts of Markham Main exinite and three Chislet high-volatile coal macerals: E, 85% C, strongly caking; V, 85% C, medium caking; I, 86% C, weakly caking.

The GL chromatograms were calibrated with 13 reference PAH containing up to five rings but retention volumes for the PAHs in the pyridine extracts corresponded to compounds predominantly of only two or three rings. GLCs of the carbon-disulphide extracts of all the nine coal/maceral groups showed a homologous series of n-alkanes (calibrated with nine n-alkanes) from C_{12} to C_{34} (12). The pyridine-extract GLC s of the Chislet and Cannock Wood macerals showed much less intense n-alkane peaks than the Markham Main macerals and less than the GLCs from any of the carbon-disulphide extracts. Extracts with high H_β and H_γ content from ^{1}H NMR had relatively high n-alkane peaks in their chromatograms. For a sample of n-alkanes, typically n=26, the methylene/methyl ratio $H_{\beta 1}/H_\gamma$ would be 8:1, whereas the highest ratios in the coal extracts are 4 to 4.5 for inertinites from Markham Main and Cannock Wood (carbon-disulphide extracts); for most samples the ratio is below three. Evidently in all the extracts alkanes must be branched and/or in short alkyl chains attached to PAH.

In all of the FDMS (available for most carbon-disulphide extracts but not for pyridine extracts), n-alkane lines are a prominent feature, especially for the inertinites from Chislet

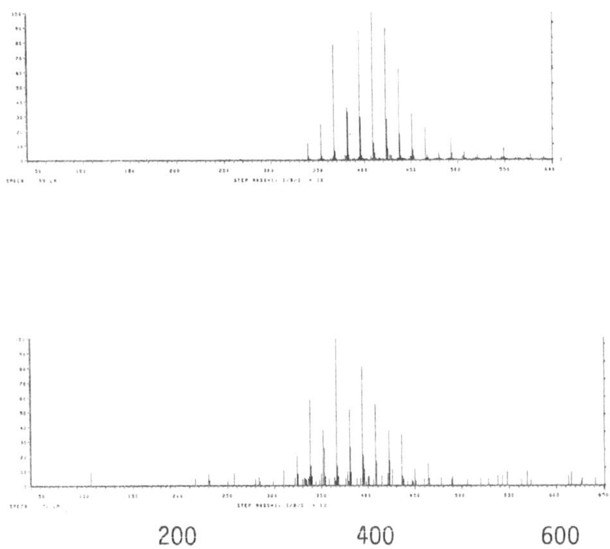

Figure 3. Parts of FDMS spectra of carbon-disulphide extracts of Markham Main coals : upper, vitrinite, MM range 100-670; lower, inertinite, MM range 100-650.

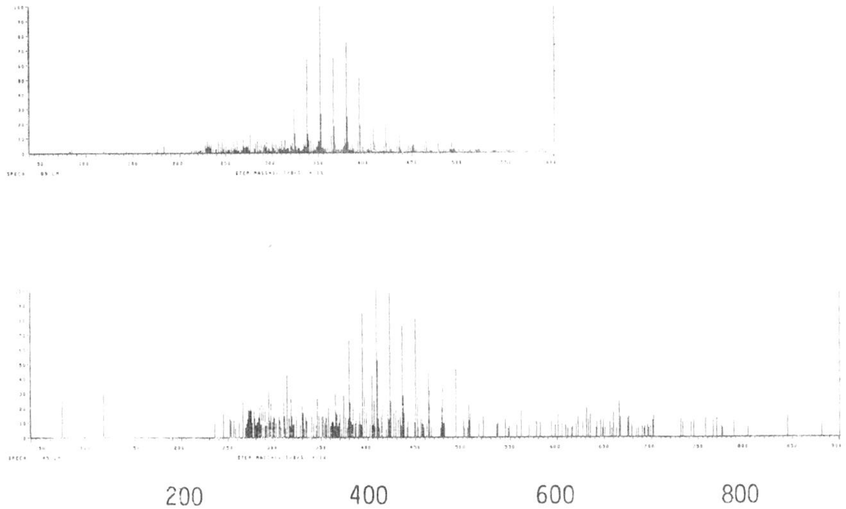

Figure 4.. Parts of FDMS spectra of carbon-disulphide extracts of (upper) Chislet inertinite (MM range 50-550) and (lower) Cannock (MM range 350-900) vitrinite (high-volatile, non-caking coal, 80% C).

and Cannock Wood (Figs. 3 and 4). Whereas GLC senses the more volatile n-alkanes, the FDMS, even those recorded at the lower temperatures, show appreciable C_nH_{2n+2} concentrations for C_{23}-C_{25}. As the source temperature is raised some fractionation can occur, so that the centre of an alkane distribution is shifted from MM 350-400 to 400 or more; some individuals appear to be as high as C_{60}. Other molecular-ion peaks appear to be as high as MM 1200, i.e. appreciably higher than sensed by EI MS and supporting the Shpol'skii spectra.

ACKNOWLEDGEMENTS

The authors thank Mr. E. Bradburn (N.C.B., Doncaster) for samples, Dr. D. E. Games (University College, Cardiff) and Mr. B. L. Gower for FD mass spectra, and Professor G. F. Kirkbright (UMIST, Manchester) and Dr. B. S. Causey for Shpol'skii spectra.

REFERENCES

1. Grimes, W.R., 1982, in M.L. Gorbaty, J.W. Larsen and I. Wender (eds) "Coal Science", Academic Press, New York, pp 21-42.

2. Bartle, K.D., and Jones, D.W., 1978, in C. Karr (ed) "Analytical Methods for Coal & Coal Products", Vol. 2, Academic Press, New York, pp 103-160.

3. Speight, J.G., and Long, R.B., 1982, in R.C. Filby, B.S. Carpenter and R.C. Ragani (eds) "Atomic & Nuclear Methods in Fossil Energy Research", Plenum Press, New York, pp 295-321.

4. Bartle, K.D., Jones, D.W., and Pakdel, H., 1982, in E. L. Fuller (ed) "Coal & Coal Products: Analytical Characterization Techniques", ACS Symposium Series 205, American Chemical Society, Washington D.C., pp 27-45.

5. Richards, D.G., Snape, C.E., Bartle, K.D., Gibson, C., Mulligan, M.J., and Taylor, N., 1983, Fuel, 62, 724-731.

6. Strausz, O.P., Rubinstein, I., Hogg, A.M., and Payzant, J.D., 1982, in R.C. Filby, B.S. Carpenter and R.C. Ragani (eds) "Atomic and Nuclear Methods in Fossil Energy Research", Plenum Press, New York, pp 409-441.

7. Jones, D.W., Pakdel, H., and Bartle, K.D., 1982, Fuel, 61, 44-52.

8. Drake, J.A.G., Jones, D.W., Causey, B.S., and Kirkbright, G.F., 1978, Fuel, 57, pp 663-666.

9. Kirkbright, G.F., and De Lima, C.G., 1976, Chem. Phys. Letters, 37, 165-171.

10. Retcofsky, H.L., and Friedel, R.A., 1970, in R.A. Friedel (ed) "Spectroscopy of Fuels", Plenum Press, New York, pp 70-89.

11. Yokoyama, S., Itoh, M., and Takaya, G., 1974, J. Chem. Soc. Japan, Chem. Ind. Chem, 4, 735-741.

12. Drake, J.A.G., Jones, D.W., Games, D.E., and Gower, J.E., *Fuel*, in the press.

600 MHZ PROTON MAGNETIC RESONANCE STUDY OF COAL-DERIVED LIQUIDS

T.Hara and N.C.Li*

Department of Chemistry, Faculty of Engineering, Yokohama National University, Yokohama, Japan 240
* Department of Chemistry, Duquesne University, Pittsburgh, PA 15219, USA

The newly developed 600 MHz ^{1}H NMR Spectrometer is used to characterize coal-derived liquids and their chromatographically separated fractions. The structural changes of the coal liquid caused by the catalytic hydroprocessing as well as the accelerated ageing were monitored by the spectrometer. The distinct and well resolved proton resonance lines in both aromatic and aliphatic regions and IR analysis have been used to identify the major compounds and compound-types. Double resonance technique has been applied for the chemical shift identification of donor protons (α-CH_2, β-CH_2) of partially hydrogenated polynuclear aromatic compounds. An NMR difference technique is applied to determine specific compositional changes in coal liquids derived under identical process conditions, but from different coal sources.

High resolution ^{1}H NMR spectrometry has been used as one of the major spectroscopic tools for chemical characterization of coal-derived materials. Based on the quantitative proton distribution data, average structural information for complex mixtures of coal-derived materials has been obtained (1-3). A higher-field NMR spectrometer (220 MHz) was applied to tar component by Bartle et al. (4). The proton resonance signals of coal-derived liquids were devided into eleven ranges of chemical shift values and the signals in each range were assigned to various types of hydrogen by the addition of pure compounds. Schweighardt et al. used a 250 MHz NMR spectrometer for characterization and to quantify hydroxyl group in coal-liquefaction oils and asphaltenes (5). Hydroxyl group in coal-derived materials was converted into trimethylsilyl ethers and proton signals in methyl groups attached to silicon were measured. The well-resolved NMR spectra enable them to char-

L. Petrakis and J. P. Fraissard (eds.), Magnetic Resonance. Introduction, Advanced Topics and Applications to Fossil Energy, 729–738.

acterize the nature of the original hydroxyl group in coal-derived materials.

In the present study, a 600 MHz spectrometer for ^{1}H NMR operating at 140.9 kgauss is used for the characterization of SRC-II liquids and their chromatographically separated subfractions (6,7). The assignment of well-resolved resonance lines becomes simpler by use of the highest magnetic field to identify major compounds and compound-types found in coal liquids. The structural changes of SRC-II liquid caused by the catalytic hydroprocessing as well as the accelerated ageing were also monitored (8). An NMR difference spectroscopic technique is used on a pair of coal-liquid samples to determine compositional change between them. Such quick and direct characterization without fractionation may help to assess the slight variation of process variables in coal-liquefaction processes.

Experimental

A liquid solvent-refined coal product (SRC-II) was made from Western Kentucky bituminous coal at the Fort Lewis SRC Pilot Plant, having an initial boiling point of 453 K and an end boiling point of 665 K. The coal liquid was fractionated using ion-exchange and silica-gel chromatography (6,7). The SRC-II Middle Distillates (boiling-point range of 450-565 K) used were from three different coals (Illinois No.6, Western Kentucky No.9/14 and Pittsburgh Seam) derived under the same process conditions (6,7).

The 600 MHz ^{1}H NMR spectrometer which operates at a magnetic field of 140.9 kgauss has been developed by Bothner-By and Dadok at Carnegie-Mellon University (9). The resolution determined by half-height width of ethanol triplet was 0.41 Hz, and signal to noise ratio for a single scan was better than 500:1. Tetramethylsilane was used as a lock signal as well as for an internal standard. Each spectrum recorded was the result of 10-100 time-average scans in $CDCl_3$ solutions. The difference spectrum between the two SRC-II Middle Distillates was obtained from the individual proton spectrum recorded under the same concentration (38.2 g/l in $CDCl_3$) and operational conditions.

Results and Discussion

The SRC-II liquid, boiling range of 453-665 K, was fractionated by solvent extraction based on its solubility in n-pentane. The pentane-soluble fraction (A) 99.3 % of SRC-II, was further fractionated by sequential treatment with anion-exchange resin, cation-exchange resin and silica-gel columns, as illustrated in Figure 1. Weight percent yields of the fractions, relative to A are included in Figure 1. The fractions were well characterized by elemental analysis, boiling-point distribution, IR spectroscopy

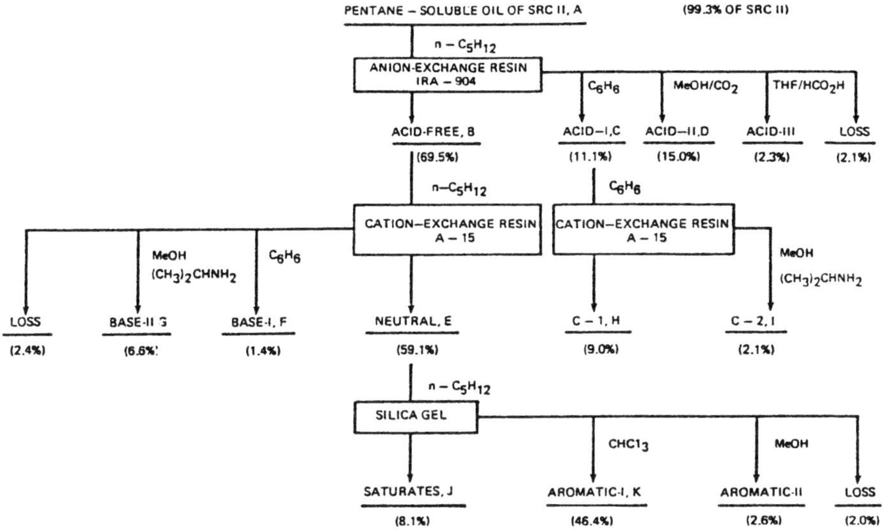

Figure 1. Chromatographic fractionation of SRC-II

and glass-capillary column GC technique (6,7). Heteroatom functionality in acidic and basic fractions were quantitatively analyzed by solution IR technique (6).

The 600 MHz NMR spectrum of Saturate Fraction J shows a triplet at 0.89 ppm (J=6.5 Hz) and a relatively sharp singlet at 1.26 ppm. The result combined with the infrared spectrum and the gas chromatogram shows that J is substantially composed of n-alkanes of high carbon numbers (C_{10}-C_{24}). Figure 2 shows an NMR spectrum of Aromatic Fraction K in the region of aliphatic protons. The resonance lines are well resolved by the use of high magnetic field. Methyl protons directly attached to aromatic ring structure appear as singlets in the region of 2.2-3.4 ppm. In the case of mono-aromatic compounds, methyl protons generally appear at higher magnetic field than 2.4 ppm. In the case of two or more condensed aromatic ring structures, methyl protons appear at lower magnetic field than 2.4 ppm. Signals in the spectrum of Fraction K which increased in their intensities by the small addition of model compounds are marked in Figure 2. It is interesting to note that methyl group is exclusively attached to β position and not to α position in the structure of naphthalenes involved in coal-derived liquids.

Methylene protons at β position to aromatic ring in partially hydrogenated polynuclear aromatic compounds appear in the region of 1.7-2.1 ppm. The multi-coupled signals centered at 1.77 ppm can be assigned as β-CH_2 protons in tetralin and its derivatives. Pentaplet signals at 2.05 ppm can be assigned as β-CH_2 protons in indan and its derivatives. These conclusions were made by observ-

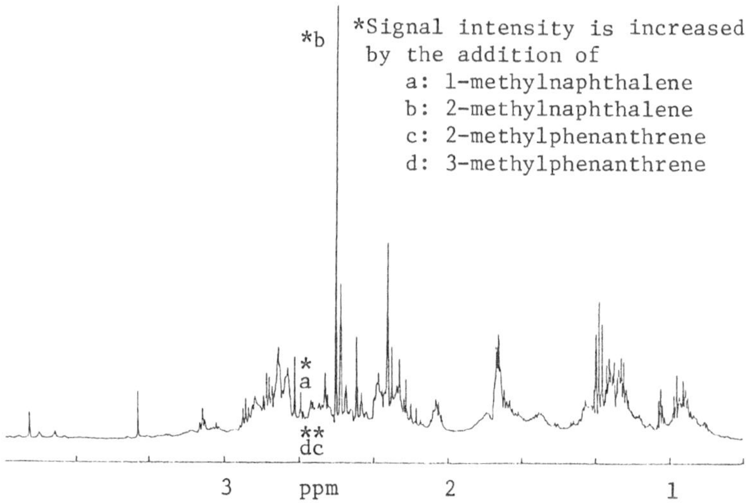

Figure 2. 600 MHz NMR Spectrum of Aromatic Fraction K from 0 to 4 ppm

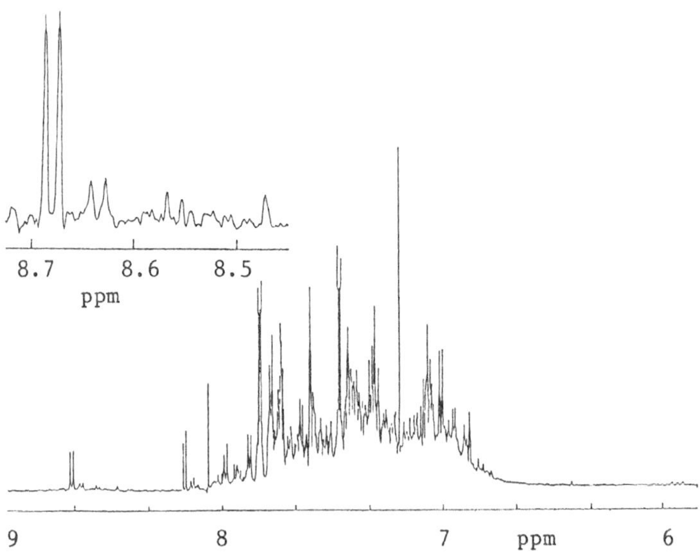

Figure 3. 600 MHz NMR Spectrum of Aromatic Fraction K from 6 to 9 ppm

ing the decoupling effects on the multiplet structure of α-CH_2 protons after irradiating at the center of β-CH_2 proton signals in tetralin, indan and their alkyl derivatives. The result of decoupling indicates the existence of several types of α-CH_2 protons which probably are located in substituted tetralin and indan (6).

These tetralin-like structures in a recycle solvent whose β-CH_2 signal is centered at 1.77 ppm play an important role in hydrogen donating behaviour in a coal liquefaction process. The concentration of such β-CH_2 protons in each heavy recycle solvent (593 K+) was kept constant through the continuous operation (number of passes=12) in the Solvolysis Coal Liquefaction Process (two stage liquefaction) using an Australian sub-bituminous coal, sponsored by the New Energy Development Organization in Japan (10). In the TSL process involving integrated hydrocracking-treatment, concentrations of partially hydrogenated phenanthrenes, pyrenes and fluoranthenes in the heavy recycle solvent decreased as a function of number of passes through the same operation.

Figure 3 shows the aromatic proton signals of Fraction K. Since more than half of the aromatic protons of Fraction K appear downfield from the benzene signal (7.37 ppm), the fraction contains a large amount of two and three condensed aromatic ring compounds. Two doublet signals (8.69 and 7.88 ppm) can be assigned to the 4,5 and 1,8 protons of phenanthrene, respectively (6,7). The small signals in the region of 8.47-8.63 ppm are also assigned to protons at the positions of 4 and 5 in methylphenanthrenes. The assignment and quantification of these signals are based either on the comparison of chemical shift values of pure compounds or on the change of spectral intensities caused by the addition of such compounds.

Figure 4 shows the aromatic proton signals of Fraction K, C, D and G. Acidic Fractions C and D, have more proton signals at higher magnetic field, because of the strong electron-donating property of phenolic OH groups. Acid-I (C) Fraction contains condensed-ring aromatic compounds such as naphthols, while Acid-II (D) has no proton signal which appears downfield from the benzene signal. This shows that Acid-II is substantially composed of mono-aromatic phenols. The spectrum in the region of aliphatic protons in Acid-II is shown in Figure 5. The intense singlets in the region of 2.20-2.40 ppm are assigned to CH_3 protons attached to mono-aromatic ring. The acidic fractions are characteristic by having high content of such CH_3 groups. The results are consistent with substantial decrease of signals between 2.2-2.4 ppm after the removal of acidic fractions from Fraction A. Two triplets at 1.20 and 0.92 ppm may be assigned to CH_3 protons in ethyl and n-propyl groups attached to aromatic rings, respectively. It is interesting to note that Acid-II (D) contains tetralinols but not indanols. As for Base-II (G) Fraction, proton signals are much more widely spread than that of Fraction K particularly at lower magnetic field, as shown in Figure 4. This may suggest that the fraction is composed of heteroaromatic compounds involving nitrogen atom in

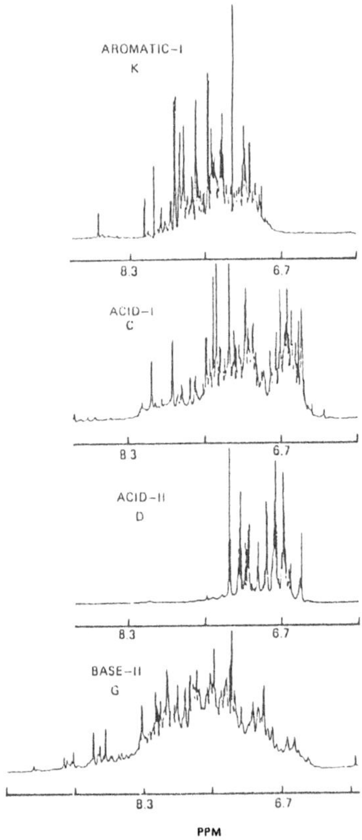

Figure 4. 600 MHz NMR Spectra of Fraction K,C,D and G

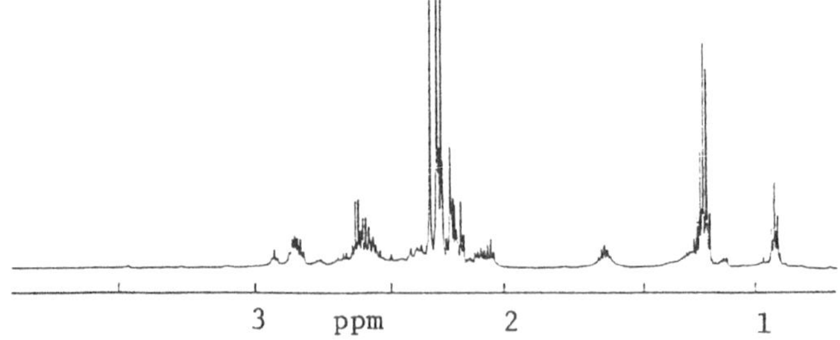

Figure 5. 600 MHz NMR Spectrum of Acid-II (D) Fraction

the ring structure. The nature and strength of basic property in Fraction G and I being both adsorbed on the cation-exchange resin, could be determined by measuring the downfield shift of OH proton signal at a constant concentration of Acid-II (D), as the concentration of G or I increased (11).

The structural change of the SRC-II liquid caused by the catalytic hydroprocessing as well as the accelerated ageing were also analyzed by the high field NMR spectrometer (3,8). The SRC-II liquid which was used for the NMR analysis being mentioned above, was hydroprocessed over a Ni-Mo/Al_2O_3 catalyst in a trickle bed reactor (3,12). The resulting upgraded oil exhibits a higher H/C atomic ratio of 1.75 and a high API° value of 33.6, compared with those of feed liquid, 1.17 and 10.1, respectively. The high-field NMR spectra of upgraded and feed oils are shown in Figure 6 in the region of aromatic protons. Though the most condensed aromatic structures are substantially disappeared in the upgraded oil, a particular compound being characteristic in the NMR signals of 8.02(singlet) and 8.13(doublet) ppm, remains rather stable in the upgraded oil in spite of the severe hydroprocessing conditions. The compound is assigned as 3-methyl-9,10-dihydrophenanthrene. The result indicates that alkyl-substituted phenanthrenes are less reactive to form perhydro derivatives, compared with non-substituted phenanthrene.

Coal-derived liquids are very susceptible to oxidative degradation by the interaction with molecular oxygen (13,8). The SRC-II Middle Distillate is rather stable against oxidative degradation and its viscosity does not increase significantly by oxygen bubbling at 333 K for three weeks. However, if oxygen is bubbled through the Middle Distillate in the presence of copper shavings, the viscosity markedly increases in four days (8). The NMR spectrum of remaining oil fraction of the aged Middle Distillate shows that aromatic protons the signals of which appear in higher magnetic field (6.40-6.75 ppm) disappear or are reduced significantly when compared with the unaged liquid (Figure 7). It is interesting to note that this aspect of the NMR spectra of unaged and aged Middle Distillate is quite similar to that of the spectra of SRC-II liquid and its chromatographically separated acid-free subfraction, also shown in Figure 7. The result indicates that acidic components (mainly phenols) react selectively in the accelerated ageing of the coal liquid to form higher molecular-weight components, which are responsible for the drastic increase in viscosity of the coal liquid.

The NMR difference spectroscopic technique was used to examine major compositional changes among the SRC-II Middle Distillates obtained under the same process conditions but different feed coals. Figure 8(a) shows the difference spectrum between two SRC-II Middle Distillates derived from Illinois No.6 and from

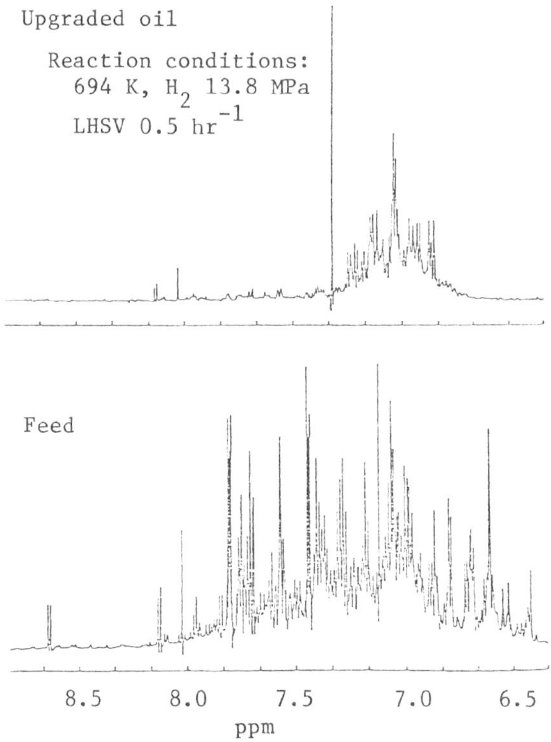

Figure 6. 600 MHz NMR Spectra of SRC-II Liquid before and after Hydroprocessing

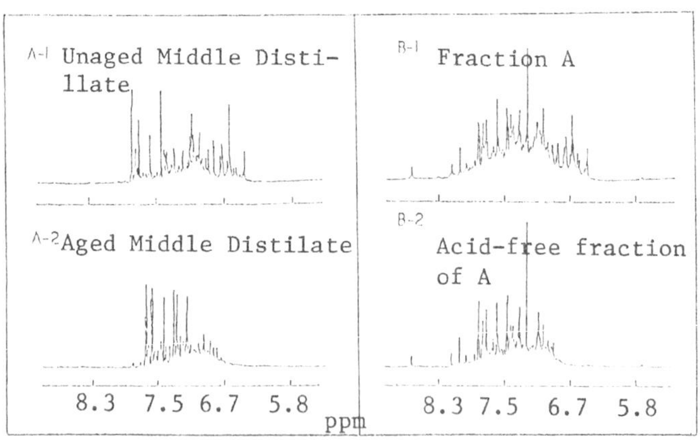

Figure 7. 600 MHz NMR Spectra of SRC-II Liquid before and after Accelerated Ageing

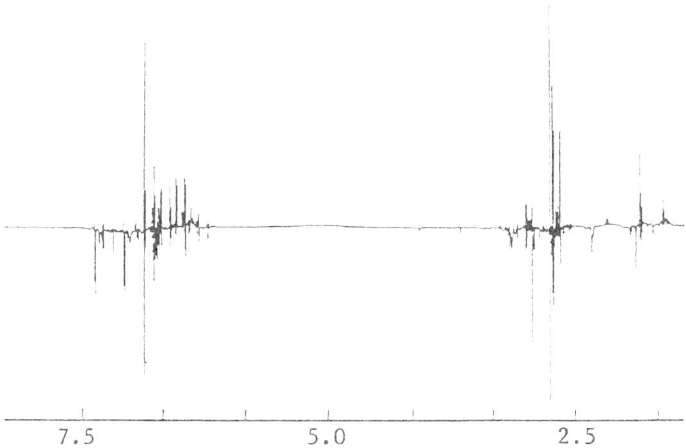

Figure 8(a). NMR Difference Spectrum between SRC-II Middle Distillates from Illinois No.6 and Pittsburgh Seam Coal

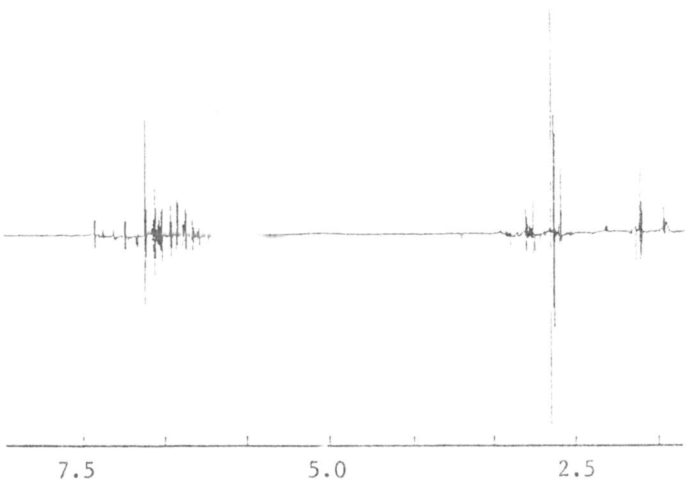

Figure 8(b). NMR Difference Spectrum between SRC-II Middle Distillates from Illiois No.6 and from Western Kentucky No.9/14 Coal

Pittsburgh Seam. The proton signals above the center line represent the types of compounds present in Illinois No.6 distillate in addition or in excess to compounds present in Pittsburgh Seam liquid. The product from Pittsburgh Seam Coal contains an excess amount of compounds whose resonance signals are displayed below the center line. The appearance of distinct resonance lines above the center line upfield to benzene and several intense singlets in the region of 2.20-2.40 ppm resemble the spectrum of Acid-II Fraction. It may be that the product from Illinois No.6 contains larger amount of alkyl-substituted mono-aromatic phenols. On the other hand, the product from Pittsburgh Seam contains larger amount of naphthalenes and tetralins.

The difference spectrum between products from Illinois No.6 and Western Kentucky is shown in Figure 8(b). In this case almost symmetrical signals can be found in both sides except for few signals. Slight changes in chemical shift of each signal may be brought about either by slight changes in structure of similar compounds in both coal-liquids, or by the presence of different compounds which affects the chemical shift of signals by intermolecular interaction. In conclusion, the chemical composition of these two coal-liquids is quite similar except for the fact that the product from Illinois No.6 contains larger amount of saturate fraction or long chain substituents.

References

1) Brown, J.K. and Ladner, W.R., Fuel,1959, 39, 87.
2) Kanda, N, Itoh, H, Yokoyama, S and Ouchi, K, Fuel, 1978, 57, 676.
3) Tewari, K.C., Hara, T., Li,N.C. and Fu, Y.C., Fuel, 1981, 60, 1137.
4) Bartle, K.D. and Jones, D.W., Fuel, 1969, 48, 21.
5) Schweighardt, F.K., Retcofsky, H.L., Friedman, S. and Hough, M., Anal. Chem., 1978, 50, 368.
6) Hara, T., Tewari, K.C., Li, N.C. and Schweighardt, F.K., 'New Approaches in Coal Chemistry', ACS Symposium Series, 169, pp. 289-318(1981).
7) Hara, T., Petrotech, 1981, 4, 925.
8) Hara, T., Jones, L., Li, N.C. and Tewari, K.C., Fuel, 1981, 60, 1143.
9) Bothner-By, A.A. and Dadok, J., Paper presented at Symposium on "Utility of 600 MHz NMR Spectroscopy", Pittsburgh, 1979.
10) Hara, T. and Ueda, T., 'Proceedings of 19th Conference on Coal Science', Tokyo, 1982, pp. 318-322.
11) Li, N.C., Tewari, K.C. and Hara, T., DOE Quarterly Report, EY-76-S-02-0063, April 1980.
12) Hara, T., Tewari, K.C., Li, N.C. and Fu, Y.C., Preprints, Div.Fuel Chem., Am.Chem.Soc., 1979, 24(3), 215.
13) Brown, F.R. and Karn, F.S., Fuel, 1980, 59, 431.

A REVIEW OF RECENT MECHANISTIC STUDIES OF THE IRON AND TIN CATALYSED HYDROLIQUEFACTION OF VICTORIAN BROWN COAL

P.J. Cassidy, W.R. Jackson, F.P. Larkins*, I.D. Watkins, P.A. Hertan, D. Rash and M. Marshall

Monash University, Clayton, Vic. 3168, Australia
*University of Tasmania, Hobart, Tas., Australia

This paper reviews our recent work in the study of the principal reactions occurring in the hydroliquefaction of Victorian Morwell brown coal and the mechanisms of iron and tin catalysts in promoting the formation of liquids during the liquefaction reaction.

THE SOLVENT-COAL INTERACTION

During the direct liquefaction process coal is finely ground and slurried with a suitable solvent. In a commercial process the solvent is a recycled distillation cut from the liquid product and it fulfills several important tasks. Firstly it is a transportation medium for the coal through the reactor. Secondly it should solvate the reacting coal particles and thirdly it should have some hydrogen donor ability to help stabilise the free radicals generated during the thermal degradation of the coal. In autoclave experiments the hydrogen donor ability of the solvent is usually the most important feature. One of the most frequently used hydrogen donors is tetralin (tetrahydronaphthalene) which can donate up to four hydrogen atoms to yield naphthalene as the stable end product by the reaction

In our experience there is little incorporation of the tetralin into the coal products and it is assumed that it acts purely as a hydrogen donor.

L. Petrakis and J. P. Fraissard (eds.), Magnetic Resonance. Introduction, Advanced Topics and Applications to Fossil Energy, 739–748.

The results in Table 1 show the effect of increasing the relative proportion of hydrogen donor solvent. The conversion increased from 50% with no tetralin (100% decalin) to 83% with 100% tetralin.

TABLE 1 The effect of increasing the hydrogen donor solvent concentration on the hydrogenation of Victorian Morwell brown coal. (405°C, 6 MPa initial H_2 pressure, 1 h at temperature)

Decalin:tetralin	3:0	2:1	1:2	0:3
Solvent:coal	3:1	3:1	3:1	3:1
		Weight % daf coal		
THF Conversion	50	64	75	83
Asphaltol	4	6	9	13
Asphaltene	6	12	18	20
Oil	20	26	29	29
C_1-C_4 HC gases	2	2	2	2
CO_2+CO+H_2O	20	19	18	19

Asphaltol THF(s)/CH_2CL_2(ins), asphaltene CH_2Cl_2(s)/X4 [hexane](ins), Oil X4[hexane](s) - s=soluble, ins=insoluble

The increased conversion appeared primarily as an increase in both oil and asphaltene yields. If the only role of the solvent was to solvate and stabilise radicals then the evidence suggests that the coal initially breaks down into asphaltene and oil "type" fragments which then repolymerise to form residue if there is insufficient available hydrogen to cap the radicals.

It is important to know whether the material that recondenses to form a residue in the absence of a hydrogen source is intractable or whether it can be re-reacted to form more oil. The product from one of our continuous reactor reactions was separated into THF insolubles, asphaltols (THF soluble/tetralin insoluble), asphaltene (tetralin soluble/petrol [hexane] insoluble) and oil (petrol [hexane] soluble) (1). Each fraction was then re-reacted in a series of batch autoclave experiments.

The results in Table 2 show that for reactions at 425°C significant conversion of the asphaltols and asphaltenes produced at 400°C was possible. In particular for the asphaltol > 95% interconversion occurred, while for the asphaltene the interconversion was > 65%. A complete range of products was formed from high oil yields to repolymerised THF insoluble material. Furthermore the THF insoluble material yielded 39% oil while hydrogenation of the oil fraction resulted in the formation of a small amount of the higher molecular weight products (< 4%) but the recovered oil contained a higher proportion of lower boiling point material. This reactivity underlines the inherent instability of these intermediate products. It can be concluded that linkage bonds generated by the recombination of radicals are not so strong that the residues are rendered intractable or there remains a number of weak linkages in the residues that were originally in the coal. This is also evidenced by the fact that bottoms or "heavy ends" recycling has been shown to be beneficial in continuous reactor systems (2,3).

TABLE 2 Product distributions from the reactions of Victorian Morwell brown coal and coal derived reactants. (1)

	Product Distribution (wt.% reactant)			
Reactant	Oil	Asphaltene	Asphaltol	THF Insoluble
Untreated coal	26	7	1	36
THF Insolubles	39	13	2	25
Asphaltol	44	28	5	13
Asphaltene	37	41	7	6
Oil	96	4	< 1	1

THE ROLE OF IRON AND TIN

Iron and tin were cation exchanged into the brown coal structure using solutions of ferrous sulphate and tin II chloride. This procedure is particularly applicable to brown coal and leads to a high dispersion of the metal throughout the coal particles. Incorporation of iron and tin led to a significant increase in the yield of methylene chloride soluble liquids (4,5,6) while the mixed catalyst of iron with a trace of tin showed a synergistic activity (5).

Total conversion yields for the temperature range 345-460°C at 6 MPa hydrogen are summarised in figure 1. The effect of increasing the reaction temperature was to increase the total conversion of all reactions irrespective of the catalyst system.

The iron-tin catalyst system was the most temperature sensitive of the four coal-catalyst systems studied. Unlike the iron and tin treated coals which were only sensitive to temperature in the range 365-405°C, the iron-tin treated coal was sensitive to temperature in the range 385°-425°C where the conversion rapidly increased from 66% to 85% (daf coal). Thus the iron-tin system was most efficient at the higher reaction temperatures up to 425°C (see Table 1, reference 6).

Below 405°C both iron and tin generated increased yields of asphaltene while there was little difference in the oil yields from the catalysed and uncatalysed reactions. However above 405°C the asphaltene yield decreased markedly and the oil yield correspondingly increased (refer Table 1, ref. 6).

Figure 1 Total conversion versus reaction temperature. Reaction conditions: 6 MPa initial H_2 pressure, 1 h at temperature 1:1 tetralin to coal ratio.

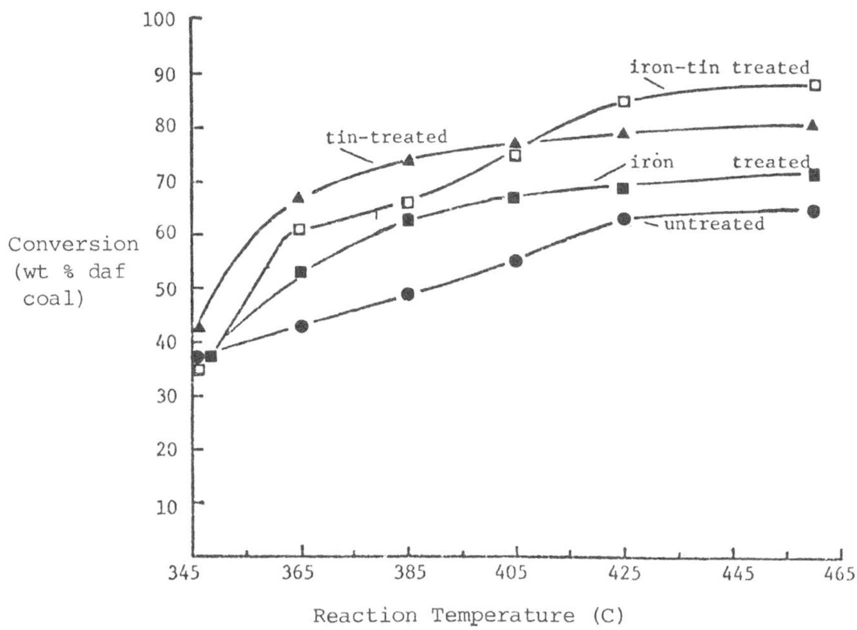

The initial hydrogen pressure dependent conversion results in the range 1-10 MPa are given in figure 2 for studies at 385°C. The tin catalysed reactions show most pressure dependence, consistent with previous findings that tin facilitates molecular hydrogen transfer during hydrogenation (7). For the untreated coal and for the other catalyst systems there is only a small pressure dependence of product yields beyond 4 MPa (6). The increased conversions with hydrogen pressure appeared as increased asphaltene yields. For example the asphaltene yield from the tin catalysed reaction was from 10% to 36% (daf coal) over the range 1-10 MPa of hydrogen.

A detailed study of the chemical constitution of the products revealed that their composition was influenced by temperature but not by hydrogen pressure or the nature of the catalyst (6). With increased reaction temperature there was a decrease in total and acidic oxygen concentrations in the asphaltenes and a corresponding increase in both aromatic content and C/H ratio. This observation is consistent with the loss of aliphatic side chains from the polycondensed ring systems and corresponds to the increased yields of hydrocarbon gases (6).

Figure 2 Total conversion versus initial hydrogen pressure. Reaction conditions: 385°C, 1 h at temperature, 1:1 tetralin to coal ratio.

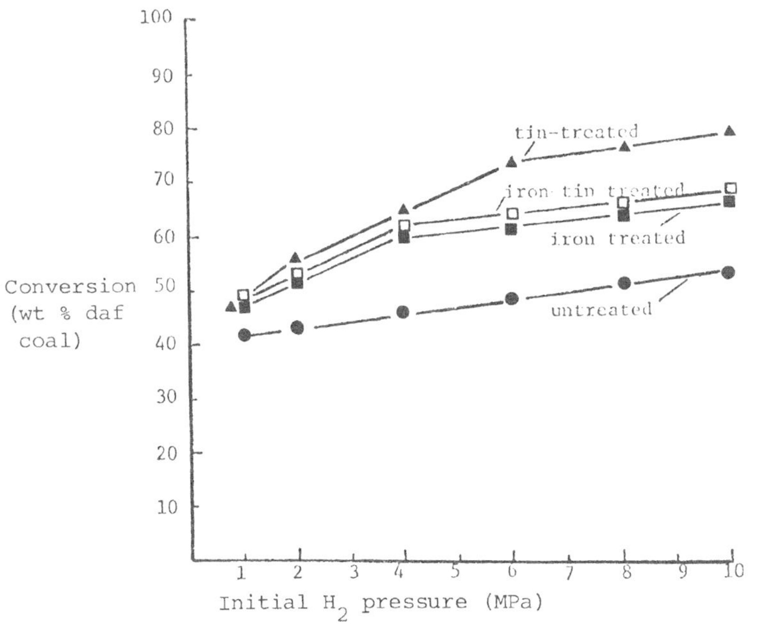

It was concluded that neither iron nor tin were acting to significantly degrade the coal structure and hence did not change the nature of the products compared to the untreated reaction. The metals appeared only to generate more or less of the same material.

In order to further clarify the activity of iron and tin in the liquefaction reaction a study of reactions of coal related model compounds with hydrogen was undertaken. Victorian brown coal is a low rank coal with up to 25% organically bound oxygen of which approximately one third is in the form of ether linkages. (8) The ether linkages are up to 28 kJ $mole^{-1}$ weaker than the equivalent carbon-carbon linkages (e.g., $PhCH_2$-OPh 221 kJ $mole^{-1}$ cf. $PhCH_2$-CH_2Ph 249 kJ $mole^{-1}$) and thus provide a potential region for bond cleavage within the brown coal structure which may be of central importance in the overall liquefaction reaction sequence.

Infrared (9) and artificial coalification (10) studies suggest that Victorian brown coal is essentially demethylated, dehydrated lignin. Lignin is formed from the polymerisation of substituted phenylpropene units such as coniferyl alcohol (HO-⟨O⟩-CH=CH-CH_2OH, with MeO on the ring). The ethers formed from the condensation of the phenylpropene units can be broadly classified into three classes depending on the type of radical formed after cleavage of the ether linkage. The classes are phenoxy ethers (PhO-$(CH_2)_n$Ph n = 1,2,3), benzyl ethers ($PhCH_2$-O-CH_2Ph) and alkyl ethers (R-O-R'). Iron and tin were added to reactions of these ethers as iron II acetate and tin metal in proportions consistent with the coal reactions.

Figure 3: The effect of iron and tin on the conversion of model ethers. (6 MPa initial H_2 pressure, 1 h at temperature)

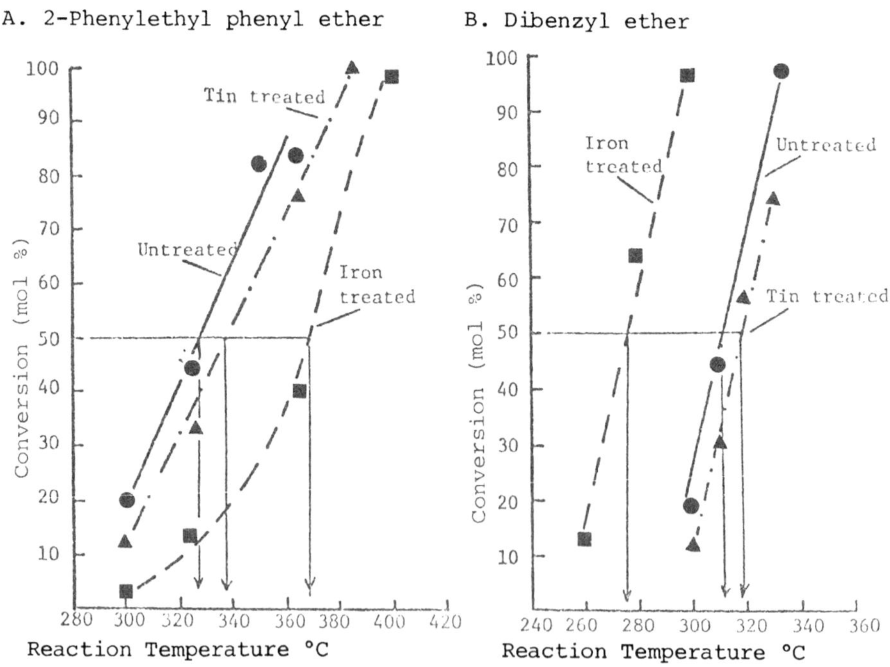

Neither iron nor tin changed the composition of the products in any of the modelsystems (11). For example there was no evidence of ring hydrogenation. However both iron and tin significantly changed the rate of disappearance of the parent compounds. In the case of the phenyl alkyl ethers (phenoxy class), both iron and tin reduced the rate of conversion. For example, for the reaction of phenylethyl phenyl ether at 325°C there was only 13% conversion in the presence of iron compared with 33% with tin and 43% without additive (see Figure 3A). This class of ether is common in lignin and thus is likely to be widely represented in brown coal structures. In contrast, addition of iron to a reaction dibenzyl ether (benzyl class) promoted the conversion while tin suppressed it relative to no additive being present (see Figure 3B). A similar effect to that with dibenzyl ether was found for the action of iron on the di-alkyl ether (alkyl class) while tin had no effect on the reaction of this class of ether.

The thermal decomposition of these ethers is believed to involve a radical chain mechanism. It is postulated that while iron may facilitate carbon-oxygen bond cleavage, as evidenced by the increased reactivity of benzyl and aliphatic ethers, when phenoxy radicals are produced they are strongly adsorbed on the catalyst surface and hence are not available to propogate the radical chain reaction. Furthermore, to account for the observed decrease in conversion, it is postulated that the phenoxy radicals generated away from the iron surface have sufficient stability to enable them to be transported to the metal surface and hence prevented from propogating the radical reaction.

It appears that tin does not have the ability to facilitate carbon-oxygen bond cleavage reactions but is able to "trap" both phenoxy and benzyl radicals that are transported from the site of reaction to the metal surface. In addition tin appears to utilise the molecular hydrogen dissociatively dissolved within its structure (12) to stabilise adsorbed coal molecules. At this time the exact reactions occurring at the iron and tin surfaces, if any, are uncertain. These may be involved in oxidation-reduction, recombination of adsorbed species or reaction with hydrogen atoms. However the net effect is a reduction in the rate of the radical propogation reactions. Such a reduction has two major consequences that can be tested in coal related reactions. Firstly, and most importantly, the rate of the thermally induced breakdown of the coal should be reduced in the initial minutes of the coal reaction. This has recently been demonstrated when iron catalysts were used in a series of short residence time studies. The results, summarised in Table 4 show that the conversion of acid washed coal after rapid heating to 405° had a conversion of 55% daf coal while the conversion of acid washed coal that had been re-exchanged with 300 mmol kg^{-1} (dry coal) of iron had a conversion of only 46%. A similar trend was found for the reaction of "as mined" coal (39% conversion) and the same coal that had been treated with a solution of iron II acetate and had a conversion of 33%. It is interesting to note that the natural inorganics of the "as-mined" coal had a very strong inherent activity to slow the rate of thermal decomposition.

TABLE 4 The effects of iron in short residence time hydroliquefaction reactions of victorian (Morwell) brown coal.

COAL TREATMENT	ACID WASHED	ACID WASH WASHED +Fe	UNTREATED COAL	UNTREATED COAL +Fe
conversion	55	46	39	33

conversion = 100 - % THF insolubles

The second effect of the reduction in the rate of radical propogation reactions is to slow down the rate of radical recombination or polymerisation. This effect has been demonstrated in hydrogenation reactions of asphaltols in the presence of the non-hydrogen donor solvent, decalin. The results are summarised in Table 5.

TABLE 5 The effect of iron and tin on the product distribution for asphaltol hydrogenation using decalin as a solvent.

	Product Distribution (wt% initial asphaltol)		
catalyst	none	iron	tin
oil	25	28	30
asphaltene	16	20	33
asphaltol	8	10	10
THF insolubles	40	30	15

In the reaction of untreated asphaltols the yield of THF insoluble "residue" was 40% by weight of the reactant asphaltol while the presence of iron reduced the residue yield to 30%. Tin had the largest effect in these reactions reducing the residue yield to 15%. The conversions from reactions of the untreated, iron and tin treated asphaltols were all 91 ± 1% and so the reduced polymerisation resulted in increased yields of oil and asphaltene.

CONCLUSION

Hydrogenation reactions of untreated Victorian brown coal indicate that it is very reactive and thermally degrades over the temperature range 345°-405°C to radical fragments that are largely oil and asphaltene in nature. Above 405°C the asphaltene fragments are further degraded by either a reduction in molecular weight or polarity to produce more oil and cracking of aliphatic side chains increases markedly the yield of hydrocarbon gases.

If the hydrogen donor solvent and hydrogen gas are unable to cap the radicals generated during reaction then radical recombination occurs forming higher molecular weight asphaltols and residue which can be thermally degraded to produce more oil and asphaltene by increasing reaction temperature to 425°C.

Adding iron, tin or a combination of both into the coal structure prior to reaction increased the yield of asphaltene but had no effect on the chemical properties of the products compared to those from the untreated reaction.

Model compound studies suggest that both iron and tin are trapping relatively stable radicals and preventing them from participating in the radical propogation reactions.

Decreasing the rate of radical propogation reactions results firstly in a decrease in the initial rate of degradation of the coal and secondly in a reduction in the rate of radical recombination (polymerisation) reactions. The combined effect is to increase the probability of radicals being stabilised by hydrogen to yield the lower molecular weight oil and asphaltene. The ability of the metal to directly utilise molecular hydrogen to cap radicals is also important. Tin is significantly more efficient than iron in this regard. At this time there is no conclusive evidence that iron and tin are catalysing any significant bond cleavage reaction. The increased oil yields from the reactions of iron, tin and iron-tin treated coals above 405°C are consistent with the thermal degradation of stable asphaltenes generated during the initial degradation of the coal.

REFERENCES

1. Larkins, F.P., Jackson, W.R., Hertan, P.A., Cassidy, P.J., Marshall, M and Rash, D., Am. Chem. Soc. (Div. Fuel Chem.) Preprints (1983), Washington meeting.

2. Sondreal, E.A., Wilson, W.G. and Stenberg, V.I., Fuel (1982), 61, pp925.

3. EDS Coal Liquefaction Process Development Phases IV-V Annual Technical Progress Report, July 1, 1979 - June 30, 1980.

4. Hatswell, M.R., Jackson, W.R., Larkins, F.P., Marshall, M., Rash, D., and Rogers, D.E., Fuel (1980), 59, pp442.

5. Marshall, M., Jackson, W.R., Larkins, F.P., Hatswell, M.R. and Rash, D., Fuel (1982), 61, pp121.

6. Cassidy, P.J., Hertan, P.A., Jackson, W.R., Larkins, F.P., and Rash, D., Fuel (1982), 61, pp939.

7. Cochran, S.J., Hatswell, M.R., Jackson, W.R., and Larkins, F.P., Fuel (1982), 61, pp831.

8. McPhail, I. and Murray, J.B., SECV Miscellaneous Report No. MR-155 31 March 1969.

9. Durie, R.A., Lynch, B.M. and Sternhell, S., Aust. J. Chem. (1960), 13(1), pp156.

10. Sternhell S., Aust. J. Chem. (1964), 17(11), pp1236.

11. Cassidy, P.J., Jackson, W.R. and Larkins, F.P., accepted for publication in Fuel.

12. Jurovtseva and Kunin, Analysis of Gases in Metals, Chapter 2, published by consultants Bureau, New York (1959).

13. Neavel, R.C., Coal Science (1982), 1, pp1.

EXTRACTION STUDY OF TURKISH LIGNITES: ÇAN, SOMA AND TUNÇBİLEK

Salih Dinçer and Esen Bolat

Boğaziçi University, Chemical Engineering Department
İstanbul, Turkey

ABSTRACT

Solvent extraction of lignites from various regions of Turkey such as Çan, Soma and Tunçbilek was studied. Effects of different solvents and particle size of lignites were examined in Soxhlet extractors. In addition, the effects of pressure, solvent to lignite ratio and reaction-extraction time on extraction yields using tetralin and N_2 gas in a batch autoclave, were studied at 325°C. Yields with tetralin in the presence of H_2 gas are compared with those in the presence of N_2 gas at 60 atm (measured at room temperature).

INTRODUCTION

Turkey has reasonably large lignite reserves which are being used either for heating purposes or for production of electricity. However, use of lignites could also be extended to the production of chemicals. In order to develop an economical conversion technology for Turkish lignites, a better understanding of the behaviour of lignites and the parameters affecting their conversion are required. There are various conversion routes that can be used for this purpose. Although it may not be the most suitable conversion technique, here, the solvent extraction and hydrogenation technique is selected. Therefore, the purpose of this work was to study the parameters affecting the lignite extraction with various solvents in Soxhlet extractor at atmospheric pressure and with a hydrogen donor solvent (tetralin) in a batch autoclave, under N_2 or H_2 pressure.

L. Petrakis and J. P. Fraissard (eds.), Magnetic Resonance. Introduction, Advanced Topics and Applications to Fossil Energy, 749–755.

1. EXPERIMENTAL

Extraction work performed here (1) can be classified as: 1) Soxhlet extraction under atmospheric pressure, 2) Autoclave extraction and hydrogenation under elevated pressure and temperature. In the Soxhlet extraction part Çan, Soma and Tunçbilek lignites have been extracted with different types of solvents and the effect of particle size on the extraction yield (given as %E/L = percentage of extract to dry lignite ratio) was observed. The autoclave extraction and hydrogenation of the same types of lignites used in Soxhlet extraction were done using tetralin as solvent. Either N_2 or H_2 gas was used as the pressurizing medium. When N_2 gas was used, the effects of pressure, solvent to lignite (S/L) ratio and reaction-extraction time on the extraction yields were observed. Heating up and cooling down times were not included in the reaction-extraction time measurements. Although for hydrogenation of coal usually temperatures around 400^oC or more are used (5), in this study a temperature of 325^oC was used. Higher temperatures could not be used because of the limitations of the autoclave. The autoclave used was a 1-liter stainless Cr-Ni-Mo steel batch autoclave (1). An Fe-Ko DIN thermocouple was used for temperature readings which are analyzed by the electronic PD/PID type controller. The autoclave is equipped with a permanent magnet revolving agitator.

The lignite samples used were obtained from MTA, Turkey. The elemental analysis of lignite samples was carried out at İ.Ü. Eczacılık Fakültesi. The H/C ratios and heating values of the lignite samples used are shown in Table 1.

Table 1. H/C ratios and lower heating values of lignite samples used

	Çan (Çanakkale)	Soma (Manisa)	Tunçbilek (Tavşanlı)
H/C	0.82	1.16	0.95
LHV (kcal/kg)	5011	4851	5551

2. RESULTS AND DISCUSSION

2.1 Soxhlet Extraction Results

The Soxhlet extraction (carried out to exhaustion) yields of Çan, Soma and Tunçbilek lignites with hexane, benzene, tetrahydrofuran and pyridine are shown in Table 2. The extracts were separated from the extracting solvent in a rotavapor and the extract amount measured. Benzene and pyridine extraction yields agree with those given in literature (4).

Table 2. % E/L of lignites

	% E/L		
	Çan	Soma	Tunçbilek
Hexane	1.1	2.0	1.8
Benzene	2.6	3.1	3.0
Tetrahydrofuran	5.3	6.5	6.2
Pyridine	9.8	9.3	11.9

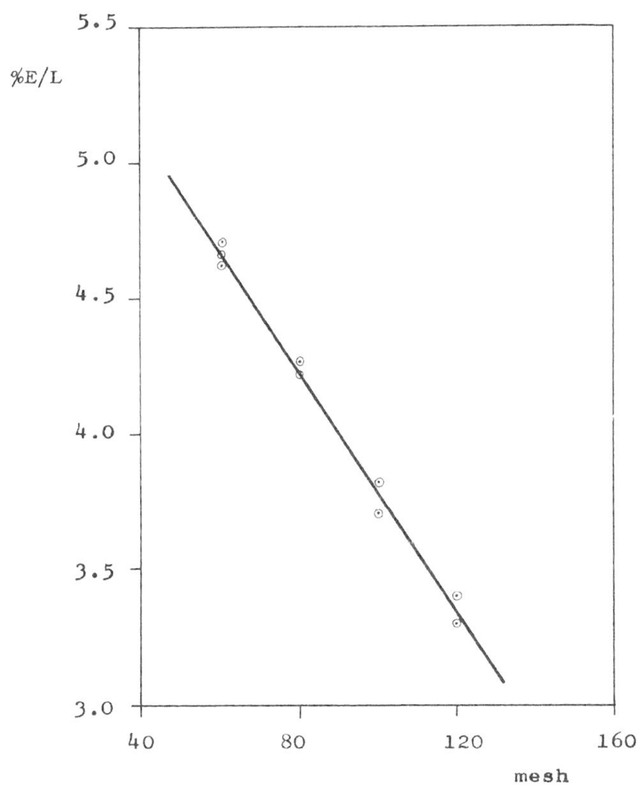

Figure 1. Effect of particle size on extract of Çan lignite using pyridine as solvent in Soxhlet extractor (Extraction not carried out to exhaustion).

Effect of lignite particle size on extract yield of Çan lignite using pyridine as solvent in Soxhlet extraction is shown in Figure 1. As seen in Figure 1, decrease in extraction yield occurs as particle size is decreased. Similar behaviour is also reported in literature (3). In this work, apart from particle size effect experiments, particle size was held at 60 mesh in all the other experiments.

2.2 Autoclave Extraction Results

The effects of S/L ratio, pressure and reaction-extraction time on extraction yields of Tunçbilek lignite samples with tetralin using N_2 gas in autoclave are shown in Figures 2, 3 and 4, respectively. From the shape of the curve in Figure 2 it can be deduced that beyond S/L ratios greater than 10:1, extraction yield will not significantly increase.

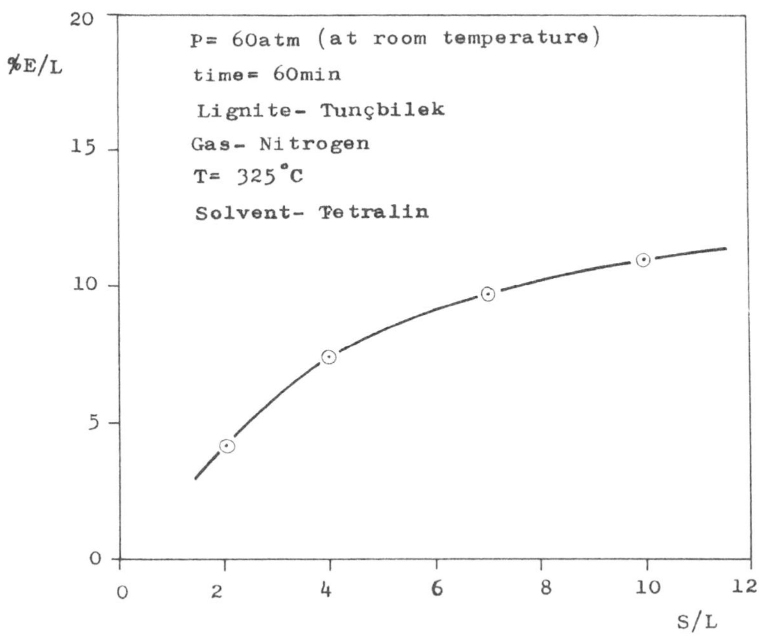

Figure 2. Effect of solvent to lignite ratio on extraction yields.

Pressure is also an important parameter in extraction. Increasing the pressure results in increased extraction yields (see Figure 3) as also reported in literature (2). Due to safety considerations, pressure was not increased further than the values given in Figure 3 but the trend in the results shows that increasing pressure beyond the pressures used in this study will increase the extraction yields further.

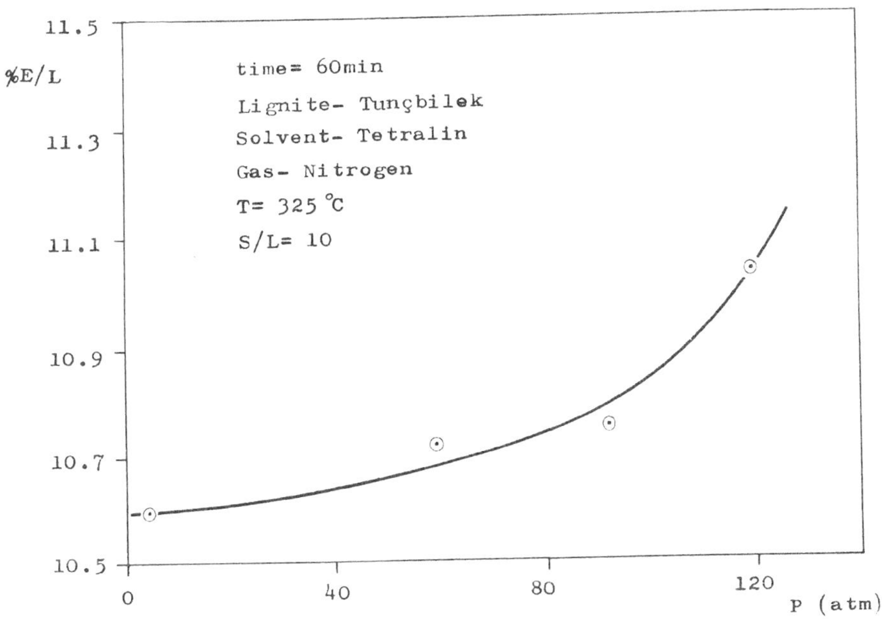

Figure 3. Effect of pressure on extraction yields (Pressures shown are at 325°C)

Reaction-extraction time was also varied and its effect on extraction yields was studied. Figure 4 shows that increasing the reaction time beyond 60 min. (excluding heating up and cooling down times) does not increase the extraction yield much; therefore in all the other runs reaction-extraction time was held constant at 60 min.

Extraction yields of Çan, Soma and Tunçbilek lignites with tetralin under N_2 and H_2 pressures are shown in Table 3.

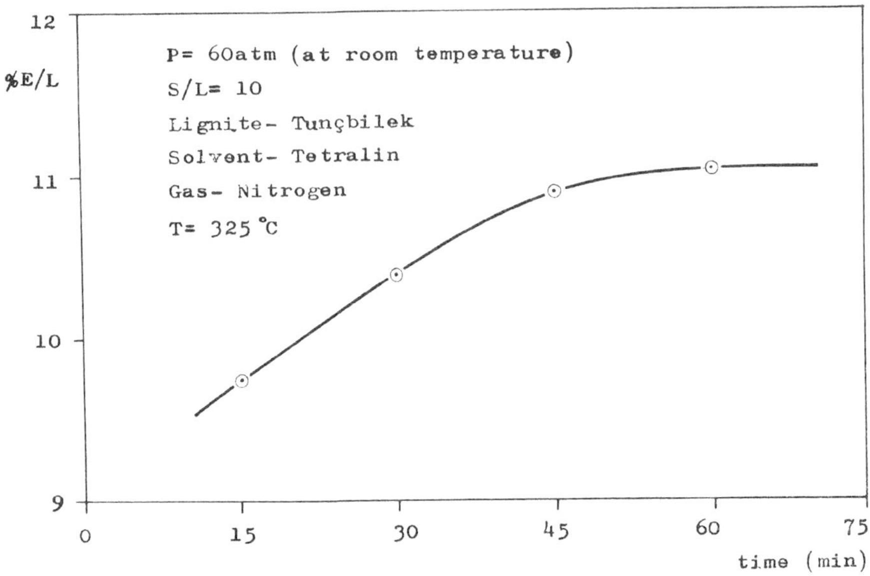

Figure 4. Effect of reaction-extraction time on yields.

Table 3. Yields of lignites with tetralin under H_2 or N_2 pressure

P = 60 atm. (at room temperature)
P = 119 atm.(at 325°C)
S/L = 10
time = 60 min.
T = 325°C

	% E/L		
	Çan	Soma	Tunçbilek
N_2 pressure	21.8	19.4	11.0
H_2 pressure	23.2	19.8	12.5

3. CONCLUSIONS AND RECOMMENDATIONS

As the pressure and reaction-extraction time both increase separately, tetralin extraction yields of Turkish lignites studied increase. From Soxhlet extraction study of Çan lignite with pyridine, it is observed that extraction yields are higher at lignite particle size of 60 mesh in the 60-120 mesh size range. At 325°C and 60 atm. (measured at room temperature), the presence of H_2 gas slightly increases the extract yield obtained with tetralin over that obtained in the presence of N_2 gas. For higher extraction yields the temperatures of 400°C or more and pressures greater than 60 atm. (measured at room temperature) should be employed.

The results obtained in this study give an idea about what to expect from Turkish lignites when they undergo extraction under various conditions. Analysis of extracts obtained was not attempted in this study. However, to better understand the quality and structures of extracts obtained, analysis of extracts by GC/MS and magnetic resonance techniques is necessary. In addition, lignite samples themselves should be analyzed by solid NMR technique so that their structures can be identified.

ACKNOWLEDGEMENTS

Authors thank for the contributions of Boğaziçi University Research Facilities and NATO Accessory Fund.

REFERENCES

1. Bolat, E., "Extraction of Lignites from Çan, Soma and Tunçbilek with Donor and Non-Donor Solvents in Soxhlet Extractor and Autoclave", 1983, MS Thesis, Boğaziçi University, p. 85.

2. Lurgi, "Express Information", 1978, Ö 1275/3, pp. 1-16.

3. Oele, A.P., Waterman, H.I., Goedkoop, M.L., van Krevelen, D.W., 1951, Fuel 30, pp. 169-178.

4. van Krevelen, D.W., "Coal", 1961, Elsevier, Amsterdam, pp. 177-200.

5. Wise, W.S., "Solvent Treatment of Coal", 1971, Mills and Boom Limited, London, pp. 7-53.

OXYGEN-17 AND SULFUR-33 NUCLEAR MAGNETIC RESONANCE SPECTROSCOPY OF ORGANOSULFUR COMPOUNDS

Slayton A. Evans, Jr.

Department of Chemistry
University of North Carolina
Chapel Hill, North Carolina 27514, U.S.A.

The ^{17}O NMR chemical shifts are obtained for a number of sulfinyl (R-S=O) and sulfonyl (R-S=O) compounds and their chemical shifts range' from δ -20 to 183 ppm. The acyclic, aliphatic, and aryl sulfoxides exhibit shifts in the narrow range of δ -20 to +20 ppm referenced to naturally abundant $^{17}OH_2$. Diastereotopic sulfonyl oxygens exhibit chemical shift nonequivalence and in the presence of $Eu(fod)_3$, the magnitude of the diastereotopism is enhanced. The ^{33}S NMR shifts of sulfones and sulfonic acids (and derivatives) occur in the range of δ 310 to 390 ppm downfield of external $C^{33}S_2$. Simple sulfonic acids and their salts afford line-widths-at-half-height of < 25 Hz (at 250 Mhz)and thus allow for relatively easy identification of similar sulfonic acids.

Our current objectives include (i) development of ^{17}O and ^{33}S NMR techniques for the efficient and rapid determination of "oxidized" forms of organosulfur compounds in coal and petroleum extracts, and (ii) determination of the bonding characteristics between sulfur and oxygen in various organosulfur materials.

OXYGEN-17 NMR

Oxygen-17 NMR spectroscopy is becoming a valuable tool for the elucidation of bonding and structural features in organosulfur molecules in spite of the extremely low natural abundance (0.037%) and sensitivity (2.91×10^{-2} times that of ^{1}H at constant field) as well as the quadrupole moment of the ^{17}O nucleus. (1-3) The ^{17}O shifts of aliphatic, acyclic sulfoxides occur between δ -20 and 20 ppm (Table 1) and the corresponding sulfones exhibit shifts

L. Petrakis and J. P. Fraissard (eds.), Magnetic Resonance. Introduction, Advanced Topics and Applications to Fossil Energy, 757–766.

between δ 120 and 183 ppm relative to external $^{17}OH_2$. This shift range is surprisingly close to water considering the bonding analogy drawn between sulfoxides and ketones [ketones exhibit shifts between δ 530-580 (4) downfield of H_2O]. Assuming that ^{17}O shifts

Table 1. ^{17}O NMR Chemical Shifts of Cyclic and Acyclic Sulfoxides and Sulfones in PPM

Sulfoxides	$\delta(^{17}O)$	Sulfones	$\delta(^{17}O)$
$CH_3S(O)CH_3$	12,19	$CH_3SO_2CH_3$	164
$CH_3CH_2S(O)CH_2CH_3$	-6	$CH_3CH_2SO_2CH_2CH_3$	140
$(CH_3)_2CHS(O)CH(CH_3)_2$	-20	$(CH_3)_2CHSO_2CH(CH_3)_2$	120
$(CH_3)_3CS(O)C(CH_3)_3$	-2	$(CH_3)_3CSO_2C(CH_3)_3$	122
$CH_3(CH_2)_2S(O)(CH_2)_2CH_3$	-6	$[CH_3(CH_2)_2]_2SO_2$	147
$[CH_3(CH_2)_3]_2SO$	-7	$[CH_3(CH_2)_3]_2SO_2$	148
$[CH_2{=}CHCH_2CH_2]_2SO$	-9	$[CH_2{=}CH]_2SO_2$	141
$C_6H_5S(O)C_6H_5$	2	$C_6H_5SO_2C_6H_5$	138
$C_6H_5S(O)CH_2C_6H_5$	-14	Thiane 1,1-dioxide	146
Thiane 1-oxide	1	Thiolane 1,1-dioxide	164
Thiolane 1-oxide	16	Thietane 1,1-dioxide	183
Thietane 1-oxide	66		

are dominanted largely by paramagnetic contributions (eq 1), we

$$\sigma_P^O = -\frac{e^2 h^2}{2 m^2 c^2 \Delta E} < r^{-3} >_{2P_O} \Sigma Q_{SO} \qquad (1)$$

we rationalize the shift range in the following way. The dipolar description of the sulfinyl linkage is of major importance and as multiple bond contributions diminish and electron density increases at the sulfinyl oxygen, the oxygen 2p orbital radius expands causing a reduction in the magnitude of the $< r^{-3} >_{2P}$ term when compared to ketones. The result is a reduction in the paramagnetic shielding contribution in sulfoxides.

The large downfield shifts for the sulfones appear consistent with the expectation that the sulfonyl S=O bonds possess more multiple bond character than the sulfinyl group. This increase in the double bond character in sulfones probably stems from the effect of "contracted" 3d orbitals of the sulfonyl sulfur which allow for a more efficient overlap with the smaller oxygen 2p orbitals.

In the acyclic sulfoxides and sulfones, the oxygens experience upfield shifts when a hydrogen atom attached to Cβ is replaced by a methyl or methylene group (i.e., γ-methyl effect).

Cδ-Cγ-Cβ-X-Cβ-Cγ-Cδ where X = S=O and $\overset{\displaystyle S=O}{\underset{\displaystyle O}{\parallel}}$

Thus, a comparison of the ^{17}O shifts for dimethyl sulfoxide, diethyl sulfoxide, and diisopropyl sulfoxide reflects γ-methyl effects of -18 and -34 ppm for diethyl sulfoxide and diisopropyl sulfoxide, respectively (or -8 to -9 ppm/methyl). By contrast, the ^{17}O shift for di-t-butyl sulfoxide is -2 ppm and only 14 ppm to higher field than that for dimethyl sulfoxide. We spectulate that steric crowding reduces the γ-methyl effect in sterically hindered systems.

We note that alkyl thiolsulfinates and thiolsulfonates exhibit ^{17}O shifts which are to lower field than their sulfoxide and sulfone counterparts (Table 2). While we are certain that

Table 2. ^{17}O NMR Chemical Shifts of Alkyl Thiolsulfinates and Thiolsulfonates in PPM

Thiolsulfinate	$\delta(^{17}O)$	Thiolsulfonate	$\delta(^{17}O)$
$CH_3S(O)SCH_3$	73	$CH_3SO_2SCH_3$	199
$CH_3CH_2S(O)SCH_2CH_3$	64	$CH_3CH_2SO_2SCH_2CH_3$	190
$(CH_3)_2CHS(O)CH(CH_3)_2$	57	$(CH_3)_2CHSO_2SCH(CH_3)_2$	183

S-S bonding interactions are important here, our final decision must await a more detailed study.

We have also discovered that the ^{17}O shift of the sulfinyl oxygen in trans-thiadecalin 1β-oxide (*1β*) is δ -11.4 with the axial sulfinyl oxygen which is 17.0 ppm to higher field than the *1α* diastereoisomer (δ 5.6) with the equatorial sulfinyl oxygen. trans-Thiadecalin 1,1-dioxide (*2*), possessing both axial and equatorial monocoordinated oxygens, exhibits ^{17}O shifts for the distereotopic sulfonyl oxygens at δ 123.9 and 138.9, reflecting a shift difference of 15 ppm. It is reasonable that the γ-methyl or methylene shielding effect is responsible for the high field shift in *1β* and we have therefore assigned the absorption at δ 123.9 in *2* to the axial sulfonyl oxygen.

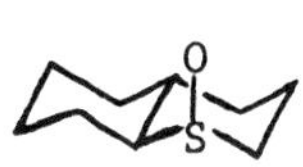

1β

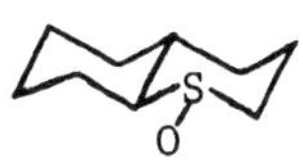

1α

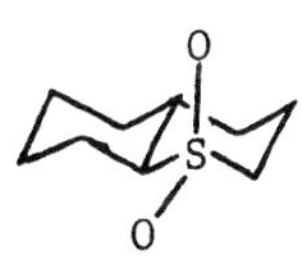

2

We have examined the influence of tris-6,6,7,7,8,8,8-heptafluoro-2,2-dimethyl-3,5-cotanedionato)europium, $Eu(fod)_3$, on the magnitude of the shift difference between the diastereotopic oxygens of the sulfonyl group in 3-isopropoxythiolane 1,1-dioxide (*3*). Several conclusions emerge from an examination of the ^{17}O NMR data (toluene solvent, 100°C) in Table 3. First, the interaction of $Eu(fod)_3$ with both sulfonyl oxygens results in shielding while the ethereal oxygen's shift is essentially invariant over the

Table 3. ^{17}O NMR Shift Data of 3-i-Propoxythiolane 1,1-dioxide in the Presence of $Eu(fod)_3$ in Toluene, 80°C

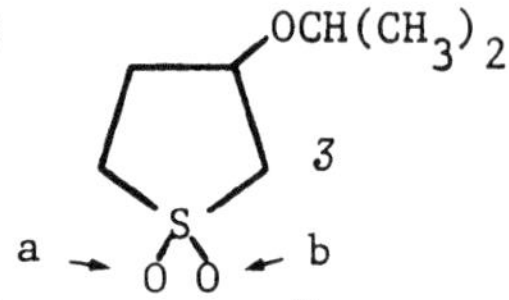

3

Conc $Eu(fod)_3$ (10^{-2}M)	δS=O (O$_a$)	$(W_{1/2})^a$	δS=O (O$_b$)	$(W_{1/2})^a$	δMe$_2$CH-O-C	$(W_{1/2})$	Δδ[b]
0	172	(163)	171	(163)	54	(272)	1.3
1.5	172	(88)	165	(108)	56	(210)	6.6
3.0	169	(91)	158	(122)	60	(217)	10.5
4.5	166	(122)	152	(135)	59	(237)	14.1
6.3	161	(-)	143	(152)	57	(-)	18.2
7.8	156	(126)	133	(163)	57	(316)	24.0

[a]The width-at-half-height in hertz. [b]The chemical shift difference between the diastereotopic sulfonyl oxygens in ppm.

concentration range of $Eu(fod)_3$ employed here. Secondly, the sulfonyl oxygen, $S(O)O_a$, with the smallest $W_{1/2}$ and the ^{17}O shift to lowest field is probably cis to the 3-i-OPr group. We arrive at this conclusion from the following analysis. The steric encumbrance caused by the 3-i-OPr substituent diminishes the coordinating ability of $Eu(fod)_3$ for the cis sulfonyl oxygen resulting in a concurrent reduction in the shielding of that oxygen. The trans sulfonyl oxygen, $S(O)O_b$, is more removed from these steric considerations. An examination of the magnitudes of the total upfield displacements for each sulfonyl oxygen makes this point clear. These results clearly show that the magnitude of the ^{17}O shift difference between diastereotopic sulfonyl oxygens can be enhanced when in the presence of $Eu(fod)_3$. In addition, the important implication is that the least sterically hindered sulfonyl oxygen will be more responsive to the shielding caused by coordinating $Eu(fod)_3$, thus allowing for possible stereochemical assignments.

As a simple illustration of the utility of ^{17}O NMR for quantitative applications, we have examined the equilibrium distribution of 2,5-dihydrothiophene 1,1-dioxide (*4*) and 4,5-dihydrothiophene 1,1-dioxide (*5*) obtained by base (HO^-/H_2O) catalyzed equilibration of dioxide *4*. The ^{17}O chemical shifts are δ 165 for *4* and δ 177 for *5* with integrated intensities corresponding to 56% *4* and 38% *5* in DMF solvent at 80°C. The ^1H NMR analysis supports the ^{17}O findings and both ^1H and ^{17}O NMR indicate the presence of a trace of hydroxy sulfone *6*. Sulfone *6* arises from conjugate addition of hydroxyl ion to sulfone *5*. (Figure 1).

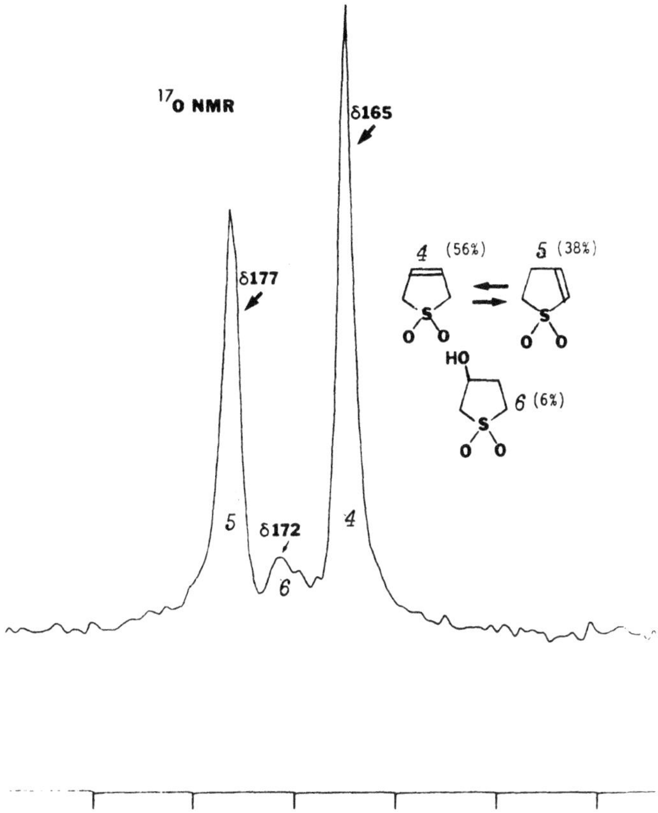

Figure 1. ^{17}O NMR of equilibrium mixture of *4*, *5*, and *6* in DMF at 80°C

SULFUR-33 NMR

The ^{33}S isotope has a spin I of 3/2 and a natural abundance of 0.76%. The ^{33}S nucleus also possesses an electric quadrupole moment and the electric field gradients at the nucleus lead to short relaxation time, thus affording broad resonance lines and low resolution. However, the short relaxation times allow for rapid pulse repetition in FT experiments and this serves to improve substantially the signal-to-noise ratio.

The widths of the ^{33}S signals particularly in sulfides may vary from 1 to 9 kHz and thus preclude accurate measurement of chemical shifts resulting from structural and electronic effects within classes of sulfides (e.g., aryl vs. alkyl). However, it is apparent that conversion of sulfides to sulfones and mercaptans to sulfonic acids affords the ^{33}S electronic symmetrization and serves to reduce the band width sufficiently to allow useful experimental distinction between similar ^{33}S nuclei in a variety of different systems (5-6).

The naturally abundant ^{33}S NMR shifts for both cyclic and acyclic aliphatic sulfones appear in the range of δ 300 to 370 ppm downfield of external CS_2 having naturally abundant ^{33}S.(Table 4).

Table 4. ^{33}S NMR Chemical Shifts of Aliphatic and Aryl Sulfones

Sulfone[a]	$\delta(^{33}S)$	$W_{1/2}$, Hz
$CH_3SO_2CH_3$	320 ($CHCl_3$); 332 (TFA)	50; 275
$CH_3CH_2SO_2CH_2CH_3$	334	70
$(CH_3)_2CHSO_2CH(CH_2)_3$	351	160
$(CH_3)_3CSO_2C(CH_3)_3$	366	160
$[CH_3(CH_2)_2]_2SO_2$	333	180
$[CH_3(CH_2)_3]_2SO_2$	336	180
$C_6H_5SO_2C_6H_5$	312	120
$(C_6H_5CH_2)_2SO_2$	330	120
Thiane 1,1-dioxide	322	50
Thiolane 1,1-dioxide	370	50
2,5-Dihydrothiophene 1,1-dioxide (*4*)	361	50
1,4-Oxathiane 1,1-dioxide	315	100

[a]All ^{33}S NMR spectra were obtained at 19.196 Mhz (22-25°C) as 20-45 w/w% chloroform solutions using CS_2 as external reference and shifts are in ppm.

In the acyclic, aliphatic sulfones, a systematic, symmetrical replacement of the hydrogens attached to the α-carbons by methyl groups causes a downfield shift of the ^{33}S nucleus. Comparisons of the chemical shifts for dimethyl sulfone (δ 320), diethyl sul-

$$C\delta-C\gamma-C\beta-C\alpha-\overset{O}{\underset{O}{\overset{\|}{\underset{\|}{S}}}}-C\alpha-C\beta-C\gamma-C\delta$$

fone (δ 334), diisopropyl sulfone (δ 351) and di-*t*-butylsulfone (δ 366) illustrate the effect of a Cα methyl substitution on the ^{33}S chemical shift. This "β-methyl effect" is nearly additive within the limits of our experimental error, averaging 7-8 ppm/Me. As the alkyl chain increases in length, the ^{33}S chemical shifts are essentially invariant indicating that alkyl substitution resulting in chain lengthening beyond the Cβ carbon exerts a minimal influence on the shift of the ^{33}S nucleus.

We also note that when dimethyl sulfone is dissolved in

trifluoroacetic acid (TFA), the ^{33}S resonance is shifted downfield by 12 ppm compared to that in chloroform solvent. It seems possible that hydrogen bonding to TFA by the sulfonyl oxygens results in the development of partial positive character on the sulfonyl sulfur and thus deshielding is the result.

We have examined the ^{33}S chemical shifts of a number of other sulfones and sulfonic acids and derivatives. The addition of

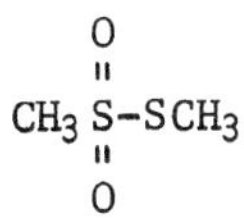

δ 328 (H_2O), $W_{1/2}$ = 25 Hz

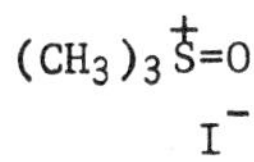

δ 310 (H_2O), $W_{1/2}$= 70 Hz

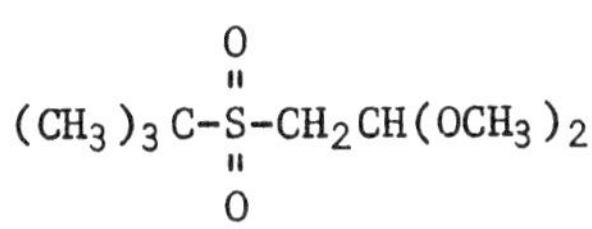

δ 350 ($CHCl_3$), $W_{1/2}$ = 270 Hz

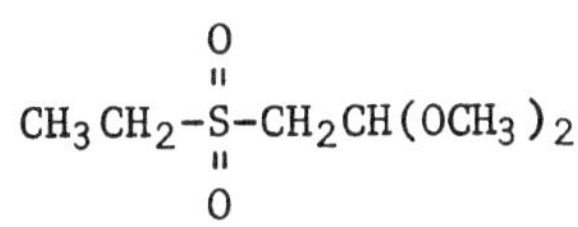

δ 335 ($CHCl_3$), $W_{1/2}$ = 110 Hz

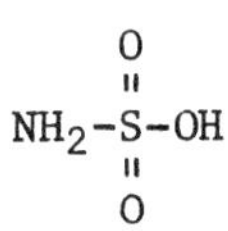

δ 328 (neat), $W_{1/2}$ = 210 Hz

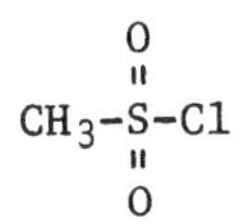

δ 341 (neat), $W_{1/2}$ = 1500 Hz

another heteroatom (i.e., O, N, S) to the sulfonyl sulfur does not appear to affect the chemical shift of the sulfonyl sulfur to any major extent. One exception is shown in Figure 2. Here, the sulfate of sodium thiosulfate is significantly deshielded compared to simple ammonium sulfate.

We have also examined the ^{33}S chemical shifts of a series of sulfonic acid and/or sodium salts. While the chemical shift differences between various structural types are small, the line widths are extremely sharp. Because of the narrow bandwidths we have been able to distinguish the components of a known mixture of sulfonic acids (Figure 3).

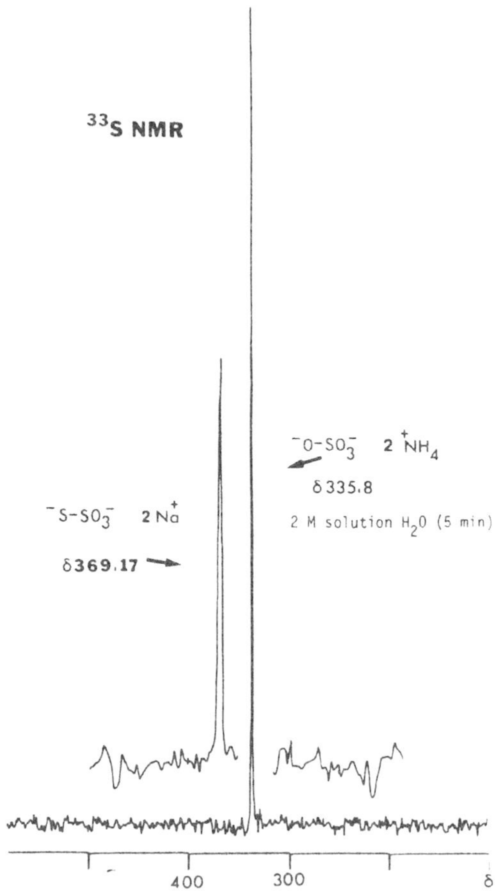

Figure 2. ^{33}S NMR of Sodium Thiosulfate and Ammonium Sulfate in H_2O.

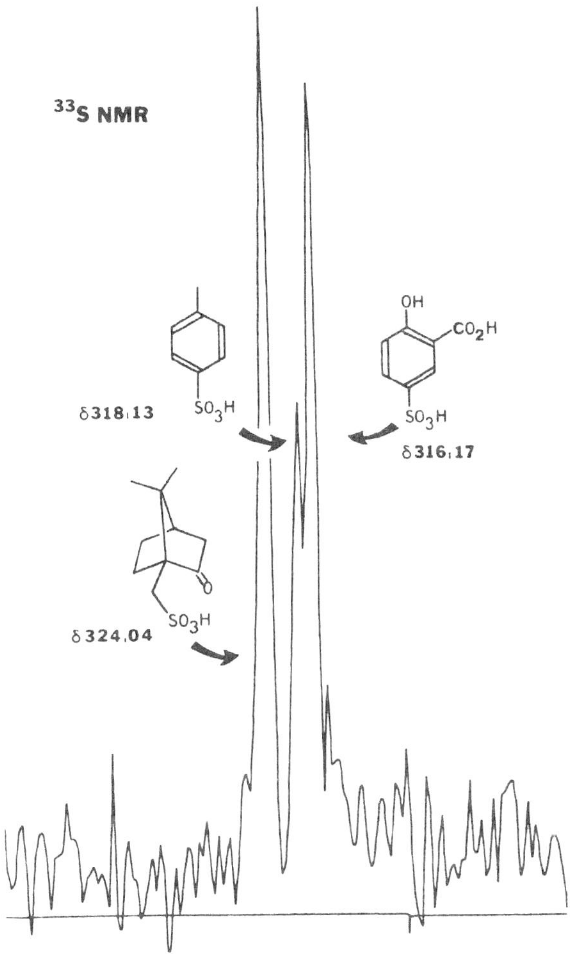

Figure 3. ^{33}S NMR of a Known Mixture of Sulfonic Acids in H_2O

Acknowledgment is made to the Department of Energy (Contract DE-A505-80ER10641) and the University of North Carolina for support of this research. The Bruker WM-250 NMR spectrometer was purchased from a grant from the National Science Foundation (CHE-79-12130).

References

(1) Block, E., Bassi, A. A., Lambert, J. B., Wharry, S. M., Andersen, K. K., Dittmer, D. L., Patwardhan, B. H., and Smith, D. J. H. 1980, J. Org. Chem. 45, pp 4807-10.

(2) Kobayashi, K., Sugawara, T., Iwamura, H. 1981, J. Chem. Soc. Chem. Commun. pp 479-480.

(3) Dyer, J. C., Harris, D. L., and Evans, Jr., S. A. 1982, J. Org. Chem. 47, pp 3660-3664.

(4) Kintzinger, J.-P. In "NMR-17. Oxygen-17 and Silicon-29", Diehl, P, Fluck, E., and Kosfeld, R., (Ed), Springer-Verlag, New York, 1981, pp 1-64.

(5) Faure, R., Vincent, E. J., Ruiz, J. M., and Lena, L. 1981, Organic Magn. Reson. 15, pp 401-403

(6) Harris, D. L., and Evans, Jr., S. A. 1982, J. Org. Chem., 47, 3355-3358.

A MULTINUCLEAR NMR STUDY OF THE INTERACTION OF CIS-PLATIN, A POTENT ANTITUMOR AGENT, WITH DNA-BASES.

Sarah G. Ward and R. Craig Taylor

Department of Chemistry, Oakland University, Rochester, Michigan 48063 USA

Cis-diamminedichloroplatinum(II), hereafter called cisplatin, is the first of a new class of compounds which exhibit potent antitumor activity in a wide variety of human neoplasms. It apparently attacks cellular DNA preventing new DNA synthesis. We have undertaken a research program aimed at elucidating the possible modes of binding of cisplatin with DNA and fragments of DNA. We wish to report here the results of a multinuclear NMR study (C-13, Pt-195, and N-15) involving the interaction of cisplatin with three nucleosides, xanthosine, inosine, and guanosine, as a first step towards understanding the binding of this compound to DNA. The NMR studies with guanosine indicate a strong interaction between cis-$Pt(NH_3)_2(H_2O)_2^{2+}$ and this nucleoside. This result is consistent with the suggestion, made by us and others, that guanosine is the primary base target in cellular DNA. Labelling cisplatin with 99% enriched $N(15)H_3$ has enabled us to further explore the DNA-base-Pt binding. The ramifications of these studies will be presented.

INTRODUCTION

The chemistry and biochemistry of platinum complexes have been the subject of considerable interest in recent years due in large part to the discovery of potent anticancer activity of cisplatin by Rosenberg and co-workers in 1969(1). Cisplatin and related platinum containing complexes are now part of the chemotherapeutic arsenal against this dreaded disease. Although the mechanism of

L. Petrakis and J. P. Fraissard (eds.), Magnetic Resonance. Introduction, Advanced Topics and Applications to Fossil Energy, 767–783.

biologic action is still the subject of extensive investigations, a great deal of evidence has been accumulated to suggest that cellular DNA is the primary target(2). Numerous hypotheses have been suggested which imply a bifunctional coordination of the bases of DNA, in particular two adjacent guanines(3). Other modes of binding are also possible and as a result the origin of the antitumor activity of cisplatin is still uncertain. A variety of spectroscopic techniques have been used to study the interaction of cisplatin with the purine and pyrimidine bases of DNA(4). It occured to us that multinuclear nmr would be an informative technique which would provide direct evidence for "platination" of specific base sites on individual DNA-bases and perhaps DNA itself. Accordingly, we wish to report the results of an nmr study involving the reaction of cisplatin with three nucleosides: inosine (I), xanthosine (X), and guanosine (G) (Figure I).

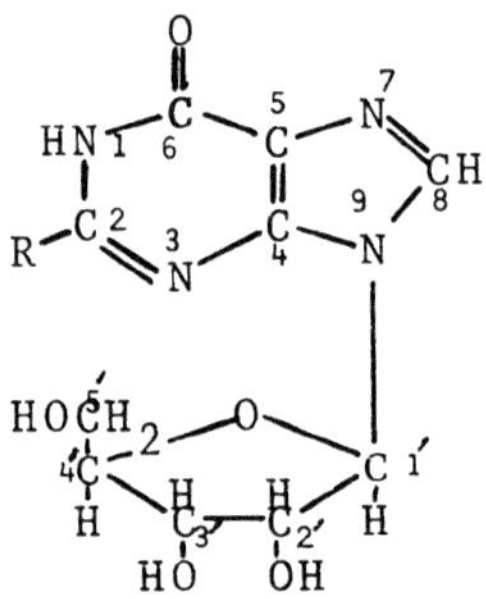

Figure I. The nucleosides; inosine(R=H), xanthosine (R=OH), and guanosine($R=NH_2$)

EXPERIMENTAL

A. Preparation of complexes.

All three nucleosides readily react with cisplatin in aqueous solution to yield cis-$[Pt(NH_3)_2N_2]Cl_2$, where N= I, X, or G(5). The $N(15)H_3$ labelled complexes were prepared in like manner from cis- $Pt(^{15}NH_3)_2Cl_2$, which in turn was prepared by reacting 99% enriched $N(15)H_4Cl$ and $N(15)H_3$ (Cambridge Isotopes) with K_2PtCl_4 according to the procedure of Kauffman and Cowan(6).

B. NMR characterization.

The NMR of the free nucleosides and the Pt complexes were recorded in H_2O and DMSO-d_6. The solutions were saturated with respect to both the free ligands and complexes. These solutions were contained in 8mm sample tubes which were inserted into 10mm tubes containing DOD as lock solvent. All spectra were recorded at 37 °C on a Varian FT-80 NMR Spectrometer equipped with a tunable broadband probe. Carbon-13 chemical shifts were referenced to external DDS, platinum-195 shifts were recorded relative to external 1M aqueous K_2PtCl_4, and nitrogen-15 shifts were relative to an external saturated solution of KNO_3. The positive shift-high frequency convention was used in accordance with IUPAC recommendations. For the spectra recorded in DMSO-d_6, the samples were run in 10mm tubes.

RESULTS AND DISCUSSION.

All three nucleosides readily react with cisplatin to produce the bis-complexes in excellent yield. All three are quite soluble, the guanosine complex exhibiting the highest aqueous solubility.

A. Cis-$[Pt(NH_3)_2G_2]Cl_2$.

(i) Carbon-13 Spectra. The C-13 proton decoupled spectrum of free guanosine in DMSO-d_6 shows the expected 10 resonances separated into two distinct groups; the five ribose resonances occuring between 61 and 87 ppm, and the purine resonances between 116 and 158 ppm. The assignments of the purine resonances are in excellent agreement with previous results(7). Further confirmation of these assignments, especially C(4), C(5) and C(8), is provided by the gated decoupled spectrum (Figure II). The $^1J_{C(8)-H}$ splitting is 213 Hz and the long range coupling with H(1') is 3.9 Hz. The $^3J_{C(5)-H(8)}$ is 11.4 Hz, whereas the splitting pattern for C(4) shows four unequally spaced lines with separations between adjacent signals ranging from 1.7 to 2.6 Hz. Presumably C(4) couples with both H(8) and H(1'). As expected, C(6) and C(2) show single resonances.

Upon reacting quanosine with cisplatin the chlorides are displaced by two quanosines to yield cis- $Pt(NH_3)_2$-G_2 Cl_2. The proton decoupled and coupled C-13 spectra of this complex are presented in Figures III and IV, respectively. Of particular interest is the C(8) resonance. Three bond platinum-195-carbon-13 couplings are observed as two satellites ($^3J_{C-Pt}$ = 42 ± 2 Hz) symmetrically placed relative to the central peak. Theoretically,

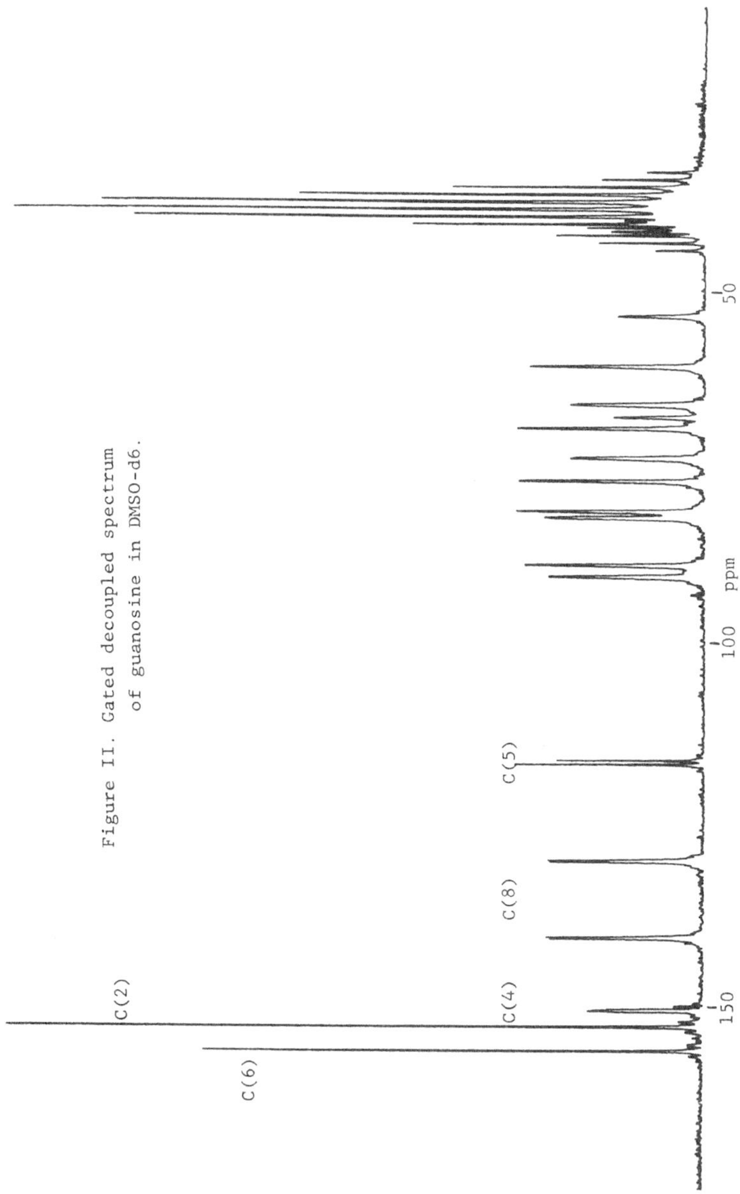

Figure II. Gated decoupled spectrum of guanosine in DMSO-d6.

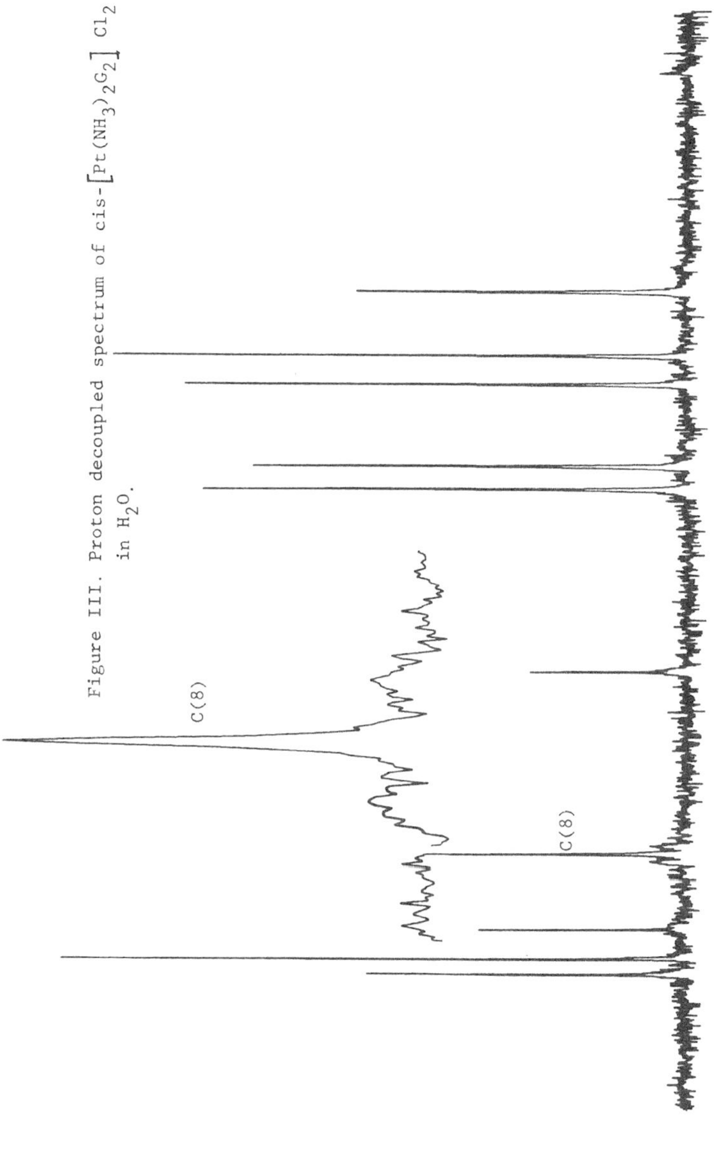

Figure III. Proton decoupled spectrum of cis-$[Pt(NH_3)_2G_2]$ Cl_2 in H_2O.

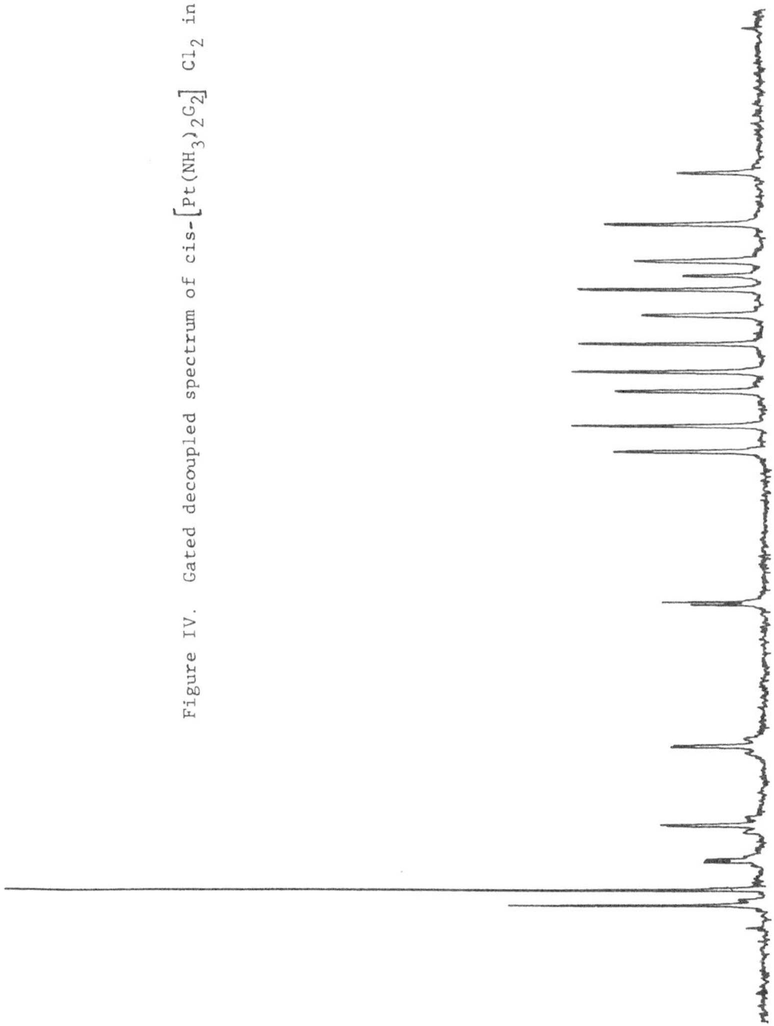

Figure IV. Gated decoupled spectrum of cis-[$Pt(NH_3)_2G_2$] Cl_2 in H_2O.

the relative peak heights should be in the ratio 1:4:1, based on the natural abundance of Pt-195 (33%). That they are not can be attributed to Pt-195 relaxation dominated by the chemical shift anisotropy mechanism (8). Figure V shows an expansion of the C(8) region of Figure IV. In addition to the observed Pt-C(8) and C(8)-H(8) splittings, long range coupling of C(8) with H(1') is evident($^3J_{CH}$ = 3.8 ± 0.1 Hz). These results point to N(7) as the site of platinum coordination, in agreement with what has been found in the solid state for both cis-[Pt-$(NH_3)_2G_2$]$Cl_{3/2}(ClO_4)_{1/2}\cdot 7H_2O$ (9) and the analogous ethylenediamine complex, cis-[Pt(en)G_2]$Cl_{3/2}I_{1/2}\cdot 2H_2O$ (10).

The C(4), C(5), and C(8) resonances show very little chemical shift upon coordination of N(7) of the guanine ring to platinum. This is somewhat surprising, although not unprecedented.

(ii) Platinum-195 and Nitrogen-15 spectra. The Pt-195 spectrum of cis- $Pt(NH_3)_2G_2\ Cl_2$ exhibits a single broad resonance ($\Delta\nu_{1/2}$ = 260 Hz) at -816 ppm relative to K_2PtCl_4. Upon substitution of $N(15)H_3$ in this complex, a triplet is observed in the proton decoupled spectrum of cis-[Pt($^{15}NH_3)_2G_2$] Cl_2 (δ = -800ppm) with a one bond N(15)-Pt(195) splitting of 309 Hz (Figure VI). This coupling has been confirmed by recording the N(15) nmr of the same sample. In the proton decoupled N-15 spectrum a triplet is observed with a one bond splitting of 309 Hz and a chemical shift of -414.4 ppm relative to external nitrate(Figure VII) in reasonable agreement with the results found by Nee and Roberts(11). The coupling constant is in the range expected for those nitrogen ligands ($N(15)H_3$) bound to platinum that are trans to nitrogen ligands (11). The proton coupled spectrum of this same complex exhibits the expected quartet pattern due to the three equivalent protons with a nitrogen-hydrogen coupling constant of 71.7 ± 1.0 Hz (Figure VIII). This constant is considerably larger than in $N(15)H_3$ itself (J = 61.2 Hz) but similar to the coupling constant for NH_4^+ (J = 73.3 Hz) (12).

Finally, both the Pt-195 and N-15 spectra for this complex are very clean,i.e., there is no nmr evidence to suggest the presence of other cis- $Pt(NH_3)_2^{2+}$ containing moieties. The complex also appears to be quite stable in aqueous solution with little or no tendency for water to displace the nucleoside. (See, however, the results for the inosine and xanthosine complexes).

(iii) Reaction of cis- $Pt(NH_3)_2G_2\ Cl_2$ with DMSO. In an attempt to increase the solubility of this complex for nmr studies and to find a suitable vehicle for the use of platinum complexes in the topical treatment of Herpes simplex viruses types 1 and 2 (*vide infra*), DMSO

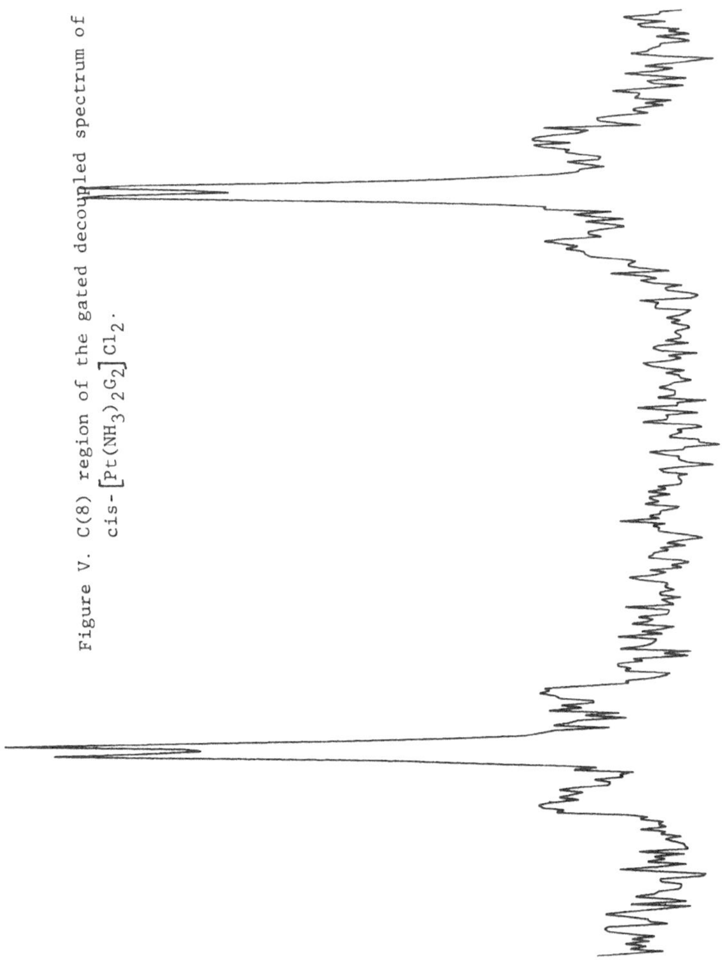

Figure V. C(8) region of the gated decoupled spectrum of cis-$[Pt(NH_3)_2G_2]Cl_2$.

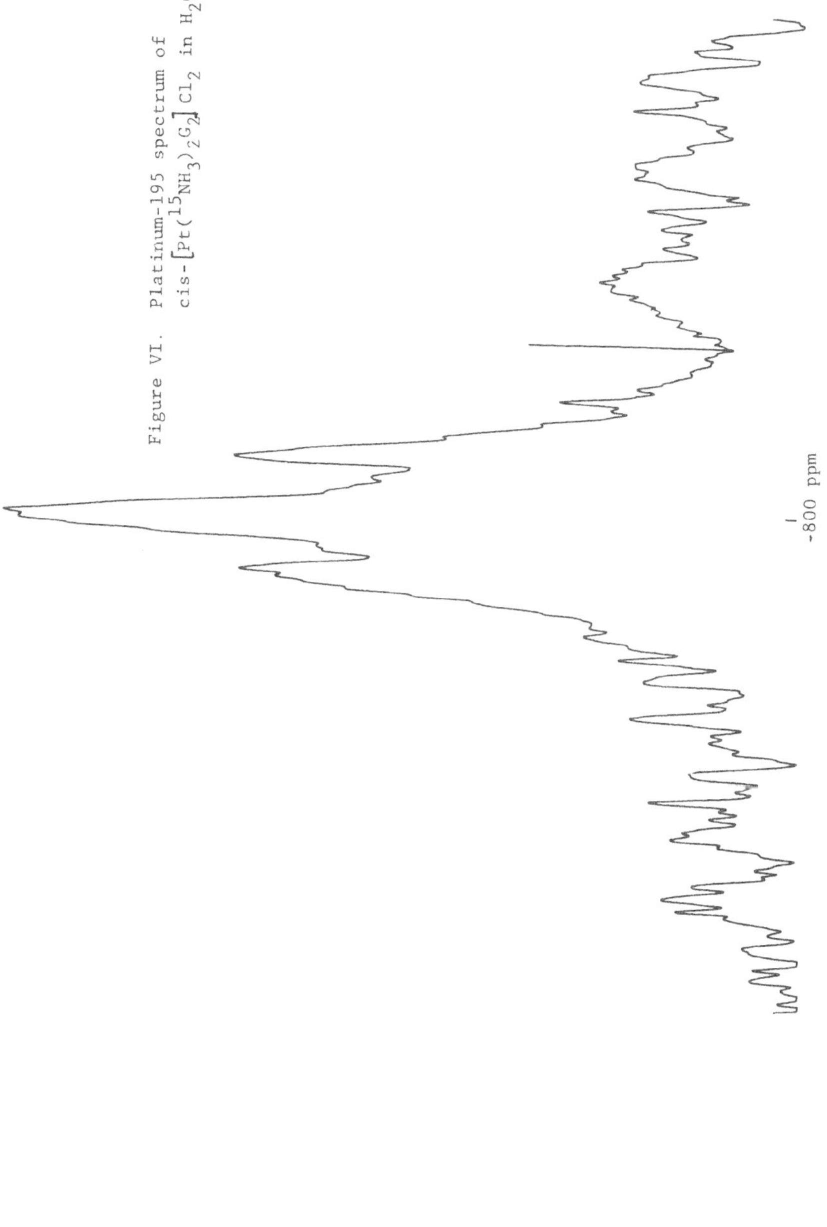

Figure VI. Platinum-195 spectrum of cis-$[Pt(^{15}NH_3)_2G_2]Cl_2$ in H_2O.

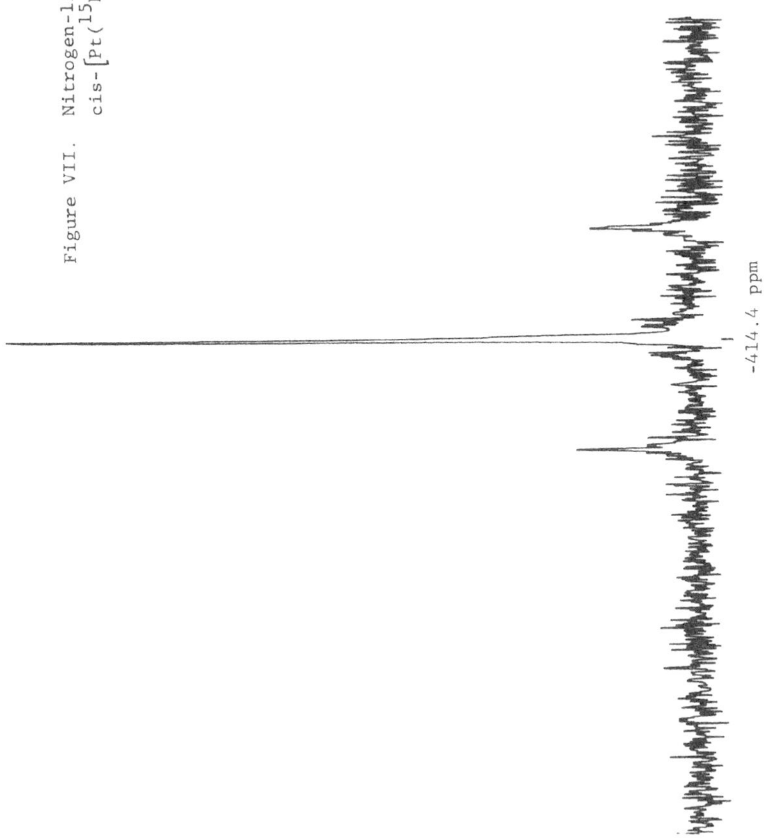

Figure VII. Nitrogen-15 nmr of cis-$[Pt(^{15}NH_3)_2G_2]Cl_2$ in H_2O.

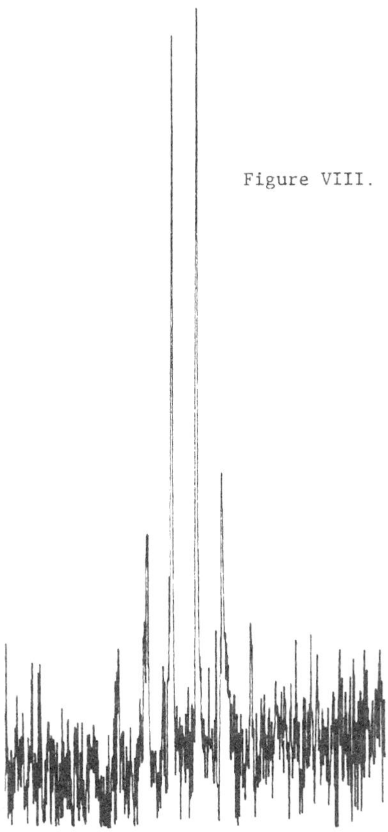

Figure VIII. Nitrogen-15 nmr of cis-[Pt($^{15}NH_3$)$_2$$G_2$] Cl_2 in H_2O. Gated decoupled spectrum.

was investigated as a potential solvent. However, the C-13 and Pt-195 spectra showed the replacement of the nucleoside by DMSO. The C-13 spectra undergo changes as a function of time. Figure IX represents what we believe to be the equilibrium position of the substitution reaction. Resonances for both free and complexed guanosine are present. The Pt-195 spectra also show two resonances, one at -816 ppm due to the guanosine complex, and one at -1027 ppm which we have assigned to the complex in which one or both of the guanosine ligands have been displaced from the coordination sphere of the platinum(II) species. These spectra also change as a function of time.

B. Cis-$[Pt(NH_3)_2I_2]Cl_2$ and cis-$[Pt(NH_3)_2X_2]Cl_2$.

(i) Carbon-13 Spectra. The carbon-13 spectra for these two complexes are quite similar to those obtained for the corresponding bis-guanosine complex. One important difference, however, is worth noting. These two complexes udergo exchange of nucleoside by water much more readily than the guanosine complex. Thus in addition to resonances which may be assigned to the bis complexes, other peaks are present which may represent other complexes such as cis-$[Pt(NH_3)_2N(H_2O)]^{2+}$ or cis-$[Pt(NH_3)_2N\text{-}Cl]^{+}$. A case in point is the xanthosine system which shows two sets of triplets for the C(8) in the proton decoupled spectrum (Figure X). The more intense triplethas been assigned to the bis-xanthosine complex based on the magnitude of its three bond carbon-platinum coupling constant of 40.7 ± 1.1 Hz which is quite similar to the result obtained for the bis-guanosine species. The weaker triplet has a three bond carbon-platinum coupling constant of 30.4 ± 0.9 Hz and has been assigned to one of the other species mentioned above. Nonetheless, the carbon-13 spectra of the xanthosine and inosine complexes are consistent with the fact that in both cases the major species present are cis-$[Pt(NH_3)_2N_2]Cl_2$ in which coordination is via N(7).

(ii) Platinum-195 and Nitrogen-15 Spectra. As pointed out above, more than one species is present in solutions prepared by dissolving cis-$[Pt(NH_3)_2N_2]Cl_2$ (N= I or X). We only observe one resonance in the Pt-195 spectrum, albeit quite broad. It is possible that the Pt chemical shifts are quite similar for the various species present and the resonance observed represents more than one complex. An alternative explanation is that the minor component(s) exhibit very broad peaks which are lost in the background noise. The platinum-195 spectra for the $N(15)H_3$ labelled complexes show the expected triplet feature with the one bond nitrogen-15-

Figure IX. Carbon-13 spectrum of reaction product of cis-$[Pt(NH_3)_2G_2]$ Cl_2 with DMSO-d_6.

Figure X. Carbon-13 nmr of cis-$[Pt(^{15}NH_3)X_2]Cl_2$ in H_2O, C(8) region only.

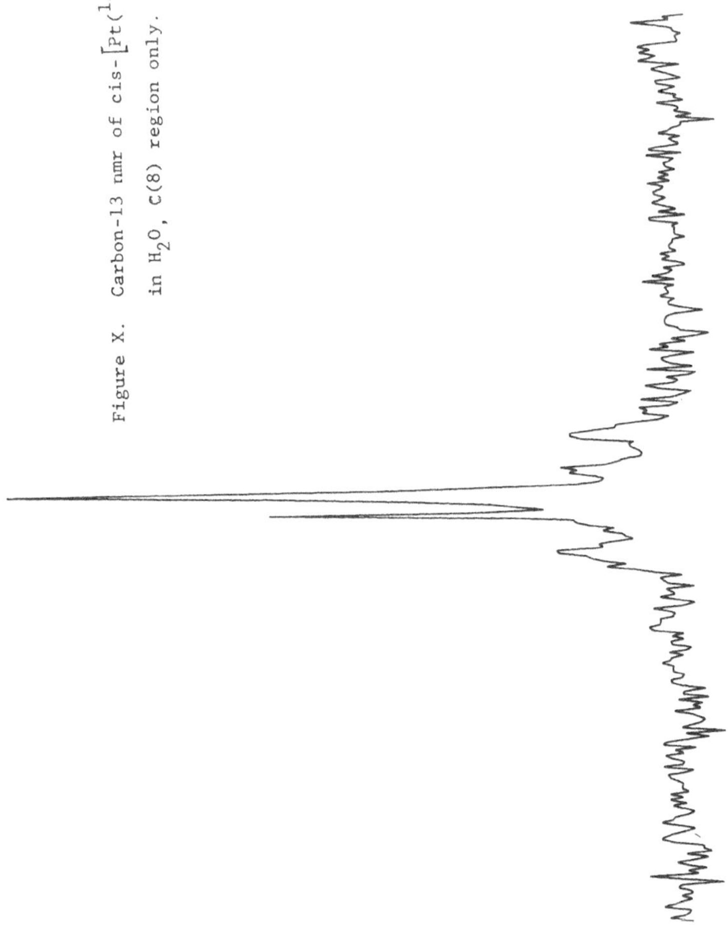

platinum-195 coupling constants of 314 and 316 Hz for the inosine and xanthosine complexes respectively.

The nitrogen-15 spectra of these species clearly suggest the presence of more than one complex (Figure XI). The chemical shifts and coupling constants for the bis-nucleoside complexes are similar to those found for the corresponding guanosine species. Minor peaks must be due to other cis-$Pt(^{15}NH_3)_2^{2+}$ moieties present in solution.

THE USE OF CISPLATIN IN THE TREATMENT OF HERPETIC LESIONS

The results presented above indicate a strong interaction between the N(7) site of guanosine and cis- Pt-$(NH_3)_2^{2+}$. This is consistent with the suggestion (13) that guanine is the probable target for "platination" of DNA. It occured to us that since herpes simplex virus types 1 and 2 are much richer in G+C content (66-68%) (14) than normal human DNA (G+C content is about 40%), cisplatin might readily interact with replicating herpes DNA viruses and prevent further DNA synthesis. In other words, cisplatin might have a salutary, efficacious effect on herpetic lesions. Accordingly, we have tried topical applications of a dilute (~1mM) aqueous solution of cisplatin on a number of volunteers who suffered from recurrent outbreaks of either herpessimplex virus type-1 (HSV-1, cold sores are a common manifestation of the virus) or varicella-zoster virus (shingles). In 14/14 cases, a regimen of 2-3 applications/day for 2-4 days after initial outbreak of the disease had the desired salutary effect. In all cases, no further outbreaks have occured. In three cases (two cases of HSV-1 and one of shingles) the volunteers have been disease free for over one year!

These encouraging results have prompted us to further explore cisplatin and related platinum containing complexes as potential antiherpes agents. We are currently studying the antiviral activity of these complexes in both tissue culture and animal models and have observed cisplatin, in particular, to be 100% effective in the systems infected with HSV-1. We have filed a patent application with the US Patent Office covering this new use of these fascinating platinum complexes.

SUMMARY

Multinuclear nmr studies of the interaction of platinum complexes with individual DNA base molecules

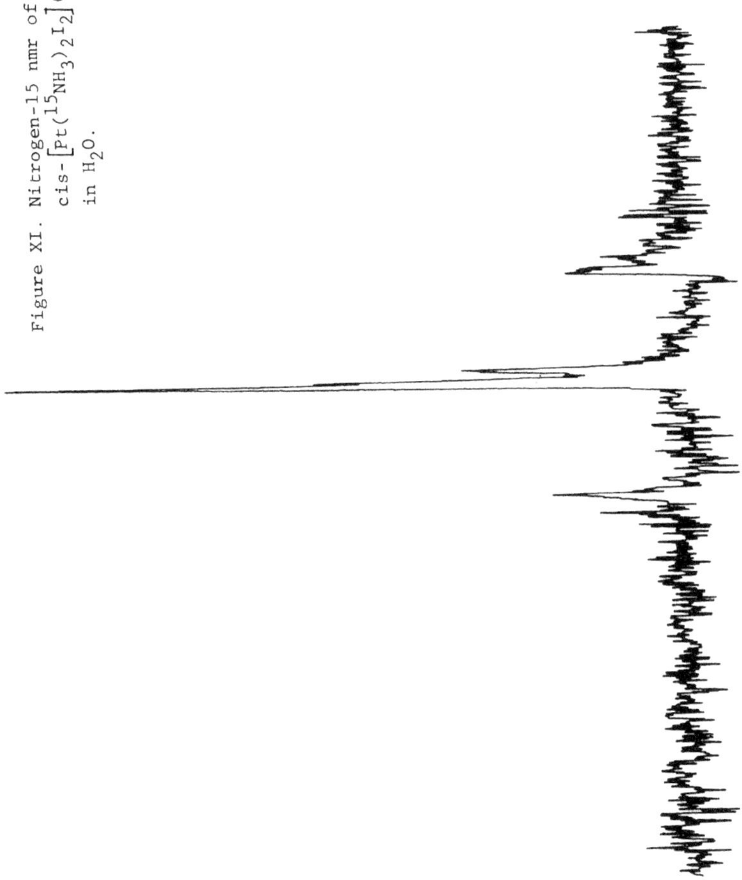

Figure XI. Nitrogen-15 nmr of cis-$[Pt(^{15}NH_3)_2I_2]Cl_2$ in H_2O.

shows a great deal of promise in elucidating the mode of binding of platinum to these ligands in solution. Whether or not this technique can be extended to the study of the "platination" of DNA itself remains to be seen.

ACKNOWLEDGMENTS

This work was supported in part by a Biomedical Research Support Grant from Oakland University. We also wish to thank NATO for providing partial support enabling us to attend this Advanced Study Institute.

REFERENCES

(1) Rosenberg,B., VanCamp,L., Trosko,J.E., and Mansour, V.H., 1969, Nature(lOndon), 222, pp. 385-386.
(2) Roberts, J.J., and Thomson, A.J., 1979, Prog.Nucleic Acid Res.Mol.Biol., 22, pp. 71-133.
(3) Kelman, A.D., and Peresie, H.J., 1979, Cancer Treat. Rep., 63, pp. 1445-1452.
(4) Nee,M. and Roberts, J.D., 1982, Biochem., 21, pp. 4920-4926, and references cited therein.
(5) Kong, P. and Theophanides, T., 1974, Inorg.Chem., 13, pp. 1167-1170.
(6) Kauffman, G.B., and Cowan, D.O., 1963, Inorg. Syn., 7, pp. 239-242.
(7) Jones,A.J., Winkley,M.W., Grant,D.M., and Robins, R. K., 1970, Proc. Nat. Acad. Sci., 65, pp. 27-30.
(8) Ismail, I.M., Kerrison, S.J.S., and Sadler, P.J., 1982, Polyhedron, 1, pp. 57-59.
(9) Cramer, R.E., and Dahlstrom, P.L., 1977, J. Clinical Hematol. and Oncol., 7, pp. 330-337.
(10) Bau, R., Gellert, R.W., Lehovec, S.M., and Louie, S., Ibid., pp. 51-62.
(11) Nee,M. and Roberts, J.D., 1982, Biochem., 21, pp. 4920-4926.
(12) Binsch, G., Lambert, J.B., Roberts, B.W., and Roberts, J.D., 1964, J.Amer.Chem.Soc., 86, pp. 5564-5570.
(13) Scovell, W.M., and Kroos, L.R., 1982, Biochem. and Biophys. Res. Comm., 104, pp. 1597-1603.
(14) Jawetz, E., Melnick, J.L., and Adelberg, E.A., 1976, Review of Medical Microbiology, 12th ed., Lange Publications, Los Altos, California 94022, Chapter 38, pp. 442-453.

RESEARCH NEEDS, OPPORTUNITIES AND PROSPECTS

M. L. Gorbaty and B. G. Silbernagel

Exxon Research and Engineering Company

INTRODUCTION

The evolution of science and the application of new scientific concepts to technology occurs by a chain reaction mechanism as indicated in Figure 1. Researchers in basic science focus on the process of understanding natural principles, with applied goals as a secondary consideration. Conversely, researchers in applied science focus on well defined problems with clear applications in mind, bringing to bear whatever knowledge is known and available to them. In the absence of communication between these groups, valuable opportunities for advancing the technology may be lost, since the goal-oriented researcher may not be aware of the implications of new scientific advances, and the researcher in basic science may not appreciate the technical significance of applications of his techniques.

The initiation of the chain reaction begins when the basic science and the applied problems become known and understood by the people involved. This process usually involves a degree of language translation and an appreciation of the dimension of the problem and the capability of basic science to contribute. In establishing this communication, there is no substitute for dialogue between the basic and applied scientists and interactions of people with multiple skills. Once this overlap is accomplished and it becomes clear how science can be used to solve the problem, new technology invariably arises. As this new technology is utilized, new problems are identified and new science areas are involved, moving both science and technology forward.

L. Petrakis and J. P. Fraissard (eds.), Magnetic Resonance. Introduction, Advanced Topics and Applications to Fossil Energy, 785–795.

The goal of this Advanced Study Institute was to bring together scientists working with advanced magnetic resonance techniques with those exploring the science of fossil energy. During the two weeks of the institute, the basic principles underlying both research areas were defined and a common language was established. Considerable discussion was also directed to the "state of the art" in each area, so that a clear picture of current capabilities was available to all participants. To aid the onset of the chain reaction and to tie this information together in a concrete way, the Advisory Board set up three subcommittees charged with defining the frontiers and opportunities in applying magnetic resonance techniques to fossil energy systems. These groups met informally throughout the two-week period and prepared short reports which were presented and discussed at the last plenary session.

FIGURE 1 - SCIENCE/TECHNOLOGY EVOLUTION: A CHAIN REACTION

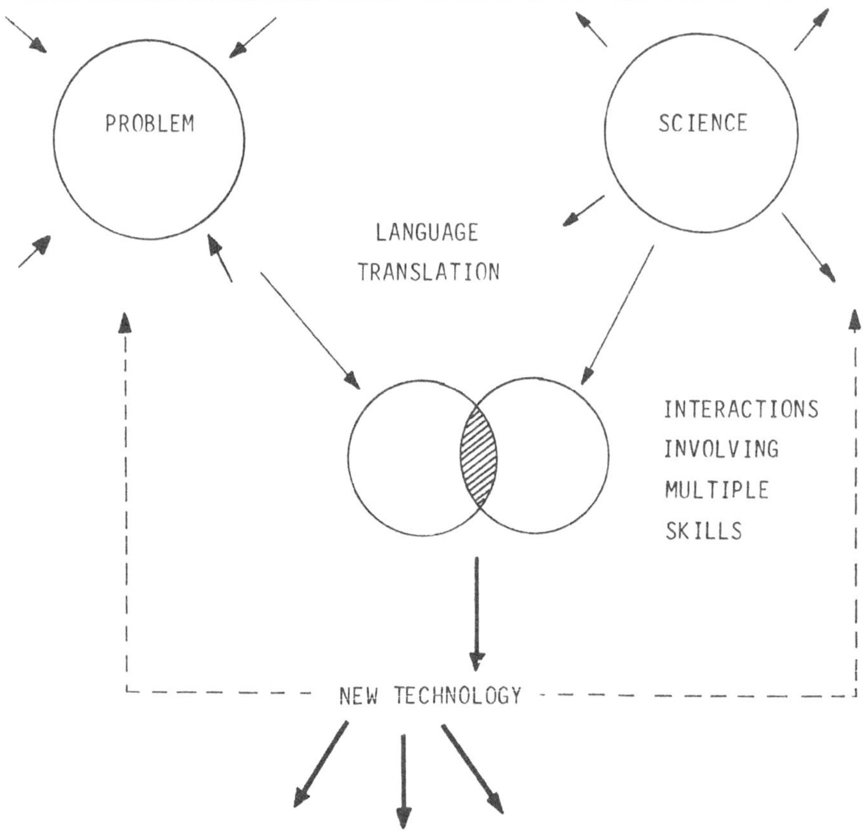

TABLE 1: ROSTERS OF STUDY SUBCOMMITTEES

I. Subcommittee on Nuclear Magnetic Resonance

*G. Maciel
F. Miknus
L. Lynch
R. Wind

II. Subcommittee on Electron Spin Resonance

*R. deBeer	J. Conard
L. Dalton	B. Silbernagel
J. Uebersfeld	D. Doetschman
H. Thomann	

III. Subcommittee on Other Resonance Spectroscopies

*P. Montano
R. Davidson
F. Kapteijn
L. Lynch
P. Cassidy

* = Chairman

TABLE 2: NMR SUBCOMMITTEE REPORT

A. Fossil Fuel Research Areas Addressable by NMR.

1. Chemical Structures of Coal and Oil Shale.
2. Coal Conversion (Liquefaction and Gasification).
3. Oil Shale Conversion.
4. Coal Burning.
5. Transportation (e.g., weathering of coal).

B. Fossil Fuel Research Opportunities -- Availability of Research Resources.

C. Information on Which NMR Should Focus.

1. Heteroatom Chemistry (e.g., N, O, S, Cl).
2. Organic Functional Group Characterization.
3. Molecular diffusion and thermal transport.
4. Catalytic behavior of heteroatom species.
5. Nature of coal surface and the interaction with small molecules.
6. Interactions of solvents with coal.
7. Anisotropies and heterogeneities which distinguish coal or kerogen from other macromolecular systems.
8. Standardized, definable parameters for characterizing coal rank.
9. Chars and cokes and their formation by combustion, pyrolysis and condensation.
10. Mesophase behavior.
11. Spatial resolution in samples (imaging).

D. Specific NMR activities of Special Merit.

1. Advanced high resolution NMR techniques (e.g., 2-dimensional NMR for liquids.
2. Standardize samples from a central coal bank.
3. Include input of geochemistry in interpretation.
4. Develop high-temperature NMR capabilities.
5. Use proton multipulse techniques for a new generation of ^{13}C NMR to elucidate structure in solids.
6. Apply dynamic nuclear polarization to ^{15}N, ^{17}O, ^{33}S, as well as ^{13}C, in solids.
7. Techniques that focus and elucidate anisotropies and domain heterogeneities in coals.
8. NMR imaging in fossil fuel samples.

TABLE 3: ESR SUBCOMMITTEE REPORT

Suggestions for future ESR/ENDOR research on coal

1. The necessity for adequate sample preparation and documentation.

2. Separation of the overlapping spectra of different radicals

3. Systematic use of transient parameters like T_{1e}, T_{2e} and hyperfine interactions to characterize radicals

4. Correlation of ESR parameters with other properties of samples

Sample Preparation: many variables reflecting sample history must be defined. An internationally recognized reference set would be desirable.

Near Term Transient Measurements:

a. Measure true T_{1e}, T_{2e} to augment steady state ESR data

b. Use ENDOR, Electron Spin Echo Envelope Modulation (ESEEM) to measure hyperfine interactions in coal

c. Resolve overlapping ESR lines on the basis of their different relaxation times.

d. Probe for possible S=1 biradicals using different sequences.

Possible Future Transient Measurements

a. Pulsed ESR/ENDOR at 70-150 GHz.

b. Photoexcitation of coal combined with pulsed ESR.

c. Fast Fourier Transform-Free Induction Decay ESR

d. Kinetics of Transient radicals during flash pyrolysis.

e. Linear Electric Field Effect studies to probe tensor properties.

Important Research Strategies

a. Chemically well-defined samples such as isolated coal macerals.

b. A standard protocol for handling samples.

c. A sample bank.

d. Use of probe molecules to elucidate surface properties oxygen physisorption, ^{17}O labeling experiments.

The three subcommittee areas were defined for convenience in terms of the three generic resonance disciplines: Nuclear Magnetic Resonance, Electron Spin Resonance, and Other Resonance Spectroscopies (Mossbauer, etc.) Subcommittees consisted of a combination of magnetic resonance and fossil fuel researchers, the committee rosters being included in Table 1. The subcommittees were asked to address the following questions:

- Where are the "frontiers" and opportunities in applying magnetic resonance techniques to fossil fuel systems?
- What are the principal difficulties associated with the application of these techniques to such chemically and physically inhomogeneous systems? What alternatives exist for overcoming these difficulties?
- What should the next steps be in studying these systems and what would be the consequences of success?

The detailed reports discussed in the next section define research needs and identify new research opportunities.

SUBCOMMITTEE REPORTS

- Nuclear Magnetic Resonance (Table 2).

Almost the complete range of fossil fuel used has at least been surveyed by NMR, from an analysis of the chemical structures of coal and oil shales, through the entire array of utilization processes and even to some very practical applications such as the weathering of coal. Basic scientists have found the problems challenging and have been ready to "dirty their hands", but the recent loss of interest in the short term commercial applications of coal conversion and shale retorting has lead to an appreciable reduction in government research support. The need for adequate, long term funding in these areas for academic researchers is an important consideration if more sophisticated tools are to be brought to bear in a systematic way.

As section C of Table 2 indicates, there are many types of information which NMR can provide, at least in principle. They deal with the chemical structure of coal, shale and such resulting products as chars, cokes, and carbonaceous mesophase materials. Also included are factors related to reactivity: molecular mobility, catalytic behavior of heteroatoms, the role of the coal surface and the way in which small molecules and solvents interact with the coal structure. One issue which recurs in the other subcommittee reports is the extent to which a microscopic, possibly NMR, parameter or set of parameters could be defined, which could give a more quantitative definition of coal rank and a better feeling for coal reactivity.

In identifying specific NMR activities of particular merit, several instrumental approaches were listed which will undoubtedly be exploited in the near future, particularly high resolution in both liquids and solids and NMR imaging, which could provide sub-millimeter spatial resolution in complex structures such as coal. Extension of dynamic nuclear polarization techniques to studying the heteroatom resonances ^{15}N, ^{17}O, and ^{33}S in solids is a greater challenge because of the relatively low density of these nuclei and the complications associated with quadrupole interactions for ^{17}O and ^{33}S. However, the basic process is understood, and particularly favorable samples can be identified for initial trial experiments. A more formidable and as yet unsolved problem is the development of high resolution NMR techniques to study solids at temperatures conventionally employed for fossil fuel conversions: 400-700°C. An additional important need is to provide well characterized, standardized samples on which these novel techniques can be attempted. In interpreting the data, a greater appreciation of the associated geochemistry would be particularly valuable.

- Electron Spin Resonance (Table 3)

The ESR considerations and concerns differ significantly from those of NMR. While NMR data in general reflect the sum of all nuclei of a given type, ESR focuses on the relatively small number of organic species which are paramagnetic. One major challenge in interpretation is to relate the ESR data to the global properties of the sample as represented by such parameters as the chemical analyses. A major concern is the introduction (or destruction) of paramagnetic species during sample handling. While careful sample handling is highly desirable for NMR samples, it is imperative for ESR samples. A second distinction between NMR and ESR techniques is that very few transient ESR experiments have been performed on fossil fuel samples. Because of the paucity of information in this area, we would like to focus primarily on potential transient ESR measurements.

A number of transient ESR experiments can be done immediately on a systematically selected sample set. Measurements of the electron spin-lattice relaxation time (T_{1e}) and the spin-spin relaxation time (T_{2e}) can be done directly, and the envelope modulation of the electron spin echo can measure the hyperfine interactions in the coal. Inequivalent radicals may be discriminated on the basis of their relaxation properties. Singlet

TABLE 4: OTHER RESONANCE TECHNIQUES REPORT

Specific Problems of Interest and Applications:

A. Fossil Fuels Characterization

1. Aromatic species (concentration and ring size distribution).

2. Bridges (functional form and distribution with respect to aromatic clusters).

3. Heteroatoms (functionality and distirbution of O, N, S).

4. Type and distribution of hydroaromatic species.

B. Coal Classification

1. An objective chemical measure of coal rank.

2. Relationship of resonance properties to reactivity.

3. Development of predictive techniques.

C. Coal Liquefaction

1. Aromatic ring size may indicate maximum quality of oil.

2. Connecting bridges indicate the kinetics which control the liquefaction reaction and guide the choice of a catalyst.

3. Different chemistry is required to break oxygen and methylene bridges.

4. Would like to assess probability of recombination after cleavage of bonds.

D. Coal Coking

- the chemistry of the plastic state is particularly important.

E. Modeling

- Microscopic chemical information will lead to more appropriate models.

F. Gasification

1. Must understand how coal structure influences pyrolysis and gasification.

2. Must understand the state of the coal surface after the onset of gasification.

G. *In Situ* Reaction Studies

1. Characterize the active catalyst.

2. Identify reaction intermediates.

3. Study reaction rates.

4. Probe effect of temperature, pressure, solvent, gas composition.

H. Promising Research Directions

1. In situ ESR to study radical reactions.

2. Multinuclear NMR for O, N, Na, Al, Si.

3. Mössbauer for Fe, Sn coordination.

4. Molecular mobility in the plastic state.

5. Spillover H_2 during catalysis.

6. Catalyst fluidity, mobility.

radicals and biradicals can be discriminated by the different pulse sequences required to maximize their spin echoes. While the number of laboratories routinely doing transient ESR is relatively small and transient ESR spectrometers are only now approaching commercial production, there is sufficient interest in these materials that detailed studies will be done in the near future.

On the longer term, these measurements will be performed at higher frequencies 70-150 GHz, vs. 10-35 GHz employed presently. More elaborate combination experiments such as photoexcitation, in situ flash pyrolysis and linear electric field effect experiments are possible if the appropriate investment in instrumental development is made. One formidable problem is the Fourier transformation of the ESR free induction decay, significantly complicated by the difficulty in fixing the phase of the transform. Recent advances in numerical analysis are promising in the resolution of this problem.

As mentioned above, a major concern is relating ESR properties to the chemistry of the coal. One particularly promising approach is to achieve a simplification by examining coal macerals of a given organic type. This allows a more direct interpretation of the chemistry and a more quantitative analysis of the resulting properties. One must also exercise considerable care in the protocol for sample preparation. For example, physisorbed oxygen can broaden steady state ESR lines and influence electron relaxation rates even when it occurs at low concentrations. Conversely, the use of paramagnetic molecules can be used as an effective probe of the coal surface, and ^{17}O labeling experiments could provide information about the nature of the coal-oxygen interaction.

- Other Resonance Techniques (Table 4)

The last subcommittee took the complementary approach by defining the chemistry of the problem first and then identifying resonance opportunities. Initially, one would like to determine the chemical structure of the organic matter in fossil fuels: the size and number of aromatic species, the bridges which connect them, associated heteroatoms (O, N, S), and the hydroaromatic species. Hydroaromatics are particularly important because they provide the source of hydrogen required in thermal reactions. One would like to use this chemical information to provide a more quantitative measure of coal rank, one that could predict conversion behavior under a variety of conditions.

Such studies on fossil fuels are directly extendable to their conversion processes. In coal liquefaction, for example, knowledge of the type of bridges would dictate the chemistry used, and the aromatic ring size of the starting coal would provide a measure of product quality. In the coking of coal, a microscopic knowledge of the chemistry of the plastic state would be most desirable. For gasification, the structure of the coal surface and how it changes upon gasification would allow a considerably more focused treatment scheme. However, to really study conversion chemistry, it should be done at reaction conditions. While ESR experiments and transient (low resolution) NMR measurements are currently being done, they are just the beginning of what should be a very profitable future research activity. The study of catalytic conversion at reaction conditions is a complex but exciting future prospect.

INDEX OF SUBJECTS

GPSR Compliance

The European Union's (EU) General Product Safety Regulation (GPSR) is a set of rules that requires consumer products to be safe and our obligations to ensure this.

If you have any concerns about our products, you can contact us on ProductSafety@springernature.com

In case Publisher is established outside the EU, the EU authorized representative is:

Springer Nature Customer Service Center GmbH
Europaplatz 3
69115 Heidelberg, Germany

Batch number: 10607216

Printed by Printforce, the Netherlands